Regional Geology Reviews

Series Editors

Roland Oberhänsli
Maarten J. de Wit
François M. Roure

More information about this series at
http://www.springer.com/series/8643

Maarten J. de Wit • François Guillocheau •
Michiel C.J. de Wit
Editors

Geology and Resource Potential of the Congo Basin

Springer

Editors
Maarten J. de Wit
AEON- Africa Earth Observatory
Network, and ESSRI - Earth
Stewardship Science Research Institute
Nelson Mandela Metropolitan University
Port Elizabeth
South Africa

François Guillocheau
Geosciences
University of Rennes
Rennes Cedex
France

Michiel C.J. de Wit
Geology
University of Pretoria
Pretoria
South Africa

ISBN 978-3-662-50951-7 ISBN 978-3-642-29482-2 (eBook)
DOI 10.1007/978-3-642-29482-2
Springer Heidelberg New York Dordrecht London

Cover illustration: Photo by Edmond Thorose
Aerial view (helicopter flying at ca. 150 m above ground) looking northeast across the southwestern Congo Basin, 215 km southwest of Kinshasa, with outliers of poorly consolidated red sandstones of the Kwango Group (Jurassic-Cretaceous) capped by a subhorizontal hardpan of Cenozoic silcrete. These outliers represent the nothernmost remnants of the Kalahari Plateau, regional peneplanation surface ("the African Surface") that is well preserved at ca. 1000 m across southern Africa.

Printed on acid-free paper

Springer is part of Springer Science+Business Media (www.springer.com)

AEON

Earth Stewardship Science Research Institute

Nelson Mandela Metropolitan University

for tomorrow

UNIVERSITÉ DE RENNES 1

Topo AFRICA

!KHURE AFRICA

The story is the African wind

Dedication

Our Congo book is dedicated to three remarkable earth scientists whose work over a span of nearly 65 years, pre- and post-independent Sub-Saharan Africa, changed the course of knowledge about the geography, geology, and natural resources of the Congo Basin and surrounding regions of Central Africa:

Lucien CAHEN (Belgium, 1912–1982), Norman Snelling (England, b. 1930), and Henri KAMPUNZU (DRC—Democratic Republic of Congo, 1952–2004)

Lucien Cahen

Norman Snelling

Henri Kampunzu

Cahen and Snelling were relatively shielded from the challenging sociopolitical and economic issues associated with colonial activities and the inevitable transition to African independence. They were able to focus mostly on what they loved: field geology and figuring out stratigraphy and tectonics using the early emergence of radiometric dating. By contrast, Kampunzu was a child of the African struggle for independence; he represented part of a new generation of African scientists freed from colonial suppression. He too loved geology and its history, and the field work needed to unravel its complexities, but he was also 'caught up' in the turbulent periods of transformation from the Belgian Congo to the Democratic Republic of Congo—DRC. Yet these three scientists formed a common bond through their scientific observations, curiosity, and writings, and the fact that they knew that they were building on the shoulders of a long line of fine, dedicated geoscientists with a love for Africa and its challenges.

The work by these three outstanding geoscientists has left a solid foundation for a modern analysis of the geology of central Africa. We hope that this book will go some way towards expressing our appreciation and recognition for their dedicated work.

Preface

This book is a collection of scientific narratives about the origin and evolution of the Congo Basin and its surrounding regions.

The Congo Basin is a vast near circular area of nearly 3.7 km^2, some 300–400 m above sea level astride the Congo River; it is surrounded by elevated regions and flanked in the east by the African Rift System where, in places, elevations of Precambrian rocks and active volcanoes stand above 5,000 m; and in the west by the Atlantic Ocean into which the Congo River delivers vast volumes of detritus and dissolved loads (Fig. 1). The Congo River is the most powerful river in Africa: during the rainy seasons over 50,000 m^3 (near 2 million ft^3) of water per second flow into the Atlantic Ocean.[1] 'The force with which the Congo River empties into the Atlantic is so great that it changes the colour of the seawater for hundreds of kilometers around' (Van Reybrouck 2010, 2014). Below sea level, gravity-driven currents and suspended sand clouds have carved an immense submarine canyon across the

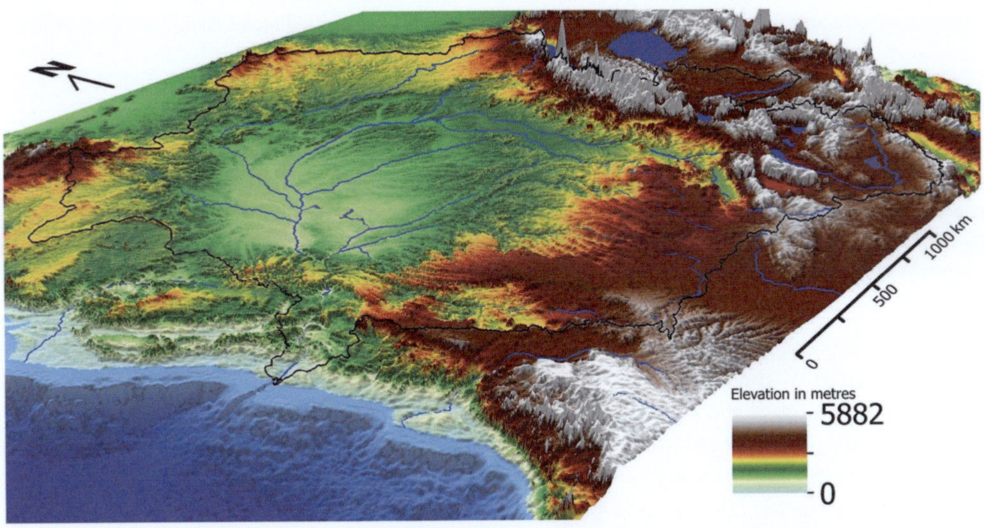

Fig. 1 Image of the Congo Basin and surrounding regions covered in this book. *Thin black outline marks* the edge of the Congo drainage system. Note that the perspective for the horizontal scale bar due to the 'depth' and orientation of the image (the image is more *diamond shape* than *square*) is correct only near the centre of the image (i.e. the km distance between the Congo River arc is within 10 % [100 km] accuracy). However, for the areas around the East African Lake District and near the Atlantic coast the accuracy is closer to within 20 %. Note also the vertical exaggeration (figure courtesy of Tyrel Flugel). See Chaps. 14 and 15 in this book for location names and greater information about the landscapes and river systems

[1] The Congo's discharge at its mouth ranges from 23,000 cubic metres per second (810,000 cu ft/s) to 75,000 m^3/s (2,600,000 cu ft/s). For a more detailed account of the Congo Basin drainage system, see Chap. 15 in this Book.

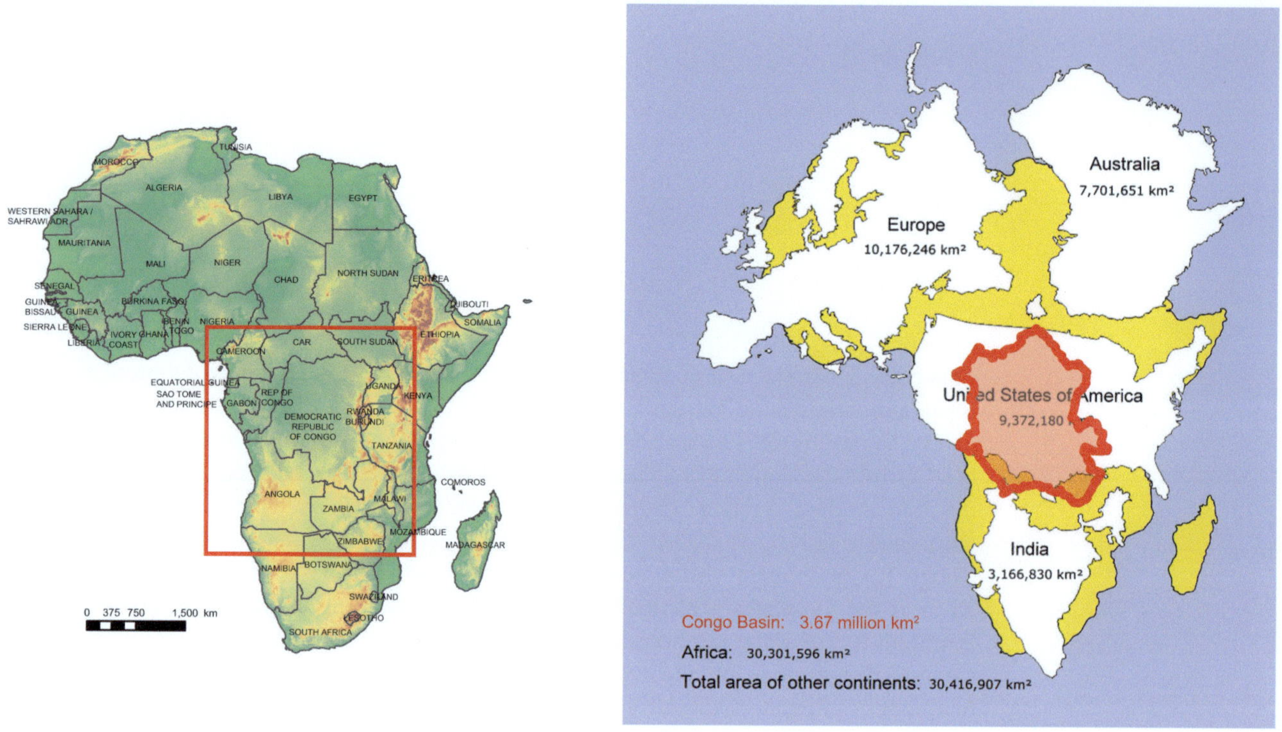

Fig. 2 (**a**) Map of Africa with country names. *Red box* shows the approximate sub-Saharan region of Central Africa that is the focus of the geophysics, geology, and geochronology of the crustal foundations of the Congo Basin (see Chaps. 1 and 2). (**b**) Relative size of Africa and the Congo Basin and, for direct comparisons to other selected nations, as plotted on an equal area projection

continental shelf through which these turbidites are transported into the oceanic abyss.

The Congo River is the second longest river in Africa, close to 5,000 km long. It emerges in the highland of the great Lake District of East Africa in the southern Hemisphere. From there, the journey to the sea snakes northward across the flatlands of the Congo Basin passing first through the savannas and swamps of Katanga, and then across the equator and through the endless equatorial forest in the northern hemisphere before it carves its way through the canyons and waterfalls of the rugged mountains of Bas-Congo to finally end up in the mangroves and the Congo Fan at the coast. It is constantly fed by alternating rainy seasons north and south of the equator. Van Reybrouck writes: ‘If the map of Congo looks like a balloon, Bas-Congo is the neck through which everything passes’.

Because it is sometimes hard to comprehend the immenseness of the Congo Basin, which occupies just under 12 % of the surface area of Africa (about 40 % of the USA; and about half of Australia or the EU), it is worthwhile to compare the size of these regions on an equal area projection (Fig. 2b). It is also true that many casual observers of Africa are challenged to name all its 50 nations, or even the 18 odd countries of central Sub-Sahara, whose livelihood is directly affected by the present drainage system of the Congo Basin, and by the underlying geology and mineral wealth all of which are explored in this book (Fig. 2a).

Port Elizabeth, South Africa
Rennes, France
Pretoria, South Africa
August, 2014

Maarten J. de Wit
François Guillocheau
Michiel C.J. de Wit

Background to a Punctuated History of Geoscience in the Congo

Quantum leaps in knowledge about geography, geology, and mineral resources of Central Africa, and in particular within the bounds of the Congo Basin, occurred across major transitions between six periods: pre-1885 (the 'Kongo'[2]); 1985–1908 (the Congo Free State); 1908–1960 (the Belgian Congo); 1960–1965 (the Republic of Congo); 1965–1997 (the Republic of Zaïre); and post-1997 (the Democratic Republic of Congo—DRC).

1. Pre-1885: The Kongo Kingdom

There is still very little documentation about the pre-imperial phase of indigenous knowledge of rocks and minerals of the Kongo. By the time of the first recorded contact with the Europeans, the Kingdom of Kongo (see footnote 2) was a developed state at the centre of an extensive trading in goods manufactured from natural geologic resources.

Substantial knowledge about the value of natural resources of the Kongo goes back to an extended period during which local wealth and industries were related to indigenous evolving technologies. From archaeology we know that hunter-gatherers from the Middle Stone Age, during the early to late Pleistocene, between ca. 230,000 and 23,000 years ago, made extensive use of iron, copper, and gold long before the Portuguese and Dutch trading phases. In excavations from this period and before there is also abundant evidence of sophisticated iron smelters, and a remarkable high density of percussion-flaked stone tools spread widely across this region associated with, for example, diamondiferous river terraces (Fig. 3; e.g. Clark 1971; Cahen 1978; Clark and Brown 2001; Chirikure 2010; Walker et al. 2014).

By the time Arabian traders reached Kongo from East Africa, gold and copper complemented ivory as sought-after commodities. Post-independence (1960-present) instabilities in the DRC have resulted in far too little emphasis on documenting and learning more about, and from, the archaeological archives of the Kongo. The sign of the times is that this is changing, undoubtedly leading to new, unimagined discoveries of our early human development (e.g. Cahen 1978; Chrétien 2000; Van Reybrouck 2010).

[2] The Kingdom of Kongo, between fourteenth and nineteenth centuries, was located in west central Africa, in what is now northern Angola, the Republic of the Congo, the western portion of the Democratic Republic of the Congo, and southernmost part of Gabon. At its greatest extent, it reached from the Atlantic Ocean in the west to the Kwango River in the east, and from the Congo River in the north to the Kwanza River in the south. Much of west Kongo became a major source of slaves for Portuguese traders and other European powers in the fifteenth to nineteenth centuries. Although not strictly the Congo Basin, we use it here for convenience in a general sense to refer to pre-imperial Central Africa west of the great lakes.

a b

Fig. 3 Stone tools from the 'Middle Stone Age' period of Lupemban Industries, dated between 23 and 230 kyr, and which are abundant across most of sub-Saharan Africa as far south as South Africa. The raw material from which the artefacts were made is a fine-grained quartzite-silcrete known as the 'grès polymorphe' of the Lower Kalahari Formation (see Chap. 10, this Book). (**a**) Hand axes recovered by divers from dredges used in diamond mining operations of GEM Diamonds in the Lubembe River near Nsumbula of the Kasai West Province, DRC (photo by MCJ de Wit). (**b**) A large variety of different hand tools from the diamondiferous Kwango River terraces, southwest DRC (photo by B Linol)

2. 1885–1908 (Congo Free State)

Phase two follows the Conference of Berlin, also known as the Congo Conference, in 1884–1885, during which European powers divided most of Central Africa between them. Its outcome, the General Act of the Berlin Conference, ushered in a period of heightened colonial activity by European powers, which eliminated or overrode most existing forms of African autonomy and self-governance. Portugal claimed the largest share of what remained of independent Kongo (now mostly Angola), whilst the onset of the Belgian invasion and takeover of the greater 'Kongo' region to the east, as private property of the Belgian King Leopold II, led to the formation of the Congo Free State.

The earliest written documentation of the geology and mineral riches in and around the Congo basin stems from this time; particularly after Jules Cornet (1865–1929) first rediscovered copper-rich ore deposits in Katanga in 1892. Cornet became the first European geoscientist whose dedicated work was translated into geologic reports, maps, and books of parts of the Congo Basin (J Cornet 1886, 1893–1894; René J Cornet 1944; Reybrouk 2010, 2014). On his return to Belgium, in 1897, he became Professor of Geology and Mineralogy at the School of Mines of the University of Mons, a post he held for 32 years, during which time he also became the mentor of the next generation of great Belgian geoscientists, and in particular Maurice Robert (1880–1958), whose first edition of 'Le Congo Physique' was dedicated to Jules Cornet (Robert 1923).

3. 1908–1960 (the Belgian Congo Colony)

During the early part of this colonial phase, Maurice Robert dominated the scientific research in geology and exploration for mineral resources throughout the Congo Basin. His prolific writings (e.g. Robert 1927; and a third edition of his 'Physical Geography of the Congo' was published 23 years later in 1946) covered a wide range of topics ranging from detailed descriptions of the early imperial expeditions around and into Africa to geology, mining and minerals, hydrogeology, climatology, landscapes and geomorphology, ecology and ecosystems, anthropology, as well as sociopolitical analysis of the Congo, and Katanga in particular. Whilst most of the literature was in French at that time, one excellent account on the geology of the Congo Basin was published in English by Veatch (1935). From the developments and knowledge gained during this pre-WWII period came a new generation of Belgian geologists and mining engineers. One of these, who at an early age had come under the mentorship of Maurice Robert, was Louis Cahen.

Lucien Simon Cahen became an active researcher during the emerging era when the world of stratigraphy was changing from being framed around crude estimate about the relative age and correlation of rocks based on fossils (or the lack thereof), to one built on precise radio-isotopic dating.

After a childhood in England, where his family relocated during WWI, Lucien, aged ten, returned to Belgium and 14 years later graduated from Free University of Brussels ('Université Libre de Bruxelles', ULB) as a mining engineer (Ingénieur Civil des Mines). In 1937, he was appointed as a geologist to the Geographical and Geological Survey of the Katanga Special Committee (Service géographique et géologique du Comité special du Katanga), a colonial Belgian institution with the mandate to develop mining activities in Katanga. He was engaged until 1941 in topographic and geologic mapping of Katanga. With the onset of WWII, Cahen and his young family were confined to the Belgian Congo, and between 1942 and 1946, Cahen served in the Belgian Artillery of Congo and the British Royal Air Forces (RAF) for which he was decorated with the Order of the British Empire. He also continued with his geologic mapping during this period in Bas-Congo.

In July 1946, Cahen took up a temporary position in the Royal Museum of Belgian Congo (now the Royal Museum of Central Africa, RMCA) on the estate of the Belgian King located in a suburb of Brussels, Tervuren. He was appointed on a permanent basis in 1947 and became its director in 1958. In 1970, after independence (see below), he was appointed also as General Director of the Institute of the Zaïre National Museums.

By the time Lucien Cahen returned to Belgium he had completed almost 8 years geologic field mapping across 30,000 km^2 of the Congo. He worked in close collaboration with the mining industry and the colonial industrial Belgian consortiums, forming a joint committee for the Geological and Mining Study of the Congolese Basin ('Syndicat pour l''Etude géologique et minière de la Cuvette congolaise') through which he successfully lobbied for funding of deep drill holes and linked geophysical surveys in the central Congo Basin. His only obligation was that he would archive all the data and cores at the Royal Museum in Tervuren.

Cahen's field observations formed the foundation of his vast number of contributions, especially his influential book 'Geology of Congo' published in 1954. This massive book of nearly 600 pages includes a synthesis of the Congo Precambrian basement ('soubassement') with the first radiometric dates from the Congo, and stratigraphic details of the overlying Phanerozoic sediments ('couverture'), landscaped into the spectacular 'chimneys' ('confins') of Katanga and Kasai, and the complex planation surfaces across the Congo Basin.

Cahen's focus turned increasingly to 'dating' of rocks and minerals, following a turning point in his professional life at the 1948 International Geological Congress in London where he learned first-hand about the pioneering work in geochronology applied to Africa by Arthur Holmes. Holmes, who had dated his first rocks from northern Mozambique to circa 1 billion years (Holmes 1918), and which were then the oldest known rocks in Africa, already had accumulated a substantial collection of radiometric dates across Africa (Holmes 1951). Lucien became hooked on the idea that geochronology was the key to unravelling the geologic complexities he had encountered in Katanga and Bas-Congo. From then on Lucien Cahen's works were dedicated to the furtherance of geochronology, and specifically the geologic history of the Congo and Africa, completing three more books, all with a focus on the geochronology of (central) Africa (Holmes and Cahen 1957; Cahen and Snelling 1966; Cahen et al. 1984).

Fig. 4 Lucien Cahen in the field with his mentor Maurice Robert, in 1938, at Lula near Mitwaba looking at pegmatites—M. Robert (*back*), A. Rollet (*left*), L. Cahen (on the *right side*—with a field notebook)—Photo by J van der Straeten (courtesy of Max Fernandez-Alonso from the archives Royal Museum of Central Africa, Tervuren, Belgium)

Cahen also worked in close association with J. Lepersonne (1909–1997; and head of the geological survey of Leopoldville [Kinshasa] from 1940 to 1945 [Liebaers 1977; Van den Audenzerde et al. 1982; Lepersonne 1983; Delhal 1988]). Later, when Lepersonne became the head of the Geology Section and Cahen the head of the Tervuren Musuem, they were famous for their early 'breakfast' discussions: '*Dans ce couple à la « Holmes-Watson », Cahen était le génie, Lepersonne la raison, ou encore le premier était l'architecte et l'autre l'entrepreneur*' (Dehal, J and Delmer A. 1999). Lepersonne is best known for his compilation of the 2nd edition of the 1:2 million 'Geological Map of Congo' (1974; the first edition was published by Cahen and Lepersonne in 1950, but this map contains distinct blank areas in west Congo without data) into which, for the first time, all the data from the Belgium Mining Consortia were incorporated.

Although not an experimentalist, Cahen's engineering background meant he could easily understand the physics of geochronology, and the important concepts of accuracy, precision, and error evaluation. In the early 1950s, he started a fruitful collaboration with E. Pioccotto of the Nuclear Physics Laboratory of the Free University of Brussels, founding one of the first isotope groups in continental Europe; and in 1951, in collaboration with T. Jacobs, a chemist at the Tervuren Museum, he published a short note providing a first U/Pb date, of 1 Ga, for the rocks of the Kibara Mountains. Later during the 1950s, he combined his geological experience and an encyclopaedic knowledge of Earth Sciences, together with his large, unique collections of African rock samples housed in Tervuren, to forge collaborations with a number of international scientists in Canada, Europe, the UK, and South Africa. The chemistry and the isotope measurements on the Congo rocks were determined, for example, in several different laboratories across the world: such as in Göttingen University (F.G. Houtermans), Toronto University, Canada (J.T. Wilson), Carnegie Institution, Washington (L.T. Aldrich), and the BPI of the University of the Witwatersrand, Johannesburg (B. Shonland). The first geochronology paper to emerge from these collaborations was published in 1954 (in French) in the Annales de la Société Géologique de Belgique entitled 'Geochronological results obtained on Congo minerals before May 1954'.

The impact of age determinations on the understanding of African Geology led the 'Association of African Colonial Geological Surveys' of the UK to assign Arthur Holmes and Lucien Cahen to evaluate and catalogue geological age determinations from African rocks, and specifically to increase the number of dated samples from central Africa. This collaborative work with Holmes resulted in the first synthesis of age determinations for Precambrian rock sequences from Africa, published separately in the UK in 1955, and in Belgium in 1956 (as 'Géochronologie africaine', a memoir by the Belgian Academy of Colonial Sciences), just before independence of the Congo (Holmes and Cahen 1955; Cahen and Holmes 1956).

4. 1960–1965 (the Republic of the Congo)

Arthur Holmes retired on the eve of independence of the Republic of Congo. He proposed that Norman Snelling of the Directorate of Colonial Geological Surveys (later the Directorate of Overseas Geological Surveys) take over the responsibilities of working with Lucien Cahen. By then Snelling was seconded to the new 'Geological Age and Isotope Research Laboratory—GAIR' at Oxford University, and Cahen was director of the Royal Museum of the Belgium Congo. At the end of this period (1965) Cahen also became a Distinguished Professor at the Free University of Brussels, where he was then planning to build his own geochronology laboratory.

Norman Snelling joined the UK Colonial Geological Surveys in 1959, and started collaboration with Lucien Cahen in 1960, after he had returned from a long geological collecting trip to the British (mainly former) African Colonies in East and southern Africa, as well as Nigeria.

Norman was nearly 20 years younger than Lucien. Trained in physics, mathematics, and with a geology degree from Manchester University, UK, he had become an experimental scientist, building up a long career in U/Pb, Rb/Sr, and K/Ar dating and design of mass spectrometers: first at the ANU in Canberra, Australia, with John Richards, and then in the K/Ar laboratory of the Geological Age Determination Group at the Geological Survey in Canada; and then from 1960 to 1995 by spells in isotope labs at Oxford and London (as Head of the Department of Isotope Geology of the British Geological Survey), and the Isotope Centre of the Indian Geological Survey, in Calcutta, India; and as a last 5-year stint, at the isotope laboratory of Complutense University, Madrid, Spain. In both of the latter laboratories he installed the first prototype K/Ar gas mass spectrometer system that he and Chris Rundle designed and requested a reputable company—V.G. Micromass—to build. Later, in Madrid, Norman and his wife, Carmen, installed a solid source Micromass VG Sector 54, and together they soon had both K/Ar and Rb/Sr dating there working continuously.

5. 1965–1997 (Republic of Zaïre)

Lucien Cahen and Norman Snelling (and their wives) developed a close long-lasting friendship and a brilliant scientific collaboration that ultimately led to their joint publications of two groundbreaking books on African Geoscience. In 1966, just after the start of era of the Republic of Zaïre, Lucien had established his own laboratory in Brussels where he could date his vast collections from the Congo: the Belgian Centre for Geochronology. In the same year their first book, 'Geochronology of Equatorial Africa' (roughly 18 degrees south to 10 degrees north), was published. By all measures a masterpiece ahead of its time. The book contains both technical information and the first real attempt at an African database management system: the first three chapters of this work outline the 'Methods of age determination', 'Interpretation of radiometric ages', and 'Terminology and Classification'. Chapters 4–16 then discuss the results from the region of Equatorial Africa, followed by a concluding Chap. 17. The book finishes with the first 'Register of Age Determinations in Africa', giving sample locations and full analytical data of over 550 age determinations—a data management system ahead of its time.

Sixteen years later, one month before Lucien's death in 1982, Lucien and Norman sealed their cooperative efforts on African geochronology by submitting to the publisher the results of their 'big project': 'The Geochronology and Evolution of Africa' (eventually published in 1984 by Oxford University Press, together with contributions and help from J. Delhal, J.R. Vail, M. Bonhomme, and D Ledent) remains to date the most valuable scientific book on African Geology. Shortly thereafter, Snelling (1983, 1985) paid tribute to his colleague and many of his other mentors and collaborators in a 'state-of-the-art' summary of geochronology and the geological record, and of his particular interest in quantifying the rates of geological processes.

By mid-stage of the then Republic of Zaïre, the radiometric database of Africa, neatly compiled and linked to a vast number of verifiable geological observations by Cahen and Snelling in 19 exhaustive chapters, had risen beyond a simple manageable database. Access to such an interrogative database for Africa has had to wait for another 30 years (e.g. Eglington et al. 2009; Perhsson et al. 2013; and http://www.earth-time.org), underscoring the mammoth task undertaken by Cahen and Snelling that was well ahead of their time.

In 1951, Lucien Cahen dated his first African rocks collected by him in Kibara, within the Congo Basin, at about 1,000 Ma (million years). Ironically, just over ten years later, in 1962, Norman Snelling dated his first Canadian rocks from the Grenville also at circa 1,000 Ma. This one billion year old age, which had also already been established in northern Mozambique by Holmes (1911), became of great interest to Henri Kampunzu, then a new rising star of African geology. By then, the origin and age of supercontinents emerged as a topic of global interest, and it just happens that the 'Kibaran' (as originally defined by Cahen and Snelling) and Grenvillian would turn out to be critical time periods for understanding the tectonic history of the supercontinent Rodinia, and the origin of Africa. Henri Kampunzu followed Cahen to become an expert on the Kibaran, and wrote extensively on the formation and break-up of this continent (e.g. Kampunzu 1998; Yoshida et al. 2003; Kampunzu et al. 1998). His major goal towards the end of his life was to produce a Geological Map of Rodinia as part of IGCP 440. As it turns out now, the 1 billion year old age related to the formation of Rodinia is prominent also in detrital zircons from all the sedimentary sequences of the Congo Basin (see Chaps. 7–11 and 13 in this book).

Ali Basira Henri Kampunzu, better known as Henri Kampunzu, was born in Bukavu in South Kivu in 1952. He was eight years old when the Republic of Congo gained independence from Belgium. He exemplifies the first post-colonial generation of leading 'Kongo' geologists to help lead Sub-Saharan Africa into a new era of Geoscience and Mining endeavours, especially after he joined Lubumbashi University in 1988.

Fig. 5 Henri Kampunzu in the field at unknown DRC location and date

The country changed its name to Republic of Zaïre by the time Henri was at high school. He attended the integrated Athénée de Bagira and specialised in biology and chemistry. Bukavu is situated on the southern edge of Lake Kivu along part of the western branch of the East African Rift System. Here, life amongst active volcanism and seismic activity was part of the norm. Along the northern edge of Lake Kivu, the Nyiragongo volcano is active, and methane gas is exploited from the lake. Quarries at Bagira, along the southwestern edge of the lake, display spectacular examples of columnar jointed basalts. Bukavu is also situated on the northeastern extremity of the Twangiza-Namoya Gold Belt. It only seems natural that Henri found passion for geology and minerals from his immediate surroundings in the African Lake District, and morphed into a leading petrologist.

Henri joined the 'Ecole des Mines' (ENM) in 1970, and then the University of Lubumbashi where he completed all his degrees within a decade, and thereafter held joint professorial appointment in Lubumbashi and the University of Aix-Marseilles in France.

His initial scientific focus was on the Neoproterozoic anorogenic alkaline complexes in the east of the DRC, along the border with Burundi, and active volcanism related to the western branch of the East African Rift System, particularly those associated with the South Kivu, Virunga, and Toro-Ankole provinces. This culminated in 1991 in the publication of 'Magmatism in Extensional Structural settings—the Phanerozoic African Plate' that he co-edited.

Whilst in Lubumbashi, he initiated a renewed drive and interest in the ore deposits of the Central African Copper Belt (e.g. Kampunzu and Lubala 1994), and in unravelling the geologic history of the complex convergence zone of the Precambrian fold belts of Katanga and surroundings, highlighted by Cahen and Snelling as unresolved tectonic enigmas. Understanding the connections between the Palaeoproterozoic Ubendian, Mesoproterozoic Kibaran, and Neoproterozoic Pan-African belts, and their relations to older cratonic blocks such as the Buenguewlu (Bangweulu) terrain, was central to Henri's research into resolving the origin and demise of the supercontinents Rodinia and Gondwana.

Kampunzu had great visions for the development of Earth Sciences in post-colonial Africa within a framework of genuine democratic principles, as illustrated by the fact that he was President of the Trade Union at the University of Lubumbashi between 1992 and 1996. But, at that stage, while he was still in the process of setting up an analytical laboratory, he left Zaïre during political hostilities.

6. 1997-Present (the Democratic Republic of Congo—DRC)

In 1996, towards the end of the era of the Republic of Zaïre, Henri Kampunzu moved to Gaborone as Professor at the University of Botswana. By then, Kampunzu had gained international recognition for his widely varied work in Katanga and sub-Saharan Africa (e.g. Cailteux et al. 2005; Ayele 2007).

Henri took on leadership roles in major large-scale IGCP projects focused on cross-border work with colleagues throughout sub-Saharan Africa and beyond (e.g. IGCP-227—Magmatism and evolution of extensional regions of the African Plate; IGCP 302—Proterozoic Belts in Central and Southwest Africa; IGCP-440—Rodinia Assembly and breakup; IGCP-450—Proterozoic Sediment-hosted Base Metal Deposits of Western Gondwana; IGCP-470—Precambrian basement of central Africa; as well as the IUGS-led 1:10 million scale map '*Geology and major ore deposits map of Africa*'; and in the USA-southern Africa Kaapvaal Craton Research Project). Henri became 'Mr Africa Geology' through time-consuming high-level administrative functions in the Geological Society of Africa, UNESCO, IPL, and ICESA (the International Commission for Earth Sciences in Africa).

These activities during the newly emerging DRC reflect Henri's deep commitment to ensuring the well-being of a new generation of responsible and well-trained African geoscientists (he supervised over 60 thesis-based graduate students at the Universities of Lubumbashi, Marseilles, and Botswana). Despite the negative and draining socio-economic conditions throughout the Congo, he devoted a lot of his time formulating research projects for Africa at different universities all over the world, and raising funds for promising young Congolese geologists in order to progress the geological knowledge of his country of birth.

Henri had an enormous impact on African earth sciences and particularly in the areas of central Africa to which he was most attached. He was ahead of his time as an African leader promoting Africa-grounded science; and he would surely have achieved his dreams to see the emergence of radiometric isotope and other laboratories across central Africa if he had lived on; and given his drive, they would have by now. Whilst Henri Kampunzu's sudden death in 2004, at the age of 52 has been an immense loss for the African geoscience community, he left a legacy in the great number of his students who have now stepped up to the plate.

Brief Overview of Early Documentation of the Geology of the Congo Basin

The first modern synthesis on the 'Geology of Congo' was published in 1954 by Cahen. This book provides the first comprehensive synthesis of the Precambrian basement ('soubassement'), including for the first time early radiometric data, and the stratigraphic details of the overlying Palaeozoic to Recent sediments ('couverture') of the Congo Basin and its erosional planation surfaces.

Following Maurice Robert (1947), Cahen identified three major episodes of deformation within the basement: (1) around 2,650 Ma, which he named the Kibalian ('plissement kibalien'); (2) around 1,200 and 1,400 Ma that he named the Kibaro-Urundian, at the end of the Kibaro-Urundian cycle; and (3) around 630 Ma, which he named the Kundelunguian, marking the end of the Katanguian cycle. A fourth episode, which he only recorded in NW and W Congo, remained undated.

Amazingly, 60 years later, a review of the dates of Precambrian basement in and around the Congo Basin generally corroborates Cahen's findings. The four major periods of deformation are now well recognised (even if more accurately dated): between 500 and 800 Ma, as the Pan African deformation (≈Kundelunguian of Cahen); between 1 and 1.4 Ga, as the Kibaran (≈ Kibaro-Urundian of Cahen); between 1.9 and 2.3 Ga as the Eburnian (≈undated deformation located in west and northwest Congo by Cahen, but now also well recognised in east Congo and Angola); and between 2 and 2.7 Ga, as unnamed orogenic events recorded in the Central Angola Mobile Belt and the West Central Africa Mobile Belt (≈ Kibalian of Cahen). Chapters 1 and 2 of this book provide more details.

Cahen also followed Robert in describing the occurrence of the late Precambrian glacial deposits associated with extensive carbonates of Bas-Congo and Katanga (tillites of the 'Grand et Petit Conglomérats'). Nearly 70 years ago, Robert (1947) had recognised these deposits as glacial, and published beautiful photographs of striated pebbles (Plate II) and stromatolites in the associated dolomitic limestones (Plate V). These sequences are today believed to represent the Sturtian ('Grand conglomérat') and Marinoan ('Petit conglomérat') periods of the Cryogenian (e.g. the Neoproterozoic 'Snow Ball Earth' times). More details are provided in Chaps. 2–6 of this Book.

A major breakthrough in understanding of the Phanerozoic geology of the Congo Basin followed the first geophysics surveys linked to two deep fully cored drill holes down to ca. 2,000 m in the centre of the basin, known as the Samba (2,038 m) and Dekese wells (1,836 m) drilled in 1959. During the 1950s, the 'Syndicat pour l'étude géologique et minière de la Cuvette congolaise' also conducted refraction and reflection seismic, gravimetric, and magnetic field surveys across the central parts of the Congo Basin. During the seismic surveys, 117 refraction stations were employed in the basin interior, and 7 reflection lines were recorded. In addition, about 6,000 gravity and magnetic measurements were made with an average station spacing of 5 km along rivers and dirt roads.

These geophysical and well data were the first available data with which to image the extent of the Congo Basin and its subsurface geometry; and major refractors provided the first regional overview of its subsurface structure. The CB appeared deeper than previously thought, with the possible existence of several thousand metres of sedimentary rocks beneath its upper Palaeozoic to Mesozoic sequences (e.g. Evrard 1960). Subsequently, during the 1970s, renewed petroleum exploration led to the acquisition of additional geophysical data and two deeper cored wells (Gilson-1, down to 4,563 m; and Mbandaka-1, to a depth of 4,350 m).

To date, these two sets of geophysical and drill core data remain the only published subsurface information available to study the structure and stratigraphy of the central Congo Basin. A very meagre basis from which to decipher the origin of this vast basin, as reflected in new analyses described in several chapters of this book.

Cahen also synthesised the late Palaeozoic sequences (Carboniferous to Permian), focusing in particular on the importance of glacial deposits, based on a decade of detailed work on these sequences by Nicolas Boutakoff (1948; and here further explored in Chap. 7, this Book). Robert and co-workers had also described tillites in the Lualaba-Lubilash Sequence of Katanga that had been mapped and named by J. Cornet more than 50 years earlier (Cornet 1893/1894). Veatch (1935) and Robert (1947) correlated these and other 'Karoo-like' sequences to the classic Gondwana sequences in India, and in particular the Karoo of South Africa, which he described in great detail, based on published observations by du Toit (1926) and other South African geologists. Robert (1947) was much influenced by Wegener, Argand, and du Toit (e.g. pages 85–90, in French), and early on recognised the importance of the Lualaba-Lubilash sequences for Gondwana correlations and their contribution to the raging debate about continental drift which had been firmly rooted by then in the stratigraphic correlations between Africa, South America, and India (e.g. du Toit 1937). Clearly, the Congo Basin sequences also have counterparts in Paraná Basin of Brazil, but this was not further elaborated on, perhaps because Robert was unable to pick sides between the fixists or mobilists. Chapter 13 of this book provides a new detailed correlation scheme between the Congo and Paraná Basins.

The Mesozoic chapter is arguably the weakest part of the Cahen's book. This reflects the general lack of knowledge about the Mesozoic due to its poor outcrop in the Congo Basin. Although the Jurassic was palaeontologically characterised for the first time in his book, this lack of data convinced Cahen of the importance of drilling in the centre of the Basin. Five years later, under his supervision, the Samba and Dekese cores were completed and logged (Cahen et al. 1959, 1960). Still, the Mesozoic sequences remain poorly dated even today, as elaborated on here in Chaps. 8 and 9.

The qualitative descriptions of the Cenozoic ('Tertiaire') sequences, including sediments in the Albert Rift, the alluvial terraces of the Congo and its tributaries, and the 'Quaternary' geology, are still largely accurate, but without reliable age constraints, including careful geomorphologic descriptive work on iconic stepped planation surfaces, volcanism, and structural features of the Virunga—Kivu rifts. Very few studies have been published subsequently on these sequences, with the exception of work in the Albert Rift. Much of that work unfortunately remains proprietary following the relatively recent exploration activities for Oil and Gas in Lake Albert.

Elsewhere, work on the uppermost sequences of the Congo Basin that can be traced far south to the Kalahari region of southern Africa has yielded new data to substantiate Cahen's work. This upper Cretaceous-Cenozoic Kalahari Group comprises largely of unconsolidated red sediments ('Sables ocres') covered by a variable thickness of hard caps of calcrete and silcrete, locally in the Congo Basin known as the 'Grès polymorphe' (the source for most of the materials used in the prehistoric stone industries that are widespread across the region). Chapters 10 and 14 elaborate on these enigmatic deposits.

Cahen's 1954 book represents the epitemy of the high-quality presentation of scientific observations and data presentation ('exposé des faits'), in the form of hand-drawn illustrations, maps, and photographs, linked to meticulously organised descriptions of each geological unit in each area, and clearly separated from interpretation and discussion ('interprétation et discussion'). It is a 'facsimile' of the rigorous and objective work of leading field geologists active in the early-mid twentieth century. It can be built on with great confidence.

Brief Overview of the Mineral Wealth of the Congo Basin (DRC)

By all accounts, the Congo is a mineral 'superpower'. Up to independence in 1960, the mineral wealth of the greater Congo Basin (here limited to the DRC) accounted for approximately two-thirds of its exports. At today's (2014) prices there is an estimated US$24 trillion worth of untapped mineral wealth in the ground (Global Business Report 2013; http://www.gbreports.com).

In 2012, the Democratic Republic of the Congo (DRC) held the largest cobalt reserves in the world, and the second largest global reserves of copper (70 million tonnes; all quantities referred to below are in metric tonnes); it controlled 55 % of global cobalt production, 25 % of the world's industrial diamonds, 14 % of its tantalum, and 3 % of its copper and tin output. Below we summarise some of the most important commodities that contribute to these statistics (Goossens 2007, 2009, 2013; AAP 2013, and other sources as shown).

Copper and Cobalt (Cu/Co)

It is estimated that since exploitation began until 2003 Katanga (Fig. 6) has produced 18 million tonnes of copper.

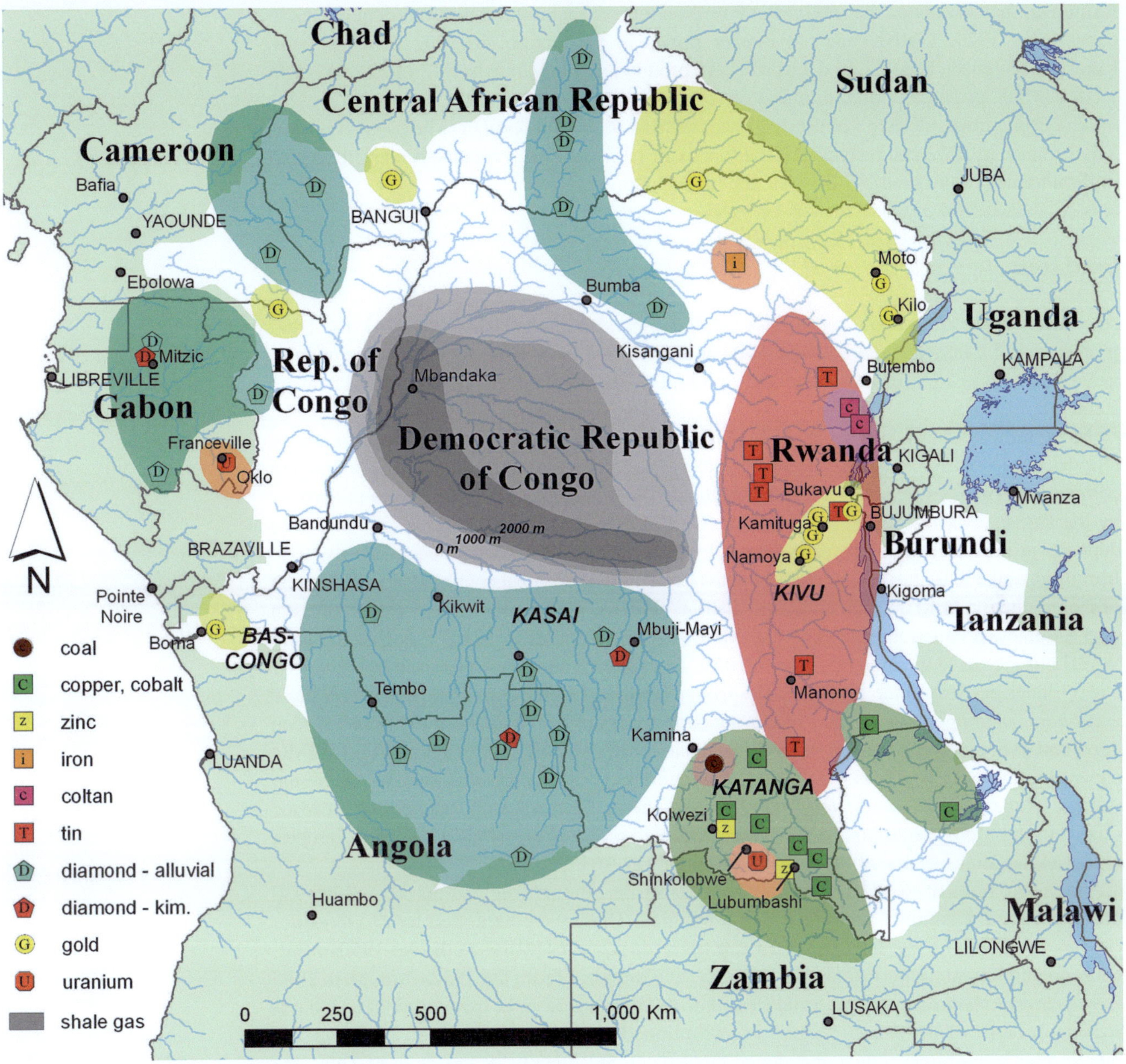

Fig. 6 Mineral Resources map of Congo Basin, showing major mineral provinces and mines of central Africa in and around the Congo Basin (compiled from numerous sources; for further details see Chap. 19 this Book)

Rich copper and cobalt deposits were discovered in Katanga by Jules Cornet nearly 125 years ago. Thereafter, *Union Minière du Haut Katanga* (UMHK), created in 1906, was granted exclusive rights to exploit the Katanga copper belt concession. Following its nationalisation, in 1967, the state-owned company, *Gécamines*, was given the monopoly to develop and mine the mineral wealth of the Katanga copper belt. The situation remained the same until 2002 when the transition government liberalised the mining industry and a new mining code was adopted by its General Assembly. Several foreign companies, often in partnership with *Gécamines*, started mining new ore deposits in addition to the existing *Gécamines* mines whose production by then had been reduced close to zero.

During the initial period, UMHK, extending the exploration work of Tanganyika Concessions Ltd (TCL), evaluated the Kambove and Etoile copper-cobalt deposits. At

Kambove, they estimated the reserves at 9 million tonnes, including 3 million tonnes at 12 % copper. By 1929, UMHK had six copper mines in production: Ruashi, Likasi-Shituru, Kambove, Kipushi, Luishia, and Lukuni, and production levels reached 136,922 tonnes. In 1928, UMHK produced 112,456 tonnes of copper metal.

The Etoile mine, close to Lubumbashi, was the first to produce copper. The production was 2,492 tonnes in 1912 and reached 27,462 tonnes in 1917. A 2008 assessment confirmed 13.5 million tonnes of ore graded 3.3 % Cu and 0.55 % Co.

Gold (Au)

According to the *Bureau de Recherches Géologiques et Minières* (BRGM 1974), the DRC produced 470 tonnes of gold between 1904 and 1972, and from 1986 until 2004 the country generated about 150 tonnes according to Gold Field Mineral Services (now Thomson Reuters).

Most of the gold has come from the Ituri region in the northeast of the Congo Basin, where alluvial gold deposits were discovered in 1895. Exploitation started in 1926 by a Belgian company, *Société des Mines d'Or de Kilo-Moto*. In 1964, the mining activities were taken over by the *Office des Mines d'Or de Kilo-Moto* (OKIMO), a state mining enterprise. Other gold deposits in the same region were exploited by *Forminière* until independence. Up to 1979, Ituri produced 330 tonnes of gold, half of which from alluvial deposits, with 90 % (300 tonnes) of the production from the Kilo and Moto districts (Fig. 6). With an annual average production of 5.6 tonnes, production from Kilo and Moto reached 8 tonnes in 1940. It declined during WWII but increased again to 8 tonnes in 1955. Thereafter production decreased to between 1 and 3 tonnes per year in the 1980s. Grades from placers and from primary deposits from these areas were running at 10–15 g/t. Total production from the Kilo and Moto districts between 1961 and 1985 was 65.9 tonnes. In recent years activities have picked up again after OKIMO signed joint venture contracts with several mining international companies such as *Anglogold Ashanti*, *Randgold*, *Mwana Africa*, and *Moto Goldmines*. Their recently opened Kibali Mine (2012) in the Moto greenstone belt, with an estimated reserve base of ca. 12 Million Ounces, and growing, ranks as one of the largest gold mines in Africa.

In the Provinces of south Kivu and northern Maniéma, alluvial gold was discovered in 1930 at Namoya and Kamituga, and in 1938 at Twangiza. These were exploited by *Minières des Grands Lacs* (MGL) and later by *Société Minière du Kivu* (SOMINKI). In the 1950s MGL subsequently found primary gold mineralisation at Twangiza. From the 1930s through to 1985 the Twangiza-Namoya gold belt produced almost 75 tonnes of gold from alluvial and hard-rock sources. In 1976, *Charter Consolidated* continued with the assessment of Twangiza, but due to other commitments then abandoned the project. SOMINKI (28.1 % state of Zaïre and 71.9 % private shareholders) restarted the exploitation until the end of the 1980s. In 1996, *Banro Corporation* signed a convention to take over the operations. After a troubled start Twangiza (100,000 ounces per annum) and Namoya are now in production, and Kamituga and Lugushwa are being evaluated.

Katanga was a relatively small gold producer. Between 1904 and 1909, for example, the Ruwe mine had produced 734 kg when mining ceased; and by 1939 Katanga produced some 777 kg.

Diamonds

It is estimated that by 2012, some 1,213 million carats had been produced from the Congo basin, 24.8 % of all diamonds produced globally, with a value of over US$ 50 billion.

The first diamond to be found in Congo Basin was in 1903 by *Tanganyika Concessions Ltd* (TCL). This led to the discovery by TCL of several barren kimberlites on the Kundelungu plateau in southeast DRC. However, the first diamond linked to economic deposits was only found in 1907, by Belgium Company *Forminière*. This developed into the discovery of the Tshikapa alluvial diamond fields that started producing in 1912. These alluvial deposits have produced in excess of 100 million carats. Diamonds around Mbuji Mayi in east Kasai (Fig. 6) were found in 1918 by geologist George Young (*Forminière*), and mining there started in 1920. In 1948 kimberlite was first described from here and it was realised then that most of the overlying 'alluvial' deposits were in fact highly weathered volcaniclastic kimberlitic materials.

Although the bulk of the DRC diamonds have come from the Mbuji Mayi deposits, these are of inferior quality and based on present market conditions are valued at between 15 and 20 US$ per carat. The best quality diamonds are found in the northern DRC and in the Kasai West Provinces around Tshikapa, which are valued at between 200 and 300 US$ per carat.

Production of diamonds in the formal sector peaked at some 20 million carats per year in the 1960s, but this has steadily declined over the years. For more detailed information, see Chaps. 16 and 17 in this Book.

Uranium

Uranium was discovered in Katanga in 1915. This Shinkolobwe uranium deposit (Fig. 6), close to Likasi, was first mined for its radium. In 1930, the Olen plant sold 30 g of radium produced from the Shinkolobwe uranium ore. In 1937, UMHK was the principal radium producer in the world and prior to World War II, uranium ore was stockpiled. These Katangan uranium ores were sold to the USA during the war and used for the construction of the first atomic bomb. In return, the Belgians were richly rewarded and granted access to nuclear technology, resulting in 1952 in a research centre in Belgium, and Africa's first nuclear reactor, in Kinshasa (Reybrouck 2010, 2014). The Shinkolobwe mine was officially closed in the early 1960s after the mine flooded. However there have been reports that uranium from the mine has been extracted by artisanal mining and sold since 1965.

Similarly of historic interests are the Uranium-rich deposits of Oklo, flanking the Congo Basin in the west, near Franceville in Gabon (Fig. 6). It was discovered by French scientists in 1972 because of unusually low ^{235}U concentrations (<0.6 % rather than 0.72 %). This is a significant difference and the only way to explain it was that the ores had reached critical conditions around 1.7 billion years ago, forming at least 17 small natural nuclear fission reactors. Most of the fission products were retained within the Okla reactor zones for 1,700 million years, and consequently became research analogues for modern radioactive waste containment technology in France.

Coal

At Luena, southeast of Kamina (Fig. 6) and about 120 km north of Tenke, coal was produced at a rate of 110,000 tonnes per year. Washed Luena coal has an ash content of about 15 % and a heating value of 5,900 kcal/kg. It has a relatively high volatile content of 35 % and is not suitable for coking. The Luena mine started in 1920 as an open pit. Four seams, varying in thickness from 0.5 to 5 m and interbedded in shales, are being mined. Reserves are about 20 million tonnes.

Other coal deposits with similar characteristics, but with much larger reserves, are known to exist in Lukuga in northern Katanga, near Lake Tanganyika. An underground mine was operated there in the 1920s by *Compagnie Géologique et Minière des Ingénieurs et Industriels Belges* (Géomines). The coal was used for electricity generation, railroads, and cement plants.

Tin (S)

Tin ores, associated with tungsten and columbium-tantalum ores, occur in the Kivu and Katanga Provinces in eastern DRC (Fig. 6). The minerals occur in alluvial deposits as well as in primary pegmatite rocks of Proterozoic age. In 1905, the Busanga mine produced 5.5 tonnes of tin. The following year this increased to 25 tonnes annually. Production at Kasongo over the same period was less than 8 tonnes per year. Although the Busanga mine was closed in 1909, it was reopened by UMHK after World War I, and by 1929, the production reached 551 tonnes of tin.

At the same time, the Géomines started the exploration for new tin deposits, which resulted in a modest exploitation of the rich tin deposits of Kilondja and Manono during World War I. The ore grades range from about 0.2–0.5 kg/m^3 of cassiterite (oxide of tin) in alluvial deposits, and between 1.0 and 1.5 kg/m^3 in primary pegmatites. Historically for each ton of tin, an average of about 25 kg of tungsten, 15 kg of columbium, and 5 kg of tantalum have been recovered as by-products.

'Coltan' (columbite) a mineral rich in columbium and tantalum, associated with cassiterite of tin deposits, was first discovered in the Kivu region in 1910. Many of the deposits are small and rarely contain more than about 1,000 tonnes of tin. In 1928, both UMHK and *Géomines* produced 1,097 tonnes of cassiterite, and in 1937, Géomines started the exportation of coltan concentrates. In the 1990s technological advances in manufacturing capacitors for mobile telephones increased the global demand for the mineral, and Coltan was mined in Kivu, Maniema, and NE Congo. Although the extraction from gravel beds was primitive, thousands of diggers have provided the markets in North America and Europe with large tonnages of Coltan especially when the price went up tenfold between 1998 and 2003. The economically mineable reserves may be in the order of 50 million m^3 of ore, containing about 40,000–50,000 tonnes of cassiterite.

Manganese (Mn)

Manganese ore was produced at Kitenge, in western Katanga, along the Angola border at 35,000 tonnes per year by the late 1980s. The manganese was mined initially from a surface deposit with about 14 million tonnes of ore reserves containing about 25 % manganese. Production halted in 1993, and about 0.6 million tonnes of ore grading at 47–50 % manganese is stockpiled.

Other Metals (Ag, Zn, PGE)

In 1934, silver production started at the Kipushi Mine (formerly Prince Leopold mine) with 105 tonnes during the first year. In 1938, UMHK exported 8,000 tonnes of roasted zinc mineral concentrates from the same deposit, and in 2006 Kipushi had an estimated 16.9 million tonnes of ore with a grade averaging 16.7 % Zn (with germanium) and 2.2 % Cu. In 1937, the *Société de Recherche Minière du Sud-Katanga* (Sud-Kat) started producing lead from the Cu–Pb–Zn Kengere mine.

Platinum and palladium were also produced by *Union Minière*. In 1936, the production was 99 kg and 391 kg, respectively. Other precious metals were recovered in Belgium, through refining Katangan copper.

Pyrochlore

Pyrochlore is a columbium ore. It occurs in carbonatites at Lueshe and Bingo, near Goma, Kivu. Proven ore reserves at Lueshe are about 2 million tonnes of weathered ore grading 2.2 % Cb_2O_5. At Bingo ore grades are as high as 6 % Cb_2O_5. The ore would be surface mineable and would make Congo an important global producer.

Energy: Oil, Gas, and Hydropower

Whilst oil and gas reservoirs have not been detected within the Congo Basin, there is, and has been, significant interest, and speculations, about its hydrocarbon potential. By the 1920s, American petroleum companies, and the Companhia de Petroleo de Angola, had funded petroleum geologist A.C. Veatch (1878–1938; US Geological Survey 1906–1910) to explore for them the western Congo and Angola (1919–1930). His wide knowledge of the region was summarised in the first comprehensive book in English on the geology of the Congo Basin (Veatch 1935). Further serious exploration for petroleum did not take place in the Congo Basin until the 1950s.

Meaningful exploration work is hampered mostly because of logistic issues associated with the need for expansive and expensive geophysics surveys. Chapter 18 of this book reviews the oil potential.

Because of the recent global developments in shale gas, we have also provided some preliminary guidelines for the potential of unconventional gas related to black shales in the Karoo sequences within the Congo Basin (Fig. 6).

Opportunities for the Congo River and its tributaries to generate hydropower are enormous The Congo River has a large flow rate, and the entire Congo Basin accounts for 13 % percent of global hydropower potential with sufficient power for all of sub-Saharan Africa's electricity needs (Weetlogs.scilogs.be; 2010). Currently there are about 40 hydropower plants in the Congo Basin (for a more detailed account of the Congo River, see Chap. 15 of this Book). One of the world's largest falls—the Inga Falls, about 200 km southwest of Kinshasa, has been recognised for its vast potential for hydroelectricity for more than 50 years. The main focus has been to complete the building of a mega-scale dam here—the Inga Dam, which has the potential to be one of Africa's largest hydroelectric power stations. So far the Inga project has spanned five decades to try and tap this massive energy potential, which, if realised, could supply up to 40,000 MW of electricity to half of Africa's 1 billion people. However, past experience has not been encouraging.

Earlier stages of the project to dam the Congo River were completed in 1972 (Inga I) and 1982 (Inga II). The combined energy produced by these projects, from a total of 14 turbines, is 2,000 MW, but, due to poor maintenance, only 25–40 % of this capacity is harvested. Construction of Inga III is scheduled to begin in 2016 and is expected to be completed by 2020. The total price tag for Inga III has been estimated between $50 billion and $80 billion, but these estimates are considered optimistic underestimates (e.g. Leslie 2005, 2014). Whilst the World Bank has recently (March 2014) approved a $73 million grant in support of this giant dam project, some economic analysts warn that foreign investors will prevent the grid from benefiting the general public and perhaps 'leave poor Africans in the dark'.

Moving Beyond the Resource Curse of the 'Kongo'

Today 98 % of the resources from the Congo Basin are exported in their raw state (AAP 2013). Thus, whilst the Congo is the world's largest exporter of cobalt, this (and its other minerals) is mostly in the form of unprocessed ore—value is added elsewhere, by the smelting industry in

Fig. 7 Typical artesian diggings along a river in the Congo Basin, in search for diamonds

China and other importing countries. The export of these and most other materials in their raw state has not fundamentally changed for more than 100 years.

The mining industry also has been a major job creator in the Congo for a long time. For example, in 1939, in Katanga alone, there were 23,705 indigenous employees and 1,064 expatriates working in the mining industry. Today there are between 500,000 and two million people involved in artisanal (Fig. 7) and small mining operations (in which women make up half the workforce), and which account for a significant share of the production of gold, diamonds, coltan, and cobalt; in the cobalt sector alone it is estimated to reach 90,000–108,000 people with a production that represents up to 2.5 % of GDP.

Yet the income share of the poorest 10 % of the people of Congo Basin, who in theory own the mineral rights, stands at ca. 2 %, whilst that of the richest 10 % equals about 35 %. The DRC then, with a per capita income of about US$1,000, classifies as 'resource rich, income poor' (AAP 2013).

Sitting at the very top of the DRC mineral value chain are some of the world's biggest mining companies; and the world's largest commodity trading company owns majority stakes in integrated copper and cobalt mines. Between 2010 and 2012, the DRC lost at least US$1.36 billion in revenues from the under-pricing of mining assets that were sold to offshore companies.

The misappropriation of resource revenues that occurs through secretive deals and the operations of offshore companies is an unconscionable blight on the lives and hopes of the people living in the Congo Basin. Perhaps therein lays the sorry tale of Africa's endemic resource curse (AAP 2013).

'The economic history of Congo is one of improbably lucky breaks. But also of improbably great misery. As a rule, not a drop of the fabulous profits trickled down to the larger part of the population. That dichotomy, that is what we call tragedy' (Van Reybroucke 2010, 2014).

Despite all its riches then, Congo suffers from the worst cases of resource disease stoked by government favours. 'Even by the standards of very poor countries, Congo scores abysmally' (AAP 2013). In 2013, the UN named Congo the least-developed country in the world, in which unaccounted private bargains, through the sales of national mines and resources, far exceed the state's annual budget in health and education. In the UN's Human Development Index—a composite measure of wealth, life expectancy, and education that covers 187 countries, the DRC is in last place. 'How can (a country) be so rich in mineral wealth and yet so poor?' (AAP 2013; Economist 2013).

In short, the Congo Basin and surroundings are endowed with a number of important mineral deposits that remain grossly undervalued, and new forms of accountability, *inter alia* through more robust and transparent accountancy of these mineral riches, are surely needed. There can be no doubt that the Congo's vast mineral resources could transform social and economic development, provided the externality costs are properly inculcated.

Certainly, the Africa Progress Panel (AAP) is convinced that Africa can better manage its vast natural resource wealth to improve the lives of the region's people (AAP 2013). Daunting as the challenges may be, there is cause for guarded optimism that this can be achieved through structures such as the Extractive Industries Transparency Initiative (EITI) that has acted as a catalyst for reform, providing governments, civil society, and foreign investors with global benchmarks for good practice and reporting in the resources industries. The DCR is now registered as an EITI candidate.

The AAP lays out some other positive opportunities through rules-based and legally enforceable multilateral regime for the global mining industry. In 2010, the United States Congress passed the Dodd–Frank Wall Street Reform and Consumer Protection Act, requiring full disclosure of payments made across the resource extraction industries. The Act also has a conflict minerals provision for tin, tantalum, tungsten, and gold (3TG) to cover minerals exported from the DRC. Similar legislation has been passed recently in the European Union.

There are also some encouraging signs that social and environmental impacts of mining now receive far greater attention than they did a decade ago. Today, most governments, donor institutions, and companies have adopted internationally recognised environmental impact assessment (EIA) and social impact assessment (SIA) tools that identify potential problems in advance.

In recent years too, there have been moves towards more stringent mandatory standards for monitoring and compliance prompted by growing concern over the role of minerals such as gold, tin, and tantalite in financing gross violations of human rights by armed groups. In the Congo, armed local militia, linked in some cases to neighbouring countries, have used mineral revenues to finance their operations. The OECD has drawn up due diligence guidance for responsible supply chains of minerals from conflict areas. The Kimberley Process Certification Scheme (KPCS) is the best-known international voluntary initiative. Founded in 2002, the KPCS brings together government, industry, and civil society to ensure that conflict diamonds and stolen diamonds do not enter the diamond markets. Under the 2010 Lusaka Declaration, governments agreed to establish a similar regional certification mechanism and database to track minerals trade, while at the same time promoting the EITI.

We welcome the recent adoption by the African Union (AU) of an African Peer Review Mechanism as the main framework for monitoring natural resources in Africa. Geoscientists can play their part in these initiatives by providing more accurate open data about the Congo's known geology and potential mineral and energy deposits. One of the aims of this book is to use the data from existing deposits to make statistical predictions about the potential for undiscovered mineral wealth throughout the region (e.g. Chap. 19, this book). Whilst still at its infancy, such studies together with socio-policy corrections can catalyse greater returns for and to the people of the Congo, and force a much needed eradication of the extreme poverty and inequality of the Congo (as clearly recorded by the Gini-index or other relative inequality measurements) if the gap between rich and poor in this part of the world is to have any chance of fundamental change (e.g. AAP 2013, Piketty 2014; Ravallion 2014; Science 2014).

What Is in This Book? A Neo-Modern Geo-Framework of the Congo Basin

The geologic history of the Congo as described in this book provides a good example of the power of recycling old data with new ideas. In part this is facilitated by decades of ongoing political instability that make it hard to collect substantial fresh geologic and geophysical data.

Moreover, recent new data that have been collected under these circumstances often remain 'proprietary'; even 'open data' gathered during World Bank- or UN-funded projects and European based surveys are 'closed'. Access to such archives is restricted and available to general public only under special dispensation. Young African scientists with bright new ideas may not benefit. In a small way this book attempts to reveal that this is not helpful for sustainable progress.

Whilst there is a trickle of new, albeit disparate, observational data that have been collected by academic researchers over the last decades, and new analytical information from archived rock samples that has been reexamined, as summarised in many of the chapters in this Book (e.g. Chaps. 3–8), there is mostly too little robust data with which to make significant advances. For example, a number of new competing ideas about the evolution of the Congo Basin (e.g. compare Chaps. 1, 6, 11–13) make it clear that even a general tectonic model for the evolution of the Congo basin, or for its underlying Precambrian basement, does not yet exist.

But there are also other approaches that have created advances. For example, a sprinkle of new radiometric dates on basement rocks described in this book can often be combined with existing geological field and core data, and carefully selected new local observations on the ground, to yield significant new ideas about regional tectonics (e.g. Chaps. 2, 7, 8, 14; and, for example, Chaps. 16 and 17, which focus on diamonds and alluvial deposits). Sometimes new and very detailed local field observations, albeit far too few, provide inertia for a rethink at a regional scale (e.g. Chaps. 4, 5, 7, 9, 16). In some instances, completely new data can be obtained using remote information and restructured into new provocative models to redesign programmes beyond the Congo Basin, involving all of Sub-Sahara Africa (Chaps. 1, 10, 14, 15, 17, and 19).

This book then provides a smorgasbord of stimulants for new ideas and programmes that could be initiated as this part of Africa aims to become a region of new knowledge development and value adding to its natural resources. But this cannot be achieved without local 'Kongo' expertise and leadership in geosciences of the kind shown by Henri Kampunzu.

Therefore, this book further aims to provide a catalyst for collaboration with equal currency and the need to transform present lingering tendencies of scientific 'colonialism' as well as the continued rampant resource exploitation, into something of more lasting value for the people of the Congo Basin and surroundings, as recommended in recent documentation (AAP 2013).

The origin of the CB has been much debated over the last decade, in part because the lack of high-resolution seismic data has precluded defining the precise depth to major crustal and mantle boundaries. Here, in *Chapter* 1, **A. Raveloson**, **A. Nyblade**, **S. Fishwick**, **A. Mangongolo, and S. Master** describe and reanalyse new and existing teleseismic data, and present a new shear wave velocity model of the upper mantle for central and southern Africa. Their inversion of Rayleigh wave group velocity measurements reveals a region of fast upper mantle velocities in the 50–100 km depth interval beneath the northwestern, central, and southern portions of the basin, and slower upper mantle velocities beneath the northeastern part of the basin. This suggests that lithospheric structure beneath the Congo Basin is not uniform, as is commonly assumed, and explores the tectonic implications for the geologic evolution for Central Africa. The authors find that late Precambrian and Phanerozoic rifts may have formed between small Archean blocks that underlie the Basin, a significant deviation from previous subsidence models of the Congo Basin that assume a large single Archean Congo craton constitutes its basement. The hard evidence for such rifts is also dealt with in Chaps. 6 and 11.

Chapter 2 reviews the Precambrian basement flanking the Congo Basin, based on new and old field geology and geochronology. **MJ de Wit and B Linol** revisit the definition of cratons and shields, and redefine the shape and size of the Congo Craton and the Congo Shield, and present a new model for the complex Precambrian to Palaeozoic tectonic history of central Africa. The chapter also highlights the preservation of the world's oldest diamondiferous kimberlites, and records the long history of episodic diamondiferous kimberlite emplacement in the region for over 2.8 billion years. They also raise the intriguing possibility that the oldest rocks of Africa, at ca. 4.0 Ga, may be preserved along exposures to the northeast of the Congo Basin, close to the border between DRC and the CRA, and that an Eburnian (ca. 2.0 Ga) mobile belt and associated plateau of Himalayan-Tibet proportions flanks the Congo Craton along the southern margin of the Congo Basin.

The next eight chapters describe, from the oldest (Neoproterozoic) to the youngest (Cenozoic), the 'sedimentary cover' that defines the 'Congo Basin'.

In *Chapter* 3, **F. Delpomdor** and **A. Préat** summarise numerous sections of Neoproterozoic carbonates and siliciclastics with tillites exposed along margins of the Congo Basin. This is an attempt to improve the poor constraints on correlations between isolated basins across the DRC and neighbouring countries. They re-emphasise that Neoproterozoic is an important Era in the Earth history because it records a complete cycle of supercontinent break-up and reassembly from Rodinia to Gondwana over a span of nearly 450 million years during which dramatic climatic fluctuations occurred. The record of these climatic fluctuations, in the form of alternating sequences of tillites and cap carbonates with stromatolites, is extremely well preserved in these late Neoproterozoic basins of Central Africa. The authors demonstrate how all the well-described super-sequences of the various isolated basins in central African may be linked using these climatic markers and then correlated with better documented sequences elsewhere in the world. For example, based on only a few detailed C-isotopic trends from a few of the sequences, the Akwokwo Tillite, the Niari Tillite, Bondo Tillite, and Petit Conglomérat all correlate well with the Sturtian glaciation. These well-preserved, albeit poorly exposed, records could become standards (particularly if drilled) for studies of global climate variations in the deep past. Sedimentary rocks younger than the Marinoan glaciation are unfortunately poorly dated around the Congo Basin, but their carbonate successions, deposited between 635 Ma and 580 Ma and possibly younger, are strikingly similar in terms of palaeoenvironments, and future chemostratigraphy will likely improve the chronostratigraphic correlations across central Africa, provided the limitations of the chemical-forensic tools are appreciated.

To emphasise this latter point in an accompanying *Chap.* 4, **F. Delpomdor**, **F. Blanpied**, **C. Virgone, and A. Préat** describe the late Mesoproterozoic to middle Neoproterozoic carbonate succession (1155–800 Ma) of the Mbuji-Mayi Supergroup, a classic late Precambrian carbonate sequence, and an important type section for Central Africa. Synthesis of microfacies analysis, sequence stratigraphy, and Fischer plots coupled with C- and O-isotopes reveal the relative importance of circulation of marine water masses and effects of meteoric or freshwater diagenesis on ramp-top carbonates during sea-level fluctuations of between 10 and 20 m. These data are the first to place a quantitative constraint on chemical stratigraphy of the late Precambrian carbonate deposits that likely covered much of the Congo Shield. To improve robust correlation of these carbonates by geochemistry, this chapter 'hammers home' that the history of early diagenesis of these carbonates must be firmly understood first.

In the following *Chap.* 5, **K. Kolo, K. Konhauser, J.-P. Prian, and A. Préat** do just that. They present detailed petrographic and SEM data and analyses of fresh Neoproterozoic stromatolitic carbonates from an old quarry in South Gabon along the western margin of the Congo Basin. Their results strongly suggest that Precambrian fungal colonisation of Neoproterozoic stromatolites also played a role in post-depositional carbonate diagenesis and alteration, and importantly, may provide early physical evidence for fungi in the rock record. This work reveals numerous per-mineralised fungal relicts, especially in the upper part of shallowing-upward evaporitic peritidal sequences. It provides a detailed evaluation of the fungal role in an attempt to better understand the mechanisms involved in the palaeo-weathering of a Neoproterozoic carbonate substrate.

In the next *Chap.* 6, **E. Kadima**, **D. Delvaux**, **M. Everaerts**, **S. Sebagenzi, and F. Lucazeau** focus on these lowermost Neoproterozoic to Cambrian succession beneath the centre of the Congo Basin and use this to better constrain the early evolution of the Congo

Basin. These authors draw on earlier work of Kadima and reinterpret the old drill core and seismic data, especially from the two fully cored, deep stratigraphic wells Gilson-1: 4,563 m and Mbandaka-1: 4,350 m) that appear to terminate in carbonates that contain intriguing evaporitic-like minerals at the base of the Congo Basin. With this information, they then correlate the successions with the carbonate-siliciclastic sequences along the peripheral margins of the Congo Basins (mentioned in the earlier chapters). The main findings of this work are the recognition of possible major Neoproterozoic (mechanical) rifts that underlie the overlying, but poorly dated, Phanerozoic cover, which they interpret as thermal subsidence sequences. This is an important contribution towards unravelling the origin of the Congo basin.

In *Chaps.* 7 *and* 8, **B. Linol and his co-workers** describe in detail the Phanerozoic sequences that overlie the above-mentioned carbonates and overlying 'Red Beds' of the Precambrian-Cambrian transition. These chapters are based on detailed field work by Linol in south west Congo basin along the Kwango River, and his reexamination of core from the two deep drill holes through the central of the Congo Basin, archived in the RMCA in Tervuren Belgium. The lithostratigraphic results are integrated with reinterpreted seismic (and other geophysical data collected in the 1980s), as well as, for the first time, with a systematic U/Pb analyses of detrital zircons throughout the cores. The chapter presents a new stratigraphic correlation across the basin that recognised three major regional unconformities in these terrestrial sequences and one possible marine incursion. The work focuses in particular on the enigmatic peak of 1,000 Ma zircons that occur throughout the Phanerozoic sequences, and highlights the desert-like conditions during deposition of the Mesozoic sequences (red beds) whose source rocks lie predominantly to the north of the Congo Basin, as determined from new and extensive palaeo-current data as well as the detrital zircon ages. Only the glacial deposits of the underlying Karoo-like sequences have a strong easterly source component. The Red Beds below this Karoo sequence are also ubiquitous across central Africa and those in the Congo basin best correlate with upper Neoproterozoic to lower Cambrian sequences elsewhere, such as in Namibia and South Africa.

EM Roberts, HA Jelsma, and TA Hegna in *Chap.* 9 describe in great detail the Mesozoic sedimentary sequences in the Kasai region, based on their lengthy field work, mapping, and drill core used during diamond exploration projects. Like the data from the Kwango River by Linol in the previous chapters, this is new and meticulously documented data. It provides a new local benchmark for terrestrial Mesozoic deposits of central Africa, and is relatively easily correlated with the data described by Kadima et al. and Linol et al. in Chaps. 6 and 8, respectively.

In the final chapter of this section of the book, *Chap.* 10, **B. Linol et al.** describe a model of dramatic collapse at around 34 Ma, of the end-Mesozoic to Cenozoic African surface comprising of a thick sequence of calcrete and silcrete overlying the near-unconsolidated Cretaceous red aeolian sands. This model, they argue, provides a link between rapid onshore erosion and a sharp increase in offshore sediment accumulation as determined from high-resolution seismic stratigraphy. In turn, this collapse model can account also for rapid change in global seawater chemistry (and in particular its palaeo-Sr-isotope record). Hitherto such changes have been contributed solely to the uplift and erosion of the Himalayas. This sudden, accelerated erosion of more than half a million km^3 of red beds across the Congo region provides a complementary, perhaps competing, model as a driver for global changes in the Oligocene, which deserves further investigations.

The reader should, however, carefully compare and contrast Chap. 10 with Chap. 14. These two chapters present radically different interpretations about the first-order topographic evolution of the Congo Basin in the mid Cenozoic. In this Chap. 10, Linol et al. link the high erosion rates to collapse of an end Cretaceous high plateau (1,500 m). In Chap. 14, Guillocheau et al. argue that this increase in erosion at that time is due to rapid uplift of the surrounding regions from a flat domain near sea level.

The next section of the book deals with large-scale, regional correlations and provides two very different interpretations of the origin of the Congo basin: one, *Chap.* 12, describes a simple long-term thermal history of the region based on geophysical data and stratigraphic vitrinite analyses; the other (in *Chaps.* 11 *and* 13) describes a complex subsidence history derived from the stratigraphic details described in *Chaps.* 7 *and* 8.

In *Chap.* 12, **F. Lucazeau, J. Armitage, and E. Kadima** provide evidence to support a model of near constant thermal relaxation of 200–250 km thick cold lithosphere after initial mechanical rifting phase in the late Precambrian (700–635 Ma), with a mild extension factor ($\beta = 1.4$). From temperatures, lithology and porosity recorded in the two deep wells in the central Congo Basin, the authors estimate the present-day surface heat flow to be 40 ± 5 mWm^2, in good agreement with a single measurement in this area of 44 mWm^2. Because the magnitude of the initial crustal thinning in their model is small, the authors argue that thermal conditions throughout the Phanerozoic were probably not very different from the present day, yielding a slow and steady Phanerozoic subsidence history for the Congo Basin. They support this with information from kimberlites, stratigraphic variations of vitrinite reflectance, and their reconstructed subsidence history. Two short-term variations in subsidence to account for maturation of vitrinite, as well as angular unconformities recognised on seismic lines, are interpreted to be due to subtle sub-lithospheric mantle instabilities and related crustal fluid activities.

By contrast, in *Chap.* 11, **B. Linol, M.J. de Wit, F. Guillocheau, C. Robin, and O. Dauteuil** argue that the Congo Basin is a complex successor basin with a multiphase history of alternating episodic subsidence/deposition and uplift/erosion throughout the Phanerozoic. They use a revised sequence stratigraphy to calculate a new first-order model for its subsidence and uplift history. Because the sedimentary sequences of this basin are largely terrestrial, they first describe and then implement a new back-stripping method especially designed for terrestrial sequences. Their analyses reveal two main episodes of subsidence: rapid subsidence during the Carboniferous-Triassic [10–20 m/Ma], followed by slower subsidence during the Jurassic-Cretaceous [5–10 m/Ma], interrupted by several short-lived episodes of exhumation during the mid Jurassic (160–180 Ma); the early Cretaceous (120–140 Ma) and the Eocene (ca. 30–50 Ma). They relate these episodic uplifts to epeirogenic activity induced by far-field geodynamic effects: First, tectonic events that led to the formation of Pangea during the late Palaeozoic, and second due to its break-up, and the further demise of Gondwana, associated with successive Large Igneous Provinces (e.g. 'Karoo', 'Paraná-Etendeka', 'Ethiopian') during the opening of the Indian and South Atlantic Oceans that effectively shaped modern Africa.

In *Chap.* 13, **B. Linol, M.J. de Wit, E.J. Milani, F. Guillocheau, and C. Scherer** present a new correlation scheme between the stratigraphic successions of the Congo Basin and its sister basins in South America and southern Africa (e.g. the Paraná and Cape-Karoo Basins) prior to the opening of the South Atlantic. This immense Central West Gondwana Basin (CWGB) allows, for the first time, accurate analysis of the evolution of a mega-basin at a supercontinental scale. In this framework, the late Palaeozoic-early Mesozoic cycle of subsidence of the CWGB is linked to long wavelength flexure of the Gondwana lithosphere during tectonic processes that formed the Mauritanian-Variscan and Cape-de la Ventana Orogens along the northwestern and southern margins of the supercontinent around ca. 300 Ma and ca. 250 Ma, respectively. Thereafter, hot and arid sedimentation across a vast Gondwana desert during the Jurassic-Cretaceous resulted in widespread deposition of northerly derived aeolian dunes, episodically covered by ash deposits from southerly derived high-stratospheric volcanic clouds. The sequences were twice affected by eruption from Large Igneous Provinces during punctuated phases of Gondwana break-up (ca. 183 Ma and 132 Ma). This shared sedimentation and climatic history of the Congo, Paraná, and Cape-Karoo Basins was finally disrupted by the early-mid Cretaceous opening of the South Atlantic Ocean and the Kalahari epeirogeny.

From then on, the Congo Basin records a history of intermittent phases of lacustrine and fluvial deposition.

The next section of the book deals with the present- and the recent palaeo-drainage systems and landscapes in and around the Congo Basin, and then a reconstruction of its geomorphic history further back in time to the Late Cretaceous, which characterises much of central and southern Africa's landforms.

First, *in Chap.* 14, **F. Guillocheau, R. Chelalou, B. Linol, O. Dauteuil, C. Robin, F. Mvondo, Y. Callec, and J.P. Colin** describe and interpret the complex landscapes of the Congo Basin and flanking topographic highs, such as the Central African Atlantic, the Cameroon and the Ubangian Highlands, as well as the East African Dome and the Kalahari and Angolese Plateaus. These authors argue that the present-day geomorphology of the Congo Basin reflects a complex differential uplift between the Congo Basin and highlands of the surrounding areas. Dates for the onset of uplift of these regions are carefully determined, and are based on detailed stratigraphy and meticulous analysis of the shapes and tilts of a spectrum of planation surfaces across central Africa, ranging from large-scale etchplains and pediplains to small-scale pediments. Nine successive regional planation surfaces are defined and dated, based on the cross-cutting relationships of these planation surfaces with dated volcanic and plutonic rocks, and on comparison with a well-dated sequence of weathering profiles and planation surfaces of West Africa. Using slope changes of the pediments, the palaeo-topography is restored and a four-stage history reconstructed from the uppermost Cretaceous to present day. Thermal subsidence of the Congo Basin terminated during the mid-Eocene (ca. 45 Ma), inconsistent with dynamic topographic models that predict continued lithospheric subsidence of the Congo Basin.

Then, in *Chap.* 15, **T. Flügel, F. Eckardt, and W. Cotterill** describe the present-day drainage patterns of the Congo River System and their Neogene evolution. Whilst the Congo Basin contains one of the world's largest fluvial systems, little is known about the basin's geomorphic evolution during the late Cenozoic. The authors of this chapter show in great detail how the present drainage pattern is influenced by several controls, such as lithology and tectonics that dominate in some regions, with zones of overlap where controls interact. The evaluation of the drainage patterns in the context of the basin's long wavelength geomorphology reveals a relative chronology of subtle events, as highlighted through detailed, high-resolution analysis of knick points and subtle rapids. Geologic timings of key events are poorly known, but evidence based on the new discipline of geoecodynamics, and data obtained through genetics and molecular clocks, constrains the ages of major river emplacements. For example, the congruence of the divergence data for the bonobos-chimpanzees, gorillas, and guenon species supports the emplacement of a modern, crescent-shaped Congo River by 1.5 Ma. This combination of geologic, geomorphic, and phylogenetic data of fishes and mammals sheds new light on the Neogene evolution of the Congo River System, and the juxtaposition of differing drainage patterns reveals a complex multistage evolution of the Congo River.

The final section of this book deals with natural mineral and energy resources.

In *Chapter* 16, **M.C.J. de Wit and E. Thorose** describe a variety of exploration techniques that have been used to evaluate alluvial diamond deposits associated with river terraces, younger river flats, and the present-day river channel of the Lower Kwango River that forms the border between Angola and the DRC. The diamonds have been derived from primary sources in the upper Kwango/Cuango Basin in Angola and have been transported and deposited down river showing a classic decrease in stone size from the source area in Angola to the confluence with the Kasai River in the DRC. It further illustrates the importance of local geomorphological influences on the concentration of alluvial diamonds.

The origin, value, and distribution of diamonds in the Congo are of great interest to geologists, explorationists, and traders alike. Here, in *Chap.* 17, **M.C.J. de Wit and H. Jelsma** document the history of diamond exploration and discoveries in the Congo, and then summarise the most important diamond mines of the DRC. They also cover the distribution and increased knowledge of the existing kimberlite fields in the DRC; and they document here, for the first time, the most recent discovery of two new clusters of kimberlites near Kabinda in the Kasai-East Province.

In *Chap.* 18, **D. Delvaux and M. Fernandez-Alonso** evaluate the petroleum potential of the Congo Basin, probably the last continental-scale sedimentary basin that is still poorly known in terms of petroleum system and related hydrocarbon resources. Whilst some studies have speculated that there may be giant oil deposits in the Congo Basin, to date no evidence for the presence of oil reservoirs has been forthcoming, despite possible potential source-rock levels, petroleum systems, and possible plays designated as targets for exploration projects in the 1950s and 1970s. The authors of this chapter present results of their recent re-analyses of the geological archives and samples stored in the collections of the Royal Museum for Central Africa. Based on their new geochemical analyses, black shales from the Neoproterozoic and early Palaeozoic sequences appear to be barren. In contrast, the Permian black shales are gas prone, whilst younger Jurassic-Cretaceous source rocks are of excellent quality (with Total Organic Carbon up to 25 %), but too immature. The authors also describe oil seeps along the Lukenie River and Lake Inongo that have been previously linked to potential oil reservoirs, but these turn out to be anthropogenic. Thus, whilst there is still no clear indication presently of active petroleum systems in the Congo Basin, this is based on very limited data. New exploration projects are needed to identify areas where Mesozoic source rocks might have reached maturity. Thus, whether the Congo Basin may become an important oil (or shale gas) producer in the future remains unanswered.

In the final *Chap.* 19 of this book, **C. Thiart and M.J. de Wit** evaluate the mineral potential of the basement rocks beneath the Congo Basin, using the data of known ore deposits across central Africa. Applying spatial statistical analyses on this data, using a 'Weight of Evidence' approach and the concept of 'Metallogenic Fingerprints', they find that the Archean areas (ca. 1.1 million km^2) beneath the Congo Basin have a strong affinity for gold and chrome, (cobalt) nickel and platinum, and a moderate affinity for tin. Inferred deformed Neoproterozoic areas beneath the Congo Basin (0.2 million km^2) have a moderate affinity for uranium and rare-earth elements, and a strong affinity for base metals, whilst the Meso-Palaeoproterozoic areas contain a moderate affinity for tin, but these rocks occupy only a very small area (<1 %).

Whilst this evaluation of the mineral potential of central Africa inculcates a significant degree of uncertainty, and whether or not ore deposits can ever be extracted beneath the Congo Basin, even along its shallow margins, remains highly speculative, this chapter highlights the enormous mineral wealth of this region. We hope that this knowledge will encourage sustained, responsible, and above all ethical stewardship of these resources for the future.

References and Citations

APP – Africa Progress Panel (2013) Equity in extractives: Stewarding Africa's natural resources for all. Switzerland, 119 pp. http://www.africaprogresspanel.org

Ayele A (2007) Obituary – In memoriam: Henri Kampunzu (1952–2004). J Afr Earth Sci 48: v–vii

Boutakoff N (1948) Les formations glaciaires et post-glaciaires fossilifères, d'âge permo-carbonifère (Karoo inférieur) de la region de Walikale (Kivu, Congo belge). Mémoire de l'Institut géologique. Université de Louvain IX (II): 214

Cahen D (1978) Vers une revision de la nomenclature de industries prehistorique de l'Afrique central. L'Anthropology (Paris) 28(1): 5–36.

Cahen L (1954) Geologie du Congo Belge. Liege, Belgium.

Cahen L, Holmes A (1956) Geochronologie africaine. Belgium Academy of Colonial Sciences, Memoir 1.

Cahen L, Ferrand JJ, Haarsma MJF, Lepersonne J, Vebeek T (1959) Description du Sondage de Samba. Annales du Musée Royal du Congo belge, Tervuren (Belgique), Série in-8, Sciences géologiques, 29, 210p

Cahen L, Ferrand JJ, Haarsma MJF, Lepersonne J, Vebeek T (1960) Description du Sondage de Dekese. Annales du Musée Royal du Congo belge, Tervuren (Belgique), Série in-8, Sciences géologiques, 34, 115p

Cahen L, Snelling NJ (1966) Geochronology of equatorial Africa. North Holland, Amsterdam

Cahen L, Snelling NJ, Delhal J, Vail J (1984) The geochronology and evolution of Africa. Clarendon Press, Oxford, 512pp

Cailteux JLH, Lubala FRT, Masaru Yoshida M (2005) Obituary – Prof. Henri Ali Basira Kampunzu. Gondwana Research (Gondwana Newsletter Section) 8: 289–291

Chirikure S (2010) Indigenous mining and metallurgy in Africa. Cambridge University Press, Cambridge.

Clark JD (1971) Human behavioural differences in southern Africa during the Later Pleistocene. Am Anthropol 73:1211–1236

Clark JD, Brown KS (2001) The twin rivers Kopje, Zambia: Stratigraphy, Fauna, and artefact assemblages from the 1954 and 1956 Excavations. J Archaeol Sci 28: 305–330

Cornet, René, J. Katanga; le Katanga avant les Belges. Brussels.

Cornet J (1886) Les gisement métallifères du Katanga. Mém. et Publ. de la Soc. des Sc. Du Hainaut, V^{e} sér., t. VIII

Cornet J (1893–1894) Les formations post-primaires du basins du Congo. Ann. Soc.géol. de Belg., t. XXI, pp. M 193-279; carte au 2.000.000

Chrétien J-P (2000) L'Afrique des grands lacs: Deux milles and d'histoire (Aubier, Paris); 2003. The great lakes of Africa: two thousand years of history (Urzone, NY)

Delhal J (1988) Remise du Prix Lucien Cahen 1987. Bull Séanc Acad R Sci Outre-Mer, 33(4): 535–536

Dehal J, Delmer A (1999) Notice sur Jacques Lepersonne. Académie royale de Belgique, Annuaire, 65–79

Du Toit AL (1926) The geology of South Africa. Oliver and Boyd, London, 444pp

Du Toit AL (1937) Our wandering continents. Oliver and Boyd, Edinburg

Economist (2013) Murky minerals. Business in the Democratic Republic of Congo. How bad is it? 18 May

Eglington BM, Reddy SM, Evans DAD (2009) The IGCP 509 database system: design and application of a tool to capture and illustrate litho- and chrono-stratigraphic information for Palaeoproterozoic tectonic domains, large igneous provinces and ore deposits; with examples from southern Africa, 27-47. In: Reddy SM, Mazumder R, Evans DAD (eds) Palaeoproterozoic supercontinents and global evolution. Geological Society, London, Special Publication 323, 300pp

Evrard P (1960) Sismique. Annales du Musée Royal du Congo belge, Tervuren (Belgique), Série in-8, Sciences géologiques, 33, 87p

Goossens PJ (2007) Phoenix rising in an uncertain world –new mining activities in Katanga. Bull Séanc Acad R Sci Outre-Mer 53(3):361–385.

Goossens PJ (2009) Mineral potential of the Democratic Republic of Congo: A geological scandal? Society of Economic Geologists Newsletter, April 2009, Number 77

Goossens PJ (2013) L'or à travers les âges, une histoire pas toujours dorée. Presses Universitaires de Liège, 306pp

Holmes A (1918) The Precambrian and associated rocks of the district of Mozambique. Quart J Geol Soc 74:31–98

Holmes A (1951) The sequence of pre-Cambrian orogenic belts in south and central Africa. In: 18th International Geol Congress, London (1948), vol 14, pp 254–269

Holmes A, Cahen L (1955) African geochronology. Results available to 1st September, 1954. Colon Geol Miner Resour Lond 5:3–39

Kampunzu AB (1998) The Mesoproterozoic Kibaran Belt System in Africa: a key for the reconstruction of Rodinia supercontinent. Gondwana Res 1: 412–414

Kampunzu AB, Kramers JD, Makutu NM (1998) Rb-Sr whole rock ages of the Lueshe, Kirumba and Numbi igneous complexes (Kivu, DRCongo) and the break-up of the Rodinia supercontinent. J Afr Earth Sci 26: 29–36

Lepersonne J (1983) Notice sur Lucien Cahen, correspondant de l'Académie. Annuaire Acad. R. Belgique, pp 57–99

Leslie J (2005) Deep water: the epic struggle over dams, displaced people and the environment. Farrar, Straus & Giroux, NY

Leslie J (2014) Large Dams Just Aren't Worth the Cost. NY Times, Sunday Review, August 24, SR5.

Liebaers H (1977) Hommage à Lucien Cahen/Hulde aan Lucien Cahen. Africa Tervuren 23:31–36

Muhongo S (1952–2004) In Memorium, Henri Kampunzu (1952–2004). Episodes 27:305–306

Piketty T (2014) Capital in the twenty-first century. Harvard University Press, Cambridge, MA, 685pp

Pehrsson SJ, Berman RG, Eglington BM, Rainbird RH (2013) Two Neoarchean supercontinents revisited: the case for a Rae family of cratons, Precambr Res 232:27–43

Ravallion M (2014) Income inequality in the developing world. Science 344: 851–855

Robert M (1923) Le Congo Physique. Brussels (with 2nd and 3rd editions in 1942 and 1946, respectively)

Robert M (1927) La Katanga Physique. Brussels

Science (2014) The science of inequality. Special issue 344:828–867

Snelling NJ (1983) Obituary. In Memoriam Lucien Cahen (1912–1982). Isotope Geosci 1:97–100.

Snelling NJ (1985) Geochronology and the geological record. Geological Society, London, Memoirs 10, 3–9.

Van den Audenaerde TD, Lepersonne J, Lebrun J, Maesen A, Delhal J (1982) In memorian *Lucien Cahen*. Africa Tervuren 28(4): 35

Van Reybrouck D (2010) Congo: Een Geschiedenis (De Bezige Bij, Amsterdam)

Van Reybrouck D (2014) Congo: The epic history of a people (HarperCollins, NY)

Veatch AC (1935) Evolution of the Congo Basin. Geol. Soc. America, Memoir 3, 183pp

Walker SJH, Lukich V, Chazan M (2014) Kathu Townlands: a high density earlier stone age locality in the interior of South Africa. PloS One 9 (7):e103436. doi: 10.1371/journal.pone.0103436

Yoshida M, Kampunzu AB, Li ZX, Watanabe T (2003) Assembly and break-up of Rodinia and Gondwana: evidence from Eurasia and Gondwana: introduction. Gondwana Res 6: 139–142

Acknowledgements

The idea of a 'Congo Book' emerged out of a special session on the 'Congo Basin: its origin, evolution and resources' at the 23rd Colloquium of African Geology (http://www.cag23.co.za), organised by the Geological Society of Africa in the University of Johannesburg, 8th–14th January 2011, during which numerous presentations and posters about Precambrian geology, recent dating and thermochronology, geophysics, structural analysis, and sedimentology of the region were presented, questioned, and further unpacked during a round-table discussion. It became clear from the interactive discourse that a compilation of the geology and natural resources of this vast region was wanted. We would like to thank all participants for their stimulated interactions. We hope that this book lives up to their expectations and that it may stimulate new research.

Financial support for this project has been provided in part by the NRF—National Research Foundation of South Africa, and the Global Change programme of the DST—Department (Ministry) of Science and Technology with their generous support of **!Khure Africa**, a collaborative Earth Systems Programme between South Africa and France coordinated through AEON at the Nelson Mandela Metropolitan University in Port Elizabeth. In France funds were made available also through the **TopoAfrica** programme coordinated in the Department of Geosciences at Rennes University.

We owe a great deal to Bastien Linol for numerous onerous tasks during the preparation of this book, and to Pierre Goossens for sharing his wide knowledge of the history of mining in the Congo and his DRC mineral information. We thank Max Fernandez-Alonso of the RMCA in Tervuren, Belgium, for general information and help with Archival Photos, and Sam Bowring (MIT, USA), Bruce Eglington (Saskatchewan, Canada), and Moctar Doucouré (AEON-ESSRI, NMMU, South Africa) for sharing their thoughts and advice. We are also keen to acknowledge our appreciation to Dörthe Mennecke-Bühler and Annett Büttner, editors of Springer-Verlag, Germany, for their efficiency, patience, and professional help to bring this book to completion.

Finally, we would like express our sincere thanks to the following referees for their peer reviews of the chapters in this book. Without their dedicated help and their time this book could not have been completed:

Fritz Achterberg, University of Ottawa, Canada
Jean Braun, University of Grenoble, France
John Bristow, Global Diamond Network, Johannesburg, South Africa
Greg Botha, Council for Geosciences, South Africa
Dominique Chardon, University of Toulouse, France
Bruce Eglington, University of Saskatchewan, Canada
Hartwig Frimmel, University of Würzburg, Germany
David Gillan, University of Mons, Belgium
Pierre Goossens, Bugeco and University of Liege, Belgium
John Gurney, Mineral Services, and University of Cape Town, South Africa

Rob Harris, Oregon State University, Corvallis, USA
Judith Kinnaird, University of the Witwatersrand, South Africa
Mike Lynn, MSA Group Services (Pty) Ltd. Johannesburg, South Africa
Tania Marshall, Explorations Unlimited, South Africa
Brendan Murphy, St Francis Xavier University, Canada
Pat Sheahan, Sheanan Literature Services, Canada
Renato Spaggiari, CNK Mining, Cameroon
Anthony Tankard, Tankard Enterprises, Calgary, Canada
John Ward, Majimba Geoconsulting, South Africa
Nicky White, Cambridge University, UK

Contents

Part I

Regional Lithosphere Framework: Crust and Mantle

The Upper Mantle Seismic Velocity Structure of South-Central Africa and the Seismic Architecture of Precambrian Lithosphere Beneath the Congo Basin

1

Andriamiranto Raveloson, Andrew Nyblade, Stewart Fishwick, Azangi Mangongolo, and Sharad Master

1.1 Introduction

The Congo Basin, covering approximately 10 % of the African continent, is one of the largest intracratonic basins on any continent, and its origin remains poorly understood (e.g. Daly et al. 1992; Downey and Gurnis 2009; Crosby et al. 2010; Kadima et al. 2011a,b). It contains up to 9 km of sedimentary rocks, which may date back to the Neoproterozoic, and its tectonic history is punctuated with major periods of compression during the late Neoproterozoic formation of Gondwana and Late Paleozoic subduction along the southern margin of Pangea. Stratigraphic information from drill holes and outcrops, seismic reflection profiles, and gravity and magnetic observations have led to a number of models for the formation of the basin, many of which begin with Neoproterozoic rifting (e.g. Daly et al. 1992; Kadima et al. 2011b) and call upon anomalous density either within or below the lithospheric mantle to sustain the long-lived (e.g. ca. 700 Ma), basin-wide subsidence (e.g. Crosby et al. 2010; Downey and Gurnis 2009).

Seismic images of mantle structure beneath the basin from continental-scale tomography place first order constraints on explanations for how the basin may have formed (e.g. Fishwick 2010; Pasyanos 2010; Priestley et al. 2008; Pasyanos and Nyblade 2007; Ritsema and van Heijst 2000), and all but one show fast, thick (i.e. > 200 km) lithosphere beneath the interior of the basin. The thick shield lithosphere imaged by most of the tomography studies underpins many of the geodynamic models explaining how the basin formed. At lithospheric mantle depths (i.e. >~40 km) the model by Pasyanos and Nyblade (2007) shows slower velocities beneath the central and eastern portions of the basin, which they suggested could indicate the presence of thinner (mobile belt) lithosphere that may have formed during the Proterozoic amalgamation of the three major cratonic blocks of the greater Congo Shield that surround the basin, the Kasai Craton to the south, the Ntem Craton to the west and the Bomu Craton to the north-east (Fig. 1.1). The suggestion that Proterozoic mobile belt lithosphere lies beneath the northern part of the basin has also been made by Master (2004), De Wit et al. (2008), and Gubanov and Mooney (2009) using geological evidence. Daly et al. (1992) and De Wit et al. (2008) showed a NW–SE trending Neoproterozoic belt running through the centre of the Congo Basin. To the NW of the basin, this belt extends into in a region with poorly dated metasedimentary rocks of Paleoproterozoic (>2080 Ma) and Mesoproterozoic (1167 Ma –ca. 1.0 Ga) age (Vicat et al. 1997), which in turn are overthrust by the south-verging nappes of the Oubanguide Belt, dated at c. 571–620 Ma (Moloto-A-Kanguemba et al. 2008; Toteu et al. 1994; De Wit et al. 2008; see also de Wit and Linol, Chap. 2, this Book).

Here we briefly review the basin's geological setting and some of the geophysical studies relevant to understanding how the basin formed, and then present a new model of upper mantle structure for central and southern Africa

A. Raveloson
Department of Geosciences, Pennsylvania State University, University Park, PA 16802, USA
e-mail: ahr12@psu.edu

A. Nyblade (✉)
Department of Geosciences, Pennsylvania State University, University Park, PA 16802, USA

School of Geosciences, University of the Witwatersrand, Johannesburg, South Africa
e-mail: andy@geosc.psu.edu

S. Fishwick
Department of Geology, University of Leicester, Leicester, LE1 7RH, UK
e-mail: sf130@leicester.ac.uk

A. Mangongolo
Council for Geosciences, Private Bag x112, Pretoria, 0001, South Africa
e-mail: amangongolo@geoscience.org.za

S. Master
School of Geosciences, University of the Witwatersrand, Johannesburg, South Africa
e-mail: Sharad.Master@wits.ac.za

M.J. de Wit et al. (eds.), *Geology and Resource Potential of the Congo Basin*, Regional Geology Reviews,
DOI 10.1007/978-3-642-29482-2_1, © Springer-Verlag Berlin Heidelberg 2015

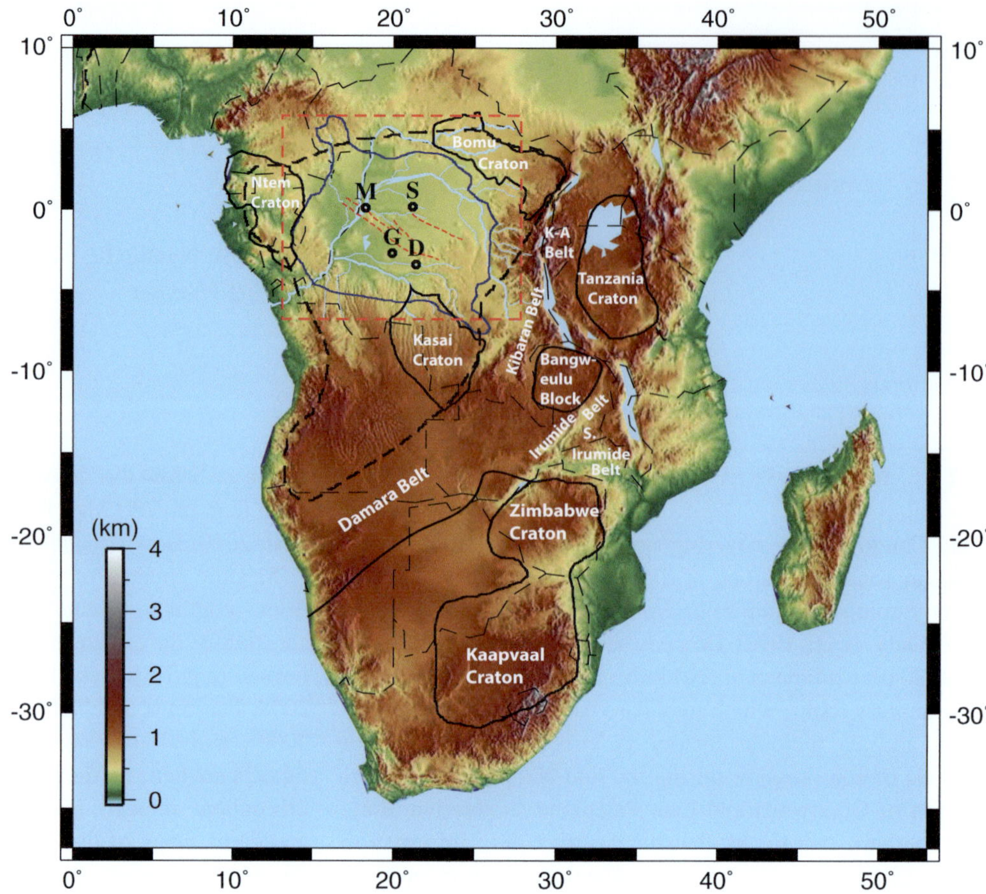

Fig. 1.1 Topographic map of central and southern Africa showing the major Archean cratons (Kaapvaal, Zimbabwe, Tanzania, Kasai, Bomu, Ntem; *solid black lines*), the greater Congo Shield (*dashed black line*), the zero thickness contour for the Mesozoic and Cenozoic sediments of the Congo Basin (*blue line*), the inferred location of Neoproterozoic rifts within the Congo Basin (*red dashed lines*), and the outline of the Bouguer gravity map in Kadima et al. (2011a) (*rectangle* outlined with *red dashed lines*). K-A Belt is the Karagwe-Ankole Belt. The location of the Samba (S), Dekese (D), Mbandaka (M), and Gilson (G) wells are shown with *black circles*

based on Rayleigh wave group velocities. By comparing similarities and differences between this model and previously published models, we gain additional insights into the architecture of lithospheric structure under the basin, which in turn we use to evaluate the proposal of a northwest-southeast trending mobile belt beneath the northern portion of the basin. Whether or not a mobile belt extends under the basin has implications not only for geodynamic models of basin formation invoking uniformly thick lithosphere beneath the basin, but also for understanding Neoproterozoic tectonism that may have initiated basin subsidence.

1.2 Geological Setting

The Congo Basin has a mean elevation of ~300–400 m above sea level and is surrounded by uplifted regions on all sides (Fig. 1.1; and see Chap. 8 for greater detail). Details of the basin fill come from four, 2–4 km deep wells (Samba, Dekese, Mbandaka, Gilson), field studies, and a number of seismic reflection profiles. Data from the wells (cores and drill logs) and the reflection profiles have been used in a number of studies (e.g. Lawrence and Makazu 1988; Daly et al. 1992; Kadima et al. 2011a,b; see Chaps. 2–4 for more details) to show that the basin contains 4 to 9 km of sediments ranging in age from Neoproterozoic to Neogene. The pre-Cretaceous sediments are thought to have been deposited during a post-rift subsidence phase (Lawrence and Makazu 1988; Daly et al. 1992; Crosby et al. 2010; Kadima et al. 2011a,b). Neoproterozoic, Late Cambrian and Permian-Triassic unconformities have been identified in seismic profiles, and the latter two have been linked to NE–SW compression that resulted in reverse faulting and the development of basement highs near the center of the basin (Daly et al. 1992). However, the validity of the ages of these unconformities has been questioned elsewhere (e.g. Linol 2013; and see Chap. 6).

A detailed description of the basin stratigraphy can be found in Kadima et al. (2011a). The four wells penetrated mostly formations of sandstones and clays and none of them reached crystalline basement. At the top of the basin is 1 km of undeformed Lower Cretaceous to Quaternary sediments deposited in fluvial, lacustrine and aeolian environments. Paleozoic and Neoproterozoic sediments outcrop along the margin of the basin, and tectonic subsidence curves from back-stripping the four wells show very slow subsidence since the Early Paleozoic compressional event (Kadima et al. 2011b), as expected for post-rift thermal relaxation of the lithosphere.

1.3 Previous Geophysical Studies

A number of geophysical techniques have been applied to the Congo Basin and surrounding regions in order to improve knowledge of the subsurface structure. Petroleum exploration of the basin led to the acquisition of the seismic reflection profiles mentioned previously, as well as a number of gravity and aeromagnetic surveys. More recently, satellite gravity missions such as the Gravity Recovery and Climate Experiment (GRACE), have provided global knowledge of the gravity and geoid fields. Interpretation of the seismic reflection data identified a series of sub-basins separated by basement highs (e.g. Lawrence and Makazu 1988; Daly et al. 1992), however a re-interpretation of the seismic profiles by Kadima et al. (2011a) suggests that salt structures may be present in the central basin and therefore the basement highs might not be as pronounced as originally thought. These various interpretations remain to be tested.

Complementing the seismic reflection data, Kadima et al. (2011a) presented results from an aeromagnetic survey over the central and northern parts of the basin and a Bouguer anomaly map for the entire basin (Fig. 1.1). The aeromagnetic survey was carried out in 1984, and shows low magnetic gradients in the central part of the basin corresponding to deep basement, and higher magnetic gradients to the northeast and southwest, where the basement is interpreted to be less deep (Kadima et al. 2011a). The gravity database used by Kadima et al. (2011a) comes from the Bureau Gravimetrique International (BGI) in Toulouse, France, and incorporates data from a variety of sources. The Bouguer anomaly map is dominated by NW-SE trending anomalies.

At a subcontinental scale the most striking feature in the gravity data is a long wavelength negative free air gravity anomaly (~ -40mGal) that coincides with the basin. Hartley and Allen (1994) proposed that this anomaly was unlikely to be supported by variations in crustal structure, and Hartley et al. (1996) suggested that the correlations between gravity anomalies and topography are indicative of convective processes in the underlying mantle. A number of recent studies have discussed the cause of the long wavelength gravity anomaly and the potential relationship to basin formation. Downey and Gurnis (2009) suggested that there must be an anomalously dense body, possibly made of eclogite, within the mantle beneath the basin. A similar interpretation involving a dense body in the lower crust or upper mantle was proposed by Braitenberg and Ebbing (2009) to explain the relationship between gravity and geoid fields for a number of intra-cratonic basins worldwide. Crosby et al. (2010) modeled the subsidence and gravity signature of the basin using a lithospheric thickness of 220 km interpreted from seismic tomography images, and argued that the long wavelength gravity signature is likely the result of recent convective drawdown beneath the basin in response to the surrounding upwellings, consistent with the mantle convection model of Forte et al. (2010). However, based on a variety of seismic models (see below), Buiter et al. (2012) conclude that the seismological results do not provide supporting evidence for a first-order role of the sub-lithospheric mantle in the recent subsidence of the Congo Basin. In contrast to these models, Kadima et al. (2011b) argue that if the effects of the low-density sediments are removed from the gravity field, then the residual gravity anomaly can be attributed to 8–10 km of relative crustal thinning beneath the deepest portions of the basin.

A review of both continental and global-scale seismic tomography models that inculcate the Congo Basin has been recently published by Buiter et al.(2012). Using simple averaging, the authors assessed uncertainties in lithospheric thickness estimates from a range of geophysical models, and also computed mean velocities, along with the associated standard deviations, to investigate the role of the sublithospheric mantle in the subsidence of the basin. The review showed generally thick (~200 km) lithosphere beneath much of the basin, but suggested that the uncertainties between the different tomographic models provide little evidence of upwelling on the flanks of the basin (Buiter et al. 2012). We briefly summarize the key features from several of the continental-scale models of Africa.

All continental-scale seismic tomography models show fast, craton-like upper mantle structures beneath some parts or else most of the Congo Basin (e.g. Fishwick 2010; Priestley et al. 2008; Ritsema and van Heijst 2000; Pasyanos and Nyblade 2007; Sebai et al. 2006). The models by Fishwick (2010) and Priestley et al. (2008) use the same basic technique, comprising waveform inversion measurements for fundamental and first four higher mode Rayleigh waves. Sebai et al. (2006) also measured and inverted phase velocities for fundamental and higher mode Rayleigh waves, in addition to Love waves. Ritsema and van Heijst (2000) modeled phase velocity measurements from fundamental mode Rayleigh waves only. In contrast, the model by Pasyanos and Nyblade (2007) is based on group velocity measurements from fundamental mode Rayleigh waves.

At the shallowest depths (100–200 km depth) displayed in the models constructed using phase velocities or waveform inversion techniques, the velocity structure is reasonably similar, with fast velocities found beneath much of the basin, but with the fastest velocities found beneath the southern portion of the basin and Kasai Craton (Fig. 1.2). In the Fishwick (2010) model (Fig. 1.2), the Ntem Craton appears as a relatively fast region compared to the Priestley et al. (2008) model. The isotropic shear wave velocity model of Sebai et al. (2006) shows similar structures to the Fishwick (2010) and Priestley et al. (2008) models across the basin, while the model of Ritsema and van Heijst (2000) shows uniformly fast upper mantle everywhere beneath the Congo Basin.

The fast velocities observed in the majority of models, are, however, not seen in the results of Pasyanos and Nyblade (2007). At upper mantle depths of 100–200 km, their model shows a region of relatively slower velocities extending from the eastern side of the basin into the interior of the basin, which is illustrated in Fig. 1.2. Faster velocities in this model are found to the south in the vicinity of the Kasai Craton, to the north beneath the Bomu Craton, and to the west beneath the Ntem Craton. As mentioned in the introduction, this is the only model that shows slower velocities beneath a substantial portion of the Congo Basin, and it is based on this result that Pasyanos and Nyblade (2007) suggested that beneath the basin there might be mobile belt lithosphere which is thinner than the lithosphere beneath the cratons.

Crustal thickness beneath the basin remains poorly known, with the majority of studies using either the global model 3SMAC (Nataf and Ricard 1996) or Crust 2.1 (Bassin et al. 2000). Pasyanos and Nyblade (2007) used surface wave tomography to estimate crustal thickness across the whole continent. For a comparison of these three crustal models, we refer the reader to Fig. 3 in Fishwick and Bastow (2011).

As discussed by Buiter et al. (2012), lithospheric thickness beneath the basin has been estimated in a number of studies. Pasyanos (2010) estimated ~130–160 km thick lithosphere compared to 160–180 km beneath parts of the Bomu Craton, 160–200 km beneath the Kasai Craton, and over 200 km thick beneath the Ntem Craton. Fishwick (2010) reported lithospheric thicknesses of >200 km beneath the southwestern and central parts of the basin, with lithosphere thinning to around 160–200 km adjacent to the margins of the basin and beneath the Bomu, Ntem and Kasai cratons. Priestley et al. (2008) reported considerably different lithospheric thicknesses from Fishwick (2010) in spite of using very similar modeling methodologies. In a more recent update, Priestley and McKenzie (2013) estimate the thickest lithosphere (>210 km) beneath the central part of the basin and thinner lithosphere (170–200 km) beneath the Ntem and Kasai Cratons. In their model, lithospheric thickness drops sharply on the northeastern margin of the basin and to the southwest of the basin.

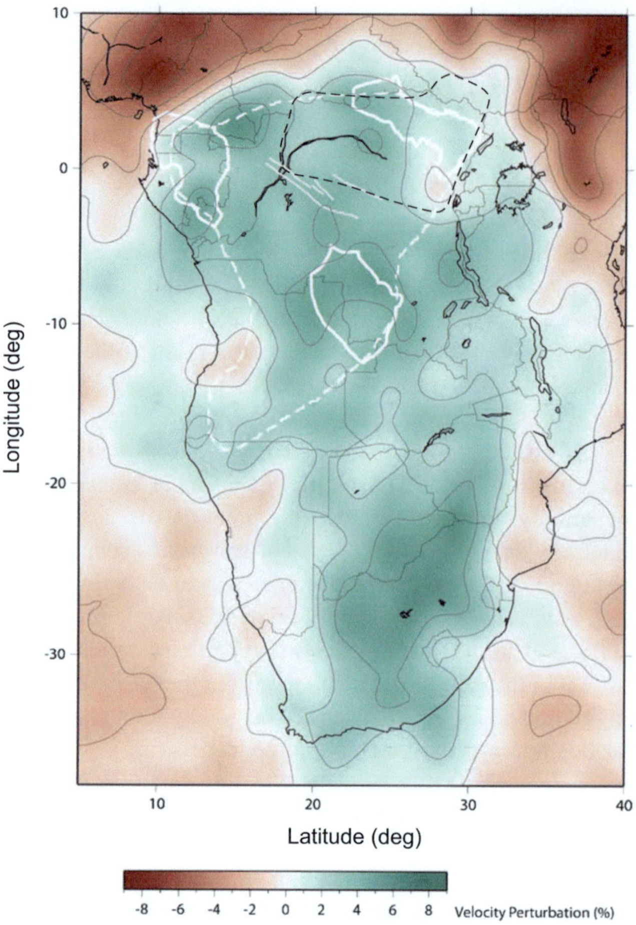

Fig. 1.2 High resolution shear velocity model at 100 km depth from Fishwick (2010). Velocities are plotted as perturbations from the reference model ak135, *green/blue shades* showing faster velocities and *red/brown* slower velocities. The outline of major Archean cratons (see Fig. 1.1 for more details) are shown with *solid white lines*, the greater Congo Shield with the *dashed white line*, the inferred location of Neoproterozoic rifts with *grey lines*, and the area of slow velocities at 100 km depth in the Pasyanos and Nyblade (2007) model with a *dashed black line*

It is very difficult to accurately assess the uncertainty on these thickness estimates. The tomographic models themselves have vertical resolution of ~25–50 km for depths shallower than ~200 km, although this is dependent on the exact periods and techniques used. Different approaches for converting velocity to thickness further challenge comparisons between the various models. It is, therefore, difficult to have confidence in variations of less than ~50 km in lithospheric thickness.

1.4 Shear Wave Tomography

To advance our understanding of upper mantle structure beneath and surrounding the Congo Basin, we have developed a new shear wave velocity model using all available open data from permanent and temporary networks, as well as from many new temporary and permanent seismic stations in eastern and southern Africa that are part of AfricaArray (http://www.africaarray.org) (Fig. 1.3). Temporary stations were deployed in Uganda, Tanzania and Zambia between 2007 and 2011 as part of the AfricaArray East African Seismic Experiment (AAEASE) and the AfricaArray Tanzania Basin Seismic Experiment (AATBSE) (Tugume et al. 2012; Mulibo and Nyblade 2013), and several permanent stations have been established since 2005 to form the AfricaArray observatory network. Data from the AfricaArray stations greatly improves ray path coverage across the Congo Basin. In our modeling approach, maps of Rayleigh wave group velocities are first constructed using event-station measurements of group velocities from fundamental mode surface waves, dispersion curves are extracted over a grid of nodes and inverted for 1D shear velocity models, and then the 1D models are combined to create a quasi-3D upper mantle shear wave velocity model of the region.

1.4.1 Rayleigh Wave Group Velocity Maps

To develop the group velocity maps, we used seismic events with magnitude greater than 4.5 and depth shallower than 100 km recorded between 1990 and 2013. Only events along the margin or within the African plate have been used to minimize the influence of structure outside the plate on our tomographic images.

Group velocity measurements for 40,337 event-station pairs were made for fundamental mode Rayleigh waves using vertical component seismograms. An example is shown in Fig. 1.4 for a magnitude 5.5 event recorded on station KMBO. The ray path coverage is illustrated in Fig. 1.5. After visual inspection for waveform quality, the instrument-corrected data were windowed to isolate the Rayleigh wave, and group velocity measurements were made using the multiple filter method (Dziewonski et al. 1969; Herrmann 1973) implemented in the program PGSWMFA (PGplot Surface Wave Multiple Filter Analysis; Ammon 1998). An example measurement is shown in Fig. 1.6 for the same data as in Fig. 1.4.

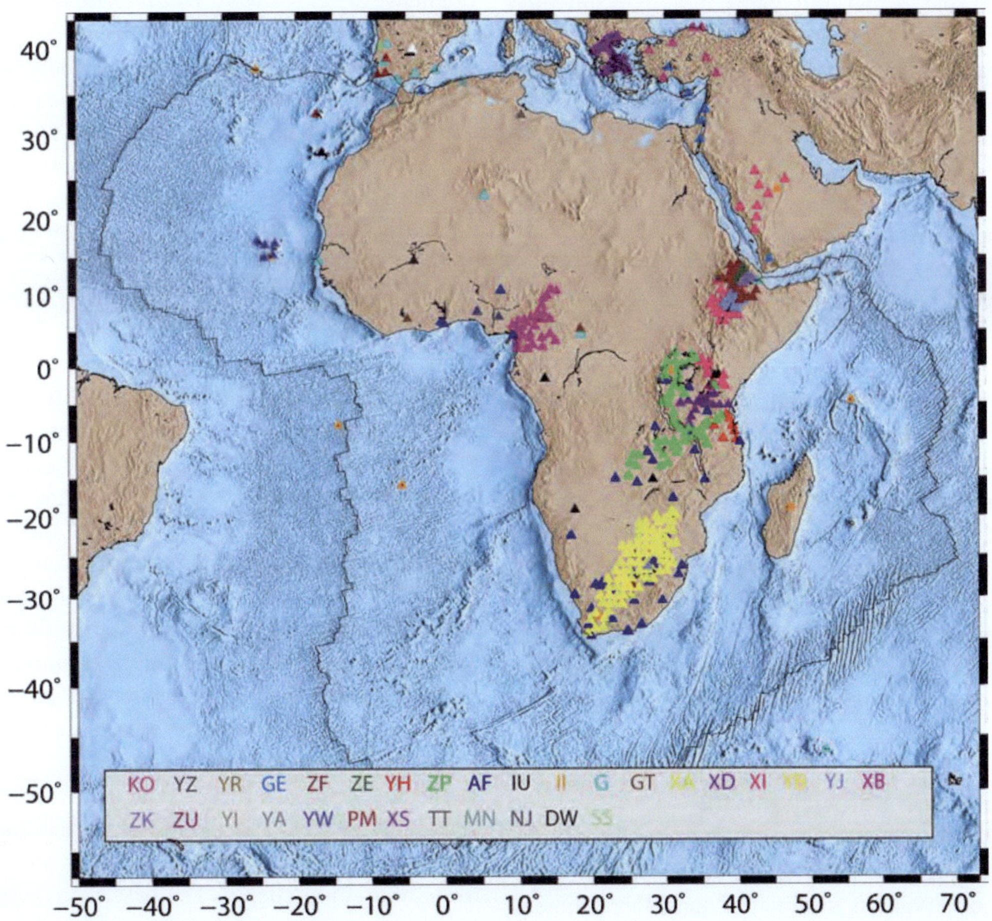

Fig. 1.3 Locations of seismic stations used for this study. Color-coded FDSN network codes are provided at the *bottom* of the figure

The ensemble of group velocity measurements at each period was inverted using a linear 2-D tomographic inversion to obtain maps of Rayleigh wave group velocities. The inversion was performed over a grid of 1° × 1° cells and the slowness for each cell was calculated using a conjugate gradient method (Paige and Saunders 1982). Damping and smoothing were used to stabilize the inversion. Changing the damping parameter did not significantly affect the results, and therefore a low damping of 0.1 was used. A smoothing value of 3,000 was used for all periods based on tradeoff curves between smoothing and travel time misfit (Fig. 1.7). Values of misfits in velocity, as defined by the expression

$$\sqrt{\frac{\sum\left(\left(dis\tan ce\left(\frac{1}{observed\ \ traveltime}-\frac{1}{predicted\ \ traveltime}\right)\right)^{2}\right)}{number\ \ of\ \ dispersion\ \ curves}}$$

are used as approximations for uncertainty in velocity and were found to be between 0.06 and 0.48 km/s.

Checkerboard tests were performed to estimate the model resolution. The velocity of the input checkers was ± 9 % of the initial model velocity, and synthetic velocities were calculated from the actual ray paths and then inverted using the procedure described above. From the recovered models, it was determined that 4° checkers were the smallest size that could be resolved over the Congo Basin (Fig. 1.8). Regions of limited resolution were excluded from further analyses, which include all regions north of 7° north latitude where there is a substantial amount of smearing between many of the checkers.

Figure 1.9 shows the maps of Rayleigh wave group velocities for periods of 20, 40, 60, 80 and 100 s. At periods greater than 100 s, the resolution degrades considerably because of limited ray path coverage, and therefore we do not use group velocities for periods greater than 100 s. At shorter periods (20 and 40 s), the most prominent features are lower velocity regions associated with the Congo and Kalahari Basins. At these periods, the Rayleigh waves still have sufficient sensitivity to the low velocity sediments to make the basin locations stand out. At periods of 60 s and higher, regions of expected thick lithosphere begin to emerge as regions of faster velocity (i.e. the Tanzania, Congo, Kaapvaal and Zimbabwe cratons) separated by slower regions.

1.4.2 Shear Wave Velocity Model

We follow the method of Park et al. (2008) for inverting the group velocity observations to obtain a quasi-three dimensional shear wave velocity model for the upper mantle. In this method, we constrain crustal structure based on a priori information, invert for a one dimensional shear wave velocity profile for grid nodes every 0.5°, and apply a Gaussian weighted smoothing parameter across the one-dimensional models to create a quasi three-dimensional shear wave velocity model.

To compute shear wave velocity profiles at each node, we follow the inversion methodology described by Julià et al. (2000). This method uses an iterative damped generalized linear least-squares approach that inverts the data with model smoothness and weighting parameters to solve for changes to a starting velocity model. The method was designed to jointly invert dispersion and receiver function data, and so to apply this joint inversion method, we designate the weight assigned to receiver function data to be equal to zero.

The starting model used for the inversions is the IASP91 (Kennett and Engdahl 1991) model. The model parameterization includes 21 layers with a 37-km thick continental crust. Within the crust, the top ten layers are each 1 km thick, the next three layers are each 2 km thick, and the bottom four layers are each 5 km thick. Ten km thick layers are used in the mantle to a depth of 700 km. In the inversion, velocities below 200 km depth are fixed to IASP91 (Kennett and Engdahl 1991) values. We include a 1 km thick layer above and below the Moho with no vertical smoothing to constrain the Moho depths, yet still allow large variation in velocity across the Moho. A linear smoothing parameter is used to prevent large and unrealistic contrasts in velocity between adjacent layers, which Rayleigh waves are not sensitive to. For the ocean basins, the model parameterization is similar to the continent but with the Moho fixed at a depth of 7 km.

A continental crustal thickness of 37 km was used everywhere because a recent study that compared crustal structure in eastern Africa to southern and central Africa using receiver functions found little variability in the thickness of Precambrian crust, which comprises most of the study region (Tugume et al. 2012). The average crustal thickness for most terrains varies between 35 and 40 km, and therefore we used a mid-range value of 37 km. Because there are no seismic estimates of crustal thickness for the Congo Basin, we used our group velocities between periods of 20 and 50 s to see if there was any evidence for substantial thinning of the crust under the basin, as argued for by Kadima et al. (2011b). Figure 1.10 shows a plot of the average group velocity curve for the regions of the basin where Kadima et al. (2011b) show crustal thickness on the order of 25–30 km, and synthetic dispersion curves calculated using a model that contains low velocity sediments in the top 5 km of the crust, an average crustal Vs of 3.7 km/s and a Moho ranging from 20 to 40 km depth. The best fit to the data is obtained for models with a Moho at 35 to 40 km depth. This result justifies using a crustal thickness beneath the basin that is

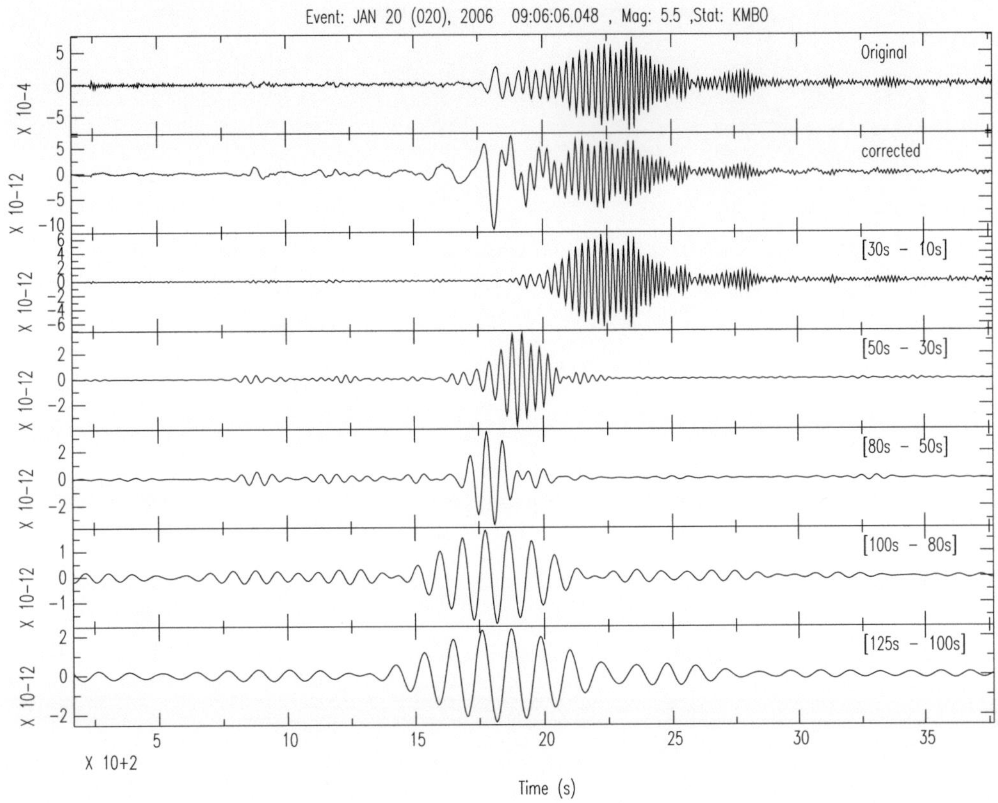

Fig. 1.4 Sample vertical component displacement seismogram for a magnitude 5.5 earthquake recorded on station KMBO shown with several bandpass filters illustrating the dispersion of the Rayleigh wave. The instrument response has been removed from the seismogram

similar to the surrounding Precambrian terranes. It also does not support the crustal thinning model of Kadima et al. (2011b).

To estimate the uncertainties in our velocity model, we follow the approach developed by Julià et al. (2005) for their inversion method, which involves repeatedly performing the inversions using a range of inversion parameters. This may be viewed as an overly simplistic approach, but it is effective in developing a sense of the range of variation in the inverted parameters given the observations and *apriori* constraints. As explained by Julià et al. (2000), more sophisticated and statistically rigorous approaches generally require that the observations meet desirable properties, such as being normally distributed, which in general are not satisfied by the data. By following the approach of Julià et al. (2005), we obtain an uncertainty of approximately 0.1–0.2 km/s for the velocity in each model layer.

The results for upper mantle depths of 50 to 90 km are shown in Fig. 1.11. Shear wave velocity structure is less well resolved at deeper depths. The shear wave velocity pattern is similar to the group velocity pattern at periods of 60 s and higher. There are several areas with faster velocities that correspond to the Archean cratons (Bomu, Kaapvaal, Kasai, Ntem, Tanzania, Zimbabwe) with slower velocities marking Proterozoic mobile belts in between. The mobile belts that exhibit slower upper mantle velocities include the Mesoproterozoic Kibaran, Karagwe-Ankole, Irumide and Southern Irumide Belts and the Neoproterozoic Damara Belt (Figs. 1.1 and 1.11). The Paleoproterozoic Bangweulu Block in northern Zambia also has lower upper mantle velocities than the Archean cratons. Within the Congo Shield (Fig. 1.1), the fastest velocities are found beneath the southern and central portions of the Congo Basin. The regions of fast velocity extending from coastal regions offshore Angola, Namibia and South Africa probably result from smearing of upper mantle structure across the continent-ocean boundary. In the resolution tests (Fig. 1.10), checkers in these regions are less well resolved than within the interior of the continent. Similarly, the region of fast upper mantle velocities in South Sudan probably results from structure in that part of the model being less well resolved than to the south.

1.5 Comparison of Models

We briefly compare the new model of uppermost mantle structure at 90 km depth (Fig. 1.11) with the model from Fishwick (2010) at 100 km illustrated in Fig. 1.2. To first order, the figures show similar structures. In both models the upper mantle beneath the Congo Shield has faster velocities,

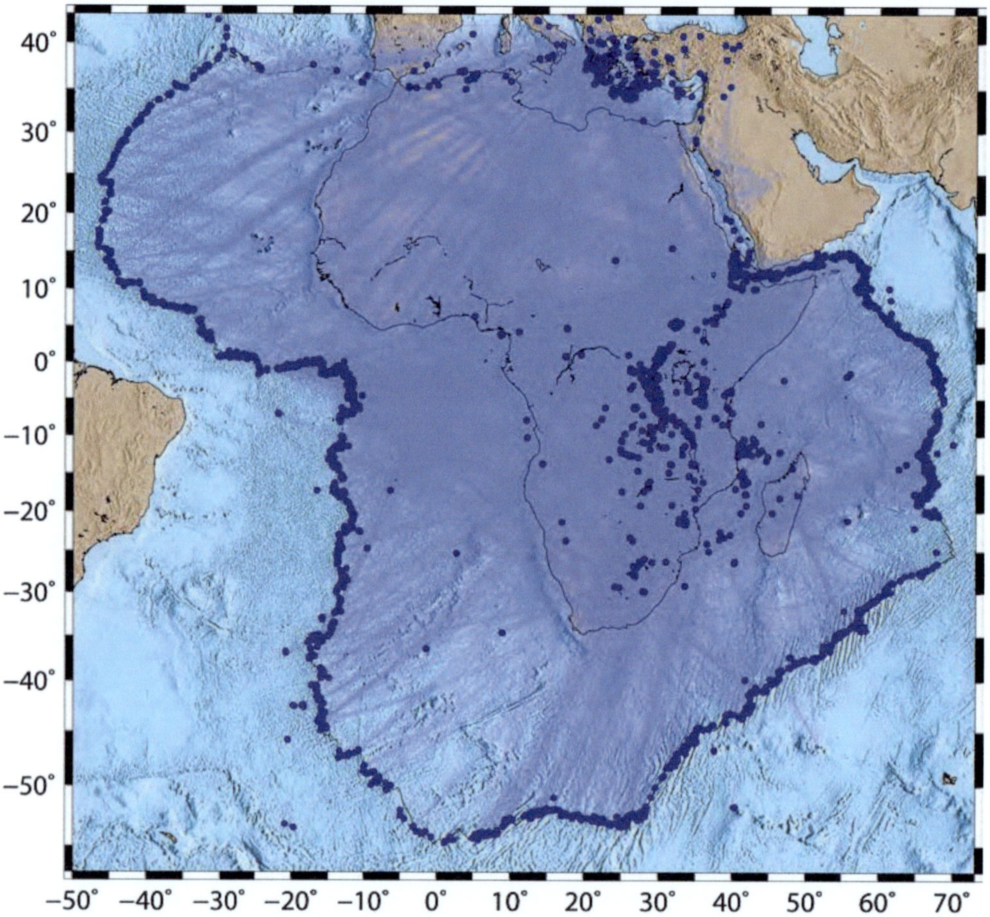

Fig. 1.5 Event-station ray paths for 20 s Rayleigh waves. *Blue dots* show event locations. Station locations are given in Fig. 1.3

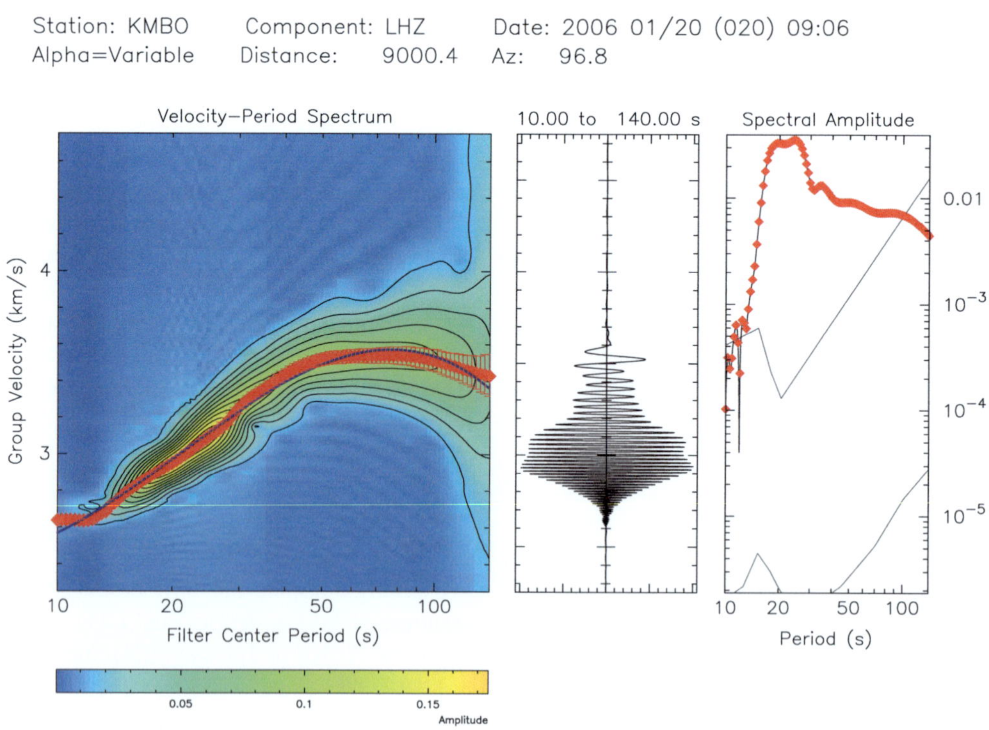

Fig. 1.6 Sample group velocity measurement for the Rayleigh wave shown in Fig. 1.4

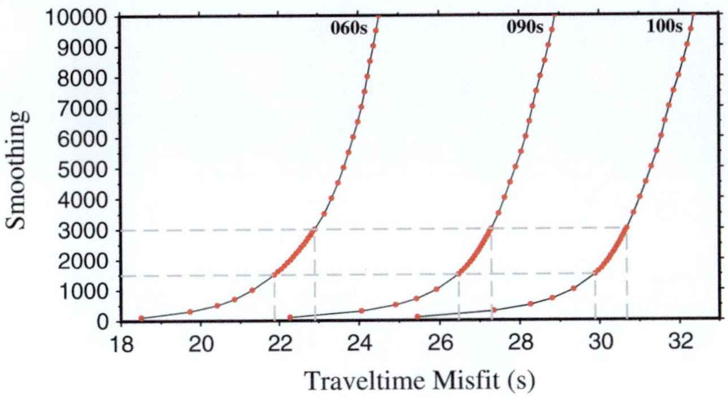

Fig. 1.7 Trade-off curves showing smoothing versus travel time misfit for 60, 90 and 100 s group velocities

and regions of slower velocities are found beneath the Proterozoic mobile belts to the east and south of the Congo Shield. In both models, the Ntem Craton has slightly slower upper mantle velocities compared to the Kasai Craton. Both models also show a region of slightly faster velocities beneath the northwestern portion of the Congo Basin to the east and northeast of the Ntem Craton, and a region of slower mantle velocities beneath the northeastern portion of the Congo Basin to the south and southwest of the Bomu Craton. The major differences between the models is that in our model the southern and central parts of the Congo Basin have the fastest upper mantle structure, whereas in the Fishwick (2010) model, the velocities in those regions are about the same or else slightly slower than beneath the Kasai Craton and the northwestern part of the basin. These differences could arise from the different data sets used or from the different inversion methodologies—the model of Fishwick (2010) shows the fastest velocities beneath the central part of the basin at slightly greater depths (125–200 km). Nevertheless, overall similar structure is obtained using two different sets of surface wave observations and modeling approaches, with both models showing velocities diminishing beneath the northeastern part of the basin. The significance of this velocity pattern for understanding the Precambrian lithospheric structure beneath the basin is discussed in the next section. We also note that the region of slower mantle velocities beneath the northeastern side of the Congo Basin is included within the region of slower mantle velocities in the Pasyanos and Nyblade (2007) model (Fig. 1.2), but the region of slower mantle velocities in that model is significantly larger than in our model or the Fishwick (2010) model.

1.6 Discussion

The seismic tomography models in Figs. 1.2 and 1.11 showing variations in shear wave velocities beneath the Congo Basin, as described in the previous section, indicate that the sub-basin lithospheric mantle may not be uniform, with the thickest lithosphere under the central and southern portions of the basin. The models can be used to evaluate whether or not a Proterozoic mobile belt extends to the NW beneath the northern portion of the Congo Basin, possibly representing a collisional zone between an Archean block to the NE and E, referred to by Master (2004) as the North-East Congo Craton (labeled the North-East Congo Shield or NECS in Fig. 1.11f) and an Archean block to the W and SW, referred to by Master (2004) as the South-West Congo Craton (labeled the South-West Congo Shield or SWCS in Fig. 1.11f). The SWCS consists of the Ntem and Angola-Kasai cratons, and the NECS consists of the Bomu Craton, the West Nile Complex in South Sudan, the northwestern Uganda Archean Complex, the Tanzania Craton and the Paleoproterozoic Bangweulu Block. The Mesoproterozoic mobile belt that Master (2004) and Gubanov and Mooney (2009) infer to occur beneath the northern part of the basin is a continuation of the Mesoproterozoic Kibaran Belt to the east (recent dating and reinterpretation of the Mesoproterozoic terrains along the eastern boundary of the Congo Basin defines the northern part of the Kibaran Belt as the Karagwe-Ankole Belt; e.g.Tack et al. 2010; Fernandez-Alonso et al. 2012).

Regardless of the name or age of the mobile belt, the mobile belt underlying the northern part of the Congo Basin in the Master (2004) interpretation formed as a result of the collision between the SWCS and the NECS. The Karagwe-Ankole Belt is in contact in the north with the Archaean granite-greenstone terrains of the Bomu Craton, but the contact between them is so poorly mapped that it appears as a blank on the 1: 2 million geological map of Lepersonne (1974). Structural features visible on satellite imagery and on radar-based topographic images (e.g. Fernandez-Alonso et al. 2012) indicate that the northern end of the Karagwe-Ankole Belt is marked by an east–west trending sinistral shear zone that disappears under the Congo Basin (Lepersonne 1978). The structural trends of the mobile belt curve and change from SW–NE to NW–SE as the shear zone is approached. In the model of Fernandez-Alonso et al. (2012), this contact would have resulted from far-field stresses associated with distant collisions in the East African Orogen. Master (2004) and Gubanov and Mooney (2009) suggested that the mobile belt continues to the WNW under the northeastern part of the Congo Basin. De Wit et al. (2008) considered this to be a Neoproterozoic mobile belt separating two Archean blocks, whereas Master (2004) considered it to be of Mesoproterozoic age, overlain by Neoproterozoic rocks. Evidence for the possible re-emergence of the mobile belt to the northwest of the Congo Basin comes from the "Serie de Liki-Bembe" (Cahen 1954; Lepersonne 1974) rocks (Master 2004). These rocks disappear to the NW in the poorly exposed

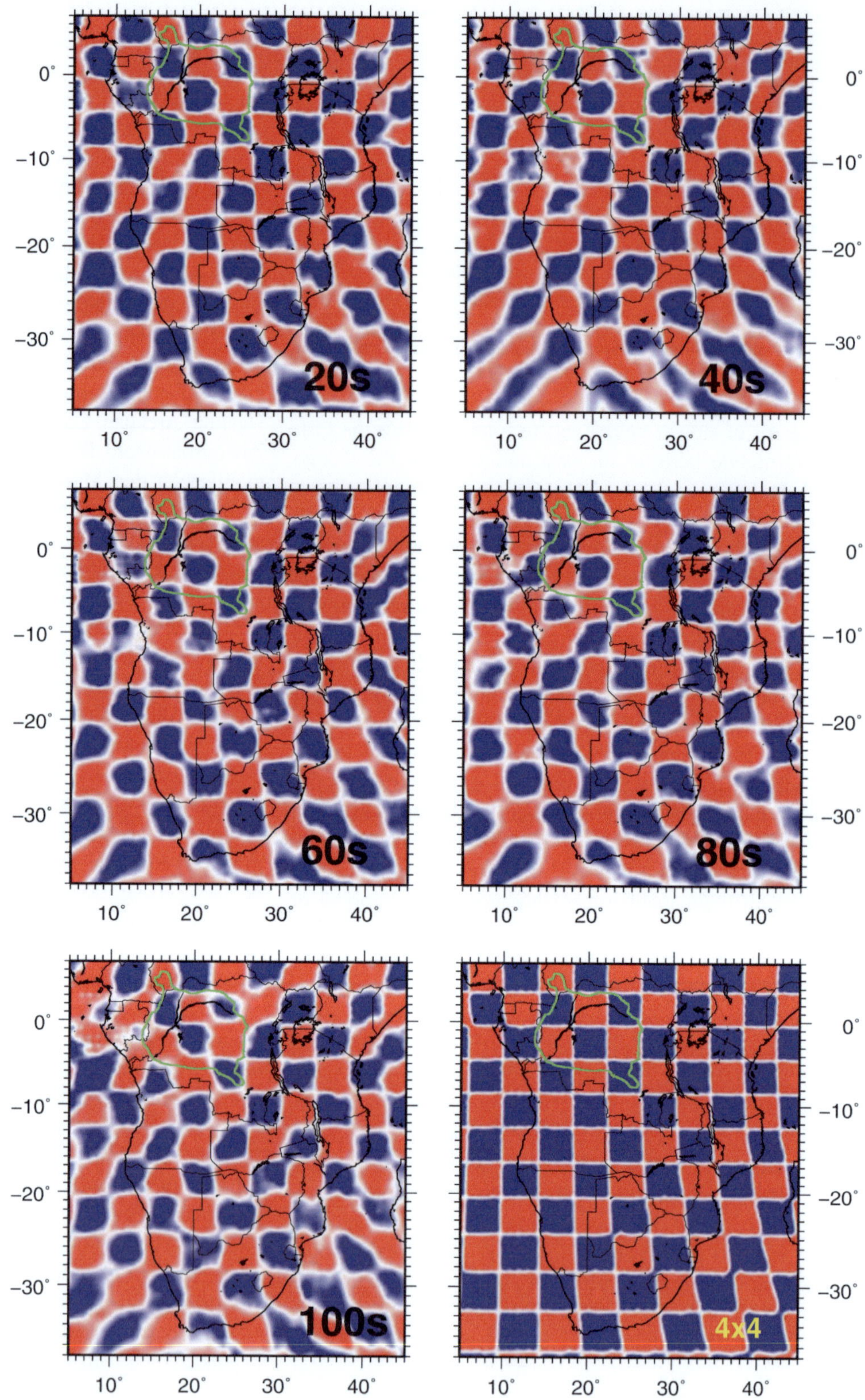

Fig. 1.8 Checkerboard resolution test results for periods of 20 to 100 s. *Green line* shows the zero thickness contour for the Mesozoic and Cenozoic sediments of the Congo Basin. *Lower right* panel shows input checker model

jungles of SE Cameroon and SW Central African Republic. In southern Cameroon, the ca. 0.6 Ga Yaounde gneisses, which represent a possible deformed passive margin of the northern Ntem Craton, have T_{DM} Sm-Nd ages of 1.6–1.0 Ga, indicating a derivation from Kibaran-aged rocks (Toteu et al. 1994). The Liki-Bembian rocks are overlain by Neoproterozoic Lindian rocks, in the same way that rocks of the classic Kibaran Belt of eastern Democratic Republic of Congo are overlain by Neoproterozoic Katangan rocks (Cahen 1954).

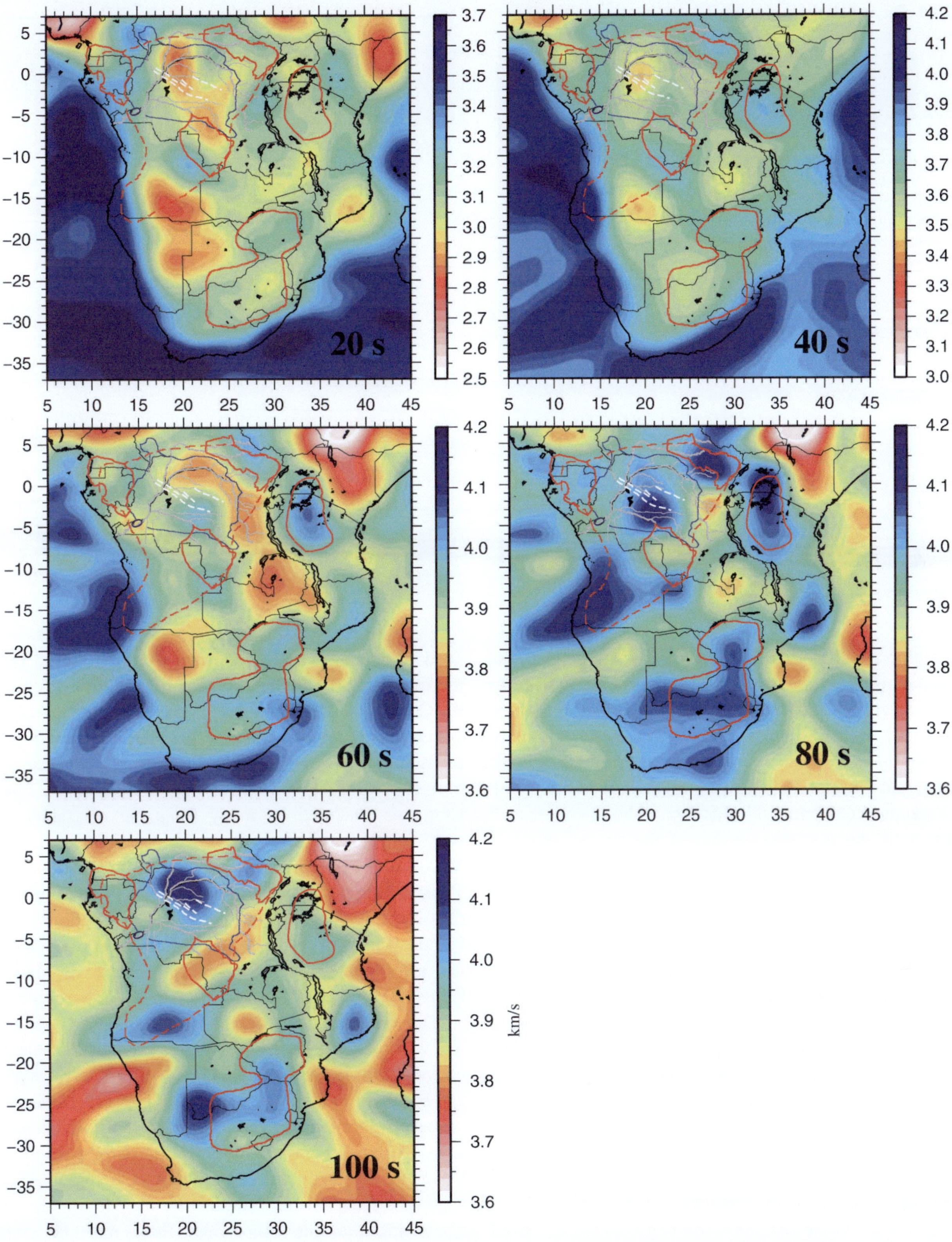

Fig. 1.9 Group velocities for fundamental mode Rayleigh waves at periods of 20–100 s. The *solid red lines* show Archean cratons, the *dashed red line* shows the outline of the greater Congo Shield, the *blue line* shows the zero thickness contour for the Mesozoic and Cenozoic sediments of the Congo Basin, and the *white lines* show rift faults as described in Fig. 1.1

Fig. 1.10 Group velocity curves for crustal structure for the portions of the Congo Basin where crustal thinning may exist (Kadima et al. 2011b). The *black line* shows the average group velocities from our model for the region of thin crust according to Kadima et al. (2011b). The *colored lines* show synthetic curves for crustal models with the Moho at the depths noted on each curve

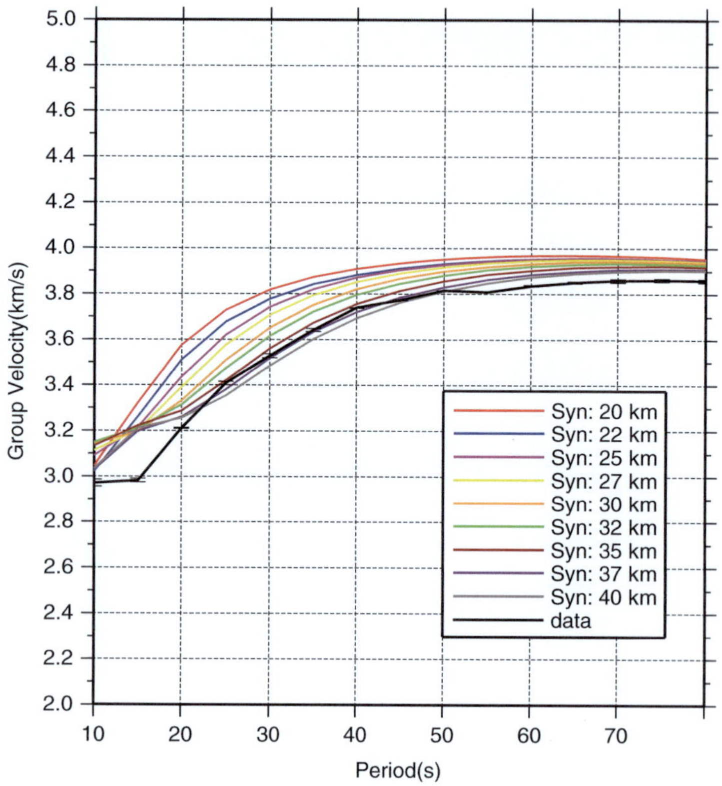

The proposal that a mobile belt may extend beneath the northern part of the Congo Basin has important implications for the origin of the basin, as discussed below. Many geodynamic models invoking density contrasts within or below the mantle lithosphere (e.g. Hartley and Allen 1994; Downey and Gurnis 2009; Crosby et al. 2010; Forte et al. 2010) assume a uniformly thick cratonic lithosphere beneath the basin. If the northern portion of the Congo Basin is underlain by Proterozoic lithosphere and is surrounded by Archean lithosphere to the east, south and north, then there could exist substantial heterogeneity in lithospheric thickness across the region. Continental-scale tomographic models typically have lateral resolution of no better than 300–500 km at upper mantle depths > 100 km, and therefore variations in lithospheric thickness associated with a Proterozoic mobile belt under the northern part of the Congo Basin may not be resolved in those models.

A second geodynamic implication of the proposal that Proterozoic lithosphere lies beneath the northern part of the Congo Basin concerns the Neoproterozoic rifting event that may have initiated basin subsidence. The location of the rifts are shown in Fig. 1.11 and are found in the region of our velocity model where the velocities begin to change from faster to slower values going from the center of the basin towards the northeast. The rifts may have formed along the northern border of the SWCS, as opposed to within the middle of a single large Archean craton, alleviating the need to explain why a Neoproterozoic rift would form in the interior of an Archean craton (Fig. 1.11f). Within both the Cenozoic East African rift system and the NE Karoo rift system there are good examples of rift basins forming along the margins of thick Archean lithosphere. For example, in northern and western Tanzania, the margin of the Tanzania Craton is fractured by many Cenozoic rift faults, and a number of half-graben basins have formed (i.e. Lake Manyara, Lake Eyasi basins). From the Carboniferous-Triassic Karoo system, a good example is the Zambezi rift that formed along the northern margin of the Zimbabwe Craton. These rifts provide plausible analogs for understanding the development of a Neoproterozoic rift under the Congo Basin within a region of cratonic lithosphere.

The tomographic models in Figs. 1.2 and 1.11 are consistent with the existence of Proterozoic lithosphere beneath the northeastern side of the Congo Basin, where lower velocities can be seen in the 50–100 km depth interval. However, the models do not support the proposal that Proterozoic lithosphere extends beneath the entire northern portion of the basin. In Figs. 1.2 and 1.11, there is a region of faster velocities beneath the northwestern side of the basin, suggesting that the lithosphere there may be cratonic. Hence, while Proterozoic lithosphere likely extends to the NW under the northeastern side of the basin, it probably does not extend all the way across the basin to southern Cameroon.

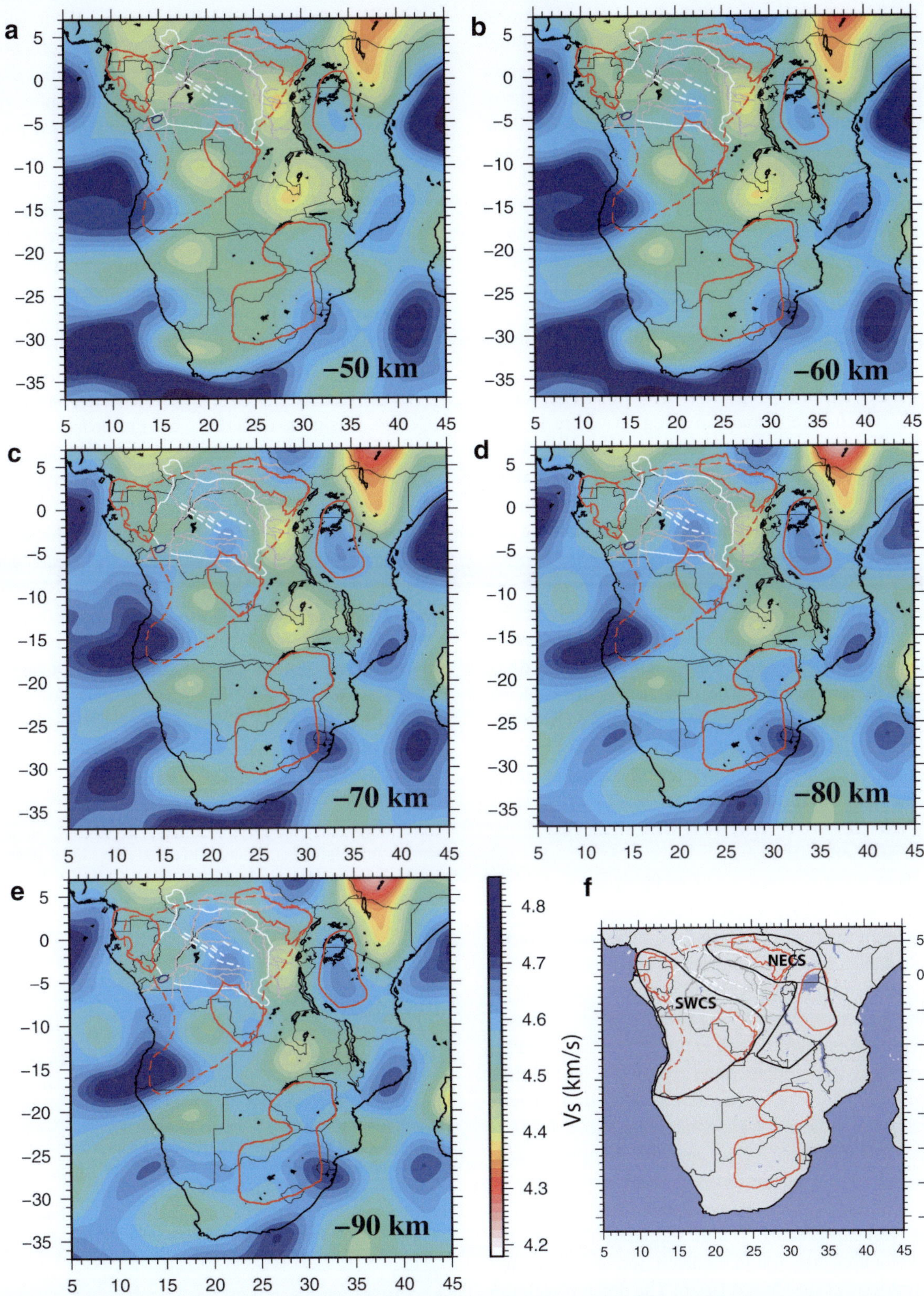

Fig. 1.11 (**a–e**) Shear wave velocities at depths of 50, 60, 70, 80 and 90 km. *Solid red lines* show Archean cratons, *dashed red line* shows the outline of the greater Congo Shield, the *solid white line* shows the zero thickness contour for the Mesozoic and Cenozoic sediments of the Congo Basin and *dashed white lines* show rift faults as described in Fig. 1.1. (**f**) *Red* and *white lines* are the same as in figures **a–e**, and the *bold*, *black lines* show the SWCS and NECS proposed by Master (2004)

The difference in shear wave velocity in the uppermost parts of the mantle between Archean and Proterozoic lithosphere upon which this interpretation is based is not necessarily well understood. It is also noted that at the depths of interest (i.e. 50–100 km) the Proterozoic mobile belts to the south of the Congo Shield are characterized by slower velocities than beneath the Archean cratons (Figs. 1.2 and 1.11). This pattern of slower shear wave velocities in the uppermost part of the mantle beneath Proterozoic terrains is a common feature of many tomographic images from Precambrian shields, although not all (Darbyshire and Eaton 2010).

Lastly, it is of interest to compare the Precambrian lithospheric architecture for the Congo Basin with the Neoproterozoic Central super-basin of Australia (e.g. Walter et al. 1995), which also formed in a region where several Archean cratons amalgamated in the Proterozoic (e.g. Betts and Giles 2006; Cawood and Korsch 2008) and where there has since been a prolonged history of deformation. Similar upper mantle structure is found beneath and surrounding this basin; at 75 km depth slow velocities are observed beneath the Proterozoic central region, while faster velocities are observed beneath the Archean cratons themselves. By 125 km depth, fast velocities are observed beneath most of the Precambrian Shield (e.g. Kaiho and Kennett 2000; Fishwick and Reading 2008). The similarity in lithospheric structure beneath the Congo Basin and central Australia suggests that the slower uppermost mantle wave speeds may reflect similar processes involved in the amalgamation and subsequent deformation of these regions.

Summary and Conclusions

A new shear wave velocity model of the upper mantle for central and southern Africa developed using an inversion of group velocity measurements from fundamental mode Rayleigh waves has been presented. The model shows several areas with faster velocities that correspond to Archean cratons (Bomu, Kaapvaal, Kasai, Ntem, Tanzanian, Zimbabwean), and areas of slower velocities marking Proterozoic mobile belts in between. The mobile belts that exhibit slower upper mantle velocities include the Mesoproterozoic Kibaran, Karagwe-Ankole, Irumide and Southern Irumide Belts, and the Neoproterozoic Damara Belt. The Paleoproterozoic Bangweulu Block in northern Zambia also has slower upper mantle velocities than the Archean cratons. Within the Congo Shield, the fastest velocities are found beneath the southern and central portions of the Congo Basin. The depth resolution of the models precludes estimating the thickness of the lithosphere.

Our model compares favorably with the tomographic model from Fishwick (2010), which is representative of a range of tomographic models using Rayleigh wave phase velocities. In both models a region of slower mantle velocities is found beneath the northeastern portion of the Congo Basin to the south and southwest of the Bomu Craton. This region of slower mantle velocities does not continue beneath the northwestern part of the Congo Basin.

This finding suggests that the lithospheric mantle beneath the Congo Basin may not be uniform, and that Proterozoic lithosphere may lie beneath the northeastern side of the basin. However, it does not support the extension of Proterozoic lithosphere beneath the entire northern portion of the basin. Many geodynamic models invoking density contrasts within or below the mantle lithosphere assume a uniformly thick cratonic lithosphere beneath the basin. If the northeastern portion of the Congo Basin is underlain by Proterozoic lithosphere, then there could exist heterogeneity in lithospheric thickness across the region. A second geodynamic implication concerns the Neoproterozoic rifting event that may have initiated basin subsidence. The proposed location of the rifts are in the region of our velocity model where the velocities begin to change from faster to slower values going from the center of the basin toward the northeast. We suggest that the rifts may have formed along the northeastern border of a southwestern Congo shield (SWCS), as defined by Master (2004), as opposed to within the middle of a greater Congo Craton, alleviating the need to explain why a Neoproterozoic rift might form in the interior of a large Archean Craton.

Acknowledgement We thank Maarten J. de Wit and an anonymous reviewer for helpful comments. This study has been funded by the National Science Foundation (Grants OISE-0530062, EAR-0440032, EAR-0824781).

References

Ammon CJ (1998) Note on seismic surface wave processing. Part I. Group velocity estimation, http://eqseis.geosc.psu.edu/~cammon/

Bassin C, Laske G, Masters G, Eos T (2000) The current limits of resolution for surface wave tomography in North America. Am Geophys Union 81:F897

Betts PG, Giles D (2006) The 1800–1100 Ma tectonic evolution of Australia. Precambrian Res 144:92–125

Braitenberg C, Ebbing J (2009) The GRACE-satellite gravity and geoid fields in analyzing large-scale cratonic or intracratonic basins. Geophys Prospect 57:559–571. doi:10.1111/j.1365-2478.2009.00793.x

Buiter SJH, Steinberger B, Medvedev S, Tetreault J (2012) Could the mantle have caused subsidence of the Congo Basin?Tectonophysics 514-517, 62-80.. doi:10.1016/j.tecto.2011.09.024

Cahen L (1954) Géologie du Congo belge. Vaillant-Caramanne, Liège, p 577

Cawood PA, Korsch RJ (2008) Assembling Australia: Proterozoic building of a continent. Precambrian Res 166:1–38

Crosby AG, Fishwick S, White N (2010) Structure and evolution of the intracratonic Congo Basin. Geochem Geophys Geosyst. 11: Q06010. doi: 10.1029/2009GC003014

Daly MC, Lawrence SR, Diemu-Tshiband K, Matouana B (1992) Tectonic evolution of the Cuvette Centrale, Zaire. J Geol Soc London 149:539–546

Darbyshire F, Eaton D (2010) The lithospheric root beneath Hudson Bay, Canada from Rayleigh wave dispersion: no clear seismological distinction between Archean and Proterozoic mantle. Lithos 120:144–159. doi:10.1016/j.lithos.2010.04.010

De Wit M, Stankiewicz J, Reeves C (2008) Restoring Pan-African-Brasiliano connections: more Gondwana control, less Trans-Atlantic corruption. In: Pankhurst RJ, Trouw R AJ, Brito neves BB, de wit MJ (eds) West Gondwana: pre-cenozoic correlations across the South Atlantic Region, vol 294. Geological Society, London, Special Publications, pp 399–412. doi: 10.1144/SP294.20

Downey NJ, Gurnis M (2009) Instantaneous dynamics of the cratonic Congo basin. J Geophys Res 114, B06401. doi:10.1029/2008JB006066

Dziewonski AM, Bloch S, Landisman M (1969) A technique for the analysis of transient seismic signals. B Seismol Soc Am 59: 427–444

Fernandez-Alonso M, Cutten H, De Waele B, Tack L, Tahon A, Baudet D, Barritt SD (2012) The Mesoproterozoic Karagwe-Ankole Belt (formerly the NE Kibara Belt): The result of prolonged extensional intracratonic basin development punctuated by two short-lived far-field compressional events. Precambrian Res 216–219:63–86

Fishwick S (2010) Surface wave tomography: imaging of the lithosphere asthenosphere boundary beneath central and southern Africa. Lithos 120:63–73

Fishwick S, Bastow ID (2011) Towards a better understanding of African topography: a review of passive-source seismic studies of the African crust and upper mantle. In: Van Hinsbergen DJJ, Buiter SJH, Torsvik TH, Gaina C, Webb SJ (eds) The formation and evolution of Africa: a synopsis of 3.8 Ga of Earth history. Geological Society, London, pp 343–371. doi:10.1144/SP357.19, Special Publications

Fishwick S, Reading AM (2008) Anomalous lithosphere beneath the Proterozoic of western and central Australia: a record of continental collision and intraplate deformation? Precambrian Res 166: 111–121

Forte AM, Quere S, Moucha R, Simmons NA, Grand SP, Mitrovica JX, Rowley DB (2010) Joint seismic-geodynamic-mineral physical modelling of African geodynamics: a reconciliation of deep-mantle convection with surface geophysical constraints. Earth Planet Sci Lett. doi:10.1016/j.cpsl.2010.03.017

Gubanov AP, Mooney WD (2009) New global geological maps of crustal basement age. Eos transactions, AGU, 90, Fall Meet.Suppl. Abstract T53B-1583

Hartley R, Allen PA (1994) Interior cratonic basins of Africa: Relation to continental break-up and role of mantle convection. Basin Res 6:95–113

Hartley R, Watts AB, Fairhead JD (1996) Isostacy of Africa. Earth Planet Sci Lett 137:1–18

Herrmann RB (1973) Some aspects of band-pass filtering of surface waves. B Seismol Soc Am 63:663–671

Julià J, Ammon CJ, Herrmann RB, Correig AM (2000) Joint inversion of receiver functions and surface-wave dispersion observations. Geophys J Int 143:99–112

Julià J, Ammon CJ, Nyblade AA (2005) Evidence for mafic lower crust in Tanzania, East Africa, from joint inversion of receiver functions and Rayleigh wave dispersion velocities. Geophys J Int 162: 555–569

Kadima E, Delvaux D, Sebagenzi SN, Tackw L, Kabeyaz SM (2011a) Structure and geological history of the Congo Basin: an integrated interpretation of gravity, magnetic and reflection seismic data. Basin Res 23:499–527. doi:10.1111/j.1365-2117.2011.00500

Kadima EK, Sebagenzi SMN, Lucazeau F (2011b) A Proterozoic-rift origin for the structure and the evolution of the cratonic Congo basin. Earth Planet Sci Lett. doi: 10.1016/j.epsl.2011.01.037

Kaiho Y, Kennett BLN (2000) Three-dimensional seismic structure beneath the Australian region from refracted wave observations. Geophys J Int 142:651–668

Kennett BLN, Engdahl ER (1991) Traveltimes for global earthquake location and phase identification. Geophys J Int 122:429–465

Lawrence SR, Makazu MM (1988) Zaire's Central Basin: Prospectivity outlook. Oil Gas J 86:105–108

Lepersonne J (1974) Map and Notice explicative de la carte géologique du Zaïre au 1/2 000 000.Direction de la Géologie, Dept. des Mines, Rép.du Zaïre, 67 pp

Lepersonne J (1978) Stucturegeologique du basin interieur du Zaire, Academie Royale des Sciences d'Outre-Mer, Bruxelles, Classe des Sciences Naturelles et Medicales, N.N. XX(2), 1–27

Linol B (2013) Sedimentology and sequence stratigraphy of the Congo and Kalahari Basins of south-central Africa and their evolution during the formation and break-up of West Gondwana. PhD thesis, Nelson Mandela Metropolitan University, 375p

Master S (2004) Archean to Neoproterozoic assembly and growth of the Greater Congo Craton. In: Geoscience Africa 2004 Conference, Abstract Volume 2, University of the Witwatersrand, Johannesburg, 12–16 July, 2004, 425–426

Moloto-A-Kanguemba GR, Trindade RIF, Monié P, Nédélec A, Siqueira R (2008) A late Neoproterozoic paleomagnetic pole for the Congo craton: Tectonic setting, paleomagnetism and geochronology of the Nola dike swarm (Central African Republic). Precambrian Res 164:214–226

Mulibo G, Nyblade A (2013) The P and S wave velocity structure of the mantle beneatheastern Africa and theAfrican superplume anomaly. Geochem Geophys Geosyst. doi:10.1002/ggge.20150

Nataf HC, Ricard Y (1996) 3SMAC: an a priori tomographic model of the upper mantle based on geophysical modeling. Phys Earth Planet Int 95:101–122

Paige C, Saunders M (1982) LSQR: an algorithm for sparse linear equations and sparse least squares. ACM Trans Math Softw 8:43–71

Park Y, Nyblade AA, Rodgers AJ, Al-Amri A (2008) S wave velocity structure of the Arabian Shield upper mantle from Rayleigh wave tomography. Geochem Geophys Geosyst 9, Q07020. doi:10.1029/2007GC001895

Pasyanos ME (2010) Lithospheric thickness modeled from long-period surface wave dispersion. Tectonophysics 481:38–50. doi:10.1016/j.tecto.2009.02.023

Pasyanos ME, Nyblade A (2007) A top to bottom lithospheric study of Africa and Arabia. Tectonophysics. 444: 27–44.

Priestley K, McKenzie D (2013) The relationship between shear wave velocity, temperature, attenuation and viscosity in the shallow part of the mantle. Earth Planet Sci Lett 831:78–91. doi:10.1016/j.epsl.2013.08.022

Priestley K, McKenzie D, Debayle E, Pilidou S (2008) The African upper mantle and its relationship to tectonics and surface geology. Geophys J Int 175:1108–1126

Ritsema J, van Heijst H (2000) New seismic model of the upper mantle beneath Africa. Geology 28:63–66

Sebai A, Stutzmann E, Montagner J-P, Sicilia D, Beucler E (2006) Anisotropic structure of the African upper mantle from Rayleigh and Love wave tomography. Phys Earth Planet Int 155:48–62

Tack L, Wingate MTD, De Waele B, Meert J, Belousova E, Griffin B, Tahon A, Fernandez-Alonso M (2010) The 1375Ma "Kibaran event" in Central Africa: Prominent emplacement of bimodal magmatism under extensional regime. Precambrian Res 180: 63–84. doi:10.1016/j.precamres.2010.02.022

Toteu SF, van Schmus WR, Penaye J, Nyobé JB (1994) U-Pb and Sm-Nd evidence for Eburnian and Pan-African high-grade metamorphism in cratonic rocks of southern Cameroon. Precambrian Res 67:321–347

Tugume F, Nyblade A, Julia J (2012) Moho depths and Poisson's ratios of Precambrian crust in East Africa: Evidence for similarities in Archean and Proterozoic crustal structure. Earth Planet Sci Lett 355–356:73–81. doi:10.1016/j.epsl.2012.08.041

Vicat J-P, Pouclet A, Nkoumbou C, Mounagué AS (1997) Le volcanismefissuralnéoprotérozoïque des séries du Djainférieur, de Yokadouma (Cameroun) et de Nola (RCA)- Signification géotectonique. CR Acad Sci, Paris, Sci de la terreet des planètes 325:671–677

Walter MR, Veevers JJ, Calver CR, Grey K (1995) Neoproterozoic stratigraphy of the CentralianSuperbasin, Australia. Precambrian Res 73:173–195

Precambrian Basement of the Congo Basin and Its Flanking Terrains

2

Maarten J. de Wit and Bastien Linol

2.1 Introduction

Africa is the world's second largest continent, with just over 30 million km^2, almost entirely underlain by Precambrian crystalline basement. The geology of this basement, particularly in central Africa, however, is generally not well-described and is poorly constrained by geochronology because of an extensive cover of Phanerozoic rocks, recent sediments, laterites and vegetation; and because of an extended period of socio-political instabilities that continues to make it difficult to explore and study this region.

Until 30 years ago, the Precambrian history and geochronology of central Africa's basement was based primarily on geologic and low resolution Rb/Sr analyses essentially carried out and synthesized by Louis Cahen and Norman Snelling, and their collaborators, in their iconoclastic book '*The Geochronology and Evolution of Africa*' (Cahen et al. 1984). Since then there has been a slow exponential increase in more precise U-Pb dates on zircons and hence improved understanding of Africa's crustal evolution, particularly in the regions of central Africa: in Gabon, Cameroon, Central African Republic (CAR) and Chad (e.g. Feybesse et al. 1998; Toteu et al. 2006, 2014; Nkoumbou et al. 2013; de Wit et al. 2014, under review); in Uganda, Democratic Republic of Congo (DRC), Rwanda and Burundi (e.g. Link et al. 2010; Tack et al. 2010; Fernandez-Alonso et al. 2012; Lawley et al. 2013, 2014); in Tanzania (e.g. Boniface et al. 2012; Kabette et al. 2012a, b; Kasanzu 2014); Mozambique (e.g. Bingen et al. 2009); in northern Zambia (e.g. Master et al. 2005; de Waele et al. 2006; Lawley et al. 2013, 2014) and in central-northeast Angola (e.g. Carvalho et al. 2000; Delor et al. 2006; Jelsma et al. 2011, 2012), amongst others.

Here we briefly review and summarize some of these results to provide a framework of the Precambrian basement underlying and flanking the Congo Basin (CB), in particular because this constitutes the upper lithosphere foundation to this very large (ca. 1.8 million km^2) and poorly understood Phanerozoic sedimentary basin that is the focus of this book.

2.2 Cratons and Shields in Africa

The African continent can be subdivided into four major Precambrian Shields that amalgamated along Neoproterozoic orogens (Fig 2.1, inset B), and each of which in turn comprises assemblages of Archean cratons further embedded within Meso- and Paleo-Proterozoic mobile belts. However, in the literature, the terms Craton and Shield are not always clearly defined. For example, geologic terrains in central Africa are often used indiscriminately and sometimes interchangeable: Congo Craton, Congo Shield and Central African Shield or Craton; Kasai Craton or Shield and Lunda Shield; Angola-Kasai Craton, Angola Craton or Shield, and NE Angolan Shield, or Cuango Shield or Craton, and southern Congo Craton; Greater Congo Shield, etc. To clarify our nomenclature here, we follow the definitions of Stankiewicz and de Wit (2013), and specifically refer to 'cratons' as Archean blocks (e.g., stabilized >2.5 Ga), and to 'shields' as stabilized, post-Archean continental domains that formed and/or amalgamated at specified times during the Proterozoic-Phanerozoic, and within which Archean cratons (or deformed Archean blocks) are embedded.

We refer collectively then to a number of relatively small Archean cratons that underlie, or in part underlie, the CB

M.J. de Wit (✉)
AEON-ESSRI (Africa Earth Observatory Network-Earth Stewardship Science Research Institute), Nelson Mandela Metropolitan University, Port Elizabeth, South Africa
e-mail: maarten.dewit@nmmu.ac.za

B. Linol
AEON-ESSRI (Africa Earth Observatory Network-Earth Stewardship Science Research Institute), Nelson Mandela Metropolitan University, Port Elizabeth, South Africa

Geological Sciences, Nelson Mandela Metropolitan University, Port Elizabeth, South Africa
e-mail: Bastien.aeon@gmail.com

M.J. de Wit et al. (eds.), *Geology and Resource Potential of the Congo Basin*, Regional Geology Reviews,
DOI 10.1007/978-3-642-29482-2_2, © Springer-Verlag Berlin Heidelberg 2015

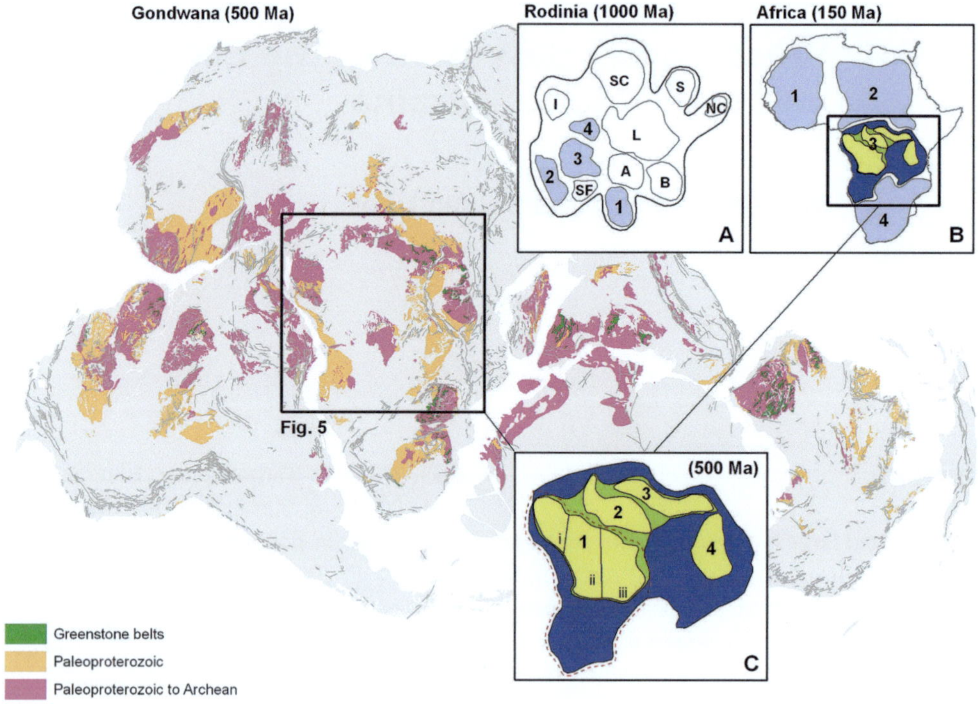

Fig. 2.1 Simplified early Precambrian map of Gondwana (GIS from de Wit et al. 1988; modified). Inset (A and B) Four major African Precambrian Shields (*pale blue*): 1 = West African Shield; 2 = Saharan Shield; 3 = Central African Shield; 4 = Southern African Shield, and other fragments of Rodinia (~1 Ga): A = Amazonia; B = Baltica; L = Laurentia; I = India; S = Siberia; SC = South China; NC = North China (simplified from Li et al. 2008 and Lindeque et al. 2011). Inset (C) crustal domains of central Africa with four cratons (*yellow*): $1_{i,ii,iii}$ = Ntem, Cuango, Kasai = SouthWest Congo Craton (SWCC); 2 = Cuvette [Central] Congo Craton (CCC); 3 = Mboumou-Uganda = NorthEast Congo Craton (NECC), forming the Congo Shield (CS, *green*), and enlarged in Proterozoic to form the Central African Shield (CAS, *dark blue*). *Red dotted outline* is the SouthWest Congo Shield (SWCS)

(e.g. Kasai, Cuango, Ntem, Bouca, [Bomu] Mboumou-Uganda) and some of which amalgamated before the end-Archean into three larger cratons: 'the Southwestern, Cuvette [or Central], and Northeastern Congo Cratons (SWCC, CCC, NECC, respectively), but the detailed tectonic history of which still remains largely unknown (Fig 2.1, inset C). We use 'Congo Shield' (CS) for a larger lithospheric region that comprises those three cratons that cannot be directly correlated beneath the present CB for lack of appropriate geophysical data, but which likely amalgamated during the Proterozoic and which all lie inboard of surrounding Pan-African (ca. 0.8–0.5 Ga), Kibaran (ca. 1.0–1.4 Ga) or Eburnian (ca. 1.9–2.3 Ga) orogenic belts and related inferred suture zones that surround the CB (Fig. 2.1, inset B; Fig. 2.2).

For simplicity, here we specifically separate the Congo Shield (CS) from potential conterminous shield areas in Brazil. Even though parts of a larger central African shield area, which we here term the Central African Shield (CAS), may have been linked to the greater São Francisco Shield (mostly referred to in the literature as a Craton) in a Rodinia framework (Fig. 2.1, inset A), there is still considerable uncertainty about the detailed timing and accretion processes that may have linked these two shields into a larger Paleoproterozic Shield in Eburnian times (circa 2.0 Ga; e.g. Pedrosa-Soares et al. 2008, and references therein). Suffice it to say that it has been argued, for example, on the basis of paleomagnetism and undeformed Mesoproterozioc sedimentary sequences found on both shields (dated between 1.7–1.8 Ga (Pedreira and de Waele 2008) that the two continental terrains may have been connected at that time (e.g. Trompette 1994), but also that the age correlations are far from detailed enough to substantiate this correlation from one continent to another (e.g. McCourt et al. 2004; Pedreira and de Waele 2008). Whilst we will briefly refer to this potential connectivity below, it is beyond the scope of this chapter to explore this in further detail here.

As early as the mid-Paleoproterozoic, the eastern and northern margins of the CS were convergent margins that experienced Eburnian subduction-obduction of ca. 2.0–2.3 Ga oceanic lithosphere, and accretion of Archean continental fragments, over a period of at least 150 Ma, between ca. 2050–1880 Ma (e.g. Boniface and Schenk 2012; Boniface et al. 2012; Nkoumbou et al. 2013; Lawley et al. 2013, 2014; and see below). By contrast, the southern margin of the CS is defined by the Central Shield Zone of Angola (cf. Carvalho et al. 2000), a wide transition zone of Eburnian granitoid magmatism and high grade tectonism that also embodies a

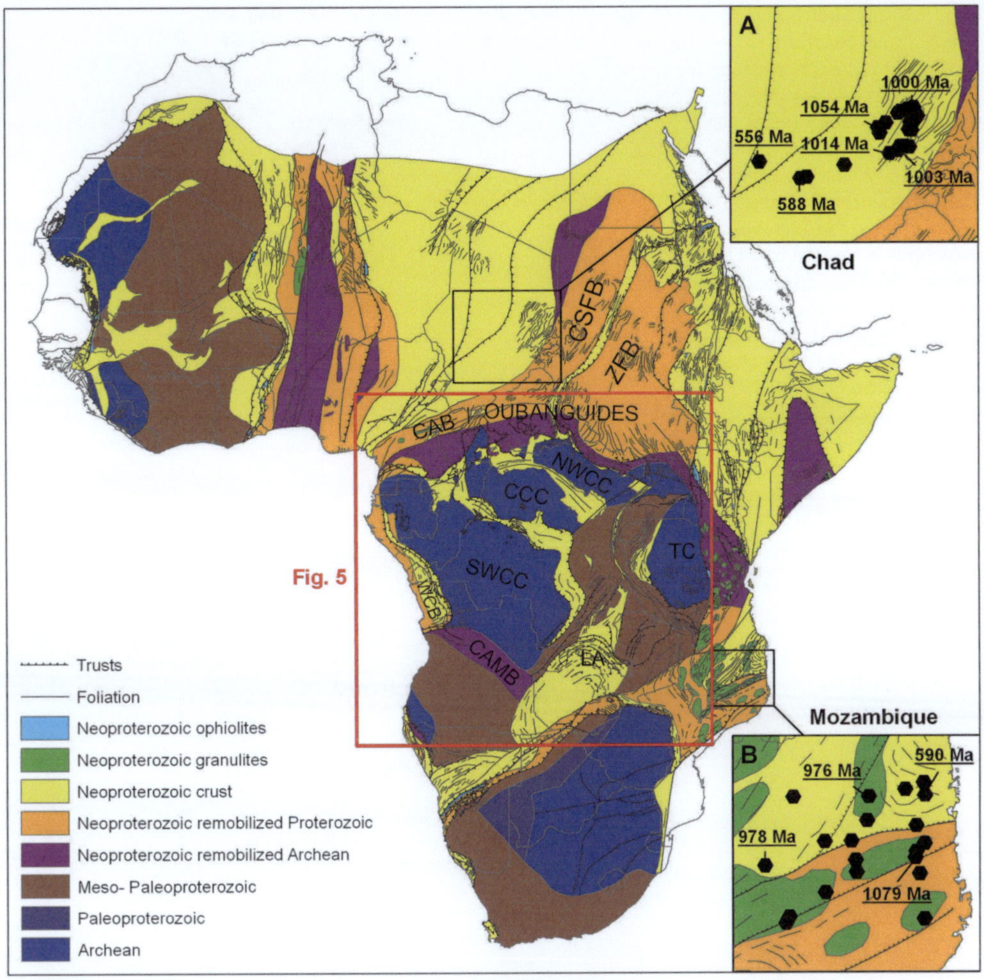

Fig. 2.2 Geological map of the Precambrian basement of Africa (GIS from de Wit et al. 2008; updated), locating the studied area in central Africa. Black rectangles detail adjacent regions with recent data: (A) in Chad (de Wit et al. under review) and (B) in Mozambique (Jamal, PhD thesis 2005)

number of remobilized Archean fragments, and which we here rename the Central Angola Mobile Belt (CAMB). This belt separates the rest of the Angola basement to the south, comprising predominantly Eburnian-Kibaran (mid-Proterozoic) crust, from the SWCC. We suggest these two blocks collided along the CAMB at ca. 2 Ga to form the 'Southwestern Congo Shield' (SWCS, as first termed by S. Masters in 2004; see Chap. 1 in this Book).

To the west of the CS lies a wide region of Proterozoic basement, including Eburnian (Ubendian), Kibaran and Pan-African segments that separate it from the Tanzanian Craton. This intervening region has a complex geologic history that is still under reconnaissance investigation, and the details of which are beyond the scope of this chapter. Suffice it to mention that an Eburnian orogenic belt/suture zone (the Rwenzori Fold Belt of Tanner, 1974), and the Buganda-Toro system, e.g. Nagudi et al. 2003, and references therein separates the Tanzanian Craton from the NWCC (Mboumou-Uganda Craton). It is not clear how this Eburnian suture links up southwards across central Africa, from Uganda and western DRC through Rwanda, Burundi, western Tanzania to Zambia and Angola. We show here two likely Eburnian terrain boundaries directly flanking the CS and the Tanzanian Craton, respectively (Fig. 2.3). This leaves the central Proterozoic block (including the Paleoproterozoic Bangwuelu Block), with occasional Archean fragments, as a separate accretionary region that first amalgamated between the CS and the Tanzanian Craton during Paleoproterozoic (Eburnian) times. This terrain was subsequently in part re-melted during the Mesoproterozoic (e.g. the 'Kibaran Event' [s.s.] in the Karagwe-Ankole region; Tack et al. 2010; Fernandez-Alonso et al. 2012) and again deformed within the Kibaran Belt (e.g. de Waele et al. 2006, 2008). Along the Ubendian Belt flanking the southwestern margin of the Tanzanian Craton, Eburnian oceanic crust was metamorphosed at low T/high P to eclogite during subduction at around 1866–1868 Ma, and was subsequently re-metamorphosed and re-deformed during Kibaran and Pan African tectonism (Boniface et al. 2012; Kabette et al. 2012a; Lawley et al. 2013, 2014). The complexity of this vast 'Kibaran Shield' is yet to be resolved.

Subsequent accretion of other crustal blocks along a number of orogenic belts surrounding the Congo Shield during the formation of Gondwana in the late Neoproterozoic-Early Cambrian created the still larger 'Central African Shield'

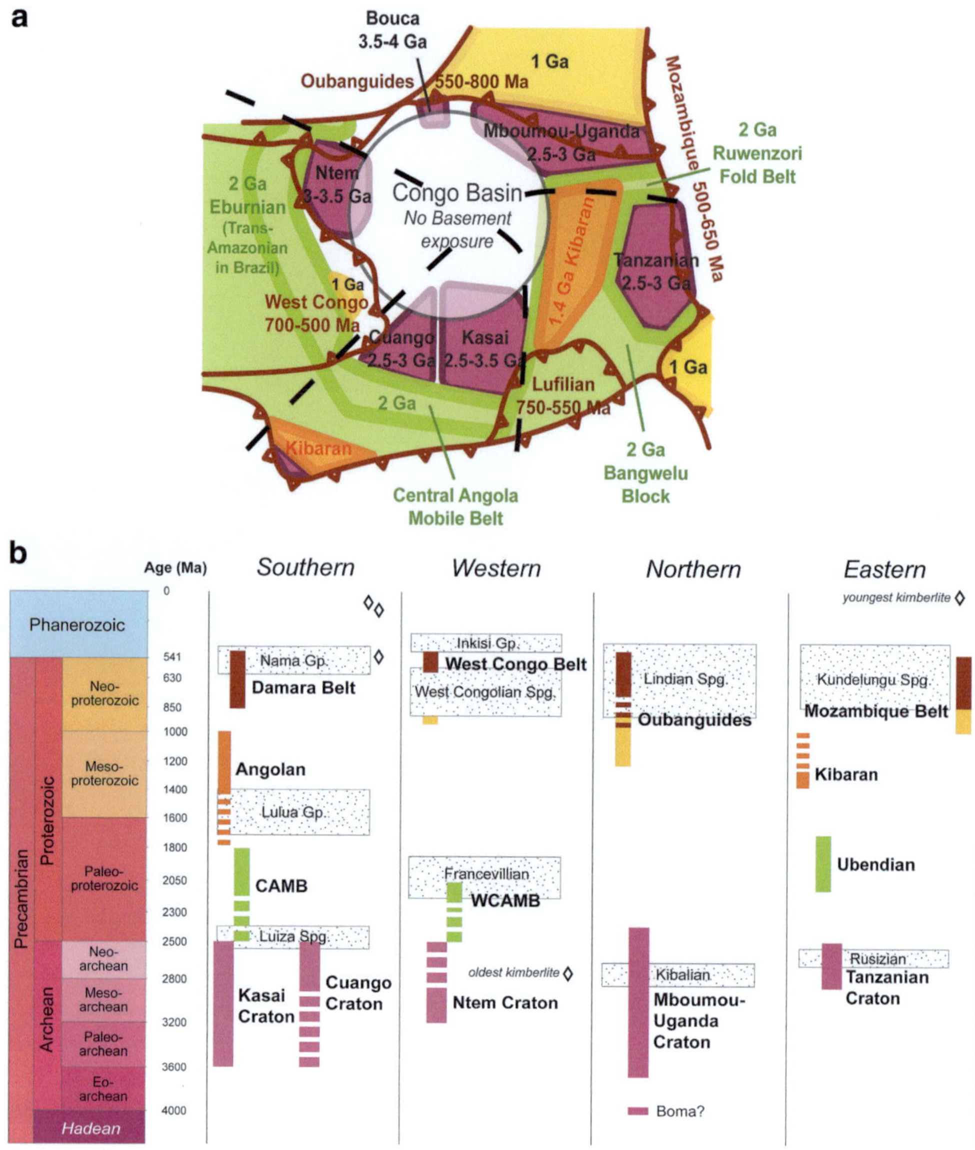

Fig. 2.3 (**a**) Sketch map of Precambrian basement surrounding the CB, showing the six Archean cratonic blocks (*purple*); Eburnian (*green*) and Kibaran terrains (*orange* and *yellow*), and tectonic fronts of Pan African fold-and-thrust belts (*brown*). Possible Pan African structures below the basin are not shown. (**b**) Time–Space diagram of Precambrian of central Africa. The four regions are shown in (**a**) (*dashed lines*) and described in detail in the text. Dotted rectangles in b = sedimentary sequences

(CAS) that stabilized by ca. 540 Ma (Fig. 2.1, inset B). CAS incorporates: the West Congo Belt of Gabon, western DRC and northwestern Angola; the Oubanguides (or sometimes called the Central African Fold Belt) of northern Gabon, Cameroon and CAR; the Mozambique Belt (as part of the southern East African Orogen) from southern Sudan, eastern Tanzania, Kenya and Mozambique; the Lufilian Belt (Arc) in Zambia and its extension into northwest Botswana and Namibia (as the Damara Belt) and from there north through the Kaoko Belt back to a link with the West Congo Belt in northwest Angola, via parts of the Braziliano Araçuaí Belt (now) of eastern Brazil (e.g. de Wit et al. 2008; Fig. 2.3). The precise connections between these Pan African Belts and their Brasiliano counterparts are still under scrutiny (e.g. Pedrosa-Soares et al. 2008; IGCP-628 project, and draft Gondwana Geological Map 2014, Renata Schmitt Personal communications, 2014).

Thereafter, this lithosphere of central Gondwana experienced prolonged and complex long-wavelength perturbations during multiple compression and then multiple rifting events as it became covered by Paleozoic and Meso-Cenozoic sedimentary basins—of which the CB is a prime and globally unique example (e.g. Linol, 2013; Linol et al., Chap. 12 of this Book). Unlike its early-linked Paraná Basin (now in Brazil), and unlike the Michigan Basin of North America to which an over-simplified history of the CB has been frequently compared (e.g. Hartley and Allan 1994), Gondwana break-up led to a CB that evolved further into a continental basin surrounded by rifts, except along its southern margin, where it is flanked by the southern (Kalahari) African high Plateau underlain by the Southern African Shield (e.g. Fig. 2.4).

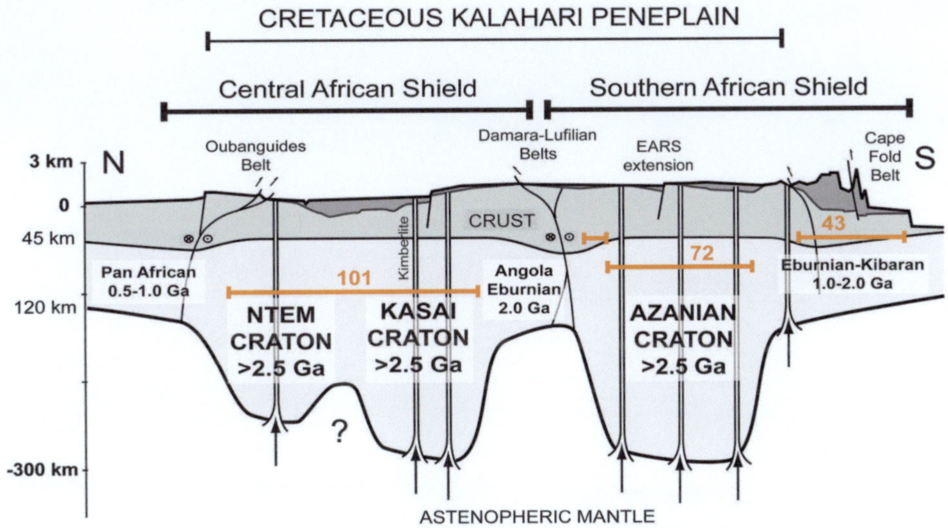

Fig. 2.4 Lithospheric model (in *pale grey*) N–S cross-section from southern Africa (Agulhas Fault) to Sahara (Chad), showing the Central and Southern African Shields capped by a sub-continental Cretaceous Kalahari peneplain. Effective elastic thicknesses of the lithosphere (Te; in km), shown in orange lines, are taken from Hartley and Allan (1994) and Doucoure and de Wit (1996), derived using the same techniques through spectral analysis

Shortly after Africa emerged out of Pangea/Gondwana, over a period of some 80 million years (ca. 200–120 Ma) that heralded the opening of the Indian and South Atlantic Oceans, the conterminous Central and Southern African Shields were uplifted during the Kalahari epeirogeny (de Wit 2007), and then peneplained to form the Kalahari Plateau (Fig. 2.4). During the end-Mesozoic to Cenozoic this plateau became covered by sands and a hard-cap of calcretes and silcretes (the Kalahari Group). Above the CB, this surface collapsed in the Eocene (Linol et al., Chap. 11, this Book), but across the Southern African Shield this surface remains mostly intact, probably because, largely, it is directly underlain by Precambrian basement rocks.

Whilst a large amount of deep local geophysical and geochemical data has enabled a firm understanding of the 3-D structure beneath the Southern African Shield (e.g. Evans et al. 2011; Bell et al. 2003; Stankiewicz and de Wit 2013, and references therein), including variations in depth to Moho and the bottom of the lithosphere, this is not the case for the CAS (e.g. Begg et al. 2009; Buiter et al., 2012; see also Raveloson et al., Chap. 1, this Book). Geophysical data is particularly scarce across the CS. Beneath the CB, there is no hard information about variations in depth to Moho or the lithosphere structure and its thickness, other than by inference from (mostly) Cretaceous diamondiferous kimberlites that have intersected the mantle lithosphere below the CS (Fig. 2.4; de Wit and Jelsma, Chap. 18, this Book); and from low resolution seismic tomography that suggest the thickness of the lithosphere beneath the cratons embedded in the CS varies between 130–200 km (e.g. Raveloson et al., Chap. 1, this Book) and is thus thick enough to contain the transition into the general lithospheric mantle stability field of diamonds (ca. >120 km). But the geophysics has not as yet tested potential lithospheric variations across the boundaries between, for example, the three central Congolese Cratons of the CS.

2.3 Regional Geology and Geochronology of Central Africa

The CB of central Africa is completely surrounded by peneplained Precambrian basement, including Archean cratonic blocks and numerous Paleo- to Meso-Proterozoic mobile belts (Figs. 2.2, 2.3, 2.4 and 2.5). This complex assemblage comprises the Kasai, Cuango, Ntem, Bouca, Mboumou-Uganda and Tanzanian Cratons (ca. 2.5–3.5 Ga; Cahen et al. 1984), the CAMB (ca. 2.0–2.3 Ga; Carvalho et al. 2000), and the West Central African Mobile Belt (WCAMB; ca. 2.0–2.5 Ga; Feybesse et al. 1998).

These oldest blocks apparently all amalgamated during Eburnian (in Africa; Trans-Amazonian in South America) orogenesis, ca. 1.8–2.2 Ga, along: an east–west suture in southern Uganda, cutting across the central Rwenzori Mountains, and collectively known as the Rwenzori Fold Belt (Tanner 1973); the Rusizian and Ubendian-Usagaran Belts in Uganda and western Tanzania (Lenoir et al. 1995; Tack et al. 2010; Link et al. 2010; Fernandez-Alonso et al. 2012; Boniface et al. 2012; Lawley et al. 2013, 2014), across central Angola (Doucoure et al. 1999; de Carvalho et al. 2000; Jelsma et al. 2012), and north–south along the South Atlantic margin where it forms the basement to the Neoproterozoic West Congo Belt and its Braziliano counterpart in Brazil (Feybesse et al. 1998; Tack et al. 2001; Toteu et al. 2001; Pedrosa-Soares et al. 2008).

We have drawn roughly defined Eburnian sutures along the CAMB and WCAMB, where major continental accretion may have occurred to consolidate the SWCS (Figs. 2.3 and 2.6). There are widespread areas in the CAMB (e.g. Central Shield Zone of Carvalho et al. 2000; Jelsma et al. 2012) where regions of Eburnian deformation, granulite-grade metamorphism and crustal melting have significantly affected Archean granite-greenstone fragments. This broad

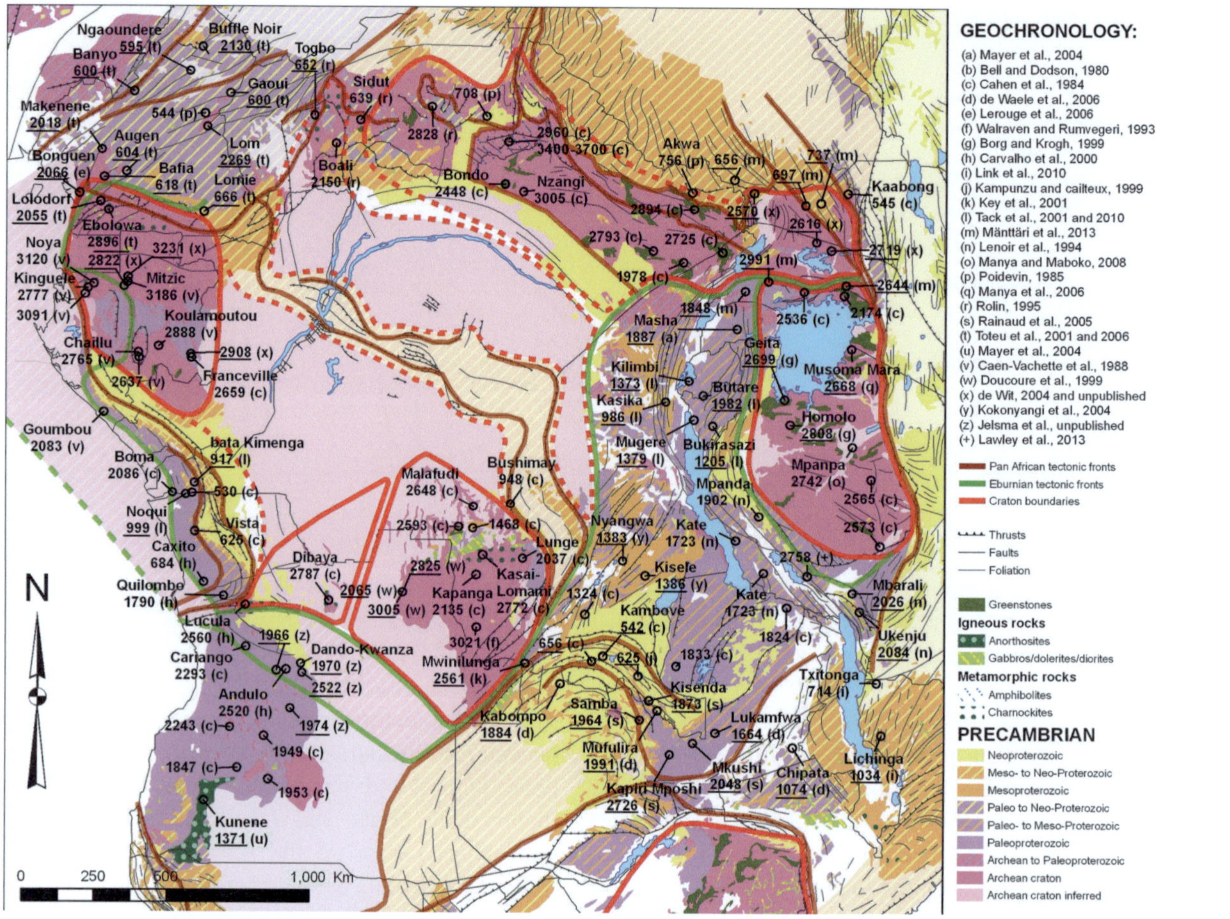

Fig. 2.5 Detailed Precambrian map of basement surrounding the CB (GIS database from de Wit et al. 1988, modified), with selected dates from the literature. *Underlined numbers* are U-Pb zircon dates, in millions of years (Ma)

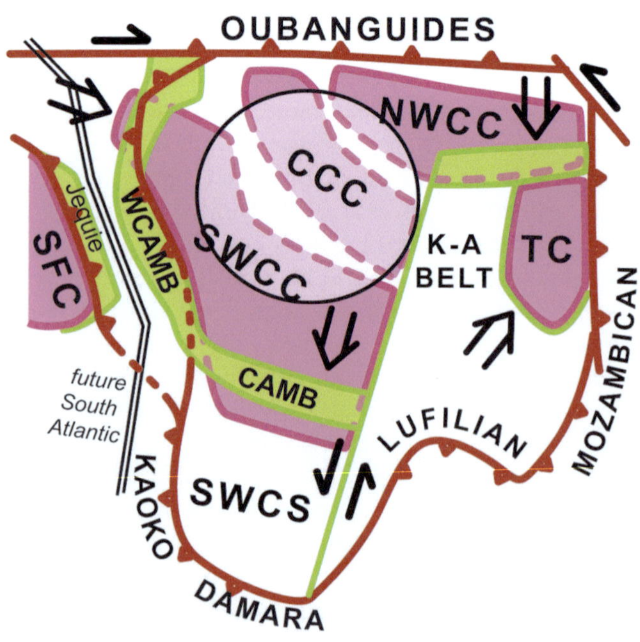

Fig. 2.6 Schematic diagram of the formation of the Central African Shield (CAS)

boundary zone (200–250 km wide) between the SWCC and the Angola region to the south may represent an Eburnian exhumed 'Alpine-Himalaya' type-zone, formed during prolonged southward under-thrusting of the Kasai Craton beneath the Angola basement, forming an Eburnian paleo-Tibet like Plateau with low pressure granulite formation and prolonged alkaline magmatism (2300–1300 Ma) and wide-spread rapakivi granites as old as 2065–2068 Ma (U/Pb zircon TIMS dates; Doucoure et al. 1999). The Angola region stretches ca. 1,000 km south towards its tectonic contact with the Kaoko-Damara Belt in Namibia, a contact boundary that contains slivers of Archean gneisses (ca. 2580–2620 Ma) within the Eburnian-Kibaran granitoids of the Angola basement (e.g. Miller 2008). Near its southern margin this Angola basement is overlain unconformably by relatively undeformed Paleo-Mesoproterozoic sedimentary rocks (Chela Group, ca. 1790 Ma McCourt et al. 2004), and intruded by the vast Mesoproterozoic mafic-anorthosite complex of Kunene at 1371 Ma (U-Pb zircon date; Mayer et al. 2004) associated with widespread red granites and syenites dated around 1400–1300 Ma (Carvalho et al. 2000).

These dates also confirm Kibaran (s.s.) mafic and granite magmatism and crustal melting along the southern margin of the SWCS, similar to the Karagwe-Ankole Belt east of the CB; both predated by Paleo-Mesoproterozoic sedimentation across a large region of the CAS, including the Kibaran Supergroup sequences flanking the western margin of the Tanzania Craton (e.g. Pedreira et al., 2008).

Extensive Archean areas north of the CAMB have also been affected locally by Eburnian tectono-metamorphic remobilisation. In parts of the Kasai and Cuango Cratons, the 2.0–2.2 Ga Mubindji Event deformed Paleoproterozoic metasediments (Luiza Group) and possibly related low-grade metabasaltic with pillow lavas (Lulua Group), which unconformably overlie Archean charnockites and enderbites (Cahen et al. 1984; Carvalho et al. 2000). Neither accurate field relationships nor precise geochronology are available to establish a more detailed geologic history.

If the Kasai, Cuango and Ntem Cratons are in fact connected beneath the CB (as the SWCC), then the flanking Eburnian high-grade mobile belt (CAMB) to the south may have stretched all the way from Zambia in the southeast (terminating along a NE-trending Eburnian-Kibaran transcurrent suture) to Gabon, where, as the WCAMB, it forms basement to the younger, overlying West Congo Belt (Figs. 2.2, 2.3 and 2.4). The westernmost tectonic margin of the (CAMB-WCAMB) Mobile Belt is exposed in Brazil between the Trans-Amazonian suture (containing the fold and thrust belt, and gold mineralisation in the Jacobina metasediments) that separates the eastern margin of the Sao Fransisco Craton from the Archean-Eburnian Jequie granulites along the east coast of Brazil (e.g. Feybesse et al. 1998; Lerouge et al. 2006).

Finally the CS was again strongly tectonized during the Pan-African (Brasiliano) orogenesis (ca. 650–540 Ma; e.g. de Wit et al. 2008 and references therein) during accretion of continental blocks along peripheral suture zones, now manifested by a number of well-studied fold-and-thrust belts, including: the West Congo, the Oubanguides (Central African Fold Belt), the Mozambique, and the Irumides-Lufilian-Damara-Kaoko Belts (Figs. 2.2, 2.3 and 2.4). These belts all verge tectonically into the CB, and provided sediment detritus to late Neoproterozoic-early Paleozoic foreland basin sequences that cover the basement rocks and/or the Cryogenian Carbonate sequences exposed along the outer margins of the CB (e.g. Delpondor and Préat, Chap. 3 this Book), and which likely also cover the basement of all of the three Congolese Cratons as shallow marine carbonates, with possibly evaporites (e.g. Kadima et al., Chap. 6, this Book; Selley et al. 2005; Jackson et al. 2014). Evaporite minerals have been recorded in these sequences where they outcrop near the southern and western margins of the CB (e.g. Mbuji-Mayi and Gabon; see Delpomdor et al., and Kolo et al., Chaps. 4 and 5, this Book, respectively), However, major salt sequences have not been observed anywhere in the deep drill holes through the central CB or along the peripheral Neoproterozoic basins (e.g. Tait et al., 2011; and, respectively, Kadima et al., and Linol et al., Chaps. 6 and 11, this Book).

Regionally, 'Redbeds' unconformably overlie the folded Neoproterozoic (carbonate) sequences of the Pan-African Belts and in cores through the CB (Kanda et al. 2011). These Redbeds can no longer be considered as only Precambrian, as frequently argued, because ca. 540 Ma detrital zircons have been found in the Redbeds of the Inkisi Group (Jelsma et al. 2011). At least the upper parts of the CB Redbeds are thus younger than earliest Cambrian; however, the general stratigraphy of these and younger Redbeds remains poorly constrained as there is no reliable chronostratigraphy to correlate them accurately (e.g. Tait et al. 2011).

It has been suggested that the CCC and SWCC amalgamated along an NW-SE trending Pan African deformation zone (Daly et al. 1992; de Wit et al. 2008; Figs. 2.2 and 2.5), but the age of this deformation, indeed even the existence of such a Neoproterozoic deformation zone is still in dispute (Linol et al., Chap. 12, this Book) and will require new high resolution reflection seismology to clarify this (e.g. Kadima et al. 2011; Kadima et al., Chap. 6, this Book). Similarly, the CCC and NECC may have amalgamated along a NW-SE trending deformation zone during the Mesoproterozoic (Kibaran) (as suggested by S. Master in 2004, see Raveloson et al., Chap. 1, this Book), or during Neoproterozoic (Pan African) as suggested by de Wit et al. (2008; see Fig. 2.2). Again there is no hard data, geologically or geophysically, to refute or substantiate these working models.

Below we briefly summarize the most reliable geology and geochronology of some of the cratons and shield regions, referred to above, surrounding the CB.

2.4 Local Geology and Geochronology of Cratonic Domains and Their Cover Sequences Surrounding the CB

The Precambrian basement of the Congo Shield and relevant parts of the CAS is described below in four sub-regions: (1) southern, (2) western, (3) northern, and (4) eastern central Africa (Fig. 2.3a, b). A detailed map with selected (most robust) radiometric dates summarized from the literature and our own unpublished data, is presented (Fig. 2.5; Table 2.1).

In almost all of the basement areas directly flanking the CB, thick (~5–10 km) sequences of Cryogenian (850–635 Ma) siliciclastics and carbonates, with excellently preserved, albeit sedimentologically complex, glacial deposits (diamictites and tillites; also sometime named 'mixites') and cap carbonates rest directly on the

Table 2.1 Selected dates of the Precambrian Central African Shield

Long		Lat		Rock name	Dates (Ma)		Method	Author
22°48′ 31″	E	7°44′42″	S	Musefu	2772	±28	Rb-Sr whole-rock	Cahen et al. (1984)
23°39′54″	E	6°8′13″	S	Bushimay basalt	948	±20	K-Ar whole-rock	Cahen et al. (1984)
17°58′11″	E	9°10′6″	S	Dibaya migmatite	2787	±164	Rb-Sr whole-rock	Cahen et al. (1984)
28°30′4″	E	2°56′38″	S	Kasika granite	986	±10	U-Pb zircon	Tack et al. (2010)
29°23′44″	E	3°29′55″	S	Mugere granite	1379	±10	U-Pb zircon	Tack et al. (2010)
29°58′51″	E	3°41′11″	S	Bukirasazi granite	1205	±19	U-Pb zircon	Tack et al. (2010)
29°40′37″	E	2°44′23″	S	Butare orthogneiss	1982	±6	U-Pb zircon	Tack et al. (2010)
29°14′1″	E	2°18′13″	S	Kilimbi Muzimu	1373	±6	U-Pb zircon	Tack et al. (2010)
13°1′21″	E	5°49′1″	S	Boma	2086	±64	Rb-Sr whole-rock	Cahen et al. (1984)
15°20′1″	E	10°37′2″	S	Cariango	2293	±43	Rb-Sr whole-rock	Cahen et al. (1984)
15°18′11″	E	9°19′0″	S	Lucala migmatite	2560	±50	Rb-Sr whole-rock	de Carvalho et al. (2000)
14°38′4″	E	9°2′36″	S	Quilombo gneiss	1790	±32	Rb-Sr whole-rock	de Carvalho et al. (2000)
13°59′48″	E	8°37′25″	S	Caxito migmatite	684	±20	Rb-Sr whole-rock	de Carvalho et al. (2000)
22°31′17″	E	6°13′29″	S		2648	±22	Rb-Sr whole-rock	Cahen et al. (1984)
22°3′18″	E	6°52′17″	S		2593	±92	Rb-Sr whole-rock	Cahen et al. (1984)
22°30′14″	E	6°55′10″	S		1468	±30	K-Ar whole-rock	Cahen et al. (1984)
24°2′40″	E	7°50′44″	S		2037	±30	U-Pb	Cahen et al. (1984)
22°36′17″	E	8°21′40″	S	Kapanga	2135	±100	Rb-Sr whole-rock	Cahen et al. (1984)
20°17′39″	E	8°54′14″	S		2825	±3	U-Pb zircon	Doucouré et al. (1999)
16°35′42″	E	11°19′40″	S	Andulo migmatite	2520	±36	Rb-Sr whole-rock	de Carvalho et al. (2000)
16°2′58″	E	14°45′13″	S		1853	± 74	Rb-Sr whole-rock	Cahen et al. (1984)
14°49′60″	E	13°8′3″	S		2243	±94	Rb-Sr whole-rock	Cahen et al. (1984)
15°54′54″	E	13°23′50″	S		1949	±30	Rb-Sr whole-rock	Cahen et al. (1984)
15°4′15″	E	14°23′36″	S		1847	±62	Rb-Sr whole-rock	Cahen et al. (1984)
21°9′59″	E	6°13′46″	N	Ippy	2828	±70	Rb-Sr whole-rock	Rolin (1995)
17°28′26″	E	5°57′30″	N	Togbo	652	±19	U-Pb zircon	Rol(1995)
13°36′2″	E	1°37′0″	S	Franceville granite	2659	±30	Rb-Sr whole-rock	Cahen et al. (1984)
13°59′57″	E	6°0′3″	N		544	±40	Rb-Sr whole-rock	Poidevin (1985)
10°51′18″	E	4°0′52″	N	Bafia metamonzodiorite	618	±7	U-Pb zircon	Toteu et al. (2006)
11°32′33″	E	4°11′56″	N	Augen metagranite	604	±3	U-Pb zircon	Toteu et al. (2006)
13°56′2″	E	8°5′6″	N	Buffle Noir amphibolite	2130	±20	U-Pb zircon	Toteu et al. (2001)
10°47′1″	E	4°52′4″	N	Makenene gneiss	2018	±9	U-Pb zircon	Toteu et al. (2001)
14°4′59″	E	5°35′1″	N	Lom metavolcanite	2269	±100	U-Pb zircon	Toteu et al. (2001)
11°47′6″	E	6°42′5″	N	Banyo charnockite	600	±10	U-Pb zircon	Toteu et al. (2001)
13°32′58″	E	7°20′12″	N	Ngaoundere granite	595	±10	U-Pb zircon	Toteu et al. (2001)
14°48′5″	E	6°39′4″	N	Gaoui granite	600	±10	U-Pb zircon	Toteu et al. (2001)
13°26′4″	E	5°52′2″	S	Noqui granite	999	±7	U-Pb zircon	Tack et al. (2001)
13°42′23″	E	5°30′47″	S	Bata Kimenga granite	917	±14	U-Pb zircon	Tack et al. (2001)
13°37′12″	E	5°49′45″	S		530	±12	Rb-Sr whole-rock	Cahen et al. (1984)
29°21′26″	E	3°4′12″	N		2894	±67	Rb-Sr whole-rock	Cahen et al. (1984)
23°27′19″	E	3°50′4″	N		2448	±28	Rb-Sr whole-rock	Cahen et al. (1984)
28°5′42″	E	1°45′8″	N		2793	±66	Rb-Sr whole-rock	Cahen et al. (1984)
32°48′4″	E	0°29′33″	N		2536	±24	Rb-Sr whole-rock	Cahen et al. (1984)
24°2′6″	E	3°36′5″	N	Nzangi gneiss	3005	±64	Rb-Sr whole-rock	Cahen et al. (1984)
29°2′20″	E	1°23′59″	N		1978	±72	Rb-Sr whole-rock	Cahen et al. (1984)
30°15′0″	E	1°43′14″	N		2725	±77	Rb-Sr whole-rock	Cahen et al. (1984)
30°43′32″	E	0°39′57″	S		1887	±245	Rb-Sr whole-rock	Cahen et al. (1984)
31°33′45″	E	8°17′30″	S	Kate granite south	1723	±41	Rb-Sr whole-rock	Lenoir et al. (1994)
34°19′49″	E	8°54′17″	S	Mbarali granite	2026	±8	U-Pb zircon	Lenoir et al. (1994)
31°24′ 21″	E	6°31′23″	S	Mpanda gneiss	1902	±73	Rb-Sr whole-rock	Lenoir et al. (1994)
32°13′4″	E	2°51′39″	S	Geita rhyolitic volcanite	2699	±9	U-Pb zircon	Borg and Krogh (1999)
32°22′52″	E	3°39′15″	S	Homolo rhyolitic volcanite	2808	±3	U-Pb zircon	Borg and Krogh (1999)
34°16′35″	E	1°17′24″	S	Musoma Mara granite	2668	±11	U-Pb zircon	Manya et al. (2006)
34°18′38″	E	4°20′12″	S	Mpanpa volcanite	2742	±27	Sm-Nd whole-rock	Manya and Maboko (2008)

(*continued*)

Table 2.1 (continued)

Long		Lat		Rock name	Dates (Ma)		Method	Author
34°54′7″	E	5°21′3″	S		2565	±40	Rb-Sr whole-rock	Cahen et al. (1984)
35°10′22″	E	7°26′16″	S		2573	±34	Rb-Sr whole-rock	Cahen et al. (1984)
34°2′55″	E	0°23′9″	N		2174	±69	Rb-Sr whole-rock	Cahen et al. (1984)
27°10′28″	E	7°54′59″	S	Nyangwa monzogranite	1383	±5	U-Pb zircon	Kokonyangi et al. (2004)
27°52′38″	E	8°22′21″	S	Kisele monzogranite	1386	±8	U-Pb zircon	Kokonyangi et al. (2004)
25°59′53″	E	9°35′9″	S		1324	±71	Rb-Sr whole-rock	Cahen et al. (1984)
28°0′24″	E	12°16′3″	S	Kinsenda trachyandesite	1873	±8	U-Pb zircon	Rainaud et al. (2005)
27°43′44″	E	12°52′32″	S	Samba metavolcanite	1964	±12	U-Pb zircon	Rainaud et al. (2005)
26°34′20″	E	10°53′35″	S	Kambove uranite	542	±10	U-Pb zircon	Cahen et al. (1984)
26°13′34″	E	11°3′38″	S		656	±20	U-Pb zircon	Cahen et al. (1984)
27°40′22″	E	11°30′33″	S		625	±5	U-Pb zircon	Kampunzu and Cailteux (1999)
32°17′20″	E	9°21′40″	S		1824	±75	Rb-Sr whole-rock	Cahen et al. (1984)
28°51′40″	E	11°11′40″	S		1833	±9	Rb-Sr whole-rock	Cahen et al. (1984)
22°52′36″	E	5°56′7″	N		708	±11	Rb-Sr whole-rock	Poidev(1985)
18°10′54″	E	5°4′26″	N	Boali	2150	±30	Rb-Sr whole-rock	Rolin (1995)
18°56′16″	E	5°48′19″	N	Sidut	639	±3	Rb-Sr whole-rock	Rolin (1995)
34°8′10″	E	3°34′57″	N	Kaabong	545	±30	Rb-Sr whole-rock	Cahen et al. (1984)
31°12′60″	E	3°34′17″	N	U03-6B	2570	±3	U-Pb zircon	de Wit (2004)
33°10′11″	E	2°2′21″	N	U03-12	2616	±1	U-Pb zircon	de Wit (2004)
13°58′25″	E	2°56′23″	N	Lomie metagabbro	666	±26	U-Pb zircon	Toteu et al. (2006)
10°22′36″	E	0°32′51″	N	Kinguele bronzite	2777	±83	Rb-Sr whole-rock	Caen-Vachette et al. (1988)
12°34′6″	E	1°15′33″	S	Koulamoutou gneiss	2888	±40	Rb-Sr whole-rock	Caen-Vachette et al. (1988)
11°35′32″	E	0°44′46″	N	Mitzic gneiss	3186	±75	Rb-Sr whole-rock	Caen-Vachette et al. (1988)
10°52′34″	E	3°19′7″	S	Goumbou granite	2083	±26	U-Pb zircon	Caen-Vachette et al. (1988)
10°5′36″	E	3°31′26″	N	Bonguen metagranodiorite	2066	±4	U-Pb zircon	Lerouge et al. (2006)
10°44′39″	E	3°15′32″	N	Lolodorf metasyenite	2055	±5	U-Pb zircon	Lerouge et al. (2006)
10°33′14″	E	0°41′24″	N	Noya charnockite	3120	±67	Rb-Sr whole-rock	Caen-Vachette et al. (1988)
11°57′49″	E	1°37′3″	S	Chaillu K-granite	2637	±33	Rb-Sr whole-rock	Caen-Vachette et al. (1988)
11°55′36″	E	1°27′13″	S	Chaillu dioritic granite	2765	±39	Rb-Sr whole-rock	Caen-Vachette et al. (1988)
10°16′49″	E	±0°22′7″	N	Kinguele gneiss	3091	±53	Rb-Sr whole-rock	Caen-Vachette et al. (1988)
11°0′0″	E	2°58′59″	N	Ebolowa charnockite	2896	±7	U-Pb zircon	Toteu et al. (1994)
13°43′43″	E	7°2′35″	S	Vista Alegre granite	625	±25	Rb-Sr whole-rock	Cahen et al. (1984)
29°18′14″	E	3°35′47″	N	Akwa mylonite	756	±40	Rb-Sr whole-rock	Poidevin (1985)
24°7′37″	E	11°7′19″	S	Mwinilunga granite	2561	± 10	U-Pb zircon	Key et al. (2001)
25°14′31″	E	11°45′22″	S	Kabompo granite	1884	±10	U-Pb zircon	de Waele et al. (2006)
28°16′23″	E	12°35′32″	S	Mufulira granite	1991	±3	U-Pb zircon	de Waele et al. (2006)
28°40′10″	E	13°57′10″	S	Kapiri Mposhi granite	2726	±36	U-Pb zircon	Rainaud et al. (2005)
29°24′13″	E	13°35′13″	S	Mkushi gneiss	2048	±6	U-Pb zircon	Rainaud et al. (2005)
30°5′18″	E	13°16′43″	S	Lukamfwa Hill granite gneiss	1664	±6	U-Pb zircon	de Waele et al. (2006)
32°29′51″	E	13°42′45″	S	Chipata porphyritic granite	1074	±3	U-Pb zircon	de Waele et al. (2006)
14°1′37″	E	15°25′21″	S	Kunene mangerite vein	1371	±3	U-Pb zircon	Mayer et al. (2004)
34°33′14″	E	9°28′36″	S	Ukenju gneiss	2084	±86	U-Pb zircon	Lenoir et al. (1994)
30°42′8″	E	7°16′54″	S	Kate granite north	1723	±41	Rb-Sr whole-rock	Lenoir et al. (1994)
22°36′53″	E	9°59′50″	S	Sandoa granodioritic granite	3021	±48	Rb-Sr whole-rock	Walraven and Rumvegeri (1993)
32°55′39″	E	8°22′19″	S		2758	±9	U-Pb zircon	Lawley et al. (2013)
35°14′10″	E	13°19′48″	S	Lichinga	1034	±14	U-Pb zircon	Link et al. (2010)
35°5′36″	E	11°42′19″	S	Txitonga	714	±17	U-Pb zircon	Link et al. (2010)
30°58′9″	E	0°30′22″	N		1848	±6	U-Pb zircon	Mänttäri et al. (2013)
34°5′37″	E	0°42′9″	N		2644	±10	U-Pb zircon	Mänttäri et al. (2013)
31°41′0″	E	1°49′38″	N		2991	±9	U-Pb zircon	Mänttäri et al. (2013)
16°44′26″	E	12°32′40″	S		1974		U-Pb zircon	Jelsma et al. (unpublished)
16°17′39″	E	11°21′58″	S		1966		U-Pb zircon	Jelsma et al. (unpublished)
17°6′55″	E	11°26′15″	S	Dando-Kwanza	2522		U-Pb zircon	Jelsma et al. (unpublished)
17°3′42″	E	11°9′7″	S	Dando-Kwanza	1970		U-Pb zircon	Jelsma et al. (unpublished)

Precambrian crystalline basement. In many places at least two glacial sequences are recognized and commonly linked to the Marinoan and Sturtian Glaciations, ca. 635–650 Ma and 720–750 Ma, respectively (e.g. Poidevin 2007; Tait et al. 2011). However, because these sections are not conterminous around the CB and because in many places Pan-African tectonism has tectonically repeated and/or strongly deformed these sequences (e.g. in the Lufilian Arc of Zambia and northwest Botswana; the Oubanguides of northern DRC, CAR and Cameroon), and because radiometric data are scarce and often imprecise, accurate correlations across the CS remain wanted (e.g. Frimmel et al. 2006; Tait et al. 2011; Delpomdor and Préat, Chap. 3, this Book).

An angular unconformity across the Neoproterozoic carbonates in central Africa is considered to represent the base of the CB (although some researchers place it beneath the deformed carbonates; e.g. Daly et al. 1992; Kadima et al. 2011). It is overlain by ubiquitous Cambrian Redbeds, which are often tentatively correlated across the entire CB; however other (similar) formations of red quartzites, conglomerates, sandstones and red siltstones are very common within the stratigraphic record of this large basin (e.g. the Triassic Haute Lueki Group; Lombard 1961) and may often have been misidentified.

2.4.1 Southern Central Africa

At the southern margin of the CB, the Precambrian basement corresponds to the Kasai and Cuango Cratons (Kasai and NE Angola Cratons of Cahen et al. 1984), which essentially comprise Meso- to Neo-Archean grey tonalitic to granodioritic gneisses (TTG), high-grade metamorphic granite gneisses (including extensive charnockites and enderbites) and migmatites (e.g. the Dibaya and Kasai-Lomami Complexes), as well as large mafic-ultramafic complexes, all intruded by widespread Neoarchean granites. Diamondiferous kimberlites of late Precambrian (582 Ma) to late Phanerozoic (110–140 Ma [Alto Cuilo]; 70 Ma [Mbuji-Mayi]; 32 Ma [Kundelungu]) intrude these two cratons (Batumike et al. 2009; Jelsma et al. 2009; de Wit and Jelsma, Chap. 17, this Book), attesting to the deep cratonic mantle-lithosphere characteristics of this region (e.g. Fig. 2.4).

2.4.1.1 The Kasai Craton

In the central core of the Kasai region of southern DRC, small regions of Mesoarchean TTG gneisses (c. Luanyi; Fig. 2.5) were dated at 3490 ± 170 Ma (Rb/Sr microcline date; in Cahen et al. 1984), and granodiorites (Kanda-Kanda gneisses; c. Sandoa) at 3021 ± 49 Ma (Pb/Pb whole-rock date; Walraven and Rumvegeri 1993). These rocks are flanked to the north and south by an extensive region of Neoarchean granulites and gabbro-norites (e.g. the charnockites and enderbites of the Kasai-Lomami Complex), dated between 2.76 and 2.89 Ga, and which, in turn, are in tectonic contact with Eburnian granites and gneisses to the south and metasediments to the north (e.g. the Lulua Group). The age of granulite metamorphism is estimated at ca. 2.78 Ga and generally referred to as the Musefu event (Cahen et al. 1984). Farther north and east are late Neoarchean granites and migmatites (the Dibaya Complex), dated between 2.58 and 2.69 Ga (Cahen et al. 1984; Delhal 1991). These granites and migmatites of the Kasai Craton may extend southeast as far as NW Zambia where granites have also been dated recently at 2561 ± 10 Ma (Key et al. 2001). Migmatisation and metamorphism of this age was referred to as the Moyo event by Cahen et al. (1984).

More recent U-Pb zircon (TIMS) analyses yielded an oldest date of 2825 ± 3 Ma on a TTG in the Lunda area of northeast Angola (and a range between 2.9–2.5 Ma in the same general region; Doucoure et al. 1999; Fig. 2.5), and of 2684–2520 Ma (SHRIMP) on biotite gneisses of the Dando-Kwanza area (Jelsma et al. 2012), suggesting an extension of the central Kasai region into northeast Angola as far south as the CAMB. Whilst a number of dates on charnockitic gneiss in the Lunda area have yielded the oldest zircon dates of the Kasai Craton yet (3.4–3.6 Ga) overprinted by metamorphic U/Pb zircon of ca. 3.25 Ga (H. Jelsma personal communication, 2012), the early geological history of the Kasai Craton is still to be resolved.

Farther south, the Angola basement is swamped by Eburnian granites (e.g. at Jamba and Cutenda, dated at 1.8 Ga by Cahen et al. 1984), and younger rapakivi granites and syenites ('red granites' dated by Rb/Sr between 1302–1411 Ma; de Carvalho et al. 2000). Recently, precise U/Pb zircon analyses re-date these granites between 2065 and 2068 Ma (U/Pb TIMS by S. Bowring, in Doucoure et al. 1999). Nevertheless, there is a great apparent range of granites and syenites from Paleo- to Meso-Proterozoic based on Rb/Sr dates and low resolution U/Pb zircon dates (between 1300–1960 Ma, and in one case as young as 1027 Ma; in Cahen et al. 1984; Carvalho et al. 2000). More precise petrology and geochronology is clearly wanted.

The Kasai Craton is overlain by a sequence of Paleoproterozoic metasediments (the Luiza Supergroup), intruded by granitoids (e.g. Kapenda and Lunge) dated between 2.0 and 2.2 Ga, and covered by younger basaltic volcanics (with pillow lavas) of the Lulua Group dated at 1.9 Ga (Delhal 1991; Andre 1994). To the NE, it is overlain by Late Mesoproterozoic carbonate platform metasediments (the Mbuji-Mayi Supergroup), with basalts in the upper part dated at 948 ± 20 Ma (K/Ar whole-rock date; in Cahen et al. 1984; see also Delpomdor et al. Chap. 4, this Book).

2.4.1.2 The Cuango Craton

In the adjacent Cuango and Malange regions of northern Angola and southern DRC (Fig. 2.5), granite gneisses and migmatites similar to the Dibaya Complex were dated near Lucala at 2560 ± 50 Ma, and at Andulo at 2520 ± 36 Ma (Rb/Sr whole-rock dates; de Carvalho et al. 2000). Thus, the reason for subdividing the Cuango from the Kasai Craton is not based on geology but because the dated (and thus exposed) regions are separated geographically by substantial distances, and two different cratonic blocks are discernable on low resolution seismic sections (e.g. Linol et al., Chap. 8, this Book).

2.4.2 Western Central Africa

The western margin of the CB is bound by a NNW-trending mobile belt, more than 1,000 km long that includes, respectively, an external and internal east-verging Eburnian Belt (WCAMB), and the east-verging Pan-African West Congo Belt with folds and thrusts of thick sequences (~10 km) of Neoproterozoic volcanic and carbonate rocks of the West Congolian Supergroup. To the north and east this Pan African belt tectonically overlies the Archean Ntem Craton and its Paleoproterozoic platform cover (Cahen et al. 1984; Feybesse et al. 1998; Tack et al. 2001; Tait et al. 2011). Farther north still, flanking the Cameroon border, the degree of Eburnian thrusting is sometimes difficult to distinguish from the Pan-African thrusts (Feybesse et al. 1998; Toteu et al. 2001, 2006 and 2014 in preparation; Kalsbeek et al. 2013; Nkoumbou et al. 2013; Ndema Mbongue et al. 2014).

2.4.2.1 The Ntem Craton

In the north-central section of the Ntem Craton, in the Mont Crystal region of northern Gabon, TTG gneisses were dated at Mitzic at 3186 ± 75 Ma and charnockites (c. Noya) at 3120 ± 67 Ma (Rb/Sr whole-rock dates; Caen-Vachette et al. 1988; Fig. 2.5). More recently, a series of TTG and charnockites (11 sites in total) close to Mitzic, also yielded an age range between 2775.2 and 3056.2 Ma; and close to Zamoho at 3231 ± 0.7 [TIMS and U/Pb zircon and monazite dates; S. Bowring, in Doucouré et al. (1999) and de Wit (2004)], and with Mesoarchean TDM dates as old as 3.2 Ga. Near the northern margin of the Ntem Craton, in southwestern Cameroon, charnockites (c. Ebolowa) have been dated at 2896 ± 7 Ma (TIMS U-Pb zircon date, Toteu et al. 1994; Nkoumbou et al. 2013). These Late Archean dates within the Ntem Craton characterize an episode of high grade metamorphism. Yet, ca. 100 km farther along to the northwest, is the Eburnian high-grade Nyong terrain of Cameroon, whose sequences are thrust across the margin of the Ntem Craton (Toteu et al. 2001, 2006; Nkoumbou et al. 2013; Ndema Mbongue et al. 2014). These Eburnian high-grade metasediments and charnockites include 2093 Ma eclogites with MORB-like chemistry that represent relics of the oldest subducted Paleoproterozoic oceanic crust that were subsequently emplaced across the northeast margin of the CS (Schenk et al. 2006; Boniface et al. 2012). Thus, subduction and obduction processes mark this northern boundary of the Ntem Craton as an Eburnian convergent margin, which has since been extensively overprinted again by the Pan-African tectonism of the Oubanguides-Sergipano Belts (e.g. Toteu et al. 1994 and 2014, in preparation; Nkoumbou et al. 2013; see below), possibly linking for the first time the CAS with the Sao Francisco Shield of Brazil (Toteu et al. 2006; Kalsbeek et al. 2013).

The geology and new dates of the Ntem Craton are consistent with the synthesis of Feybesse et al. (1998) of a complex Archean Craton comprising Mesoarchean granite-greenstone-BIF and high grade terrains dated between ca. 3.2 and 2.9 Ga, and a subsequent Neoarchean history of widespread granites and granodiorite intrusions followed by tectono-metamorphism at ca. 2.7–2.8 Ga, and then Late Archean granite activity.

Near Mitzic, the Ntem Craton is locally cut by a substantial number of Archean meta-kimberlites, emplaced at 2848-2862 Ma (U-Pb zircon; LA-ICP-MS) and metamorphosed ($^{40}Ar/^{39}Ar$) at ca. 2615–2550 Ma (at amphibolite grade) and ca. 2000–1940 Ma (at greenschist grade). These are the world's oldest known diamondiferous kimberlites; and together with the chemistry of these kimberlites, attest to the presence of Archean depleted mantle lithosphere beneath the SWCC at that time (Henning et al. 2003). However, since no younger diamondiferous kimberlites have yet been discovered on the Ntem Craton, it is not known if the Archean mantle lithosphere here has retained its original structure and thickness.

To the south, the Ntem Craton also includes extensive younger Archean granites of the Chaillu Massif, dated between 2.63 and 2.76 Ga (Caen-Vachette et al. 1988), very similar to that of the Kasai and Cuango Cratons (e.g. the Moyo event of Cahen et al. 1984).

The Archean basement of the Ntem Craton is tectonically and unconformably overlain in the west by the Eburnian metavolcanic sedimentary rocks of the Oguoué orogenic domain (2235–2040 Ma) and to the east by the Francevillian platform/foreland basin sequences (2150–1780 Ma), respectively. Metasediments of the Francevillian Supergroup (Mossman et al. 2005; Ossa Ossa et al. 2013), include black shale and uranium-rich deposits with the Eburnian natural nuclear reactors at Oklo, dated at 2050 ± 30 Ma (U-Pb uraninite date, Cahen et al. 1984).

2.4.2.2 The West Congo Belt

Precambrian basement to the West Congo Belt includes Eburnian-age granites (c. Goumbou) dated at 2083 ± 26 Ma

(U-Pb zircon date; Caen-Vachette et al. 1988), and migmatites dated near Boma at 2086 ± 64 Ma (Rb/Sr whole-rock date; in Cahen et al. 1984; Fig. 2.5). This Paleoproterozoic basement is locally intruded by younger granites (c. Noqui) dated at 1.0 Ga, and capped by 5–6 km thick basalts and rhyolites of the Zadinian and Mayumbian Groups, dated between 912 ± 7 Ma and 920 ± 8 Ma (U-Pb zircon dates; Tack et al. 2001). These early Neoproterozoic magmatic sequences are, in turn, unconformably overlain in the eastern part of the belt by relatively thick (~5 km) carbonate platform and foreland metasediments of the West Congolian Group (Frimmel et al. 2006; Delpomdor et al., Chap. 3; Kolo et al., Chap. 5, this Book). Recent U-Pb and Ar/Ar geochronology within these sequences shows that Pan-African orogenesis occurred during two main episodes, at ca. 490 Ma and 540 Ma (Monié et al. 2012). This Early Cambrian (Tardive) phase of deformation is very similar to that of the Araçuaí Orogen in eastern South America (e.g. da Silva et al. 2005).

The deformed Neoproterozoic West Congolian Group includes in the lower part two formations of diamictites (the Lower and Upper Mixtites), linked to the Sturtian (710–750 Ma) and Marinoan (635–650 Ma) glaciations, respectively. This is only supported, however, with Sr-isotope dates of 575 Ma from overlying cap carbonates (Poidevin 2007). This sequence terminates upward with molasse-type deposits (the Mpioka Subgroup) unconformably overlain by southward prograding fluvial-delta red sandstones and conglomerates of the Cambrian Inkisi Group that might best be interpreted to reflect the final exhumation and erosion of the West Congolian orogenic Pan-African mountains (Alvarez et al. 1995; Tack et al. 2001; Tait et al. 2011).

2.4.3 Northern Central Africa

In northern Central Africa, the vast E–W trending and S-verging fold-and-thrust belt of the Oubanguides is more than 3000 km long, extending from Cameroon (where it has sometimes been called the Central African Fold Belt), and farther west into northeast Brazil as the Sergipano Belt), through the CAR, to northeastern DRC and Uganda, where it is tectonically thrust across the Bouca and Mboumou-Uganda Cratons (Fig. 2.5). No diamondiferous kimberlites have been discovered across this region.

To the North of the Oubanguides lies a region in the Sahara often referred to as the Sahara Metacraton, but which in fact as far north as Darfur mostly comprises Mesoproterozoic basement, with extensive Kibaran granitoids (ca. 1000–1100 Ma) deformed during the Pan-African orogenesis at ca. 625 Ma along the Central Saharan and Zalingai Fold Belts (CSFB and ZFB, respectively; Fig. 2.2, inset A; de Wit et al. 2005 and 2014, under review).

2.4.3.1 The Mboumou-Uganda Craton

The central section of this craton (also referred to as the Congo-Uganda Block by Tait et al., 2011), in northeastern DRC, is predominantly a classic low-grade granite greenstone terrain (de Wit and Ashwal 1997). The oldest TTG gneisses (c. Nzangi), dated at 3005 ± 64 Ma (Rb/Sr whole-rock date, in Cahen et al. 1984), are intruded by undeformed tonalites (e.g. Moto and Kilo) dated between 2.7 and 2.8 Ga, and relatively younger granites dated between 2.4 and 2.6 Ga (Cahen et al. 1984). These Neoarchean dates are similar to those of the Kasai and Cuango Cratons. However, high grade mafic gneisses (Bomu and Monga) that define the northwestern margin of this craton have been dated at 3.4–3.7 Ga (Cahen et al. 1984); and in one report the Boma rocks were speculated to be ca. 4.0 Ga, based on a Rb/Sr errorchron; Lavreau and Ledent 1975), which would make these the oldest mafic gneisses in Africa and some of the oldest in the world. Clearly new precise geochronology of these Boma gneisses is sorely needed.

The eastern extremity of the Mboumou-Uganda Craton contains Neoarchean high-medium grade TTG and paragneisses exposed over a wide area of north-central Uganda. This is supported by more recent (TIMS) U-Pb zircon dates of 2570 ± 3 Ma and 2616 ± 1 Ma obtained from basement rocks in central Uganda (S. Bowring, in de Wit 2004, unpublished), and several others between ca. 2550–2650 Ma (Link et al 2010; Mänttäri et al. 2013). In two other localities, flanking either side of the Uganda-DRC border, Mesoarchean dates have been recorded (2991 and 3079 Ma, respectively, Mänttäri et al. 2013). The easternmost boundary of this craton is defined by NW-SE trending Neoproterozoic mylonite shear zones (of which the Aswa shear zone is the best known, and dated around 700–680 Ma; S. Bowring, in de Wit 2004, unpublished).

2.4.3.2 The Oubanguides Belt

The northern margin of the CAS and the southern margin of the Central Sahara Shield is marked by the poorly studied, E–W striking Oubanguides, an orogenic belt that stretches from Cameroon through the CAR and southernmost South Sudan to northeast Uganda (Poidevin 1985; van Schmus et al. 2008; de Wit et al. 2008; Toteu et al. 1994, 2001, 2006, and 2014 in preparation). High-grade granitic gneisses and charnockites are widespread, some of which are Neoarchean in age (U/Pb zircon dates between 2.56–2.64 Ga; Toteu et al. 1994, 2001, 2006, and 2014 in preparation) whilst most others are Neoproterozoic (U/Pb zircon dates; Toteu et al. 1994, 2001, 2006, and 2014 in preparation). Many K/Ar and Rb/Sr whole rock dates on paragneisses and associated batholitic granitoids, and U/Pb dates on their zircons, cluster in the range of ~800–1200 Ma and between 480–655 Ma. The latter ages relate to dextral strike-slip and south-directed thrusting of the Oubanguides over the Archean rocks and its overlying Cryogenian sequences (Poidevin 1985; Toteu et al. 2001, 2006, and 2014, in preparation). Farther south scarce Archean dates mark the diffuse tectonic transition between the northern margin of the Archean Bouca and Mboumou Cratons and the allochthonous southern margin of the Oubanguides (Cahen et al. 1984; Poidevin 1985; Toteu et al. 2014, in preparation). Apart from its western extension in Cameroon (e.g. Nkoumbou et al. 2013), this tectonic transition has not been studied with modern structural and radiometric analysis in the central section of the belt.

In Cameroon and western CAR, the Oubanguides comprise both Eburnian and Pan-African thrust sheets with high grade allochthons of Archean (e.g. 3072 Ma), Eburnian (2372 Ma) and Pan-African juvenile volcanogenic rocks (1100–625 Ma; Nkoumbou et al. 2013; Toteu et al. 2006; Toteu et al. 2014, in preparation; de Wit et al. 2014 under review) emplaced across a ~3 km thick Neoproterozoic platform sequences known as the Lindian Supergroup (Lepersonne 1974; Poidevin 1985). Pan-African tectonism has been dated, for example at Akwa at 756 ± 40 Ma, at Boukouma at 708 ± 11 Ma (Rb/Sr whole-rock dates, in Poidevin 1985), near Sidut at 639 ± 3 Ma (U-Pb zircon dates, in Rolin 1995) and in southwestern Cameroon between 640–600 Ma (Nkoumbou et al. 2013). This is comparable with gabbros (c. Lomie) dated at 666 ± 26 Ma in the western part of the belt, and slightly younger monzodiorites in western Cameroon (c. West of Bafia) dated at 619 ± 4 Ma (U-Pb zircon dates, Toteu et al. 2006). More recent dates from western and central CAR includes extensive granite activity between 640–660 Ma, with magmas derived from melting of Kibaran crust (1020 Ma), as well as Archean granitoids with Neoproterozoic overprints; and ca. 800 Ma metasediments with metamorphic overprints at ca. 650 Ma (LA-ICP-MS; K. Drost and M. de Wit, unpublished; Toteu et al. 2014 in preparation). In general, the Pan-African dates of this northern region of central Africa seems to represent orogenic events that apparently young westward across this vast and understudied belt; but this is poorly constrained.

The deformed Lindian Supergroup comprises carbonate metasediments (the Ituri and Lokoma Groups) including dolomites with spectacular stromatolites (Plate 5; Robert 1946). This sequence is dated at 730–755 Ma (Rb/Sr dates, Poidevin 2007) and includes a tillite (the Akwokwo Tillite) that may represents either the Sturtian or Marinoan glaciation. These sequences are overlain by about 1,500 m thick red quartzites, schists, and arkoses (the Galamboge, Alolo, and Banalia Formations) of the late Neoproterozoic-lower Paleozoic Aruwimi Group (Lepersonne 1974). Based on lithostratigraphy, the uppermost Banalia Redbeds unit is now considered to be Cambrian in age (Tait et al. 2011).

2.4.4 Eastern Central Africa

Along the eastern margin of the CB, the 300 km wide, NNE-trending Kibaran Belt (s.s.) separates the Tanzanian Craton from the CS (Figs. 2.2, 2.3 and 2.5), and is tectonically overlain to the south by the late Neoproterozoic (Pan African) Lufilian-Damara Belts.

2.4.4.1 The Tanzanian Craton (TC)

The Tanzanian Craton (>3.8–2.6 Ga) comprises a complex amalgamation of Archean blocks (seven super-terranes and twelve or so terranes as defined by Kabette et al. 2012a) in which Neoarchean granitic gneisses and a number of classic greenstone belts predominate; but Mesoarchean dates have also been sporadically reported (Borg and Shackleton, 1997; Kabette et al. 2012a, b; Kasanzu 2014).

In the northern section of the Tanzanian Craton (TC), volcanics of the Nyanzian and Kavirodian Supergroups have been dated, for example at Homolo (2808 ± 3 Ma) and at Geita (2699 ± 9 Ma; U-Pb zircon dates; Borg and Krogh 1999; Fig. 2.5), and near Mpampa at 2742 ± 27 Ma (Sm/Nd whole-rock date; Manya and Maboko 2008). These dates represent a significant episode of volcanism within the TC that clearly differs from that of all Congolese Cratons.

In the central TC, Meso- to Paleoarchean TTG gneisses have been reported (Kabette et al. 2012a), and basement to juvenile greenstone rocks in the Central Tanzania Region includes orthogneisses with enclaves of ~ 3600 Ma fuchsitic quartzite with detrital zircons as old as 4013 Ma (Kalsbeek et al. 2013).

In the south, the TC includes late Neoarchean gneisses and migmatites of the Dodoman Complex dated at 2573 ± 34 Ma (Rb/Sr whole-rock date; in Cahen et al. 1984), and granites (e.g. Singida) dated at 2535 ± 30 Ma (Rb/Sr whole-rock date; Bell and Dodson 1981), and between 2660 to 2850 (U/Pb Zircon SHRIMP, Kabette et al. 2012a).

The southern and western margins of the TC are poorly defined and includes a number of remobilized Archean endogenous and exogenous terranes within a broad zone, known as the Ubendian-Usagaran Belt, of long-lived Proterozoic tectonism that lasted over a period of ca. 300 million years between 2.1–1.8 Ga (Boniface et al. 2012; Lawley et al. 2013, 2014). The zone includes sheared Paleoproterozoic (Eburnian) metasediments of the Rusizian Supergroup dated between 1.9 and 2.1 Ga, and younger granitoids dated around 1.7–1.8 Ga, that form the Ubendian Belt (Cahen et al. 1984; Lenoir et al. 1995; Lawley et al. 2013, 2014 and references therein). The complex series of allochthonous terranes of the Ubendian Belt are in turn overlain by Mesoproterozoic metasediments that, in turn, were deformed during the Kibaran Orgogeny. Due to lack of reliable geophysical data, there is no agreement about the origin of the allochthonous fragments in the Ubendian Belt, and whether or not the TC lithosphere may originally have extended as far south as the Bangweulu Block (e.g. Lawley et al. 2013 and references therein; and see below).

The eastern margin of the TC comprises mainly Archean blocks remobilized during the Neoproterozoic at high grades within the central zone of the Pan African Mozambique belt (e.g. Fritz et al. 2013; Kabette et al. 2012a; Fig. 2.2).

Across the Tanzanian Craton, repeated diamondiferous kimberlite activity (at 189 Ma [Mwadui] and 53 Ma [Nzoga]) and young Quaternary volcanoes with magmatic rocks characteristic of kimberlite (c. Iqwisi), as well as geochemistry of lithospheric mantle xenoliths, attest to a cratonic lithosphere with a thickness in excess of 120 km (e.g. Dawson 1994; Rudnick et al. 1998; Begg et al. 2009; Brown et al. 2012a, b).

2.4.4.2 The Kibaran Belt (s.s.)

Across most of eastern DRC, Rwanda, and Burundi, Precambrian basement is widely intruded by Mesoproterozoic granites and these represent the type area of the famous ca. 1.0 Ga Kibaran Belt defined by Cahen et al. (1984). Several generations of these granites were recently re-dated, ranging between 986 ± 10 Ma and 1383 ± 17 Ma, but the majority of the granites are ca. 1375 Ma in age (U-Pb zircon dates; Tack et al. 2010). In the southern part of the belt, granites (e.g. Kisele and Nyanga; Fig. 2.5) have been dated similarly at 1385 Ma (Kokonyangi et al. 2004). Thus, this 1370–1385 Ma age characterizes the main episode of Mesoproterozoic magmatism and crustal melting in eastern central Africa, the original type area of the Kibaran (as defined by Cahen et al. 1984, but now referred to as the Kibaran Karagwe-Ankole Belt). Whether the general term Kibaran should be applied to the younger, near 1.0 Ga, rocks elsewhere in Africa (e.g. the Irumide, Mozambican, Namaquan belts) remains an as yet unanswered question. This is a debate that has not yet been resolved (e.g. Tack et al. 2010; Fernandez-Alonso et al. 2012), and is similar to earlier misunderstandings about the Pan African, which was originally defined around ca. 500 ± 50 Ma (Kennedy 1964), but is presently mostly used for rocks in the range between 500–800 Ma (e.g. see discussion in de Wit et al. 2001).

In the type area, the Kibaran Belt is unconformably overlain by late Mesoproterozoic to Neoproterozoic platform metasediments of the Bukoban Supergroup, with basalts in the upper part dated at 815 ± 14 Ma (K/Ar whole-rock date, in Cahen et al. 1984).

2.4.4.3 The Lufilian Arc

In Zambia and southeastern DRC, basement consists of Paleo- to Meso-Proterozoic granite gneisses of the Bangweulu Block, dated at ~2050 Ma, and with a widespread volcanic arc sequence (c. Lubufu) that yield U-Pb SHRIMP zircon ages ranging between 1874 ± 8 Ma and 1980 ± 7 Ma and granites (e.g. Mufulira; Fig. 2.5) dated at ca. 2 Ga (de Waele and Mapani 2002; Rainaud et al. 2005). This deformed (Eburnian) basement is locally intruded by younger granitoids at 1.6 Ga and between 1000–1050 Ma (c. Serenge; de Waele et al. 2008). It is tectonically overlain by the Lufilian Arc: a convex Pan-African northward-verging fold and thrust belt with two principal tectonic nappes that contain about a 10 km thick Neoproterozoic sequence of metasediments, known as the Katanga Supergroup (formerly Kundelungu Supergroup; Lepersonne 1974), and famous in Zambia and DRC for its high-grade Cu-Co mineral deposits. The deformation style is believed to have been influenced by evaporites, since the structures resemble salt-tectonism (e.g. Jackson et al. 2003; Selley et al. 2005). However, no salt deposits are preserved, and if this tectonic interpretation is correct, the evaporites have all been dissolved during the Pan African deformation and related fluid activity, perhaps inducing the base-metal mineralization (c.f. El Desouky et al., 2010, and references therein).

The Lufilian tectonism has been dated between about 700 Ma and 530 Ma, similar to that of the Damara Belt of southern Namibia (Kampunzu and Cailteux 1999; Miller 2008), and to which it is likely linked via northwestern Botswana (de Wit 2010). Here, in northwest Botswana, a thick Neoproterozoic sequence of sediments (ca. <541–750 Ma) comprises medium-grade metamorphosed and deformed Cryogenian rocks, including diamictites and substantial carbonates interbedded with intermittent BIF. These metasediments overly a complex granitoid Archean and Proterozoic basement (recently dated at ca. 2.6 and 2.0 Ga, respectively; U/Pb zircon LA-ICP-MS; de Wit 2012; Unpublished Report for Tsodilo Resources Ltd.). Any affiliation with the Archean block below the Lufilian Fold Belt remains unknown.

In the Lufilian Arc of southeastern DRC and northern Zambia, the Katanga (formely Kundelungu) Supergroup comprises Cryogenian carbonate platform metasediments (the Roan and Nguba Groups), which locally include volcanics dated at 760 ± 5 Ma (U-Pb zircon date, in Master et al. 2005), and with two glacial conglomerates (Grand and Petit Conglomerates) basically attributed to the Sturtian and Marinoan glaciations (e.g. Selley et al., 2005; Poidevin 2007). At the top, the Supergroup terminates with a relatively thick (~5 km) Redbed foreland basin sequence (e.g. Wendorff 2003), in turn unconformably overlain by about 1 km thick red siltstones of the Biano Group (Master et al. 2005). The latter may be correlated to the (Cambrian) upper Nama Group covering the Damara Belt of Namibia (e.g. Grotzinger and Miller 2008), and the Cambrian Redbeds across the CS discussed earlier (the Inkisi and Banalia sequences).

2.4.4.4 The Mozambique Belt

The Mozambique Belt of northeastern Mozambique flanks the eastern margins of the CAS and the Southern African Shield at Africa's southernmost extent of the Pan-African East African Orogen. Details of this orogen are beyond the scope of this chapter, but the interested reader is referred to Fritz et al. 2013, and references therein). Suffice it to mention that the Mozambique Belt yielded the first 1.0 Ga Kibaran (Mozambican) dates on orthogneisses and high grade granulites in Africa (Holmes 1918, 1951), an age which has since been recorded by Rb/Sr analyses (e.g. Cahen et al. 1984), and more recently confirmed by precise zircon and monazite dating (e.g. Jamal 2005; Bingen et al. 2009; Fritz et al. 2013). For example, the Mozambican orthogneisses of the Unango and Marrupa complexes date between ca. 946 Ma and 1062 Ma, overprinted by the extensive Pan-African high grade granulite-facies metamorphism with charnockitization between 700–525 Ma (Fig. 2.2, Inset B; Jamal 2005). These lower crustal rock types and ages have strong affinities with similar rocks of southern India and Madagascar. All these regions are similarly overprinted by extensive Pan-African deformation dated between 530–620 Ma. Whilst in detail the Kibaran/Pan-African history is complex (Jamal 2005; Bingen et al. 2009; Fritz et al. 2013), there is justified reason here for mentioning these Kibaran/Pan-African dates of the Mozambique Belt, because in many of the Phanerozoic sequences of the CB, detrital zircon analyses invariably yield both abundant ca. 1.0 Ga and 0.6 Ga dates (Linol et al., Chaps. 7 and 8, this Book). To date, the source of this bimodal detrital zircon population has been controversial. Since the age of crystalline rock-types of the Mozambique Belt is similar to parts of the Oubanguides flanking the northern margin of the CAS, and the regions to the north (in Central Sahara), the detritus may have been sourced from either north of the CB (Fig. 2.2, Inset A; e.g. Linol et al. 2014, submitted) or southeast of the CB (Fig. 2.2, Inset B; e.g. Roberts et al., 2012). Only when integrated with sedimentology data (paleocurrent directions) can a conclusive interpretation for a northerly source of many CB sequences be verified (see Linol et al., Chaps. 8, 9 and 13, this Book).

2.5 Summary and Conclusions

The CB of central Africa is underlain and completely surrounded by Precambrian basement that spans a history of up to 3.8 Ga, including Archean cratonic blocks, with rocks along its northeastern edge possibly as old as 4.0 Ga (and detrital zircons of similar age on the Tanzanian Craton), separated and surrounded by numerous Paleo- to Neo-Proterozoic mobile belts. Diamondiferous kimberlites span a range of almost 3.0 Ga billion years. This complex assemblage inculcates peripheral (outcropping) Archean nuclei: the Kasai, Cuango, Ntem, Bouca and Mboumou-Uganda cratonic blocks that constitute three larger Congolese Cratons largely unexposed beneath the CB: the SouthWest-, the Cuvette- or Central- and the NorthEast- Congo Cratons (SW-, C- and NE -CC) and that in turn, together, constitute the Congo Shield (CS). Subsequently these three aggregated along the Central Angola Mobile Belt (CAMB; ca. 2.0–2.3 Ga; Carvalho et al. 2000) and the West Central African Mobile Belt (WCAMB; ca. 2.0–2.5 Ga, Feybesse et al. 1998) to form the SouthWestern Congo Shield (SWCS). In the east, the CS enlarged further to form the Central African Shield (CAS) during the Proterozoic along Eburnian, Kibaran and Pan-African orogenic belts (Fig. 2.6), but the details of the accretion processes of these continental domains remain uncertain. The working model we provide here serves only to open discussions and encourage renewed field work and radiometric dating across this region, as that catalysed by Arthur Holmes, William Kennedy, Luiz Cahen and Norman Snelling in the last century. There is much work to be undertaken before a robust model for the formation of the CAS can emerge.

The basement beneath Phanerozoic CB is clearly complex. The oldest Archean cratons may have amalgamated in the Proterozoic, although the precise age of amalgamation is in dispute, and may even have occurred in the Archean. Parts of these cratons are exposed along the northern, southern and eastern margins of the CB, where they are flanked by Eburnian and Pan-African sequences; and the Tanzanian Craton (with its remobilised Archean crust to the west, within the northern Mozambique belt), which collectively is referred to here as the CAS.

The southern margin of the CS is defined by the Central Shield Zone of Angola, a wide transition zone of Eburnian granitoid magmatism and high-grade tectonism that also embodies a number of Archean fragments, and which we have named the CAMB, inferred to represent a deep, exhumed Himalayan-style Eburnian orogenic belt. This belt separates the rest of the Angola basement to the south, comprising predominantly Eburnian-Kibaran (mid-Proterozoic) Tibetan-like crust, from the SWCC. These two blocks collided along the CAMB at ca. 2 Ga to form the SWCS.

Between the Tanzanian Craton and the CS, a complex Eburnian-Kibaran tectonic history includes large scale melting of Eburnian crust between 1.3–1.5 Ga. Near the southern margin of the Angola basement, the vast Mesoproterozoic mafic-anorthosite complex of Kunene associated with widespread red granites and syenites dated around 1.3–1.4 Ga Ma confirm Kibaran (s.s.) mafic and granite magmatism and large scale crustal melting along the southern margin of the CS that may link along a continental shear zone to the type Kibaran of the Karagwe-Ankole Belt (Fig. 2.6). Subsequent accretion of other crustal blocks, along a number of orogenic belts surrounding the CS in the Mesoproterozoic, including the Tanzanian Craton during the formation of Rodinia, and Gondwana in the late Neoproterozoic-Early Cambrian, created the still larger CAS that stabilized by ca. 540 Ma.

The Precambrian rock sequences were regionally peneplained during and following formation and exhumation of the Pan-African orogenic Mountain systems that lasted until at least 540 Ma, and possibly 490 Ma (e.g. de Wit et al. 2001); and the unconformities were covered by marine carbonate-quartzite platform rocks as early as the Cryogenian and, in turn unconformably overlain by Cambrian Redbeds. These fluvial-lacustrine and aeolian Redbeds, which generally characterize hot and arid continental environments, form the basal units of the CB, recording marked extreme climate changes across the CB (and the CAS) during the Neoproterozoic-Cambrian transition.

In summary, the basement of the CB has a highly varied geological history with a complex mantle lithosphere that dates back to the Early Archean. Whilst detailed geophysics and mantle petrology across the CAS are mostly lacking, the variable ages of diamondiferous kimberlites, or the lack thereof, from its varied cratonic blocks illustrates that a simple 'one-fits all' lithospheric mantle structure cannot be used to model the evolution of the CB.

Acknowledgments MdW would like to thank many colleagues from De Beers for facilitating and participation during a decade of his many long field escapes into Africa between 1995 and 2005; and BL into the CB in 2008. Both of us are grateful to Tsodilo Resources for supporting similar trips to Botswana. MdW acknowledges Sam Bowring (MIT) for precise U/Pb zircon dates (most of which still remain unpublished, but we are trying to do something about that) and Moctar Doucouré for his calm and diplomatic Africa advice and wisdom during our African adventures. We thank Bruce Eglington, Brendan Murphy and Mike de Wit for critical reviews. Some of this work was funded through the DST/NRF to AEON and Inkaba yeAfrica/!Khure Africa programs. This is AEON contribution number 134 and Inkaba yeAfrica number 107.

References

Alvarez P, Maurin JC, Vicat JP (1995) La Formation de l'Inkisi (Supergroupe Ouest-congolien) en Afrique centrale (Congo et Bas-Zaire): un delta d'âge Paléozoique comblant un bassin en extension. J Afr Earth Sci 20(2):119–131

Andre L (1994) Age Rb-Sr Proterozoique inferieur du Magmatism continental du Groupe de la Lulua (Kasai, Zaire): ses implications géodynamiques. Ann Soc Geol Belg 116(1):1–12

Batumike JM, Griffin WL, O'Reilly SY, Belousova EA, Pawlitschek M (2009) Crustal evolution in the central Congo-Kasai Craton, Luebo, D.R. Congo: insights from zircon U–Pb ages, Hf-isotope and trace-element data. Precambrian Res 1–2:107–115

Bell K, Dodson MH (1981) The geochronology of the Tanzanian shield. J Geol 89(1):109–128

Bell DR, Schmitz MD, Janny PE (2003) Mesozoic thermal evolution of southern African mantle lithosphere. Lithos 71:273–287

Begg GC, Griffin WL, Natapov LM, O'Reilly SY, Grand SP, O'Neill CJ, Hronsky JMA, Poudjom Djomani Y, Swain CJ, Deen T, Bowden P (2009) The lithospheric architecture of Africa: seismic tomography, mantle petrology, and tectonic evolution. Geosphere 5:23–50

Bingen B, Jacobs J, Viola G, Henderson IHC, Skår Ø, Boyd R, Thomas RJ, Solli A, Key RM, Daudi EXF (2009) Geochronology of the Precambrian crust in the Mozambique belt in NE Mozambique, and implications for Gondwana assembly. Precambrian Res 170: 231–255

Boniface N, Schenk V, Appel P (2012) Paleoproterozoic eclogites of MORB-type chemistry and three Proterozoic orogenic cycles in the Ubendian Belt (Tanzania): Evidence from monazite and zircon geochronology, and geochemistry. Precambrian Res 192–195:16–33

Boniface N, Schenk V (2012) Neoproterozoic eclogites in the Paleoproterozoic Ubendian Belt of Tanzania: evidence for a Pan-African suture between the Bangweulu Block and the Tanzania Craton. Precambrian Res 208–211:72–89

Borg G, Shackleton RM (1997) The Tanzania and NE Zaire cratons. In: de Wit MJ, Ashwal LD (eds) Greenstone belts. Clarendon Press, Oxford, pp 608–619

Borg G, Krogh T (1999) Isotopic age data of single zircons from the Archean Sukumaland Greenstone Belt, Tanzania. J Afr Earth Sci 29(2):301–312

Brown RJ, Manya S, Buisman I, Sparks RSJ, Field M, Stuart FM, Fontana G (2012a) Physical volcanology, geomorphology, and cosmogenic 3He dating of the youngest kimberlite volcanoes on earth (the Holocene Igwisi Hills volcanoes, Tanzania). Abstract 10th IKC, Bangalore

Brown RJ, Manya S, Buisman I, Fontana G, Field M, Mac Niocaill C, Sparks RSJ, Stuart FM (2012b) Eruption of kimberlite magmas: physical volcanology, geomorphology and age of the youngest kimberlite volcanoes known on earth (the Upper Pleistocene/ Holocene Igwisi Hills volcanoes, Tanzania). Bull Volcanol 74: 1621–1643

Buiter SJH, Steinberger B, Medvedev S, Tetreault JL (2012) Could the mantle have caused subsidence of the Congo Basin? Tectonophysics 514–517:62–80

Caen-Vachette M, Vialette Y, Bassot J-P, Vidal P (1988) Apport de la géochronologie isotopique à la connaissance de la géologie gabonaise. Chroniques de la recherche minière 491:35–54

Cahen L, Snelling NJ, Delhal J, Vail J (1984) The geochronology and evolution of Africa. Clarendon, Oxford, p 512

Carvalho H, Tassinari C, Alves PM, Guimarães F, Simões MC (2000) Geochronological review of the Precambrian in western Angola: links with Brazil. J Afr Earth Sci 31:383–402

Daly MC, Lawrence SR, Diemu-Tshiband K, Matouana B (1992) Tectonic evolution of the Cuvette Centrale, Zaire. J Geol Soc 149 (4):539–546

Delhal J (1991) Situation géochronologique 1990 du Precambrian du Sud-Kasai et de l'Ouest-Shaba. Rapport annuel du Musée Royal de l'Afrique centrale, Tervuren (Belgique), Département de Géologie et de Minéralogie, pp 119–125

Delor C, Lafon JM, Rossi P, Cage M, Pato D, Chevrel S, Lê Metour J, Matukov e Sergeev DS (2006) Unravelling precambrian crustal growth of Central West Angola: In: 21th colloquium of African geology, Moçambique, Abstracts, pp 40–41

da Silva LC, McNaughton NJ, Armstrong R, Hartmann LA, Fletcher IR (2005) The Neoproterozoic Mantiqueira Province and its African connections: a zircon-based U-Pb geochronologic subdivision for the Brasiliano/Pan-African systems of orogens. Precambrian Res 136:203–240

Dawson JB (1994) Quaternary kimberlitic volcanism on the Tanzania Craton. Contributions Mineral Petrology. 116, 473M85

de Waele B, Mapani B (2002) Geology and correlation of the central Irumide belt. J Afr Earth Sci 35:385–397 de Waele B, Liégeois J-P, Nemchin AA, Tembo F (2006) Isotopic and geochemical evidence of proterozoic episodic crustal reworking within the Irumide belt of south-central Africa, the southern metacratonic boundary of an Archaean Bangweulu Craton. Precambrian Res 148:225–256

de Waele B, Johnson SP, Pisarevsky SA (2008) Paleoporterozoic to Neoproterozoic growth and evolution of the eastern Congo Craton: its role in the Rodinia puzzle. Precambrian Res 160:127–141

de Wit MJ (2004) African Craton Project: North Africa and Uganda. Unpublished CIGCES Internal Report, and 9th Annual Compte-Rendu to De Beers Cons.Mines Ltd.

de Wit MJ (2007) The Kalahari Epeirogeny and climate change: differentiating cause and effect from core to space. S Afr J Geol 110(2–3):367–392

de Wit MJ (2010) Unpublished report for Tsodilo Resources Ltd; http://www.tsodiloresources.com

de Wit MJ, Doucoure M (1995) A focus on the Central African Shield. Unpublished CIGCES Internal Annual Report for Anglo American Corp Ltd and De Beers Cons.

de Wit MJ, Ashwal L (eds) (1997) Greenstone belts. Oxford University Press, Oxford, UK, p 830

de Wit MJ, Jeffery M, Berg H, Nicolayson LO (1988) Geological Map of sectors of Gondwana reconstructed to their position ~150 Ma (with explanatory notes), scale 1: 1.000.000. American Association of Petroleum Geologist, Tulsa

de Wit MJ, Bowring SA, Ashwal LD, Randrianasolo LG, Morel VPI, Rambeloson RA (2001) Age and tectonic evolution of Neoproterozoic ductile shear zones in southwestern Madagascar, with implications for Gondwana studies. Tectonics 20:1–45

de Wit MJ, Bowring S, Dudas F, Kamga G (2005) The great Neoproterozoic central Saharan arc and the amalgamation of the north African Shield. In: GAC-MAC-CSPG-CSSS conference, Halifax, Canada. Abstracts with programs, p43

de Wit MJ, Stankiewicz J, Reeves C (2008) Restoring Pan-African-Brasiliano connections: more Gondwana control, less Trans-Atlantic corruption. In: Pankurst RJ, Trouw RAJ, Brito Neves BB, de Wit MJ (eds) West Gondwana: Pre-Cenozoic correlations across the South Atlantic region, vol 294. Geological Society of London, London, pp 399–412, Special Publications

de Wit MJ, Bowring S, Buchwaldt R, Dudas F, MacPhee D, Tagne-Kamga G, Dunn N, Salet AM, Nambatingar D (2014). Proterozoic crust of the Central Sahara Shield. Gondwana Res (under review)

Doucoure M, de Wit MJ (1996) Effective elastic thickness of the continental lithosphere in South Africa. J Geophys Res 101(B5): 11,291–11,303

Doucoure M, de Wit MJ, Bowring S (1999) African craton project: a focus on the West African shield, with new data from the Congo Craton, Central African shield. Unpublished CIGCES Annual Report for De Beers Africa Exploration

El Desouky HA, Muchez P, Boyce AJ, Schneider J, Cailteux JLH, Dewaele S, Albrecht von Quadt A (2010) Genesis of sediment-hosted stratiform copper–cobalt mineralization at Luiswishi and Kamoto, Katanga Copperbelt (Democratic Republic of Congo). Miner Deposita 45:735–763

Evans RL, Jones AG Garcia X, Muller M, Mark Hamilton M, Evans S, Fourie CJS, Spratt J, Susan Webb S, Jelsma H, Hutchins D (2011) Electrical lithosphere beneath the Kaapvaal craton, southern Africa. J Geophys Res. 116: BO 4105

Fernandez-Alonso A, Cutten H, De Waele B, Tack L, Tahon A, Baudet D, Barritt SD (2012) The Mesoproterozoic Karagwe-Ankole Belt (formerly the NE Kibara Belt): The result of prolonged extensional intracratonic basin development punctuated by two short-lived far-field compressional events. Precambrian Res 216–219:63–86

Feybesse JL, Johan V, Triboulet C, Guerrot C, Mayaga-Mikolo F, Bouchot V, Eko N'Dong J (1998) The West Central African belt: a model of 2.5-2.0 Ga accretion and two-phase orogenic evolution. Precambrian Res 87:161–216

Frimmel H, Tack L, Basei M, Nutman A, Boven A (2006) Provenance and chemostratigraphy of the Neoproterozoic West Congolian Group in the Democratic Republic of Congo. J Afr Earth Sci 46(3):221–239

Fritz H, Abdelsalam M, Ali KA, Bingen B, Collins AS, Fowler AR, Ghebreab W, Hauzenberger CA, Johnson PR, Kusky TM, Macey P, Muhongo S, Stern RJ, Viola G (2013) Orogen styles in the East African Orogen: A review of the Neoproterozoic to Cambrian tectonic evolution. J Afr Earth Sci 86:65–106

Grotzinger JP, Miller RM (2008) Nama Group. In: Miller RM (ed) The Geology of Namibia, in three volumes. Windhoek, Geological Survey of Namibia, pp 229–272

Hartley AP, Allan PA (1994) Interior cratonic basins of Africa. Relation to continental break-up and role of mantle convection. Basin Res 6:95–113

Henning A, Kiviets G, Kurzlaukis S, Barton E and Mayaga-Mikolo F (2003) Early Proterozoic metamorphosed kimberlites from Gabon. In: Long Abstract, 8th International Kimberlite conference, Victoria, Canada

Holmes A (1918) The Precambrian and associated rocks of the district of Mozambique. Q J Geol Soc 74:31–98

Holmes A (1951) The sequence of pre-Cambrian orogenic belts in south and central Africa, vol 14. In: 18th Int. Geol Congress, London, 1948, pp 254–269

Jackson MPA, Warin ON, Woad GM, Hudec MR (2003) Neoproterozoic allochhonous salt tectonics during the Lufilian orogeny in the Katanga Copperbelt, central Africa. Geol Soc Am Bull 115: 314–330

Jamal DL (2005) Crustal studies across selected geotransects in NE Mozambique: Differentialting between Mozambican (-Kibaran) and Pan Africa events, with implications for Gondwana studies. Unpublished PhD Thesis, University of Cape Town, 365 pp

Jelsma HA, Barnett W, Richards S, Lister G (2009) Tectonic setting of kimberlites. Lithos 112S:155–165

Jelsma HA, Perrit SH, Amstrong R, Ferreira HF (2011) Shrimp U-Pb zircon geochronology of basement rocks of the Angolan Shield, Western Angola. Abstract, 23rd CAG, Johannesburg, 8th–14th January

Jelsma H et al (2012) Kimberlites from Central Angola: a case study of exploration findings. In: 10th International Kimberlite conference, extended abstract reg. no 101KC-42

Kadima E, Delvaux D, Sebagenzi SN, Tack L, Kabeya SM (2011) Structure and geological history of the Congo Basin: an integrated interpretation of gravity, magnetic and reflection seismic data. Basin Res 23(5):499–527

Kabette JM, McNaughton NJ, Groves DI, Mruma AH (2012a) Reconnaissance SHRIMP U–Pb zircon geochronology of the Tanzania Craton: evidence for Neoarchean granitoid–greenstone belts in the Central Tanzania Region and the Southern East African Orogen. Precambrian Res 216–219:232–266

Kabette JM, Gorves DI, McNaughton NJ, Mruma AH (2012b) A new tectonic and temporal framework for the Tanzanian Shield: implications for gold metallogeny and undiscovered endowment. Ore Geol Rev 48:88–124

Kalsbeek F, Ekwueme BN, Penaye J, de Souza ZS, Thrane K (2013) Recognition of Early and Late Neoproterozoic supracrustal units in West Africa and North-East Brazil from detrital zircon geochronology. Precambrian Res 226:105–115

Kampunzu AB, Cailteux J (1999) Tectonic Evolution of the Lufilian Arc (Central Africa Copper Belt) during Neoproterozoic Pan African Orogenesis. Gondwana Res 2(3):401–421

Kanda NV, Mpiana C, Cibambula E, Fernandez-Alonso M, Delvaux D, Kadima E, Delpomdor F, Tahon A, Dumont P, Hanon M, Baudet D, de Waele S, Tack L (2011) The 1000 m thick Redbeds sequence of the Congo River Basin (CRB): a generally overlooked testimony in

Central Africa of post-Gondwana amalgamation (550 Ma) and pre-Karoo break-up (320 Ma). Abstract, 23rd CAG, Johannesburg, 8th–14th January

Kasanzu C (2014) Dating the unroofing and cooling histories of the Archaean Tanzania Craton, Eastern Africa: using a combination of apatite fission-track and (U-Th)/He thermochronometric techniques. Unpublished PhD thesis, Univ. of Cape Town

Kennedy WQ (1964) The structural differentiation of Africa in the Pan-Africa (±500 Ma) tectonic episode. Leeds University Research Institute African Geological Annual Report 8, 48–49

Key RM, Liyungu AK, Njamu FM, Somwe V, Banda J, Mosley PN, Armstrong RA (2001) The western arm of the Lufilian Arc in NW Zambia and its potential for copper mineralization. J Afr Earth Sci 33:503–528

Kokonyangi J, Armstrong R, Kampunzu A, Yoshida M, Okudaira T (2004) U-Pb zircon geochronology and petrology of granitoids from Mitwaba (Katanga, Congo): implications for the evolution of the Mesoproterozoic Kibaran belt. Precambrian Res 132 (1–2):79–106

Lavreau J, Ledent D (1975) Etat actuel de l'étude géochronologique du complex amphibolitique et gneissique du Bomu (Zaire et République Centrafricaine). Rapp. Annu. Mus. R.Afr. centr., Tervuren (Belg.), Dept. Geol. Min., pp. 123–141

Lawley CJM, Selby D, Condon DJ, Horstwood M, Millar I, Crowley Q, Imber J (2013) Lithogeochemistry, geochronology and geodynamic setting of the Lupa Terrane, Tanzania: Implications for the extent of the Archean Tanzanian Craton. Precambrian Res 231:174–193

Lawley CJM, Selby D, Condon D, Imber J (2014) Palaeoproterozoic orogenic gold style mineralization at the Southwestern Archaean Tanzanian cratonic margin, Lupa Goldfield, SW Tanzania: Implications from U–Pb titanite geochronology. Gondwana Research 26:1141–1158

Lenoir JL, Liegeois J-P, Theunissen K, Klerkx J (1995) The Palaeoproterozoic Ubendian shear belt in Tanzania: geochronology and structure. J Afr Earth Sci 19(3):169–184

Lepersonne J (1974) Carte géologique du Zaïre au 1: 2.000.000. Kinshasa, République du Zaïre: Direction de la Géologie/Musée Royal de l'Afrique centrale, Tervuren (Belgique)

Lerouge C, Cocherie A, Toteu SF, Penaye J, Milési J-P, Tchameni R, Nsifa EN, Mark Fanning C, Deloule E (2006) Shrimp U-Pb zircon age evidence for Paleoproterozoic sedimentation and 2.05Ga syntectonic plutonism in the Nyong Group, South-Western Cameroon: consequences for the Eburnian-Transamazonian belt of NE Brazil and Central Africa. J Afr Earth Sci 44(4–5):413–427

Li ZX, Bogdanova SV, Collins AS, Davidson A, De Waele B, Ernst RE, Fitzsimons ICW, Fuck RA, Gladkochub DP, Jacobs J, Karlstrom KE, Lu S, Natapov LM, Pease V, Pisarevsky SA, Thrane K, Vernikovsky V (2008) Assembly, configuration, and breakup history of Rodinia: A synthesis. Precambrian Res 160:179–210

Lindeque A, de Wit MJ, Ryberg T, Weber M, Chevallier L (2011) Deep crustal profile across the Southern Karoo Basin and Beattie magnetic anomaly, South Africa: an integrated interpretation with tectonic implications. S Afr J Geol 114:265–292

Link K, Koehn D, Barth MG, Aanyu K, Foley SF (2010) Continuous cratonic crust between the Congo and Tanzania blocks in western Uganda. Int J Earth Sci (Geol Rundsch) 99:1559–1573

Linol B (2013) Sedimentology and sequence stratigraphy of the Congo and Kalahari Basins of south-central Africa and their evolution during the formation and break-up of West Gondwana. PhD Thesis, Nelson Mandela Metropolitan University, 375p

Linol B, de Wit MJ, Barton E, de Wit MJC, Guillocheau (2014) U-Pb detrital zircons dates and source provenance analysis of Phanerozoic sequences of the Congo Basin, Central Gondwana (accepted at Gondwana Research)

Lombard AL (1961) La série de la Haute Lueki (partie orientale de la cuvette congolaise). Bulletin de la Société belge de Géologie, de Paléontologie et d'Hydrologie 70:65–72

Mänttäri I, Westerhof P, Härmä P, Kigereigu F, Koistinen T, Kuosmanen E, Lahaye Y, Lehtonen MI, Mäkitie H, Manninen T, Pokki J, Virransalo P (2013) Mesoarchean to Neoproterozoic evolution of north Uganda: evidence from new U-Pb rock ages. Abstract, 24rd CAG, Addis Ababa, 8th–14th January

Manya S, Maboko MAH (2008) Geochemistry and geochronology of Neoarchaean volcanic rocks of the Iramba-Sekenke greenstone belt, central Tanzania. Precambrian Res 163(3–4):265–278

Master S, Rainaud C, Armstrong R, Phillips D, Robb L (2005) Provenance ages of the Neoproterozoic Katanga Supergroup (Central African Copperbelt), with implications for basin evolution. J Afr Earth Sci 42(1–5):41–60

Mayer A, Hofmann A, Sinigoi S, Morais E (2004) Mesoproterozoic Sm–Nd and U–Pb ages for the Kunene Anorthosite Complex of SW Angola. Precambrian Res 133(3–4):187–206

McCourt S, Armstrong RA, Kampunzu AB, Mapeo RB, Morais RBM (2004) New U-Pb SHRIMP ages for zircons from the Lubango region, SW Angola: insights into the Proterozoic evolution of South-Western Africa. In: Geoscience Africa, 2004. Geological Society of South Africa, pp 438–439

Miller R (2008) The geology of Namibia. In: Upper Palaeozoic-Cenozoic, vol 3. Geological Survey, Windhoek, Namibia

Monié P, Bosch D, Bruguier O, Vauchez A, Rolland Y, Nsungani P, Buta Neto A (2012) The Late Neoproterozoic/Early Palaeozoic evolution of the West Congo Belt of NW Angola: geochronological (U-Pb and Ar-Ar) and petrostructural constraints. Terra Nova 24(3): 238–247

Mossman DJ, Gauthier-Lafaye F, Jackson SE (2005) Black shales, organic matter, ore genesis and hydrocarbon generation in the Paleoproterozoic Franceville Series, Gabon. Precambrian Res 137: 253–272

Nagudi B, Koeberl C, Kurat G (2003) Petrography and geochemistry of the Singo granite, Uganda, and implications for its origin. J Afr Earth Sci 36(1–2):73–87

Ndema Mbongue JL, Ngnotue T, Ngo Nlend CD, Nzenti JP, Cheo Suh E (2014) Origin and Evolution of the Formation of the Cameroon Nyong Series in the Western Border of the Congo Craton. J Geosci Geomat 2(2):62–75

Nkoumbou C, Barbey P, Yonta-Ngouné C, Paquette JL, Villiéras F (2013) Pre-collisional geodynamic context of the southern margin of the Pan-African fold belt in Cameroon. J Afr Earth Sci. http://dx.doi.org/10.1016/j.jafrearsci.2013.10.002

Ossa Ossa F, El Albani A, Hofmann A, Bekker A, Gauthier-Lafaye F, Pambo F, Meunier A, Fontaine C, Boulvais P, Pierson-Wickmann A-C, Cavalazzi B, Macchiarelli R (2013) Exceptional preservation of expandable clay minerals in the ca. 2.1 Ga black shales of the Francevillian basin, Gabon and its implication for atmospheric oxygen accumulation. Chem Geol 362:181–192

Pedreira AJ, de Waele B (2008) Contemporaneous evolution of the Palaeoproterozoic-Mesoporterozoic sedimentary basins of the São Fransisco-Congo Craton. In: Pankurst RJ, Trouw RAJ, Brito Neves BB, de Wit MJ (eds) West Gondwana: Pre-Cenozoic correlations across the South Atlantic region. Geological Society of London, London, pp 33–48, Special Publications

Pedrosa-Soares AC, Alkmin FF, Tack L, Noce CM, Babinski M, Silva LC, Martins-Neto MA (2008) Similarities and differences between Brazilian and African Counterparts of the Neoproterozoic Araçuaí-West Congo orogen. In: Pankurst RJ, Trouw RAJ, Brito Neves BB, de Wit MJ (eds) West Gondwana: Pre-Cenozoic correlations across the South Atlantic region. Geological Society of London, London, pp 153–172, Special Publications

Poidevin JL (1985) Le Proterozoic supérieur de la Républic Centrafricaine. *Annales du Musée Royal de l'Afrique centrale, Tervuren (Belgique),* Serie in-8, Sciences Geologiques, 91, 75p

Poidevin JL (2007) Stratigraphie isotopique du strontium et datation des formations carbonatées et glaciogéniques néoprotérozoiques du Nord et de l'Ouest du craton du Congo. C R Geosci 339(3–4): 259–273

Rainaud C, Master S, Armstrong RA, Robb LJ (2005) Geochronology and nature of the Palaeoproterozoic basement in the Central African Copperbelt (Zambia and the Democratic Republic of Congo), with regional implications. J Afr Earth Sci 42(1–5):1–31

Robert M (1946) Le Congo physique. Troisième edition. H. Vaillant-Carmanne, Liége, 449 pp

Rolin P (1995) Carte tectonique de la Republique centrafricaine, au 1:1.500.000. BRGM

Rudnick RL, McDonough FM, Richard J, O'Connell RJ (1998) Thermal structure, thickness and composition of continental Lithosphere. Chem Geol 145:395–411

Schenk V, Boniface N, Loose D, 2006 Paleoproterozoic subduction zones at the margins of the Tanzania and Congo Cratons: evidence from eclogites with MORB-type chemistry in the Usagaran—Ubendian Belts of Tanzania and the Nyong Complex of Cameroon. In: IGCP 509 (Abstract), Brazil, pp 102–103

Selley D, Broughton D, Scott R, Hitzman M, Bull S, Large R, McGoldrick P, Croaker M, Pollington N, Barra F (2005) A new look at the geology of the Zambian Copperbelt. Econ Geol 100th Anniversary Volume:965–1000

Stankiewicz J, de Wit M (2013) 3.5 billion years of reshaped Moho, southern Africa. Tectonophysics 609:675–689

Tack L, Wingate MTD, Lie J (2001) Early Neoproterozoic magmatism (1000-910 Ma) of the Zadinian and Mayumbian Groups (Bas-Congo): onset of Rodinia rifting at the western edge of the Congo craton. Precambrian Res 110:277–306

Tack L, Wingate MTD, de Waele B, Meert J, Belousova E, Griffin B, Tahon A, Fernandez-Alonso M (2010) The 1375 Ma "Kibaran event" in Central Africa: Prominent emplacement of bimodal magmatism under extensional regime. Precambrian Res 180(1–2): 63–84

Tanner PWG (1973) Orogenic cycles in East Africa. Bull Geol Soc Am 84:2839–2850

Tait J, Delpomdor F, Preat A, Tack L, Straathof G, Nkula VK (2011) Neoproterozoic sequences of the West Congo and Lindi/Ubangi Supergroups in the Congo Craton, Central Africa. In: Arnaud E, Halverson GP, Shields-Zhou G (eds) The geological record of neoproterozoic glaciations, vol 36. Geological Society of London, Memoirs, London, pp 185–194

Toteu SF, Van Schmus WR, Penaye J, Nyobe JB (1994) U-Pb and Sm-N evidence for Eburnian and Pan African high-grade metamorphism in cratonic rocks of southern Cameroon. Precambrian Res 67(3–4): 321–347

Toteu SF, Van Schmus WR, Penaye J, Michard A (2001) New U-Pb and Sm-Nd data from north-central Cameroon and its bearing on the pre-Pan African history of central Africa. Precambrian Res 108(1–2):45–73

Toteu SF, Fouateu RY, Penaye J, Tchakounte J, Mouangue ACS, Van Schmus WR, Deloule E, Stendal H (2006) U-Pb dating of plutonic rocks involved in the nappe tectonic in southern Cameroon: consequence for the Pan-African orogenic evolution of the central African fold belt. J Afr Earth Sci 44(4–5):479–493

Toteu SF, de Wit MJ, Bowring S, Drost K (2014) New geochronology and geologic map to facilitate correlations along the southern margin of the Oubanguides, from Cameroon to CAR, and the transition from this PanAfrican Orogenic belt to the Ntem Craton

Trompette R (1994) Geology of Western Gondwana (2000-500 Ma). Pan African-Brasiliano aggregation of South America and Africa. A.A. Balkema, Rotterdam

Van Schmus WR, Oliviera EP, Silva Filho AF, Toteu SF, Penaye J, Guimares IP (2008) Proterozoic links between the Borborema Province, NE Brazil, and the Central African Fold Belt. In: Pankurst RJ, Trouw RAJ, Brito Neves BB, de Wit MJ (eds) West Gondwana: Pre-Cenozoic correlations across the South Atlantic region, 294th edn. Geological Society of London, London, pp 69–100, Special Publications

Viegas LGF, Archanjo CJ, Hollanda MHBM, Vauchez A (2014) Microfabrics and zircon U–Pb (SHRIMP) chronology of mylonites from the Patos shear zone (Borborema Province, NE Brazil). Precambrian Res 243:1–17

Walraven T, Rumvegeri BT (1993) Implications of whole-rock Pb-Pb and zircon evaporation dates for the early metamorphic history of the Kasai craton, Southern Zaire. J Afr Earth Sci 16(4):395–404

Wendorff M (2003) Stratigraphy of the Fungurume Group: evolving foreland basin succession in the Lufilian fold-thrust belt, Neoproterozoic- Lower Palaeozoic, Democratic Republic of Congo. S Afr J Geol 106(1):17–34

Part II

Basin Sedimentology and Stratigraphy: Neoproterozoic-Cambrian; Paleozoic-Early Mesozoic; Mesozoic; Cenozoic

3 Overview of the Neoproterozoic Sedimentary Series Exposed Along Margins of the Congo Basin

Franck Delpomdor and Alain Préat

3.1 Introduction

The Neoproterozoic (1000–542 Ma) is an important Era in the Earth history, because it records a complete cycle of supercontinent break-up and re-assembly from Rodinia to Gondwanaland, during which major climatic fluctuations occurred, e.g. several recurring ice ages as part of the Snowball Earth hypothesis (Kirschvink 1992; Hoffman et al. 1998; Hoffman and Schrag 2002) and the emergence of Metazoans (Knoll 1992; Narbonne 2010; Erwin et al. 2011). The Central African Shield consists of several tectonic blocks of Archean age amalgamated during the Mesoproterozoic, whilst the margins of the Congo Shield (CS) consist of mid-Neoproterozoic back-arc type rifts that briefly opened and closed during this supercontinental cycle (e.g. De Waele et al. 2008a, b; Begg et al. 2009; see also de Wit and Linol 2014).

In the light of this general tectonic configuration, we aim here to synthetize the sedimentology, geochemistry and paleogeography of the Neoproterozoic sequences flanking the CS. The series belong to: (i) the West Congo Supergroup that extends along the western margin of the CS from Gabon in the north to northern Angola through the Democratic Republic of the Congo (DRC) and the Republic of the Congo (RC); (ii) the Lindi Supergroup of the Lindi/Ubangui and Fouroumbala-Bakouma exposed along the northern margin of the CS in the Central African Republic (CAR) and DRC; (iii) the Itomwe and Malagarazi supergroups exposed on the eastern margin of the CS; (iv) the Katanga and Mbuji-Mayi supergroups along the southern margin of the CS.

3.2 Structural Regional Tectonic Framework

The CS (Fig. 3.1) comprises several Archean nuclei welded during the Eburnean orogeny approximately between 2.1–1.8 Ga (Pinna et al. 1996; Noce et al. 2007; De Waele et al. 2008a, b; Delor et al. 2008; Begg et al. 2009; de Wit and Linol 2014), and forming a coherent block throughout Late Paleoproterozoic and Mesoproterozoic times (Linol et al. 2014). The late Mesoproterozoic period (1.3–1.0 Ga) is marked by the assembly of the Rodinia supercontinent (Piper 1976; Bond et al. 1984; McMenamin and McMenamin 1990; Dalziel 1991; Hoffman 1991; Karlstrom et al. 1999). However, the exact configuration of various elements, e.g. the CS, surrounding Rodinia is still not clear (Dalziel 1997; Piper 2000; Sears and Price 2000). Nonetheless, palaeomagnetic data suggested that the CS was independent of Rodinia in the late Mesoproterozoic (Pisarevsky et al. 2003) and joined only its south-eastern margin, facing Laurentia at c. 1.0 Ga (Hoffman 1991; Li et al. 2004, 2008).

During this time-span, the southern and eastern margin of the CS was characterized by the Kibaran orogen (1.4–1.0 Ga) related to the accretion of the CS and Tanzania-Bangweulu-Kalahari cratons (Pinna et al. 1996; Fernandez-Alonso et al. 2012), and the amalgamation of microcontinental belts (Southern Mozambique, Sinclair zone, Gordonia, Namaqua) between the proto-Zambezi orogen and the northern Kalahari Craton, ending with a collision along the Irumide orogen at c. 1.02 Ga (De Waele et al. 2003; De Waele 2005; Johnson et al. 2005; Begg et al. 2009). In the southeastern margin of the CS, the fragmentation of Rodinia started around 880–850 Ma with the emplacement of magmatic rocks (Tack et al. 1984; Porada and Berhorst 2000; Mbede et al. 2004; Armstrong et al.

F. Delpomdor (✉)
Université libre de Bruxelles, Department of Earth Sciences and Environment, Unit of Biogeochemistry and Modeling of the Earth System, Avenue F.D. Roosevelt, 50 CP160/2 1050, Brussels, Belgium
e-mail: franck.delpomdor@ulb.ac.be

A. Préat
Université libre de Bruxelles, Department of Earth Sciences and Environment, Unit of Biogeochemistry and Modeling of the Earth System, Brussels, Belgium

M.J. de Wit et al. (eds.), *Geology and Resource Potential of the Congo Basin*, Regional Geology Reviews,
DOI 10.1007/978-3-642-29482-2_3, © Springer-Verlag Berlin Heidelberg 2015

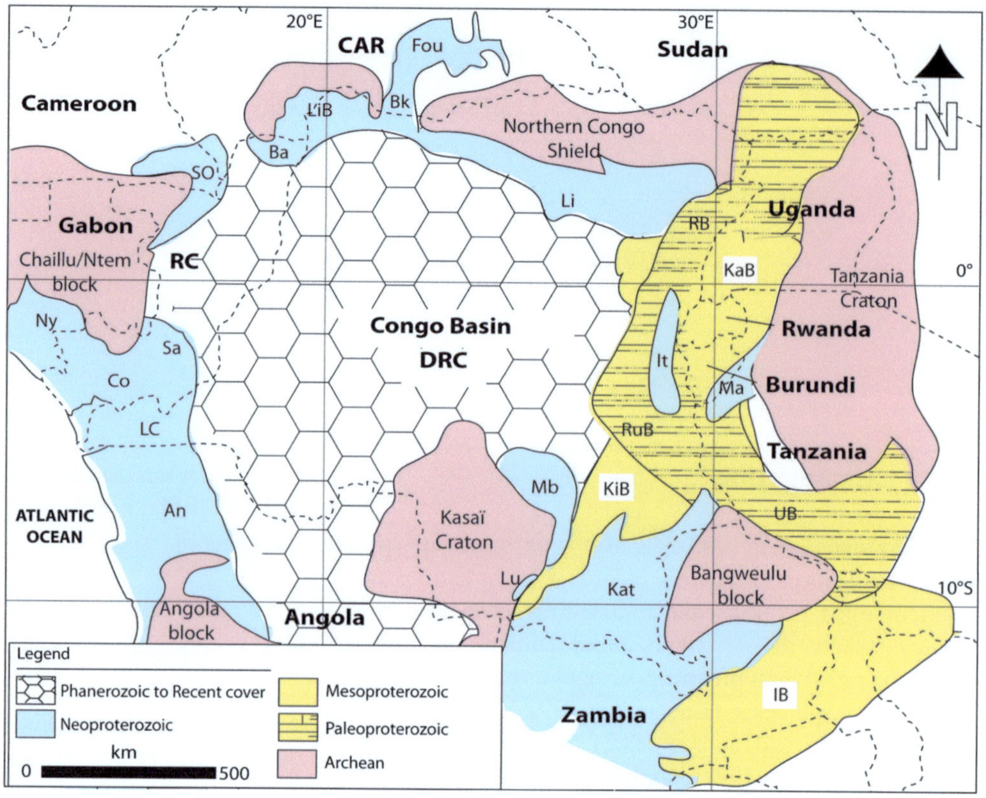

Fig. 3.1 Simplified geological map of Sub-Saharan Africa showing the Neoproterozoic basins on and around the Congo Shield (modified after De Waele et al. 2008a, b). *LC* Lower Congo Basin, *Co* Comba Basin, *Sa* Sangha Basin, *Ny* Nyanga-Niari Basin, *An* Angola Basin, *SO* Sembé-Ouesso Basin, *Ba* Ubangui Basin, *LiB* Likki-Bembé Basin, *Bk* Bakouma Basin, *Fou* Fouroumbala Basin, *Li* Lindi Basin, *It* Itombwe Basin, *Ma* Malagarazi-Bukoban Basin, *Mb* Sankuru-Mbuji-Mayi-Lomami-Lovoy Basin, *Lu* Luamba Group, *Kat* Katanga Basin, *RB* Ruwenzorian Belt, *KaB* Karagwe-Ankole Belt, *RuB* Ruzisian Belt, *KiB* Kibaran Belt, *UB* Ubendian Belt, *IB* Irumide Belt

2005; Kampunzu et al. 2007), related to the opening of the Mozambique Ocean (Li et al. 2008).

Around 600 Ma, West Gondwanaland was coherent with the assembly of Amazonia, West Africa and Congo–São Francisco (Maurin 1993; Trompette 1994; Brito-Neves et al. 1999; Cordani et al. 2003), initiating the earlier Pan-African deformations of the Araçuaí-West Congo Belt (e.g. Pedrosa-Soares et al. 2001, 2007; Alkmim et al. 2006, 2007; Pedrosa-Soares and Alkmin 2011). The convergence of East and West Gondwanaland was initiated between approximately 650 Ma and 515 Ma with the closure of the Mozambique Ocean (Boger and Miller 2004; Jacobs and Thomas 2004). In Central Africa, the Pan-African tectonic episode was related to a c. 560–550 Ma collision between the western margin of the Congo and the São Francisco shields (Maurin 1993; Trompette 1994; Brito-Neves et al. 1999; Cordani et al. 2003); between the Kasai-(Angola)-Kalahari cratons, CS and the Tanzania Craton along a southeast-northwest trending suture linking up the southern Mozambique Belt with the Araçuai-West Congo Belt (Porada and Berhorst 2000; Alkmim et al. 2006, 2007; Pedrosa-Soares et al. 2001, 2007; Pedrosa-Soares and Alkmin 2011); at c. 620– 610 Ma collision between the Central African Mobile Zone and the northern margin of the Congo and São-Francisco shields (Pin and Poidevin 1987; Toteu et al. 2006; Oliveira et al. 2006; Bueno et al. 2009); and between the northern CS and the Saharan Metacraton (Abdelsalam et al. 2002). Recently, Monié et al. (2012) reported two deformational Pan-African events in the hinterland domain of the Angola Basin with recrystallisation ages of c. 540 Ma, coeval with the age of 490 ± 3 Ma (Torquato 1974; Carvalho et al. 2000).

3.3 Neoproterozoic Stratigraphy

3.3.1 The Western Margin of the CS

The West Congo Belt (WCB) is exposed, on 1,300 km, along the western margin of the CS. It comprises a thrust-fold-belt with top-to-(north-)east transport direction from the hinterland domain in the west and grades into a foreland domain in the east with decreasing regional metamorphism from amphibolite facies (in the west) to unmetamorphosed rocks (in the east) (Frimmel et al. 2006).

3.3.1.1 The Sangha-Comba-Lower-Congo basins, DRC (Fig. 3.2A)

Lithostratigraphically, the West Congo Supergroup (Fig. 3.2B) is divided into, from oldest to youngest, the rift-related volcanoclastic 999 ± 7 Ma Zadinian and volcano-sedimentary ± 910 Ma Mayumbian groups deposited on a ± 2.1 Ga polymetamorphic Kimezian basement, and the sedimentary West Congolian Group (Tack et al. 2011). This latter (originally described by Cahen 1978) is subdivided into four subgroups: (i) the 1,650 m-thick

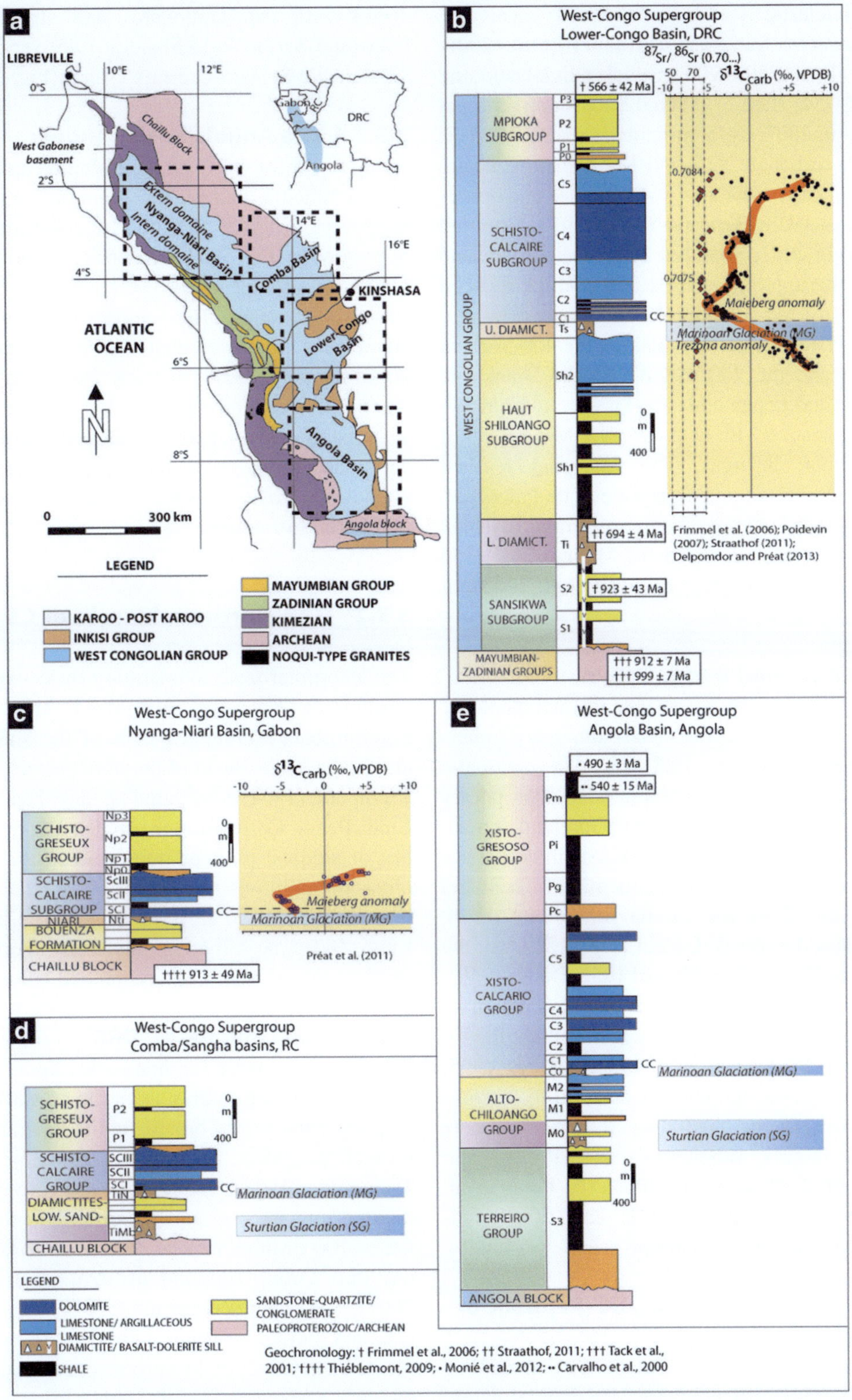

Fig. 3.2 Western margin of the Congo Shield. (A) Sketched geological map of the West Congo Belt (modified after Frimmel et al. 2006). (B) Synthetic stratigraphic log combined with C and Sr isotopic curves of the West Congolian Group in the DRC (modified after Tait et al. 2011). Note strontium isotope ratios of the Haut-Shiloango Subgroup ranging between 0.7068–0.7072 (Frimmel et al. 2006; Poidevin 2007) and of the C3 to C5 formations of the Schisto-Calcaire Subgroup ranging between 0.7074 to 0.7084 suggesting deposition, respectively, between 800 Ma and 650 Ma and between 635 and 575 Ma (Frimmel et al. 2006; Poidevin 2007; Delpomdor and Préat 2013) and the negative shift of ± 13 ‰, consistent, in absolute values, with the negative swing of the Trezona anomaly (Halverson et al. 2007). (C) Synthetic stratigraphic log combined with C and Sr isotopic curves of the West Congolian Group in Gabon (modified after Thiéblemont et al. 2009). Note slightly more negative $\delta^{13}C$ in the cap carbonate reported as typical 'Marinoan' cap ranging form −1 ‰ to −5 ‰ (Préat et al. 2011). (D) Synthetic stratigraphic log of West Congolian Group in the RC (modified after Alvarez 1995). (E) Synthetic stratigraphic log of West Congolian Group in Angola. *cc* cap carbonates; *SG* Sturtian Glaciation, *MG* Marinoan Glaciation

continental siliciclastic and rift-related volcanic 980–920 Ma Sansikwa Subgroup (Frimmel et al. 2006; Straathof 2011); (ii) the 800–700 m-thick siliciclastic and carbonate 800–650 Ma Haut-Shiloango Subgroup (Frimmel et al. 2006; Poidevin 2007; Delpomdor and Préat 2013); (iii) the 1,000-m-thick carbonate 575–550 Ma Schisto-Calcaire Subgroup including cap carbonates, stromatolites and oolites (Delhaye and Sluys 1923; Bertrand-Sarfati 1972; Frimmel et al. 2006; Poidevin 2007; Straathof 2011; Delpomdor and Préat 2013); and the approximately 1,000 m-thick red-bed siliciclastic 566 ± 42 Ma Ma Mpioka Subgroup (Frimmel et al. 2006). Up to two diamictite units, i.e. the 400 m-thick Lower Diamictite and the 200 m-thick Upper Diamictite formations (Cahen and Lepersonne 1981), were interpreted, respectively, as Sturtian (U-Pb age: 694 ± 4 Ma; Straathof 2011) and Marinoan in ages. The overlying red-bed siliciclastic Inkisi (Sub)Group (I) is reported late Paleozoic, altough the youngest detrital zircon grain yielded Neoproterozoic U-Pb ages of 851 ± 18 Ma and 558 ± 56 Ma, respectively (Straathof 2011; Frimmel et al. 2006), or Permian (Alvarez 1995) in ages.

3.3.1.2 The Nyanga-Niari Basin, Gabon

The Nyanga-Niari Basin comprises two distinct geologic domains from west to east, i.e. the metasedimentary hinterland domain deposited on c. 1000–850 Ma magmatic Mabouiné and Mayumba complexes, and the poorly constrained unmetamorphosed sedimentary foreland domain (Thiéblemont et al. 2009). The hinterland domain comprises the Loukoula Group, fluvio-glacial sedimentary M'beia and metasedimentary 713 ± 49 Ma Louila formations (*idid*). The unmetamorphosed internal domain (Fig. 3.2C), i.e. the Nyanga syncline, is divided into (Gérard 1958; Dadet 1969): (i) the 100 m-thick fluvial Bouenza Formation resting on an Archean basement (Prian 2008); the 0 m to 8 m-thick fluvio-glacial Niari Group (Béchennec et al. 1981; Alvarez 1995; Prian et al. 2009a); (iii) the 400 m-thick dominantly carbonate Schisto-Calcaire Group including a cap carbonate ('Dolomie de Bongolo'), lithoherms, stromatolites and oolites (Alvarez 1992; Prian et al. 2009a, b; Préat et al. 2010, 2011; Delpomdor and Préat 2013); and (iv) the red-bed siliciclastic Schisto-Gréseux Group.

3.3.1.3 The Sangha-Comba basins, RC (Republic of the Congo)

Lithostratigraphically, the West Congo Supergroup is divided into, from oldest to youngest: (i) approximately 400 m-thick 'Diamictites and Lower Sandstones' Group deposited directly on the Neoproterozoic Mayumbian Complex in the western basin (Mayumbe) or on an Archean-Paleoproterozoic basement in the eastern basin (Niari); (ii) unconformable > 600 m-thick carbonate Schisto-Calcaire Group (Babet 1935; Scolari 1965; Dadet 1969; Alvarez 1992) with cap carbonates and stromatolitic bioherms (Bertrand-Sarfati and Milandou 1989); and (iv) red-coloured siliciclastic Schisto-Gréseux Group (Fig. 3.2D).

3.3.1.4 The Angola Basin, Angola

Lying unconformably on the Lulumba and Uonde Eburnean metasediments, and on the granitoids and gneisses of the northwest Angola, the poorly constrained West Congo Supergroup (Fig. 3.2E) comprises from the youngest to oldest (Stanton et al. 1963): (i) 2,400 m-thick siliciclastic Terreiro Group; (ii) approximately 1,250 m-thick siliciclastic and carbonate Alto-Chiloango Group; (iii) approximately 2,000 m-thick carbonate sequence of the Xisto-Calcario Group starting by 10–80 m-thick diamictitic succession of the Upper Tilloid Formation capped by a cap carbonate unit; (iv) and 3,250 m-thick exclusively siliciclastic Xisto-Gresoso Group subdivided into M'Pioka and Inkisi formations.

3.3.2 The Northern Margin of CS

The Neoproterozoic sedimentary rocks exposed on northern margin of CS, are recognized in the Ubangui and Fouroumbala-Bakouma basins of the south-western CAR, and in the Lindi Basin of the northeastern DRC (Fig. 3.3A). Trompette (1994) considered all these basins to be part of the Lindi Basin. Correlation is problematic due to limited number of studies, poor or non-existent age constraints, poor outcrop conditions and accessibility. Diamictite units have been recognized in all three basins. The best studied is the Lindi Basin described in detail by Verbeek (1970), the stratigraphy of which is summarized below.

3.3.2.1 The Lindi Basin, DRC

The Neoproterozoic Lindian rocks, up to 2,500 m in thickness, rest unconformably upon the crystalline Archean and/or Early Proterozoic basement and have undergone little or no metamorphism. As summarized by Verbeek (1970) and Trompette (1994), the Lindi Supergroup (Fig. 3.3B) is divided into three groups, from base to top, as follows: (i) the > 100 m-thick marine carbonates (Daly et al. 1992) of the Ituri Group reported as terminal Riphean (Poidevin 2007); (ii) the unconformable >500 m-thick siliciclastic Lokoma Group starting with the 40 m-thick Akwokwo Tillite capped locally by carbonaceous argillites interbedded of cap carbonate lenses; and (iii) the 1,500 m-thick siliciclastic Aruwimi Group, interbedded with oolitic limestones and dolomitic lenses. The overlying >1,000 m-thick red-bed siliciclastic Banalia Group, which apparently conformably overlies the Aruwimi Group, is correlated with the Inkisi Group of the Lower Congo region (Tait et al. 2011).

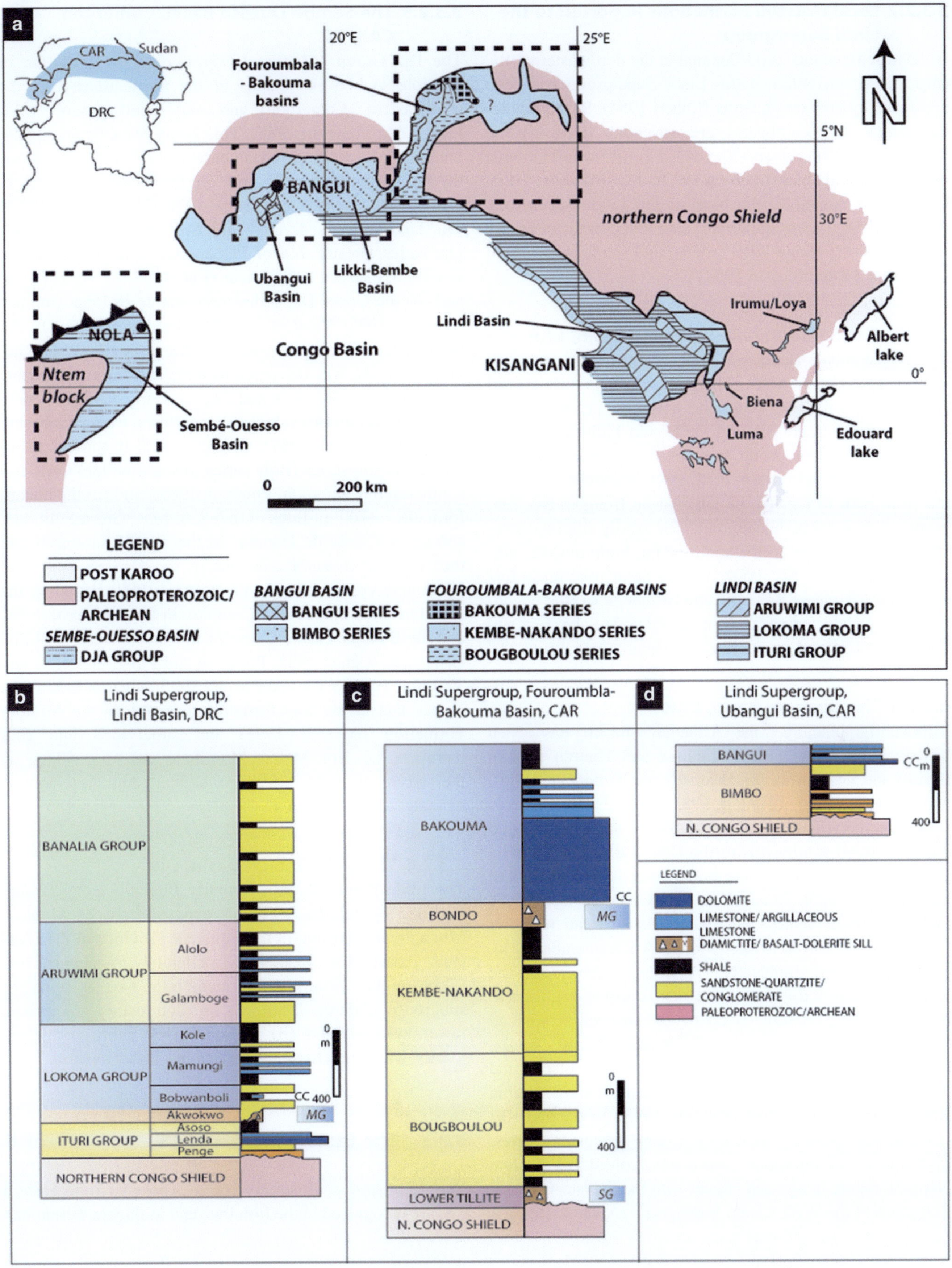

Fig. 3.3 Northern margin of the Congo Shield. (A) Geological sketch map of the Lindi Basin (modified after Verbeek 1970; Poidevin 1985). (B) Stratigraphic logs of the Lindi Supergroup in the DRC (modified after Verbeek 1970). (C and D) Stratigraphic logs of the Lindi Supergroup in CAR (modified after Poidevin 1985). *cc* cap carbonates, *SG* Sturtian Glaciation, *MG* Marinoan Glaciation

3.3.2.2 Local Isolated Formations Reported to the Lindi Supergroup

Isolated sedimentary series located in the northeastern DRC (Fig. 3.3A) attributed to the Lindi Supergroup has been separated by erosion or faults (Cahen 1954). Stratigraphic-equivalent Ituri carbonates are observed along (i) the Niangara and Gada rivers (Meyer 1953); (ii) near Yambuya on the Aruwimi river (Mathieu 1922); and (iii) Buta, Poko, Gwane and Amadi to the base of the Lindi Supergroup (Cahen 1954). Two important diamictitic sequences are observed in the Haute Ibina (Biena) and south of Irumu, in the Homas Mount (Cahen 1954).

3.3.2.3 The Fouroumbala-Bakouma Basin, CAR

The Fouroumbala-Bakouma Basin is located in the south central region of the CAR. Three main sedimentary groups (Fig. 3.3C), with two diamictite horizons, i.e. the Lower Tillite in the extreme SW and the Bondo Tillite at the base of the Bakouma Subgroup, have been recognized (Bigotte and Bonifas 1968; Poidevin et al. 1981; Poidevin 1985) and are from base to top: (i) the siliciclastic Bougboulou Subgroup, with a cap carbonate unit with locally silicified stromatolitic carbonates in the south of the Fouroumbala sub-basin (Mestraud 1952, 1953), overlies an angular unconformity on the Archean Bangui-Ketté basement; (ii) the unconformable deltaic and neritic Kembé-Nakando Subgroup; and (iii) the post-glacial fluvio-marine Bakouma Subgroup.

3.3.2.4 The Ubangui Basin, CAR

Lithostratigraphically, the Ubangui Basin has been described by Babet (1935), Legoux and Hourcq (1943), Gérard and Gérard (1952), Bessoles and Trompette (1980) and Poidevin (1976, 1979a, b, 1985, 2007). Neoproterozoic sedimentary rocks exposed in the SW of the CAR, around Bangui extend southwards to the Congo River and overlie metasedimentary rocks of the Nola Group that are thought to be Paleoproterozoic in age. The Nola Group is intruded by undated dolerite dykes, but yielded Pan-African resetting $^{40}Ar–^{39}Ar$ age of 571 Ma (Moloto-A-Kenguemba et al. 2008). Lithostratigraphically, the Neoproterozoic of the Ubangui Basin is divided as follows, from the youngest to the oldest, (Fig. 3.3D): (i) up to 200 m-thick tectonized 'Kembe conglomerates'; (ii) 130 m-thick fluvial 'Bimbo sandstones' passing eastwards into the fluvio-lacustrine Kembé-Nakando sandstones of the Fouroumbala-Bakouma Basin in the eastern CAR (Alvarez 1995, 2000); and (iii) 'Bangui carbonates', including a cap carbonate series, rhythmites and microbial limestones, considered as late Neoproterozoic in age and coeval with the Schisto-Calcaire Subgroup of the West Congo Supergroup (Alvarez 1995).

3.3.2.5 The Sembé-Ouesso Basin, Cameroon, CAR, RC

The Dja Group (Alvarez 1995) records the evolution of multiple graben-like basins in the northeastern RC, the southeastern Cameroon and the southwestern CAR (Fig. 3.3A; Poidevin 1985). This group rests, by an angular unconformity on the Archean basement of the Chaillu block in the northeastern RC and on the Ntem Complex in Cameroon (Lasserre and Soba 1976; Vicat et al. 1997). The Dja Group is divided into two series: (i) the Lower Dja series, presumably of Paleoproterozoic (Poidevin and Pin 1986; Censier 1989; Vicat et al. 1997) or Tonian ages; and (ii) the Upper Dja series, presumably of Ediacaran age (Alvarez 1995; Vicat et al. 1997; Poidevin 2007). In the northeastern RC, the Lower Dja series, previously Sembé-Ouesso series, was divided into three sub-series (Chochine 1950; Sonet 1958; Grès and Legras 1967; Jaunatre 1969), and grouped into two sub-series as follows, from youngest to oldest (Vicat et al. 1997): (i) the $\pm$1,500 m-thick Schisto-Quartzitic sub-series (Bolé series in southwestern CAR and southeastern Cameroon; Poidevin 1985); and (ii) the unconformable $\pm$700 m-thick 1167–850 Ma 'Gréso-Pélitique' sub-series ('Grès de Douma' in the Sembé sub-basin and the 'Grès de Séka-Nzoulabout in the Ouesso sub-basin; Vicat and Vellutini 1987; Vicat et al. 1997) including the $\pm$100 m-thick Dja Tillitic Complex in the northeastern RC and the Bandjia Tillitic Complex in southwestern CAR and Cameroon (Cahen 1982; Poidevin and Pin 1986; Vicat and Vellutini 1987; Vicat and Pouclet 1995; Alvarez 1995). The Upper Dja series was formerly renamed as the Mintom Formation (Alvarez 1995) and subdivided into four members; i.e. Kol, Métou, Momibolé and Atog Adjap; on the basis of field, sedimentological and geochemical evidences, by Caron et al. (2010).

3.3.2.6 The Likki-Bembé Basin, DRC

The Likki-Bembé Group, formerly Bembe-Ugangi Group, renamed by Aderca (1950), is exposed between Mobayi-Mbongo and Libenge in the Equateur Province of the DRC (Fig. 3.3A). The weakly to moderately metamorphized Likki-Bembé Group is divided into two systems or subgroups (Aderca 1950): (i) the dominantly siliciclastic lower systems; and (ii) the quartzites, schists, phyllads locally with calcschists and limestones upper systems.

3.3.3 The Eastern Margin of CS

Neoproterozoic sedimentary rocks exposed on the eastern CS, are recognized in the Itombwe and Malagarazi-Bukoban

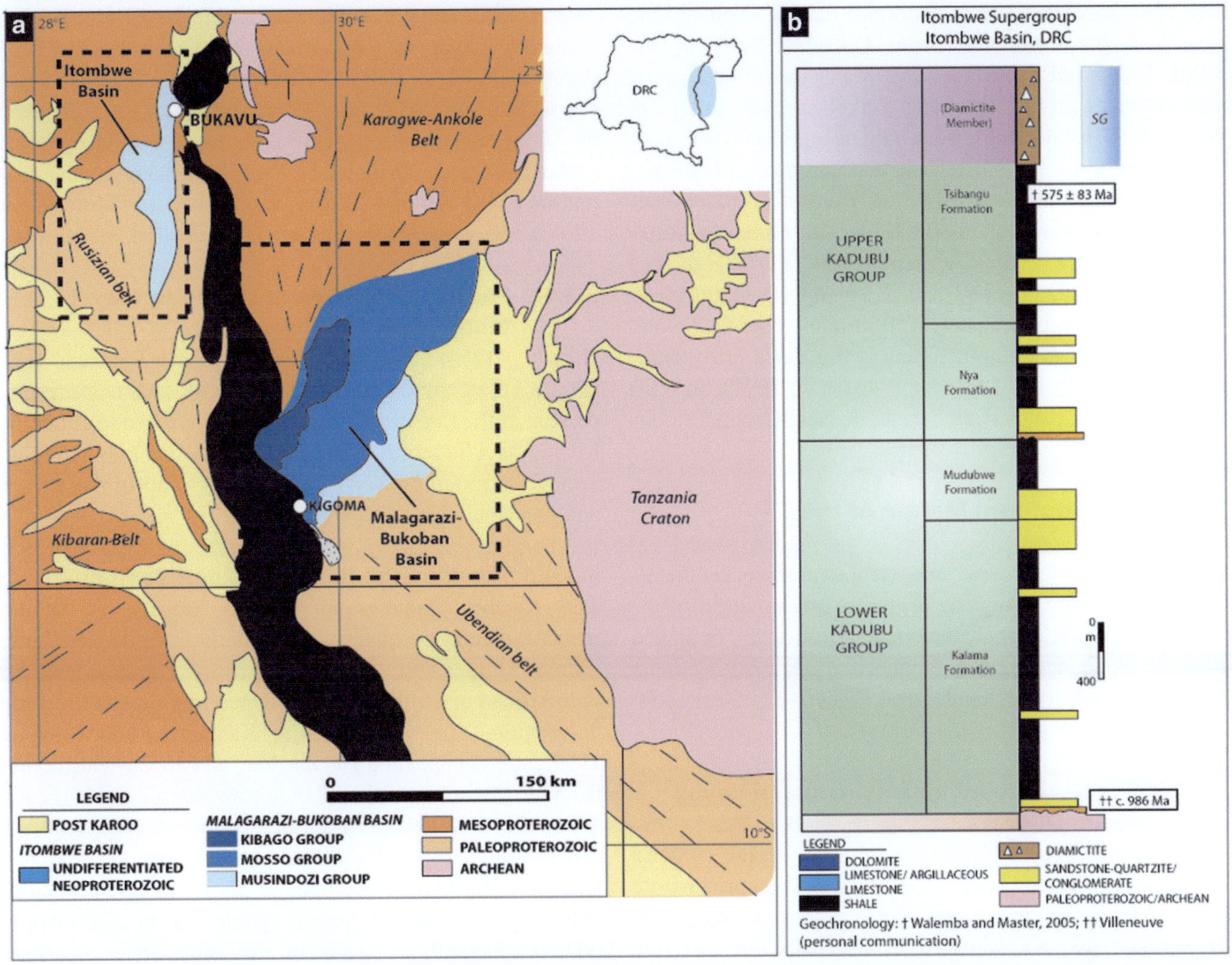

Fig. 3.4 Eastern margin of the Congo Shield. (A) Geological sketch map of the Itombwe and Malagarazi-Bukoban basins. (B) Synthetic stratigraphic log of the Itombwe Supergroup in the DRC (modified after Walemba and Masters 2005). *SG* Sturtian Glaciation

basins in the DRC and Burundi-Tanzania successively (Fig. 3.4A). The Itombwe Supergroup is affected by metamorphism from greenschist facies to weakly wrinkled metamorphosed (Villeneuve 1976a, b, 1977, 1979; Walemba 2001), while the Malagarazi-Bukoban Supergroup are unmetamorphosed (Tack 1995). Deformation in the Itombwe and Malagarazi-Bukoban basins may be kinematically linked with strike-slip reactivation of the Kibaran and Karangwe-Ankole belts and/or Ubendian Belt during Pan-African orogen (Theunissen et al. 1992; Fernandez-Alonso and Theunissen 1998).

3.3.3.1 Itombwe Basin, DRC

The Itombwe Supergroup is lithostratigraphically divided into (Fig. 3.4B; Villeneuve 1976a, b, 1977, 1979; Cahen et al. 1979): the Lower Kadubu and Upper Kadubu groups. The Lower Kadubu Group rests unconformably on a metamorphic and crystalline Burundian-Ruzizian basement, and consists of approximately 3,700 m-thick siliciclastic rocks subdivided into three formations, i.e. Kigogo, Kalama and Mudubwe. The Upper Kadubu Group has two formations, i. e. Nya-Kasiba and Tsibangu. This Group comprises approximately 8,000 m-thick siliciclastic rocks including at the top >800 m-thick diamictitic sequences of the Diamictite Member (Walemba and Masters 2005). The age of the Itombwe Supergroup is estimated between 1310 ± 40 and 648 ± 12 Ma (Villeneuve 1976a, b, 1979; Cahen et al. 1979; Waleffe 1988), and between 1020 ± 50 and 575 ± 83 Ma (Walemba and Master 2005). Fernandez-Alonso et al. (2012) proposed a maximum age of approximately 750 Ma for the base of the Itombwe Supergroup.

3.3.3.2 The Malagarazi-Bukoban Basin, Burundi, Tanzania

The Malagarazi-Bukoban Basin (Fig. 3.4A) is located along the boundary between the Archean Tanzania Craton to the east, the Mesoproterozoic Karagwe-Ankole Belt to the west and the Paleoproterozoic Ubendian Belt to the south. Lithostratigraphically, the Malagarazi Supergroup, stratigraphic-equivalent to the Bukoban Supergroup of Tanzania, is divided into, from base to top (Tack 1995): (i) the 290–890 m-thick siliciclastic and carbonate Musindozi Group, resting unconformably on Burundian basements, with minor intrusions of dolerite feeder sills of the Nyaganza Basalts (Waleffe 1965; De Paepe et al. 1991) emplaced between 709 ± 2 Ma and 815 ± 14 Ma (Deblond et al. 2001); (ii) the approximately 150 m-thick siliciclastic and carbonate Mosso Group including 100 m-thick amygdaloidal basalts at 815 ± 14 Ma (De Paepe et al. 1991); and (iii) the 865 m-thick red-bed siliciclastic Kibago Group.

3.3.4 The Southern Margin of the CS

The late Mesoproterozoic and Neoproterozoic sediments outcrop in two large regions of the southern margin of the CS, each with its own characteristics (Robert 1956; Raucq 1957, 1970; François 1974; Wendorff 2000, 2005; Cailteux et al. 2005, 2007; Delpomdor and Préat 2013; Delpomdor et al. 2013).

3.3.4.1 The Sankuru-Mbuji-Mayi-Lomami-Lovoy Basin, DRC

The Sankuru-Mbuji-Mayi-Lomami-Lovoy Basin (SMLL; Fig. 3.5A) is located between the Archean-Paleoproterozoic Kasai Craton to the north and west, and along the Mesoproterozoic Kibaran Belt to the south and east (see also Delpomdor et al. 2014). The Mbuji-Mayi sedimentary sequence is weakly or no affected by regional metamorphism (Polinard 1935; Wasilewsky 1954; Cahen 1954; Raucq 1970). The Mbuji-Mayi Supergroup (Fig. 3.5B) is divided as follows, from oldest to youngest (Raucq 1957, 1970): (i) 550 m to 3,000 m-thick siliciclastic 1174 ± 22 Ma to 882.2 ± 8.8 Ma BI group (Cahen 1954; Holmes and Cahen 1955; Delhal et al. 1966, 1989; Delpomdor et al. 2013) unconformably overlying the approximately 3.0–2.5 Ga metamagmatic rocks of the Dibaya Complex to the north (Delhal et al. 1976; Cahen et al. 1984; Delhal 1991; Key et al. 2001; Batumike et al. 2009); and (ii) 1,000 m-thick carbonate ± 810 Ma BII group (Delpomdor et al. 2013) including stromatolitic bioherms (Bertrand-Sarfati 1972) with local shales. Two important masses or bodies of basalts are reported in the basin: (i) basaltic lavas, they yielded an altered age of 948 ± 20 Ma (Cahen et al. 1974) and overlie the BII group at the confluence of the Mbuji-Mayi and Sankuru rivers (Cahen et al. 1984); (ii) doleritic sills emplaced, at 888.2 ± 8.8 Ma (Delpomdor et al. 2013), close to the contact between the BI and BII groups in the Lomami area; and (iii) undated sills of dolerite with large extensions of amygdaloidal lavas within the BI group along Kiankodi and Lovoy rivers, (Cahen and Mortelmans 1947). Along the Kibaran Belt in the southern part of the basin, i.e. the Kabele and Kabenga Conglomerates, with more than 50 % of clastic material derived from the Mbuji-Mayi carbonates, were attributed to the Grand Conglomérat Formation of the Katanga Supergroup (Cahen and Mortelmans 1947).

3.3.4.2 Local Luamba Group Linked to the Mbuji-Mayi Supergroup

Ninove (1954) described, in the Lueo-Lubilash area (Fig. 3.1), an approximately 800 m-thick sedimentary succession of the Luamba Group which they linked to either the Mbuji-Mayi Supergroup or as they suggested may represent a medium group between the Lubudi Group of the Kibara and the Mbuji-Mayi Supergroup.

3.3.4.3 The Katanga Basin, DRC and Zambia

The Katanga Supergroup consists largely of tabular metasedimentary sequences that extend from the DRC into Zambia (Fig. 3.6A). These sequences are affected by the Pan-African or Lufilian north-directed folding and thrusting deformations (c. 620–570 Ma; Kampunzu and Cailteux 1999), stretching over a 700 km-long and 50 km-wide region from the Mwinilunga district in northwestern Zambia (Brock 1961; Steven 2000), east-northeastwards through Kolwezi and Likasi (DRC), and southeastwards to Bwana Mkubwa (Zambia) and Lonshi (Cailteux et al. 2007). In thrust-fold-belt, regional metamorphism grades from amphibolite prehnite-pumpellyite facies in the outer zones to lower greenschist facies toward the inner zones (Ramsay and Ridgeway 1977; Lefebvre and Patterson 1982). Lithostratigraphically, the Katanga Supergroup (Fig. 3.6B) is divided, from oldest to youngest (Cailteux et al. 2007) into: (i) the Roan Group; (ii) the Nguba Group (formerly Lower Kundelungu); and (iii) the Kundelungu Group (formerly Lower Kundelungu). Wendorff (2000, 2005) proposed an alternative interpretation of the stratigraphy, with absence of the Lower Roan in the DRC. This subject forms a matter of intense discussion (e.g. Cailteux et al. 2005) beyond the scope of this chapter. The 880–750 Ma Roan Group (Master et al. 2005) is subdivided into four subgroups (Cailteux et al. 2007): (i) the 'Roche Argilo-Talqueuse Subgroup' (R.A.T.; undeterminated thickness) reported stratigraphically to the ± 880 Ma Mindola Subgroup in Zambia (Armstrong et al. 2005); (ii) the > 400 m-thick carbonate Mines Subgroup correlated with the Kitwe Subgroup in Zambia; (iii) the >1,000 m-thick carbonate Dipeta Subgroup equivalent to the Kirilabombwe Subgroup in Zambia (Cailteux 1994; Cailteux and Kampunzu 1995); and (iv) the 600–800 m-thick carbonate and siliciclastic ± 760 Ma Mwashya Subgroup (formerly Upper Mwashya; Armstrong 2000; Liyungu

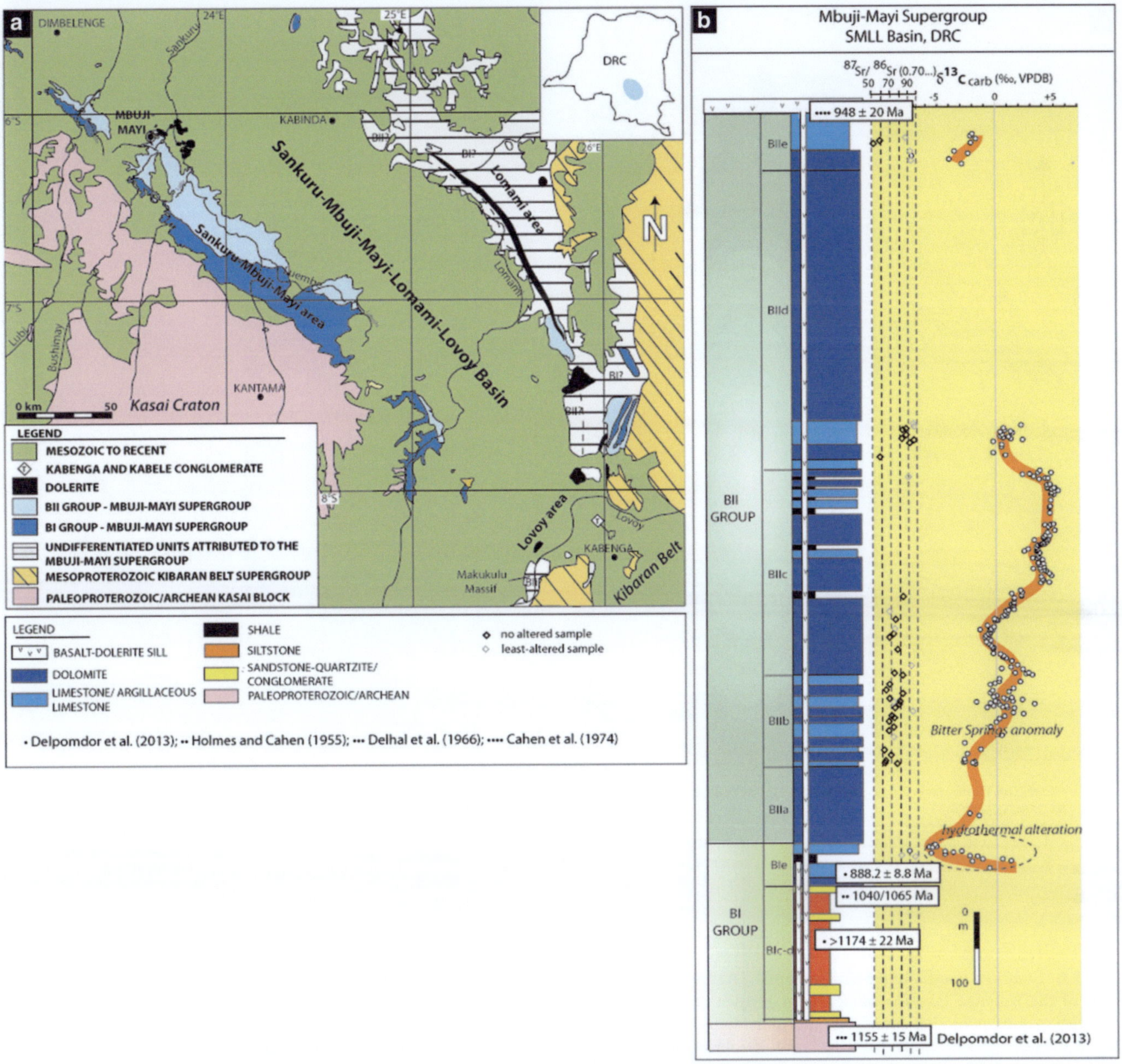

Fig. 3.5 Southern margin of the Congo Shield. (A) Sketched geological map of the Sankuru-Mbuji-Mayi-Lomami-Lovoy Basin (modified after Raucq 1957, 1970). (B) Synthetic stratigraphic log combined with C and Sr isotopic curves of the Mbuji-Mayi Supergroup in the DRC (Modified after Raucq 1957, 1970). Note the $^{87}Sr/^{86}Sr$ ratios ranging between 0.7069 and 0.7071 and the negative $\delta^{13}C$ shifts of 4.0 ‰ and 6.1 ‰ in the BIIb and BIIc transition, which are reported with the Bitter Springs anomaly around c. 810 Ma (Delpomdor et al. 2013)

et al. 2001; Key et al. 2001; Barron et al. 2003) including mafic volcaniclastic rocks (Lefebvre 1973; Cailteux 1983; Cailteux et al. 2003; Kabengele et al. 2003) coeval with the gabbroic bodies emplaced within rocks of the Dipeta Subgroup (DRC) and Bancroft-Kanwangungu formations (Mendelsohn 1961; Lefebvre 1975; Cailteux et al. 2007). Cailteux et al. (2007) adopted a similar lithostratigraphic scheme for the Upper Katanga Supergroup, i.e. the Nguba and Kundelungu groups, between the DRC and Zambia. The ± 2,000 m-thick Nguba Group comprises a shaly carbonate succession becoming siliciclastic at the top, and is subdivided into (i) the Muombe Subgroup including the diamictitic 765–735 Ma 'Grand Conglomérat' or Mwale Formation (Key et al. 2001; Wendorff and Key 2009) correlated with the global Sturtian glacial event (Cailteux et al. 2007); and (ii) the carbonate and siliciclastic Bunkeya Subgroup. The Kundelungu Group is further subdivided into: (i) the Gombela Subgroup including the glacigenic

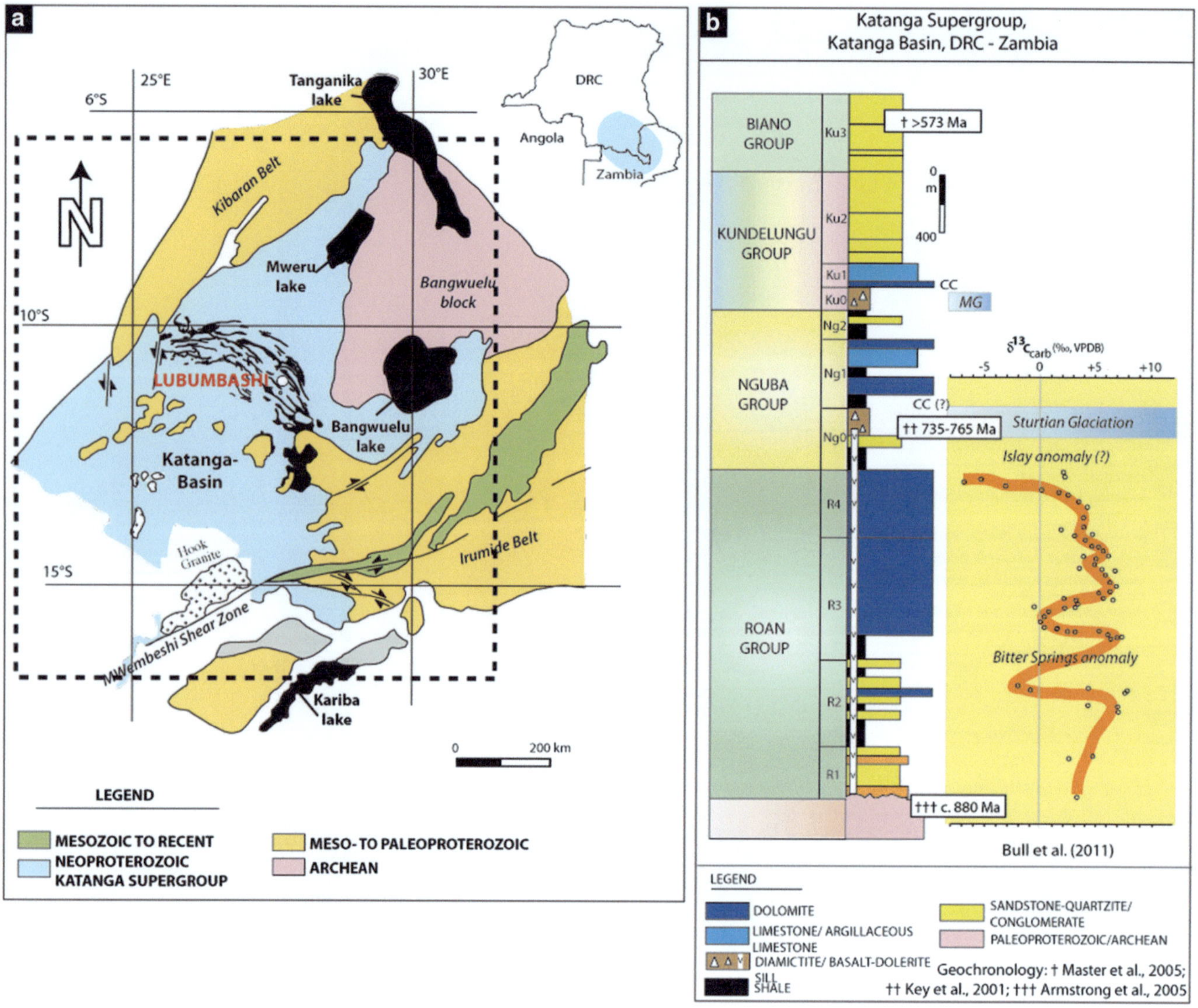

Fig. 3.6 Southern margin of the Congo Shield. (A) Geological sketch map of the Katanga Basin (modified after Heijlen et al. 2008). (B) Synthetic stratigraphic log combined with C and Sr isotopic curves of the Katanga Supergroup in the DRC and Zambia (modified after Cailteux et al. 2005). Note the declines in $\delta^{13}C$ estimated at 6.7 ‰ in the Lower Roan Group, followed by intervals with strong $\delta^{13}C$ enrichments, attributed to the Bitter Spring negative excursion and a second negative $\delta^{13}C$ shift at the top of the Mwashia Group, reported to the Sturtian glacial events (Bull et al. 2011)

diamictites of the 'Petit Conglomérat' or Kyandamu Formation and the cap carbonates of the Lusele Formation; (ii) the carbonate, then siliciclastic Ngule Subgroup; and (iii) the siliciclastic ± 573 Ma Biano Subgroup (Master et al. 2005).

3.4 Discussion and Conclusion

3.4.1 Depositional settings

Since the Mesoproterozoic Eon, the CS region was subjected to intermittent episodes of extensional activity resulting in long-lived shallow cratonic basins, which were subsequently deformed by short-lived phases of compression during the Neoproterozoic (e.g. Fernandez-Alonso et al. 2012). The Itombwe and Malagarazi–Bukoban supergroups accumulated in passive intraplate or cratonic sag-type basins (Mbede 1991), and the Mbuji-Mayi Supergroup in a failed-rift intracratonic basin (Kadima et al. 2011; Delpomdor et al. 2013). These successions are interpreted as molasse deposits related to a tectonic reactivation of the Kibaran and/or Karangwe-Ankole mountains (Fernandez-Alonso et al. 2012), or as rift deposits during the first phases of Rodinia break-up. Along the western margin of the CS, the Zadinian and Mayumbian groups, and the time-equivalent Mabouiné and Mayumba complexes, represent a continental rift sequence related to the break-up of Rodinia (Tack et al. 2001; Thiéblemont et al. 2009), while the West Congolian Group appears to represent passive margin siliciclastic and carbonate platform deposits preserved in the foreland domain of the WCB. The Sansikwa Subgroup represents continued pre-glacial continental rifting (Frimmel et al. 2006). The time-equivalent Roan and BII groups are interpreted as lacustrine or lagoonal systems (Cailteux et al. 2007; see also Delpomdor et al. 2014), while the Roan Group from Zambia appear to be marine. The Musindozi and Mosso groups correspond, respectively, to continental and lacustrine deposition, and the Kibago Group to shallow-water molasse sediments in a foreland basin. The Sturtian-equivalent glaciogenic and associated deposits (well constrained at 715–680 Ma from IUGS Terminal Neoproterozoic Subcommission 2005) accumulated in the most part of basins, except in the Malagarazi-Bukoban Basin. In the western margin of the CS, the Bouenza Group represents a fluvio-lacustrine and pre-glacial environment (Alvarez 1995), and the upper part of the Haut-Shiloango Subgroup is interpreted as a open-marine carbonate unit recording a transgression on a passive margin, following a rift stage (Frimmel et al. 2006). The time-equivalent Ituri Group represents lagoonal or shallow marine platform type systems (Daly et al. 1992). A second global glacial event, coeval with the Marinoan glaciation (well constrained at 660–635 Ma from IUGS Terminal Neoproterozoic Subcommission 2005) is marked by several diamictites in each Neoproterozoic basins (Tait et al. 2011), except in the SMLL Basin (Delpomdor et al. 2013). The Schisto-Calcaire (Sub) Group and Bangui Group record a transgressive marine system including distal to proximal outer shelf, offshore barriers and lagoonal facies submitted to hypersaline shallow subtidal and supratidal conditions, with lagoonal setting colonized by stromatolites and cyanobacterial mats (Alvarez 1992; Préat et al. 2010). Locally, continental and fluvial environments are observed at the top of these groups (Alvarez 1995; Lepersonne 1973). The Upper Dja Series bas been recently reinterpreted as Gaskiers-equivalent lacustrine or lagoonal systems in a graben-like syn-rift basin, with limited sediment input and episodic marine invasion (Caron et al. 2010). The Mpioka Subgroup, Schisto-Gréseux Group, Aruwimi and Kundelungu groups accumulated in continental, lacustrine to fluvio-deltaic semi-arid environments, and all are interpreted as Pan African-related late-orogenic molasse deposits (Frimmel et al. 2006; Cailteux et al. 2007; Heijlen et al. 2008). In the light of more recent works, however, the Inkisi Group and Biano Group are considered to be red-bed sequences of post-Pan African and pre-Karoo age (Tait et al. 2011; see also de Wit and Linol 2014).

3.4.2 Regional Correlations

Available stratigraphic and geochronological data support regional correlation (Fig. 3.7). The late Mesoproterozoic–early Neoproterozoic BI Group has no stratigraphic equivalent in Central Africa. Nevertheless, the 'Gréso-Pélitique' unit of the Lower Dja series is estimated between 1167 Ma and 850 Ma in age (Vicat et al. 1997), but no correlation to date is possible with the BI Group of the Mbuji-Mayi Supergroup. The 1000–910 Ma Zadinian and Mayumbian groups are compatible in terms of age with the Mabouiné and Mayumba complexes in the hinterland domain from Gabon (Thiéblemont et al. 2009). However, the upper part of the Mayumba Complex, dated at 904 $\pm$ 6 Ma (U-Pb ages on zircon from gabbro) may be correlated with the 912 $\pm$ 7 Ma Sansikwa Subgroup of the DRC (Frimmel et al. 2006). The lack of age constraints does not allow precise correlations between the Sansikwa Subgroup, Loukoula and Terreiro groups and stratigraphic references. Nonetheless, on the basis of lithological observations, the Lower Diamictite Formation of the DRC and Angola may be Sturtian in age, and correlated with the M'Beia Formation, reported to the Lower Tilloid Formation by Dévigne (1959).

In the southern and eastern margins of the CS, the isolated Itombwe and Malagarazi-Bukoban supergroups also cannot be correlated due the lack of age constraints. However, the emplacement of intrusive sills of basalts in the Musindozi and Mosso groups, dated between ca. 815 $\pm$ 14 Ma and 709 $\pm$ 2 Ma (De Paepe et al. 1991; Deblond et al. 2001) may be correlated with the 888.2 $\pm$ 8.8 Ma dolerite sills at the BI/BII Group transition or with the second igneous sequences overlying the Mbuji-Mayi Supergroup (Delpomdor et al. 2013); and with the magmatic rocks (Kafue rhyolites [879 Ma]; Nchanga Granite [877 Ma]; Lusaka Granite [865 M]) and the 760–735 Ma Mwashya Subgroup (Key et al. 2001) of the Katanga Supergroup. The BII Group time-equivalent to the Upper Mines and Dipeta subgroups of the Katanga Supergroup (Delpomdor et al. 2013) is supported by C-isotopic and geochronological data (Cahen et al. 1974; El Desouky et al. 2010; Bull et al. 2011; Delpomdor et al. 2013). The Kabele and Kabenga conglomerates, considered as coeval with the Grand Conglomérat Formation (Cahen and Mortelmans 1947), may be correlated with the Sturtian-equivalent Tshibangu Formation and the Bandjia Tillitic Complex.

In Gabon, the Louila Formation in the internal domain, is a lateral equivalent unit of the Bouenza Formation (Thiéblemont et al. 2009), and is a correlative of the Haut-Shiloango Subgroup (Chevallier et al. 2002), consistent with a post-Sturtian deposition for the latter. Previously, it was suggested that the base of the Lindi Supergroup, i.e. the Ituri

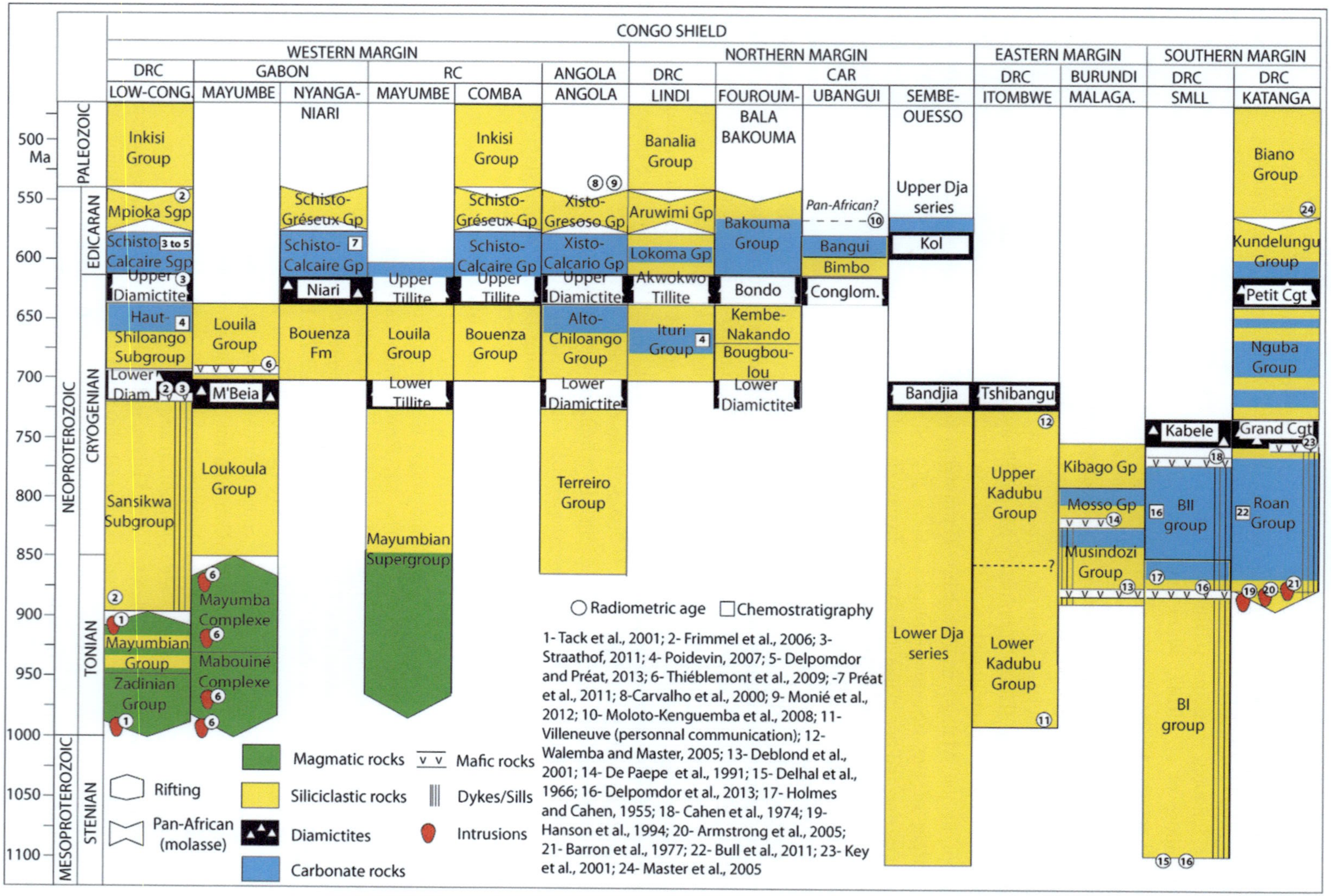

Fig. 3.7 Stratigraphic correlation between the Neoproterozoic sedimentary basins in Central Africa. Correlations based on Sr isotopic data, radiometric age constraints and revised lithological relationships

Group, is post-Sturtian in age and correlated with the Haut-Shiloango Subgroup (Daly et al. 1992; Trompette 1994). The lack of time constraints makes it difficult to establish detailed stratigraphic correlations. Nonetheless, on the basis of the lithological evidences and the stratigraphic position of the Lower Diamictite Formation from the CAR, the Ituri Group can be correlated with the Bougboulou and Kembé-Nakando series. The Akwokwo Tillite is correlated with the Sturtian-equivalent Lower Diamictite Formation (Poidevin 2007). However, recent lithological observations show that cap carbonates are ubiquous of the post-Marinoan event, except for the Grand Conglomérat Formation (Master and Wendorff 2011). Accepting the 'Snowball Earth hypothesis' with a synchronous deposition of diamictites, the Niari Tillite, Upper Diamictite, Akwokwo Tillite, Bondo Tillite, and Petit Conglomérat Formation are coeval in age (Tait et al. 2011). All post-Marinoan sedimentary rocks are poorly constrained in Central Africa. However, the carbonate successions are comparable in terms of paleoenvironments and C-isotopic trends, and are thus likely to have been deposited between 635 Ma and 580 Ma (Frimmel et al. 2006; Poidevin 2007; Straathof 2011). Nevertheless, Caron et al. (2010) estimated that a minimum age for the succession is 580 Ma in the Upper Dja series. Poidevin (2007) and Tait et al. (2011) concluded that the Upper Bakouma Series is correlated with the Aruwimi Group and the Mpioka (Sub-) Group of the West Congo Supergroup. Finally, the Inkisi Group, with a maximum depositional late Neoproterozoic age, is correlated with the Banalia and Biano groups.

Conclusion

Based on the present knowledge of the Neoproterozoic basinal litho-stratigraphy exposed along the margins of the Congo Shield, an attempt of chronostratigraphic correlations has been established between the isolated basins in the Democratic Republic of the Congo and neighboring countries. We conclude that: (i) the lithostratigraphy of each sedimentary basin may be used for intrabasinal and interregional correlations; (ii) all the Supergroups may be potentially broadly synchronous based on widespread climatic markers, represented by glacial deposits. However, such correlations remain inadequate because of the lack of precise such as biostratigraphy, sequence stratigraphy, and palaeomagnetism. New chemostratigraphic and geochronological data for the Neoproterozoic sedimentary series of Central Africa are thus sorely needed.

Acknowledgments The manuscript has been greatly improved following reviews and suggestions of H Frimmel and M. de Wit. We would like to thank M. Rottesman for her critical encouragement.

References

Abdelsalam MG, Liégeois JP, Stern RJ (2002) The Saharan Metacraton. J Afr Earth Sci 34(3–4):119–136

Aderca B (1950) Etude pétrographique et carte géologique du district du Congo-Ubangi (Congo belge). Mémoire de l'Institut Royal Colonial Belge, Sciences Naturelles et Médicales, in-8°, t. XVIII, fasc. 4, 3–65

Alkmim FF, Marshak S, Pedrosa-Soares AC, Peres GG, Pereira Cruz SC, Whittington A (2006) Kinematic evolution of the Araçuai-West Congo orogen in Brazil and Africa: nutcracker tectonics during the Neoproterozoic assembly of Gondwana. Precambrian Res 149: 43–64

Alkmim FF, Pedrosa-Soares AC, Noce CM, Cruz SCP (2007) Sobre a evolução tectônica do Orógeno Araçuaí-Congo Ocidental. Geonomos 15:25–43

Alvarez P (1992) Répartition de la sédimentation dans le golfe Protérozoïque supérieur du Schisto-Calcaire au Congo et Gabon. Implications en Afrique centrale. Palaeogeogr Palaeoclimatol Palaeoecol 96:281–297

Alvarez P (1995) Evidence for a Neoproterozoic carbonate ramp on the northern edge of the Central African Craton: relations with Late Proterozoic intracratonic troughs. Geol Rundsch 84:636–648

Alvarez P (2000) A quantitative method for the study of non-fossiliferous clastic formations: pre-Pan-African sandstones from central Africa and the northern Democratic Republic of Congo [ex-Zaïre]. J Afr Earth Sci 31(2):263–284

Armstrong RA (2000) Ion microprobe (SHRIMP) dating of zircons from granites, granulites and volcanic samples from Zambia. Unpublished Report, Australian National University, PRISE Job No. A99-160, Canberra

Armstrong RA, Master S, Robb LJ (2005) Geochronology of the Nchanga Granite, and constraints on the maximum age of the Katanga Supergroup, Zambian Copperbelt. J Afr Earth Sci 42:32–40

Babet V (1935) Esquisse géologique provisoire de la région comprise entre Bangui et la frontière du Cameroun. Chronique Minière Coloniale de Paris 38:160–164

Barron JW, Broughton DW, Armstrong RA, Hitzman MW (2003) Petrology, geochemistry and age of gabbroic bodies in the Solwezi area, northwestern Zambia. In: Contributions presented at the 3rd IGCP-450 Conference, Proterozoic Sediment-hosted Base Metal Deposits of Western Gondwana; Conference and Field Workshop Lubumbashi 2003, Lubumbashi, Democratic Republic of Congo, pp 75–77

Batumike JM, Griffin WL, O'Reilly SY, Belousova EA, Pawlitschek M (2009) Crustal evolution in the central Congo-Kasaï Craton, Luebo, D.R. Congo: insights from zircon U-Pb ages, Hf-isotope and trace-element data. Precambrian Res 170:107–115

Béchennec F, Hutin G, Martel-Jantin B (1981) Cartographie géologique et prospection générale. Mission Nyanga 1980. Région de Tchibanga. Rapport du Bureau de Recherches Géologiques et Minières 81 LIB 004

Begg GC, Griffin WL, Natapov LM, O'Reilly SY, Grand SP, O'Neill CJ, Hronsky JMA, Poudjom Djomani Y, Swain CJ, Deen T, Bowden P (2009) The lithospheric architecture of Africa: Seismic tomography, mantle petrology, and tectonic evolution. Geosphere 5(1):23–50

Bertrand-Sarfati J (1972) Stromatolites columnaires de certaines formations carbonatées du Précambrien supérieur du bassin congolais (Bushimay, Lindien, Ouest-Congolien). Annales du Musée Royal de l'Afrique Centrale, Tervuren, Belgique, Série in-8 - n° 74, 45 pp

Bertrand-Sarfati J, Milandou R (1989) Mécanismes de croissance des stromatolites géants infralittoraux, Protérozoïque supérieur de la République Populaire du Congo. Bull Soc Geol Fr 8(5):1185–1192
Bessoles B, Trompette R (1980) Géologie de l'Afrique: la chaîne panafricaine « zone mobile d'Afrique Centrale (partie sud) et zone mobile soudanaise. Mémoire du Bureau de Recherches Géologiques et Minières 92:398
Bigotte G, Bonifas G (1968) Faits nouveaux sur la géologie de la région Bakouma. Chroniques des Mines et de Recherches Minières 36(370):43–46
Boger SD, Miller JMcL (2004) The terminal suturing of Gondwana and the onset of the Ross-Delamerian Orogeny: the cause and effect of a Late Cambrian reconfiguration of plate motions. Earth Planet Sci Lett. 219:35–48
Bond GC, Nickeson PA, Kominz MA (1984) Breakup of a supercontinent between 625 and 555 Ma: new evidence and implications for continental histories. Earth Planet Sci Lett 70:325–345
Brito-Neves BB, Campos-Neto MC, Fuck RA (1999) From Rodinia to Western Gondwana: an approach to the Brasiliano-Pan African cycle and orogenic collage. Episodes 22:155–199
Brock BB (1961) The structural setting of the Copperbelt. In: Mendelsohn F (ed) The geology of the Northern Rhodesian Copperbelt. Macdonald, London, pp 81–89
Bueno JF, Oliveira EP, McNaughton NJ, Laux JH (2009) U-Pb dating of granites in the Neoproterozoic Sergipano Belt, NE-Brazil: implications for the timing and duration of continental collision and extrusion tectonics in the Borborema Province. Gondwana Res 15(1):86–97
Bull S, Selley D, Broughton D, Hotzman M, Cailteux J, Large R, McGoldrick P (2011) Sequence and carbon isotopic stratigraphy of the Neoproterozoic Roan Group strata of the Zambian Copperbelt. Precambrian Res 190(1–4):70–89
Cahen L (1954) Géologie du Congo Belge. Liège, Imprimerie H. Vaillant-Carmanne, S.A., 577 pp.
Cahen L (1978) La stratigraphie et la tectonique du Supergroupe Ouest-Congolien dans les zones médiane et externe de l'orogénèse Ouest-Congolien (Pan-African) au Bas-Zaïre et dans les régions voisines. Annales du Musée Royal de l'Afrique Centrale, Tervuren, in 8°, Sciences Géologiques, 83, 150 pp
Cahen L (1982) Geochronological correlation of the Late Precambrian sequences on and around the stable zones of Equatorial Africa. Precambrian Res 1:73–86
Cahen L, Lepersonne J (1981) Proterozoic diamictites of Lower Zaire. In: Hambrey, MAHW (ed) Earth's pre-pleistocene glacial record. Cambridge University Press, Cambridge, pp 153–157
Cahen L, Mortelmans G (1947) Le système de la Bushimaie au Katanga. Bulletin de la Société Belge de Géologie 56:217–253
Cahen L, Ledent D, Snelling NJ (1974) Données géochronologiques dans le Katangien inférieur du Kasaï oriental et du Shaba nord-oriental (République du Zaïre). Musée Royal de l'Afrique Centrale, Tervuren, Rapport Annuel, Département de Géologie et des Mines, 59–70
Cahen L, Ledent D, Villeneuve M (1979) Existence d'une chaîne plissée Protérozoïque Supérieur au Kivu oriental (Zaïre). Données géochronologiques relatives au Supergroupe de l'Itombwe. Bulletin de la Société Belge de Géologie 88(1):71–83
Cahen L, Snelling NJ, Delhal J, Vail JR (1984) Geochronology and evolution of Africa. Clarendon, Oxford, 512 pp
Cailteux JJ (1983) Le Roan Shabien dans la région de Kambove (Shaba-Zaïre). PhD Thèse, Université de Liège, unpublished
Cailteux J (1994) Lithostratigraphy of the Neoproterozoic Shaba type (Zaïre) Roan Supergroup and metallogenesis of associated stratiform mineralization. J Afr Earth Sci 19:279–301
Cailteux J, Kampunzu AB (1995) The Katangan tectonic breccias in the Shaba province (Zaïre) and their genetic significance. In: Wendorff M, Tack L (eds) Late Proterozoic Belts in Central Africa. Musée Royal de l'Afrique Centrale, Tervuren, pp 63–76
Cailteux JLH, Kaputo AK, Kampunzu AB (2003) Structure, lithostratigraphy and Cu–Co mineralization of the Mines Subgroup at Luiswishi, central Africa Copperbelt. In: Cailteux J (ed) Proterozoic sediment-hosted base metal deposits of Western Gondwana, Lubumbashi, pp 103–107
Cailteux JLH, Kampunzu AB, Lerouge C, Kaputo AK, Milesi JP (2005) Genesis of sediment-hosted stratiform copper–cobalt deposits, central African Copperbelt. J Afr Earth Sci 42:134–158
Cailteux JLH, Kampunzu AB, Lerouge C (2007) The Neoproterozoic Mwashya–Kansuki sedimentary rock succession in the central African Copperbelt, its Cu–Co mineralization, and regional correlations. Gondwana Res 11:414–431
Caron V, Ekomane E, Mahieux G, Moussango P, Ndjeng E (2010) The Minton Formation (new): sedimentology and geochemistry of a Neoproterozoic, paralic succession in south-east Cameroon. J Afr Earth Sci 54(4):367–385
Carvalho H, Tassinari C, Alves PH, Guimaraes F, Simoes MC (2000) Geochronological review of the Precambrian in western Angola: links with Brazil. J Afr Earth Sci 31:383–402
Censier C (1989) Dynamique sédimentaire d'un système fluviatile diamantifère mésozoïque. La Formation de Carnot (République Centrafricaine). Thèse de l'Université de Bangui, 591 pp, unpublished
Chevallier L, Makanga JF, Thomas RJ (2002) Carte géologique de la République gabonaise, 1: 1 000 000. Notice explicative. Council for Geoscience, South Africa
Chochine N (1950) Carte géologique de reconnaissance de l'Afrique Equatoriale Française à l'échelle du 1/500 000 (feuille Makoukou Est) avec notice explicative. Gouvernement Genéral de l'Afrique Equatoriale Française, Direction des Mines et de Géologie, Brazzaville, 16 pp
Cordani UG, Brito-Neves BB, D'Agrella MS, Trindade RIF (2003) Tearing-up Rodinia: the Neoproterozoic paleogeography of South American cratonic fragments. Terra Nova 15:343–349
Dadet P (1969) Notice explicative de la carte géologique de la République du Congo-Brazzaville au 1/500 000. Mémoire du Bureau de Recherches Géologiques et Minières 40:103
Daly MC, Lawrence SR, Diemu-Tshiband K, Matouana B (1992) Tectonic evolution of the Cuvette Centrale, Zaïre. J Geol Soc Lond 149:539–546
Dalziel IWD (1991) Pacific margins of Laurentia and East Antarctica-Australia as a conjugate rift pair: evidence and implication for an Eocambrian supercontinent. Geology 19:598–601
Dalziel IWD (1997) Neoproterozoic-Paleozoic geography and tectonics: review, hypothesis and environmental speculation. Bull Geol Soc Am 109:16–42
De Paepe P, Tack L, Moens L, van de Velde P (1991) The basic magmatism of the upper Proterozoic in southeast Burundi. In: Departement de Géologie et de Minéralogie (eds), Musée Royal de l'Afrique Centrale, Tervuren, Belgique, rapport annuel 1989-1990, pp 85–104
De Waele B (2005) The Proterozoic geological history of Irumide belt, Zambia. PhD thesis, Curtin University of Technology, Perth, Australia, 358 pp., unpublished
De Waele B, Wingate MTD, Mapani BSE, Fitzsimons ICW (2003) Untying the Kibaran knot: a reassessment of Mesoproterozoic correlations in southern Africa based on SHRIMP U-Pb data from the Irumide belt. Geology 31:509–512
De Waele B, Johnson SP, Pisarevsky SA (2008) Palaeoproterozoic to Neoproterozoic growth and evolution of the eastern Congo Craton: its role in the Rodinia puzzle. Precambrian Res 160:127–141
de Wit MJ, Linol B (2014) Precambrian Basement of the Congo Basin and Flanking Terranes. In: De Wit M, Guillocheau F (eds) The

geology and resource potential of the Congo Basin. Springer, Heidelberg

Deblond A, Punzalan LE, Boven A, Tack L (2001) The Malagarazi supergroup of southeast Burundi and its correlative Bukoba supergroup of northwest Tanzania: Neo- and Meso-Proterozoic constraints from Ar–Ar ages of mafic intrusive rocks. J Afr Earth Sci 32:435–449

Delhal J (1991) Situation géochronologique 1990 du Précambrien du sud-Kasaï et de l'Ouest-Shaba. Musée Royal de l'Afrique Centrale, Tervuren (Belgique). Rapport Annuel 1989–1990:119–125

Delhal J, Lepersonne J, Raucq P (1966) Le Complexe sédimentaire et volcanique de la Lulua (Kasaï). Annales du Musée Royal de l'Afrique Centrale, Tervuren (Belgique). Sciences Géologiques 51:106 pp

Delhal J, Ledent D, Torquato JR (1976) Nouvelles données géochronologiques relatives au complexe gabbro-noritique et charnockitique du bouclier du Kasaï et à son prolongement en Angola. Ann Soc Geol Belg 99:211–226

Delhal J, Deutsch S, Snelling NJ (1989) Datation du Complexe sédimentaire et volcanique de la Lulua (Protérozoïque inférieur, Kasai, Zaïre). Musée Royal de l'Afrique Centrale, Tervuren (Belgique), Département de Géologie et de Minéralogie. Rapport Annuel 1987–1988:93–99

Delhaye F, Sluys M (1923-4 and 1928-9) Observations ayant servi a l'élaboration de l' "Esquisse géologique du Congo occidental". Etude du Système Schisto-Calcaire. 1er - 3e mémoire. Annales de la Société Géologique de Belgique, 1923-4, C. 50-91 et 1928-9, C. 69–114

Delor C, Theveniaut H, Cage M, Pato D, Lafon J-M, Bialkowski A, Roig J-Y, Neto A, Cavongo M, Sergeev S (2008) New insights into the Precambrian Geology of Angola: basis for an updated lithochronological framework at 1:2 000 000 scale. 22nd Colloquium African Geology (CAG22), 04–06.11.2008, Hammamet, Tunisia, Abstracts volume, pp 52–53

Delpomdor F, Préat A (2013) Early and late Neoproterozoic C, O and Sr isotope chemostratigraphy in the carbonates of West Congo and Mbuji-Mayi supergroups: a preserved marine signature? Palaeogeogr Palaeoclimatol Palaeoecol 389:35–47

Delpomdor F, Linneman U, Boven A, Gartner A, Travin A, Blanpied C, Virgone A, Jelsma H, Préat A (2013) Depositional age, provenance, tectonic and paleoclimatic settings of the late Mesoproterozoic – middle Neoproterozoic Mbuji-Mayi Supergroup, Democratic Republic of Congo. Palaeogeogr Palaeoclimatol Palaeoecol 389: 4–34

Delpomdor F, Blanpied C, Virgone A, Préat A (2014). Sedimentology and sequence stratigraphy of the Late Precambrian carbonates of the Mbuji-Mayi Supergroup in the Sankuru-Mbuji-Mayi-Lomami-Lovoy Basin (Democratic Republic of Congo). In: De Wit M, Guillocheau F (eds) The geology and resource potential of the Congo Basin. Springer, Heidelberg

Dévigne JP (1959) Le Précambrien du Gabon occidental en Afrique Occidentale Française et régions limitrophes. Thèse, Clermont-Ferrand, 19 avril 1958 et Bulletin Division des Mines et Géologie de l'Afrique Equatoriale Française 11, 315 pp, unpublished

El Desouky HA, Muchez P, Boyce AJ, Schneider J, Cailteux JLH, Dewaele S, von Quadt A (2010) Genesis of sediment-hosted stratiform copper-cobalt mineralization at Luiswishi and Kamoto, Katanga Copperbelt (Democratic Republic of Congo). Miner Deposita 45:735–763

Erwin DH, Laflamme M, Tweedt SM, Sperling EA, Pisani D, Peterson KJ (2011) The Cambrian Conundrum: early divergence and later ecological success in the early history of animals. Science 334(6059): 1091–1097

Fernandez-Alonso M, Theunissen K (1998) Airborne geophysics and geochemistry provide new insights in the intracontinental evolution of the Mesoproterozoic Kibaran Belt (Central Africa). Geol Mag 135(2):203–216

Fernandez-Alonso M, Cutten H, De Waele B, Tack L, Tahon A, Baudet D, Barritt SD (2012) The Mesoproterozoic Karagwe-Ankole Belt (formerly the NE Kibaran Belt): The result of prolonged extensional intracratonic basin development punctuated by two short-lived far-field compressional events. Precambrian Res 216–219:63–86

François A (1974) Stratigraphie, tectonique et minéralisations dans l'arc cuprifère du Shaba (République du Zaïre). In: Bartholomé P (ed) Gisements Stratiformes et Provinces Cuprifères. Société Géologique de Belgique, Liège, pp 79–101

Frimmel HE, Tack L, Basei MS, Nutman AP, Boven A (2006) Provenance and chemostratigraphy of the Neoproterozoic West Congolian Group in the Democratic Republic of Congo. J Afr Earth Sci 46: 221–239

Gérard G (1958) Carte géologique de l'Afrique Equatoriale Française au 1/2 000 0000 avec notice explicative. Brazzaville, Direction des Mines et Géologie de l'Afrique Equatoriale Française, 198 pp, 4 feuilles

Gérard G, Gérard J (1952) Stratigraphie du Précambrien d'Oubangui-Chari occidental (Afrique Equatoriale Française) et essai de corrélations avec les territoires voisins. XIXème Congrès Géologique International, Alger, Association des Services Géologiques Africains, pp 145–153

Grès M, Legras M (1967) Mission Ouesso. Rapport de synthèse (1964-1967). Rapport du Bureau de Recherches Géologiques et Minières, 70 RME 039

Halverson GP, Dudas FO, Maloof AC, Bowring SA (2007) Evolution of the Sr-^{87}/Sr-^{86} composition of Neoproterozoic seawater. Palaeogeogr Palaeoclimatol Palaeoecol 256(3–4):103–129

Heijlen W, Banks DA, Muchez P, Stensgard BM, Yardley BWD (2008) The nature of mineralizing fluids of the Kipushi Zn–Cu deposit, Katanga, Democratic Republic of Congo: quantitative fluid inclusion analysis using laser ablation ICP-MS, and bulk crush-leach methods. Econ Geol 103:1459–1482

Hoffman PF (1991) Did the breakout of Laurentia turn Gondwanaland inside-out? Science 252:1409–1412

Hoffman PF, Schrag DP (2002) The Snowball Earth hypothesis: testing the limits of global change. Terra Nova 14:129–155

Hoffman PF, Kaufman AJ, Halverson GP, Schrag DP (1998) A Neoproterozoic Snowball Earth. Science 281:1342–1346

Holmes A, Cahen L (1955) African geochronology. Colon Geol Min Resour 5(1):3–38

Jacobs J, Thomas RJ (2004) A Himalayan-type indentor-escape tectonic model for the southern part of the Late Neoproterozoic/Early Paleozoic East African/Antarctic Orogen. Geology 32:721–724

Jaunatre A (1969) Rapport synthèse géologique Ouesso. Rapport de la Commission économique pour l'Afrique

Johnson SP, Rivers T, De Waele B (2005) A review of the Mesoproterozoic to Early Paleozoic magmatic and tectonothermal history of Central Southern Africa: Implications for Rodinia and Gondwana. J Geol Soc Lond 162:433–450

Kabengele M, Mashala T, Loris NBT (2003) Geochemistry of the Lower Mwashya pyroclastic rocks in the Likasi-Kambove area (D.R. Congo). In: Contributions presented at the 3rd IGCP-450 Conference, Proterozoic Sediment-hosted Base Metal Deposits of Western Gondwana; Conference and Field Workshop Lubumbashi 2003, Lubumbashi, D.R. Congo, pp 69–74.

Kadima E, Delvaux D, Sebagenzi SN, Tack L, Kabeya M (2011) Structure and geological history of the Congo Basin: An integrated interpretation of gravity, magnetic and reflection seismic data. Basin Res 1–29

Kampunzu AB, Cailteux J (1999) Tectonic evolution of the Lufilian Arc during Neoproterozoic Pan African orogenesis. Gondwana Res 2:401–421

Kampunzu AB, Makutu M, Rocci G, Kramers J, Pineau F, Louaradi I, Tembo F (2007) Neoproterozoic Alkaline and Carbonatite Magmatism along the Western Rift in Central-Eastern Africa: break-up of Rodinia Supercontinent and reconstruction of Gondwana. Gondwana Res 1:155–156

Karlstrom KE, Harlan SS, Williams ML, McClelland J, Geissman JW, Ahall K-I (1999) Refining Rodinia: geologic evidence for the Australia-western US connection in the Proterozoic. Geol Soc Am Today 9(10):1–7

Key RM, Liyungu AK, Njamu FM, Somwe V, Banda J, Mosley PN, Armstrong RA (2001) The western arm of the Lufilian Arc in NW Zambia and its potential for copper mineralization. J Afr Earth Sci 33:503–528

Kirschvink J (1992) Late Proterozoic low-latitude glaciation: the Snowball Earth. In: Schopf J, Klein C (eds) The proterozoic biosphere. Cambridge University Press, Cambridge, pp 51–52

Knoll AH (1992) Biological and geochemical preludes to Ediacaran radiation. In: Lipps, JH, Signor PW (Eds) Origin and Early evolution of the Metazoa. Topics in geobiology, vol 10, Plenum, New York, pp 53–84.

Lasserre M, Soba D (1976) Age libérien des graniodiorites et des gneiss à pyroxènes du Cameroun méridional. Bulletin du Bureau de Recherches Géologiques et Minières 2:17–32

Lefebvre JJ (1973) Présence d'une sédimentation pyroclastique dans le Mwashya inférieur du Shaba méridional (ex-Katanga). Ann Soc Geol Belg 96:197–218

Lefebvre JJ (1975) Les roches ignées dans le Katangien du Shaba (Zaïre). Le district du cuivre Annales de la Société Géologique de Belgique 98:47–73

Lefebvre JJ, Patterson LE (1982) Hydrothermal assemblage of aluminium serpentine florencite and kyanite in the Zaïrian Copperbelt. Ann Soc Geol Belg 105:51–71

Legoux P, Hourcq V (1943) Esquisse géologique de l'Afrique Equatoriale Française. (Notice explicative de la carte géologique de l'Afrique Equatoriale Française à 1/3.500.000). Bulletin de Service des Mines de l'Afrique Equatoriale Française.

Lepersonne, J., 1973. Carte géologique à l'échelle 1/200000. Notice explicative de la feuille Ngungu (degré carré S6/14 = SB 33.9). République Démocratique du Congo, Département Mines Direction du Service Géologique

Li ZX, Evans DAD, Zhang S (2004) A 90° spin on Rodinia: possible causal links between the Neoproterozoic supercontinent, superplume, true polar wander and low-latitude glaciation. Earth Planet Sci Lett 220:409–421

Li ZX, Bogdanova SV, Collins AS, Davidson A, De Waele B, Ernst RE, Fitsimons ICW, Fuck RA, Gladkochub DP, Jacobs J, Karlstom KE, Lu S, Natapov LM, Pease V, Pisarevsky SA, Thrane K, Vernikovsky V (2008) Assembly, configuration, and break-up history of Rodinia: A synthesis. Precambrian Res 160(1–2):179–210

Linol B, de Wit MJ, Guillocheau F, Robin C, Dauteuil O (2014) Multiphase Phanerozoic subsidence and uplift history recorded in the CB – A complex successor basin. In: De Wit M, Guillocheau F (eds) The geology and resource potential of the Congo Basin. Springer, Heidelberg

Liyungu AK, Mosley PN, Njamu FM, Banda J (2001) Geology of the Mwinilunga area. Rep Geol Surv Zambia 110:36

Master S, Wendorff M (2011) Neoproterozoic glaciogenic diamictites of the Katanga Supergroup, Central Africa. In: Arnaud E, Halverson GL, Shields-Zhou G (eds) The Geological Record of Neoproterozoic Glaciations, vol 36, Geological Society, London, Memoir., pp 173–184

Master S, Rainaud C, Armstrong RA, Phillips D, Robb LJ (2005) Provenance ages of the Neoproterozoic Katanga Supergroup (Central African Copperbelt), with implications for basin evolution. J Afr Earth Sci 42:41–60

Mathieu, F.F., 1922. Note sur les calcaires de l'Itimbiri, de l'Uele et de l'Aruwimi. Annales de la Société Géologique de Belgique, Publication Relative au Congo Belge, XLIV, pp. c25-33.

Maurin J-C (1993) La chaîne panafricaine ouest-congolienne : corrélation avec le domaine Est-brésilien et hypothèse géodynamique. Bull Soc Geol Fr 764:51–60

Mbede EI (1991) The sedimentary basins of Tanzania reviewed. J Afr Earth Sci 13:291–297

Mbede, E.I., Kampunzu, A.B., Armstrong, R., 2004. Neoproterozoic Inheritance during Cainozoic rifting in the Western and Southwestern branches of the East African Rift System: evidence from Carbonatite and Alkaline intrusions. The East African Rift System: Geodynamics, Resources and Environment, United Nations Conference Centre, Addis Ababa, Ethiopia, June 20–24, Extended Abstracts volume, p. 142.

McMenamin MAS, McMenamin DLS (1990) The emergence of Animals: The Cambrian Breakthrough. Columbia University Press, New York, 217 pp

Mendelsohn F (1961) Ore genesis: summary of the evidence. In: Mendelsohn F (ed) The geology of the Northern Rhodesian Copperbelt. Macdonald, London, pp 130–146

Mestraud J-L (1952) Formation du socle en Oubangui-Chari central. XIXème Congrès Géologique International, Alger, Association des Services Géologiques Africains, pp 155–162

Mestraud, J.-L., 1953. Notice explicative sur la feuille Bangassou Ouest. Carte géologique de reconnaissance au 1/500.000. Publication de la Direction des Mines et de Géologie de l'Afrique Equatoriale Française.

Meyer A (1953) Une formation à *Collenia* dans la région de Niangara. Bulletin de la Société Belge de Géologie, LXII, pp 213–216

Moloto-A-Kenguemba GR, Trindade RIF, Monie P, Nédelec A, Siqueira R (2008) A late Neoproterozoic paleomagnetic pole for the Congo Craton: Tectonic setting, paleomagnetism and geochronology of the Nola dike swarm (Central African Republic). Precambrian Res 164(3–4):214–226

Monié P, Bosch D, Bruguier O, Vauchez A, Rolland Y, Nsungani P, Buta NA (2012) The Late Neoproterozoic/Early Palaeozoic evolution of the West Congo Belt of NW Angola: geochronological (U-Pb and Ar-Ar) and petrostructural constraints. Terra Nova 00:1–10

Narbonne GM (2010) Evolutionary biology: When life got big. Nature 470:339–340

Ninove G (1954) Observations inédites. In: Cahen L (ed) Géologie du Congo Belge. Liège, Imprimerie H. Vaillant-Carmanne, S.A., pp. 577

Noce CM, Pedrosa-Soares AC, da Silva LC, Armstrong R, Piuzana D (2007) Evolution of polycyclic basement complexes in the Araçuai Orogen, based on U-PB SHRIMP data: Implications for Brazil-Africa links in Paleoproterozoic time. Precambrian Res 159:60–78

Oliveira EP, Toteu SF, Araujo MNC, Carvalho MJ, Nascimento RS, Bueno JF, McNaughton N, Basilici G (2006) Geologic correlation between the Neoproterozoic Sergipano belt (NE Brazil) and the Yaounde belt (Cameroon, Africa). J Afr Earth Sci 44(4–5):470–478

Pedrosa-Soares AC, Alkmin FF (2011) How many rifting events preceded the development of the Araçuai-West Congo orogen? Geonomos 19(2):244–251

Pedrosa-Soares AC, Noce CM, Wiedemann CM, Pinto CP (2001) The Araçuaí-West Congo orogen in Brazil: An overview of a confined orogen formed during Gondwanland assembly. Precambrian Res 110:307–323

Pedrosa-Soares AC, Noce CM, Alkmim FF, Silva LC, Babinski M, Cordani U, Castañeda C (2007) Orógeno Araçuaí: síntese do conhecimento 30 anos após Almeida 1977. Geonomos 15:1–16

Pin C, Poidevin JL (1987) U-Pb Zircon Evidence for a Pan-African Granulite Facies Metamorphism in the Central-African-Republic - a New Interpretation of the High-Grade Series of the Northern Border of the Congo Craton. Precambrian Res 36(3–4):303–312

Pinna P, Cocherie A, Thiéblemont D, Feybesse J-L, Lagny P (1996) Evolution géodynamique du Craton Est-Africain et déterminisme gîtologique. Geodynamic evolution and metallogenic controls in the East-African Craton (Tanzania, Kenya, Uganda). Bureau des Recherches Géologiques et Minières, Chroniques des Recherches Minières 525:33–43

Piper JDA (1976) Palaeomagnetic evidence for a Proterozoic supercontinent. Philos Trans R Soc Lond A280:469–490

Piper JDA (2000) The Neoproterozoic supercontinent: Rodinia or Palaeopangea? Earth Planet Sci Lett 176:131–146

Pisarevsky SA, Wingate MTD, Powell CM, Johson S, Evans DAD (2003) Models of Rodinia assembly and fragmentation. Geol Soc Lond Spec Publ 206(1):35–55

Poidevin J-L (1976) Les formations du Précambrien supérieur de la région de Bangui (R.C.A.). Bulletin de la Société Géologique de Belgique 7, XVIII, 4, 999–1003.

Poidevin J-L (1979a) Echelle stratigraphie des formations précambriennes de Centrafrique (R.C.A.). Résumé 10ème Colloque de Géologie Aficaine, Montpellier, p. 12.

Poidevin J-L (1979b) Les basaltes et dolérites Précambrien supérieur de la région de Bakouma (Empire Centraficain). 7ème Réunion Association Sciences de la Terre, Lyon, p. 374.

Poidevin J-L (1985) Le Protérozoïque supérieur de la République Centrafricaine. Annales du Musée Royal de l'Afrique Centrale, Tervuren, Belgique, Sciences Géologiques 91:75

Poidevin J-L (2007) Stratigraphie isotopique du strontium et datations des formations carbonatées et glaciogéniques néoprotérozoïques du Nord et de l'Ouest du Craton du Congo. Compt Rendus Geosci 339: 259–273

Poidevin J-L, Pin C (1986) 2 Ga U-Pb zircon dating of Mbi graniodiorite (Central African Republic) and its bearing on the chronology of the Proterozoic of Central Africa. J Afr Earth Sci 5–6:581–587

Poidevin J-L, Alabert J, Miauton JD (1981) Géologie des séries du Précambrien supérieur de la région de Bakouma (République Centrafricaine). Bulletin du Bureau de Recherches Géologiques et Minières, Section IV 4:313–320

Polinard E (1935) La géographie physique de la region du Lubilash, de la Bushimaie et de la Lubi vers le 6e parallèle sud. Institut Royal du Congo Belge, Section Sciences Naturelles et Médicales, Mémoire in-4°, t. IV, fasc. 1.

Porada H, Berhorst V (2000) Towards a new understanding of the Neoproterozoic–Early Paleozoic Lufilian and Northern Zambezi Belts in Zambia and Democratic Republic of Congo. J Afr Earth Sci 30:727–771

Préat A, Kolo K, Prian J-P, Delpomdor F (2010) A peritidal evaporite environment in the Neoproterozoic of South Gabon (Schisto-Calcaire Subgroup, Nyanga Basin). Precambrian Res 177(3–4): 253–265

Préat A, Prian JP, Thiéblemont D, Mabicka Obame R, Delpomdor F (2011) Stable isotopes of oxygen and carbon compositions in the Neoproterozoic of South Gabon (Schisto-Calcaire Subgroup, Nyanga Basin): Are cap carbonates and lithoherms recording a particular destabilization event after the Marinoan glaciation? J Afr Earth Sci 60(4):273–287

Prian J-P (2008) Notice géologique et ressources minérales de la carte de N'Dendé à 1/200 000. Document provisoire, carte géologique du Gabon, 79 pp.

Prian J-P, Thiéblemont D, Préat A, Gujou GC, Simo Ndounze S, Ekogha H (2009a) Notice explicative de la carte géologique de la République du Gabon à 1/200000, feuille Ndendé. Editions DGMG, Ministère des Mines, du Pétrole, des Hydrocarbures. Libreville. Document provisoire, carte géologique du Gabon, 79 pp.

Prian J-P, Thiéblemont D, Préat A, Walemba A, Simo Ndounze S, Gujou GC, Ekogha H, Kassadou AB (2009b) Carte géologique de la République du Gabon à 1/200000, feuille Ndendé. Editions DGMG, Ministère des Mines, du Pétrole, des Hydrocarbures. Libreville. Document provisoire, carte géologique du Gabon, 79 pp.

Ramsay CR, Ridgeway J (1977) Metamorphic patterns in Zambia and their bearing on problems of Zambian tectonic history. Precambrian Res 4:321–337

Raucq P (1957) Contribution à la reconnaissance du Système de la Bushimay. Annales du Musée Royal du Congo Belge (Tervuren), Série 8, vol.18, 427 pp.

Raucq P (1970) Nouvelles acquisitions sur le système de la Bushimay (Répubique Démocratique du Congo). Annales du Musée Royal de l'Afrique Centrale, Tervuren, Belgique, Série in-8° - n° 69.

Robert M (1956) Géologie et Géographie du Katanga, y compris l'étude des ressources et de la mise en valeur. Edition M. Hayez, Bruxelles, 621 pp

Scolari G (1965) Etude géologique du bassin du Niari oriental (République du Congo Brazzaville) et des minéralisations Cu Pb-Zn. Mémoire du Bulletin de Recherches Géologiques et Minières 35:219

Sears JW, Price RA (2000) New look at the Siberian connection: no SWEAT. Geology 28:423–426

Sonet J (1958) Carte géologique de reconnaissance de l'Afrique Equatoriale Française à l'échelle du 1/500 000 (feuille Ouesso) avec notice explicative, Gouvernement Général de l'Afrique Equatoriale Française. Direction des Mines et de Géologie, Brazzaville, 24 pp

Stanton WI, Schermerhorn LJG, Korpershoek HR (1963) The West Congo System. Angola Serviços Geologico de Minas Boletim, Angola 8:69–78

Steven NM (2000) A Shaba-type Cu–Co(–Ni) deposit at Luamata, West of the Kabompo Dome, northwestern Zambia. Explor Min Geol 9:277–287

Straathof GB (2011) Neoproterozoic Low Latitude Glaciations: An African Perspective. Ph.D. Thesis University of Edinburgh, 285 pp., unpublished.

Tack L (1995) The Neoproterozoic Malagarazi Supergroup of SE Burundi and its equivalent Bukoban Supergroup in NW Tanzania: a current review. In: Wendorff M, Tack L (eds) Late Proterozoic Belts in Central and Southwestern Africa. Annales du Musée Royal de l'Afrique Centrale, Sciences Géologiques, pp 121–l 29.

Tack L, De Paepe P, Deutsch S, Liégeois JP (1984) The alkaline plutonic complex of the Upper Ruvubu (Burundi): geology, age, isotopic geochemistry and implications for the regional geology of the Western Rift. In: Klerkx J, Michot J (eds) African geology. A Volume in Honour of L. Cahen, Musée Royal de l'Afrique Centrale, Tervuren, Belgium, pp 91–114

Tack L, Wingate MTD, Liégeois J-P, Fernandez-Alonzo M, Deblond A (2001) Early Neoproterozoic magmatism (1000-910 Ma) of the Zadinian and Mayumbian Groups (Bas-Congo), onset of Rodinia rifting at the western edge of the Congo Craton. Precambrian Res 110:277–306

Tack L, Fernandez-Alonso M, Tahon A, De Waele B, Baudet D, Dewaele S (2011) The "Kibaran belt" of central Africa: What's in a name? 23rd Colloquium of African Geology (CAG23), Johannesburg, South Africa, 8–14 January 2011. Abstracts volume, p. 376.

Tait J, Delpomdor F, Préat A, Tack L, Straathof G, Kanda Nkula V (2011) Neoproterozoic sequences of the West Congo and Lindi/ Ubangi Supergroups in the Congo Craton, Central Africa. In: Arnaud E, Halverson GL, Shields-Zhou G (eds) The geological record of neoproterozoic glaciations. Geological Society, London, Memoir 36, pp 185–194

Theunissen K, Lenoir J-L, Liégeois J.-P, Delvaux D, Mruma A (1992) Major Pan-African imprint in the Ubendian belt of SW Tanzania: U–Pb on zircon geochronology and structural context. Comptes Rendus de l'Académie des Sciences 314 Séries II, 1355–1362

Thiéblemont D, Castaing C, Billa M, Bouton A, Préat A (2009) Notice explicative de la carte géologique et des ressources minérales de la République gabonaise à 1/1000000. Programme Sysmin 8 ACP GA 017, Ministère des Mines, du Pétrole, des Hydrocarbures. Direction Générale des Mines et de la Géologie, 384 pp.

Torquato J (1974) Geologia do sudoeste de Moçâmedes e suas relações com a evolução tectónica de Angola. Tese de Doutoramento não publicada. Instituto de Geociências, Universidade de Sao Paulo, Brasil, unpublished.

Toteu SF, Penaye J, Deloule E, Van Schmus WR, Tchameni R (2006) U-Pb dating of plutonic rocks involved in the nappe tectonic in southern Cameroon: consequence for the Pan-African orogenic evolution of the Central African fold belt. J Afr Earth Sci 44(4–5): 479–493

Trompette R (1994) Geology of Western Gondwana (2000-500 Ma). Pan-African-Brasiliano aggregation of South America and Africa. Balkema, Rotterdam, 350 pp

Verbeek T (1970) Géologie et lithostratigraphie du Lindien (Précambrien supérieur du nord de la République Démocratique du Congo). Annales du Musée Royal de l'Afrique Centrale série n°8, 66, 309 pp.

Vicat J-P, Pouclet A (1995) Nature du magmatisme lié à une extension pré-panafricaine: les dolérites des bassins de Comba et de Sembé-Ouesso (Congo). Bull Soc Geol Fr 166–4:355–364

Vicat J-P, Vellutini P-J (1987) Sur la nature et la signification des dolérites du bassin précambrien de Sembé-Ouesso (République Populaire du Congo). Precambrian Res 37:57–69

Vicat J-P, Pouclet A, Nkoumbou C, Semé Mouangué A (1997) Le volcanisme fissural néoprotérozoïque des séries du Dja inférieur, de Yokadouma (Cameroun) et de Nola (RCA) – Signification géotectonique. Comptes Rendus de l'Académie des Sciences de Paris 325:671–677

Villeneuve M (1976a) In: X- Réunion de travail des 1, 2, et 3 septembre 1975: La géologie des terrains précambriens voisins du Fossé tectonique occidental, spécialement dans les régions de part et d'autre de la partie sud de Lac Kivu et du Nord du Lac Tanganyika au Kivu, au Rwanda et au Burundi. Musée Royal de l'Afrique Centrale, Tervuren (Belgique), Département de Géologie et Minéralogie, Rapport Annuel 1975, pp 143–170

Villeneuve M (1976b) Mise en évidence d'une discordance angulaire majeure dans les terrains précambriens au nord du flanc oriental du 'synclinal de l'Itombwe' (région du Kivu, Zaïre). Comptes Rendus de l'Académie des Sciences, Paris 282, Section D, 1709–1712.

Villeneuve M (1977) Précambrien du Sud du lac Kivu (région du Kivu, Zaïre). Etude stratigraphique, pétrographique et tectonique. Thèse 3e cycle, Faculté des Sciences St-Jérôme, Université d'Aix-Marseille, 195 pp., unpublished.

Villeneuve M (1979) Etude photogéologique du secteur Précambrien de Luemba (Sud Kivu, Zaïre). La partie méridionale du 'Synclinal de l'Itombwe' et son substratum. Ann Soc Geol Belg 101:47–52

Waleffe A (1965) Etude géologique du Sud-Est du Burundi (Régions du Mosso et du Nkomal). Annales du Musée Royal de l'Afrique Centrale, Sciences Géologiques, Séries ln-8°, 48.

Waleffe A (1988) Etude photogéologique du Synclinorium de l'Itombwe et des régions avoisinants au sud du 3ème parallèle sud (Kivu, Zaïre). Bulletin de la Société Belge de Géologie 97(2): 211–221

Walemba KMA (2001) Geology, geochemistry, and tectono-metallogenic evolution of Neoproterozoic gold deposits in the Kadubu area, Kivu, Democratic Republic of Congo. Ph.D. thesis, University of the Witwatersrand, Johannesburg, South Africa, 491 pp., unpublished.

Walemba KMA, Master S (2005) Neoproterozoic diamictites from the Itombwe Synclinorium, Kivu Province, Democratic republic of Congo: Palaeoclimatic significance and regional correlations. J Afr Earth Sci 42:200–210

Wasilewsky I (1954) Exploration en profondeur des formations du Système de la Bushimaie (Bakwanga, Kasai). Mémoire de l'Institut de Géologie, Université de Louvain, t. XIX, fasc. II, pp. 145-176.

Wendorff M (2000) Genetic aspects of the Katangan megabreccias: Neoproterozoic of Central Africa. J Afr Earth Sci 30:703–715

Wendorff M (2005) Evolution of Neoproterozoic–Lower Palaeozoic Lufilian arc, Central Africa: a new model based on syntectonic conglomerates. J Geol Soc Lond 162:5–8

Wendorff M, Key RM (2009) The relevance of the sedimentary history of the Grand Conglomérat Formation (Central Africa) to the interpretation of the climate during a major Cryogenian glacial event. Precambrian Res 172(1–2):127–142

Sedimentology and Sequence Stratigraphy of the Late Precambrian Carbonates of the Mbuji-Mayi Supergroup in the Sankuru-Mbuji-Mayi-Lomami-Lovoy Basin (Democratic Republic of the Congo)

4

Franck Delpomdor, Christian Blanpied, Aurelien Virgone, and Alain Préat

4.1 Introduction

The late Mesoproterozoic to middle Neoproterozoic times heralded extraordinary climatic and biological changes deeply linked to global tectonics (e.g. Kaufman et al. 1993, 1997; Halverson et al. 2007). These dramatic climatic changes, global carbon cycling and atmospheric oxygen budget (Knoll et al. 1986; Derry et al. 1992; Knoll 1992; Des Marais 1994; Strauss 1997; Hoffman et al. 1998; Canfield 1999) are recorded globally by chemostratigraphic fluctuations of C, Sr and S isotopic compositions in the worldwide late Mesoproterozoic (Stenian, i.e. 1200–1000 Ma; Fairchild et al. 1990; Des Marais et al. 1992; Knoll et al. 1995; Podkovyrov et al. 1998; Canfield 1999; Kah et al. 1999; Bartley et al. 2001), and mid- (Cryogenian, i.e. 850–635 Ma) to late (Ediacaran, i.e. 635–541 Ma) Neoproterozoic series (Derry et al. 1989; Kaufman and Knoll 1995; Asmeron et al. 1991; Halverson et al. 2005, 2007; Kaufman et al. 2006; Tewari and Sial 2007). The origin of these large global isotope fluctuations is still being debated. In spite of the abundance of late Mesoproterozoic to middle Neoproterozoic sedimentary rocks chemostratigraphically studied with stable isotope geochemistry to establish regional or global inter-basinal correlations (Melezhik et al. 2001; Shields and Veizer 2002; Halverson et al. 2005, 2010), their depositional environment remains poorly constrained.

Phanerozoic carbonate rocks may reflect short-term marine environmental changes with high frequency sea-level fluctuations as well as long-term patterns of sedimentary evolution by the formation of carbonate buildups, because they are controlled by biogenic production variations as well as by sea-level fluctuations, accommodation, and sedimentation rates (Flügel 2004). This is also especially relevant in the Precambrian, because most of the carbonate sediments were directly precipitated with microbial assistance (Day et al. 2004). Precambrian sedimentary successions have recently been discussed within the context of sequence stratigraphy (Catuneanu et al. 2005; Martins-Neto 2009). This approach interprets the sedimentary changes as a response to the interaction occurring between authigenic (i.e. the sedimentology and facies analyses), and allogenic processes, (i.e. sedimentation including climate, tectonics and sea level changes) regardless of scale and age of the strata (Vail et al. 1977; Miall 1997; Posamentier and Allen 1999; Catuneanu 2006). The largest sequences (Vail et al. 1977; Krapez 1996), recognized as first-order, last between 200 to 400 Myrs, and are subdivided into second-order (10–100 My), third-order (1–10 My), and fourth-order (0.01–1 My) sequences etc. (e.g. Vail et al. 1977; Miall 2000). The durations of these sequences are related to global tectonic processes (first to third orders) and orbital controls (fourth to fifth orders) with possible superimposed eustatic fluctuations. These sea level fluctuations may be revealed by Fisher plots, a popular tool in cyclostratigraphy to graphically illustrate deviations from the average thickness of a sedimentary cycle (Fischer 1964). Fischer plots provide graphically and illustrate such thickness variations, and they help identify also longer-term trends and patterns. The thickness variations in the stacking patterns of meter-scale (or more) may reflect third-order changes in accommodation space. Because Fischer plots have often been criticized as being too subjective, and speculative, statistical analyses should be carried out, and ideally based upon more than 30–50 cycles to install confidence in the models (Saddler et al. 1993). Fischer plots, combined with the analysis of cycles, are useful and upward changes in facies can be examined in the context of upward changes in cycle thickness through accommodation cycles. Fischer plots can also

F. Delpomdor (✉) • A. Préat
Université libre de Bruxelles, Department of Earth Sciences and Environment, Unit of Biogeochemistry and Modeling of the Earth System, Avenue F.D. Roosevelt 50, CP160/2, 1050 Brussels, Belgium
e-mail: franck.delpomdor@ulb.ac.be; apreat@ulb.ac.be

C. Blanpied • A. Virgone
Total Exploration and Production, Avenue Larribau, 64018 Pau Cedex, France
e-mail: Christian.Blanpied@total.com; aurelien.virgone@total.com

M.J. de Wit et al. (eds.), *Geology and Resource Potential of the Congo Basin*, Regional Geology Reviews,
DOI 10.1007/978-3-642-29482-2_4, © Springer-Verlag Berlin Heidelberg 2015

contribute to the discussion of the origin of parasequences. One expects non-random patterns for those produced by orbital forcing (allocyclicity; Goldhammer et al. 1990). Were the cyclicity was produced by sedimentary processes, for example of tidal-flat progradation (autocylicity), then the Fischer plots should display more random patterns of cycle thickness and one would not expect the plots to correlate over any significant lateral distances. Limitations of the use of these plots concern principally 'missed beats' that occur where oscillations in relative sea level have not been preserved in the sedimentary strata through, for example emergence, so that the complete cycle history is not recorded. Fischer plots can also pick up noise where autocyclic processes, such as tidal flat progradation, or periodic faulting, overprint a eustatic signature.

The purpose of our work is to report advances in the understanding of the tectono-sedimentary evolution of the Sankuru-Mbuji-Mayi-Lomami-Lovoy Basin in terms of paleoecology and sea level variations, recorded in the late Mesoproterozoic—middle Neoproterozoic carbonate successions of the Mbuji-Mayi Supergroup in the Democratic Republic of the Congo (DRC), and thereby contribute to a better understanding of global Late Precambrian climate fluctuations.

4.2 Geological Setting

The Sankuru-Mbuji-Mayi-Lomami-Lovoy (SMLL; Fig. 4.1) Basin is located between 6°S and 8°S latitude and 23°E and 26°E longitude (Kasai-Oriental region, DRC), flanking the Archean-Paleoproterozoic Kasai Craton to the southwest, the Mesoproterozoic Kibara Supergroup to the east (Fig. 4.1). The Mbuji-Mayi sedimentary sequence is undisturbed and only weakly affected by regional metamorphism. The strata have a maximum dip of 3° in the west and between 20 and 45° in the southeastern parts of the SMLL Basin (Cahen 1954). The Mbuji-Mayi Supergroup (Fig. 4.1, Table 4.1) is divided, from oldest to youngest, into siliciclastic BI and carbonate BII groups (Raucq 1957, 1970). In the Sankuru-Mbuji-Mayi area, the lower siliciclastic series of the Mbuji-Mayi Supergroup (BI group) is ~500 m thick, and consists of five subgroups: BIb, BIc, BId, BIE (the latter only visible in the Kafuku region) and BIe. The BIa subgroup is missing in the Sankuru-Mbuji-Mayi area, but has been observed near the Makululu and Kiankodi villages in the southern part of the SMLL Basin. Detailed descriptions of this subgroup have been given by Cahen and Mortelmans (1947). It consists, from base to top, of ~1,500 m-thick red quartzites and shales with an interbedded pink chert horizon. The BII group consists of a ~1,000 m-thick carbonate successions embedded of thin levels of organic-rich shales, and subdivided into five subgroups: BIIa, BIIb, BIIc, BIId and BIIe. Basic igneous rocks have been identified as (i) basaltic lavas overlying the BII Group at the confluence of the Mbuji-Mayi and Sankuru Rivers (Cahen et al. 1984); (ii) dolerite sills emplaced within the succession close to the BI and BII contact in the Lomami area; and (iii) within the BI group along Kiankodi and Lovoy Rivers, with sills of dolerite (Cahen and Mortelmans 1947). Along the Kibaran Belt in the southern part of the SMLL Basin (Fig. 4.1), polymictic conglomerates (Kabele and Kabenga conglomerates) with more than 50 % of clasts material derived from the Mbuji-Mayi carbonates, were correlated with the Grand Conglomérat Formation of the Katanga Supergroup (Cahen and Mortelmans 1947).

4.3 Geochronology (Table 4.1)

In the north, oldest siliciclastic BI Group rests unconformably on 2648 ± 22 Ma migmatitic gneisses with granitic to tonalitic compositions of the Dibaya Complex (Delhal et al. 1976). In the south, the BI Group overlies the 1155 ± 15 Ma Kibara Supergroup (Delhal et al. 1966). An upper limit to the age of deposition of the Mbuji-Mayi Supergroup is provided a K–Ar date of 1152 ± 15 Ma (Delhal et al. 1989, recalculated to 1118 ± 15 Ma using the decay constants from Steiger and Jäger 1977) on biotite, pyroxene and amphibole from E–W trending syenodiorite dykes that outcrop in the eastern part of the Lulua Complex. Concordant U-Pb dates on detrital zircon grains from the BId Subgroup yield and age of 1174 ± 22 Ma, which is consistent with the maximum K–Ar age of 1152 ± 15 Ma (Delpomdor et al. 2013a). The transition between the BI and BII groups, constrained by $^{207}Pb/^{206}Pb$ dates on presumed syngenetic galena, yielded 1040 Ma and 1065 Ma (Cahen 1954; Holmes and Cahen 1955). Recent data from $\delta^{13}C$ and $^{87}Sr/^{86}Sr$ chemostratigraphy on BII carbonate samples reveal primary marine signals that are coeval with the Bitter Springs negative anomaly around c. 810 Ma (Delpomdor et al. 2013a). Amygdaloid basaltic pillow lavas that overlie the Mbuji-Mayi Supergroup at the Sankuru-Mbuji-Mayi confluence, yield an ^{40}Ar–^{39}Ar date of 928 ± 20 Ma. ^{40}Ar–^{39}Ar dating on dolerite located in the eastern part of the SMLL basin yield and age of 888.2 ± 8.8 Ma (Delpomdor et al. 2013a).

4.4 Sampling and Analytical Methods

4.4.1 Sampling

Drillcores were studied for this investigation. They are situated within the Sankuru-Mbuji-Mayi area between the Lubi and Luembe Rivers (Fig. 4.1). The cores are stored in

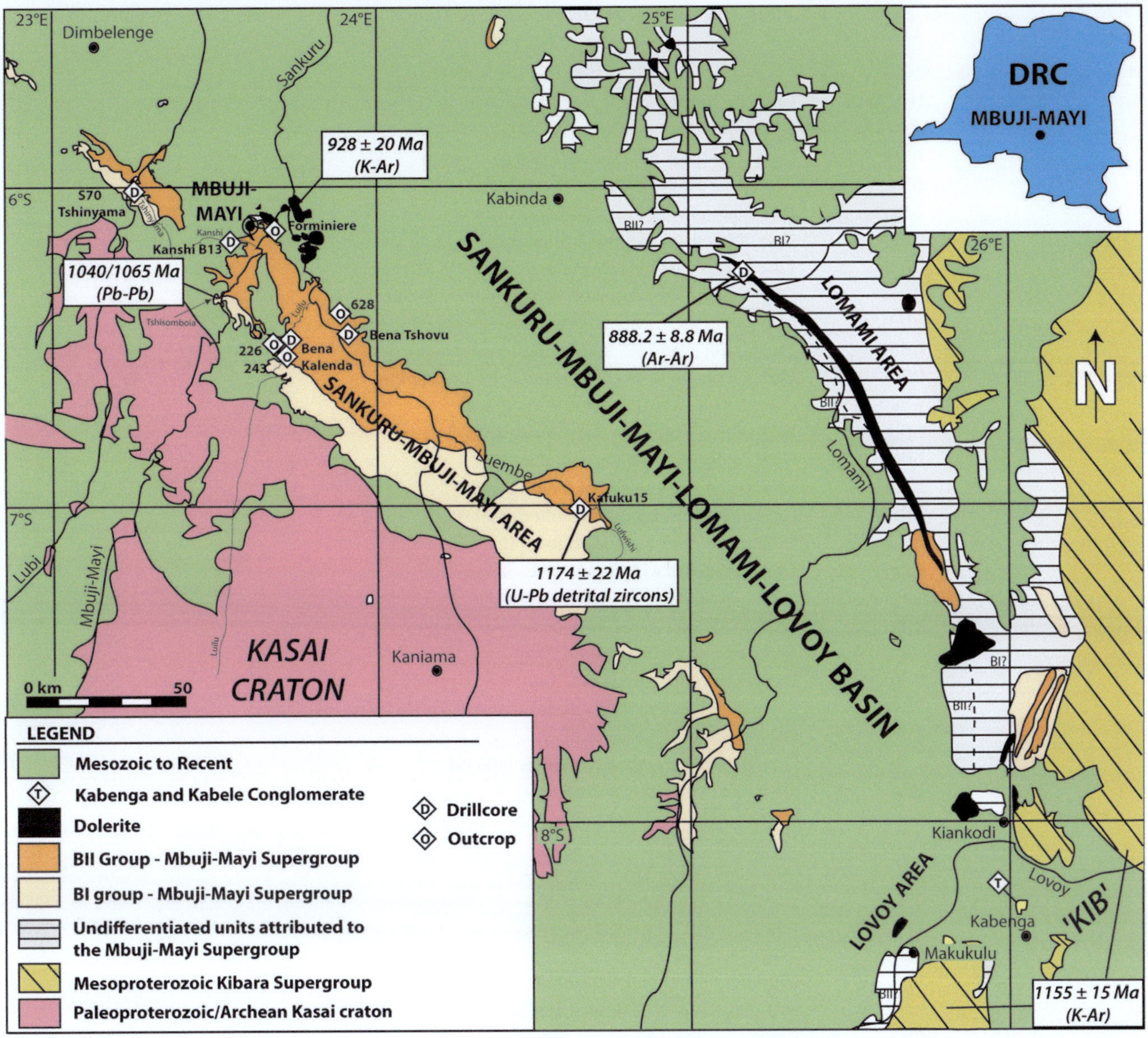

Fig. 4.1 Simplified geological map of the Sankuru-Mbuji-Mayi-Lomami-Lovoy Basin in the Kasai-Oriental region (Democratic Republic of the Congo). Also shown are location of drillcores (S70 Tshinyama, B13 Kanshi, Bena Kalenda, Bena Tshovu, Kafuku 15) and outcrops (stops 226, 243, 628, 602). In the south, the BI Group overlies the 1155 ± 15 Ma Kibara Supergroup (Delhal et al. 1966). Concordant U-Pb dates on detrital zircon grains from the BId Subgroup yielded 1174 ± 22 Ma, which are consistent with the maximum K–Ar age of 1152 ± 15 Ma (Delpomdor et al. 2013a). The transition between the BI and BII Groups, constrained by $^{207}Pb/^{206}Pb$ dates on presumed syngenetic galena, yielded 1040 Ma and 1065 Ma (Cahen 1954; Holmes and Cahen 1955). Amygdaloid basaltic pillow lavas at the Sankuru-Mbuji-Mayi confluence, $^{40}Ar–^{39}Ar$ date yield 928 ± 20 Ma, and overlie the Mbuji-Mayi Supergroup. $^{40}Ar–^{39}Ar$ dating on dolerite located in the eastern part of the SMLL basin have 888.2 ± 8.8 Ma dates. (Delpomdor et al. 2013a). Although no reliable data on the deepening direction have been reported, Raucq (1970) suggested a general southward deepening of the basin

the Geological Department of the Royal Museum for Central Africa (RMCA) in Tervuren (Belgium). A detailed lithological description of these drillcores is provided by Raucq (1970). The total studied drillcores (Fig. 4.2) is ~525 m. It intersects stratigraphic levels from the BIb to BIId subgroups. Here, we use the lithological descriptions of Raucq (1957, 1970) for the siliciclastic series (BI group), but for the carbonate series we updated the descriptions using a new nomenclature to reflect the degree of diagenetic alteration. 750 samples of carbonates from split core (10 cm in diameter) at sampling intervals of ±0.3 m were collected in the BIe subgroup (S70 Tshinyama and Kafuku 15 drillcores), BIIb Subgroup (Bena Kalenda and B13 Kanshi drillcores), BIIc Subgroup (B13 Kanshi drillcore), and BIId Subgroup (Bena Tshovu drillcore). In addition, 33 samples from outcrops intersecting the BIIa (stops 226 and 243) and

Table 4.1 Simplified stratigraphy of the Mbuji-Mayi Supergroup in the north Sankuru-Mbuji-Mayi-Lomami-Lovoy Basin, Kasai-Oriental region, DRC (after Raucq 1970)

Mbuji-Mayi Supergroup, DRC			
Group	Subgroup	Formation	Lithostratigraphy
	948 ± 20 Ma (Cahen et al. *1974,* 1984*)*		
	BIIe	$BIIe_5$	Grey limestones with stromatolites (17.5 m)
		$BIIe_4$	Brecciated grey limestones, stromatolites (24 m)
		$BIIe_3$	Pink limestones (11 m)
		$BIIe_2$	Grey limestones, grey to pink dolostones (15 m)
		$BIIe_1$	Pink to grey shaly to calcareous dolostones (13.5 m)
	BIId	$BIId_3$	Grey cherty dolostones with cherts, pink shaly dolostones (343 m)
		$BIId_2$	Grey dolostones, shaly dolostones and dolomitic limestones (42 m)
		$BIId_1$	Shales with cherts (15 m)
	BIIc	$BIIc_8$	Dark shales, green to purple shaly dolostones (24 m)
		$BIIc_7$	Grey dolostones with stromatolites (5 m)
BII		$BIIc_6$	Dark shales, grey shaly dolostones (27 m)
		$BIIc_5$	Grey dolostones with stromatolites (25 m)
		$BIIc_4$	Dark shales, grey shaly dolostones (12.5 m)
		$BIIc_3$	Grey dolostones with stromatolites (88 m)
		$BIIc_2$	Dark shales, grey shaly dolostones (21 m)
		$BIIc_1$	Grey dolostones with stromatolites (89 m)
	± 810 Ma (Delpomdor et al. 2013a*)*		
	BIIb	$BIIb_5$	Dark cherty dolostone, shaly dolostones (25 m)
		$BIIb_4$	Shaly calcareous dolostones (23 m)
		$BIIb_3$	Polygenetic conglomerates (19 m)
		$BIIb_2$	Shaly dolostones and calcareous dolostones, often brecciated (39 m)
		$BIIb_1$	Grey cherty dolostones (9 m)
	BIIa		Grey dolostones with stromatolites (105 m)
	BIe	BIe_2	Grey-pink feldspathic psammites (1 m), grey shaly dolostones (26.5 m)
		BIe_1	Grey to pink dolostones with galena (25.5 m)
	888.2 ± 8.8 Ma (Delpomdor et al. 2013a*)*		
	1040–1065 Ma (Cahen 1954*; Holmes and Cahen* 1955*; Raucq* 1957*)*		
	(BIE)	BIE_2	Locally visible in the Kafuku area—red shales, shaly psammites, grey calcareous dolostones (28 m)
		BIE_1	Locally visible in the Kafuku area—pink to brown shales, grey dolostones (31.5 m)
BI	BId	BId_2	Quartzites (308 m)
		BId_1	Red sandy psammites, sometimes dolomitic
	BIc	BIc_2	Red shales and shaly psammites (180 m)
		BIc_1	Conglomerates
	BIb		Basal conglomerate, alternations of sandstones and conglomerates (17 m)
	BIa		Only visible in south SBLL basin—shales and quartzites (500 to 1,500 m)
	1152 ± 15 Ma (Delhal et al. 1989)		
	1155 ± 15 Ma (Delhal et al. 1966)		
	1174 ± 22 Ma (Delpomdor et al. 2013a)		
	1310 ± 25 Ma (Cahen et al. 1984)		

BIIe (Forminière and 'stop 628' samples; Fig. 4.2) Subgroups were studied from the collection of the RMCA.

4.4.2 Microfacies Analysis

Microfacies descriptions are based on the recognition of grain and sediment types, sedimentary structures and textures, using 788 selected thin sections from drillcores and outcrop samples. Our carbonate microfacies are named after the depositional fabric according to the classification of carbonate rocks (Dunham 1962; Embry and Klovan 1972; Sibley and Gregg 1987) and take account of sedimentary structures and/or fossils, e.g. wavy laminated dolomudstones, oolitic grainstones, and stromatolites. Each sample, in thin section, was etched with Alizarine red S and potassium ferricyanide for recognition of carbonate minerals (for more details, see Delpomdor and Préat 2012).

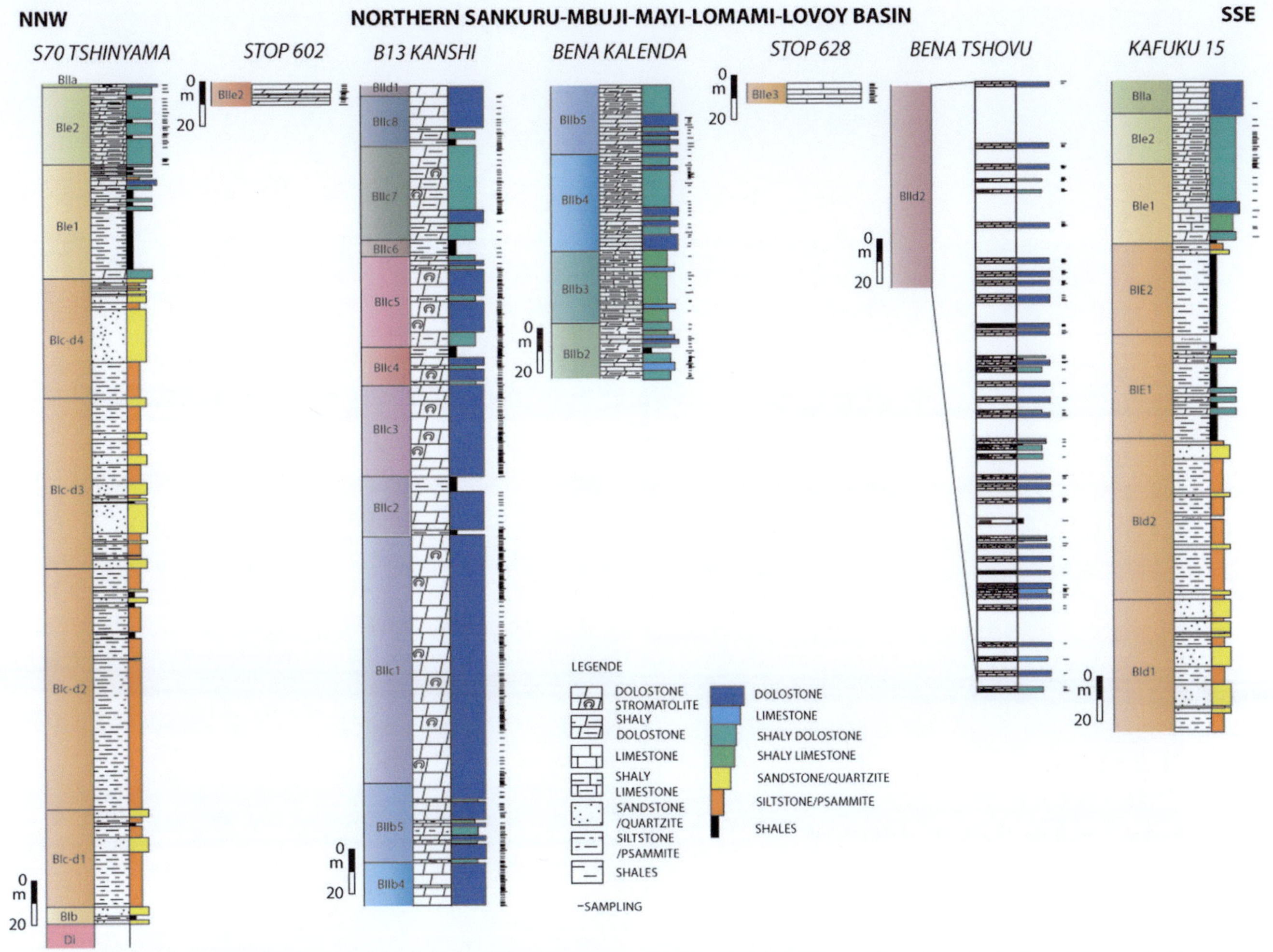

Fig. 4.2 Facies logs of studied drillcores (S70 Tshinyama, B13 Kanshi, Bena Kalenda, Bena Tshovu, Kafuku 15) and outcrop sections (stops 602 and 628) in the northern Sankuru-Mbuji-Mayi-Lomami-Lovoy Basin, showing the succession of formations of the Mbuji-Mayi Supergroup. See Fig. 4.1 for location, the cores and samples from outcrops are stored in the Royal Museum for Central Africa (Tervuren, Belgium) since 1960

4.4.3 Sequence Stratigraphy

The sequence stratigraphy of the studied Mbuji-Mayi series is based on the recognition/definition of elementary sequences of fifth order rhythms/cycles resulting from the interpretation of our lithologic curves based on lithofacies (clastic sediments) and microfacies (carbonates) analysis. Applying systematically the sequence standard defined in the Mbuji-Mayi series, 133 fifth-order 'sequences' or elementary parasequences have been identified from BIb to BIIe Subgroups (Delpomdor and Préat 2012); a parasequence is a stratigraphic unit here defined as 'a relatively conformable succession of genetically related beds or bedsets bounded by a flooding surface' (following Van Wagoner 1995).

Fischer graph (Fischer 1964) plots display cumulative departure from mean thickness against cycle number so that thicker-than-average cycles display as a plot with a positive slope and thinner-than-average cycles a plot with a negative slope. In our study, the base of a cycle was identified by the initial backstepping of a less restricted or an open marine facies over a more restricted facies. The backstepping is related to a relative increase in water depth with increased water circulation in a 'protected' lagoon or in restricted area subjected to open marine conditions. When sea level falls, the restricted facies is deposited.

4.5 Results

4.5.1 Microfacies Analyses

We recognized 11 microfacies whose vertical successions from MF1 to MF11 constitute a standard sequence integrating sea-level, energy, and salinity variations. Detailed descriptions of these microfacies and their organization in the standard sequence are available in Delpomdor and Préat (2012). This sequence encompass a distal to proximal open-marine outer ramp (MF1-MF4; Fig. 4.3a–e), a

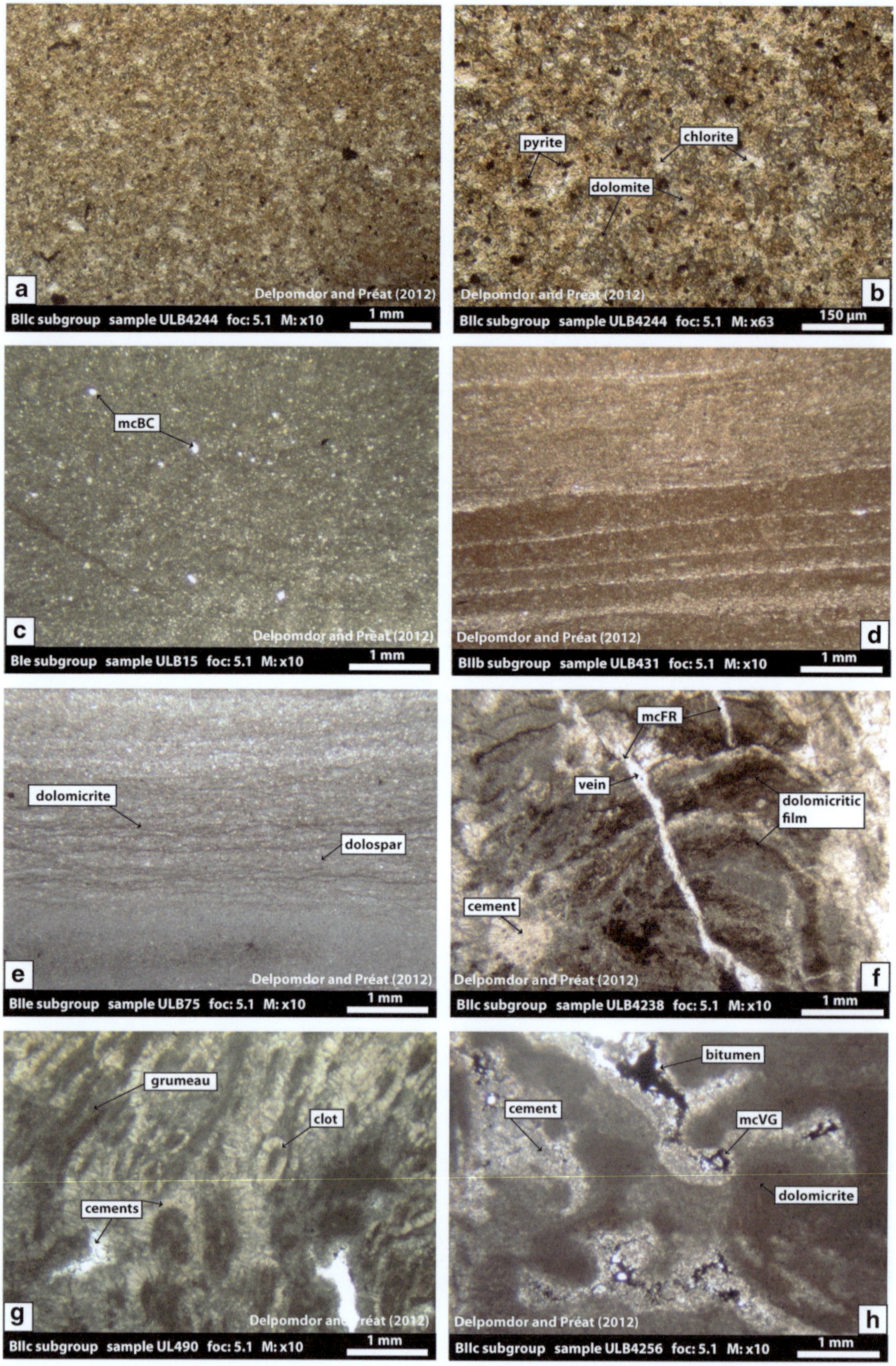

Fig. 4.3 Basinal to lower-upper subtidal marine environment—*Marine basinal environment*: (**a**) fine-grained shales. MF1, BIIc Subgroup, sample ULB4244, depth: −278.00 m, B13 Kanshi drillcore; (**b**) framboidal pyrites and elongate flakes or laths of chlorite-illite

reefal middle ramp (MF5; Fig. 4.3f–h), a restricted back-reef inner ramp (MF6-MF9; Fig. 4.4a–e), closed lacustrine ponds (MF10; Fig. 4.4f–h) and carbonate-evaporite transitional environments (MF11; Fig. 4.5a–h). The major characteristics of these microfacies are briefly described in Table 4.2.

4.5.2 Sequence Stratigraphy

We present a synthesis of the sequence stratigraphic framework of late Mesoproterozoic—middle Neoproterozoic series in the Mbuji-Mayi Supergroup (Fig. 4.6) based on the recognition/ definition of elementary sequences or fifth order rhythms/cycles resulting from the interpretation of our lithologic curve. The recognized parasequences, identified as shallowing-upward aggrading or prograding lobes from marine to near-shore settings present a bimodal distribution of their thicknesses, the most abundant being <5 m-thick on average, the others >10 m-thick on average (Delpomdor and Préat 2012).

Fischer plots are extremely useful for comparing and for assessing variations in long-term sea level. The Fischer plots were only used with cycles that show full regression to intertidal-supratidal facies, since it is only then that one can be confident that accommodation space has been completely filled. In contrast, Fischer plots of subtidal cycles may not necessarily reflect the full amount of available accomodation space and in our case for the 'first' MF1–MF5 cycle or standard parasequence that is dominated by subtidal facies. The 'second' MF6-MF11 cycle or standard parasequence is associated with upper subtidal-supratidal environments in very shallow water. The general setting of both standard sequences can be considered as subtidal-peritidal.

4.5.2.1 Mbuji-Mayi Fischer Plots

Our Fischer curve (Fig. 4.6) starts with the clastic BIb Subgroup and the clastic/carbonate lower part of the BIc-d Subgroup (in the northern SMLL Basin with the S70 Tshinyama drillcore and in the central SMLL Basin with the Kafuku 15 drillcore) and presents a strong positive slope with succession of parasequences 1–9 (in the S70 Tshinyama drillcore). Overlying parasequences (10–42) are thinner, ranging from 1 m to ± 10 m (average thickness is 3.6 m) except for two of them (parasequence 23 is 25 m-thick; and parasequence 30 is 16.4 m-thick) that are six times thicker than the average cycle thickness. The Fischer curve reveals three 'cyclic' packages (parasequence sets I, II and III), showing a slight thinning upward trend.

Parasequence thicknesses (43 to 57) in the BIIb Subgroup (Bena Kalenda drillcore) fall in the same metric range of thickness than the previous parasequences of the BIc-d/BIe Subgroups. Their average thickness is 4.3 m (but for thicker parasequences 48 and 49, 15.9 and 20.7 m-thick, respectively) and the Fischer curve is stabilized without significant thickness variations of parasequences, but for thicker parasequences 48 and 49 (15.9 and 20.7 m-thick, respectively).

The BIIc Subgroup in the B13 Kanshi drillcore contains 50 parasequences (55 to 104) recording shallowing-upward sedimentation. As for previous parasequences, a bimodal distribution of thicknesses is the rule with a 4.8 m-thick for 'thinner' parasequences (37 parasequences, not including > 10 m-thick parasequences) and 16.8 m for 'thicker' parasequences (13 parasequences, not including <>10 m-thick parasequences). The average parasequence thickness is 7.9 m. Parasequences 55 to 67 are similar to parasequences described at the top of the Bena Kalenda drillcore and have a similar average thickness (4 m-thick excluding >10 m-thick parasequences 61 and 64).

Despite the discontinuous profile of the Bena Tshovu drillcore in the BIId Subgroup, parasequence facies contents and thicknesses (on average 3.1 m-thick for parasequences 105–117) are similar to those of the second part of the B13 Kanshi drillcore. Moreover, the upper part of the BIId Fischer plot (parasequences 115–117) has a slight positive slope. The following parasequences in the BIIe subgroup are in the same range as the previous ones (i.e. 3.9 m-thick, excluding >10 m thick parasequences).

4.5.2.2 Parasequence Thicknesses

To determine the magnitude of relative sea level change needed to produce observed metre-scale cyclicity, we now

Fig. 4.3 (continued) minerals in a shaly matrix. MF1, BIIc Subgroup, sample ULB4244, depth: −278.00 m, B13 Kanshi drillcore. *Marine lower subtidal environment (outer ramp)*: (**c**) *grey/brown*, fine- to medium-grained dolomitic matrix. McBC—Micrometric intercrystalline porosity. MF2, BIe Subgroup, sample ULB15, depth: −88.30 m, S70 Tshinyama drillcore; (**d**) submillimetric-thick laminae of *dark grey/brown* microcrystalline organic rich dolomite and *light grey/ light brown* medium-grained organic-poor dolomite MF3, BIIb subgroup, sample ULB431, depth: −420.00 m, B13 Kanshi drillcore; (**e**) submillimetric-scale, continuous wavy laminites of *dark grey/brown* organic-rich and *light grey/light brown* organic-poor dolomite. MF4, BIIe subgroup, sample ULB75, Handspecimen 628a. *Marine upper subtidal environment (middle ramp)*: (**f**) organic-rich dolomicritic films in columnar stromatolite. *McFR* micrometric-scale fracture porosity. MF5, BIIc Subgroup, sample ULB4238, depth: −281.40 m, B13 Kanshi drillcore; (**g**) superimposed linear and coalescent dolomicritic clots and grumeaux in a dolomicritic matrix. MF5, BIIc Subgroup, sample ULB490, depth: −365.80 m, B13 Kanshi drillcore; (**h**) replacive drusy dolomites in pore-filling and space-filling interparticules and vugs. Notice dark amorphous bitumen in the cavities. *McVG* micrometric-scale vuggy porosities. MF5, BIIc Subgroup, sample ULB4256, depth: −262.40 m, B13 Kanshi drillcore

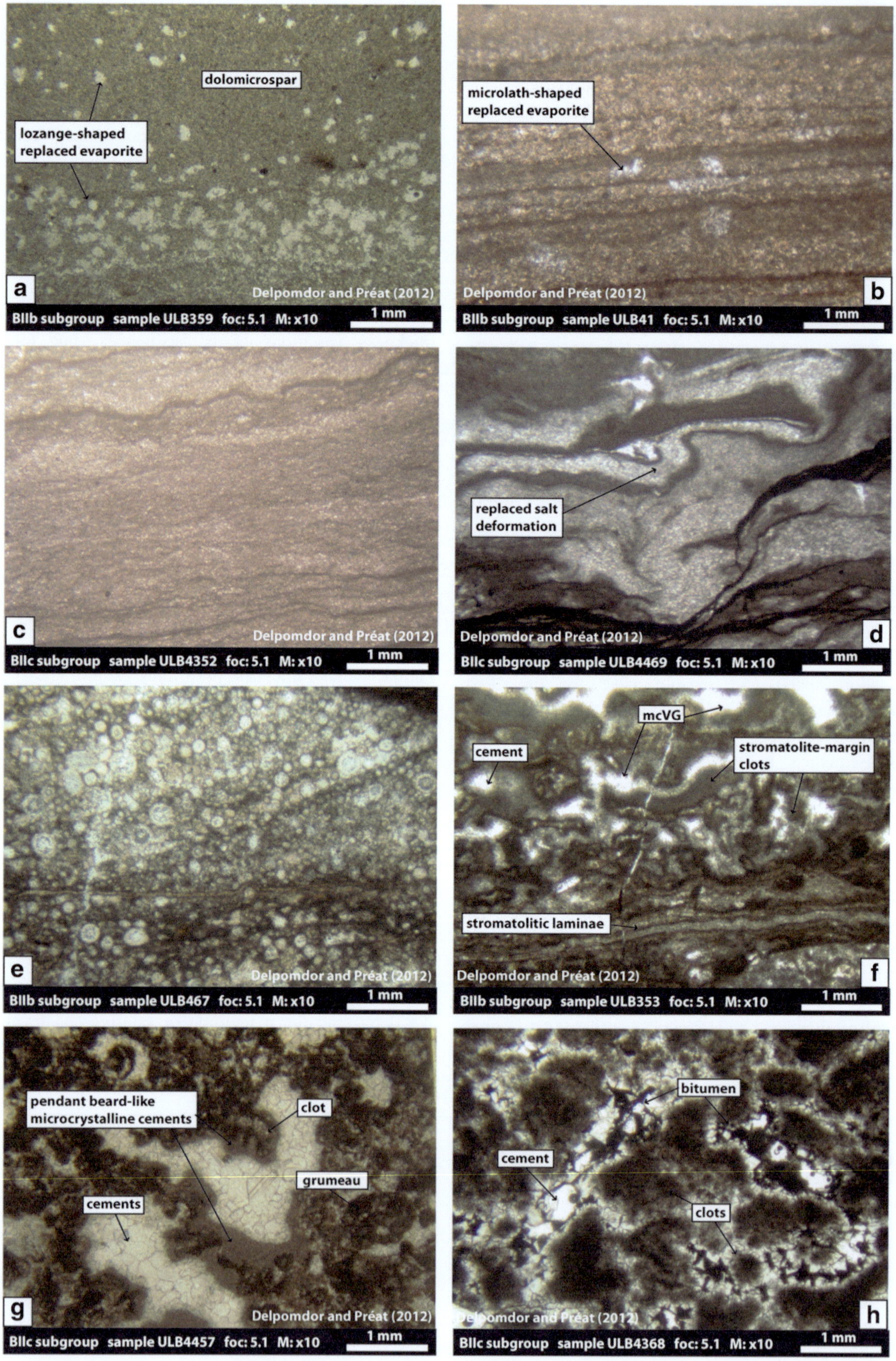

Fig. 4.4 Evaporative upper subtidal-supratidal zones in marine and lacustrine environments—*Evaporative marine environments (inner ramp)*: (**a**) *dark/grey* fine-grained dolomitic matrix with randomly oriented replaced thin microlath- and lozenge-shaped sulfate replaced

turn to the environmental data (mainly sedimentology) and thicknesses provided by elementary parasequences. The distribution of cycle thicknesses of Mbuji-Mayi carbonates follow a bimodal pattern with 'thin' peritidal cycles and 'thick' subtidal cycles. The former are from ≈ 1 m (or less) to ≈ 6 m (or more) in thickness, with an average cycle thickness of 3.4 m (S70 Tshinyama drillcore, $n = 35$); 3.5 m (Kafuku 15 drillcore, $n = 31$), 4.3 m (Bena Kalenda drillcore, $n = 13$); 3.1 m (Bena Tshovu drillcore, $n = 12$); 4.8 m (B13 Kanshi drillcore, $n = 37$). This leads to an average thickness of $\pm$ 4 m for peritidal cycles. On the other hand the subtidal cycles range from ≈ 10 m to ≈ 20 m in thickness, with an average cycle thickness of 17.7 (S70 Tshinyama drillcore, $n = 7$); 16.3 m (Kafuku 15 drillcore, $n = 4$); 4.3 m (Bena Kalenda drillcore, $n = 18$); 16.8 m (B13 Kanshi drillcore, $n = 13$).

4.6 Interpretation and Discussion

4.6.1 Paleoenvironmental Reconstruction (Fig. 4.7)

The shale microfacies (MF1) consists of organic-rich benthic microbial mats (0.16–0.69 % in TOC; Delpomdor and Préat 2012) formed at the interface between a quiet basin and a distal outer ramp, with sporadic sedimentary influx of detrital minerals and planktonic matter deposited in deeper parts or in small depressions under anaerobic and dysaerobic conditions. The transition from the distal outer ramp to the middle ramp is recorded in the MF2-MF5 evolution. MF2 and MF3 represent lower subtidal dolomitic muds deposited by suspension, probably below storm wave base as no storm-generated sedimentary structures were observed. MF4 represents therefore the shallowest water depth in the subtidal facies zone as suggested by the preserved microbial filaments or remnants of microbial mats. This suggests that light may have been a limiting factor for the growth of phototrophic microbial communities near the base of the photic zone (Batten et al. 2004). Absence of shallow-water structures (cross-laminations, mudcracks) and other indicators of shallow-water flows (ripples or trough-cross bedding) are consistent with deposition in relatively deep water (tens to hundreds of meters).

The transition between the deep- and shallow-water microfacies is highlighted by crinkly laminated dolomudstones associated with stromatolitic bioherms (MF5) consisting of muddy-dominant non-branching and branching columnar stromatolites, developed in upper subtidal and lower intertidal middle-ramp zones (Schröder et al. 2005). This interpretation is supported by the nature of stromatolites: *Conophyton*- and *Kasaia*-forms are characteristics of the subtidal zone (Bertrand-Sarfati 1972a), while *Baicali*- and *Tungussia*-forms developed in shallow subtidal and intertidal zones (Hoffman 1976; Bertrand-Sarfati 1972b). The importance of carbonate muds surrounding the stromatolitic carbonates of the Mbuji-Mayi Supergroup suggests that these microbial forms were built in quiet water environments without emersion (Bertrand-Sarfati 1972a). The absence of storm wave and shallow-water structures, point to quiet water deposition in relatively deep water (tens to hundreds of meters) on the Mbuji-Mayi outer ramp. The middle-ramp is bracketed offshore by the storm wave base and inshore by the fairweather wave base (Burchette and Wright 1992).

A similar context with stromatolite accumulation on slumps and deep-water environments has been reported elsewhere in Mesoproterozoic reefs of the Victor Bay Formation near the Stathcona River on the northern Baffin Island, Canada (Narbonne and James 1996). However, the abundance of evaporite casts in our inner-ramp microfacies (MF6 to MF9) suggests that Mbuji-Mayi bioherms composed a barrier, allowing development of a restricted environment, probably lagoonal, with episodic evaporitic precipitation. Desiccation cracks and fenestrae associated with bioherms, point to episodic peritidal environment from shallow upper subtidal to interdidal zones with high-energy sandy carbonate

Fig. 4.4 (continued) by dolomite crystals. MF6, BIIb subgroup, sample ULB359, depth: −45.50 m, Bena Kalenda drillcore; (**b**) submillimetric-thick laminae of fine-grained organic-rich dolomicritic matrix and light medium-grained organic-poor dolomicrosparitic matrix. MF7, BIIb Subgroup, sample ULB41, depth: −435.80 m, B13 Kanshi drillcore; (**c**) submillimetric-scale wavy laminae of dark micritic and *white* to *light grey* dolomitic laminae. MF8, BIIc Subgroup, sample ULB4352, depth: −214.00 m, B13 Kanshi drillcore; (**d**) small folded structures of deformations in wavy laminae during the displacive growth of evaporitic minerals. MF8, BIIc subgroup, sample ULB4469, depth: −118.10 m, B13 Kanshi drillcore; (**e**) tightly packed, fine-grained *dark grey/brown* dolomitic matrix, often silicified, with tangential ooids (ooid dolopackstone). Notice the wedge-shaped cross-lamination with planar to irregular layers inclined (average 20–30°) like internal foresets. MF9, BIIb Subgroup, sample ULB467, depth: -388.80 m, B13 Kanshi drillcore.—*Lacustrine environment*: (**f**) dark planar to undulate dolomicritic with thick laminations (*bottom*), overlain by dark dolomicritic stromatolitic margin clots (at the *top*). *McVG* micrometric vuggy porosities. MF10a, BIIb Subgroup, sample ULB353, depth: −47.00 m, Bena Kalenda drillcore; (**g**) rounded to irregular anastomosing patches of stromatolite-margin clots, selvages and 'grumeaux' filled by *drusy white* dolomite cements. Notice the pendant beard-like microcrystalline cements. MF10b, BIIc subgroup, sample ULB4457, depth: −128.20 m, B13 Kanshi drillcore; (**h**) discontinuous microbial clots of stromatolites filled by drusy dolomitic cement and bitumen. MF10c, BIIc Subgroup, sample ULB4368, depth: −207.80 m, B13 Kanshi drillcore

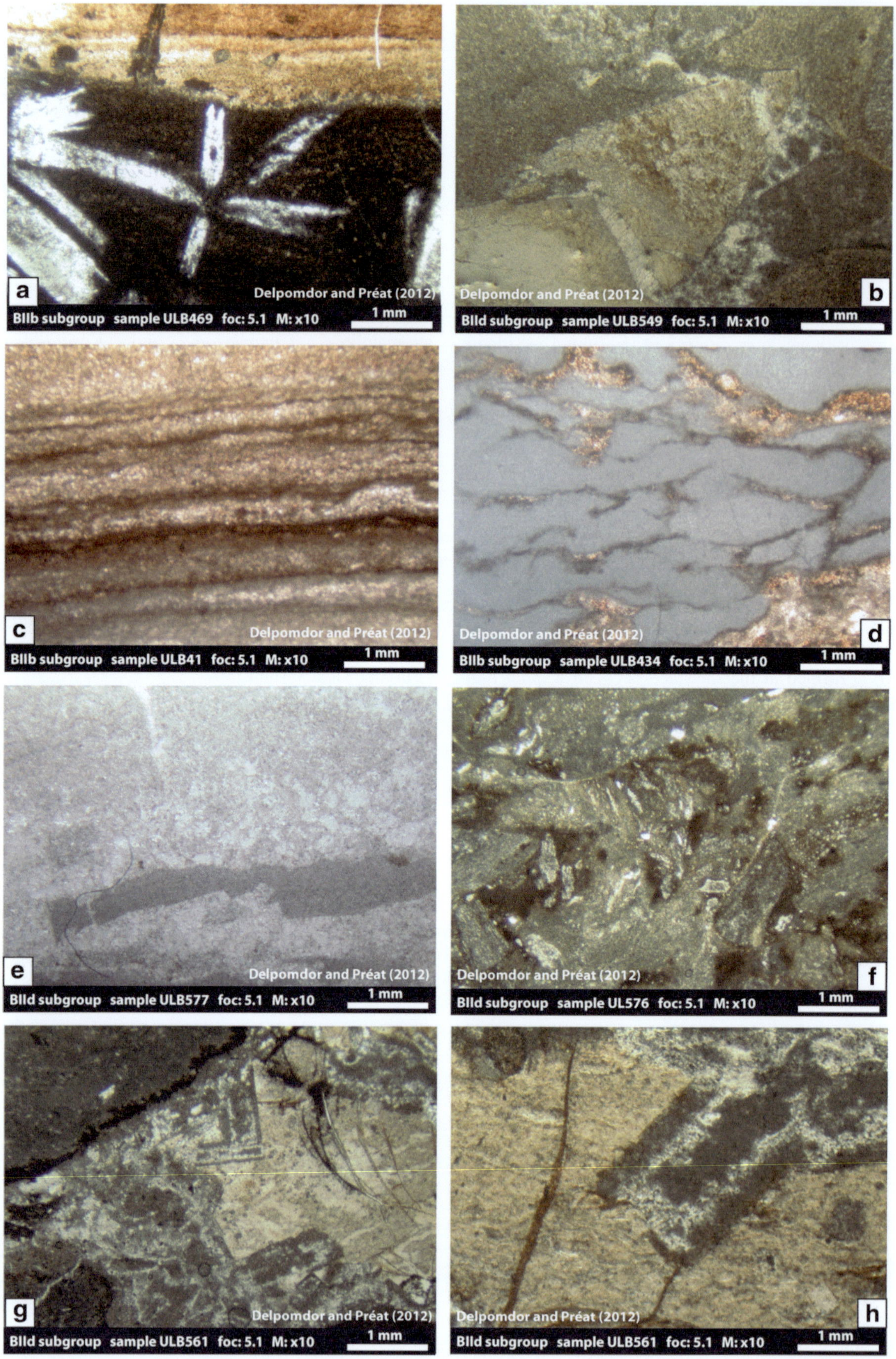

Fig. 4.5 Sabkha environments—*Early diagenesis*: (**a**) rosette-like structure of replaced sulfate by dolomitic crystals in a dark homogeneous dolomudstone. MF11a, BIIb subgroup, sample ULB469, depth: −394.00 m, B13 Kanshi drillcore; (**b**) *arrow-shaped*

shoals (MF9) under restricted conditions. Association of microbially laminated or stratiform carbonates, desiccation cracks and muddy intraclasts are frequently attributed to peritidal origins in ancient successions (Grotzinger 1986; Pratt et al. 1992). Along the inner-ramp, ephemeral supratidal lacustrine ponds, seasonally flooded by marine water and followed by precipitation of evaporitic minerals, characterize organic-rich carbonate mud (MF10a) during dry season. Buildups of cyanobacterial mats and stratiform stromatolites (MF10b) occurred preferentially in the shallow-margin of the lacustrine ponds (Freytet and Plaziat 1972), and finally during wet season coated pisoids along a lacustrine shore resulted from aggregation of stromatoclasts and reworked ooids (MF10c). Vadose or meteoric cements, suggesting subaerial exposures and temporarily humid periods affected these microfacies (Delpomdor and Préat 2012). Increasing concentration of evaporation may result in deposition of subtidal aragonitic muds, intertidal microbial mats and precipitation of gypsum, halite in the sabkha domains (MF11).

4.6.2 Sea Level Fluctuations in the SMLL Basin

The general thickening-upward trends of both parasequences and parasequences sets (from ± 20 m to ± 40 m) in the BIb and BIc-d subgroups point to a retrogradational stacking pattern, progressively dominated by subtidal deposition with interbedded shales. Parasequence sets I, II and III (including elementary parasequences 1–9, Fig. 4.6 and Delpomdor and Préat 2012), containing shaly levels, are characterized by a deepening-upwards trend highlighting a transgressive systems tract (TST). However, deposition of the BIc-d Subgroup is characterized by a relatively thick, single shallowing-upward parasequence as resurgent carbonate production was able to outpace relative sea level change succeeding the previous deepening phase.

The long-term sea level is relative constant and parasequences are thinner in the upper part of the BIc-d subgroup than in its lower part. Facies are similar along the unit and point to a constant water depth during deposition. Carbonate supratidal facies (BIIb subgroup) progressively replaced clastic and mixed clastic/carbonate facies after depth shallowed. This evolution could record the change from an early highstand system succeeding one (HST). The available accommodation space generated in the upper part of the BIe subgroup declined, producing a slight 'long term' negative slope in the Fischer curve.

The two thicker parasequences (48 and 49) create to a short positive slope in the curve that could record an aggradational step within the general progradational trend, highlighted by shallowing parasequences with intertidal and supratidal caps and thin calcrete levels. Oolitic shoals disappeared during this unstable phase and are only associated with stabilized portion of the Fischer curve. The two thicker parasequences at the beginning of HST record partly underfilled accomodation spaces due to a short deepening event related to sea-level fluctuations or temporary tectonic activation. This HST probably overlies a transgressive systems tract, consistent with stromatolitic buildups in the non-studied BIIa subgroup. Stromatolites are well developed in the lower part of next subgroup (BIIc).

A major process or facies and parasequence thickness change occurs in parasequences 68–72. Here, stromatolite buildups and shaly dolomudstones record a major shift from previous inner ramp setting to an outer/middle ramp system. The change is recorded in very thick parasequences (22.4 m-thick on average) typical of deepening-upward sediments (shales) overlain by slight shallowing-upward facies of stromatolitic boundstones. This suggests a transgressive systems tract (TST) with deposition in deeper water at a depth in which the former carbonate 'factory' was no longer able to return to state of optimum carbonate production. The deepening, or 'drowning', was probably not very significant as carbonate deposition never ceased. Overlying parasequences have thicknesses more-or-less near the average BIIc parasequence thickness with an evolution toward thinner thicknesses along this stratigraphic interval. The general evolution then is the one of a highstand systems tract (HST), locally interrupted by aggradational thicker parasequences (95–97), as previously recorded in the BIIc Subgroup. Distribution of different parasequences or cycle-types (aggradational vs. progradational) along the BIIc Subgroup therefore provides evidence of the variation of relative

Fig. 4.5 (continued) replaced gypsum by silica in a medium-grained dolomicritic matrix. MF11a, BIId Subgroup, sample ULB549, depth: −55.15 m, core 41, Bena Tshovu drillcore; (**c**) bedded and laminated replaced pale anhydrite by dolomitic crystals in brown/grey and pale laminated dolomite matrix. MF11b, BIIb Subgroup, sample ULB41, depth: −435.80 m, B13 Kanshi drillcore; (**d**) bedded light nodules of replaced anhydrite by dolomitic cements in a dolomitic matrix. MF11c, BIIb subgroup, sample ULB434, depth: −417.40 m, B13 Kanshi drillcore; (**e**) packed aggregates of tiny lath-shaped crystals (20–50 μm long and 5–10 μm wide) of replaced anhydrite forming a massive structure. MF11d, BIId Subgroup, sample ULB577, depth: −43.15 m, Bena Tshovu drillcore; (**f**) grain-supported breccia with angular dolomitic clasts and replaced anhydrite (bladed crystals). MF11e, BIId Subgroup, sample ULB576, depth: −43.00 m, Bena Tshovu drillcore; (**g**) silicified pyramidal hopper structures in a silicified massive evaporite. MF11f, BIIc subgroup, sample ULB561, depth: −47.20 m, Bena Tshovu drillcore; (**h**) filled mimetic molds of halite crystals in loosely packed aggregates of tiny lath-shaped silicified crystals of anhydrite. MF11f, BIId subgroup, sample ULB561, depth: −47.20 m, Bena Tshovu drillcore

Table 4.2 Detailed descriptions of carbonate microfacies

Lithofacies/ microfacies	Occurence and thickness	Sedimentary structures	Components and microfacies	Environment
MF1: shale	Thickness: decametric to metric beds interbedded of carbonates; top gradual to MF2	Submillimetric laminated structures	Fine-grained matrix chlorite-illite clay minerals, often dolomitic, and rich in organic matter	Basinal to lower subtidal zone
MF2: homogenous dolomudstone	Lowermost carbonate unit of the BII Group; thickness ± 10–60 cm; top gradual to MF3	Clean structureless dolomudstone; rare lamination; rare erosion surface	Fine to medium crystalline dolomite (5–15 cm); xenotopic to hypidiotopic mosaic texture; rare rounded silt-fine dolomite (>15 μm); peloidal texture	Lower subtidal zone
MF3: planar dololaminite	Thickness ± 10–30 cm; to gradual to MF4; near base intercalations of MF2 with sharp or gradual contacts	Millimetric laminated dark and light dolomite; sharp or gradual contacts	Alternation of submillimetric-thick flat laminated dolomudstone with (i) dark fine-grained crystalline organic-rich dolomite (5–10 μm) and (ii) light medium-grained organic-poor dolomite (10–15 μm); (i) equigranular xentopic to hypidiotopic texture; (ii) hypidiotopic texture; between the laminations, pseudomorphs of sulphate	Lower subtidal zone
MF4: wavy dololaminite	Thickness: decimetric bedsets often associated with MF2–3; gradual contact to MF5	Submillimetric to millilmetric smooth to crinckly laminated dolomite; in-situ brecciation	Submillimetric-scale couplets with (i) dark organic-rich and (ii) light dolomites; (i) smooth to crinckly laminae of equigranular xenotopic to hypidiptopic texture (5–10 μm); (ii) hypidiotopic dolomite crystals (10–15 μm); poorly preserved dolomicritic films in (i); interlaminated 'pallisadic' dolomitic crusts with reminescent pseudomorphs of sulphate; crystalline and diffuse pseudomorphs of sulphate	Shallow subtidal zone
Middle ramp				
MF5: columnar stromatolite	Plurimetric-scale bioherms (>100 m thick); interbeds of MF1 to MF4	LLH- and SH-type stromatolites; millimetric to metric-sized domes; unbranched columns like type-*Conophyton* stromatolite or branching columnar stromatolite like *Kasaia-Baicalia* and *Tungussia;* transitional to dololaminites (MF3 and MF4); often syn-sedimentary or early diagenetic breccia; rare desiccation cracks	Submillimetric-scale streaky structure with (i) dolomicritic films; (ii) clots and (iii) selvages and grumeaux; (i) xenotopic to hypidiotopic mosaic structure; (ii) coalescent spindle-shaped and patchy dolomicritic matrix coated by fibrous and radial cements; (iii) coalescent irregular patches of dolomicrite; replacive evaporites; strongly recrystallized dolomite; dolomitized herringbone cements	Bioherms on an upper subtidal zone
Lithofacies/ microfacies	Occurence and thickness	Sedimentary structures	Components and microfacies	Environment
Inner ramp				*Evaporative environment*
MF6: evaporative massive dolomudstone	Basal unit of inner ramp; to gradual to MF5; centimeter- to decimeter-thick beds	Clean structureless dolomudstone; replacive evaporite crystals; rare lamination; rare erosion surface	Fine crystalline dolomite (5–10 μm); equigranular xenotopic to hypidiotopic mosaic texture; replacive evaporites	Restricted subtidal zone
MF7: evaporative laminated dolomudstone	Centimetric to decimetric thick; top gradual to MF6	Similar to MF3 and MF4; millimetric-scale lamination, even planar or wavy, often crinckly; replacive evaporites; gradual contacts	Alternation of submillimetric-thick (i) dark fine-grained crystalline organic-rich dolomite (5–10 μm) and (ii) light medium-grained organic-poor dolomite (10–15 μm); (i) equigranular xenotopic to hypidiotopic texture,	Restricted shallow subtidal to intertidal zone

(continued)

Table 4.2 (continued)

Lithofacies/ microfacies	Occurence and thickness	Sedimentary structures	Components and microfacies	Environment
			(ii) hypidiotopic texture; possible mudcracks; fenestral cavities; between the laminations, replacive evaporites and evaporite cements	
MF8: stratiform stromatolite	Centimeter- to decimeter-scale beds associated with MF7-MF9; microfacies in an inner ramp and lacustrine closed-lake (salt lake); sharp contact with MF9	Stratiform, LLH-S stromatolites, millimeter to decimeter-sized domes and massive beds of mats; gradual transition with MF7, as well as sharp to MF9; fenestral and thomboidal texture; often syn-sedimentary or early diagenetic brecciation; desiccation cracks; often silicification	Alternation of submillimetric laminations of (i) dark micrite and (ii) light dolomicrospar; (i) smooth to crinckly laminae of equigranular xenotopic to hypidiotopic texture (5–10 μm); (ii) hypidiotopic dolomite crystals (10–15 μm); dolomicritic films, clots and grumeaux; interlaminated 'pallisadic' dolomite crusts; replacive evaporites and evaporite cements	Restricted shallow intertidal zone
MF9: oolitic grainstone	Thickness: centimetric beds; interbeds of MF7; association with MF8	Massive; planar cross-lamination, low angle through cross-lamination; graded; rare erosion surface	sand-sized oolites and oncoids with micritic rim; ovoid with microcrystalline dolomite (5–10 μm) anc concentric tangential cortex; rare aggregate grains; microfacies associated with bioclasts; evaporite crystals and cements; often silicified	Restricted high-energy shallow intertidal zone including sandy carbonate shoal
Lacustrine pond (closed-lake)				*Lacustrine environment*
MF10: lacustrine dolomite	Metric-thick beds compiled to MF6, MF7, MF8; centimeter-scale interbeds of MF9	Clean structureless dolomudstone, laminated dolomustone, stratiform stromatolite; abundant fenestral and thromboidal texture; brecciation; desiccation cracks; recrystallized microfacies; silicification	Microfacies similar to MF6-MF7 and MF8; peloidal texture filled by dolomitic vadose cements; rare rendant beard-like cements	Closed-lake with probable perennial pond submitted semi-arid and warm conditions
				Early diagenesis
MF11: evaporite	Millimeter- and centimeter-thick interbeds associated to MF6 untill MF10; abundant in MF10	Isolated evaporite casts, nodular, bedded or massive skeletal halite casts; slumping; deformation; brecciation	Prismatic and hemipyramidal crystals with dolomicritic rim; laminar chevron-like texture; rare halite crystals; replacive evaporites by dolomite or silica	Evaporative environment

sea level change magnitude. The first half part of the unit records greater relative sea-level fluctuations from deeper subtidal outer-ramp to shallower subtidal middle-ramp settings (parasequences 68–80) than second half part from restricted environments to intertidal and supratidal domains (parasequences 81–104). This change is also recorded in average thickness cycle ranging from <5 m in first half and >5 m in second one. This model distinguishes an early highstand systems tract (first half) and a late highstand systems tract (second half). Their Fischer plots are also similar and suggest that the studied BIId subgroup is the continuation of the previous late highstand systems tract. One has to note the second appearance of numerous levels of oolites.

The parasequences 115–117 indicate that aggradational processes are superimposed to progradational ones. Relative deepening, if at all, was limited since most of these parasequences are capped by supratidal facies with thin calcretes and evaporites. Subsequent parasequences in the BIIe subgroup display shallowing from deep subtidal to shallow subtidal. Previously overfilled accomodation space of the BIIc-BIId Subgroups is now replaced by underfilled accomodation space, since no intertidal or supratidal caps are present. This leads to recognition of early highstand systems tracts, which probably followed a transgressive systems tract between BIId and BIIe subgroups. No cores or material across this transitional stratigraphic interval are available for study. Reported biohermal buildups

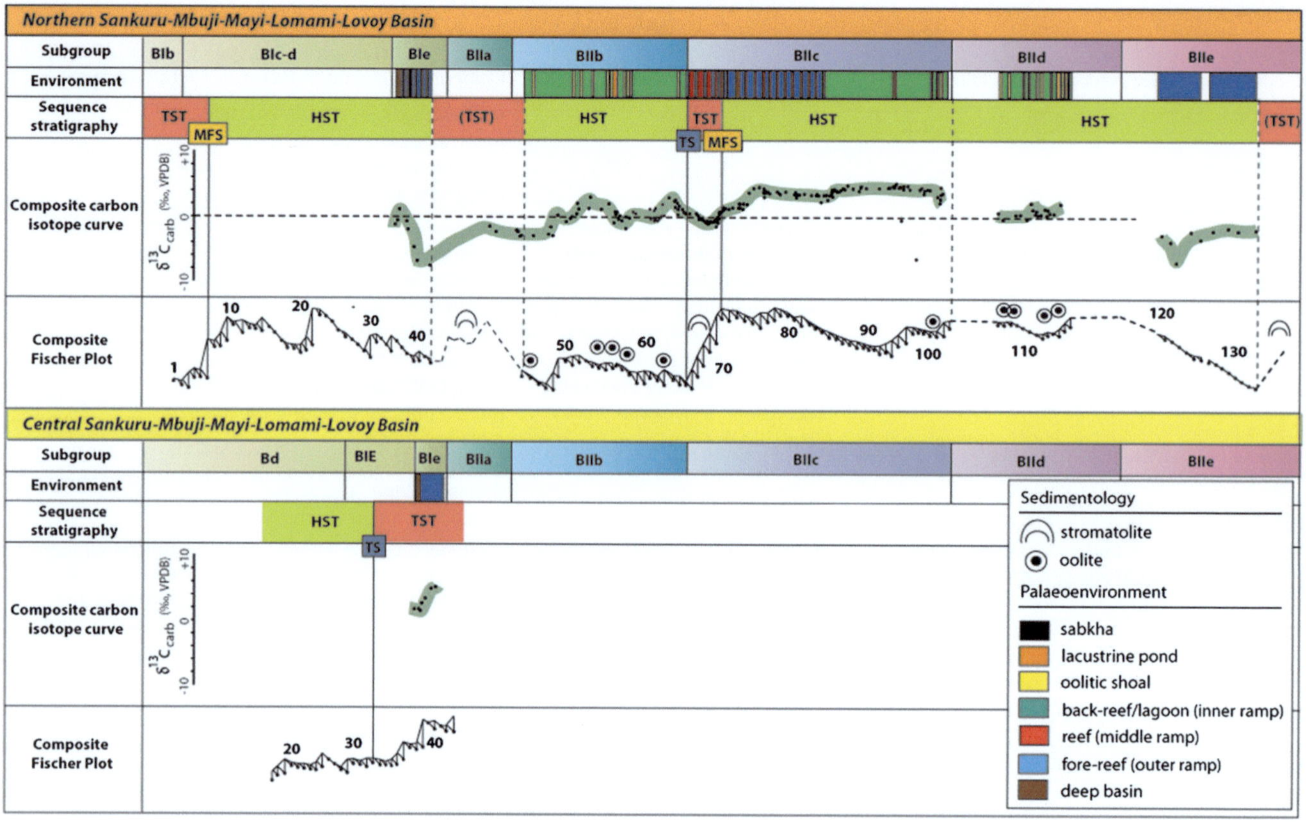

Fig. 4.6 Composite Fischer plot, sequence stratigraphy and $\delta^{13}C_{carb}$ curves in the Mbuji-Mayi Supergroup of the northern (and central) Sankuru-Mbuji-Mayi-Lomami-Lovoy Basin. Elementary parasequences in BII carbonates are delimited by the microfacies standard sequence (MF1–MF11) and by the lithofacies standard sequence (lf1–lf3) in the mixed clastic-BI group (Delpomdor and Préat 2012). The shallowing upward lithofacies succession is the following: lf1, shales; lf2, psammites; lf3, sandstones and quartzites or conglomerates (Based on a total of 133 elementary parasequences [fifth order]). Abbreviations: *TST* Transgressive Systems Tract, *HST* Highstand Systems Tract, *TS* Transgressive Surface, *MFS* Marine Flood Surface

(Bertrand-Sarfati 1972a) could indicate that a new transgressive systems tract will replace the BIIe HST.

Finally a comparison of clastic and mixed clastic/carbonate sedimentation in part of the northern (and central) SMLL Basin is revealing. The Fischer plot of the Kafuku 15 drillcore displays a regular succession of 30 plurimetric shallowing-upward clastic/carbonate parasequences (on average 4.1 m-thick excluding four >10 m-thick parasequences). The curve is stabilized in the BId Subgroup supporting a HST. Parasequences thicken progressively, with overlying lithostratigraphic units starting with the BIE subgroup pointing to a TST in an outer ramp setting. Comparing the S70 Tshinyama (northern basin) and Kafuku 15 drillcores (central basin) separated by a distance of 190 km, it appears that Fischer curves are different with a positive slope (Kafuku) and a negative one (Tshinyama). In this case, mixed clastic/carbonate sedimentation is evidently diachronic and related to a slow transgression across the craton. In addition, the BIE Subgroup is missing in the S70 Tshinyama drillcore, suggesting that differential subsidence characterized different parts of the basin.

4.6.3 Paleodepth

We estimate that the clastic and carbonate sequences of the Mbuji-Mayi Supergroup have an average thickness of ± 4 m for peritidal cycles, and ± 17 m for subtidal cycles. Using approximate depositional depths of microfacies (or lithofacies lf1 to lf3 in siliciclastics and mixed siliciclastics/carbonates of the S70 Tshinyama and Kafuku 15 drillcores), we thus suggest that 'peritidal' cyclicity' within the shelf (or ramp with MF6-MF10/11 succession) records relative sea level fluctuations of a maximum of 4 m, and that most fluctuations probably ranged between 1 to 4 m. Precise relative sea level fluctuations are more difficult to determine in the 'deeper' subtidal cycles, because facies could have been deposited in a wider range of water depths (example of shales and stromatolites), rarely reaching shallow waters (see MF1-MF5 sequence). It is probable, therefore, that fluctuations with a magnitude of 10–20 m occurred on the ramp, but whether the depositional environments were able to record these fluctuations is not easy to determine. Both cycle-types (thin and thick) are well expressed on the

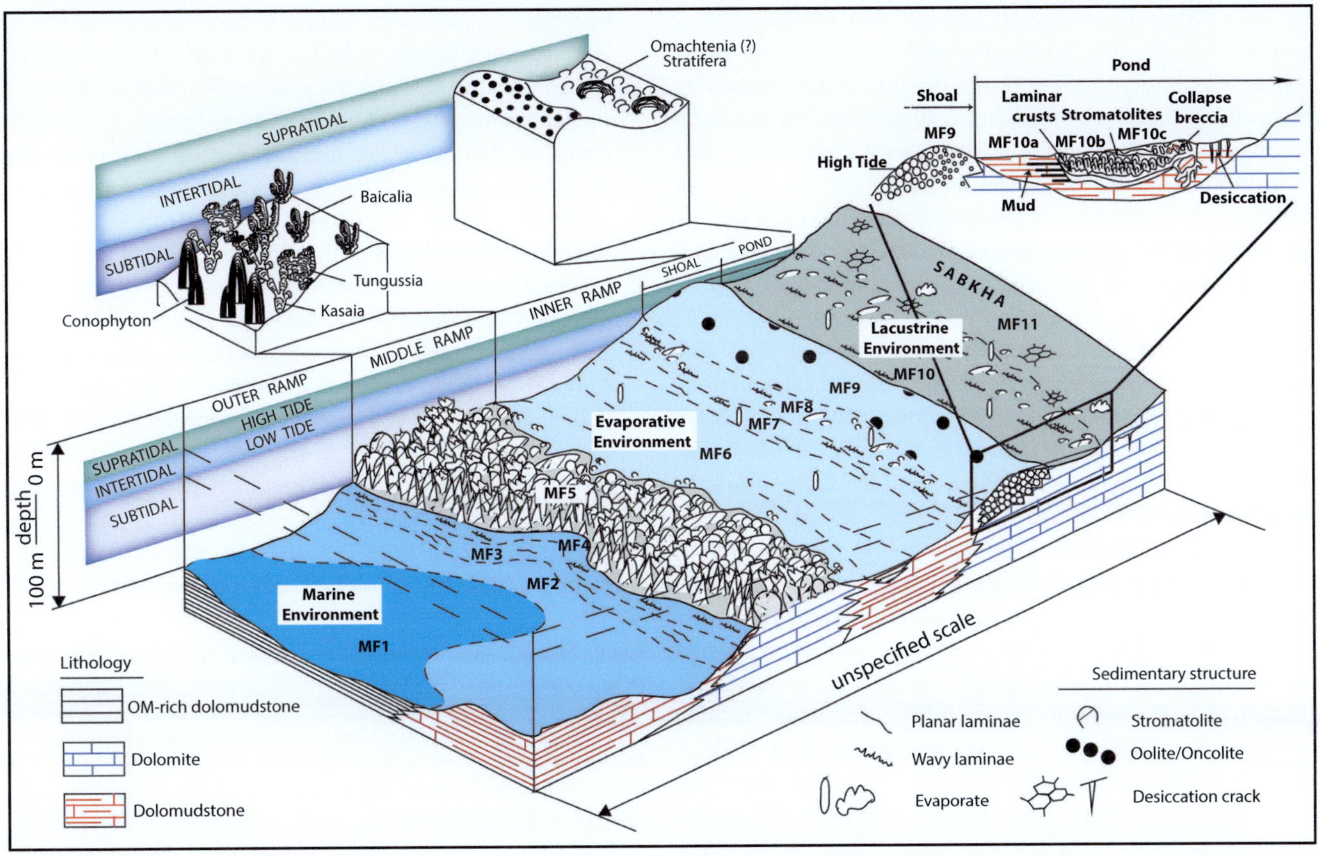

Fig. 4.7 Schematic reconstitution of depositional environments of the carbonate Mbuji-Mayi ramp. Idealized 3D-model illustrating the distribution of microfacies (MF1–MF11) from the outer-, middle- and inner-ramp system, with evaporative lacustrine ponds and sabkha. No vertical and horizontal scales. Ecological distribution of stromatolites is also shown in the middle ramp zone

Fischer plots: thin cycles characterize a stabilized curve with slight falling limbs, while thick cycles dominate the rising limbs of the plot. This suggests that fluctuations in eustacy had a strong control on the cyclicity

4.6.4 Sequence Stratigraphy Versus Carbon Isotope Curve

Combining the Fischer plots with our $\delta^{13}C$ isotopic data (Delpomdor et al. 2013a) displays alternations of $\delta^{13}C$ positive–negative shifts in tune with cycle-thickness variations that reflect third-order and possible fourth-order changes in accommodation space available. The BII carbonates exhibit $\delta^{13}C$ and $\delta^{18}O$ isotopic trends towards higher values, reflecting open-marine signatures episodically subjected to evaporitic conditions, and towards lower values reflecting early meteoric and burial signatures (Delpomdor and Préat 2013). Immenhauser et al. (2003) attempted to explain local variations of C and O isotope shifts in the Late Carboniferous Atokan series in northern Spain as a result of chemical changes of marine water masses in ramps affecting carbonates during early diagenesis. In contrast, Allan and Matthews (1982) argued that the low isotopic values of carbonates in the ramp are mainly related to (i) influence of ^{18}O-depleted early meteoric fluids and a ^{13}C-depleted soil-zone CO_2 (overprinting heavier marine values during a sea level fall and (ii) effects of 'sea-water ageing' with oxidation of organic matter in CO_2, which increase of bicarbonate concentrations in shallow poorly circulated sea water masses (e.g. Holmden et al. 1998). In our study, the relationship between cycle-thickness variations reflecting third-order changes, and $\delta^{13}C$ variations in the Mbuji-Mayi carbonate Supergroup, are linked (Fig. 4.6). We have shown that transgressive sequences highlighted transgressive systems tracts (TST) while 'relative' regressive sequences developed during 'late' stages of highstand systems tracts (HST). Two trends are identified (i) 'long term' negative slopes in the Fischer curve, with negative $\delta^{13}C$ excursions; and (ii) 'long term' positive slopes associated with positive $\delta^{13}C$ excursions. These trends suggest that (i) during transgressive systems tracts (TST) circulation of ramp-top marine water-masses was enhanced, and effects of meteoric diagenesis on ramp-top carbonates were strongly reduced, causing an isotope shift towards higher marine values; and (ii) during highstand systems tracts (HST) circulation of ramp-top marine water masses is reduced, and effects of meteoric diagenesis on ramp-top carbonates much increased, leading to a negative $\delta^{13}C$ shift (e.g. Immenhauser et al. 2003). Furthermore, the effects of meteoric or freshwater diagenesis, observed within REE + Y distributions (Delpomdor et al. 2013b) coincides with negative slopes of the Fischer curve and negative $\delta^{13}C$ excursions.

Similar carbon isotopic ratios that co-vary with third- and fourth-orders have already been documented in numerous Phanerozoic successions. For example, Wynn and Read (2007) show that carbon (and oxygen) ratios of Mississippian ramp-slope muds tend to co-vary at the scale of fourth-order sequences. There is a 1 to 2 ‰ decrease in $\delta^{13}C_{carb}$ in the maximum flooding zone relative to the low-stand, followed by a positive shift upward in the high-stand. Positive carbon isotope ratios reflect increased productivity due to increased global circulation associated with transgressive carbonates in the Mississippian. Our 100 m stromatolitic biohermal succession in the BIIb/BIIc transition is also transgressive (TST) and shows a positive shift of 8 ‰ in $\delta^{13}C_{carb}$ while a decrease of 4 ‰ in $\delta^{13}C_{carb}$ occurs at the end of the HST. Late HST is also characterized in the Mississippian carbonates by more negative values related to the influx of light carbon by riverine waters. Our late HST at the top of the BIIc and BIId Subgroups are characterized by lacustrine environments influenced by riverine input as shown by REE + Y analyses (Delpomdor et al. 2013b; Delpomdor and Préat 2013).

Conclusion

The Mbuji-Mayi sequence (Sankuru-Mbuji-Mayi-Lomami-Lovoy Basin), now dated between ca. 1155 Ma – 800 Ma, consists of approximately 550 m of clastic (BI group) and 1,000 m of stromatolitic carbonates (BII group). Our study, based on microfacies analysis, sequence stratigraphy, and Fischer plots combined with carbon and oxygen isotopes, indicates that the succession was deposited in 'deep' shaly and stromatolitic subtidal environments and 'very shallow' subtidal/supratidal marine and non-marine carbonate/evaporitic environments. Microfacies analysis shows that carbonate succession, i.e. BIe to BIIe Subgroups, consists of strata accumulated on a carbonate ramp, and sequence stratigraphy indicates that the sedimentation was cyclic in the inner ramp, represented by plurimetric 'thin' peritidal cycles (±4 m on average,) recording relative sea level of a maximum of 4 m, with fluctuations in the range around 1 m to 4 m. These shallow water depths, and the abundance of (photosynthesizing) cyanobacteria, indicate that the water column was oxygenated. The subtidal outer/middle ramp facies are also cyclic with 'thick' subtidal cycles characterized by an average thickness of ± 17 m. Accurate relative sea level fluctuations are difficult to assess in this 'deeper' environment since the facies could have been deposited in a wider range of shallow water and did not fill completely the accomodation space. The likely magnitude for such sea level fluctuations was around 10 to 20 m. Despite no reliable data on the deepening direction have been recently reported, we have to mention that Raucq (1970) suggested a general southward deepening of the basin.

In conclusion, the carbonate succession of the Mbuji-Mayi Supergroup is a good example of sedimentary and geochemical records in the late Mesoproterozoic to middle Neoproterozoic (1150–800 Ma), and in particular preserves well the local variations of carbonate $\delta^{13}C$ that reflect the relative importance of circulation of ramp-top marine water masses and effects of meteoric or freshwater diagenesis on ramp-top carbonates.

Acknowledgements Financial support from the TOTAL PhD scholarship (grant ULB/TOTAL-FR00003322) is gratefully acknowledged. The Royal Museum for Central Africa (Tervuren, Belgium) is thanked for providing samples of the Mbuji-Mayi Supergroup. The manuscript has been greatly improved following reviews and suggestions by M.J. de Wit and H. E. Frimmel. We would like to thank M. Rottesman for her critical encouragement.

References

Allan JR, Matthews RK (1982) Isotope signatures associated with early meteoric diagenesis. Sedimentology 29:797–817

Asmeron Y, Jacobsen S, Knoll AH, Butterfield NJ, Swett K (1991) Strontium isotope variations of Neoproterozoic seawater: implications for crustal evolution. Geochim Cosmochim Acta 55:2883–2894

Bartley JK, Semikhatov MA, Kaufman AJ, Knoll AH, Pope MC, Jacobsen SB (2001) Global events across the Mesoproterozoic-Neoproterozoic boundary: C and Sr isotopic evidence from Siberia. Precambrian Res 111:165–202

Batten KL, Narbonne GM, James NP (2004) Paleoenvironments and growth of early Neoproterozoic calcimicrobial reefs: platformal Little Dal Group, northwestern Canada. Precambrian Res 133:249–269

Bertrand-Sarfati J (1972a) Stromatolithes columnaires de certaines formations carbonatées du Précambrien supérieur du bassin congolais (Bushimay, Lindien, Ouest-Congolien). Annales du Musée Royal de l'Afrique Centrale, Tervuren, Belgique, Série in-8 - n° 74, 45pp

Bertrand-Sarfati J (1972b) Stromatolithes columnaires du Précambrien supérieur du Sahara Nord-Occidental—inventaire, morphologie et microstructure des laminations, corrélations stratigraphiques. Thèse de Doctorat d'Etat, Université des Sciences et Technologies de Montpellier et du Centre national de la recherche scientifique. Centre de Recherche des Zones arides, France No. 14, unpublished

Burchette TP, Wright VP (1992) Carbonate ramp depositional systems. Geology 79:3–57

Cahen L (1954) Extension et âge d'une minéralisation Cu, Pb, Zn en Afrique centrale et australe. Bulletin de la Société belge de Géologie, Hydrologie, Paléontologie 63:89–100

Cahen L, Mortelmans G (1947) Le système de la Bushimaie au Katanga. Bulletin de la Société belge de Géologie, Hydrologie, Paléontologie 56:217–253

Cahen L, Snelling NJ, Delhal J, Vail JR (1984) Geochronology and evolution of Africa. Clarendon, Oxford, 512 pp

Canfield DE (1999) A new model for Proterozoic ocean chemistry. Nature 396:450–453

Catuneanu O (2006) Principles of sequence stratigraphy. Elsevier, Amsterdam, 375 pp

Catuneanu O, Martins-Neto MA, Eriksson PG (2005) Precambrian sequence stratigraphy. Sediment Geol 176:67–95

Day ES, James NP, Narbonne GM, Dalrymple RW (2004) A sedimentary prelude to Marinoan glaciation, Cryogenian (Middle Neoproterozoic) Keele Formation, Mackenzie Mountains, northwestern Canada. Precambrian Res 133:223–247

Delhal J, Lepersonne J, Raucq P (1966) Le Complexe sédimentaire et volcanique de la Lulua (Kasaï). Annales du Musée Royal de l'Afrique Centrale, Tervuren 51:106

Delhal J, Ledent D, Torquato JR (1976) Nouvelles données géochronologiques relatives au complexe gabbro-noritique et charnockitique du bouclier du Kasaï et à son prolongement en Angola. Ann Soc Geol Belg 99:211–226

Delhal J, Deutsch S, Snelling NJ (1989) Datation du Complexe sédimentaire et volcanique de la Lulua (Protérozoïque inférieur, Kasaï, Zaïre). Musée Royal de l'Afrique Centrale, Tervuren (Belgique), Département de Géologie et de Minéralogie. Rapport Annuel 1987–88:93–99

Delpomdor F, Préat A (2012) Hydrocarbon reservoir potential of Neoproterozoic carbonates in the Mbuji-Mayi Supergroup (Sankuru-Bushimay area), Democratic Republic of Congo: stratigraphy, sedimentology, geochemistry, petrophysics. ULB/Total-FR00003322, 276pp (unpublished, confidential document)

Delpomdor F, Préat A (2013) Early and late Neoproterozoic C, O and Sr isotope chemostratigraphy in the carbonates of West Congo and Mbuji-Mayi supergroups: a preserved marine signature? Palaeogeogr Palaeoclimatol Palaeoecol 389:35–47

Delpomdor F, Linneman U, Boven A, Gartner A, Travin A, Blanpied C, Virgone A, Jelsma H, Préat A (2013a) Depositional age, provenance, tectonic and paleoclimatic settings of the late Mesoproterozoic – middle Neoproterozoic Mbuji-Mayi Supergroup, Democratic Republic of Congo. Palaeogeogr Palaeoclimatol Palaeoecol 389:4–34

Delpomdor F, Blanpied C, Virgone A, Préat A (2013b) Paleoenvironments in Meso-Neoproterozoic carbonates of the Mbuji-Mayi Supergroup (Democratic Republic of Congo) – Microfacies analysis combined with C-O-Sr isotopes, major-trace elements and REE + Y distributions. J Afr Earth Sci 88:72–100

Derry LA, Keto LS, Jacobsen SB, Knoll AH, Swet K (1989) Sr isotopic variations in Upper Proterozoic carbonates from Svalbard and East Greenland. Geochim Cosmochim Acta 53:2331–2339

Derry LA, Kaufman AJ, Jacobsen SB (1992) Sedimentary cycling and environmental change in the Late Proterozoic: evidence from stable and radiogenic isotopes. Geochim Cosmochim Acta 56:1317–1329

Des Marais DJ (1994) Tectonic control of the crustal organic carbon reservoir during the Precambrian. Chem Geol 114:303–314

Des Marais DJ, Strauss H, Summons RE, Hayes JM (1992) Carbon isotope evidence for the stepwise oxidation of the Proterozoic environment. Nature 359:605–609

Dunham RJ (1962) Classification of carbonate rocks according to depositional texture. In: Ham WE (ed) Classification of carbonate rocks, vol 1, American Association of Petroleum Geologists Memoir., pp 108–121

Embry AF, Klovan JE (1972) Absolute water depth limits of late Devonian paleoecological zones. Geol Rundsch 61:672–686

Fairchild IJ, Marshall JD, Bertrand-Sarfati J (1990) Stratigraphic shifts in carbon isotopes from Proterozoic stromatolitic carbonates (Mauritania): influences of primary mineralogy and diagenesis. Am J Sci 291-A:46–79

Fischer AG (1964) The Lofer cyclothems of the Alpine Triassic. In: Merriam DF (ed) Symposium on cyclic sedimentation, vol 1, State of Geological Survey of Kansas Bulletin., pp 107–149

Flügel E (2004) Microfacies analysis of limestones, analysis interpretation and application. Springer, Berlin, 976 pp

Freytet P, Plaziat J-C (1972) Les constructions algaires continentales stromatolithiques, examples pris dans la Crétacé supérieur et le Tertiaire de France et d'Espagne du Nord. International Congress of Geology, Montréal

Goldhammer RK, Dunn PA, Hardie LA (1990) Depositional cycles, composite sea level changes, cycle stacking patterns, and the hierarchy of stratigraphic forcing: examples from platform carbonates of the Alpine Triassic. Geol Soc Am Bull 102:535–562

Grotzinger JP (1986) Cyclicity and paleoenvironmental dynamics, Rocknest platform, northwest Canada. Geol Soc Am Bull 97:1208–1231

Halverson GP, Hoffman PF, Schrag DP, Maloof AC, Rice AHN (2005) Towards a Neoproterozoic composite carbon isotope record. Geol Soc Am Bull 117:1181–1207

Halverson GP, Dudas FO, Maloof AC, Bowring SA (2007) Evolution of the $^{87}Sr/^{86}Sr$ composition of Neoproterozoic seawater. Palaeogeogr Palaeoclimatol Palaeoecol 256:103–129

Halverson GP, Hurtgen MT, Porter SM, Collins AS (2010) Neoproterozoic-Cambrian Biogeochemical Evolution. In: Gaucher C, Sial AN, Frimmel HE, Halverson GP (eds) Neoproterozoic-Cambrian Tectonics, vol 16, Global change and evolution: a focus on South Western Gondwana. Developments in Precambrian geology., pp 351–365

Hoffman HJ (1976) Precambrian microflora, Belcher Island, Canada: significance and systematic. J Paleontol 50:1040–1073

Hoffman P, Kaufman A, Halverson G, Schrag D (1998) A Neoproterozoic Snowball Earth. Science 281:1342–1346

Holmden C, Creaser RA, Muehlenbachs K, Leslie SA, Bergström SM (1998) Isotopic evidence for geochemical decoupling between ancient epeiric seas and bordering oceans: Implications for secular curves. Geology 26:567–570

Holmes A, Cahen L (1955) African geochronology. Colo Geol Miner Res 5:3–38

Immenhauser A, Della Porta G, Kenter JAM, Bahamonde J (2003) An alternative model for positive shifts in shallow-marine carbonate $\delta^{13}C$ and $\delta^{18}O$. Sedimentology 50:953–959

Kah LC, Sherman AB, Narbonne GM, Kaufman AJ, Knoll AH, James NP (1999) $\delta^{13}C$ isotope stratigraphy of the Mesoproterozoic Bylot Supergroup, Northern Baffin Island: Implications for regional lithostratigraphic correlations. Can J Earth Sci 36:313–332

Kaufman AJ, Knoll AH (1995) Neoproterozoic variations in the C-isotopic composition of seawater: stratigraphic and biogeochemical implications. Precambrian Res 73:27–49

Kaufman AJ, Jacobsen SB, Knoll AH (1993) The Vendian record of Sr and C isotopic variations in seawater: implications for tectonics and paleoclimate. Earth Planet Sci Lett 120:409–430

Kaufman AJ, Knoll AH, Narbonne GM (1997) Isotopes, ice ages, and terminal Proterozoic earth history. Proc Natl Acad Sci 95:6600–6605

Kaufman AJ, Jiang G, Christie-Blick N, Banerjee DM, Rai V (2006) Stable isotope of the terminal Krol platform in the Lesser Himalayas of northern India. Precambrian Res 147(1–2):156–185

Knoll AH (1992) Biological and geochemical preludes to Ediacaran radiation. In: Lipps JH, Signor PW (eds) Origin and early evolution of the Metazoa. Topics in geobiology, vol 10. Plenum, New York, pp 53–84

Knoll A, Hayes J, Kaufman A, Swett K, Lambert I (1986) Secular variation in carbon isotope ratios from Upper Proterozoic successions of Svalbard and east Greenland. Nature 321:832–837

Knoll AH, Kaufman AJ, Semikhatov MA (1995) The carbon isotopic composition of Proterozoic carbonates: Riphean successions from Northwestern Siberia (Anabar Massif, Turukhansk uplift). Am J Sci 295:823–850

Krapez B (1996) Sequence-stratigraphic concepts applied to the identification of basin-filling rhythms in Precambrian successions. Aust J Earth Sci 43:355–380

Martins-Neto MA (2009) Sequence stratigraphic framework of Proterozoic successions in eastern Brazil. Mar Petrol Geol 26:163–176

Melezhik VA, Gorokhov IM, Kuznetsov AB, Fallick AE (2001) Chemostratigraphy of Neoproterozoic carbonates: implications for 'blind dating'. Terra Nova 13:1–11

Miall AD (1997) The geology of stratigraphic sequences. Springer, Berlin, 433 pp

Miall AD (2000) Principle of sedimentary basin analysis, 3rd edn. Springer, New York, 616 pp

Narbonne GM, James NP (1996) Mesoproterozoic deep-water reefs from Borden Peninsula, Arctic Canada. Sedimentology 43:827–848

Podkovyrov VN, Semikhatov MA, Kuznetsov AB, Vinogradov DP, Kozlov VI, Kislova IV (1998) Carbonate carbon isotopic composition in the Upper Riphean Stratotype, the Karatau Group, Southern Urals. Stratigr Geol Correl 6:319–335

Posamentier HW, Allen GP (1999) Siliciclastic sequence stratigraphy: concepts and applications. Society and economic paleontologists and mineralogists, Concepts in sedimentology and paleontology 7, p 210

Pratt BR, James NP, Clinton AC (1992) Peritidal carbonates. In: Walker RG, James NP (eds) Facies models: response to sea level change. Geological Association of Canada, Canada, p 409

Raucq P (1957) Contribution à la reconnaissance du Système de la Bushimay. Annales du Musée Royal du Congo Belge (Tervuren), Série 8, vol 18, 427 pp

Raucq P (1970) Nouvelles acquisitions sur le système de la Bushimay (République Démocratique du Congo). Annales du Musée Royal de l'Afrique Centrale, Tervuren, Belgique, Série in-8°- n° 69

Saddler PM, Osleger DA, Montanez IP (1993) On the labelling, length and objective basis of Fischer plots. J Sediment Petrol 63:360–368

Schröder S, Grotzinger JP, Amthor JE, Matter A (2005) Carbonate deposition and hydrocarbon reservoir development at the Precambrian-Cambrian boundary: The Ara Group in South Oman. Sediment Geol 180:1–28

Shields G, Veizer J (2002) Precambrian marine carbonate isotope database: Version 1.1. Geochem Geophys Geosyst. 3. doi: 10.1029/2001GC000266

Sibley DF, Gregg JM (1987) Classification of dolomite rock textures. J Sediment Res 57:967–975

Steiger RH, Jäger E (1977) Subcommission on geochronology: Convention on use of decay constants in geo- and cosmochronology. Earth Planet Sci Lett 126:359–362

Strauss H (1997) The isotopic composition of sedimentary sulfur through time. Palaeogeogr Palaeoclimatol Palaeoecol 132:97–118

Tewari VC, Sial AN (2007) Neoproterozoic-Early Cambrian isotopic variation and chemostyratigraphy of the Lesser Himalaya, India, Eastern Gondwana. Chem Geol 237:64–88

Vail PR, Mitchum RM Jr, Todd RG, Widmier JM, Thompson S III, Sangree JB, Bubb JN, Hatlelid WG (1977) Seismic stratigraphy and global changes of sea-level. In: Payton CE (ed) Seismic stratigraphy—applications to hydrocarbon exploration, vol 26, American Association of Petroleum Geologists Memoir., pp 49–212

Van Wagoner JC (1995) Sequence stratigraphy and marine to nonmarine facies architecture of foreland basin strata, Book Cliffs, Utah, USA. In: Van Wagoner JC, Bertram GT (eds) Sequence stratigraphy of Foreland Basin Deposits-Outcrop and Subsurface Examples from the Cretaceous of North America, vol 64, American Association of Petroleum Geologists Memoir., pp 137–223

Wynn TC, Read JF (2007) Carbon-oxygen isotopes signal of Mississippian slope carbonates, Appalachians, USA: A complex response to climate-driven fourth-order glacio-eustacy. Palaeogeogr Palaeoclimatol Palaeoecol 256:254–272

Probable Fungal Colonization and Carbonate Diagenesis of Neoproterozoic Stromatolites from South Gabon, Western Congo Basin

5

Kamal Kolo, Kurt Konhauser, Jean-Pierre Prian, and Alain Préat

5.1 Introduction

Fungi are able to colonize any number of rock surfaces in their efforts to extract nutrients and trace metals for their metabolism. Their filaments, called hyphae, physically exploit grain boundaries, cleavages and cracks to gain access to new mineral surfaces, and in the process, they cause several alteration features, ranging from simple surface roughing by etching and pitting to selective mineral dissolution and cavity formation to extensive physical disintegration of the minerals (see Konhauser 2007 for details). Simultaneously, all exposed mineral surfaces become covered in EPS, which serves to retain water and fuel hydrolysis reactions (Welch et al. 1999). Through the release of organic acids, such as oxalic acid or citric acid (Richter et al. 2007), mineral dissolution is accelerated because the acids dissociate and release protons that can attack minerals directly by complexing with ligands at the minerals surface. Deprotonated organic anions (e.g., oxalate, citrate) indirectly affect dissolution rates by complexing with metal cations in solution, thereby lowering the mineral's saturation state (e.g., Bennett et al. 1988). These interactions not only result in the slow alteration of the primary mineral surfaces, but frequently they induce the formation of secondary mineral phases, such as Ca- and Mg-oxalate and calcite (Gadd 1999; Verrecchia 2000; Chen et al. 2000; Burford et al. 2003a; Hoffland et al. 2004) or the so-called desert varnish comprising Fe- and Mn-oxides (Krumbein and Jens 1981; Grote and Krumbein 1992). Finally, extreme bioweathering can even form a new diagenetic "mycogenic rock fabric" (Burford et al. 2003b).

As weathering agents, fungi have played a particularly important role in the alteration of carbonate rocks: the biodeterioration of carbonate monuments and buildings (Hoffland et al. 2004; Sterfinger and Krumbein 1997), bioerosion of corals and sediment particles (Vogel et al. 2000; Golubic et al. 2005), and the accumulation of carbonate-sourced metals (Sterflinger 2000; Gadd 2007), are just a few examples. The large quantities of oxalic acid produced by fungi can also react with carbonate host rocks to yield Ca-oxalates crystals or re-precipitation of Ca-minerals in the form of calcretes (Verrecchia 2000). Indeed, it has been suggested that fungi are probably at the origin of much the calcium carbonate accumulation in paleosols and $CaCO_3$ enrichment of surficial sediments throughout the Phanerozoic (Verrecchia et al. 2003; Cardon and Whitbeck 2007). For instance, paleosols in the Lower Carboniferous of South Wales contain needle-fibre calcite as coatings on sediment grains and rhizocretions (Wright 1986,). The fibres were probably formed by the calcification of fungal hyphae. Esteban and Klappa (1983) illustrated fungal hyphae in a Pleistocene caliche hardpan from Spain. A well-developed biogenic structure (sparmicritization), related to the activity of fungi and algae is reported by Kahle (1977) on the Pleistocene Miami Limestone which has been transformed into calcareous crusts. Part of the spar-micritization was caused by boring of sparry calcite cement by fungi, followed by in situ calcification. Fossilized fungal hyphae and spores have also been observed in the Upper Devonian of the Rocky Mountains (Canada), in the Lower Carboniferous of northern France and in the Cretaceous of Central Italy by Préat

K. Kolo (✉)
Faculty of Science and Engineering, Department of Petroleum Geosciences, Soran University, Erbil-Soran, Iraq
e-mail: kamal.kolo@soran.edu.iq

K. Konhauser
Department of Earth and Atmospheric Sciences, University of Alberta, Edmonton, AB, Canada
e-mail: kurtk@ualberta.ca

J.-P. Prian
Bureau Recherches Géologiques et Minières, 3 Av Claude Guillemin, BP 36009, 45060 Orléans, Cedex2, France
e-mail: jp.prian@brgm.fr

A. Préat
Département des Sciences de la Terre et de l'Environnement, Université Libre de Bruxelles, CP160/02, Av. F.-D. Roosevelt, 50, 1050 Bruxelles, Belgium
e-mail: apreat@ulb.ac.be

M.J. de Wit et al. (eds.), *Geology and Resource Potential of the Congo Basin*, Regional Geology Reviews,
DOI 10.1007/978-3-642-29482-2_5, © Springer-Verlag Berlin Heidelberg 2015

et al. (2003). The fungi are systematically associated with calcrete levels at the top of thin shallowing-upward evaporitic tidal sequences, and their roles in biomineralization were dominant as indicated by particular coated grains, bridging grains, rhizoconcretions, sparmicritization, micritization, perforations and secondary calcite formation from excreted fungal Ca-oxalates.

Despite their importance, it is surprising that only a few detailed studies have been undertaken with the view of assessing carbonate alteration features induced by fungal growth. For instance, Burford et al. (2003a) investigated the role of fungi in the transformation of Paleozoic limestones and dolostones, showing that fungal dissolution of the carbonate substrates led to the formation of new microfabrics, such as polymorphic growth patterns with mineralized hyphae and crystalline material (Na plate-like and Ca-blocky crystals) adhering to their external walls. Although the precise mechanisms involved in the formation of these new 'mycogenic' fabrics were not described, it seems that both metabolism-dependent and metabolism-independent processes played integral roles. Kolo et al. (2007) described the formation of small pits and alveolar structure related to the dissolution of dolomite crystals by fungi of Carboniferous aged dolostones, and showed that the pits were occupied by a dense network of fungal hyphae penetrating the dolomitic substrate and circumventing the grain boundaries through dissolution paths.

There has similarly been a paucity of study directed to observing fossilized fungal communities in the ancient rock record, in part due to a bias directed at the more common and easily recognizable land plants (Taylor et al. 2005) and the difficulties at resolving whether fungal forms in ancient sediments can be reasonably considered "fungi" from one or more of the four classical phyla: Ascomycota, Basidiomycota, Chytridiomycota, and Zygomycota which comprise the so-called true "Fungi" (Cavalier-Smith 1987). At present, the oldest definitive fungal fossils are associated with lichen-like symbiosis is reported from the Doushantuo Formation dated at 600 Ma (Yuan et al. 2006). Butterfield (2005) re-interpreted the presence of *Tappania* in the Mesoproterozoic Roper Group of Australia and extended the record of putative fungi to 1430 Ma. Along with other Proterozoic acritarchs exhibiting fungal-like characteristics (e.g. *Trachyhystrichosphaera, Shuiyousphaeridium, Dictyosphaera, Foliomorpha*), this author considered that there is a case to be made for an extended and relatively diverse record of Proterozoic fungi. However, these reports were later discounted (Cavalier-Smith 2006), leaving us with a reasonably confident fungal phylogeny (Glomalean fungi: Redecker et al. 2000) dating back to ca. 600 Ma. Berbee and Taylor (2001) also suggested such a Neoproterozoic age for this colonization. Compared to this, nearly two decades ago, the oldest reliable fossil fungal finding was from the Devonian Rhynie chert (Stubblefield and Taylor 1988). Other recent studies describe zygomycetes fungal forms from macerations of Late Riphean (Neoproterozoic) raising thus again the issue of when Fungi first appear on Earth (Hermann and Podkovyrov 2006). Given this lack of a reliable Fungi fossil record, and identification problems, the answer is perhaps better resolved through molecular genetics., Based on protein fungal sequence analysis, Heckman et al. (2001) extended all major fungal group lineages back to ca. 1 Ga, and inferred a fungal colonization of Earth deep into the Precambrian. Nevertheless, compelling fossil evidence to corroborate this hypothesis has not been produced to date (Krings et al. 2013).

Finally, little attention has been given to fossilized fungal communities or to their role in carbonate diagenesis (Krumbein 1972); and studies of fungal activity in carbonate sediments older than the Quaternary are relatively limited in number. Our study of the Neoproterozoic of South Gabon revealed numerous per-mineralized fungal relicts: sporangia, sporangiophores, columellae, zygosporangia, suspensors, dichotomous hyphae and spores in the upper part of shallowing-upward evaporitic peritidal sequences. It provides a detailed evaluation of the fungal role in an attempt to better understand the mechanisms involved in the paleo-weathering of a Neoproterozoic carbonate substrate. As documented below, comparison of our observations with in vitro experiments allow us to define an eogenetic sequence driven by fungal invasion and colonization of a Neoproterozoic substrate. This finding, corroborated by comparison with present day fungal-mineral surface interactions, also bears upon the role of microbial activity in the shaping of the late Neoproterozoic Earth surface.

5.2 Geological Setting

The Nyanga synclinorium is an important geotectonic unit of southern Gabon, consisting of a 250 km long deformed basinal structure. It formed during the West Congolian orogeny, which forms a vast Neoproterozoic (Pan-African) metamorphic belt stretching from Angola, in the south, some 2,000 km northwards to Gabon (Thiéblemont et al. 2009; Préat et al. 2011a; and de Wit and Linol, Chap. 2, this Book). The Gabonese part of the belt, averaging 100 km in width, runs parallel to the Atlantic coast and terminates in northern Gabon. Among the five major lithostratigraphical units of the Gabonese part of the West Congolian orogen, the West Congo Supergroup outcrops in the Nyanga synclinorium (Gérard 1958) and consists of the succession of several informal units. The 'Schisto-Calcaire Group' directly overlies the 'Niari Formation' (or 'Niari Tillite'), which, based on chronostratigraphic and sedimentological studies

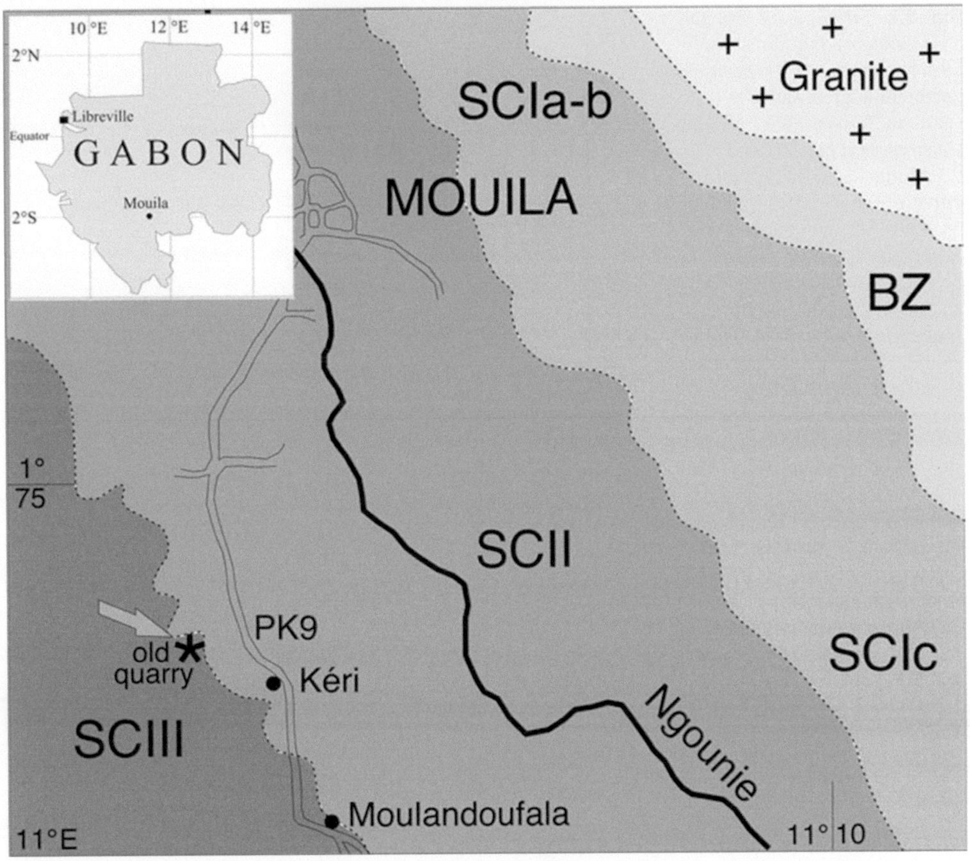

Fig. 5.1 General geological map of the studied area and location of Mouila city where the old Mouila quarry is situated, exposing the carbonate Neoproterozoic ('Nsc3' or 'SCIII') sampled for the study. Bz is for Bouenza 'Group', SCIa-b, SCIc and SCII for 'NSc1a-b, Nsc1c and Nsc2 formations of the Schisto-Calcaire Subgroup (modified from Thiéblemont et al. 2009; Préat et al. 2010)

in Namibia, has a proposed maximum age of 564 Ma (Saylor et al. 1998), and therefore likely correlates with the Gaskiers, or possibly another Ediacaran glaciation. However, the precise age of the formations of the West Congo Supergroup is still unknown.

The 'Schisto-Calcaire Subgroup' is predominantly a carbonate sequence with four formations ('Nsc1 to Nsc4'), and an overall maximum thickness of about 650 m. The lowermost three formations are calcareous to dolomitic shales with an uppermost sandy shale-siltstone unit with interbedded limestones. For our study, we collected fresh samples (standard thin sections and slabs were prepared from freshly cut samples after removing 2–3 cm from the outer surfaces) from formation Nsc3, which is exposed in the old Mouila quarry in South Gabon (Fig. 5.1).

5.2.1 Microfacies of the 'Nsc3 Formation'

In the old Mouila quarry, 20 m of thin to medium-bedded homogeneous dolo-mudstones interstratified with massive laminar peloidal and stromatolitic greyish dolomites are exposed (Figs. 5.2 and 5.3a,b). The rocks are partly silicified and contain stratiform black chert nodules, a few centimeters thick, between the beds and thin siliceous impregnations perfectly moulding the stromatolite laminae (Fig. 5.3c,d). Clays are not abundant and consist of subcentimetre-scale thicknesses. The stromatolites are hemispherical (height of 10–20 cm, lenghts up to 50 cm) and typically laterally linked (respectively LLH and SH, Logan et al. 1964, Fig. 5.3c,d). The facies are fine grained, mud-supported, greyish to slightly whitish and particularly homogeneous on the field (Fig. 5.3a,b).

Sixty samples were collected to ascertain the lithology and sedimentary features of formation Nsc1. Detailed imaging of the samples was performed using a JEOL Model JSM-6400 Scanning Electron Microscope (SEM) with a Pioneer Si–Li crystal X-ray detector (EDX) operated at a specific resolution of 138 eV. The samples were gold or carbon coated and mounted on metal stubs. The software used for spectral interpretation was a Voyager Version 3.5 from Thermo-Noran. High resolution images (1.2–3.0 nm at 15 and 1 kV, respectively) were further obtained with a JEOL JSM–7000F Field Emission Scanning Electron Microscope. Thin sections and rock slabs were sputter coated with gold-palladium and mounted on metal stubs.

The analyses enabled the recognition of 5 major 'diagenetic' dolomitic microfacies (MF1–5), whose succession (1–5) constitutes a standard sequence of shallowing-upward sedimentation and a corresponding increase in post-depositional diagenetic events related to the influence of hypersaline groundwaters in a sabkha-like environment

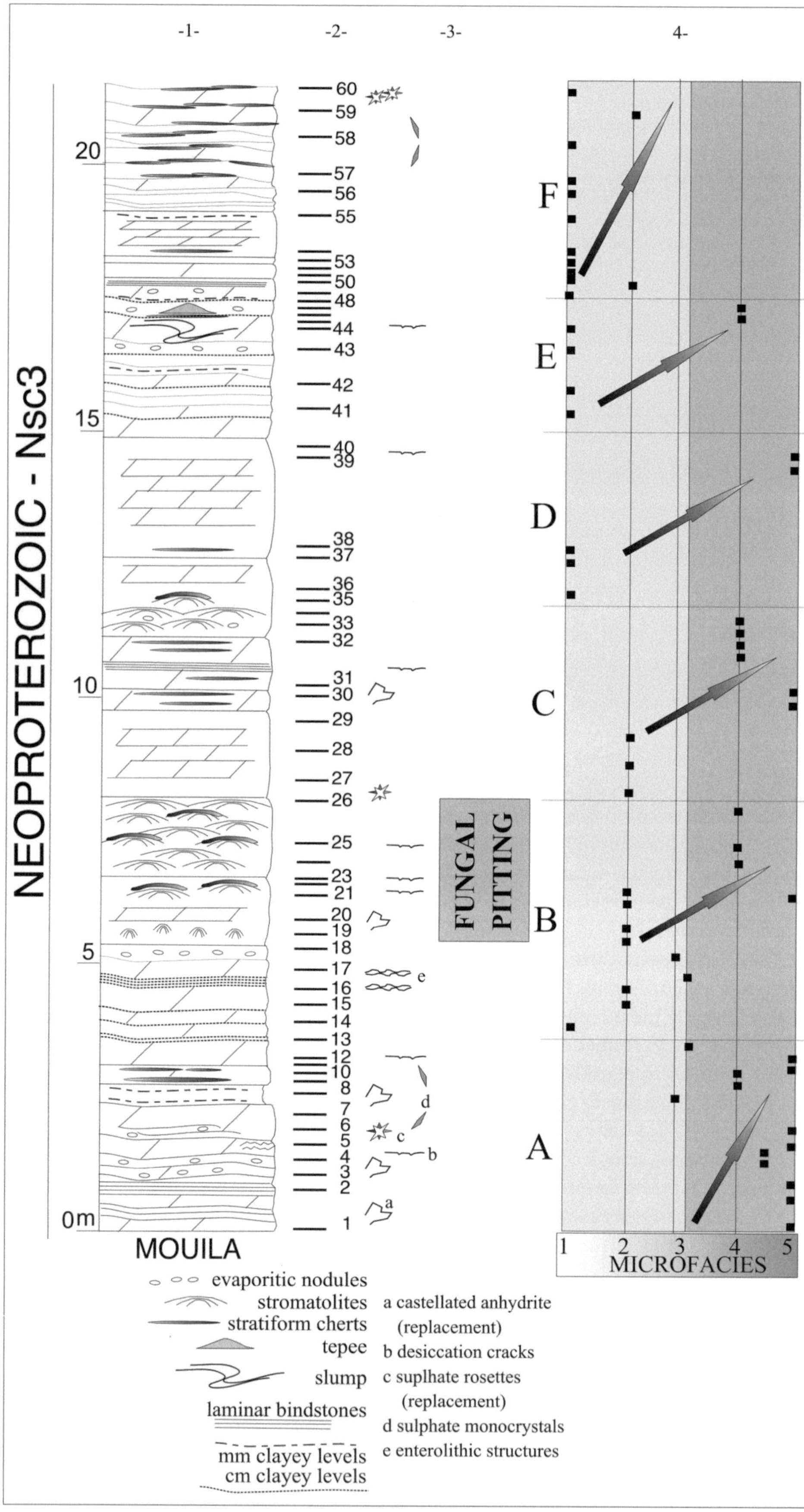

Fig. 5.2 Stratigraphy (column 1), lithology and sedimentary structures (column 2), position of sedimentological samples (column 3), position of samples illustrated on plates 1 and 2 (column 4) and lithological curve of the microfacies (column 5) of the Mouila quarry (Préat et al. 2010). The *arrows* inside column 5 indicate regressive shallowing-upward metric sequences (A–F), from shallow subtidal (microfacies 1 and 2) to supratidal and sabkha environments (microfacies 3–5). Samples 19 and 26 contain partly preserved cyanobacteria in the dolomicritic matrix (Fig. 5.4a) and well-preserved fungal hyphae in small-sized pits (Fig. 5.4b–d). The matrix of samples 22 and 23 is strongly replaced by homogeneous dolomicrosparitic crystals. See text for explanations and Fig. 5.3g, h

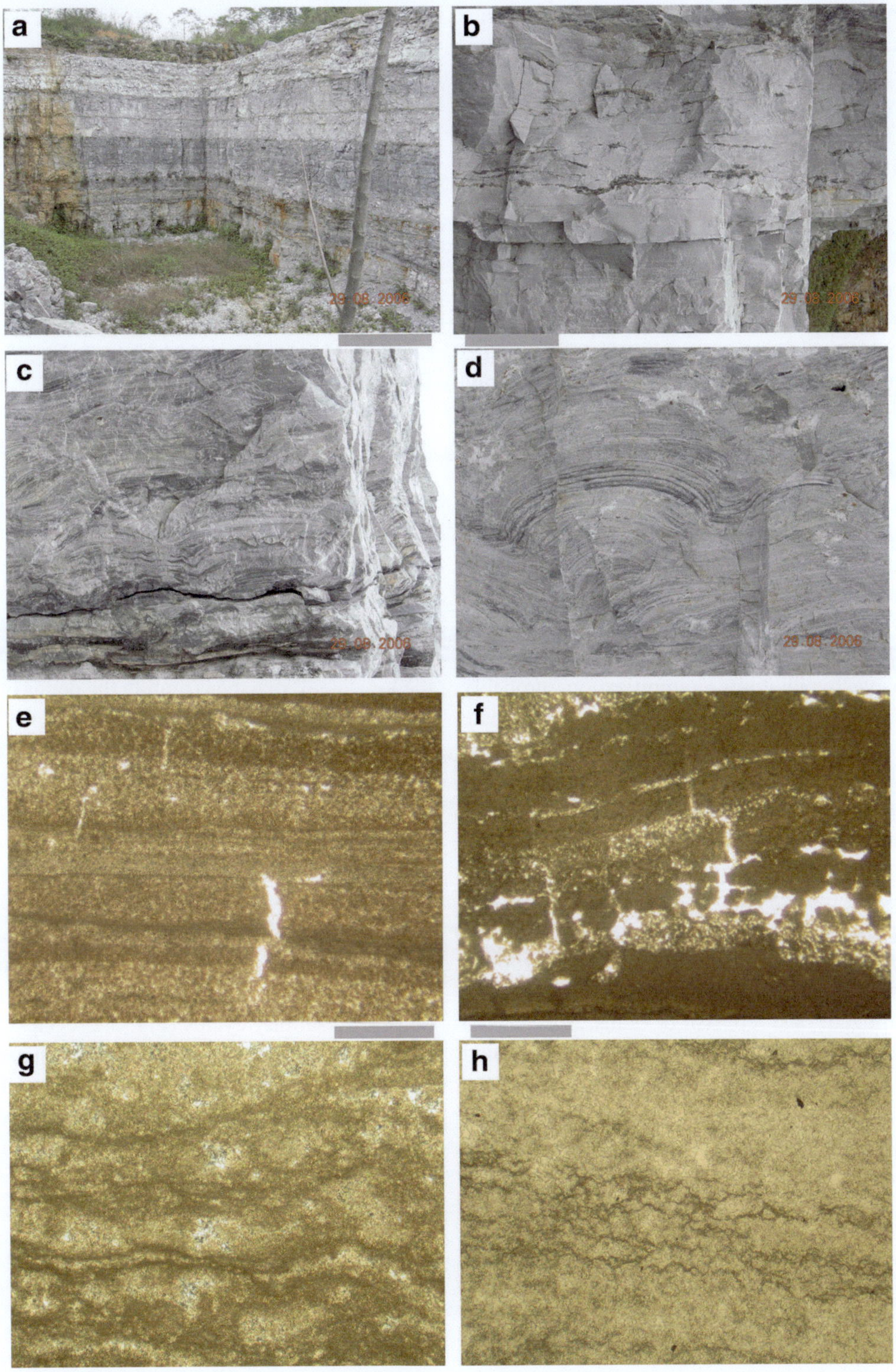

Fig. 5.3 (**a**) Mouila old quarry (South of Gabon). The picture shows the lower level of the series (10 m) belonging to the 'Nsc3' formation (Schisto-Calcaire Subgroup, Neoproterozoic). The level is composed of thin to medium-bedded homogeneous dolomudstones interstratified with stromatolitic greyish dolostones. The first massive stromatolitic layer occurs at 6 m from the base. The height of the front is 10 m (photo 1503/2006/ap). (**b**) Thin-bedded and laminated cyanobacterial greyish dolomudstones. Thin irregular undulating stratiform chert layers form slightly discontinuous *blackish* interstratified levels. Some 'ripples' are superimposed on slightly curved stromatolites. The *top* of the photograph is a thicker blackish chert level. *Lower part* of the massive stromatolitic layer (thickness of 60 cm), Mouila old quarry, Gabon (photo 1497/2006/ap). (**c**) Clusters of closely packed decimetric-scale greyish domal stromatolites interstratified with regular laminar (millimetric) thin cyanobacterial dolostones. Stratiform stromatolites are roughly parallel to layer orientation. Some flanking stromatolites are developed on sloped layers and are also interstratified with laminar dolostones. The base of the figure shows blackish chert levels. Other

(Fig. 5.2) (Préat et al. 2010). One of the primary facies consists of flat-laminated (Fig. 5.3b) to low domal stromatolite columns (Fig. 5.3c,d), the latter having laminae that form overlapping domes, with younger laminae truncating against older ones and leaving no intercolumn space. Laminae are produced by alternation of organic-rich and organic-poor horizons, with some individual laminae traceable over a few centimetres. The organic-rich horizons are generally richer in pyrite framboids (<1 μm up to 10 μm). The brighter layers contain micropeloidal micrite or clotted mudstone draping over dense mat layer. Mat constructors in thin section are poorly preserved morphologically (filamentous mat ghosts are present), but are still visible on the SEM despite the dolomitization process (Fig. 5.4b). In the outcrop, this 'cryptomicrobialite' is composed of submillimeter-scale white and gray micritic laminae couplets (Fig. 5.3e). Smooth flat laminated dolostone (Fig. 5.3f) associated with disrupted fenestral and crinkled fabrics are common. The latter typically exhibit near horizontal sheet-cracks associated with vertical and step-like thin mudcracks isolating micritic lumpy patches (Fig. 5.3f).

The major diagenetic alteration of the facies consists of a thin pervasive hypidiotopic dolomitization, probably related to episodes of anhydritization since sulphate microenterolithes developed inside the mats. Consequently, the former greyish microlaminar micritic sediment is progressively replaced by a relatively fine-grained homogeneous whitish dolo-microsparite, which may still contain thin discontinuous microbial mat relicts (Fig. 5.3g,h). At the beginning of the replacement process the dolomite is a mimetic fabric-preserving dolomite with crystal size varying between micrite and microsparite (<50 μm). As mentioned above, some evaporite minerals remain present in the matrix (i.e., not dolomitized) and consist of laths, rosette-like aggregates, enterolithic small nodules and castellated crystals (*sensu* Clark 1980) which often grow inside the mat levels. 'Elephant skin texture' with micropinnacles and net-like structures (Gerdes et al. 2000) are common in the facies associated with the stromatolitic layers. Silica is the last diagenetic phase observed.

5.2.2 Paleoenvironmental Interpretation

Micritic, millimeter-scale laminae interstratified with organic-rich thin horizons (benthic microbial mats, Fig. 5.4a) indicate initial deposition in a tidal-flat environment (Purser 1973; Hardie 1977; Sellwood 1986). The soft peloidal mud contains wavy and discontinuous lenticular laminae. Crinkled fenestral laminae (Fig. 5.3f), being either flat or domal, even (Fig. 5.3e) or pinching (Fig. 5.3f), are probably related to cyanobacterial mats. Despite strong diagenetic overprinting (microsparitization, dolomitization), slightly altered bacterial filaments are still observable by SEM imaging of the stromatolite laminae (sample MOU26, Fig. 5.4a).

The sediment laminae of the Mouila facies are typically disrupted by mudcracks and sheet-cracks a few millimeters to a few centimeters long associated with irregular small-sized fenestrae. Similar characteristics are observed today in the low 'algal' marshes fringing the ponds of channeled belts at Andros Island (Hardie and Ginsburg 1977), particularly along the backslope of the levees and the beach-ridge washovers where very shallow (millimeter range), closely spaced (around 1 cm) mudcracks are present. The cracking process may be quickly stopped by rapid growth of cyanobacterial colonies (microstromatolites) giving incomplete mudcracks as those present in Fig. 5.3f.

Sedimentological evidence also reveals that the Mouila series consists of a succession of plurimetric-thick shallowing-upward sequences which correspond to early diagenetic salinity cycles (Fig. 5.2, cycles A–F,) with well-developed upper parts related to subaerially exposed mudflats in a marginal marine sabkha. The cycles start with open marine subtidal-intertidal sedimentation in association with stromatolites (MF1 and MF2) and grade into

Fig. 5.3 (continued) discontinuous thinner silicified zones are present in the stromatolites (see Fig. 5.3d) and in the laminar dolostones. Massive stromatolitic layer (1 m thick), Mouila old quarry, Gabon (photo 1458/2006/ap). (**d**) Domal hemispherical stromatolites (LLH Logan's type, 1964) with very *thin blackish* chert layering. Stromatolitic *greyish* and *whitish* laminae are gently convex, without pronounced asymmetry in the domes. They are only interrupted when a slope is encountered. Massive stromatolitic layer, wide of the stromatolitic dome is 16 cm, Mouila old quarry, Gabon (photo 1467/2006/ap). (**e**) Smooth flat laminated dolostone composed of the alternation of millimetric fine-grained well-sorted peloidal laminar dolopackstone and thinner homogeneous dolomustone. The laminae are rather parallel at this scale of the microphotograph but pinch out laterally at at pluricentimetric scale. Small-sized vertical mudcracks cut several laminae. Sample MOU26, Mouila old quarry, Gabon (photo 0400/2006/ap). The sample has been taken in the domal stromatolite of Fig. 5.3c. (**f**) Crinkled fenestral laminar dolostone of the same type as Fig. 5.1e. The laminae are wavy, some contain sediment 'clots' or irregular peloids, particularly near the fenestral fabric. Sample MOU26, Mouila old quarry, Gabon (photo 0401/2006/ap). The sample has been taken in the domal stromatolite of Fig. 5.3c. (**g**) Irregularly dolomicrosparitized mudstone with remnants of thin layers of homogeneous blackish dolomudstone of the same type of those illustrated in Fig. 5.3e, f. The dolomicrospar is greyish and displays a patchy distribution, it contains slightly recrystallized dolomicrite matrix. Sample MOU22, Mouila old quarry, Gabon (photo 0417/2006/ap). (**h**) Strongly dolomicrosparitized mudstone with very thin uneven remnants of blackish dolomudstone. The dolomicrospar is coarser than in Fig. 5.3g and more whitish. Sample MOU23, Mouila old quarry, Gabon (photo 0409/2006/ap). Scale bars = 1 mm (**h**) and 400 μm (Fig. 5.3f, g)

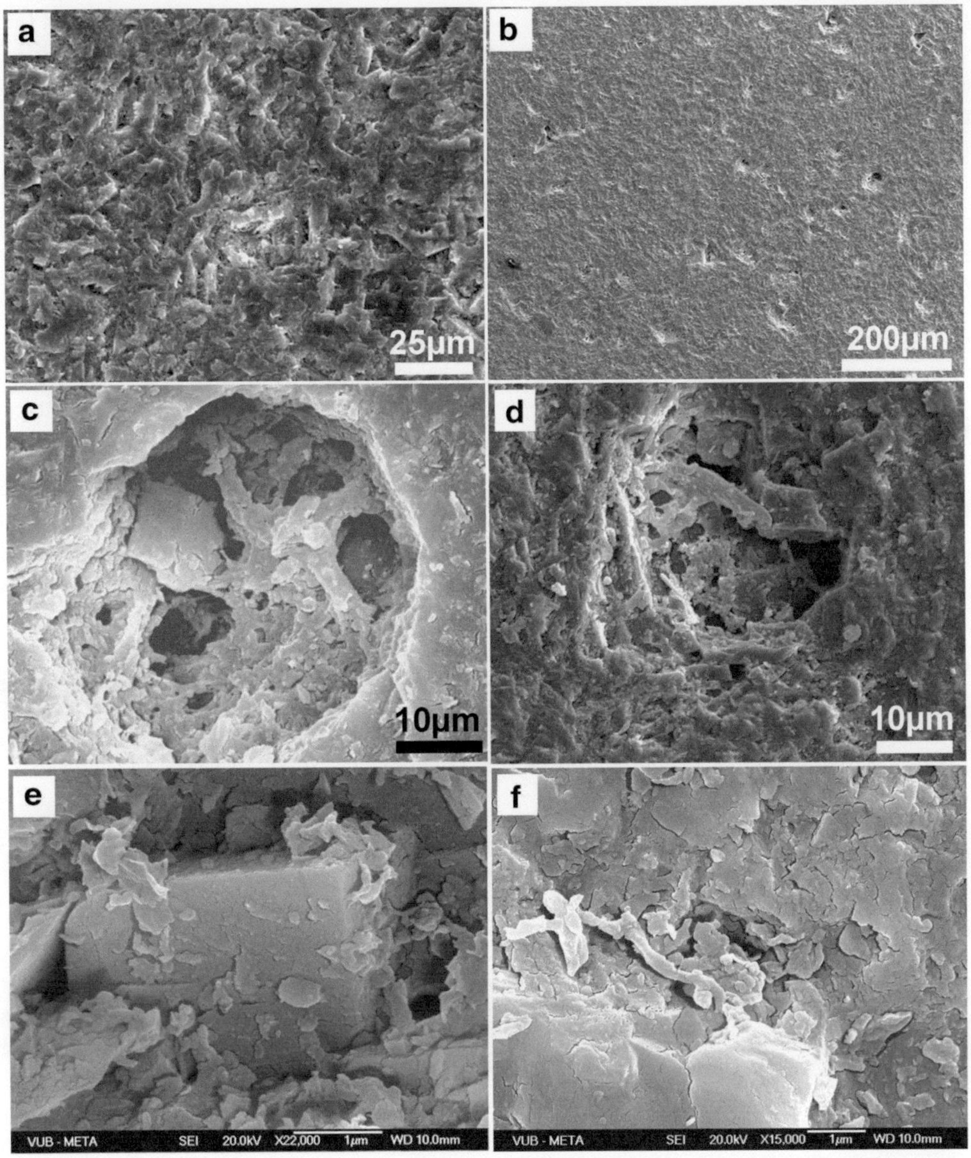

Fig. 5.4 (**a**) Filamentous dichotomous bacterial (probable cyanobacteria) ranging from 2 up to 5 μm in diameter. They are partly destroyed by a fine-grained dolomicropsar. The photograph has been taken in a homogeneous dolomudstone layer (see Fig. 5.3f). Sample MOU26, Mouila old quarry, Gabon (photo mou26-63/2007/ap). (**b**) Surface of sample MOU26 (*thin section*) under the SEM showing numerous inframillimetric irregular to subrounded pits or 'cavities'. Sample MOU26, Mouila old quarry, Gabon (photo mou26-8/2007/ap). The sample was taken in the domal stromatolite of Fig. 5.3c before any previous treatment (acid attack, coloration). (**c**) Small-sized (~30 μm) rounded pit in a dolomicrosparitized mudstone. The pit is filled with various microbial filaments forming a mesh containing well-crystallized minerals (see Fig. 5.3e). The filaments are slightly curved, some are dichotomous or rod-shaped, and the larger have diameters varying from 0.5 up to 1 μm. The dolomicrospar is irregular, varies in size between 1 and 3 μm and covers partly destroys the filaments. Thinner and shorter filaments (0.1–0.2 μm in diameter) are also present. Sample MOU19, Mouila old quarry, Gabon (photo mou19-17/2007/ap). (**d**) Small-sized (~25 μm) pit in a dolomicrosparitized mudstone (see Fig. 5.4a). Same filaments as previous figure with diameters ranging from 0.25 to 1 μm. Dolomicrospar is more regular and consists of small-sized (1–5 μm) well crystallized rhombs growing from a dolomicritic matrix. Thinner (0.1 or less micron in diameter) are associated with the larger filaments. They exhibit a discrete *barrel-shaped cells*. Some filaments are engulfed in the dolomicrospar. Sample MOU26, Mouila old quarry, Gabon (photo mou26-20/2007/ap). (**e**) Quadratic dolomite crystal probably derived from a calcium oxalate crystal associated with a few irregular microbial filaments (see upper corners of the crystal) with diameters around 0.1 μm. The crystal is long of 2.5 μm and has grown in a pit similar to those illustrated in Fig. 5.4b. (**f**) Thin filament with *septate* appearance similar to hyphae and diameters <1.0 μm likely enveloped by EPS material. Other thicker filaments (flattened?) appear well-embedded in the pit walls. Sample MOU19, Mouila old quarry, Gabon (photo mou19-21/2007/ap)

evaporitic supratidal conditions or subaerial exposition (MF–MF5) with progressive replacement of primary evaporitic minerals by dolomite (dolomicrosparite). The diagenetically altered upper parts of the cycles are related to subaqueous deposition of muds associated with desiccation and/or intrasediment precipitation of evaporitic minerals from groundwater brines, similar to some modern sabkha evaporites. Samples MOU19, 22 and 23 (Fig. 5.2) are strongly affected by evaporite brines leading to dolo-microsparitization of the microenterolithes and sample MOU26 is characterized by desiccation (Figs. 5.2 and 5.3e, f). These samples containing "fungi" (samples MOU19, 23 and 26) come from the upper part of such a diagenetic salinity cycle overprinted on a stromatolitic layer (cycle A, MOU19) and cycle B, MOU22, 23, 26 Fig. 5.2) (Préat et al. 2010, 2011a, b).

5.3 Detailed Petrography of Nsc3

5.3.1 Diagenetic Alteration

Both dolomicrite and dolomicrosparite replace (at an infra-millimeter-scale) the microbial laminae and developed progressively from the cyanobacteria which are partly or totally mineral-enveloped and still recognizable: they form a 3D-network, some are dichotomic, the average diameter is between 2–5 μm and their minimal length is 20 μm (Fig. 5.4a). Dolomite crystals are greyish (abundant micritic inclusions), xenototopic to hypidiotopic and approach sizes up to 50 μm. Larger whitish hypidiotopic crystals (50–100 μm) are associated with the replacement of former sulfate crystals and irregular fenestrae. The association of fine-grained dolomite with mudcracks and sheet-cracks (disrupted flat laminar lamination), together with their very fine grain-size and the presence of former sulfates, suggest that dolomite is a secondary mineral phase, most likely precipitated from hypersaline waters during the dry season.

Petrographic and SEM study reveal abundant subrounded (circular to oval-shaped) and irregular pits (quasi-rectangular) in the first stromatolitic level (samples MOU19 and MOU26, Figs. 5.2 and 5.4b). They range between 5 and 50 μm and are spatially arranged into either single pits or a network pattern. The latter is formed from boundary-connected subrounded-rounded pits that form a honeycomb or alveolar structure containing both colonizing fungal material and neominerals (see Sect. 4.1 below). Generally, single pits show three major features. The first is the presence of an elevated mineral "collar" or ring (originally probably Ca- or Mg-oxalates) around the pits' circumference that is composed of authigenic minerals. The second is the deposition of authigenic minerals inside the pits in a process which we are tempted to call "nesting". The minerals are dolomite in the form of rhombohedra (Fig. 5.4c,d), but also as quadratic crystals possibly related to former oxalates (Fig. 5.4e). The third, as discussed in the section below, is the colonization of the pits by invading fungal hyphae. Indeed, it is our contention that the pits are dissolution cavities related to fungal colonization of the original mineral substrate.

The colonizing microbes display various morphologies, more irregular than the ones of the cyanobacteria seen in the matrix, with non-septate (Fig. 5.4c) or septate-like (Fig. 5.4f) thin filaments similar to hyphae, and diameters <1.0 μm. Other filaments are more regular and have diameters varying between 0.25 to 1.0 μm (Fig. 5.4c). The filaments are associated and engulfed with what we presume to be fossilized EPS material (Figs. 5.4c–f and 5.5a,b). Some spherical bodies (diameter around 1.0 μm) are visibly adhering to the fossilized EPS (Fig. 5.5c–d). These spheres are richly encrusted with sub-micron sized rounded crystals that collectively yield the framboidal shapes characteristic for authigenic pyrite. However, microprobe analyses reveal that the spheres are entirely dolomite in composition. These spheres could represent fungal spores encrusted with mineral crystals. A similar observation has also been made under laboratory conditions (Kolo and Claeys 2005). The EPS is systematically desiccated (Figs. 5.4c–e and 5.5a–c) yielding strands reminiscent of actual microbial filaments. Very fine (<1 μm) aggregates of minerals or clusters adhering to the strands are frequently observed in the cavities (Fig. 5.4c–f).

5.3.2 Evidence of Fungal Colonization

The probable 'dolomitic' spores, the dolomitic prismatic quadratic and tetragonal crystals, the clusters of very fine crystals along the filaments and on the former larger crystals inside the cavities and the abundance of the thin filaments (<1.0 μm in diameters) in the pits suggest that the pits were formed through the activity of ancient fungi. Figures 5.6a–d show completely mineralized and well embedded forms in the rock matrix that can be attributed to fungal vegetative parts such as sporangia, sporangiophores and hyphae. Careful study of these images reveals a dense fungal colonization of the sediments, especially as shown in Fig. 5.6c–e. In Fig. 5.6e, f several individual fungal sporangia and sporangiophores can be clearly seen. The black color of fungal parts is attributed to organic content. The low contrast in the backscattered image (Fig. 5.6f) emphasizes the embedding of the fungal forms within the matrix and the uniform elemental composition. On close examination, the blackish sporangial area in the lower-mid part of Fig. 5.6e reveals that it is actually harboring three superimposed sporangial bodies that make intersecting circles. The lower one of these circles shows very fine and continous zig-zag wavy ornamentation likely representing the ancient ornamental spines or ridges at the perimeter of the sporangial wall. These are comparable to present day spine ornamentation on the sporangia of some Mucorales fungi (Alexopoulos and Mims 1979; Moore-Landecker 1991).

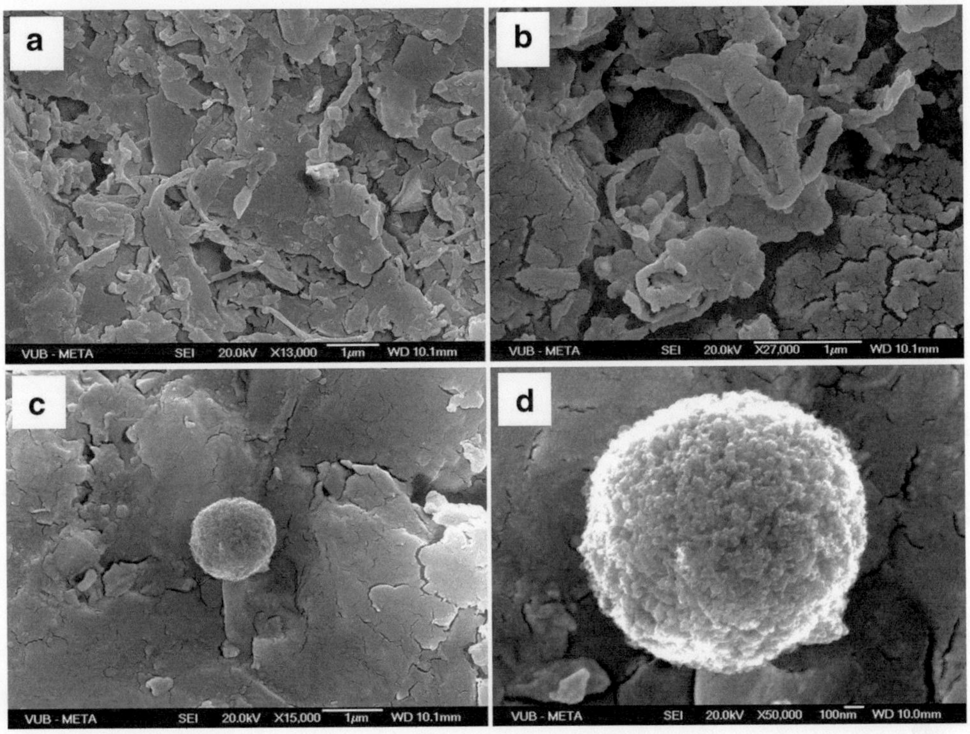

Fig. 5.5 (**a**) Very small-sized curved filaments (diameters around 0.1–0.2 μm) irregular embedded in platelet-like anhedral dolomite crystals (possibly former EPS substances). (**b**) Irregular polygonal networks of microcracks are omnipresent on the dolomite as well on the filaments. Sample MOU19, Mouila old quarry, Gabon (photo mou19-40/2007/ap). (**c, d**) Dolomitic sphere (diameter of 1 μm) inside platelet-like dolomicrospar with irregular microcracks. The *sphere* is composed of subrounded microcrystals of dolomite (10 nm in diameter, Fig. 5.5d). Sample MOU19, Mouila old quarry, Gabon (photo mou19-36 and 33 respectively/2007/ap). SEM photomicrographs. Bar scale: as indicated

Interestingly, some specific features of fossilized reproduction parts of fungi can be observed also. Figure 5.7a shows the remains of sporangial forms connected to their sporangiophores, and the black protrusions to the right suggest a typical form of spore dispersion. A second smaller sporangium lies next to the first structure. The first sporangium reveals two concentric circles separated by about a 8–10 μm thick zone. The inner circle is visibly continuous (diameter ~30 μm) and joins the sporangiophore. This structure and configuration actually reveal a fungal columella. Furthermore, Fig. 5.7b shows a richly colonized substrate (not all traces shown) where an external black perimeter of the sporangium suggests the sporiferous region of the sporangium with a probable attached spore mass. The traces of the ancient fungal parts are sometimes outlined by the dissolution figures, which relates the visible pitting on the Neoproterozoic substrates, at least in these instances, to fungal activity. Figure 5.7c shows another fungal-related shape and setting where possible sexual reproduction organs involving the formation of a zygosporangium and suspensors are observed. They display an anatomical morphology comparable to modern fungal sexual reproductive cycle and the production of zygosporangia, zygospores and suspensors, e.g., in the Zygomycetes (Alexopoulos and Mims 1979; Moore-Landecker 1991; Kendrick 2000).

A negative and enhanced image (Fig. 5.7d) of the above reveals the well-defined and preserved morphology and contours of these fungal structures that undoubtly appear well-embedded in the mineral matrix of the dolomitic substrate. This strongly suggests a pre-lithification synsedimentary process and early diagenesis. Moreover, these fungal structures are very similar to fungal parts retrieved from the Late Riphean Neryuenskaya Formation of southeastern Siberia (Hermann and Podkovyrov 2006), which represent the oldest reported fossil fungi.

Although the poor preservation caused by dolomitization and per-mineralization of fungal relicts do not allow for more detailed descriptions, and did not allow for extraction, the well-contoured fungal features, substrate dissolution, pit forms and colonization when compared with modern fungi (e.g. Mucorales), and their patterns of colonization and interaction with mineral surfaces, especially carbonates (Kolo and Claeys 2005; Kolo et al. 2007), nonetheless suggests strong similarities between modern and ancient fungi.

5.4 Neoproterozoic and Modern Fungal Diagenesis: Analogous Patterns

We have attempted to experimentally demonstrate (detailed experimental procedure *in* Kolo and Claeys 2005; Kolo et al. 2007) that fungal interaction with dolomite crystals produce characteristic colonization patterns, dissolution, pitting, and neomineral formation “nesting” as observed in our Neoproterozoic strata. For the experimental work, thin

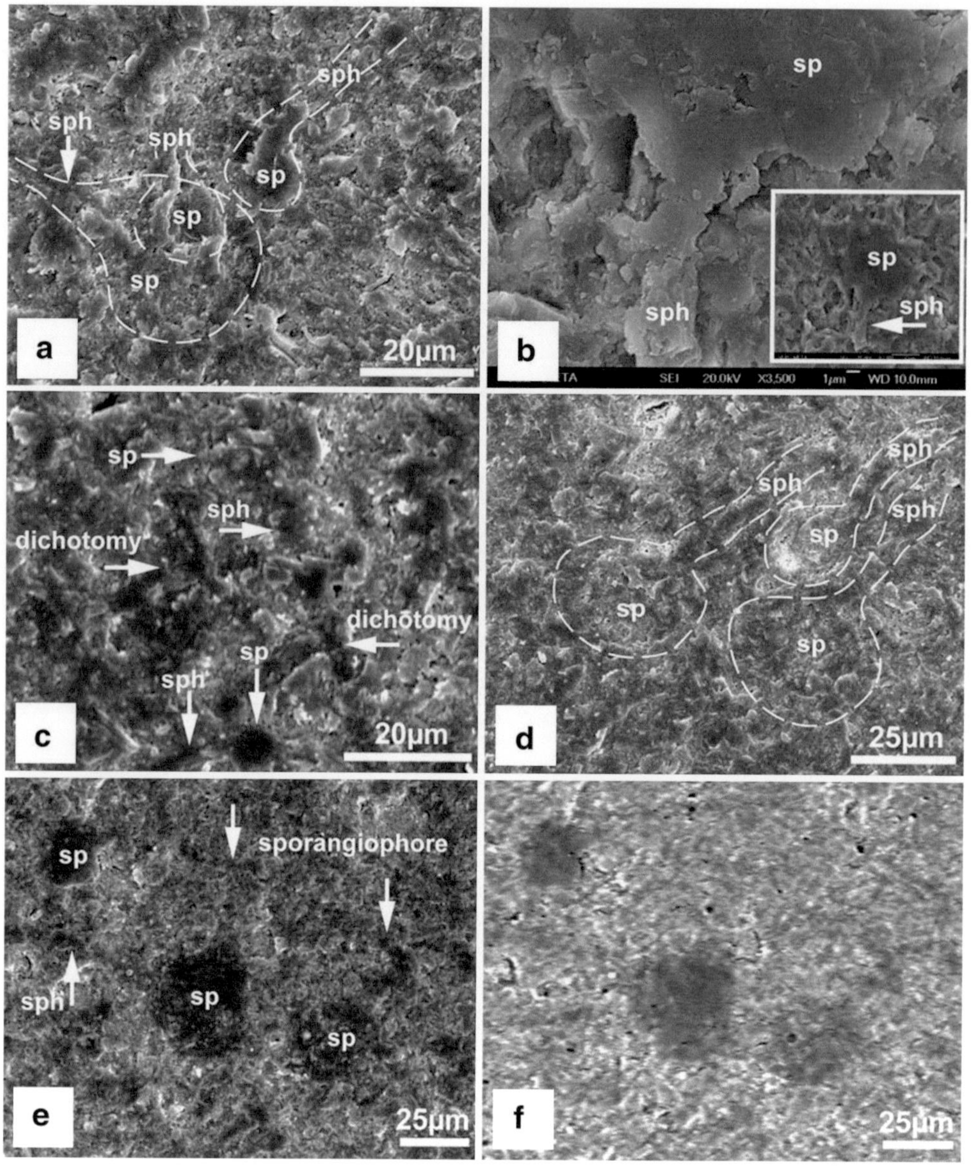

Fig. 5.6 Permineralized embedded relicts and ghost traces of fungal remains in the Neoproterozoic section of Mouila, Gabon depicting typical fungal structures that colonized the substrates. Pictures are from samples MOU26 (Fig. 5.6a, c, d, mou26-15, 26-23, 26-9/2007/ap), MOU 19 (Fig. 5.6b, 19-2/2007/ap) and MOU22 (Fig. 5.6e, f, 22-23, 22-24/2007/ap) (**a**) Per-mineralized (now dolomitic) sporangi(a)um (sp) appear attached to their sporangiophores (sph). In the *lower-left* appear two superposed sporangia. The *dashed lines* contour the visible morphology of these forms. (**b**) A visible per-mineralized single sporangium (sp) attached to its sporangiophore (sph). The inset figure shows the entire contour of this reproduction structure and its complete embedding in the rock matrix. (**c**) Various sporangia and sporangiophores showing blackish color probably related to organic content. In the lower part of the figure is a typical sporangial and sporangiophore shape of fungi. Dichotomy is also visible. (**d**) Here, together with Fig. 5.6a–c, the pits' shapes can be at least partly related to fungal remains, especially the sporangia from where the oval and circular shapes of pits are seemingly inherited. The detailed scrutiny of these figures reveals the rich colonization of these deposits by fungi. (**e**) Normal SEM image showing similar fungal remains compared to secondary backscattered image. (**f**) where the traces of the fungal remains are deeply imprinted in the substrates

sections (4 × 2 × 0.5 cm) and rock slabs from Carboniferous dolomites of the Terwagne Formation (Viséan, Bocahut quarry at Avesnes-sur-Helpe, northern France, in Mamet and Préat 2005) were used as substrates for fungal interaction. All samples were examined at the end of the experimental work by FE-SEM, SEM and EDX.

5.4.1 Honeycomb Dissolution Pattern

The Neoproterozoic thin sections show secondary porosity that is developed after dolomite crystal dissolution. Rhombic and quadratic pore spaces are still discernable in many instances (Fig. 5.8a) and are typically arranged in a

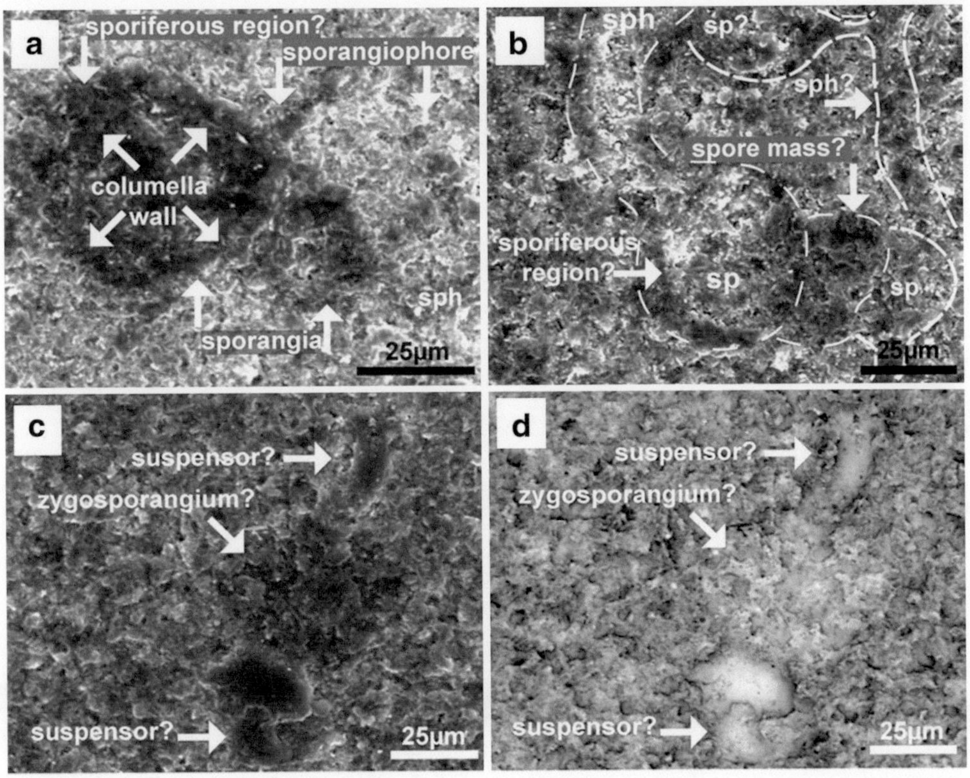

Fig. 5.7 SEM photomicrographs of relicts of fungal structures from Neoproterozoic section of Mouila quarry, Gabon. Sp = sporangi(a)um, sph = sporangiophore. Bar scale as indicated. Images are from samples MOU22 (Fig. 5.7a–d 22-20, 22-16, 22-18, 22-17/2007/ap) (**a**) Fungal remains showing sporangia (*black rounded bodies*), sporangiophores (sph), and a probable sporiferous region. In the larger sporangium are seen two concentric perimeters, the internal one (delineated by *white arrows*) depicts the columella. (**b**) Richly colonized substrate (not all traces shown) where sporangia are showing an external black perimeter depicting a sporiferous region and a probable spore mass attached to it. The traces of the ancient fungi are clearly outlined by the later dissolution diagenesis. (**c**) The visible shapes, compared to modern and ancient fungal analogues (e.g. *in*: Alexopoulos and Mims 1979; Hermann and Podkovyrov 2006), depict zygosporangium and suspensors. (**d**) For further demonstration, here is an artificially enhanced negative image from Fig. 5.7c. The permineralized structures shown in Fig. 5.7c are visibly outlined. Note how well the structures are immersed in the matrix, indicating a syn-sedimentary process

honeycomb pattern. The pore space boundaries contouring the shape of dissolved crystals have a filamentous appearance with fine crystal aggregations (Fig. 5.8a). This arrangement strongly resembles the pattern of interaction of modern fungi with Carboniferous dolomites (Fig. 5.8b). Experimentally, fungal invasion of carbonate substrata have been shown to selectively occur by hyphal penetration along grain boundaries (Sterflinger 2000; Kolo et al. 2007). This stage is followed by active microbial dissolution of the crystals through organic acid generation, creating hollow dolomite crystals or whole rhombic-quadratic and roundish pore spaces with fungal hyphae as boundaries and authigenic mineral deposition (biominerals such as Ca- and Mg-oxalates: weddelite, whewellite and glushinskite). These hollow dolomite crystals, where only boundaries are preserved, are diagenetically different from dolomite crystals with hollow centres and preserved rims (Vahrenkamp and Swart 1994; Feldmann and McKenzie 1997; Jones 2005) that may have precipitated on minute particles or metastable material andwas subsequently dissolved or from bacterially-formed dumbell-shaped hollow-core dolomite crystals (Cavagna et al. 1999).

5.4.2 Intracavity Biomineralization in Natural and Experimentally-Weathered Dolomites

Figure 5.9a shows a circular pit surrounded by a visible elevated mineral "collar" that was also observed in some thin sections of Neoproterozoic strata. Tiny crystals and also per-mineralized filaments litter the interior and exterior of the pit. The mineral "collar" consists of a mixture of per-mineralized filamentous material (probably with EPS material) and attached crystals. Similar diagenetic features (Fig. 5.9b) comprising pits formed by dissolution of

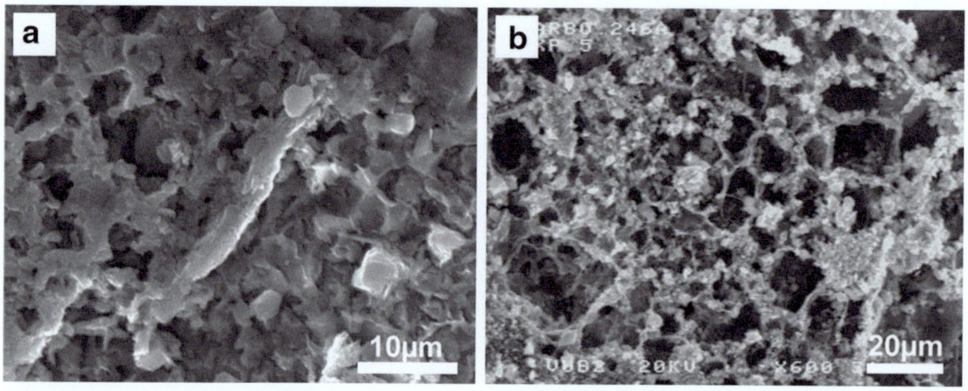

Fig. 5.8 SEM photomicrographs from samples MOU26 (Fig. 5.8a mou26-2/2006/ap) and sample carb246A, showing comparative honeycomb-alveolar structures produced by fungi on dolomitic substrates. Bar scale as indicated. (**a**) An ancient and naturally produced one from the Neoproterozoic of Mouila quarry, Gabon and (**b**) an experimentally produced structure from the Carboniferous of the Bocahut quarry in France. Both figures share common features, such as quadratic-rhombic pore space after the dissolution of dolomite crystals and the formation of a filamentous-EPS mat cover on old crystal boundaries associated with new crystal deposition on those boundaries, suggesting a similar colonization pattern and diagenetic process. In Fig. 5.8a the original fungal material (filaments and EPS) are all permineralized but the old honeycomb-alveolar structure is still well preserved

dolomite crystals, an elevated mineral "collar" composed of mineral authigenesis (here Ca-oxalates—weddellite and whewellite), fungal hyphae and EPS material lining the pits were experimentally produced by fungal interaction with a dolomite substrate, indicating these diagenetic features are a characteristic of fungal interaction with carbonate substrata and generally also a part of the honeycomb structures demonstrated above. Figures 5.9c, d reveal more details on the above-mentioned similarities. In both cases, the filaments are occasionally encrusted with fine crystals (Fig. 5.9c) and usually form a lining on the pit wall. Fungal hyphae are known to form envelopes of Ca-oxalates crystals that partially or completely engulf the hyphae especially under Ca-rich conditions (Gadd 1999; Verrecchia 2000; Kolo and Claeys 2005). This crystal adherence to fungal hyphae is also visible in both cases.

"Nesting" is a term we use here to describe the deposition of fungally produced biominerals, mainly Ca–Mg-oxalates or even calcite, in the partially to totally dissolved crystals of dolomite substrata (Fig. 5.9e, f) following fungal colonization and growth. In our experiments, these biominerals are much finer (1–3 μm) than the original substrate crystal size and typically are represented by various forms of Ca-oxalates: prismatic, tetragonal bipyramidal, and rhombic (Fig. 5.9f). Such fine crystals are also observed in the Neoproterozoic pits (Fig. 5.9e). The formation of these metal-oxalates is largely attributed to the reaction of oxalic acid, excreted by fungi, and the high availability of Ca^{2+} and Mg^{2+} in the growth environment (Gadd 1999).

5.5 Discussion

5.5.1 Neoproterozoic-Aged Colonization and Weathering

The Neoproterozoic carbonate stromatolites of the Mouila series were originally composed of a magnesian micritic mud colonized by benthic cyanobacterial mats in a shallow tidal depositional system. Then, through a combination of microbially-induced biomineralization of fine- to medium-grained dolomicrospar, coupled with copious EPS excretion, the original muddy sediment was transformed into domal stromatolites, similar to those occurring today in evaporitic carbonate sabkha-like environments (Walter 1976; Grotzinger and Knoll 1999). These microbialites seemed to have resisted erosion as evident by the preservation of the overall morphology and internal features of the stromatolites (see below). As suggested by the typical shallowing-upward sequences, this semi-lithified to progressively well-lithified sediment experienced periodic or episodic desiccation (as revealed by a 'polygonal pattern of cracks' of the mud in thin sections and the EPS under the SEM) coupled with evaporitic salty brine invasion leading to gypsum and other evaporitic minerals being interstratified within the mats. Cyanobacteria were progressively destroyed and cementation was the dominant process. During this period of subaerial conditions, fungi were able to colonize the substrate and drive carbonate diagenesis. The most striking result of their activity was the formation of the circular-oval-shaped pits in

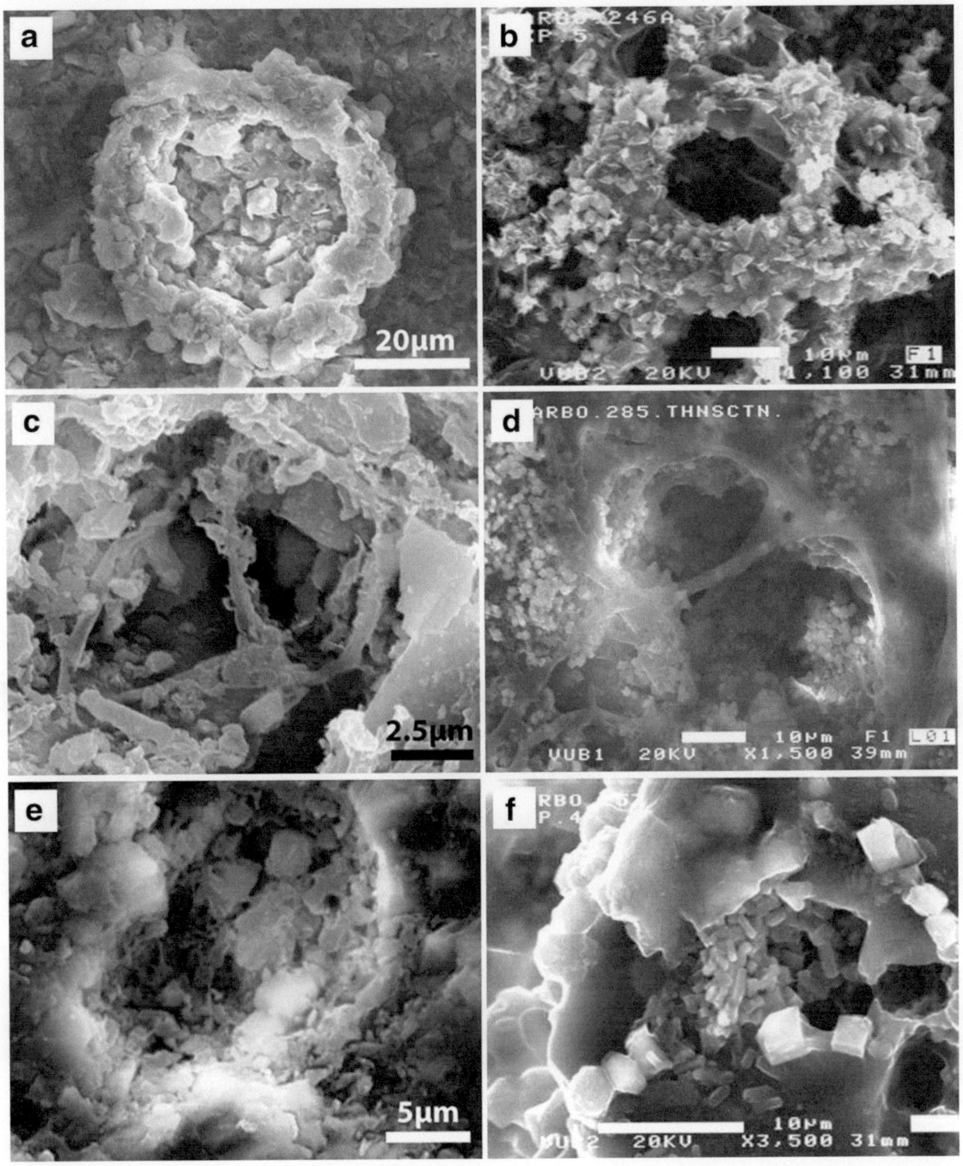

Fig. 5.9 Images showing comparative patterns of fungally produced mineral deposition "nesting" and colonization of pits as fossil and permineralized fungal relicts in the Neoproterozoic of Mouila section, Gabon (Fig. 5.9a,c,e) and of pits from *in vitro* experiments (Fig. 5.9b,d, f) on Carboniferous dolomite of the Bocahut quarry in France. Pictures are from samples MOU22 (Figure a, mou22-22/2006/ap), MOU 19 (Fig. 5.9c,e, 19-17, 19-38/2006/ap) and samples carb246A, 285 and 257 (Fig. 5.9a,b). Elevated mineral "collar" formation surrounding a pit. In (**a**) the pit appears filled with mineral crystals (dolomite) and the mineral collar is a mixture of crystals and per-mineralized filaments. In (**b**) the deep pit reveals similar mixture of neominerals (here Ca- and Mg-oxalates) and fungal hyphae. (Fig. 5.9c, d) Colonization pattern of formed pits by fungi appears similar especially the lining of inner walls of the pits. Note how the filaments' surface in (**c**) is rough and show many blebs and attachments. These are crystal aggregates adhering to their surface, a typical fungal phenomenon as is also shown in Fig. 5.9a. In (**d**) the colonizing fungi have already produced a large quantity of crystals inside and outside the pit (**e**, **f**) Showing further the "nesting" of minerals by fungal interaction with the substrates. In Fig. 5.9e fine crystals and filaments are littering the pit's bottom as well as the walls, while in Fig. 5.9f a typical pit made in a dolomite crystal is filled by fungally bio-mineralized prismatic and bi-pyramidal crystals of the mineral weddellite

the stromatolitic levels. The fungal relicts are well-embedded in the rock matrix and show homogenous early diagenetic character which indicates their penecontemporaneous nature with lithification, i.e., the Neoproterozoic.

There are several lines of evidence that support our contention that the pits formed while the sediment was already lithified and that this rock constituted a good substrate for fungal colonization:

1. The pits are hosted in a substrate formed from uncompacted sediment. The grains (flat pebbles, microbreccia, lumps, and aggregates) and the fenestral cavities are undeformed, and they do not exhibit any

interpenetration, because the pseudomorphs of sulfates or the microenterolithic levels have not collapsed. These replacements were probably formed nondisplacively in an enclosed volume of lithifying muddy sediment. Dolomite and also fine- to coarse-grained silica have completely replaced these original voids, thus maintaining their original shapes without deformation. The process was probably rapid, as no mechanical compaction and fracturing are observed. This early replacement-cementation (dolomicrospar) prevented late fracturation due to overpressuring in response to burial.

2. The mudcracks and sheet-cracks are well preserved, undeformed and filled with a dolomicrospar (and sometimes silica) having the same size (generally very-fine to fine-grained) as the dolomicrospar that replaced the primary carbonate mud. Boundaries separating peloid and muddy laminae are quite sharp and of constant thickness inside a particular laminar structure. No interpenetration of the different laminae is observed. Detailed fabric preservation of primary cracks suggests that dolomitization and silicification occurred early in the diagenetic history; they do not cross the cracks.
3. Evaporite facies display micro-slumped or contortion structures without any signs of compaction. The tiny contorted levels are folded, keeping their uniform original thickness (<200 μm for the thinnest).
4. The pits and fungi are confined to the same stratigraphic levels in a stromatolite horizon. In the field, the laminar structure consists of irregular bands and lenses of dark and light carbonate mudstone. The bands constitute sets with uniform thickness.
5. The spheres attributed to spores are dolomitized at a nanoscale level and embedded in the dolomicrospartic matrix.
6. The pits are invaded by thin hyphae associated at a very small scale (<100 μm) with former quadratic crystals which were probably primary oxalates (identical to our experiments). Those oxalates were then dolomitized through either the cycling of dolo-microspar or simply via interaction with near-coeval seawater or seawater-derived fluids.
7. Excellent fabric retention of highly-soluble evaporate phases (rosettes, swallow-tails, laths, microenterolithes, and nodules) during dolomitization indicate that dolomitization had occurred under hypersaline conditions. These conditions are also suggested by ^{18}O enrichment of the facies constituting the upper part of the shallowing-upward salinity sequences where the fungi developed. Oxygen isotopes ($\delta^{18}O$) (Préat et al. 2010), have values ranging from −5.1 to −1.3 ‰, recording stabilization in normal Neoproterozoic marine water (Veizer et al. 1992). Lighter values indicate continued evaporation and upwards percolation of underlying pore waters, ruling out an increase in temperature (during burial) and an early or late influence of meteoric fluids. In this context, late-stage diagenetic alteration resulting from large-scale convection of marine, meteoric or hydrothermal waters during burial can be dismissed. The dolomicrospar is therefore neomorphic and replaced the original fine-grained (dolo)micrite without a dissolution phase.

Collectively, these points lead one to infer that the primary carbonate muds were rapidly lithified by dolomitization associated with evaporitive marine or coeval marine waters. Under such conditions, fungi were able to inhabit this stressful environment and subsequently played an important role in the pit formation in specific or particular interstratified levels (here a stromatolite layer).

5.5.2 Fungal Colonization and the Pit Formation Hypothesis

The different depth levels of these the fungally-generated solution pits suggest a progressive process of pit formation through incipient, moderate and advanced pitting stages (Fig. 5.10a–d), possibly caused by different stages of microbial colonization-diagenesis. The incipient pit stage (Fig. 5.10a, b) has a shallow quasi-circular/oval form, visibly corresponding to a bioweathered, fragmented, micropitted, and decolorized original mineral surface compared to the surroundings where some fine-grained authigenic minerals had started to precipitate. Interestingly, some incipient pits visibly show fungal form morphology (Fig. 5.10a) that suggests a sporangium and sporangiophore relicts. This morphological resemblance between fungal parts and pits' leads us to assume a cause and effect process.

In moderately developed pits, colonization by fungal forms and mineral authigenesis can already be observed (Fig. 5.10c) relating the two processes and suggesting an interaction of fungally induced biochemical and biomechanical factors with the mineral surface. The circular/oval shape of the pits suggests therefore an inherited form after the fungal parts (e.g., sporangia, as depicted by Fig. 5.10a) or by selective fungal attack on certain sites of weakness on the mineral surface (e.g., Fig. 5.10c, grain boundaries giving rise to irregular pit shapes). Selective fungal attack that produced alveolar-type mineral structure and pitting has already been shown to occur with dolomite and limestone (Kolo et al. 2007).

At the advanced stage, the pits have clear 3-D forms, display visible inner walls, depth and variable diameters (Fig. 5.10d). Inside the pits, a dense network of colonizing fungal hyphae are interwoven with the pit's inner walls as well as the bottom which was coated with EPS (Fig. 5.10d).

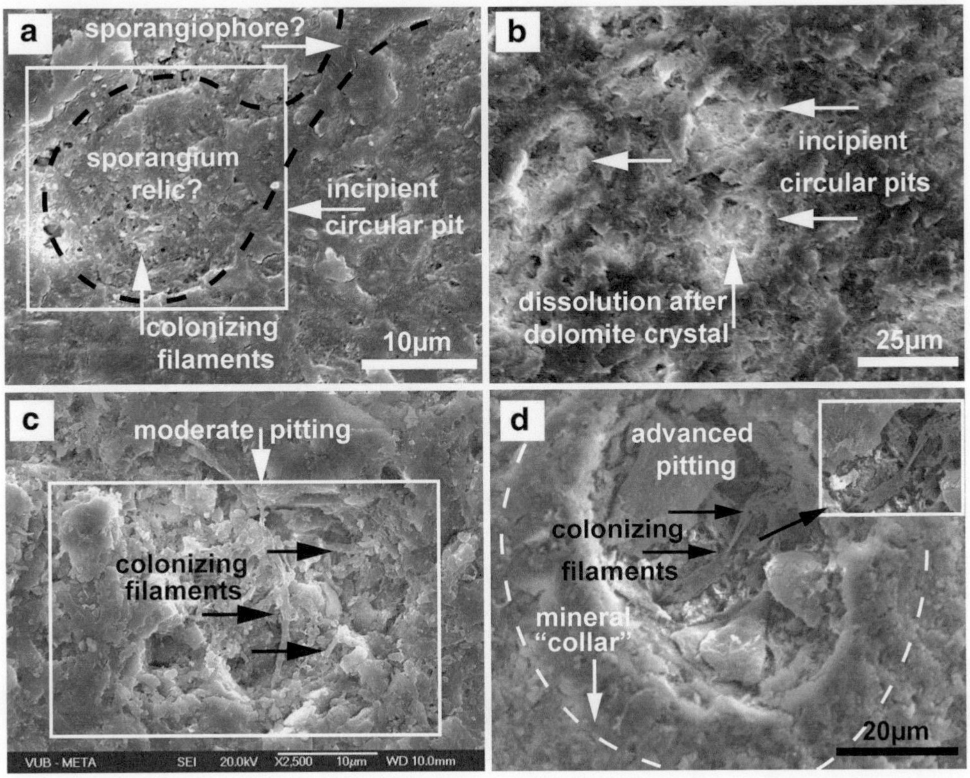

Fig. 5.10 Images showing three stages of pit formation in the Neoproterozoic section of Mouila quarry, Gabon. SEM and FE-SEM photomicrographs (from thin sections). Bar scale as indicated. *Black* and *white strokes* are for contrast only. Pictures are from samples MOU26 (Fig. 5.10a, mou26-14/2007/ap) and MOU 19 (Fig. 5.10b,c, d, mou19-7, 19-14, 19-49/2007/ap). (**a**) Incipient pitting with quasi circular shape displaying many small pits (*black spots* <1µm) and fragmented surface lower than the surrounding. The surface is bioweathered. In the centre of the pit, very small colonizing filaments (~1 µm) can be observed. The whole pit displays the form of a fungal sporangium attached to its sporangiophore. (**b**) Shows similar several incipient pits. The *lighter colored* areas are characteristic of pit formation. Interestingly, is also visible the slightly elevated mineral "collar" of the pits. (**c**) Moderately formed pit, with shallow depth but visibly deeper than the incipient ones. In this stage, colonization by microbial filaments (fungal?) is important and associated with small neomineral formation (*white crystal* aggregates in the centre). The colonizing filaments occupy the centre of the pit as well as the border of pit wall. (**d**) The advanced pitting stage creates well-developed pits with well-defined contours, formation of elevated mineral "collar" around the pit as well as mineral deposition inside the pit, and rich colonization by microbial filaments associated sometimes with polysaccharides film

Sometimes, these fungal hyphae form a 'lining' to the inner walls. Authigenic minerals related to fungal activity litter the pits and are visibly attached to fungal material as well. The latter minerals are distinguishable by their light color and inherited crystal forms. Originally, the minerals, now dolomitic, were probably Ca- or Mg-oxalates (weddellite, whewellite or glushinskite). The spatial distribution of these mineral forms is visibly restricted to encrusting the fungal hyphae at the inside of the pit and to the "collar" area surrounding the pit. This limited distribution is consistent with fungal activity. Furthermore, the colonization of the pits by fungi displays a complex pattern (Fig. 5.10d and inset) that highlights the invasive behavior of fungal hyphae into the pit walls and the mineral matrix that resulted in micro-fragmentation of the pit's wall through both mechanical dislocation and chemical dissolution in addition to mineral precipitation.

The mechanisms underpining the three-stages of pit formation are schematically represented by two scenarios (Fig. 5.11) that could have worked separately or in combination. The first mechanistic scenario involves fungal colonization of the Neoproterozoic substrate through hyphae-stolons and rhizoids invasion of the semi-lithified sediment surface, penetrating along grain boundaries. This process would have resulted in grain dissolution by fungal organic acids exudates (mainly oxalic), increasing the pore space and, with continuing colonization, development of pits with authgenic minerals precipitating on their external perimeter and within their inner walls. These pits themselves become the target of new invasion by exploratory fungal hyphae. This scenario is shown to occur in experimental studies (Kolo et al. 2007). In a calcium-magnesium rich environment, the formation of oxalate biominerals (e.g. weddelite whewellite and glushinskite) would have been

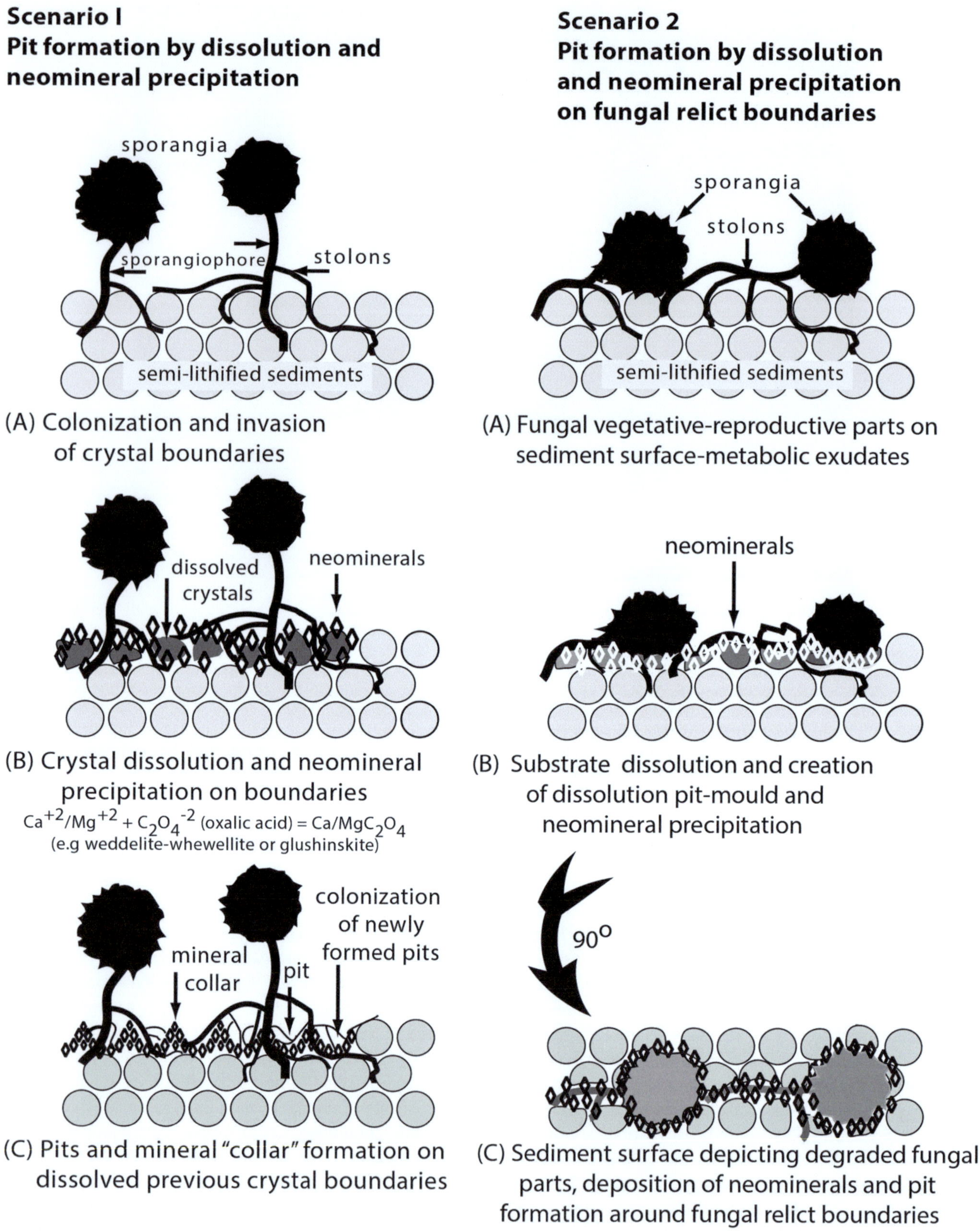

Fig. 5.11 Schematic figure showing a proposed two-pronged mechanism, for the three-stage pit formation, which could have worked separately or in combination. Scenario (1) involves fungal colonization of the Neoproterozoic substrate through hyphae-stolons and rhizoids invasion of the semi-lithified sediment surface penetrating it through grain boundaries resulting into grain dissolution, biominerals precipitating (mainly Ca/Mg-oxalates) and pits formation. The pits themselves become the target of new invasion by fungal hyphae. Scenario (2) envisages fungal vegetative parts (e.g sporangia, sporangiophores) spreading across the colonized substrate surface and with metabolic exudates released in the growth environment causing substrate dissolution, biomineral precipitation and molding the fungal parts within the substrate

Fig. 5.12 (**a**, **b**) Images of flat laminated mats associated with stacked microstromatolitic laminae characterized by partially microsparitized thin organic-rich layers giving stratiform light layers. Thin section comes from a domal stromatolite (height 50 cm, wide 65 cm)

significant. The second scenario implies that fungal vegetative and reproductive parts (e.g., sporangia, sporangiophores) could spread across the colonized substrate surface and metabolic exudates that were released in the growth environment would have caused substrate dissolution, authigenic mineral precipitation and cementation of fungal parts within the substrate.

5.6 Implications

It is our contention that we have described here one of the earliest physical records of fungi, and that these organisms having inhabited the upper supratidal part of a shallowing-upward carbonate sequence. We also show how the fungi impacted the main petrophysical characteristics of the rock. Despite this importance and stratigraphic distribution, fungi are rarely reported in ancient series in the literature. This is particularly the case in the Precambrian of West Africa, where numerous stromatolites have been described in great details (Amard and Bertrand-Sarfati 1997). Clusters of closely packed meter-scale ellipsoid to upward expanding cone-shaped bioherms several meters (up to 5 m wide and 3 m thick, Fig. 5.12e–h) developed relief of several meters above the top of laminar microbial bindstone and small-sized LLH stromatolites associated with collapse breccia containing anhydrite relicts (Fig. 5.12). The biohermal level is 15 m-thick and belongs to the post-Marinoan Neoproterozoic SCIc unit (Schisto-Calcaire Group) recognized in the Niari Basin (the Republic of Congo) by Alvarez and Maurin (1991). The Niari Basin extends over more than 75,000 km^2 and is mainly constituted by two depressions, the Niari depression in the Republic of Congo and Nyanga depression in Gabon where the Mouila quarry is located. Study of the microbial contents (cyanobacteria and fungi) of the SCIc stromatolites is in progress (Yannick Callec and Alain Préat) and is focused on the lamina microstructure forming irregular bands and lenses of grey and light carbonate mud. They are associated with early diagenesis related to replacement by evaporitive brines (Fig. 5.12a–d).

The Mouila sediments were probably partly or totally lithified during early diagenesis through pervasive dolomitization in hypersaline brines allowing pits to be formed. In this very shallow environment (backslope of the levees and beach-ridge washovers) exposure was probably high, with very dry conditions proving favourable to fungal colonization. Numerous mudcracks seen in thin sections or under the SEM support this interpretation. Cyanobacteria are partly or entirely destroyed by the dolomicropar. This contrasts with the fungal hyphae which are reasonably well-preserved and intimately associated with the dolomicrospar and the dolomitized EPS that constituted an integral component of the original microbial mats.

Evidence for ancient life typically exists within sedimentary environments, where microbial mats and colonies of filamentous, coccoid or rod-shaped microbes have been found in Early Archean strata such as in cherts of the Pilbara and Barberton greenstone belts (Westall 2005). As fungi are increasingly pushed deeper into the Precambrian, their role in early Earth processes is also increasingly linked to two major events: the "Snowball Earth" and the rise of oxygenation in the Neoproterozoic (Heckman et al. 2001; Canfield 2005; Kennedy et al. 2006). How fungi may have impacted terrestrial weathering, and to what effect this may have played a role in the broader evolution of the Earth system remains unclear. We feel that this work takes a step further into the deep past by describing how fungal relicts within the Neoproterozoic Mouila series points towards their colonization and diagenesis of shallow sediments at the time.

Acknowledgements Kamal Kolo would like to thank Prof. Philippe Claeys, Department of Geology/Vrije Universiteit Brussels for supporting the experimental work in this study. We also thank the Department of Metallurgy/Vrije Universiteit Brussels for kindly giving access to their SEM and FE-SEM laboratories. The fieldwork was done under the terms of the SYSMIN program (Eighth Fonds Européen de Développement, BRGM-CGS-SANDER-MRAC). KOK would like to thank the Natural Sciences and Engineering Research Council of Canada for continued support. The authors thank Dr Yannick Callec, BRGM (Bureau Recherches Géologiques et Minières, Orléans, France) for guiding Alain Préat on the field in the Niari area during dry season (September 2012). We thank Prof. David Gillan for a comprehensive review which helped improve the MS.

Fig. 5.12 (continued) interstratified in strongly deformed, slumped evaporitic laminated dolomudstones (pictures **c** and **d**). Sample CB9, outcrop MAD8122-Yannick Callec, Republic of Congo, photo cb9252 and 9253/ap/2013). (**c**, **d**). Salt migration (slump, microenterolithe, folding -c, tepee -d) in a dolo-microsparitized mudstone with remnants of organic-rich microbial laminae. Same ouctrop as previous pictures, sample CB10 (50 cm above CB9), photo cb9258 and 9269/ap/2013). (**e**) Flat to slightly domal stromatolites switched between irregularly-laminated microbial dolomudstones. Outcrop MAD0165-Yannick Callec, Republic of Congo, photo P1170717/ap/2012). (**f**) Massive stromatolitic 'table' reef (height 2 m) bordered by recent tufa deposits, (**g**) Concentric sheet stromatolitic bioherm, (**h**) Stacked patch reef units flanked by intraclastic (angular stromatolitic chips) dolopackstones on both sides.(f–g–h) : same ouctcrop as (**e**), respectively photos P1170726/P1170734/ P1170744ap/2012). The stromatolites constitute a 15 m-thick level interstratified in well-bedded dolomudstones and ooid-pisoid dolopackstones and dolograinstones

References

Alexopoulos CJ, Mims CW (1979) Introductory mycology, 3rd edn. Wiley, New York

Alvarez P, Maurin JC (1991) Evolution sédimentaire et tectonique du bassin protérozoïque supérieur de Comba (Congo): stratigraphie séquentielle du Supergroupe Ouest-Congolien et modèle d'amortissement sur décrochements dans le contexte de la tectonogenèse panafricaine. Precambrian Res 50:137–171

Amard B, Bertrand-Sarfati J (1997) Microfossils in 2000 Ma old cherty stromatolites of the Franceville Group, Gabon. Precambrian Res 81:197–221

Bennett PC, Melcer ME, Siegel DI, Hassett JP (1988) The dissolution of quartz in dilute aqueous solutions of organic acids at 25°C. Geochim Cosmochim Acta 52:1521–1530

Berbee ML, Taylor JW (2001) Fungal molecular evolution: gene trees and geological time. In: McLaughin DJ, McLaughin EG, Lemke PA (eds) The Mycota VIIB systematic and evolution. Springer, Berlin, pp 229–245

Burford EP, Fomina M, Gadd GM (2003a) Fungal involvement in bioweathering and biotransformation of rocks and minerals. Miner Magaz 67:1127–1155

Burford EP, Kierans M, Gadd GM (2003b) Geomycology: fungi in mineral substrata. Mycologist 17:98–107

Butterfield NJ (2005) Probable proterozoic fungi. Paleobiology 31 (1):165–182

Canfield D (2005) The early history of atmospheric oxygen: Hommage to Robert A Garrels. Annu Rev Earth Planet Sci 33:1–36

Cardon Z, Whitbeck JL (eds) (2007) The rhizosphere. an ecological perspective. Elsevier Academic Press, New York, 212 pp

Cavagna S, Clari P, Martire L (1999) The role of bacteria in the formation of cold seep carbonates: geological evidence from Monferrato (Tertiary, NW Italy). Sediment Geol 126:253–270

Cavalier-Smith T (1987) The origin of fungi and pseudofungi. In: Rayner AM, Brasier CM, Moor D (eds) Evolutionary biology of the fungi. Cambridge University Press, Cambridge, pp 339–353

Cavalier-Smith T (2006) Cell evolution and Earth history: stasis and revolution. Philos Trans R Soc B: Biol Sci 361(1470):969–1006

Chen J, Blume HP, Beyer L (2000) Weathering of rocks induced by lichen colonization, a review. Catena 39(2):121–146

Clark DN (1980) The diagenesis of Zechstein carbonate sediments. In: Füchtbauer H, Peryt T (eds) The Zechstein Basin with emphasis on carbonate sequences. Stuttgart, E, Schweizerbart'sche Verlagsbuchandlung, Contribution to sedimentology, vol 9, pp 167–203

Esteban M, Klappa CF (1983) Subaerial exposure environment. In: Scholle P, Bebout DG, Moore CH (eds) Carbonate depositional environments, vol 33, American Association Petroleum Geolologists Memoir., pp 2–54

Feldmann M, McKenzie JA (1997) Messinian stromatolite-thrombolite associations, Santa Pola, SE Spain: an analogue for the Palaeozoic? Sedimentology 44:893–914

Gadd GM (1999) Fungal production of citric and oxalic acid: importance in metal speciation, physiology and biogeochemical processes. Adv Microb Physiol 41:47–92

Gadd GM (2007) Geomycology: biogeochemical transformations of rocks, minerals, metals and radionuclides by fungi, bioweathering and bioremediation. Mycol Res 111(1):3–49

Gérard G (1958) Carte géologique de l'Afrique Equatoriale Française au 2 000 000 avec notice explicative. Brazza. Direction Mines et Géologie. Afrique Equatoriale Française, 198 pp, 4 feuilles

Gerdes G, Krumbein W, Noffke N (2000) Evaporite microbial sediments. In: Riding R, Awramik S (eds) Microbial sediments. Springer, Berlin, pp 196–208

Golubic S, Gudrun R, Le Campion-Alsumard T (2005) Endolithic fungi in marine ecosystems. Trends Microbiol 13(5):229–235

Grote G, Krumbein WE (1992) Microbial precipitation of manganese by bacteria and fungi from desert rock and rock varnish. Geomicrobiol J 10:49–57

Grotzinger JP, Knoll AH (1999) Stromatolites in Precambrian carbonates: evolutionary mileposts and environmental dipsticks? Ann Rev Earth Planet Sci 27:313–358

Hardie LA (ed) (1977) Sedimentation on the modern carbonate tidal flats of Northwest Andros Island, Bahamas, vol 22. John Hopkins University Studies in Geology, Baltimore, 202 pp

Hardie LA, Ginsburg RN (1977) Layering: the origin and environmental significance of lamination and thin bedding. In: Hardie LA (ed) Sedimentation on the modern carbonate tidal flats of Northwest Andros Island, Bahamas, vol 22. John Hopkins University Studies in Geology, Baltimore, pp 50–12

Heckman DS, Geiser DM, Brooke RE, Rebecca L, Stauffer NL, Kardos S, Blair H (2001) Molecular evidence for the early colonization of land by fungi and plants. Science 293(5532):1129–1133

Hermann T, Podkovyrov V (2006) Fungal remains from the Late Riphean. Paleontolog J 40:207–214

Hoffland ET, Kuyper W, Wallander H, Plassard C, Gorbushina AA, Haselwandter K, Holmström S, Landeweert R, Lundström US, Rosling A, Sen R, Smits MM, van Hees PAW, van Breemen N (2004) The role of fungi in weathering. Front Ecol Environ 2(5):258–264

Jones B (2005) Dolomite crystal architecture: genetic implications for the origin of the tertiary Dolostones of the Cayman Islands. J Sediment Res 75:177–189

Kahle CF (1977) Origin of subaerial Holocene calcareous crusts: role of algae, fungi and sparmicritization. Sedimentology 24:413–435

Kendrick B (2000) The fifth kingdom, 3rd edn. Focus Publishing, R. Pullins Company, Newburyport, MA

Kennedy M, Droser M, Mayer LM, Pevear D, Mrofka D (2006) Late precambrian oxygenation; inception of the clay mineral factory. Science 311(5766):1446–1449

Kolo K, Claeys P (2005) In vitro formation of Ca-oxalates and the mineral glushinskite by fungal interaction with carbonate substrates and seawater. Biogeosciences 2:277–293

Kolo K, Keppens E, Préat A, Claeys P (2007) Experimental observations on fungal diagenesis of carbonate substrates. J Geophys Res Biogeosci 112, GO1007

Konhauser K (2007) Introduction to geomicrobiology. Blackwell Publishing, Malden, MA, 425 pp

Krings M, Taylor TN, Dotzler N (2013) Fossil evidence of the zygomycetous fungi. Persoonia 30:1–10

Krumbein WE (1972) Rôle des microorganismes dans la genèse, la diagenèse et la dégradation des roches en place. Revue Ecol Biol Sol 9:283–319

Krumbein WE, Jens K (1981) Biogenic rock varnishes of the Negev Desert (Israel): an ecological study of iron and manganese transformation by cyanobacteria and fungi. Oecologia 50:25–38

Logan BW, Rezak R, Ginsburg RN (1964) Classification and environmental significance of algal stromatolites. J Geol 72:68–83

Mamet B, Préat A (2005) Sédimentologie de la série viséenne d'Avesnes-sur-Helpe (Avesnois, Nord de la France). Geol Belgica 8:91–107

Moore-Landecker E (1991) Fundamentals of the fungi, 3rd edn. Prentice Hall, Englewood Cliffs, New Jersey, USA

Préat A, Kolo K, Mamet B, Gorbushina AA, Gillan DC (2003) Fossil and subrecent fungal communities in three calcrete series from the Devonian of the Canadian Rocky Mountains, Carboniferous of Northern France and Cretaceous of Central Italy. In: Krumbein WE, Paterson DM, Zavarzin GA (eds) Fossil and recent biofilm. A natural history of life on Earth. Kluwer Acad. Publ., pp 291–306

Préat A, Kolo K, Delpomdor F (2010) A peritidal evaporate environment in the Neoproterozoic of South Gabon (Schisto-Calcaire Subgroup). Precambrian Res 177:253–265

Préat A, Prian JP, Thiéblemont D, Obame RM, Delpomdor F (2011a) Stable isotopes of oxygen and carbon compositions in the Neoproterozoic of South Gabon (Schisto-Calcaire Subgroup, Nyanga Basin): are cap carbonates and lithoherms recording a particular destabilization event after the Marinoan glaciation? J Afr Earth Sci 60:273–287

Préat A, Delpomdor F, Kolo K, Gillan D, Prian JP (2011b) Stromatolites and cyanobacterial mats in peritidal evaporitive environments in the Neoproterozoic of Bas-Congo (Democratic Republic of Congo) and South Gabon. In: Seckbach J, Tewari VC (eds) Stromatolites: interaction of microbes with sediments, Series: cellular origin, life in extreme habitats and astrobiology. Springer, Heidelberg, pp 43–63. doi:10.1007/978-94-007-0397

Purser BH (ed) (1973) The Persian Gulf. Holocene carbonate sedimentation and diagenesis in a shallow epicontinental Sea. Springer, Berlin, Heidelberg, New York, 471 pp

Redecker D, Kodner R, Graham LE (2000) Glomalean fungi from the Ordovician. Science 289(5486):1920–1921

Richter DD, Oh NH, Fimmen R, Jason J (2007) The rhizosphere and soil formation. In: Cardon Z, Whitbeck JL (eds) The rhizosphere. An ecological perspective. Elsevier Academic Press, pp 179–200

Saylor BZ, Kaufman AJ, Grotzinger JP, Urban F (1998) A composite reference section for terminal Proterozoic strata of Southern Namibia. J Sedim Res 68(6):1223–1235

Sellwood BW (1986) Shallow-marine carbonate environments. In: Reading HG (ed) Sedimentary environments and facies. Blackwell Science Publ, Oxford, pp 283–234

Sterflinger K (2000) Fungi as geologic agents. Geomicrobiol J 17:97–124

Sterfinger K, Krumbein WE (1997) Dematiaceous fungi as a major agent for biopitting on Mediterranean marbles and limestones. Geomicrobiol J 14:219–230

Stubblefield SP, Taylor TN (1988) Recent advances in palaeomycology. New Phytol 108:3–25

Taylor TN, Hass H, Kerp H, Krings M, Hanlin RT (2005) Perithecial ascomycetes from the 400 million year old Rhynie chert: an example of ancestral polymorphism. Mycologia 97:269–285

Thiéblemont D, Castaing C, Billa M, Bouton P, Préat A (2009) Notice explicative de la Carte géologique et des Ressources minérales de la République gabonaise à 1/1 000 000, Editions DGM, Ministère des Mines et du Pétrole, des Hydrocarbures, Libreville, 381 pp

Vahrenkamp VC, Swart PK (1994) Late Cenozoic sea-water generated dolomites of the Bahamas: metastable analogues for the genesis of ancient platform dolomites, In: Purser BH, Tucker ME, Zenger DH (eds) Dolomites, a volume in honour of Dolomie, vol 21, Special Publication. International Association of Sedimentologists, pp 133–153

Verrecchia EP (2000) Fungi and Sediments. In: Riding R, Awramik S (eds) Microbial sediments. Springer, Berlin, pp 68–75

Verrecchia EP, Loisy C, Braissant O, Gorbushina AA (2003) The role of fungal biofilm and networks in the terrestrial calcium carbonate cycle. In: Krumbein WE, Paterson DM, Zavarzin GA (eds) Fossil and recent biofilm. A natural history of life on Earth. Kluwer Ac. Publ., pp 363–369

Veizer J, Plumb KA, Clayton RN, Hinton RW, Grotzinger JP (1992) Geochemistry of Precambrian carbonates: V. Late Paleoproterozoic seawater. Geochim Cosmochim Acta 56:2487–2501

Vogel K, Gektidis M, Golubic S, Kiene WE, Radtke G (2000) Experimental studies on microbial bioerosion at Lee Stocking Island, Bahamas and One Tree Island, Great Barrier Reef, Australia: implications for paleoecological reconstructions. Lethaia 33:190–204

Walter MR (ed) (1976) Stromatolites. Developments in sedimentology, vol 20. Elsevier, Amsterdam, Oxford, New York, 790 pp

Welch SA, Barker WW, Banfield JF (1999) Microbial extracellular polymers and plagioclase dissolution. Geochim Cosmochim Acta 57:2939–2948

Westall F (2005) Early life on Earth: the ancient fossil record, astrobiology: future perspectives. pp 287–316

Wright VP (1986) The role of fungal biomineralization in the formation of Early Carboniferous soil fabrics. Sedimentology 33:831–838

Yuan X, Xiao S, Taylor TN (2006) Lichen-like symbiosis 600 million years ago. Science 308(5724):1017–1020

Neoproterozoic to Lower Paleozoic Sequences of the Congo Shield: Comparisons Between the Congo and Its Peripheral Basins

6

Étienne Kadima Kabongo, Damien Delvaux, Michel Everaerts, Mwene Ntabwoba Stanislas Sebagenzi, and Francis Lucazeau

6.1 Introduction

The Congo Basin (CB) is the largest and the least explored African continental sedimentary basin. Outcrop scarcity, intense weathering and dense forest constitute obstacles to undertake direct geologically-based exploration programmes, highlighting the importance of airborne and ground-based geophysics. In the 1950s, the 'Syndicat pour l'étude géologique et minière de la Cuvette congolaise' conducted refraction and reflection seismic, gravimetric and magnetic field surveys across the central parts of the CB (Evrard 1957). During the seismic surveys, 117 refraction stations were employed in the basin interior, and 7 reflection lines were recorded along rivers (Evrard 1960). In addition, about 6,000 gravity and magnetic measurements were made with an average station spacing of 5 km along rivers and dirt roads (Jones et al. 1960). Moreover, two fully-cored stratigraphic wells down to ca. 2,000 m had been drilled in the 1950s, reaching red sandstones at the bottom, that were interpreted as pre-Carboniferous (Cahen et al. 1959, 1960).

Using these geophysical and well data, Evrard (1957, 1960) and Jones et al. (1960) first imaged the structure and extent of the CB and its subsurface geometry. The major refractors mapped provided the first regional overview of the basin subsurface structure. The CB appeared deeper than previously thought, with the possible existence of several thousand meters of sedimentary rocks beneath its upper Paleozoic–Mesozoic sedimentary sequences.

Additional geophysical surveys were subsequently conducted by various petroleum companies. Exxon-Texaco shot 2,900 km of seismic reflection profiles in 1974–1976, and Japan Oil National Corporation (JNOC) performed an aeromagnetic survey and gravity measurements around Kisangani in 1984; and two additional wells, more than 4,000 m deep, were drilled (Esso Zaire 1981a, b). The interpretation of this data was first presented in an ECL (1988) report by Petrozaïre, of which a short summary was published by Lawrence and Makazu (1988). More detailed interpretations of the seismic reflection profiles and wells were presented as synthetic stratigraphic columns and a tectonic model by Daly et al. (1992), in which early subsidence of the basin was interpreted to reflect Neoproterozoic processes of rifting and thermal relaxation, followed by two regional contractional deformation phases that inverted the subsiding basin during the early Paleozoic and the late Paleozoic. The first inversion was attributed to late stages of the Pan-African orogeny in Central Africa, and the second inversion phase to far field intraplate stresses generated at the southern margin of Gondwana during the formation of the Cape Fold Belt. Daly et al. (1992) also speculated on the possible role of evaporites in enhancing deformation process at the deepest levels of the basin. Subsequently, Kadima et al. (2011a) further constrained the interpretation of some of the seismic profiles using 2D gravity and magnetic models, and delineated heterogeneous crust beneath the sedimentary sequences of the basin consistent with the interpretation that the central part of the basin is an intracratonic rift inverted during compressional tectonics facilitated by the presence of evaporites and salt tectonics.

Despite a renewed interest over the last decade, there are still many unanswered questions regarding the nature of the crystalline basement, the Neoproterozoic and Phanerozoic stratigraphy, as well as the structure of the CB. The

É.K. Kabongo (✉) • M.N.S. Sebagenzi
Laboratoire de Géophysique et Géodynamique, Département de Géologie, Université de Lubumbashi, Lubumbashi, DR Congo
e-mail: ekadimak@gmail.com; ssebagenzi@yahoo.fr

D. Delvaux • M. Everaerts
Geodynamics and Mineral Resources, Royal Museum for Central Africa, 3080 Tervuren, Belgium
e-mail: damien.delvaux@africamuseum.be; michel.everaerts@africamuseum.be

F. Lucazeau
Dynamique des Fluides Géologiques, Institut de Physique du Globe de Paris/Sorbonne, Paris Cité, UMR CNRS 7154, 1 rue Jussieu, 75005 Paris, France
e-mail: lucazeau@ipgp.fr

M.J. de Wit et al. (eds.), *Geology and Resource Potential of the Congo Basin*, Regional Geology Reviews,
DOI 10.1007/978-3-642-29482-2_6, © Springer-Verlag Berlin Heidelberg 2015

crystalline basement comprises Archean to Mesoproterozoic igneous and metamorphic rocks of the Congo Shield (see de Wit and Linol, Chap. 2, this Book). Scarce outcrops, limited geological data and stratigraphic age constraints still prevent a detailed reconstruction of the Neoproterozoic history of the CB. Nevertheless, many authors have attempted to decipher the Neoproterozoic geological evolution of Central Africa and the CB in particular (Cahen 1954; Evrard 1960; Lawrence and Makazu 1988; Daly et al 1991, 1992; Kadima 2011; Kadima et al. 2011a, b; Linol 2013; see also Chaps. 2 to 5, this Book).

Here, we re-examine the largely hidden Neoproterozoic to lower Paleozoic sequences of the CB near the centre of the Congo Shield and compare its evolution with the surrounding and partially exposed marginal sedimentary basins.

6.2 Neoproterozoic to Lower Paleozoic Sequences of the Congo Shield Flanking the CB

In Central Africa, the Neoproterozoic geological evolution is marked by a succession of major events initiated by the break-up of the Rodinia supercontinent and opening of rift-basins (e.g. West Congolian, Sembé-Ouesso, Mbuji-Mayi, Katanga, Zambian basins), and terminating by the amalgamation of Gondwana and the filling of associated foreland basins (e.g. de Waele et al. 2008). Traces of these events can be observed within the Congo Shield (sensu Stankiewicz and de Witt 2013) and particularly in marginal sedimentary basins surrounding the CB (Fig. 6.1a), such as the West Congo, Sembé-Ouesso, Bangui and Liki-Bembian basins to the West and Northwest; the Fouroumbala-Bakouma and Lindian basins to the North and Northeast; the Itombwe and Malagarazi-Bukoban basins to the East; the Katanga, Zambian and Mbuji-Mayi basins to the Southeast (Fig. 6.1b).

Studies based on shear-wave tomography (Crosby et al. 2010; Priestley et al. 2008; Ritsema and van Heijst 2000), admittance models (Downey and Gurnis 2009; Hartley and Allen 1994; Hartley et al. 1996; Pérez-Gussinyé et al. 2009), kimberlite data (Crosby et al. 2010; Batumike et al. 2009), subsidence modelling (Kadima et al. 2011b; Kadima 2011) and heat flow estimations (Kadima et al. 2011b; see also Lucazeau et al., Chap. 12, this Book) suggest that the CB and surrounding Lindian and Mbuji-Mayi basins are underlain by a thick mantle lithosphere (200 km or more), with an equivalent elastic thickness of ca. 100 km (for further details see Raveloson et al., Chap. 1, this Book).

Evidence of Rodinia break-up as main mechanism of formation for some of the sedimentary basins surrounding the CB is supported by the age of magmatic intrusions observed into these basins (Vicat and Vellutini 1987; Vicat et al. 1989; Key et al. 2001; Tack et al. 2001; Johnson et al. 2007; de Waele et al. 2008). A brief summary of the stratigraphy of these marginal basins is given below, in clockwise order as depicted in Fig. 6.1:

1. The West Congo basin contains rhyolites and felsic volcanoclastics dated at 950 Ma (Tack et al. 2001) and intruded by granitoids (e.g. in the West Congo Supergroup and the Sangha-Comba Aulacogen). These volcano-magmatic series have been considered to represent a continental rift sequence related to the break-up of Rodinia and the opening of the Adamastor Ocean with the deposition of passive margin sediments (Alvarez 1995a, b; Tack et al. 2001). These sequences are overlain by the Ediacaran-Cryogenian West Congo Group comprising a succession of siliciclastics interbedded with diamictites and carbonates (Frimmel et al. 2006; see also Delpomdor et al., Chap. 3, this Book).
2. The Sembé-Ouesso basin exposed in the Republic of the Congo (Congo-Brazzaville) contains sequence of schist and quartzite, overlain by a pelitic and sandstone sequence, and in turn by the Dja tillite, estimated at ca. 950 Ma (Vicat and Vellutini 1987). Dolerites intrude all these sequences; and their emplacement is interpreted to be related to crustal extension leading to basin subsidence in a failed rift (Vicat and Vellutini 1987; Vicat et al. 1989; Vicat and Pouclet 1995).
3. The Fouroumbala-Bakouma and Bangui basins in the Central African Republic (CAR), the Liki-Bembian, Lindian, Itombwe syncline in the Democratic Republic of Congo (DRC), as well as the Malagarazi-Bukoban basin in Burundi and Tanzania all have been recognized to be Neoproterozoic in age, based on limited radiometric dating and lithostratigraphic correlations (Alvarez 1995a, b, 1999; Poidevin et al. 1980/1981; Poidevin 1985, 2007; Deblond et al. 2001). The age of the Bakouma Formation was first estimated at ca. 840 ± 50 Ma (Poidevin 1996), although Alvarez (1999) obtained a whole rock Rb-Sr date for these rocks of ca. 683 ± 11 Ma. Magmatic intrusions have been documented, in particular in the Fouroumbala-Bakouma, Bangui, Itombwe and Malagarazi-Bukoban basins (Waleffe 1988; Poidevin 1976, 1979; De Paepe et al. 1991; Tack 1995; Deblond et al. 2001). Amygdaloid lava belonging to the Malagarazi Supergroup has been dated at 795 ± 7 Ma (Deblond et al. 2001). Syenite, dated at 700 ± 50 Ma, intrude sediments filling the Itombwe basin (Villeneuve 1983).
4. The Katangan and Zambian basins, now part of the Lufilian (Arc) fold-and-thrust belt, developed during two magmatic events between 880 Ma and 750 Ma (Porada 1989; Armstrong et al. 2005; Johnson et al. 2005), and were subsequently deformed during the Pan-

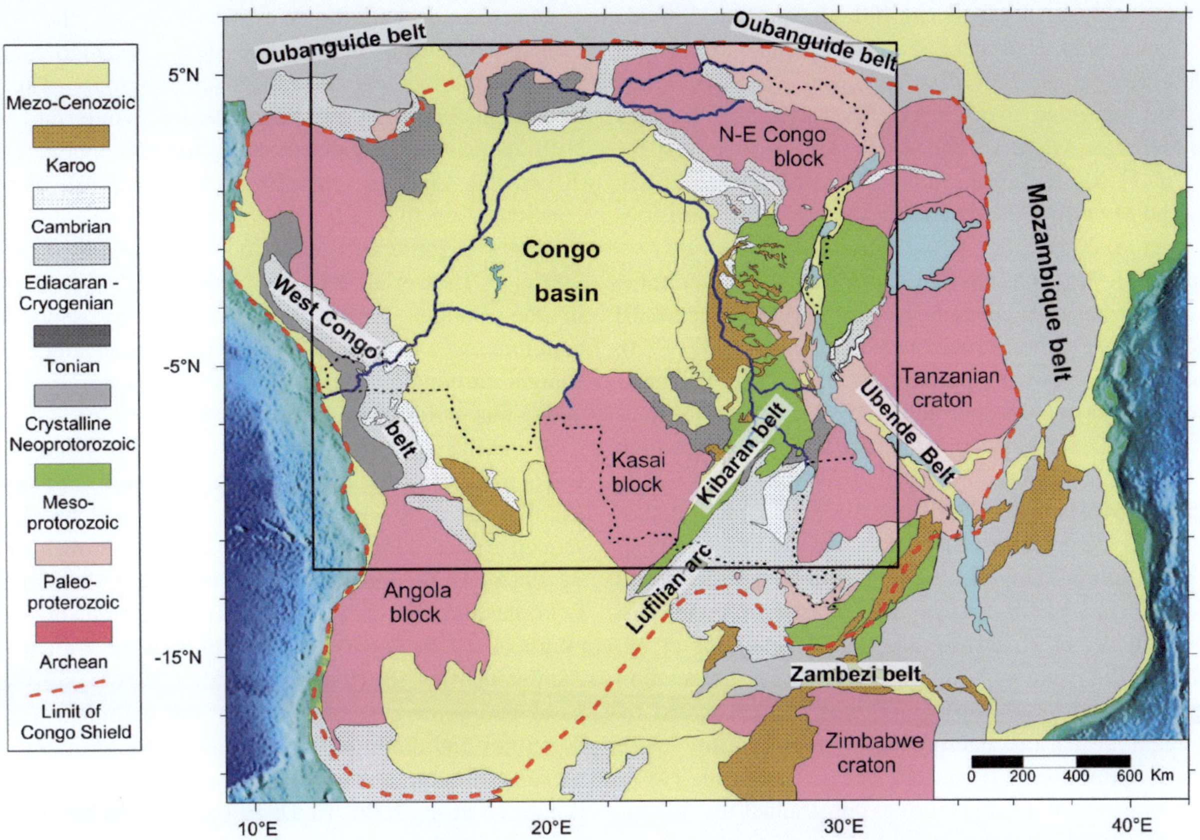
Mezo-Cenozoic
Karoo
Cambrian
Ediacaran - Cryogenian
Tonian
Crystalline Neoprotorozoic
Meso-protorozoic
Paleo-proterozoic
Archean
Limit of Congo Shield
Oubanguide belt
Oubanguide belt
N-E Congo block
Congo basin
West Congo belt
Kasai block
Kibaran belt
Tanzanian craton
Ubende Belt
Mozambique belt
Angola block
Lufilian arc
Zambezi belt
Zimbabwe craton
0 200 400 600 Km
5°N
-5°N
-15°N
10°E
20°E
30°E
40°E

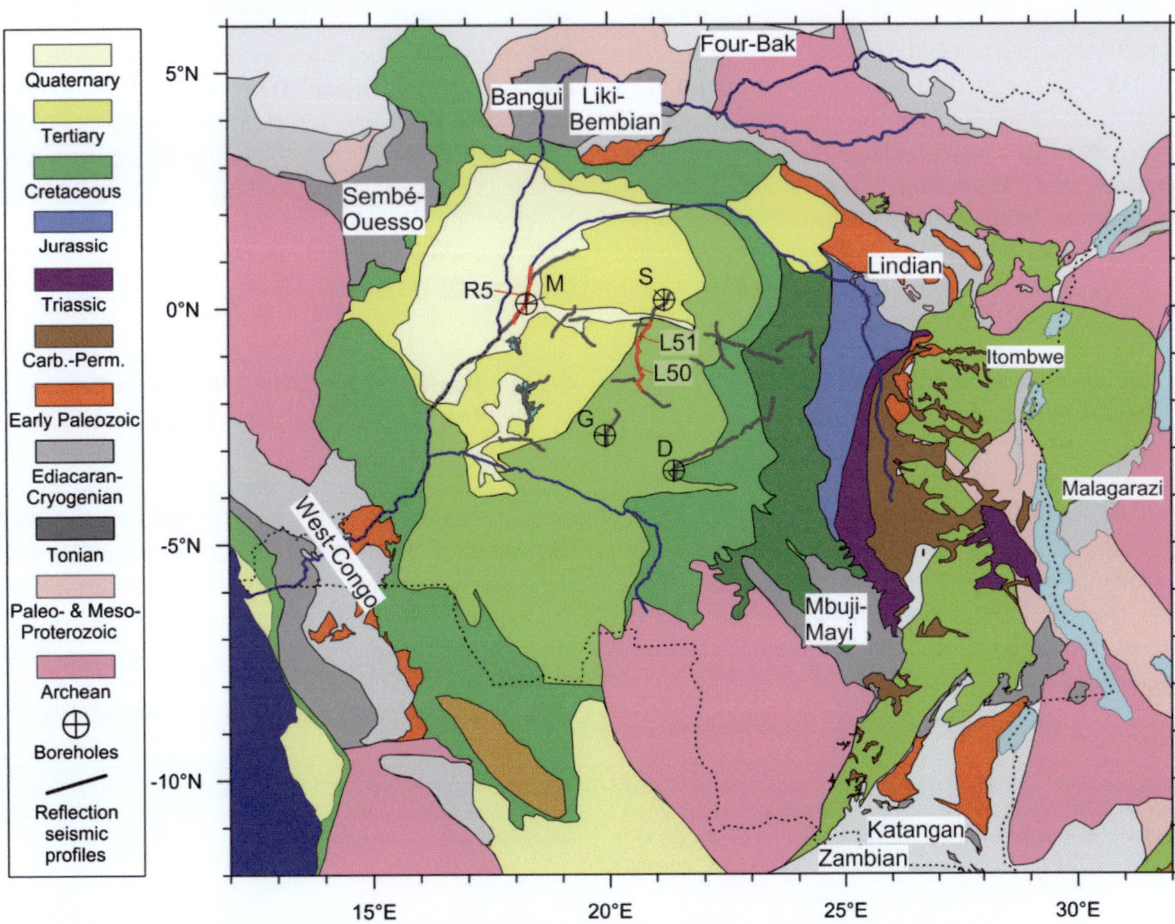
Quaternary
Tertiary
Cretaceous
Jurassic
Triassic
Carb.-Perm.
Early Paleozoic
Ediacaran-Cryogenian
Tonian
Paleo- & Meso-Proterozoic
Archean
Boreholes
Reflection seismic profiles
Four-Bak
Bangui
Liki-Bembian
Sembé-Ouesso
Lindian
R5
M
S
L51
L50
G
D
Itombwe
Malagarazi
West-Congo
Mbuji-Mayi
Katangan
Zambian
5°N
0°N
-5°N
-10°N
15°E
20°E
25°E
30°E

African orogen, between ca. 700 Ma and ca. 530 Ma (Cailteux 1994; Hanson et al. 1994; Porada and Berhorst 2000; Key et al. 2001; De Waele et al. 2008). These basins comprise the Roan, Mwashya, Nguba and Kundelungu Groups. The Roan Group overlies the 877 ± 11 Ma Nchanga granite (Armstrong et al. 2005), and the Mwashya Group contains 765 ± 5 Ma intrusions (Key et al. 2001).

5. The Mbuji-Mayi basin is a SE–NW trending intracratonic failed-rift basin. It contains a lower clastic sequence BI and an upper carbonate sequence BII (Raucq 1957, 1970; Cahen et al. 1984; Delpomdor, et al. 2013a). The presence of dolerite intrusions and pillow lavas were taken as evidence of extensional magmatism during the deposition of the Mbuji-Mayi sediments (Raucq 1957, 1970; Cahen et al. 1984). Isotopic data from the Mbuji-Mayi carbonates and the presence of pseudomorphs of anhydrite and gypsum filling veins and fractures reflect deposition and early diagenesis in marine and evaporitic conditions (Delpomdor, et al. 2013b; see also Delpomdor et al., Chap. 4, this Book). The minimum age of sedimentation for the Mbuji-Mayi Supergroup was first estimated by amygdaloid basalts interpreted as capping the entire sequence and dated at ca. 940 Ma (Cahen et al. 1984). Delpomdor et al. (2013a) further constrained the deposition of this Group between 1174 ± 22 Ma and ca. 800 Ma; and the carbonate BII Group is now dated at 760–820 Ma by Carbon and Strontium chemical stratigraphy (Delpomdor et al., Chap. 4, this Book) and the sliciclastic BI Group is either older than 880 Ma or aged between 880 Ma and 850 Ma. Following this second hypothesis, the Mbuji-Mayi Supergroup would be coeval with the Roan Group in the Katanga and Zambian basins, reflecting a similar early Neoproterozoic extension event.

6.2.1 Neoproterozoic Siliciclastic and Carbonate Sequences

The Neoproterozoic basins surrounding the CB have consistent (similar) stratigraphic successions. In most of them, a carbonate unit, of variable thickness, overlies a basal siliciclastic sequence comprising silts, sandstones and conglomerates. The Neoproterozoic carbonates are in turn overlain by relatively thick and persistent siliciclastic sequences.

For example, in the West-Congo basin, the Schisto-Calcaire Subgroup overlies the Haut Shiloango Subgroup and is overlain by the Mpioka Subgroup. The Haut Shiloango Subgroup comprises predominantly carbonates, with increasing siliciclastics upwards (Delpomdor et al. 2014; and Chap. 4, this Book). The Mpioka Subgroup consists of up to 1,000 m of molasse-like sequences comprising conglomerates, sandstones and argillites (Alvarez 1995a, b; Tack et al. 2001; Frimmel et al. 2006). Similarly, in the Bangui basin, the basal Kembe sequence of conglomerates is overlain by the Bimbo sandstones and the Bangui carbonates. Here, no clastic sequence is observed above this carbonate sequence (Poidevin 1976).

A similar succession is observed in the Fouroumbala-Bakouma basin (Alvarez 1995a, b; Tait et al. 2011). In the Lindian basin, the Ituri Group consists of a basal clastic sequence containing conglomerates and sandstones, overlain by thick oolitic limestones and dolomites. Above it, the Lokoma and Aruwimi Groups are predominantly clastic (Verbeek 1970; Poidevin 2007). In both the Katangan and Zambian basins, the Roan Group (880–750 Ma) consists of a basal conglomerate, siliciclastics and carbonates (mainly dolomites and dolomitic shales) that unconformably overly Mesoproterozoic Kibaran formations (Cailteux et al. 2005; Batumike et al. 2006; El Desouky et al. 2008). The Roan Group is overlain by the predominantly clastic and volcano-sedimentary rocks of the Mwashya the Nguba Groups (750–620 Ma). The latter contain a basal tillite known as the "Grand Conglomerat" (Lepersonne 1974), overlain by a thick carbonate sequence including limestones, dolomites, shales, dolomitic shales and interbbeded silts and sandstones (Cailteux et al. 2005; Batumike et al. 2007). The overlying Kundelungu molasses-like sequence (620–570 Ma) starts with a second glacial horizon, known as "Petit Conglomerat" (Lepersonne 1974).

6.2.2 RedBeds and the Transition Between the Neoproterozoic and Paleozoic

Throughout Central Africa, the transition between the upper Neoproterozoic and the earliest Paleozoic sedimentary sequences is poorly constrained because the transition sequences are non-fossiliferous siliciclastic redbeds. Throughout the CB and its surrounding basins, relatively thick sequences of lithologically similar red sandstones (known as the 'Redbeds') have long been correlated and

Fig. 6.1 (Continued) (**a**) Geological setting of the Congo Shield with the Congo Basin in its center, surrounded by Archean cratonic blocks and Proterozoic mobile belts. *Rectangle* shows contour of (**b**). (**b**) Geological map of the Congo basin and surrounding marginal Neoproterozoic basins with location of the 2 stratigraphic wells (Samba and Dekese) and the two exploration wells (Mbandaka-1 and Dekese-1), indicated by their *initials*

assigned to the late Precambrian (Cahen et al. 1960; Verbeek 1970; Lepersonne 1974), although some researchers recognized parts could also be early Paleozoic in age (Evrard 1957). Subsequently many authors have tried to define the Neoproterozoic-Paleozoic transition in the basins surrounding the CB on the basis rather of structural criteria in Katanga and West Congo; and on the basis of lithological correlations with the Lindian Supergroup (Kampunzu and Cailteux 1999; Porada and Berhorst 2000; Master et al. 2005; Tack et al. 2008; Tait et al. 2011).

In West-Congo and Katanga, where these Redbeds occur along the tectonic fronts of the West-Congo and Lufilian fold belts, and from there extend towards the CB into their foreland, structural relations with reference to a supra-regional late Pan-African unconformity have been used to distinguish the red sandstones as both pre- and post-dating Pan-African deformation. In the West-Congo belt, for example, the Pan-African deformation has been dated at 566 ± 42 Ma, the age of metamorphism of a dolerite sill that intrudes the Haut Shiloango Subgroup and the Inkisi Subgroup redbeds (Alvarez 1999; Tack et al. 2001; Frimmel et al. 2006). The Inkisi Subgroup is therefore considered as of latest Neoproterozoic to early Paleozoic age (Alvarez et al. 1995c; Tack et al. 2008; Tait et al. 2011). Recently, Monié et al. (2012) determined that the West Congo belt of Angola underwent two main deformation events of amphibolite grade at c. 540 and 490 Ma, followed by tectonically assisted exhumation.

Similarly, in the Katanga basin, the sub-horizontal siliciclastic sequences of the Biano (Plateaux) Subgroup (uppermost subgroup of the Kudelungu Group) discordantly overlie the folded Katanga units and are considered therefore to be post Pan-African in age (Batumike et al. 2007; Kampunzu and Cailteux 1999). The Lufilian Arc formed during collision between the Congo and Kalahari Shields between 650 and 530 Ma, with deformation peaking at ca. 550 Ma (Porada 1989; Hanson et al. 1993; Kampunzu and Cailteux 1999; Porada and Berhorst 2000; John et al. 2004; Frimmel et al. 2006). Detrital muscovites from the Biano Subgroup have a maximum Ar/Ar age of 573 ± 5 Ma at the top of the Katanga Supergroup (Master et al. 2005), while the minimum age of the Biano Subgroup is estimated as younger than 540 Ma (Kampunzu and Cailteux 1999). This suggests that at least some of the Biano sequences may have been deposited during the Cambrian (Kipata 2013).

In the sub-horizontal intracratonic Lindian Supergroup along the northern margin of the CB, no clear angular discordance has been observed within the Aruwimi Group, but the Banalia red sandstones that form the upper part of this group, have been correlated lithostratigraphically with the Inkisi Group in West-Congo and the Biano (Plateaux) Subgroup in Katanga (Alvarez et al. 1995; Tack et al. 2008; Tait et al. 2011), but there are no direct age constraints.

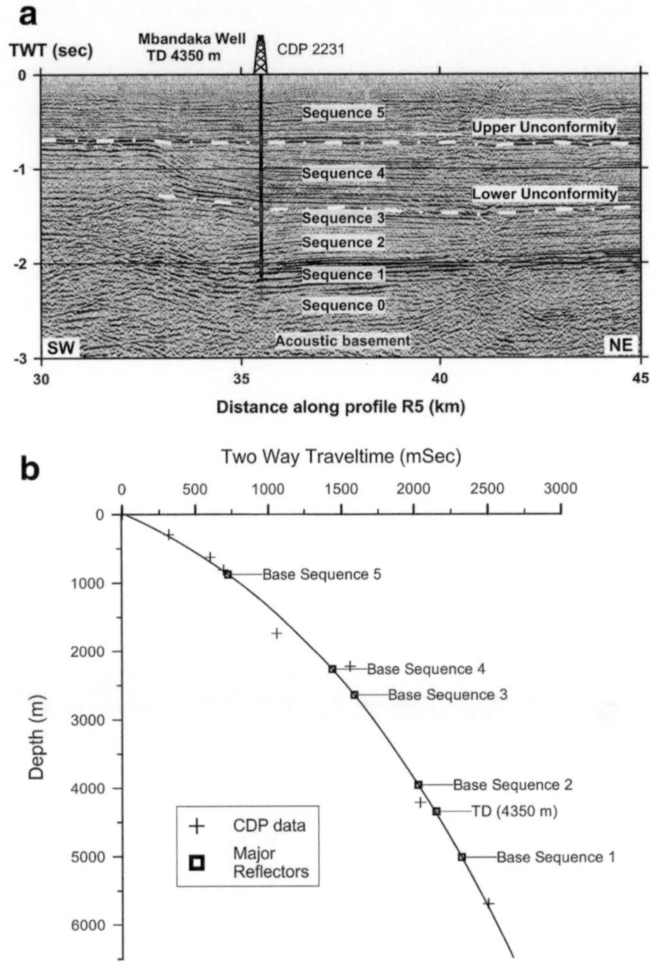

Fig. 6.2 (**a**) Seismic-stratigraphy of sequence on a section of seismic profile R5, showing six depositional sequences, two unconformities and the calibrated trace of Mbandaka-1 well as identified in this chapter. TWT: Two Way Travel-time (TWT in seconds). *Vertical black line*: position of the well, *red crosses*: base of the seismic sequences (corresponding depth in Table 6.2a). CDP 2231: Common Depth Point location. (**b**) Time–Depth curve obtained from the CDP location 2231 along line R05, near the Mbandaka-1 projected trace well. The curve shows the bottom limit of the different seismic sequences and the well TD (4,350 m). Crosses: CDP data from Table 6.1a. The depth–time curve was constructed by a polynomial interpolation of the CDP data. Empty squares: depth–time plot of the base of the seismic sequences from data of Table 6.2a

Clearly throughout the Congo Shield, robust correlations between RedBeds sequences require more precise chronostratigraphy.

6.3 Structure and Stratigraphy of the Deep CB Based on Geophysics and Well Data

Since most of the CB is not accessible to direct observations, geophysical investigations can complement the surface geology observations described above. Here we summarise interpretations of the seismic and gravity data to further unravel the early history of the CB.

Whilst a distinct seismic reflector recognized below the CB has been interpreted to record Pan-African orogenesis (Daly et al. 1991, 1992; Kadima et al. 2011a), it only locally resolves a tectonic unconformity (for example on lines L50 and L51 in the centre of the basin; Figs. 6.3 and 6.4: marked as 'Lower unconformity'). More often the reflector marks a conformable contact (see for example, line R5 calibrated against the Mbandaka-1 well; Fig. 6.2a).

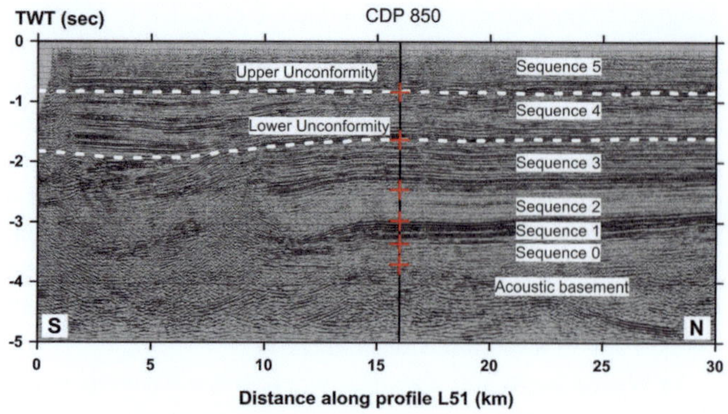

Fig. 6.3 Seismic stratigraphy of sequences identified on a section of profile L51 in the CB and linked further north to the Samba well (Fig. 6.1b) and two major unconformities as interpreted in this chapter. The *first four* sequences beneath the lower unconformity are deformed on the *left* (SW) side of the profile. *TWT* Two Way Travel-time (in seconds). *Vertical black line*: position of CDP 850, *red crosses*: base of the seismic sequences (corresponding depth in Table 6.2b)

6.3.1 Well Data

The original lithostratigraphic scheme of the central CB was established from logging the two fully cored stratigraphic wells: Samba and Dekese (Cahen et al 1959, 1960) and the two exploration wells: Mbandaka-1 and Gilson-1 (Esso Zaire 1981a, b) (location on Fig. 6.1). The Jurassic to Cretaceous sequences that form the uppermost 700–1,200 m are relatively well described and dated by biostratigraphy (Colin 1994; Linol 2013; Linol et al., Chap. 8, this Book). Below, the Permo-Carboniferous sequences of the Lukuga Group were recognized in the Dekese well on the basis of their fossil plants and spores, and their glacial to peri-glacial characteristics. These were not encountered in the Samba well but suspected in the Mbandaka-1 and Gilson-1 wells. In the latter two deeper wells, these sequences overlie siliclastics with dolomites (some with stromatolites) assigned to the Neoproterozoic, although the transition between the Precambrian and Paleozoic is poorly defined (see detailed lithostratigraphic subdivision of the four wells in Linol et al., Chap. 7, this Book). The Mbandaka-1 borehole stops at the depth of −4,350 m in a soft basement, interpreted to be halite (Esso Zaire 1981a).

6.3.2 Velocity Structure Based on Seismic Refraction Data

Evrard (1957) defined for the first time the P-wave velocity structure of the CB, based on the results of a refraction seismic survey calibrated with surface and well data at Samba and Dekese. Whilst Evrard (1957) stressed that this velocity structure is based on the physical characteristics of the sedimentary rocks and does not directly reflect the stratigraphy, the following synthesis seems robust: P-wave velocities <3,600 m s^{-1} characterize the Mesozoic-Cenozoic sequences whilst velocities of ca. 3,900 m s^{-1} are typical for the middle to upper Paleozoic sequences. Higher velocities, between 4,200 and 4,600 m s^{-1}, likely represent the late Neoproterozoic to lower Paleozoic RedBeds, whilst velocities > 5,000 m s^{-1} should represent the Neoproterozoic or older sediments and possibly crystalline basement.

6.3.3 Seismic-Stratigraphic Sequences

Using the seismic profiles acquired by Exxon-Texaco in 1974–1976 (location on Fig. 6.1), ECL (1988) and Daly et al. (1992) distinguished six seismic-stratigraphic sequences (or "Supersequences") above the acoustic crystalline basement, each bounded by regional unconformities. These sequences are re-described below (using the same numbering as in Daly et al. 1992) with reference to the seismic profile R5 shot along the Congo River and passing close to the Mbandaka-1 well (Figs. 6.1b, 2a). They are also illustrated on the land seismic profile L51 whose northern extremity approaches the Samba well (Figs. 6.1b, 3).

On the R5 profile (Fig. 6.2a), the depth-time curve is presented for CDP (Common Depth Point) location 2,231, measured closest to the Mbandaka-1 well (Fig. 6.2b). The depth-time curve was constructed by a polynomial interpolation of the CDP data (Table 6.1a). This allowed calculating the depth of the base of the different seismic-stratigraphic sequences and their thicknesses, as identified on the seismic line R5 (Table 6.2a). For reference, the bottom (TD) of the Mbandaka-1 well at 4,350 m deep correspond to a TWT time of 2,139 ms. Similarly, the depth to the base of the sequences identified on Line L51 have been calibrated for CDP 850 (Fig. 6.3, Tables 6.1a and 6.2b).

Except for the uppermost sequence (5), the ages of all sequences is poorly defined. Two major unconformities are identified and traced in most of the seismic profiles

Table 6.1 CPD data

(a) For location 2231 along the line R5, close to the Mbandaka-1 well									
Times (ms)	1	316	601	691	1,058	1,555	2,041	2,500	5,000
VRMS (m/s)	1,756	1,859	2,085	2,345	3,282	3,865	4,133	4,555	8,000
Depth (m)	1	294	623	786	1,621	2,824	4,023	5,422	15,000
(b) Location 850 along line R51									
Times (ms)	0	625	750	1,050	1,450	1,850	2,225	2,975	5,000
VRMS (m/s)	2,000	2,500	2,690	3,145	3,565	4,075	4,480	4,950	6,000
Depth (m)	0	701	999	1,609	2,506	3,615	4,758	7,059	14,422

Table 6.2 Depth for the base of seismic sequences and their thickness, obtained after calibration using the CDP data

(a) Location 2231 along the line R5, close to the Mbandaka-1 well						
Seismic sequences	5	4	3	2	1	0
Depth (m)	877	2,270	2,844	3,960	4,350	[a]
Thickness (m)	877	1,393	370	1,320	390	[a]
(b) Location 850 along line R51						
Seismic sequences	5	4	3	2	1	0
Depth (m)	1,120	2,990	5,410	7,070	8,290	9,400
Thickness (m)	1,120	1,770	2,420	1,660	1,220	1,210

[a]Base of the sequence No defined

(relatively distant one from another). However their age control is also poor (especially for the lower one) and it is possible that there are more than 2 regional unconformities (e.g. Daly et al. 1992 identified 6 unconformities).

6.3.3.1 Sequence 0

Along several profiles, a reflection-free seismic pattern that corresponds to the acoustic basement is interpreted to represents crystalline basement. The latter is overlain by a low-amplitude, discontinuous and transparent seismic patterns forming the first sequence (Sequence 0 of Daly et al. 1992). The reflectors diverge towards the centre of the CB, suggesting tilting during deposition; and the sequence is thicker in the centre of the basin than at its border. Sequence 0 is apparently discontinuous, with rapid variations in thickness and its base generally is not well-imaged. Sequence 0 was not penetrated by any cored wells. It could represent siliciclastic sediments (essentially conglomeratic) if correlated with the Liki-Bembian Group outcropping to the NW of the basin (e.g. Daly et al. 1992) and may thus represents the basal Precambrian sedimentary unit of the CB. Alternatively, it may also be correlated with the lower Neoproterozoic clastic sequence (BI) of the Mbuji-Mayi basin (Delpomdor et al. 2013a, b; and Chap. 4, this Book) and the Roan Group of the Katangan and Zambian basins.

The base of sequence 0 is not identified in the R05 seismic profile, but suspected in profile L51, at ~ TWT 3,710 ms, which is calibrated using the CDP 850 data at ~9,400 m depth, for a thickness of ~ 1,100 m (Fig. 6.3; Tables 6.2a, b).

6.3.3.2 Sequence 1

Sequence 1 has a layered seismic pattern characterized by highly continuous and divergent reflectors. In the Mbandaka-1 well its top occurs at ca. 3,960 m, and its bottom at 4,350 m. From this bottom upward to 4,133 m depth, the sequence is represented by a 217 m thick succession of dark-grey calcareous silty shales that grade into argillaceous and dolomitic limestones, with interbedded salt crystals and sparse anhydrite at the base (Esso Zaire 1981a). In the Gilson-1 well, it is represented between 4,503 m depth and the bottom of the well (4,665 m) by alternating beds of sandstone, siltstone, shale and dolomite, with massive dolomite at the base (Esso Zaire 1981b). Sequence 1 is correlated with the Ituri carbonates of the Lindian Supergroup of Verbeek (1970) and represents the M9 sequence of Linol (2013) in the Mbandaka-1 well, and the G10 sequence of Linol (2013) in the Gilson-1 well. It could possibly be correlated also with the carbonate sequence (BII) of the Mbuji-Mayi basin (Delpomdor et al. 2013a, b). Using the CDP 2231 data, its thickness at the position of the Mbandaka-1 well would be ~ 670 m. In profile L51, at CDP 850, Sequence 1 is ~1,220 m thick with its base at a depth of ~ 8,290 m (Fig. 6.3, Table 6.2b).

6.3.3.3 Sequence 2

A second transparent seismic pattern with discontinuous reflectors and one or two strong to medium continuous reflectors in the middle overlies Sequence 1. In the R5 profile (Fig. 6.2a, b; Table 6.2b), it corresponds to an interval

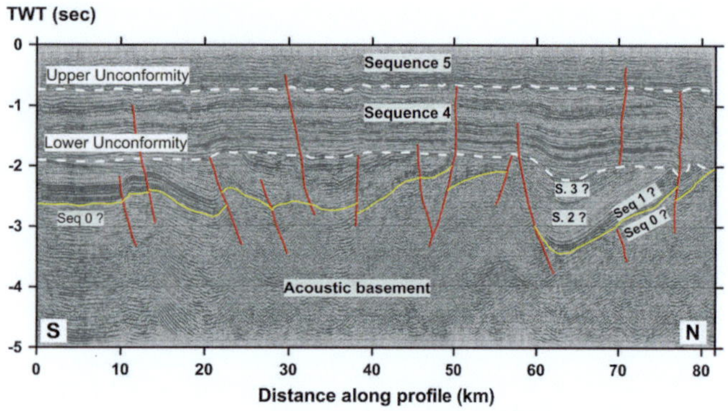

Fig. 6.4 Seismic stratigraphy and structural interpretation of N–S seismic line L50 in the center (Fig. 6.1b). The lower unconformity has been deformed during or after deposition of Sequence 5 by reactivation of underlying structures. *Yellow line*: folded and faulted reflector at the base of strongly reflecting sequence 2. *TWT* Two Way Traveltime (Seconds)

between 2,644 m and 3,960 m depth (1,316 m thick). It also broadly corresponds to conglomeratic and sandstone sequences M9 and M8 of Linol (2013). The transparent character is well seen in the seismic profile R51 (Fig. 6.3), where at CDP 850, it is ca. 1,660 m thick and its base is at ca. 7,070 m deep (Table 6.2b).

6.3.3.4 Sequence 3

Sequence 3 has a banded seismic pattern characterized by slightly divergent and strongly to moderately continuous reflectors. An angular unconformity observed along the deformed part of the basin constitutes the upper boundary of the sequence, which defines the lower major regional unconformity and can be detected on most of the seismic profiles (Daly et al. 1992; Kadima 2011; Kadima et al. 2011a, b). The angular character of this unconformity is well defined in the southern part of profile L51 (Fig. 6.3), and in the central part of profile L50 (Fig. 6.4). In profile R5 (Fig. 6.2a), it appears as a conformable contact and is therefore more difficult to identify. In the R5 profile, at the location of the Mbandaka-1 well, Sequence 3 appears between ~2,270 and ~2,640 m deep (Table 6.2) and is correlated in the Mbandaka-1 well with the red-orange conglomerates of the M7 sequence of Linol (2013). This sequence is observed in line L51 at the position of CDP 850, between ~5,410 and ~2,990 m (~2,420 m thick; Table 6.2b). We tentatively correlate this Sequence 3, together with the underlying Sequence 2, with similar (predominantly clastic) sequences in the marginal basins: the molassic Mpioka Group in West Congo (Frimmel et al. 2006), the predominatly clastic Kundelungu and Nguba Groups in the Katanga basin (Cailteux et al. 2007), and the dominantly clastic Aruwimi Group in the Lindian basin (Verbeek 1970).

6.3.3.5 Sequence 4

Overlying the lower regional unconformity, a parallel to subparallel seismic pattern marked by moderate to low continuous reflectors constitutes the fifth sequence (e.g. Sequence 4 of Daly et al. 1992). Its upper boundary is another angular unconformity that constitutes the second (or upper) regional unconformity in the CB. Well logs show that Sequence 4 is mostly siliciclastic, comprising red sandstones and conglomerates. According to the interpretation of seismic profile R5 (Fig. 6.2a), this sequence is 1,390 m thick in the region of the Mbandaka-1 well, between depths ca. 877 m and ca. 2,270 m. It corresponds to the M5 and M6 sequences of Linol (2013, and assigned to the Triassic). On CDP 850 along line 51, it is ca. 1,870 m thick, between ca. 1,120 m and ca. 2,990 m deep.

6.3.3.6 Sequence 5

The sixth and youngest seismic sequence (Sequence 5) consists of parallel highly continuous reflectors grading upwards to wavy moderate to strong reflectors. The lower boundary of this sequence corresponds to the upper regional unconformity as defined above. Well logs and outcrop data show that it contains essentially continental clastic formations ranging in age from the Late Jurassic to the Neogene. It appears undeformed on the seismic profiles and has a relatively constant thickness over the entire CB: 715 m at Dekese (Cahen et al. 1960), 1,167 m at Samba (Cahen et al. 1959), 844 m at Mbandaka and 998 m at Gilson (Esso Zaire 1981a; Linol 2013), and ~1,120 m thick along line L51 at CDP 850.

To directly compare their seismic-stratigraphic interpretations of the CB, Kadima et al. (2011a) regroup the first four sequences of Daly et al. (1992) into a lower seismic unit (Unit A), contained beneath the lower major unconformity. This Unit A is locally affected by tectonic deformation. By correlation with the depocenters surrounding the CB and considering that this lower unconformity could represent the end of Pan-African deformation, the seismic Unit A likely represents the equivalent the Neoproterozoic sedimentary sequences flanking the CB. Sequence 4, which is bound between the two regional unconformities, is named seismic Unit B, and possibly represents the Paleozoic and early Mesozoic (e.g. Triassic; see below). In this scheme, the uppermost seismic Unit C would represent the Jurassic-Cretaceous and Cenozoic.

6.3.4 Tectonism

The Neoproterozoic sequences of the seismic Unit A as defined above have been affected by Pan-African deformation. Some of those structures apparently have been reactivated also by late Paleozoic-early Mesozoic deformations. In contrast, the lower-middle Paleozoic red sandstones at the base of Unit B, together with the overlying Carboniferous-Permian diamictites and black shales, have been deformed only by the late Paleozoic-early Mesozoic deformation. This is well illustrated by the N-trending seismic profile L50 in the centre of the basin (Fig. 6.4) by structures such as flexures, anticlines, synclines and thrust faults, although it would be difficult to distinguish these from syn-depositional features related to slumping and glacial processes, or even salt tectonics. The first deformation event occurred prior to the deposition Sequence 4 and the second affects it, mainly by reactivating the structures underlying the lower unconformity. Each deformation event is characterized by an erosional phase followed by renewed sedimentation with by top-lap structures. These deformational events are interpreted as representing intraplate compressional tectonic episodes related to far-field tectonic stresses generated by Pan African collisional tectonics flanking the margin of the Congo Shield for the first event, and at the margin of Gondwana (during the formation of the Cape Fold Belt at ca. 250 Ma) for the second one (e.g. Daly et al. 1992; Kadima et al. 2011a; Kipata et al. 2013. However, if the Sequence 4 also includes the Triassic, then the second deformation would be post-Triassic and therefore cannot be related to the Cape Fold Belt and would require a different interpretation of the far field stresses (Linol 2013; see also Linol et al., Chap. 11, this Book).

6.3.5 Gravity Anomalies and Modelling

The CB is associated with a large-scale and pronounced negative free-air gravity anomaly (Jones et al. 1960; Sandwell and Smith 1997; Tapley et al. 2005). Initiation of the CB is frequently related to the development of a NW–SE Neoproterozoic rift and a large part of its subsequent subsidence to post-rift thermal relaxation (e.g. Kadima et al. 2011a, b; Buiter et al. 2012).

The large scale and amplitude of the gravity anomaly has led various researchers of the CB to propose very different geodynamic models to account for subsidence of the CB. Hartley and Allen (1994) and Hartley et al. (1996), for example, suggested the negative Bouguer gravity could be due to the combined effect of lower density sediments in the basin and higher density material in the lithospheric mantle to isostatically compensate for the sediments; and indeed, Downey and Gurnis (2009) show numerically that the topography and negative free-air gravity data over the basin can be explained by a high-density body within the deeper lithosphere.

By contrast, Heine et al. (2008), suggested that subsidence might be due to dynamic topography, consistent with Sahagian (1993) and Burke and Gunnell (2008), who prefer models in which the CB acquired its modern shape through mantle (plume) driven uplift of swells surrounding the basin.

Others prefer a model in which recent subsidence of the CB is controlled by a late phase of the ca. 700 Ma post-rift thermal subsidence following the Neoproterozoic extension (e.g. Armitage and Allen 2010; Crosby et al. 2010; Kadima et al. 2011b), while Buiter et al. (2012) suggests that the observed negative gravity anomaly across the CB is mainly due to the thick (up to 9 km) sedimentary units in the basin, and that the sub-lithospheric mantle structures did not apparently play a role in the recent subsidence of the basin.

To further test a simple long lived thermal history for the CB, Kadima et al. (2011b) backstripped the effects of the CB sediments and noted a residual NW–SE positive and narrow gravity anomaly across the central CB, which they interpreted as the remaining crustal thinning associated with the Neoproterozoic rift that initiated the CB. The paleo-rift is aligned along the Mbandaka-1 and Dekese positive Bouguer anomalies of Jones et al. (1960) and Kadima et al. (2011a). Assuming that isostasy is governed by crustal necking and flexural response to sediment loads, Kadima et al. (2011b) obtained a best fit to the residual gravity with a necking depth of 10 km and an equivalent lithospheric elastic thickness of 100 km. Consequently, the linear anomaly is interpreted as an old tectono-thermal heritage of the initial Neoproterozoic rifting, when denser mantle associated with Moho uplift invaded the necking zone and has remained in place ever since (Kadima et al. 2011b). 2D coupled gravity and magnetic models were constructed across this elongated NW-trending positive residual anomaly, along the seismic lines constrained by well data, to test the hypothesis for a rifting process with extensional magmatism activity prior to the basin formation (Kadima et al. 2011a). The main results are as follows:

- Modeling of two SW–NE profiles, one along the Congo River and passing close to the Mbandaka-1 well and one passing through the Dekese well is consistent with a lateral change in basement density along the trend of the profile, suggesting that the elongate positive NE-tending residual gravity anomaly zone could correspond to a deep crustal discontinuity injected by mafic magma during the Neoproterozoic rifting.
- The modelling is also consistent with (but does not prove) the presence of evaporite sequences in some of the deeper units in lateral continuity with evaporite showings

observed at the bottom of the Mbandaka-1 and Gilson-1 wells. The modelled low susceptibility values of some layers suggest that they could contain a significant proportion of salt.

- The poorly defined seismic facies that led to the previous interpretations of crystalline basement uplift is shown to be also consistent with paleo-salt tectonics.

6.4 Discussion

The Neoproterozoic period in Central Africa was marked by crustal extension with the opening of rift basins during the initial development of the CB near the centre of the Congo Shield. This was accompanied by the formation and preservation of Neoproterozoic basins surrounding the CB. Lithostratigraphic, structural, and paleo-environmental similarities have been established between all these basins (Tait et al. 2011; Delpomdor 2013; Delpomdor et al., Chap. 4, this Book). The sediments were subsequently affected by compressional tectonics related to the Pan African and late Paleozoic-early Mesozoic orogens.

The data discussed here suggest to us rather that the Neoproterozoic CB was initiated during an intracratonic rifting process, possibly related to the break-up of the Rodinia Supercontinent. Such extensional setting may also have initiated the development of the other Neoproterozoic basins now preserved along the periphery of the Congo Shield. Mafic magmatism associated with this regional rifting event is observed in the West Congo, Sembé-Ouesso, Bangui, Fouroumbala-Bakouma, Itombwe, Malagarazi-Bukoban, Roan, Kundelungu and Mbuji-Mayi basins (Tack et al. 2001, 2010; Vicat et al. 1989; Hanson et al. 1994; Johnson et al. 2005; Key et al. 2001; Kampunzu and Cailteux 1999; Batumike et al. 2009; Delpomdor et al. 2013b). Within the CB, there is no evidence to confirm or exclude similar magmatic activity, but 2-D modelling suggests that a linear NW-SE positive residual gravity anomaly in the centre of the basin may be associated with mafic intrusions into the crustal basement. Both Evrard (1957) and Kadima et al. (2011a, b) attributed the short wavelength gravity anomalies observed near Dekese and Mbandaka-1 to such intrusions into the upper crust.

The lithostratigraphic successions in all the Neoproterozoic basins developed across the Congo Shield show first-order similarities, with in their lower parts carbonate-rich sediments overlain by siliciclastic sequences that include tillites (for more details see Delpomdor et al., Chap. 4, and de Wit and Linol, Chap. 2, this Book). In most of the peripheral basins, the carbonate units overlie a basal conglomeratic sequence. As shown by the deep well logs (Esso Zaire 1981a, b) and previous seismic analysis, the deepest part of the CB may contain similar units. A massive carbonate formation with evaporite showings (the "carbonate-evaporite" sequence of ECL 1988; Lawrence and Makazu 1988; Daly et al. 1992), was reported from the lower Mbandaka-1 well and defined as Sequence 1 (670–1,200 thick). The carbonates are considered as late Neoproterozoic, Ediacaran-Cryogenian (e.g. Daly et al. 1992; Kadima et al. 2011a; Linol 2013; Delpomdor, et al., Chap. 4, this Book) by lithostratigraphic correlation with and between the outcropping sequences of carbonates in the peripheral basins. Beneath this carbonate unit, the basal siliclastic sequence (Sequence 0) tends to thicken towards the centre of the CB (to up to ca. 1,100 m). Overlying the carbonate unit is a predominance of siliciclastic sequences with intercalation of thin carbonate layers, defined in the two deep exploration wells and in the seismic profiles as Sequences 2 and 3 (1,700 to 4,100 m), and here attributed to the late Neoproterozoic-early Paleozoic.

In the marginal West Congo, Lindian and Katangan basins, recent observations suggest that the uppermost clastic sequences (the Inkisi, Biano and Banalia Groups) are not affected by Pan-African deformations and therefore may be early- to mid-Paleozoic in age (Alvarez et al. 1995; Master et al. 2005; Batumike et al. 2006; Tack et al. 2008; Kampunzu and Cailteux 1999; Kampunzu et al. 2009; Cailteux et al. 2007; Tait et al. 2011).

In the CB, seismic Sequences 1 to 4 have been locally affected by a compressional tectonism event responsible for inversion structures such as folds and thrusts (Daly et al. 1992 and Kadima et al. 2011a). An erosional unconformity separates Sequence 4 from Sequence 5. We consider that the first four sequences form the late Neoproterozoic-early Paleozoic sedimentary record of the CB, with the lower major erosional surface represented by a Pan African unconformity.

Conclusion

A series of basins formed in Central Africa at the periphery of the Congo Shield and the large CB formed near its centre during the early Neoproterozoic by rifting and mafic magmatism related to the break-up of Rodinia. The early phase of the CB developed during this time along a NW-SE elongated structure corresponding to the Mbandaka-Dekese axis, intruded by mafic material that locally densified the crust; and was filled subsequently by up to 9 km of Neoproterozoic to Cenozoic sediments following long term (ca. 700 million years) thermal- and density-driven subsidence. Indirect observations suggest that the first seismic-stratigraphic sequence in the CB (Sequence 0) may consist of syn-rift clastics, with slightly divergent reflectors, followed by the deposition of carbonates, with an undetermined amount of evaporites (Sequence 1), and then two siliciclastic sequences with thin carbonate layers and tillites interbedded (Sequences 2 and 3). These four sequences are grouped in the seismic

Unit A of Kadima et al. (2011a), which is more than 2,360 m thick in the Mbandaka-1 well and up to 6,400 m thick along Line L51. They all have been affected by contractional deformation interpreted to be related to Pan-African orogenesis, causing basin inversion. They are therefore considered as of pre- to syn-tectonic relative to this event and are thus of Neoproterozoic age.

A regional unconformity truncates these structures and forms the base of Sequence 4 (1,400–1,900 m thick). This sequence comprises undated red sandstones at the base, followed by glacial to peri-glacial sediments of Late Carboniferous—Permian age. It is delineated at the top by a second regional tectonic unconformity and forms the seismic Unit B of Kadima et al. (2011a). This unit is less deformed than the four deeper sequences, and mainly by the reactivation of earlier structures in Unit A. The overlying Sequence 5 (Seismic Unit C of Kadima et al. 2011a) contains essentially continental formations attributed to the Upper Jurassic-Cenozoic and is between 800 m to 1,200 m thick.

Acknowledgements This work was supported by the Frame Agreement program MRAC-DGCD under project S1_RDC_Geodyn_-UNILU. E. Kadima and S. Sebagenzi gratefully acknowledge this program and the Royal Museum for Central Africa for coordinating this project. C. Blanpied is thanked for further discussion on the interpretation of seismic profiles. M. de Wit and B. Linol are thanked for the critical review of several versions of this Ms.

References

Alvarez P (1995a) Les facteurs de contrôle de la sédimentation du Supergroupe Ouest-congolien (Sud-Congo). Rampe carbonatée et activité biologique au Protérozoïque supérieur. Implication en Afrique centrale (Congo, Zaïre, Gabon, Cameroun, Guinée Equatoriale, Centrafrique). Document BRGM, 239, 273 p

Alvarez P (1995b) Evidence for a Neoproterozoic carbonate ramp on the northern edge of the Central African Craton: relations with Late Proterozoic intracratonic troughs. Geol Rund 84:636–648

Alvarez P, Maurin J-C, Vicat J-P (1995) La Formation de l'Inkisi (Supergroupe Ouest-congolien) en Afrique centrale (Congo et Bas-Zaïre): un delta d'âge Paléozoïque comblant un bassin en extension. J Afr Earth Sci 20:119–131

Alvarez P (1999) La transition Précambrien-Cambrien en Afrique Centrale. Approche intégrée: Paléoenvironnements et données paléontologiques. Habilitation à diriger des recherches. Université de Poitiers, 1999, 307 p

Armitage JJ, Allen PA (2010) Cratonic basins and the long-term subsidence history of continental interiors. J Geol Soc London 167: 61–70

Armstrong RA, Master S, Robb LJ (2005) Geochronology of the Nchanga Granite and constraints on the maximum age of the Katanga Supergroup, Zambian Copperbelt. J Afr Earth Sci 42: 32–40

Batumike MJ, Cailteux JLH, Kampunzu AB (2007) Lithostratigraphy, basin development, base metal deposits, and regional correlations of the Neoproterozoic Nguba and Kundelungu rock successions, central African Copperbelt. Gondwana Res 11(3):432–447

Batumike J, Griffin W, O'Reilly SY, Belousova EA, Pawlitschek M (2009) Crustal evolution in the central Congo-Kasaï Craton, Luebo, D.R. Congo: insights from zircon U-Pb ages, Hf-isotope and trace-element data. Precambrian Res 170(1–2):107–115

Batumike MJ, Kampunzu AB, Cailteux J (2006) Petrology and geochemistry of the Neoproterozoic Nguba and Kundelungu Groups, Katangan Supergroup, southeast Congo: implications for provenance, paleoweathering and geotectonic setting. J Afr Earth Sci 44:97–115

Buiter SJH, Steinberger B, Medvedev S, Tetreault JL (2012) Could the mantle have caused subsidence of the Congo Basin? Tectonophysics 514–517:62–82

Burke K, Gunnell Y (2008) The African erosion surface: a continental-scale synthesis of geomorphology, tectonics, and environmental change over the past 180 million years. Geol Soc Am Mem 201

Cahen L (1954) Géologie du Congo belge. Vaillant-Caramanne, Liège, 577p

Cahen L, Ferrand JJ, Haarsma MJF, Lepersonne J, Verbeek Th (1959) Description du Sondage de Samba. Ann. Mus. Roy. Congo belge, Tervuren (Belgique), série in-8, Sci Géol 29, 210 pp

Cahen L, Ferrand JJ, Haarsma MJF, Lepersonne J, Verbeek Th (1960) Description du Sondage de Dekese. Ann. Mus. Roy. Congo belge, Tervuren (Belgique), série in-8, Sci Géol 34, 115 pp

Cahen L, Snelling NJ, Delhal J, Vail JR, Bonhomme M, Ledent D (1984) Geochronology and evolution of Africa. Clarendon, Oxford, 512 p

Cailteux J (1994) Lithostratigraphy of the Neoproterozoic Shaba-type (Zaïre) Roan Supergroup and metallogenesis of associated stratiform mineralisation. J Afr Earth Sci 19:279–301

Cailteux J, Kampunzu AB, Lerouge C, Kaputo AK, Milesi JP (2005) Genesis of sediment-hosted stratiform copper-cobalt deposits, Central African Copperbelt. J Afr Earth Sci 42:134–158

Cailteux J, Kampunzu AB, Lerouge C (2007) The Neoproterozoic Mwashya-Kasuki sedimentary rock succession in the central African copperbelt, its Cu-Co mineralisation, and regional correlations. Gondwana Res 11:414–431

Colin JP (1994) Mesozoic-Cenozoic lacustrine sediments of the Zaïre Interior Basin. In: Gierlowski-Kordesch E, Kelts K (eds) Global geological record of lake basins, vol 1. Cambridge University Press, pp 31–36

Crosby AG, Fishwick S, White N (2010) Structure and evolution of the intracratonic Congo Basin. Geochem Geophys Geosyst 11: Q06010

Daly MC, Lawrence SR, Diemu-Tshiband K, Matouana B (1992) Tectonic evolution of the Cuvette Centrale, Zaire. J Geol Soc London 149:539–546

Daly MC, Lawrence SR, Kimun'a D, Binga M (1991) Late Paleozoic deformation in central Africa: a result of distant collision? Nature 350:605–607

Deblond A, Punzalan LE, Boven A, Tack L (2001) The Malagarazi supergroup of southeast Burundi and its correlative Bukoba supergroup of northwest Tazania: Neo- and Meso-proterozoic constraints from Ar-Ar ages of mafic intrusive rocks. J Afr Earth Sci 32: 435–449

Delpomdor F (2013) Sedimentology, geochemistry and depositional environments of the 1175-570 Ma carbonate series, Sankuru-Mbuji-Mayi-Lomami-Lovoy and Bas-Congo basins, Democratic Republic of Congo - New insights into Late Mesoproterozoic and Neoproterozoic glacially- and/or tectonically-influenced sedimentary systems in equatorial Africa. PhD thesis, ULB, Brussels, 497 p

Delpomdor F, Blanpied C, Virgone A, Préat A (2013a) Paleoenvironments in Meso-Neoproterozoic carbonates of the Mbuji-Mayi Supergroup (Democratic Republic of Congo) - Microfacies analysis combined with C-O-Sr isotopes, major-trace elements and REE+Y distributions. J Afr Earth Sci 88:72–100

Delpomdor F, Linnemann U, Boven A, Gärtner A, Travin A, Blanpied C, Virgone A, Jelsma H, Préat A (2013b) Depositional age, provenance, and tectonic and paleoclimatic settings of the late Mesoproterozoic–middle Neoproterozoic Mbuji-Mayi Supergroup, Democratic Republic of Congo. Palaeogeogr Palaeoclimatol Palaeoecol 389:35–47

Delpomdor F, Kant F, Préat A (2014) Neoproterozoic uppermost Haut-Shiloango Subgroup (West Congo Supergroup, Democratic Republic of Congo): Misinterpreted stromatolites and implications for sea-level fluctuations before the onset of the Marinoan glaciation. J Afr Earth Sci 90:49–63

De Paepe P, Tack L, Moens L, van de Velde P (1991) The basic magmatism of upper Proterozoic in southeast Burundi. Musée royal de l'Afrique Centrale, Tervuren, Belgique, rapport annuel 1989–1990:85–104

De Waele B, Johnson SP, Pisarevsky SA (2008) Paleoproterozoic to Neoproterozoic growth and evolution of the eastern Congo Craton: its role in the Rodinia puzzle. Precambrian Res 160:127–141

Downey NJ, Gurnis M (2009) Instantaneous dynamics of the cratonic Congo Basin. J Geophys Res Solid Earth 114, B06401

ECL (1988) Hydrocarbon potential of Cuvette Centrale (Republic of Zaire). Exploration Consultants Limited, Cellule Technique Pétrolière, Pétrozaire, unpublished report, 41pp. + figures, tables, appendices and enclosures.

El Desouky HA, Muchez P, Dewaele S, Boutwooe A, Tyler R (2008) Postorogenic Origin of the Stratiform Cu Mineralization at Lufukwe, Lufilian Foreland, Democratic Republic of Congo. Econ Geol 103:555–582

Esso Zaire SARL (1981a) Geological completion report. Mbandaka-1. Unpublished internal report

Esso Zaire SARL (1981b) Geological completion report. Gilson-1. Unpublished internal report

Evrard P (1957) Les recherches géophysiques et géologiques et les travaux de sondage dans la Cuvette congolaise. Académie Royale des Sciences Coloniale., Sc. Techn. Bruxelles, VII(1), 62 p

Evrard P (1960) Sismique. (Résultat scientifique des missions du Syndicat pour l'étude géologique et minière de Cuvette congolaise). Ann Mus royal Afrique centrale, Tervuren (Belgique), série in-8, Sci Géol, 33, 87 p

Frimmel H, Tack L, Basei M, Nutman A, Boven A (2006) Provenance and chemostratigraphy of the Neoproterozoic West Congolian Group in the Democratic Republic of Congo. J Afr Earth Sci 46: 221–239

Hanson RE, Wardlaw MS, Wilson TJ, Mwale G (1993) U-Pb zircon ages from the Hook granite massif and Mwembeshi dislocation: constraints on Pan-African deformation, plutonism, and transcurrent shearing in central Zambia. Precambrian Res 63:189–206

Hanson RE, Wilson TJ, Munyanyiwa H (1994) Geologic evolution of the Neoproterozoic Zambezi orogenic belt in Zambia. J Afr Earth Sci 18:135–150

Hartley RW, Allen PA (1994) Interior cratonic basins of Africa: relation to continental break-up and role of mantle convection. Basin Res 6:95–113

Hartley R, Watts AB, Fairhead JD (1996) Isostasy of Africa. Earth Planet Sci Lett 137:1–18

Heine C, Müller RD, Steinberger B, Torsvik TH (2008) Subsidence in intracontinental basins due to dynamic topography. Physics of the Earth and Planetary interoirs. Recent advances in computational geodynamics: theory. Num Appl 1–4:252–264

John T, Schenk V, Mezger K, Tembo F (2004) Timing and PT evolution of whiteschist metamorphism in the Lufilian Arc - Zambezi belt orogen (Zambia): implications for the assembly of Gondwana. J Geol 112:71–90

Johnson SP, Rivers T, De Waele B (2005) A review of the Mesoproterozoic to Early Paleozoic magmatic and tectonothermal history of Central Southern Africa: Implications for Rodinia and Gondwana. J Geol Soc London 162:433–450

Johnson SP, De Waele B, Tembo F, Evans D, Iizuka T, Tani K (2007) Geochronology of the Zambezi Supracrustal sequence, southern Zambia: a record of Neoproterozoic divergent processes along the southern margin of the Congo Craton. J Geol 115:355–374

Jones L, Mathieu PL, Strenger H (1960) Gravimétrie: Les résultats scientifiques des missions du syndicat pour l'étude géologique et minière de la Cuvette Congolaise et travaux connexes. Ann. Mus. Roy. Congo belge, Tervuren (Belgique), série in-8, Sci Géol, 36, 46 pp

Kadima EK (2011) Contribution géophysique à la connaissance du bassin de la Cuvette congolaise. Modélisation de la structure sédimentaire, Mécanisme de subsidence et structure de la lithosphère sous-jacente. PhD thesis, University of Lubumbashi, 278 p

Kadima E, Delvaux D, Sebagenzi MN, Tack L, Kabeya M (2011a) Structure and geological history of the Congo Basin: an integrated interpretation of gravity, magnetic and reflection seismic data. Basin Res 23:499–527

Kadima EK, Sebagenzi SMN, Lucazeau F (2011b) A Proterozoic-rift origin for the structure and evolution of the cratonic Congo Basin. Earth Planet Sci Lett 304:240–250

Kampunzu AB, Cailteux J (1999) Tectonic evolution of the Lufilian Arc during Neoproterozoic Pan African orogenesis. Gondwana Res 2:401–421

Kampunzu AB, Tembo F, Matheis G, Kapenda D, Huntsman-Mapila P (2009) Geochemistry and tectonic setting of mafic igneous units in the Neoproterozoic Katangan Basin, Central Africa: implication for Rodinia break up. Gondwana Res 3:125–153

Key RM, Liyungu AK, Njamu FM, Somwe V, Banda J, Mosley PN, Armstrong RA (2001) The western arm of the Lufilian Arc in NW Zambia and its potential for copper mineralisation. J Afr Earth Sci 33(3/4):503–528

Kipata ML (2013) Brittle tectonics in the Lufilian foldand-thrust belt and its foreland. An insight into the stress field record in relation to moving plates (Katanga, DRC). PhD Thesis, KULeuven, 130 p

Kipata ML, Delvaux D, Sebagenzi MN, Cailteux J-J, Sintubin M (2013) Brittle tectonic and stress field evolution in the Pan-African Lufilian arc and its foreland (Katanga, DRC): from orogenic compression to extensional collapse, transpressional inversion and transition to rifting. Geol Belg 16(1–2):001–017

Lawrence SR, Makazu MM (1988) Zaire's Central basin: prospectivity outlook. Oil Gas J 86(38):105–108

Lepersonne J (1974) Carte géologique du Zaïre au 1/2.000.000 et notice explicative. République du Zaïre, Direction de la Géologie, Kinshasa & Mus. Roy. Afr. Centr., Tervuren

Linol B (2013) Sedimentology and sequence stratigraphy of the Congo and Kalahari basins of South-Central Africa and their evolution during the formation and breakup of West-Gondwana. PhD thesis Nelson Mandela Metropolitan University, South Africa, 370 p

Master S, Rainaud C, Armstrong RA, Phillips D, Robb LJ (2005) Provenances ages of the Neoproterozoic Katanga Supergroup (Central African Copperbelt), with implications for basin evolution. J Afr Earth Sci 42:41–60

Monié P, Bosch D, Bruguier O, Vauchez A, Rolland Y, Nsungani P, Buta Neto A (2012) The Late Neoproterozoic/Early Palaeozoic evolution of the West Congo Belt of NW Angola: geochronological (U-Pb and Ar-Ar) and petrostructural constraints. Terra Nova 24:238–247. doi:10.1111/j.1365-3121.2012.01060

Pérez-Gussinyé M, Metois M, Fernandez M, Vergés J, Fullea J, Lowry A (2009) Effective elastic thickness of Africa and its relationship to other proxis for lithospheric structure and surface tectonics. Earth Planet Sci Lett 287:152–167

Poidevin J-L (1976) Les formations du Précambrien supérieur de la région de Bangui (R.C.A.). Bulletin de la Société Géologique de France 18(4):999–1003

Poidevin J-L (1979) Les basaltes et dolérites Précambrien supérieur de la région de Bakouma (Empire centrafricain) 7ème Réunion Annuelle des Sciences de la Terre., Lyon, 374

Poidevin J-L (1985) Le Protérozoïque supérieur de la République Centrafricaine. Ann. Mus. royal Afrique centrale, Tervuren (Belgique), série in-8, Sci Géol, 91, 75p.

Poidevin J-L, Alabert J, et Miauton J-D (1980/1981) Geologie des series du Precambrien superieur de la region de Bakouma (Republique Centrafricaine). Bull. BRGM (2eme ser.) IV(4): 311–318

Poidevin J-L (1996) Réponse. Un segment de rampe carbonatee d'age proterozoi'que supérieur au Nord du craton d'Afrique centrale (Sud-Est de la Republique centrafricaine). J Afr Earth Sci 23, 2

Poidevin J-L (2007) Stratigraphie isotopique du strontium et datations des formations carbonatées et glaciogéniques néoprotérozoïques du Nord et de l'Ouest du craton du Congo. Comptes Rendus Geoscience 339:259–273

Porada H (1989) Pan-african rifting and Orogenesis in Southern to Equatorial Africa and Eastern Brazil. Precambrian Res 44:103–136

Porada H, Berhorst V (2000) Towards a new understanding of the Neoproterozoic-Early Paleozoic Lufilian and northern Zambezi belts in Zambia and the Democratic Republic of Congo. J Afr Earth Sci 30:727–771

Priestley K, Mckenzie D, Debayle E, Pilidou S (2008) The Africa upper mantle and its relationship to tectonics and surface geology. Geophys J Int 175(3):1108–1126

Raucq P (1957) Contribution à la connaissance du Système de la Bushimay. Ann. Mus. Roy. Afr. Cent., Tervuren (Belgique), série in-8, Sci Géol, 18, 427 p

Raucq P (1970) Nouvelles acquisitions sur le système de la Bushimay (République démocratique du Congo). Ann. Mus. Roy. Afr. Cent., Tervuren (Belgique), série in-8, Sci Géol, 69, 156 p

Ritsema J, van Heijst H (2000) New seismic model of the upper mantle beneath Africa. Geology 28(1):63–68

Sahagian DL (1993) Structural evolution of African basins: stratigraphic synthesis. Basin Res 5:41–54

Sandwell DT, Smith WHF (1997) Marine gravity anomaly from Geosat and ERS 1 satellite altimetry. J Geophys Res 102:10–039

Stankiewicz J, de Witt M (2013) 3.5 billion years of reshaped Moho, southern Africa. Tectonophysics 609: 675–689

Tack L (1995) The Neoproterozoic Malagarazi supergroup of SE Burundi and its equivalent Bukoban Supergroup in NW Tanzania: a current review. In: Wendorff M, Tack L (eds) Late Proterozoic belts in Central and Southwestern Africa. Annales du Musée royal de l'Afrique Centrale, Sciences Géologiques, pp 121–129

Tack L, Wingate MTD, Liégeois J-P, Fernandez-Alonzo M, Deblond A (2001) Early Neoproterozoic magmatism (1000–910 Ma) of the Zadinian and Mayumbian Groups (Bas-Congo), onset of Rodinia rifting at the western edge of the Congo Craton. Precambrian Res 110:277–306

Tack L, Delvaux D, Kadima E, Delpomdor F, Tahon A, Dumont P, Hanon M, Fernandez-Alonso M, Baudet D, Dewaele S, Cibambula E, Kanda Nkula V, Mpiana Ch (2008) The 1000 m thick Redbeds sequence of the Congo River Basin (CRB): a generally overlooked testimony in Central Africa of post-Gondwana amalgamation (550) and pre-Karoo break-up (320 Ma). 22nd Colloquium African Geology (CAG22), Hammamet, Tunisia, Abstracts volume, 4–6 November 2008, pp 86–88

Tack L, Wingate MTD, De Waele B, Meert J, Belousova E, Griffin A, Tahon A, Fernadez-Alonso M (2010) The 1375 Ma "Kibaran event" in Central Africa: prominent emplacement of bimodal magmatism under extensional regime. Precambrian Res 180:63–84

Tait J, Delpomdor F, Préat A, Tack L, Straathof G, Kanda Nkula V (2011) Neoproterozoic sequences of the West Congo and Lindi/ Ubangi Supergroups in the Congo Craton, Central Africa. In: Arnaud E, Halverson GP, Shields-Zhou G (eds) The geological record of Neoproterozoic Glaciation, vol 36. Geol Soc, London, Mem, pp 185–194

Tapley B, Ries J, Bettadpur S, Chambers D, Cheng M, Condi F, Gunter B, Kang Z, Nagel P, Pastor R, Pekker T, Poole S, Wang F (2005) GGM02-An improved Earth gravity field model from GRACE. J Geod 79:467–478

Verbeek T (1970) Géologie et lithologie du Lindien (Précambrien Supérieur du nord de la République Démocratique du Congo). Ann. Mus. Roy. Afr. Cent., Tervuren (Belgique), série in-8, Sci Géol, 66, 311 pp

Vicat J-P, Pouclet A (1995) Nature du magmatisme lié à une extension pré-panafricaine: les dolérites des bassins de Comba et de Sembé-Ouesso (Congo). Bulletin de la Société Géologique de France 166(4):355–364

Vicat J-P, Vellutini P-J (1987) Sur la nature et la signification des dolérites du bassin précambrien de Sèmbe-Ouesso (République du Congo). Precambrian Res 37:57–69

Vicat J-P, Gioan P, Albouy Y, Cornacchia M, Giorgi L, Blondin P (1989) Mise en évidence, sur la bordure ouest du craton du Congo, de fosses d'effondrement d'âge Protérozoique supérieur, masqués par les formations phanérozoïques de la cuvette du Zaïre. Comptes Rendus de l'Académie des Sciences, série II A 309: 1207–1213

Villeneuve M (1983) Les sillons tectoniques du Precambrien superieur dans l'Est du Zaire. Comparaisons avec les directions du rift est-africain. Bull Centres Rech Explor-Prod Elf-Aquitaine 7(1): 163–174

Waleffe A (1988) Etude photogéologique du synclinorium de l'Itombwe et des régions avoisinantes au Sud du 3è parallèle Sud (Kivu, Zaire). Bulletin de la Société belge de Géologie 97(2): 211–221

7 Paleogeography and Tectono-Stratigraphy of Carboniferous-Permian and Triassic 'Karoo-Like' Sequences of the Congo Basin

Bastien Linol, Maarten J. de Wit, Erika Barton, Francois Guillocheau, Michiel C.J. de Wit, and Jean-Paul Colin†

7.1 Introduction

The Congo Basin (CB) of central Africa, also known as the 'Cuvette centrale', was first systematically explored geologically during the 1950s when this region was part of a Belgian colony. This early exploration included extensive field mapping, regional magnetic and seismic surveys (e.g. Jones et al. 1959; Evrard 1960), and drilling of two boreholes (Samba-1 and Dekese-1), each ca. 2 km deep (Cahen et al. 1959, 1960). This work defined the structural setting and stratigraphy of the CB, culminating in the completion of the Geological Map of Zaire (Lepersonne 1974), now the Democratic Republic of Congo (DRC). All research materials from this exploration period are archived in the Royal Museum for Central Africa (RMCA), in Tervuren, Belgium.

Subsequently, during the 1970s, renewed petroleum exploration led to the acquisition of additional geophysical data and the drilling by 'Esso-Zaire' of two deeper wells (Gilson-1 and Mbandaka-1), each ca. 4.5 km in depth (Esso-Zaire 1981a, b). This data has largely remained proprietary, and only short summaries of this work were published (Lawrence and Makazu 1988; Daly et al. 1991, 1992). Since then, very little new geologic or geophysical data has been acquired across the CB (e.g. Kadima et al. 2011; Sachse et al. 2012; Linol 2013). For that reason, the stratigraphy of this unusually large (ca. 1.8 million km^2) sedimentary basin remains very poorly characterized, preventing a more modern basin analysis. Today, there is still controversy about the basic definition of its major depositional sequences (i.e. supersequences) and the age of their bounding unconformities.

Here, new field observations are integrated with re-examination of the old literature (in French) and the available seismic- and well-data to resolve some of these major stratigraphic uncertainties, focusing on the 'Karoo-like' (Carboniferous-Permian and Triassic) sequences of the CB. In addition, U-Pb dating of detrital zircons from core samples of two of the deep boreholes characterizes the source provenances for these sediments and helps to constrain the paleogeography of central Africa during the late Paleozoic and early Mesozoic.

7.2 Geological Setting

The CB is surrounded by Precambrian basement and peripheral Pan African fold-and-thrust belts of Neoproterozoic carbonate and siliclastic platform sequences (Fig. 7.1; see

†Author was deceased at the time of publication.

B. Linol (✉) • E. Barton
AEON-ESSRI (African Earth Observatory Network – Earth Stewardship Science Research Institute), Nelson Mandela Metropolitan University, Port Elizabeth, South Africa

Geological Sciences, Nelson Mandela Metropolitan University, Port Elizabeth, South Africa
e-mail: bastien.aeon@gmail.com; erikabarton@yahoo.com

M.J. de Wit
AEON-ESSRI (African Earth Observatory Network – Earth Stewardship Science Research Institute), Nelson Mandela Metropolitan University, Port Elizabeth, South Africa
e-mail: maarten.dewit@nmmu.ac.za

F. Guillocheau
Géosciences-Rennes, UMR 6118 Université de Rennes 1 – CNRS, OSUR, Université de Rennes 1, Campus de Beaulieu, 35042 Rennes Cedex, France
e-mail: francois.guillocheau@univ-rennes1.fr

M.C.J. de Wit
Delrand Resources Pty Ltd, Toronto, Ontario, Canada
e-mail: mdewit@delrand.com

M.J. de Wit et al. (eds.), *Geology and Resource Potential of the Congo Basin*, Regional Geology Reviews,
DOI 10.1007/978-3-642-29482-2_7,

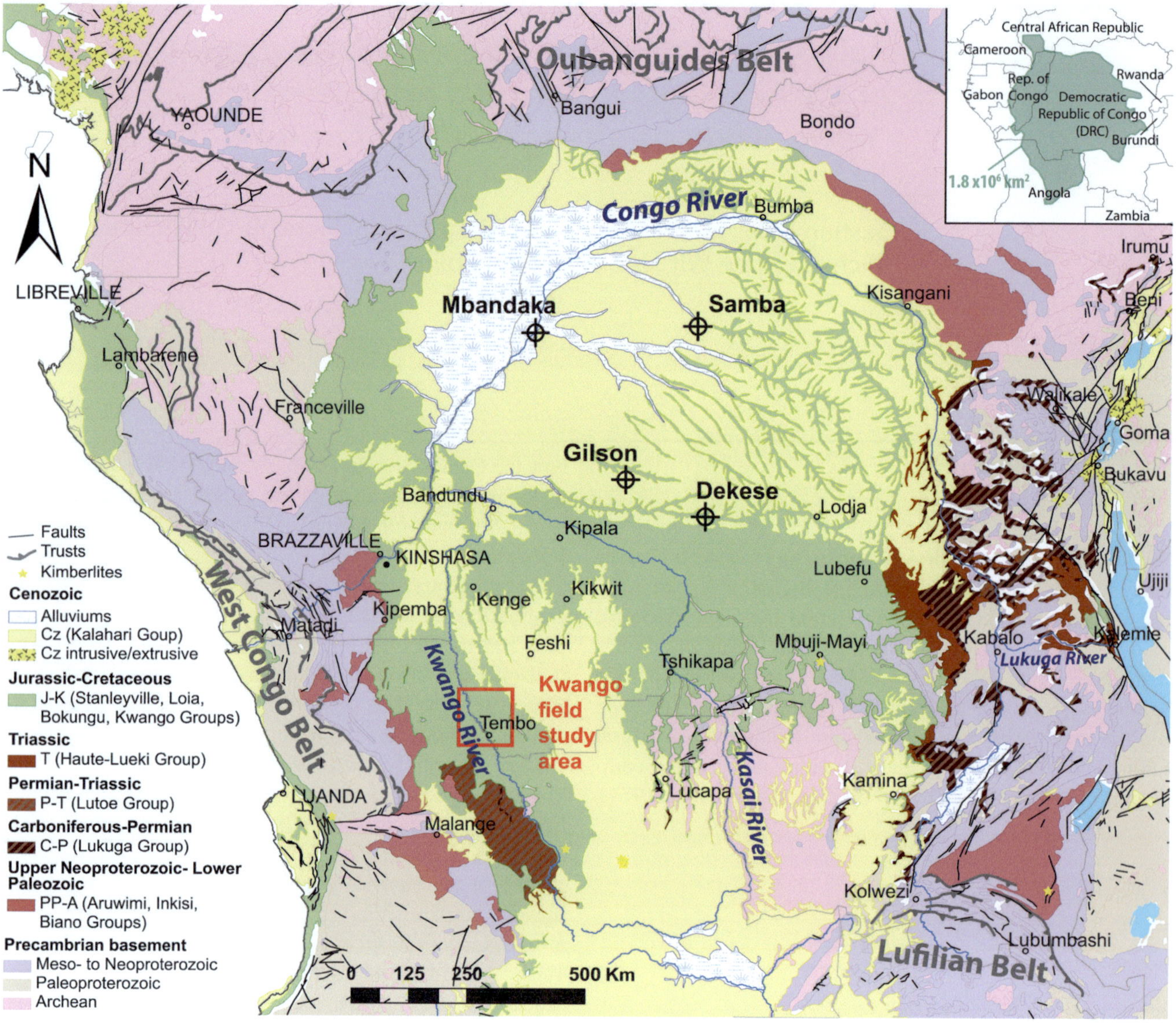

Fig. 7.1 Simplified geological map of the CB with location of the studied four deep boreholes and the field area in the Kwango Valley (*red box*). Precambrian geology is from de Wit et al. (1988), and Phanerozoic sedimentary cover compiled from various national geological maps (e.g. Lepersonne 1974; de Carvalho 1981; Desthieux 1995; Rolin 1995)

Chaps. 2 and 3, this Book). Along the margins of the basin these Pan African mobile belts are overlapped by widespread upper Neoproterozoic to lower Paleozoic 'red-beds' (e.g. Lepersonne 1978; Poidevin 1985), and Carboniferous to Triassic sequences that contain a rich Gondwanan flora, similar to those from southern Africa, Madagascar and India (Bose and Kar 1978; Cahen and Lepersonne 1978). In the CB, these successions are peneplained and in turn covered by Jurassic-Cretaceous and Cenozoic sediments (see Chaps. 8, 9 and 10, this Book).

The lowermost, upper Neoproterozoic to lower Paleozoic succession of red-beds comprises the Inkisi, Aruwimi and Biano (formerly Upper Kundelungu) Groups, each about 1,000–1,500 m thick (Lepersonne 1974; see also Chap. 6, this Book), unconformably overlying the West Congo, Oubanguides and Lufilian fold-and-thrust belts (Fig. 7.1). These sequences, mainly of quartzitic and conglomeratic red sandstones and siltstones, were deposited with regional paleocurrents to the south (Alvarez et al. 1995; Master et al. 2005), and all are considered to post-date the Pan African orogens (ca. 530–650 Ma; de Wit et al. 2008). Their stratigraphy, however, is poorly constrained by field observations and to date there are no reliable age constraints (Tait et al. 2011).

The overlying Carboniferous to Triassic sequences in the CB correspond to the Lukuga and Haute Lueki Groups that outcrop mostly in eastern DRC (Figs. 7.1 and 7.2). Along the southwestern margin of the basin, in northern Angola, their stratigraphic equivalents correspond to the Lutoe and Cassange Groups (de Carvalho 1981). These sequences

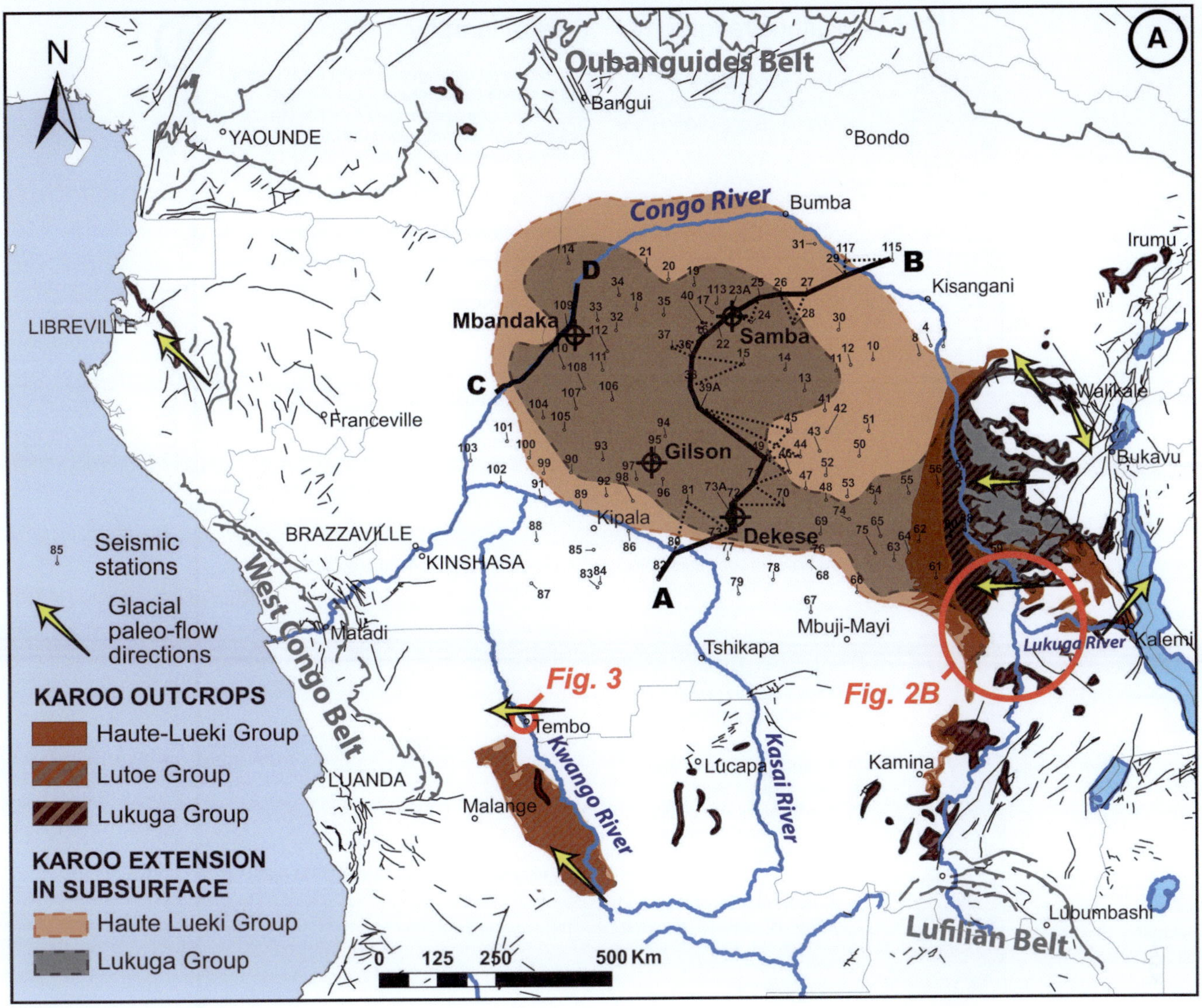

Fig. 7.2 (continued)

comprise glacial and periglacial, fossiliferous deposits with black shales that pass upward into arid continental sediments (Cahen and Lepersonne 1978; Cahen 1981), tracking the characteristic climatic evolution of central Gondwana during the late Paleozoic and early Mesozoic (e.g. Veevers et al. 1994; Milani and de Wit 2008). Similar Gondwana sequences, in east Africa (in Kenya and Tanzania; e.g. Kreuser 1984; Wopfner and Diekmann 1996), Madagascar (e.g. Rakotosolofo et al. 1999) and southern Africa (Johnson et al. 1996, 2006) are generally referred to as the Karoo Supergroup.

7.2.1 The Lukuga Group

The Lukuga Group is between about 300 m and 600 m thick along the eastern margin of the CB, best preserved within large west-facing paleo-glacial valleys (Fig. 7.2). Other glacial relics, also deposited with predominant paleocurrents oriented to the west, are found sporadically along the southern margin of the basin, in the Kasai and Kwango regions of southern DRC and northern Angola (Asselberghs 1947; Lepersonne 1951; Rocha-Campos 1976), and along the western and northern margins of the CB, in Gabon (the Agoula Series; e.g. Mounguengui et al. 2002) and Central African Republic (CAR) (the Mambere Formation; e.g. Rolin 1995).

The type-section of the Lukuga Group in eastern DRC is divided into two subgroups (Fig. 7.2B):

1. The Lower Subgroup comprises two formations of tillites: 'the lower and upper, mainly glacial beds' (Cahen and Lepersonne 1978), separated by black shales. It overlies a major unconformity on Precambrian

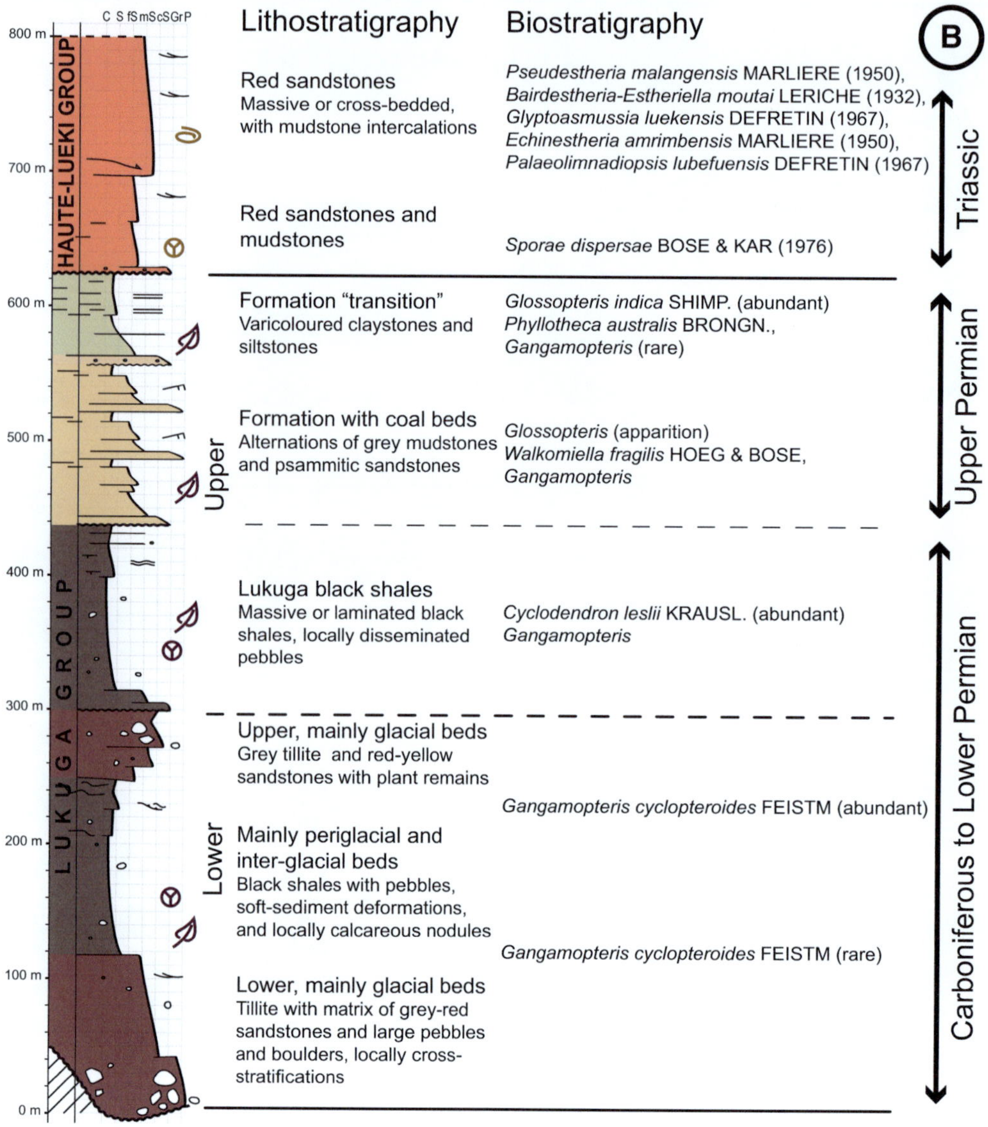

Fig. 7.2 (**A**) Map of Carboniferous-Permian and Triassic sequences of the CB, showing their vast extent in subsurface (as inferred from borehole and seismic data), and (**B**) stratigraphic type-section along the eastern margin of the basin (reconstructed from descriptions summarized in Cahen and Lepersonne (1978) and Cahen (1981)); see Figs. 7.4 and 7.10 for lithologic and biostratigraphic symbols. Seismic profiles A–B and C–D (*thick black lines*) are re-interpreted in Fig. 7.6 and 7.8, respectively

basement in the east (e.g. the Kibaran Belt), and is estimated to be between 100 m to 300 m thick in total, mainly based on the field studies of Fourmarier (1914), Jamotte (1932) and Boutakoff (1948).

2. The Upper Subgroup is known in more detail because of coal exploration and mining (Cahen and Lepersonne 1978). It includes 120 m thick black shales ('the Lukuga Black Shale'), overlain by a 20 m to 125 m thick formation of sandstones and mudstones with thin coal beds, and then a 'Transitional Formation' of varicolored mudstones, between 30 m and 65 m thick. The top of this sequence is an erosion surface, covered by a conglomerate bed at the base of the Haute Lueki Group.

The Lukuga Group is dated by paleobotany from the Carboniferous to the Upper Permian (Bose and Kar 1978). This compares well with other records of the late Paleozoic glaciation of Gondwana, such as the Dwyka and Ecca Groups of southern Africa (Johnson et al. 2006; Isbell et al. 2008). The lowermost glacial sediments are difficult to date accurately, however, in southern South Africa uppermost beds (the Waaipoort Formation) underlying the Dwyka Group contain plant remains no older than Middle Tournaisian (350 Ma) and have dropstones and soft sediment deformation structures of glacial origin (Streel and Theron 1999; Opdyke et al. 2001). Thus, the onset of the glaciation in southern (and central) Africa is likely to be

Early Carboniferous (e.g. Milani and de Wit 2008; Montañez and Poulsen 2013). The termination of the glaciation is constrained by palynostratigraphy to the Early Permian (e.g. Visser 1995; Modie and Le Hérissé 2009), and with geochronology on zircons from tuff horizons within the lowermost Ecca Group (the Prince Albert Formation), dated at 288 Ma (Bangert et al. 1999).

7.2.2 The Haute Lueki Group

The overlying Haute Lueki Group is between 50 m and 200–500 m thick in southeastern DRC, best preserved in NNW-trending Cenozoic rift valleys (e.g. Luama), west of Lake Tanganyika (Fig. 7.2). Outcrops are relatively poorly described (Cahen 1954; Lombard 1961) and generally divided into two:

1. A lower subgroup of grey-brown fine sandstones and siltstones with red mudstones, up to 150 m in thickness.
2. An upper subgroup of cross-bedded red sandstones, between 150 m and 300 m thick.

The Haute Lueki Group is dated locally in its lower part by palynology as Lower Triassic (Bose and Kar 1976). Ostracods and phyllopods collected from this group are common in the Cassange Group of Angola (Cahen 1981), which also yields Triassic fish fossils (e.g. Antunes et al. 1990). Thus, all these sequences of the CB are time-equivalent to the Beaufort Group of southern Africa (Catuneanu et al. 2005; Johnson et al. 2006).

7.3 New Stratigraphic Basin Analysis

Below new field observations are described from the Kwango region of the southwest CB, and the old seismic- and well-data in the center of the basin are re-examined to facilitate regional correlations of the 'Karoo-like' (Carboniferous-Permian and Triassic) sequences. This is then complemented with U-Pb dates of detrital zircons from core samples of two of the deep boreholes (Samba-1 and Dekese-1) to constrain the ages and the source provenances of these sediments.

7.3.1 Field Observations

Fieldwork was conducted (in 2008) along the Kwango River of the southwest DRC to document the structure and stratigraphy of exposed rocks and sediments in this remote region (Figs. 7.1 and 7.2A for location; Linol 2013). Here, Archean bedrock (the Dibaya Complex; Cahen et al. 1984) outcrops locally within waterfalls and rapids near Tembo, forming a major nick-point 150 m high (Fig. 7.3). This bedrock comprises mainly grey granites and migmatites (dated to >2.6 Ga; Cahen et al. 1984), strikes E-W, and dips 30–40° to the north to disappear in the CB beneath red sandstones of the Kwango Group (see Chap. 8, this Book).

Within the bottom half of the Kwango nick-point, a lowermost succession of well-bedded purplish-red conglomerates and siltstones was mapped along a 5 km long section, onlapping to the south and to the north across the crystalline bedrock (Fig. 7.3). This succession, maximum 30 m in thickness, comprises superimposed thick bars (1–3 m) of massive and cross-bedded conglomerates that grade upward into red siltstones with intercalations of micro-conglomerates (e.g. Fig. 7.3A). Pebble imbrications and cross-stratifications indicate river bed-load deposition with a general paleocurrent to the west (e.g. Fig. 7.3B). These conglomerates also contain boulder-size dropstones (e.g. Fig. 7.3C) that suggest a proximal glacial environment (Eyles and Eyles 1992). Thus, this lowermost fluvial-glacial succession can be correlated confidently to the other glaciogenic deposits of the Lukuga Group (Asselberghs 1947), and attests to the vast extent of ice-sheets across a paleo-relief located in east and south-central Africa during the Carboniferous (Dwyka) glaciation of Gondwana.

Field observations in the Kwango Valley show that this succession is tilted 5–15° to the south and to the west, and is unconformably overlain by tabular formations of white and red sandstones of the Jurassic-Cretaceous Kwango Group (Chap. 8, this book for details). This is consistent with the other descriptions of the Lukuga Group along the eastern margin of the CB (e.g. Jamotte 1932), but further field studies are needed to precisely characterize this angular unconformity regionally.

7.3.2 Seismic- and Well-Data

The two different seismic and drilling methods completed in the 1950s and in the 1970s represent the only subsurface dataset available to study the structure of the central CB.

7.3.2.1 Borehole Locations and Depths

The four deep boreholes drilled in the center of the CB (Figs. 7.1 and 7.2A for locations) were examined using the original core and cutting descriptions, the well logs (Cahen et al. 1959, 1960; Esso-Zaire 1981a, b), and by re-logging the cores that are stored at the RMCA museum in Belgium (Linol 2013). Figure 7.4 presents the four new reconstructed litho-stratigraphic sections: Samba, Dekese, Gilson and Mbandaka, together with overlapping seismic refraction data.

The Samba-1 borehole (0°09′45″N; 21°15′10″E) is located along the Maringa River, in the northeastern part of the central CB. It cuts through 1,167 m thick subhorizontal, red to green, bedded sandstones and mudstones (Units S1 to S5; Fig. 7.4), and then 871 m thick red quartzitic sandstones

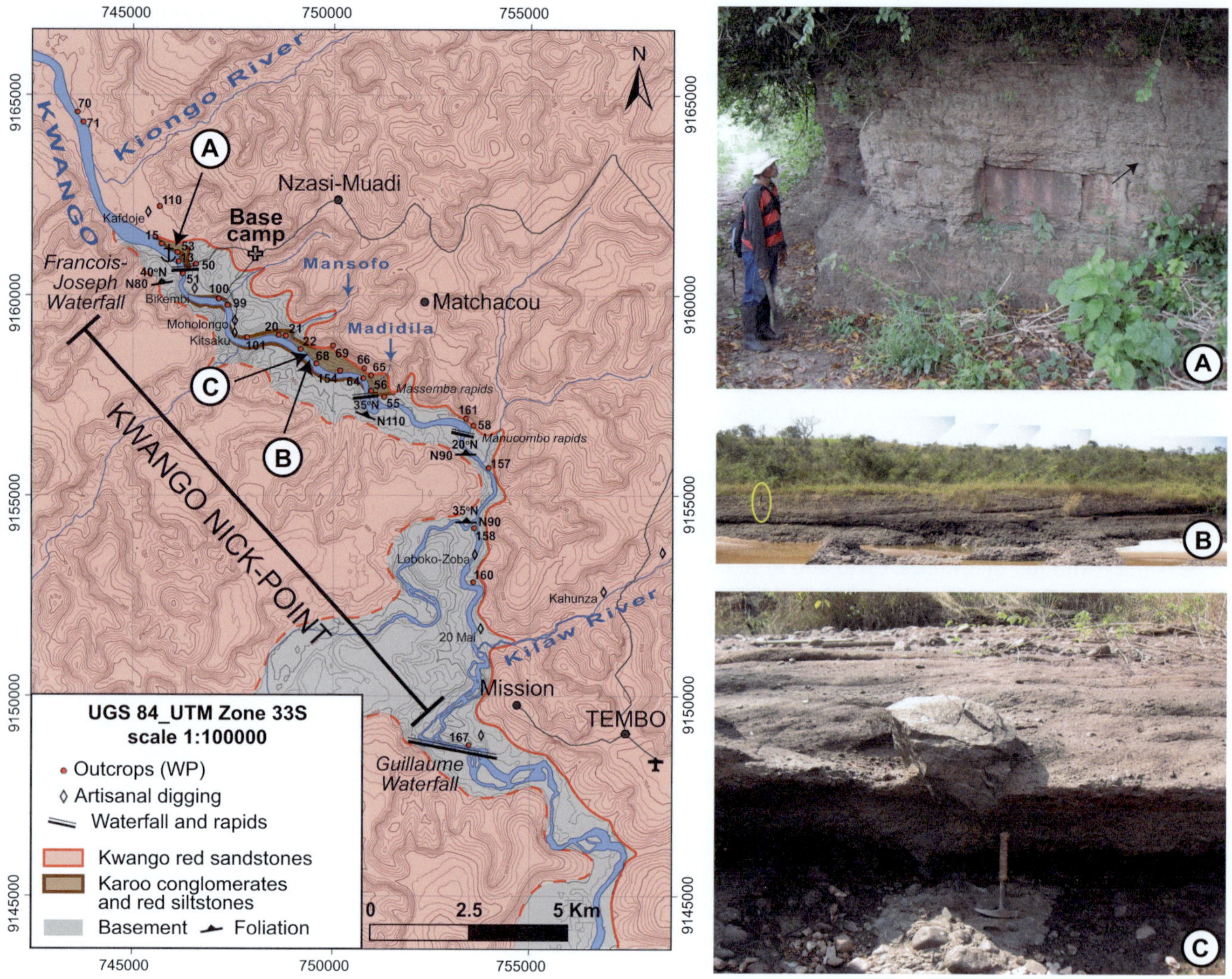

Fig. 7.3 Field map with location of outcrops and photos of fluvial-glacial conglomerates and red siltstones in the area surrounding the Kwango nick-point, between Tembo and Nzasi-Muadi, in southwestern DRC. (**A**) Red siltstones including lenses of micro-conglomerates (*arrow*) at the Port of Nzasi-Muadi (WP13). (**B**) Large-scale cross-bedded conglomerates with paleocurrents to the west (WP22); note person for scale (*yellow ellipse*). (**C**) Example of large boulder dropstone (WP22); hammer for scale

(Unit S6). The hole ends at a depth of −2,038 m, and does not reach the base of the quartzites.

The Dekese-1 borehole (3°27′26″S; 21°24′28″E) is located about 400 km to the south of Samba, in the south-central part of the basin. It cuts through 755 m thick subhorizontal, predominantly red sandstones (Units D1 to D6), and then 962 m thick soft-deformed (slumped) black shales and diamictites (Units D7 to D9). This section also rests on a basement of red quartzitic sandstones (Unit D10). The hole ends at a depth of −1,956 m, and does not reach the base of the quartzites (Fig. 7.4).

The Mbandaka-1 borehole (0°02′30″S; 18°14′10″E) is located along the Congo River, in the northwestern part of the central CB. It cuts through 844 m thick red to green bedded sandstones and mudstones (Units M1 to M4), and then 3,147 m thick red-brown quartzitic sandstones (Units M5 and M6), red-orange conglomerates (Unit M7) and grey-brown quartzitic and conglomeratic sandstones interbedded with siltstones (Units M8 and M9; Fig. 7.4). This section overlies carbonated siltstones and dolomites (Unit M10), and the hole stops at a depth of −4,343 m in a relative soft basement, which was interpreted to be halite (Esso-Zaire 1981a), as the deepest core recovered (at the depth of −4,147 m) shows large nodules of anhydrite (Fig. 7.5).

The Gilson-1 borehole (02°44′10″S; 19°54′30″E) is located about 350 km to the southeast of Mbandaka, in the south-central part of the basin. It cuts through 998 m thick interbedded red sandstones with mudstones (Units G1 to G5), and then 2,269 m thick red sandstones (Unit G6), dark brown siltstones (Unit G7) and red-brown mudstones and conglomeratic sandstones (Units G8 to G10). This section overlies 1,229 m thick interbedded sandstones with limestones (Unit G11) and dolomites (Unit G12). The hole

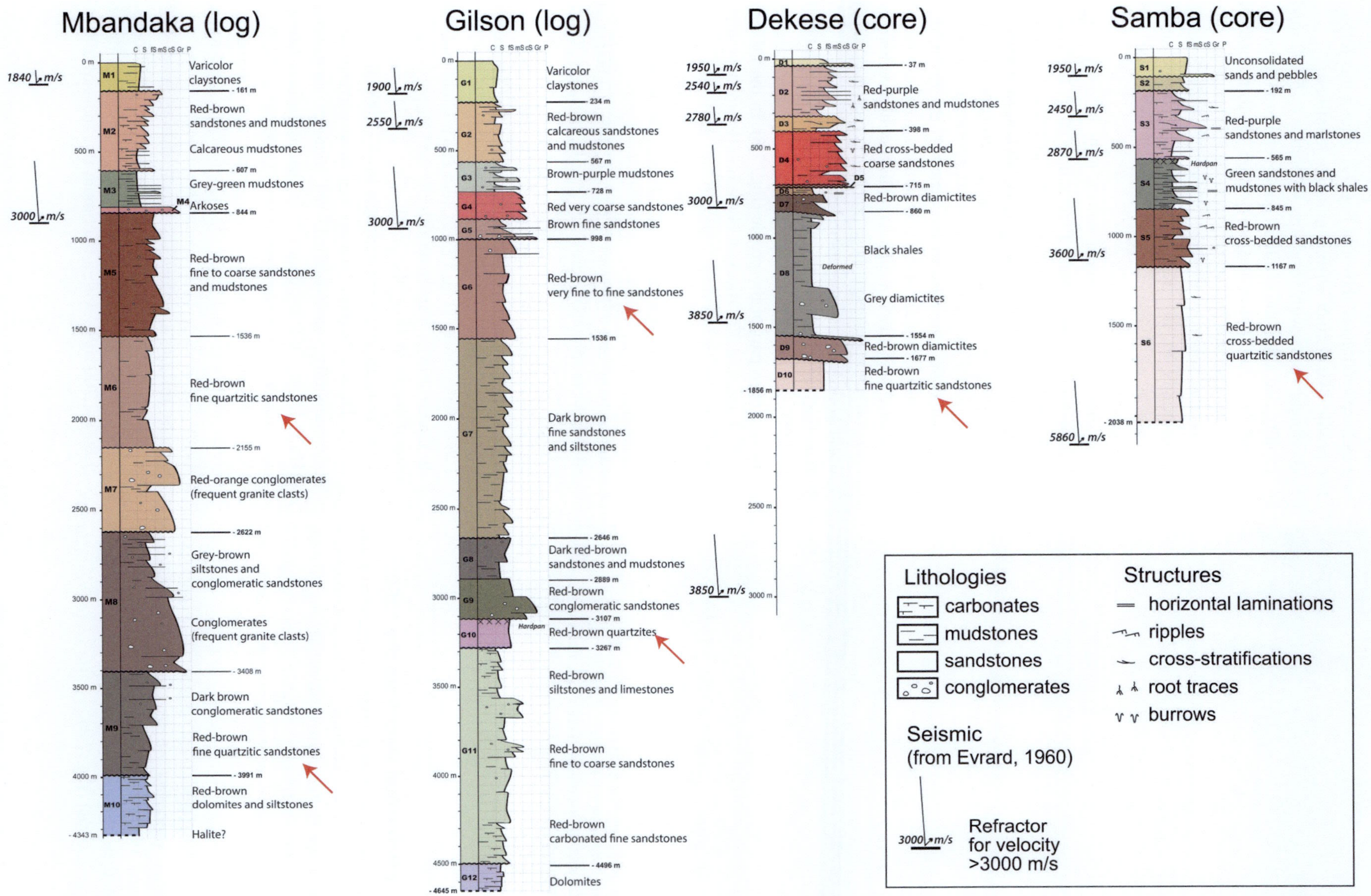

Fig. 7.4 Revised stratigraphic sections from the four deep boreholes drilled in the center of the CB in the 1950s and 1970s (Figs. 7.1 and 7.2A for locations). Note that the main geological contacts are also identified in seismic refraction analyses (Fig. 7.6). *Red arrows* show the different formations of red quartzites

Fig. 7.5 Photo of drill core (No. 8) from the depth interval −4,139 m to −4,147 m in the Mbandaka-1 borehole. The cut face of the core shows thinly laminated (stromatolithic?) dolomitic fine sands and dark shales with large nodules of anhydrite. The core is 90 mm in diameter

terminates at a depth of −4,645 m in the dolomites (Fig. 7.4).

7.3.2.2 Seismic Refraction Data from the 1950s

The old (1952–1956) seismic refraction surveys essentially determined the sediment thicknesses at different locations (e.g. 111 stations) across the entire central CB (Fig. 7.2A). The depths of the different seismic discontinuities (reflectors) were calculated at each seismic station, and because the sedimentary packages were considered largely tabular and the horizontal velocities much higher than the vertical velocities, the calculated depths were corrected using an arbitrary reduction of 10 % on the observed velocities (Evrard 1960). Comparison of results with the well-data shows that all the main geological contacts were systematically determined (Fig. 7.4).

An 850 km long, N–S reference profile across the Samba and Dekese sections is shown in Fig. 7.6. Along this refraction profile five distinct seismic units are identified:

1. Basement rocks with very high velocities (>5,200 m/s) rapidly rise along the southern part of the profile, from a depth of −3,000 m at Dekese (station 72) to a depth of −400 m (station 82) over a distance of 250 km (Fig. 7.6). Along the southern margin of the CB, this bedrock corresponds to the Kasai and NE Angola Cratons (Cahen et al. 1984), and here named the Kasai and Kwango Highs (Fig. 7.7).
2. A marked seismic contrast with a high velocity range (4,200–4,600 m/s) defines a tectonic high in the center of the profile (between stations 39A and 49), and progressively rises northward to the surface (station 115) where it links with outcropping upper Neoproterozoic quartzitic sandstones of the Aruwimi Group (Lepersonne 1974). In the four deep boreholes in the center of the basin, this seismic discontinuity also corresponds to the tops of quartzitic sandstones, which have distinct high velocities (>4,200 m/s). Although these lithologies (e.g. quartzites) occur at different stratigraphic levels (see Fig. 7.4).
3. A distinct seismic unit with intermediate velocities (3,600 to 3,900 m/s) occurs in a limited zone, 170 km wide, of the southern part of the profile (Fig. 7.6). In the Dekese section (station 72), this unit corresponds to 694 m thick black shales with diamictites (Unit D8), dated by palynostratigraphy to the Permian and attributed to the Lukuga Group (Cahen et al. 1960; Boulouard and Calandra 1963).
4. An overlying unit of lower velocities (2,400 to 3,600 m/s) stretches continuously across the profile, between 300 m (station 73) to maximum 1,000 m thick at Samba (station 23A). In the four deep boreholes, it is dated biostratigraphically to the Upper Jurassic and middle Cretaceous (Cahen et al. 1959, 1960; Grekoff 1960; Defretin-Lefranc 1967; Maheshwari et al. 1977; Colin 1981; Colin and Jan du Chêne 1981). This main seismic unit displays major lateral variation along the profile (Fig. 7.6). In the south the seismic velocities are relatively low (~3,000 m/s), and in the Dekese section (station 72) these correspond to predominantly cross-bedded red coarse sandstones (Units D2 to D5). By contrast, in the north their lateral equivalents have higher velocities (~3,600 m/s), which in the Samba section (station 23A) correspond to alternating green mudstones and black shales with lesser sandstones (Units S3 to S5). This discernible change in seismic velocity reflects a major N-S lateral change in the late Mesozoic lithostratigraphy (see also Chap. 8, this Book).
5. A thin uppermost seismic unit of low velocity range (1,800–2,200 m/s), with a maximum thickness of 200 m

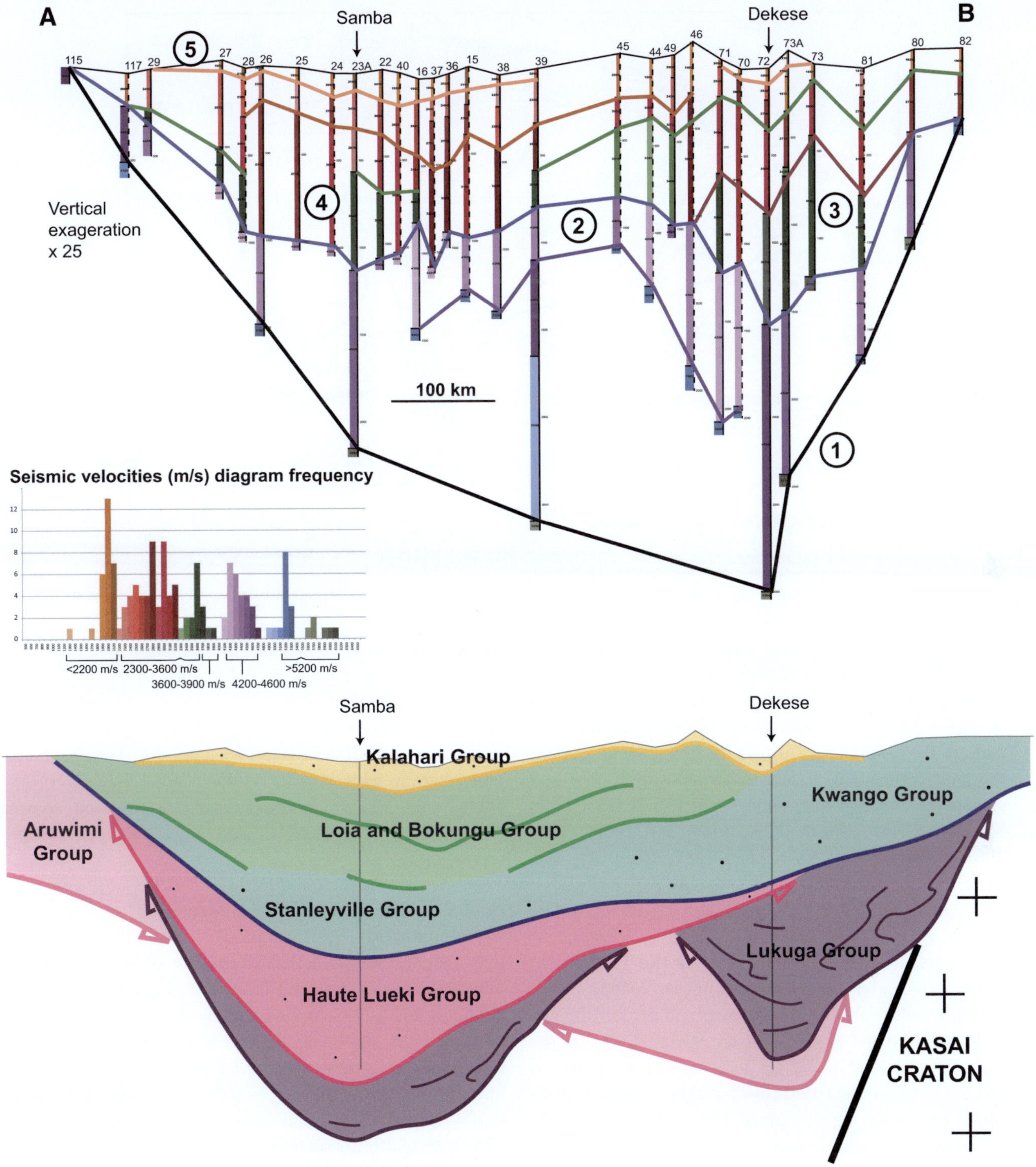

Fig. 7.6 N–S seismic refraction profile (data are from Evrard 1960), 850 km long across the Samba and Dekese sections (Fig. 7.2A for location). The *upper* profile shows five distinct units separated by seismic discontinuities: (1) basement rocks (>5,200 m/s); (2) quartzitic sandstones (4,200–4,600 m/s); (3) black shales and diamictites (3,600–3,900 m/s); (4) poorly consolidated sandstones and mudstones (3,000 and 3,300–3,600 m/s); and (5) uppermost unconsolidated material (<2,200 m/s). *Lower figure* is a schematic interpretation (*arrows* pointing up represent cross-cutting relationships)

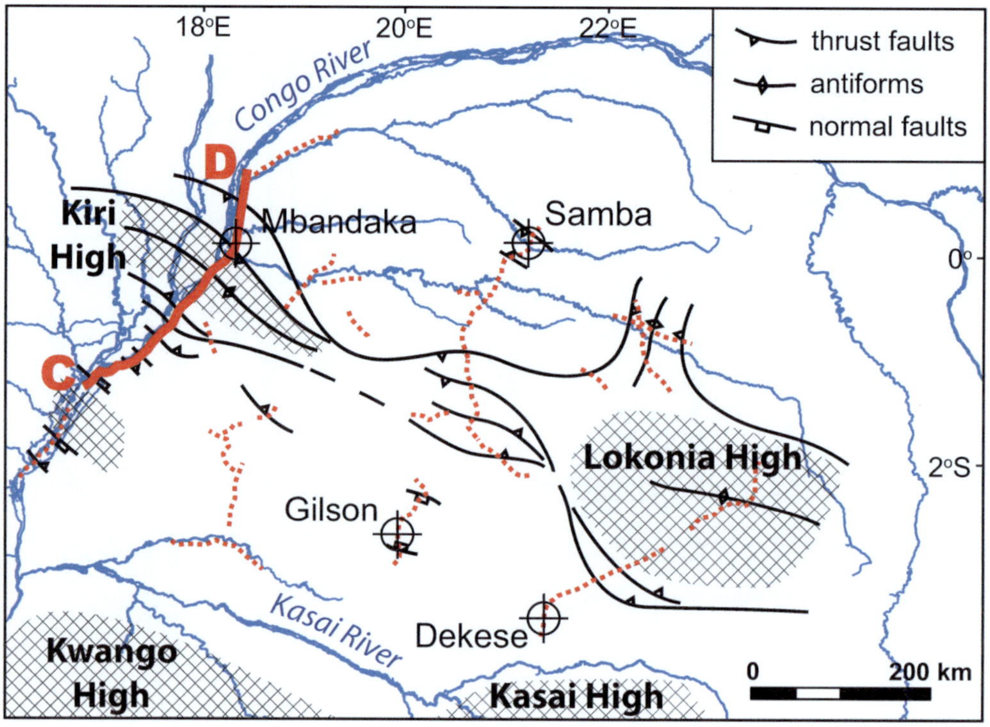

Fig. 7.7 Structural sketch map of the central CB with location of seismic reflection lines and the four deep boreholes (modified from Daly et al. 1992). The profile C–D (*thick red line*) is re-interpreted in Fig. 7.8. The structures may represent a large-scale NW–SE 'flower structure', or an extensional tilt (see text for further explanation)

at station 22, occurs both in the south and in the northern parts of the profile (Fig. 7.6). In the boreholes (Fig. 7.4), this unit corresponds to unconsolidated sandstones and claystones (<2,200 m/s), designated to the (Cenozoic) Kalahari Group (Cahen et al. 1959, and 1960; Colin 1981; Colin and Jan du Chêne 1981; Chap. 10, this Book).

7.3.2.3 Seismic Reflection Data from the 1970s

Between 1974 and 1976, petroleum exploration work acquired 36 reflection lines along rivers and roads, discontinuously between the Congo and the Kasai Rivers in the center of the basin (Fig. 7.7 for locations). The first interpretation of these data described a series of two NW-trending basement highs, named the Kiri and Lokonia Highs (Lawrence and Makazu 1988; Daly et al. 1991, 1992). Along the seismic profiles (e.g. Fig. 7.8), truncations of folded sedimentary packages along the flanks of these highs clearly indicate compressive deformation. Accordingly, the highs were interpreted to represent east- and west-verging fold-and-thrust structures ('flower structure' or 'pop-up'; Daly et al. 1991). More recent studies (Crosby et al. 2010; Kadima et al. 2011) have proposed that these reflection data may also be interpreted as a series of rifts and horsts bounded by normal faults, and further suggested the probable influence of salt tectonic associated with deep lying Neoproterozoic evaporites. Thus, whilst the principal structural framework of the CB has been presented as a reticulate fault pattern with transfer faults accommodating offsets between the Kiri and Lokonia Highs, the available seismic data are of poor quality and would require modern reprocessing.

A NE-SW reflection profile, 170 km long along the Congo River and across the Mbandaka section is re-interpreted in Fig. 7.8. This profile shows three major seismic unconformity-bounded sequences, or supersequences, which overlie a 'presumed basement of complex structure' (Daly et al. 1992) capped locally by distinct reflectors of very high amplitude. In the Mbandaka section, these reflectors correspond to massive dolomites grading upward into interbedded brown sandstones with limestones (Unit M10) that can be correlated to Pan African deformed carbonate platforms surrounding the CB, such as for example the upper Neoproterozoic West Congolian Group along the western margin of the basin (Tack et al. 2001; Frimmel et al. 2006; Chap. 3, this book). Along the seismic profile (Fig. 7.8), the carbonates are truncated by a regional erosion surface (interpreted by Daly et al. 1992 as an early Pan African unconformity). This first angular unconformity (U1; Figs. 7.8 and 7.9), located at a depth of −3,991 m in the Mbandaka-1 borehole, is considered to represent the base of the Phanerozoic CB, above which the following three supersequences are recognized:

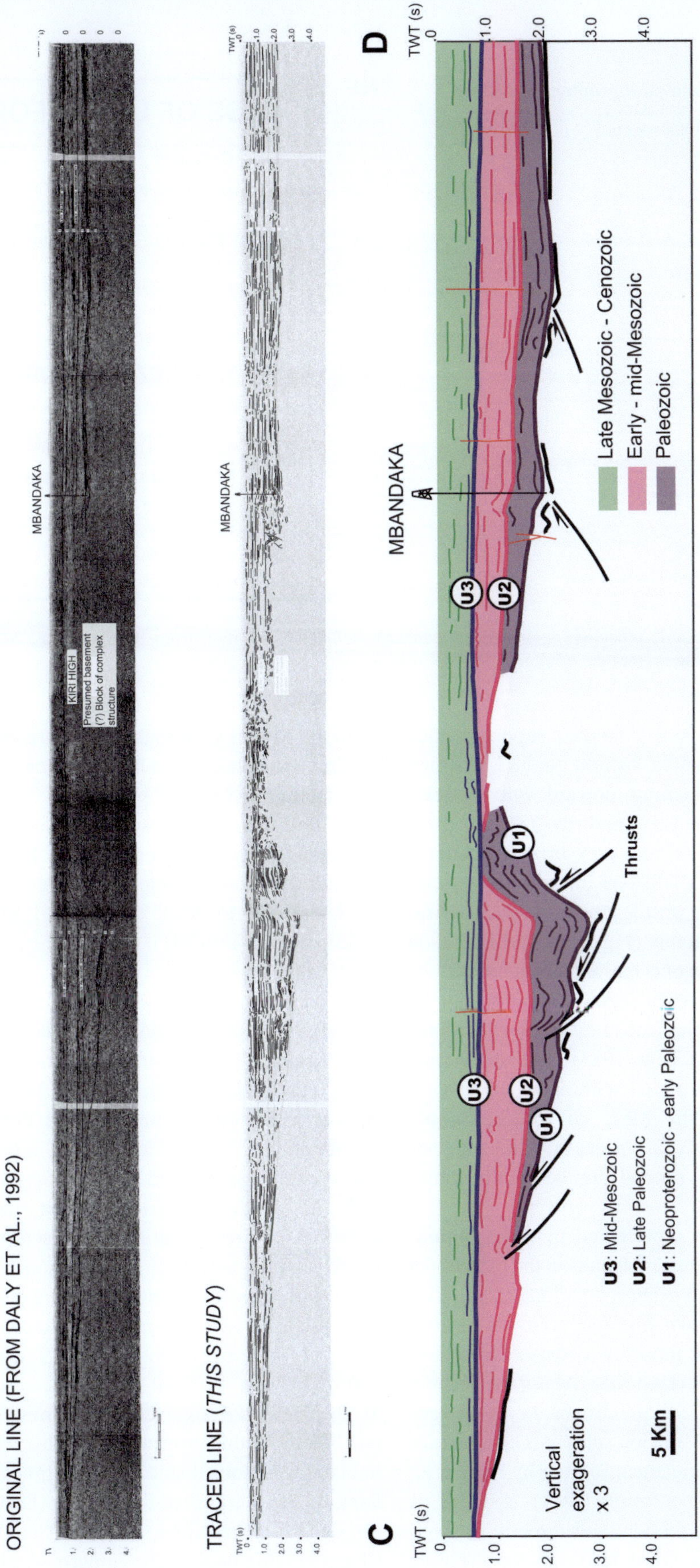

Fig. 7.8 Re-interpreted NE–SW seismic reflection profile (data are from Daly et al. 1992), 170 km long along the Congo River and across the Mbandaka section (Figs. 7.2A and 7.7 for location). This profile shows three seismic supersequences bounded by regional unconformities (U1, U2, and U3). The time-depth relationship at borehole Mbandaka is shown in Fig. 7.9

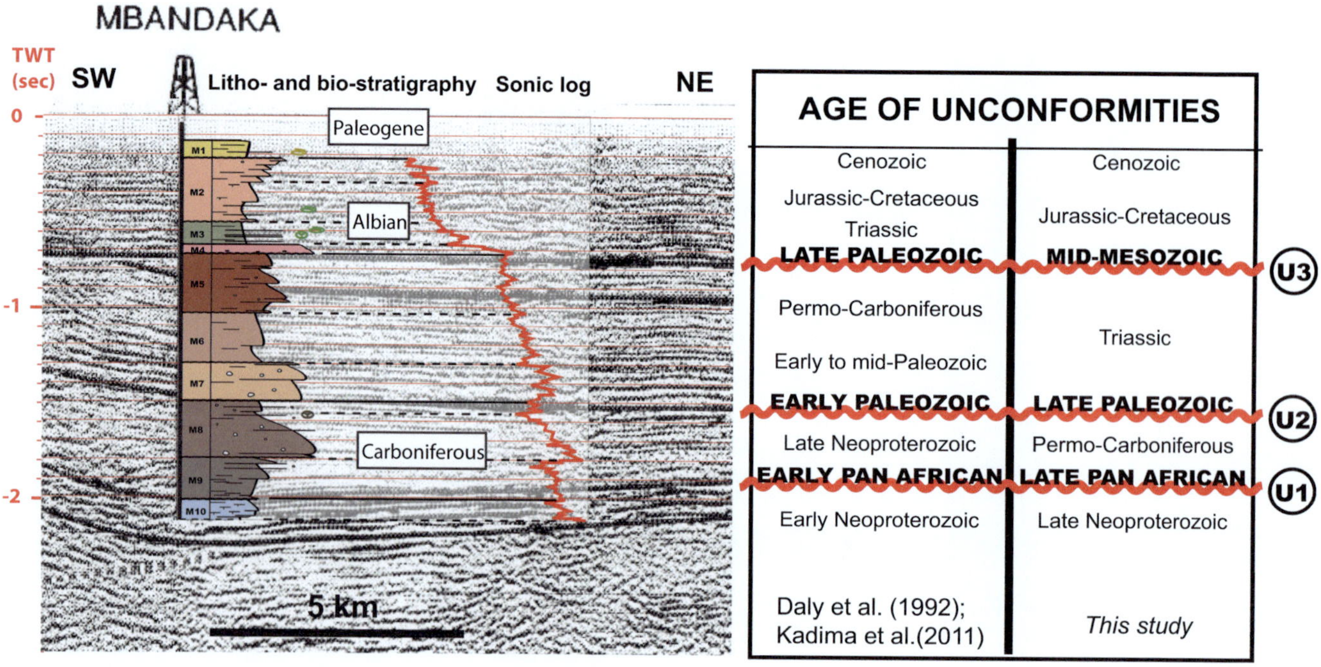

Fig. 7.9 Re-calibration of the seismic reflection against the Mbandaka well logs (modified from Kadima et al. 2011), and ages of major unconformities. Biostratigraphic data is from the original study of Colin and Jan du Chêne (1981)

1. A lowermost, thick and weakly banded supersequence corresponds in the Mbandaka section to 583 m thick red-brown siltstones coarsening upward into conglomeratic sandstones (Unit M9) and to 786 m thick conglomerates fining upward into grey-brown conglomeratic sandstones with dark siltstone intercalations (Unit M8). Along the seismic profile these two lithostratigraphic units are conformable (Fig. 7.8), and in unison they are truncated by a second regional erosion surface (U2; interpreted as an early Paleozoic unconformity by Daly et al. 1992). A palynological study of the Mbandaka section (Colin and Jan du Chêne 1981) indicates an age younger than mid-Carboniferous within the underlying grey-brown siltstones (Unit M8), on a core-sample between −2,805 m and −2,825 m depth. This suggests a late Paleozoic age for this second angular unconformity (U2; Figs. 7.8 and 7.9).
2. A thick middle succession represented by a discontinuously banded supersequence corresponds in the Mbandaka section to 467 m thick red-orange conglomerates (Unit M7), 619 m thick red-brown fine quartzitic sandstones (Unit M6) and 692 m thick red-brown coarse sandstones with siltstone intercalations (Unit M5). Along the seismic profile (Fig. 7.8), these units are truncated by a very conspicuous, third regional erosion surface (U3) that was interpreted as a late Paleozoic unconformity by Daly et al. (1992). However, this unconformity is at a depth of −844 m in the Mbandaka section, and directly covered by 40 m thick red arkoses (Unit M4) and fossiliferous grey-green mudstones (Unit M3) dated by biostratigraphy to the Albian-Cenomanian (Colin and Jan du Chêne 1981). This rather suggests a late Mesozoic age for the underlying unconformity (U3; Figs. 7.8 and 7.9).
3. A third and uppermost supersequence corresponds in the Mbandaka section to 683 m thick red arkoses and green mudstones coarsening upward into carbonated sandstones (Units M2 to M4), and to 161 m thick varicoloured claystones (Unit M1), dated to the middle Cretaceous and Paleogene, respectively (Colin and Jan du Chêne 1981).

In summary, the seismic reflection data from the 1970s reveals a Phanerozoic development of the CB punctuated by two regional episodes of deformation that previously were interpreted by Daly et al. (1992) to be related to: (1) the Pan African collisions during the end Precambrian and early Cambrian (~530–650 Ma), and (2) a possible phase of distant contraction induced by development of the Cape Fold Belt in southern Africa during the Permian-Triassic (~250 Ma). A recalibration of the seismic reflection against the Mbandaka well logs (i.e. Integrated Travel Time of the Sonic Log), tied to the litho- and bio-stratigraphy (Fig. 7.9), shows that the two lower, Carboniferous-Permian and Triassic supersequences (Units M5 to M9) unconformably overlie deformed Pan African carbonate-siliclastic rocks (Unit M10). These two supersequences are separated by a regional late Paleozoic unconformity (U2) that possibly relates to far-field deformations resulting of multiple collisional processes flanking the margins of Gondwana (Daly et al. 1992; Trouw and de Wit 1999; Newton et al. 2006; Dabo et al. 2008; see Chap. 11, this book). The uppermost unconformity (U3) is overlain by Jurassic-Cretaceous and Cenozoic sediments

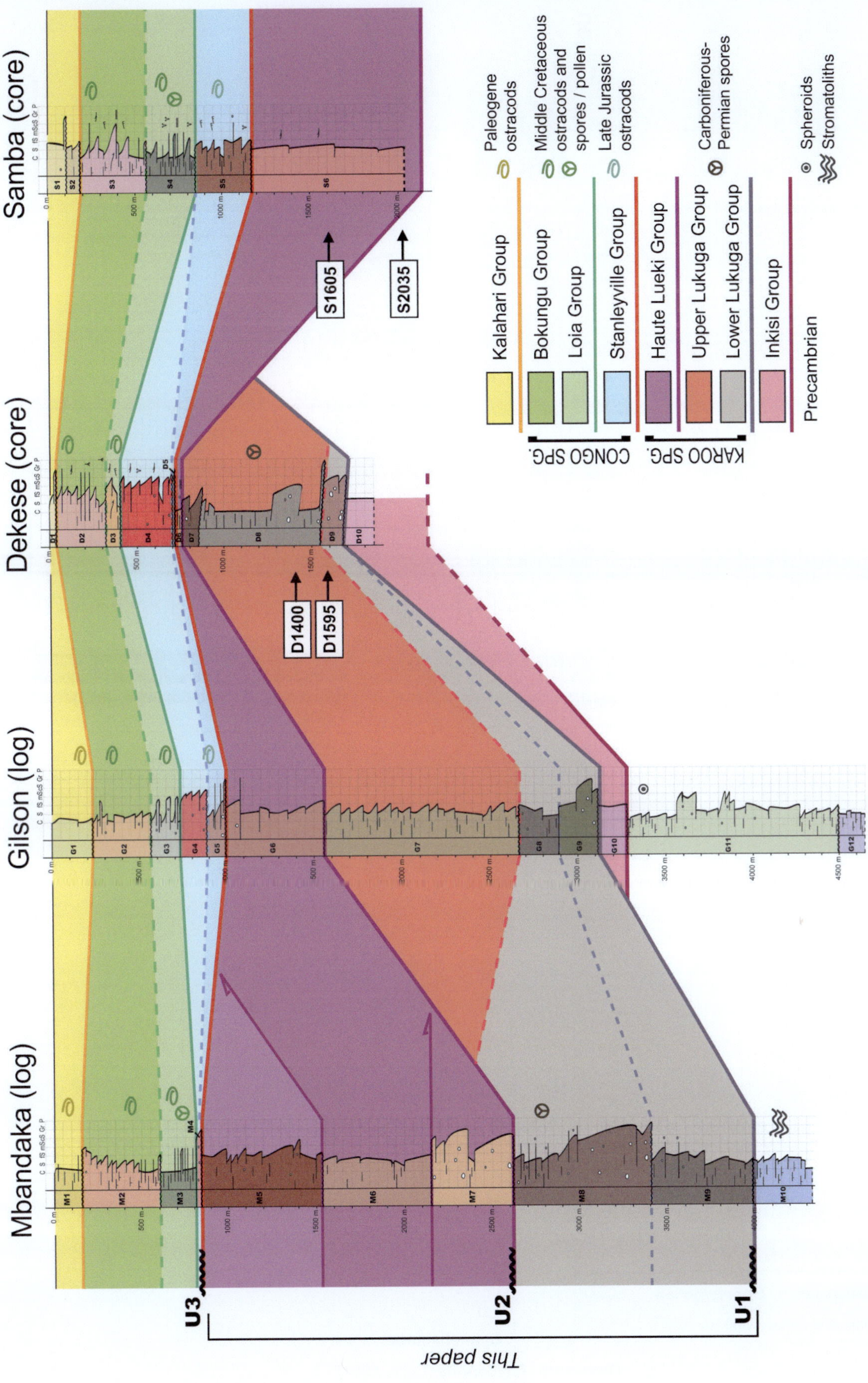

Fig. 7.10 Litho- and bio-stratigraphic correlations of the Gilson, Mbandaka, Samba and Dekese sections, with location of core-samples with dated detrital zircons

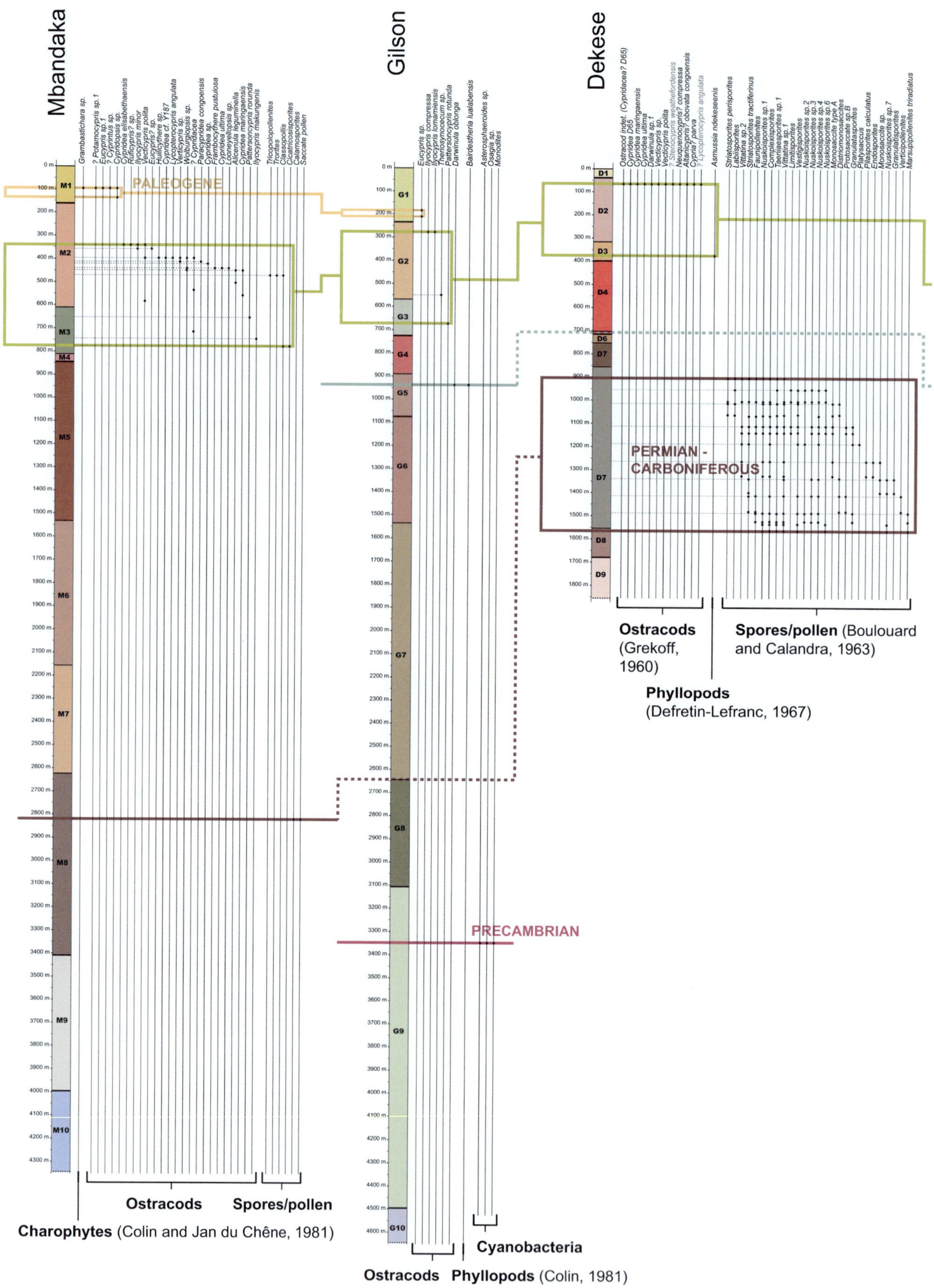

Fig. 7.11 (continued)

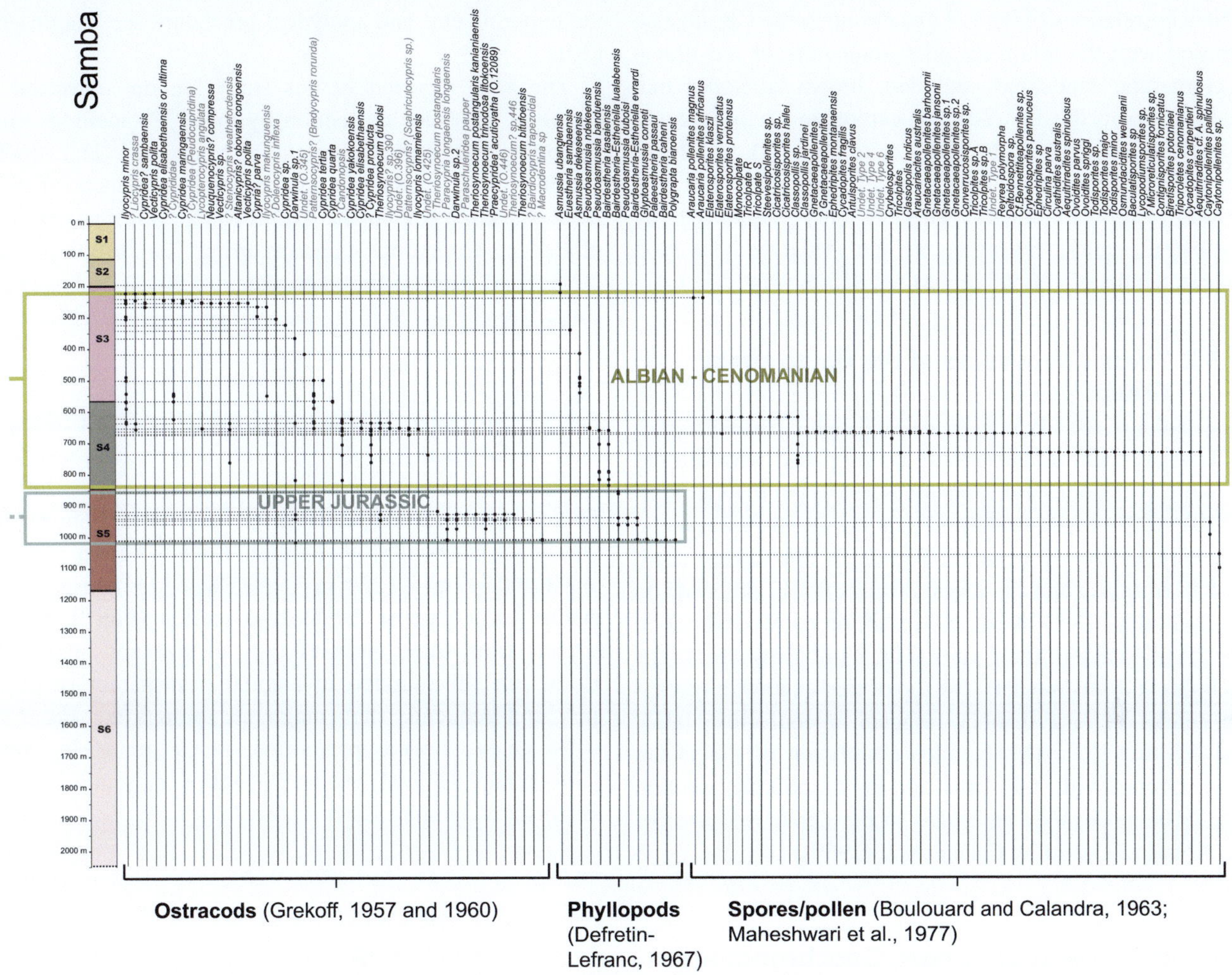

Fig. 7.11 Revised biostratigraphic data of the Gilson, Mbandaka, Samba and Dekese sections

(Units M1 to M4), and is ascribed to the late Mesozoic (Fig. 7.9). It is here postulated to be related to intracontinental deformation associated with the initial period of break-up of Gondwana, ca. 160–180 Ma (see also Chap. 13, this Book).

7.3.3 Seismic-, Litho- and Bio-Stratigraphic Correlations

Based on seismic-, litho- and bio-stratigraphic analyses, new correlations of the Carboniferous-Permian and Triassic ('Karoo-like') sequences are proposed between the type-sections of the Lukuga and Haute-Lueki Groups along the eastern margin of the CB (Fig. 7.2B), and the four deep boreholes in the center of the basin (Fig. 7.10). The available biostratigraphic data of the four borehole-sections is revised in Fig. 7.11.

7.3.3.1 Carboniferous-Permian Sequences of the Lukuga Group

In the lower part of the Dekese section (Fig. 7.10), unusually thick and slumped diamictites and black shales (Units D7 to D9), totaling 962 m in thickness, were correlated to the Lukuga Group by Cahen et al. (1960). In this section, palynology within the Unit D8 (Fig. 7.11), between depths of −905 m and −1,565 m, suggests an age ranging from Lower to Upper Permian with the predominance of species: *Nuskoïsporites*, *Vertigisporites* and *Vittatina* (Boulouard and Calandra 1963). These results from the 1960s would need to be re-examined in the light of more recent palynostratigraphic works from south-central and northern Africa (e.g. Modie and Le Hérissé 2009; Gonzalez et al. 2011). Nevertheless, the thick sequence of black shales (Unit D8) in the CB can be correlated confidentially with the distinctive (organic-rich) Permian black shales of the Prince Albert and Whitehill Formations of the Lower Ecca Group in the main Karoo Basin of southern Africa (Johnson et al. 2006).

New correlations of the Lukuga Group in the CB are also proposed with the relatively thick sequences of red-brown conglomeratic sandstones and dark brown siltstones that directly overlie the Pan African carbonate rocks in both, the Gilson (Units G7 to G9, in total 1,571 m thick; Fig. 7.10) and the Mbandaka sections (Units M8 and M9, in total 1,369 m thick). This is only supported by palynostratigraphy, however, in the Mbandaka section between depths of −2,805 m and −2,825 m (Fig. 7.11).

7.3.3.2 Triassic Sequences of the Haute Lueki Group

In the center of the CB, new lithostratigraphic correlations of the Triassic Haute Lueki Group are proposed with the red quartzitic sandstones and mudstones overlying the Carboniferous-Permian black shales and dark brown siltstones in the Dekese (Unit D6, 40 m thick; Fig. 7.10), the Gilson (Unit G6, 538 m thick) and the Mbandaka sections (Units M5 to M7, in total 1,778 m thick). No fossil, or spores and pollen were found in any of these units (Fig. 7.11). However, in the boreholes and seismic data, the top of this quartzitic succession marks a very distinct regional unconformity (U3; Figs. 7.8 and 7.9). This enables correlation of the Haute Lueki Group with the Unit S6 of the Samba section (Fig. 7.10), previously attributed to the upper Neoproterozoic Aruwimi Group by Cahen et al. (1959).

7.4 U-Pb Detrital Zircons Geochronology of the Samba and Dekese Cores

U-Pb geochronology of detrital zircons collected from the Samba and Dekese cores (Fig. 7.10 for sample locations) is used to further constrain the stratigraphy and characterize the source provenances of these Carboniferous-Permian and Triassic sequences filling the CB, which in turn help to reconstruct the basin's paleogeography (e.g. Fedo et al. 2003).

255 zircons were analyzed from the Lukuga and Haute Lueki Groups (Tables 7.1 and 7.2; see Linol 2013 and Linol et al. 2014 for details), with two samples from each group (about 50–80 analyses per sample). U-Pb dating was done using AEON's high resolution Multi-Collector Inductively Coupled Plasma Mass Spectrometer (MC-ICP-MS), coupled to a UP 193 solid-state laser system (New Wave Research) and a desolvation nebulizer system (DSN-100, Nu Instruments). The Nu Plasma MC-ICP-MS is equipped with a special collector block allowing for simultaneous detection of ion signals from masses ^{238}U to ^{203}Tl. U and Tl isotopes are measured on Faraday cups, while Pb isotopes are measured on ion counters. For more information about the experimental and analytical procedure see Biggin et al. (2011).

The ages reported in this study are not corrected for common Pb, since correction by the ^{204}Pb method cannot be applied due to the isobaric interference of ^{204}Hg, which is a contaminant of the He gas carrying the ablated material from the laser cell to the MC-ICP-MS. All analyses were plotted on Concordia diagrams using Isoplot 3.0 (Ludwig 2003), and most concordant dates (>80 % and <110 % conc.) are compared in frequency age-diagrams (Fig. 7.12).

7.4.1 Samples D1400 and D1595 (The Lukuga Group)

Two samples: D1400 and D1595 were taken from diamictites of the Carboniferous-Permian Lukuga Group, at depths of −1,400 m and −1,595 m (Units D8 and D9), respectively, in the lower part of the Dekese section (Fig. 7.10). The zircons from these samples vary in size and shape. The grains are sub-rounded to angular, transparent or smoky grey, and frequently show magmatic zonings (e.g. Fig. 7.13).

74 analyses were performed on 69 grains in sample D1400, and 83 analyses on 79 grains in sample D1595. Results are summarized in Table 7.1.

7.4.2 Samples S1605 and S2035 (The Haute Lueki Group)

Two samples: S1605 and S2035 were taken from red quartzitic sandstones of the Triassic Haute Lueki Group at depths of −1,605 m and −2,035 m (Unit S6), respectively, in the lower part of the Samba section (Fig. 7.10). The zircons from this unit are generally small (100–250 μm), elongated, transparent or reddish, and with crystal-shapes that indicate relatively limited transport (e.g. Fig. 7.14). The grains show magmatic zoning and some xenotime overgrows.

64 analyses were performed on 59 grains in sample S1605, and 60 analyses on 48 grains in sample S2035. Results are summarized in Table 7.2.

7.4.3 Synthesis and Sources Provenance Analysis

In all samples of the Lukuga and Haute Lueki Groups, four main age-populations were found: 1) early Paleoproterozoic to Archean, 2) mid-Paleoproterozoic (Eburnian), 3) early Neoproterozoic to Mesoproterozoic (Kibaran), and 4) Cambrian to late Neoproterozoic (Pan African), although there

Table 7.1 Summary of U-Pb detrital zircon dates from samples D1400 and D1595 (the Lukuga Group)

Age-population	U-Pb Dates ($^{206}Pb/^{207}Pb$)	Notes
1. Mesoarchean	One (oldest) zircon dates at 2925 ± 8 Ma	
2. Early Paleoproterozoic to Neoarchean	10 zircons date between 2.4 Ga and 2.7 Ga, with 9 concentrated between 2566 ± 17 Ma and 2690 ± 16 Ma	Indicate some contributions from Neoarchean aged-terrains. These dates are similar to dated basement rocks in southern DRC and northern Angola (Delhal 1991; Jelsma et al. 2011), western Tanzania and Uganda (Link et al. 2010), and Gabon (Caen-Vachette et al. 1988)
3. Paleoproterozoic (*Second most common age-population*)	48 zircons date between 1.8 Ga and 2.1 Ga, with 21 restricted between 1852 ± 17 Ma and 1898 ± 17 Ma, and 17 between 2006 ± 17 Ma and 2048 ± 10 Ma	Overlap the conventional Eburnian period (e.g. Cahen et al. 1984). Such dates are common within the Ruwenzori and Ubendian Belts in Uganda and western Tanzania (Lenoir et al. 1995; Nagudi et al. 2003), and within the (Kimezian) basement of the West Congo Belt in western DRC and its large counterpart in eastern Brazil (Tack et al. 2001; Pedrosa-Soares et al. 2008)
4. Mesoproterozoic	8 zircons date between 1.2 Ga and 1.6 Ga, with 5 bracketed between 1371 ± 18 Ma and 1421 ± 16 Ma	Coincide with the main peak of magmatism of the Kibaran Belt in eastern DRC, Rwanda and Burundi (Kokonyangi et al. 2004; Tack et al. 2010)
5. Early Neoproterozoic to late Mesoproterozoic (*Most common age-population*)	49 zircons date between 902 ± 27 Ma and 1181 ± 29 Ma	Indicate large contributions from late Mesoproterozoic aged-terrains. These dates are common with the Oubanguides, in particular the central Sahara Belt in CAR and Chad (de Wit et al. 2014, in press), and the Mozambique Belt (Jamal et al. 1999; Bingen et al. 2009)
6. Late Neoproterozoic	30 zircons date between 544 ± 21 and 865 ± 25 Ma	Indicate large contributions from Pan African aged-terrains, such as the West Congo, Oubanguides, Mozambique and Lufilian Belts

Table 7.2 Summary of U-Pb detrital zircon dates from samples S1605 and S2035 (the Haute Lueki Group)

Age-population	U-Pb dates ($^{206}Pb/^{207}Pb$)	Notes
1. Early Paleoproterozoic	One (oldest) zircon dates at 2470 ± 10 Ma	
2. Paleoproterozoic	6 zircons date between 1952 ± 11 Ma and 2103 ± 5 Ma	Indicate relatively small contributions from Eburnian aged-terrains. Such dates are common in Uganda and western Tanzania (Lenoir 1995; Link et al. 2010), and in Angola and western DRC (de Carvalho et al. 2000; Tack et al. 2001)
3. Mesoproterozoic	15 zircons date between 1.2 Ga and 1.5 Ga, with only one concordant date at 1399 ± 65 Ma	Kibaran Belt (ca. 1.4 Ga) was not a major source, possibly largely covered by sediments
4. Early Neoproterozoic to late Mesoproterozoic (*Second most common age-population*)	36 zircons date between 940 ± 9 Ma and 1143 ± 15 Ma	Indicate large contributions from late Mesoproterozoic aged-terrains, dinstinctively younger than the 1.4 Ga Kibaran type-area. Potential source areas have been reported in CAR and Chad (de Wit et al. 2014 in press), and in Mozambique (Jamal et al. 1999; Bingen et al. 2009)
5. Late Neoproterozoic (*Most common age-population*)	40 zircons date between 579 ± 14 Ma and 817 ± 15 Ma	Correspond to tectonic episodes of the Pan African orogens, such as within the Oubanguides that include multiple structures active between 850 Ma and 550 Ma (Poidevin 1985; Rolin 1995; Toteu et al. 2006)

are some differences in their proportions that apparently reflect changes in provenance. Figure 7.15 shows a simplified map of basement ages around the CB to evaluate possible source regions for these detrital zircons.

7.4.3.1 Early Paleoproterozoic to Archean (2.4–2.7 Ga and 2.9 Ga)

Early Paleoproterozoic to Archean dates are more common in the samples (D1400 and D1595) from the Lukuga Group (8 grains) in the Dekese section compared to those from the Haute Lueki Group (only 1 grain) in the Samba section (S1605 and S2035). This suggests greater (more proximal) derivations from early Precambrian sources to the glacial sediments in the Dekese section, or that these sources became progressively covered during deposition of the Haute Lueki Group. These 2.4–2.9 Ga sources most likely correspond to the Kasai and Tanzanian Cratons located immediately to the south (~200 km) and to the east (~800 km) of the Dekese section, respectively. This provenance is also supported by the E-W orientations of Carboniferous glacial valleys (Fig. 7.15).

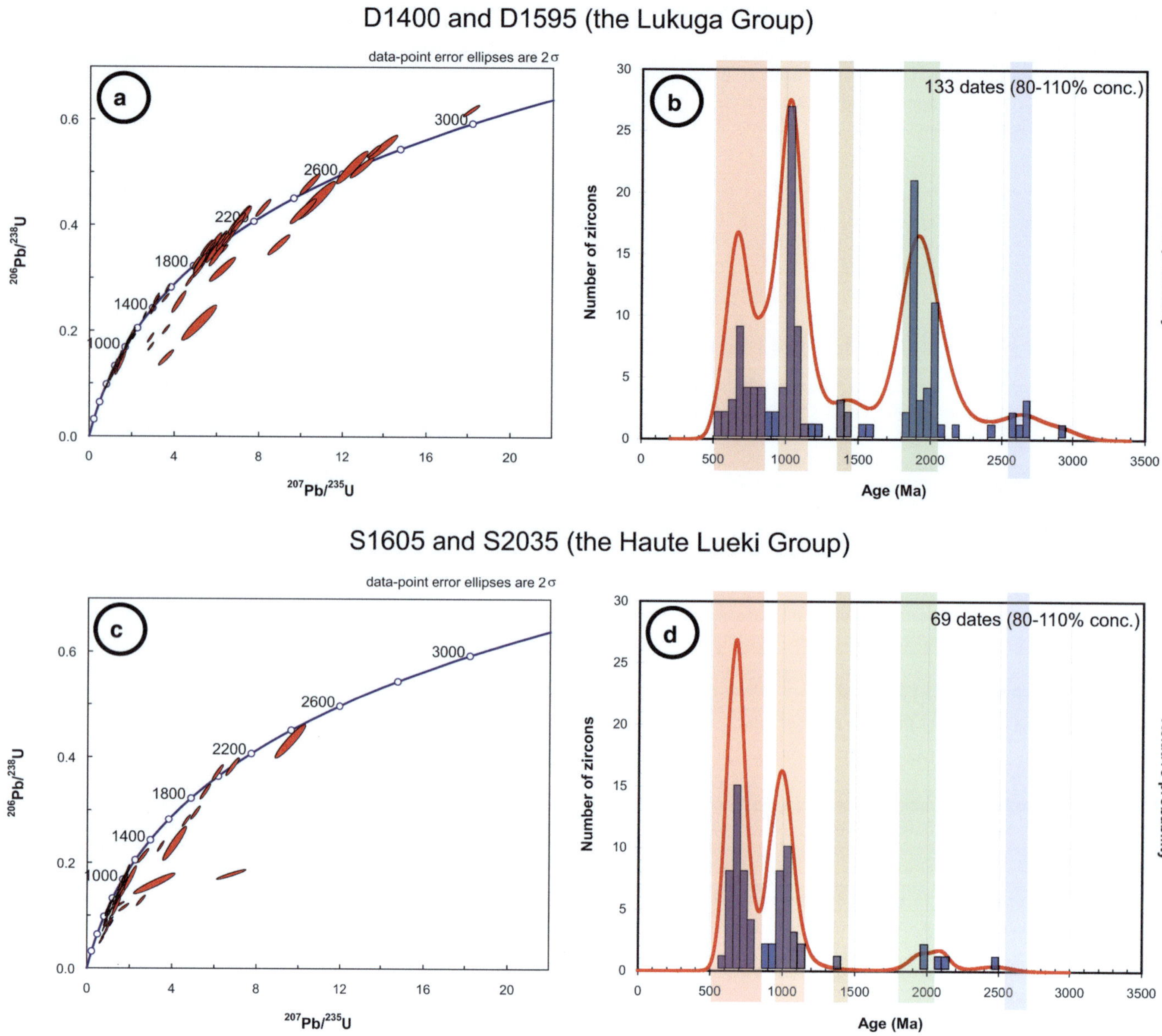

Fig. 7.12 U-Pb Concordia- and frequency age-diagrams of detrital zircons from core-samples of (**a** and **b**) the Carboniferous-Permian Lukuga Group, and (**c** and **d**) the Triassic Haute Lueki Group, in the lower parts of the Samba and Dekese sections (Fig. 7.10 for sample locations). Main age-populations are *highlighted*

7.4.3.2 Mid-Paleoproterozoic (1.7–2.2 Ga)

Mid-Paleoproterozoic (1.7–2.2 Ga) dates are better represented in the samples from the Dekese section (44 grains) than those from the Samba section (6 grains). This suggests a greater contribution of Eburnian-age rock sequences to the Lukuga Group in the Dekese section, such as found within the Ruwenzori and Ubendian Belts in east Africa (e.g. Lenoir et al. 1995; Link et al. 2010). These Paleoproterozoic dates also overlap with similar age basement rocks along the eastern margin of the CB, in Angola (De Carvalho et al. 2000), western DRC, Gabon and Cameroon (Caen-Vachette et al. 1988; Tack et al. 2001), and large terrains of the same age (Transamazonian) in northeastern Brazil (e.g. Toteu et al. 2001; Lerouge et al. 2006). However, a provenance from westernmost Gondwana (Brazil) is not supported by the westward paleocurrents for the Lukuga Group (Fig. 7.15).

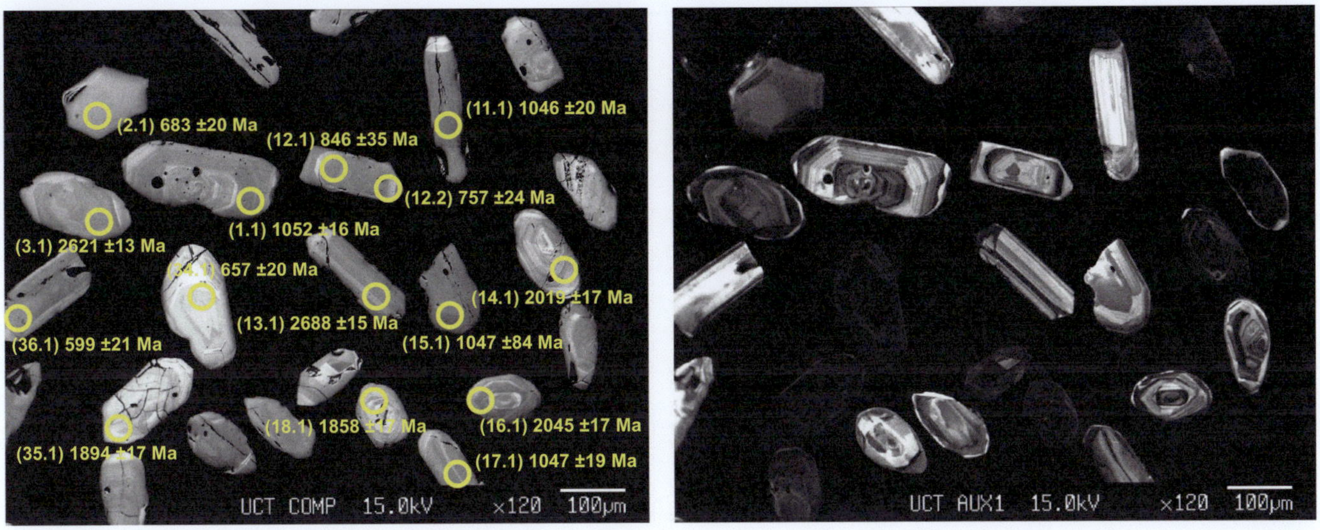

Fig. 7.13 Example of back-scattered electron (*left* photo) and cathodoluminescence (*right* photo) images of detrital zircons from sample D1595 (the Lukuga Group), with location of laser spots and U-Pb results

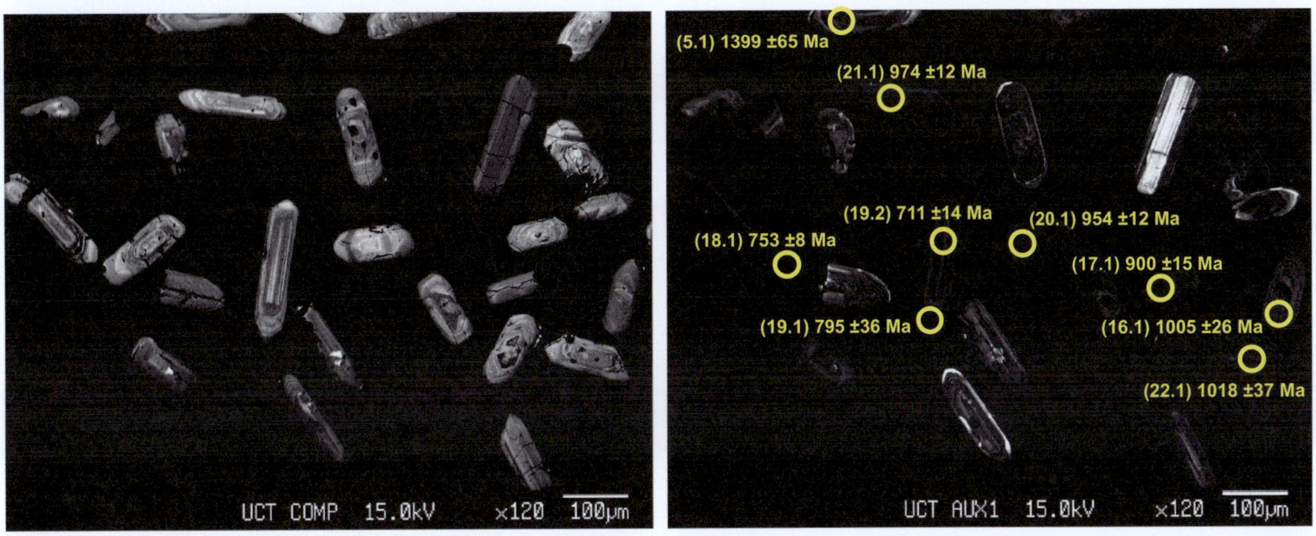

Fig. 7.14 Example of back-scattered electron (*left* photo) and cathodoluminescence (*right* photo) images of detrital zircons from sample S2035 (the Haute Lueki Group), with location of laser spots and U-Pb results

7.4.3.3 Early Neoproterozoic to Mesoproterozoic (850– 1400 Ma)

Early Neoproterozoic to Mesoproterozoic dates are abundant in all the samples (Fig. 7.12), indicating important and prolonged contributions from Kibaran-age sources to the CB. Although only in the samples from the Lukuga Group, 5 zircons dated at 1.37–1.42 Ga coincide with the Kibaran type-area along the eastern margin of the basin, recently dated at 1375 Ma (Tack et al. 2010). Most other dates are restricted between 950 and 1100 Ma (Figs. 7.12b, d). The largest potential primary sources for this 1 Ga zircon age-population are within the Oubanguides to the north, in particular the central Sahara Belt in CAR and Chad (e.g. the North African Shield; de Wit et al. 2014 in press), and on the opposite side of the Atlantic within the Brasiliano (Araguaia and Aracuai) Belts of northeastern Brazil (e.g. Dos Santos et al. 2010). These northern provenances for the CB are supported by other recent detrital zircons studies (Frimmel et al. 2006; Jelsma et al. 2011) that also found dominant 1 Ga age-populations within the upper Neoproterozoic to early Paleozoic West Congolian and Inkisi Groups along the western margin of the basin. The Pan

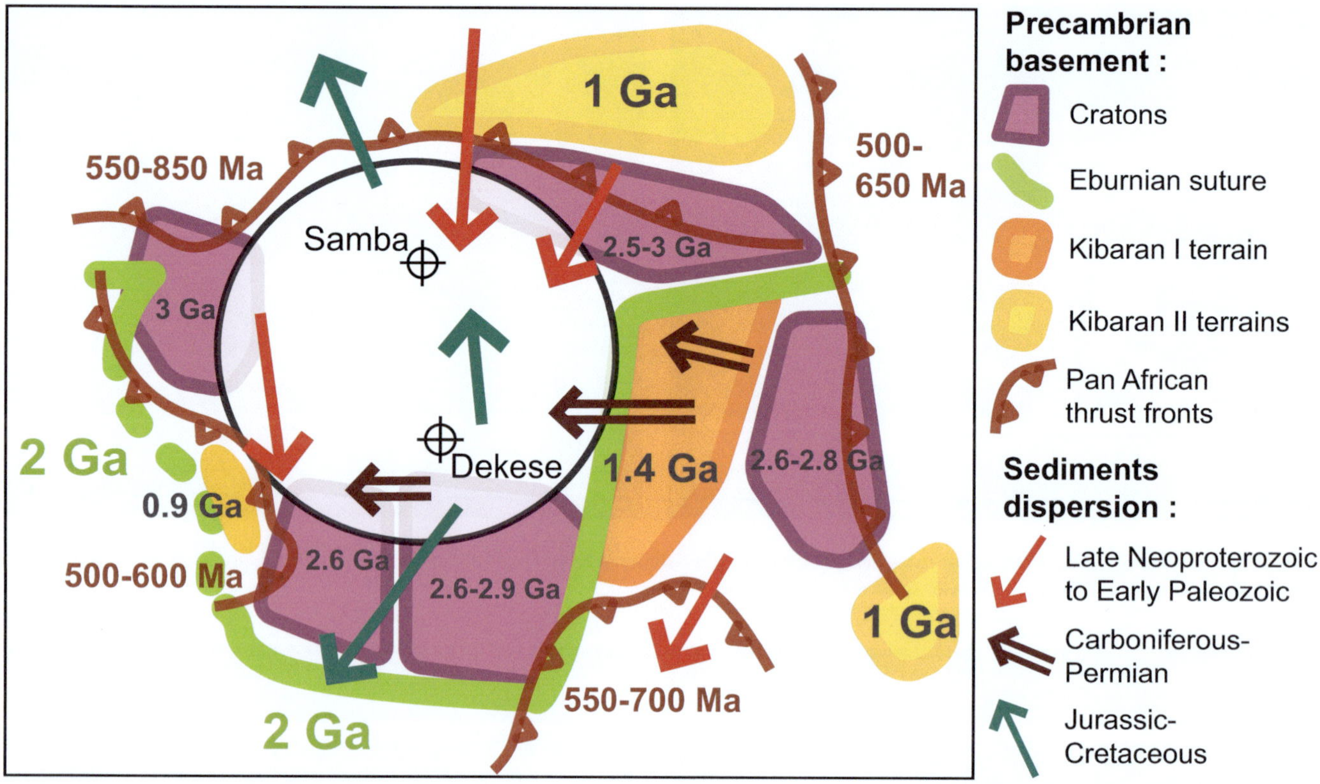

Fig. 7.15 Simplified map of Precambrian basement ages surrounding the CB (see also Chap. 2, this Book), showing main sediment dispersal directions

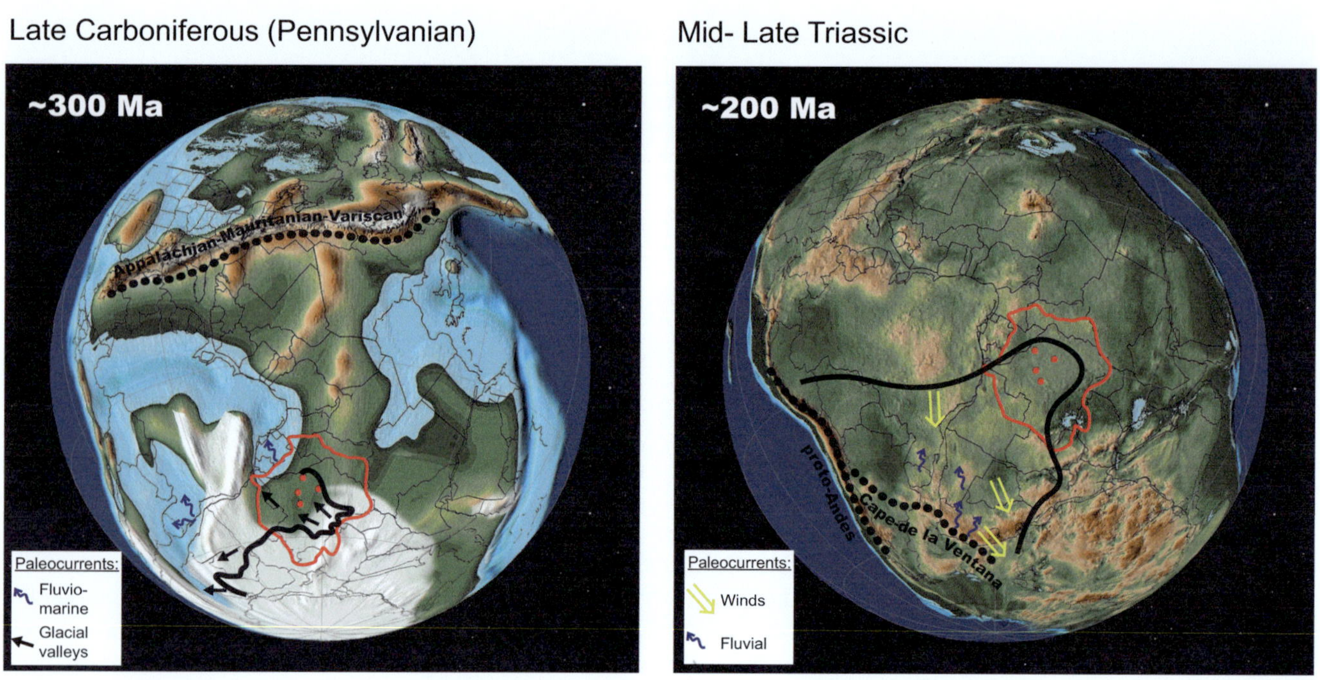

Fig. 7.16 Schematic paleogeographic reconstructions (*left*: Late Carboniferous; *right*: Middle- Late Triassic) showing the extension of depositional sequences (*thick black line*) and paleocurrent directions (*arrows*) in central West Gondwana, superimposed on paleo-globes from Moore and Scotese (2013). *Red dots* and outline are the borehole locations and modern Congo drainage basin, respectively

African mobile belts and their overlying sedimentary cover surrounding the CB, thus could have constituted important secondary sources for the sediments in the center of the basin (Fig. 7.15).

7.4.3.4 Cambrian to Late Neoproterozoic (500–850 Ma)

Cambrian to late Neoproterozoic dates are relatively abundant in all the samples (Figs. 7.12b, d). This large peak between 500 and 850 Ma indicates important Pan African contributions to the CB. However, because Pan African fold-and-thrust belts and associated molassic sequences with this age-range completely surround the basin, at this stage it is not possible to differentiate between these widely different source terrains (Fig. 7.15). Sediment dispersal directions favor Pan African sources located to the east (e.g. the Lufilian and Mozambique Belts) and to the north (e.g. the Oubanguides Belt). Also, the increase in abundance of this 500–850 Ma zircon age-population in the samples from the Lukuga Group (22 grains) to the Haute Lueki Group (36 grains) suggests progressive concentration by sediment recycling during the early Mesozoic.

Conclusion

The four new stratigraphic sections reconstructed from the deep boreholes drilled in the center of the CB in the 1950s and 1970s, all show at their lower part great thicknesses of conglomerates, quartzitic sandstones and red siltstones (e.g. 3 km at Mbandaka) overlying deformed Pan African carbonate rocks (Fig. 7.10). The lowermost sequences locally include dark siltstones and black shales dated by palynostratigraphy to the Carboniferous-Permian in the Dekese and Mbandaka sections (Boulouard and Calandra 1963; Colin and Jan du Chêne 1981), and attributed to the glacial-periglacial Lukuga Group. In the center of the basin, this first supersequence is between 900 and 1,600 m thick, truncated at the top by a regional erosion surface (U2), and is in turn overlain by 900–1,800 m thick red quartzitic sandstones and siltstones of the Triassic Haute-Lueki Group. The angular unconformity separating these two groups is equivalent to the main erosion surface that separates the lower and upper Karoo successions in southern Africa, and thus may be linked to large-scale deformation across the entire interior of Gondwana during the late Paleozoic Mauritanian-Variscan (ca. 275–325 Ma; Dabo et al. 2008) and Cape-de la Ventana (ca. 245–278 Ma; Newton et al. 2006) orogens along its northwestern and southern margins, respectively (Fig. 7.16; Scotese 2014). The overlying major unconformity (U3) across these Carboniferous to Triassic sequences of the CB is now re-assigned to the Jurassic, and believed to be related to the initial period of break-up of Gondwana during the opening of the Indian Ocean (see Chap. 13, this Book).

U-Pb detrital zircons geochronology from these sequences characterizes the evolution of source provenances for the CB during the late Paleozoic and early Mesozoic. In the lower diamictites of the Lukuga Group in the Dekese section, abundant detrital zircons of 1.85–2.05 Ga and subordinates of 1.37–1.42 Ga indicate largely dominant contributions from Eburnian and Kibaran sources in east Africa, as supported by the west-facing paleo-glacial valleys along the margins of the basin (Fig. 7.16). An elevated paleo-topography in east Africa during the Carboniferous-Permian is also consistent with recent thermochronology in Tanzania (Kasanzu 2014) that shows ca. ~7 km of exhumation of the Tanzanian Craton from 460 Ma to 220 Ma. In the overlying Haute Lueki Group in the Samba section, more dominant detrital zircons of 950–1100 Ma and 500–800 Ma derived from north-central Africa, as shown by the crystal shapes of the zircons that indicate limited transport (Fig. 7.14). In contrast, input from the Kibaran Belt of eastern DRC is almost absent in all these samples, suggesting that this vast region stopped acting as a main source for the CB, possibly being covered by sediments or subsided below sea level during the Mesozoic.

Acknowledgments We acknowledge funding through the Inkaba yeAfrica and !Khure Africa programs, supported by the DST/NRF of South Africa. We thank Anthony Tankard and an anonymous referee for critical reviews that improved the chapter. This is AEON contribution number 127 and Inkaba yeAfrica contribution number 97.

References

Alvarez P, Maurin JC, Vicat J-P (1995) La Formation de l'Inkisi (Supergroupe Ouest-congolien) en Afrique centrale (Congo et Bas-Zaire): un delta d'âge Paléozoique comblant un bassin en extension. J Afr Earth Sci 20(2):119–131

Antunes MT, Maisey JG, Marques MM, Schaeffer B, Thomson KS (1990) Triassic fishes from the Cassange Depression (R.P. de Angola). Universitade de Lisboa, Ciencias da Terra, Numero Esp, pp 1–64

Asselberghs A (1947) Caractères glaciaires des couches de base du Système du Karoo dans la vallée du Kwango. Bulletin de la Société Belge de Géologie, de Paléontologie et d'Hydrologie, Tome LVI, fascicule 56(1–2):58–62

Bangert B, Stollhofen H, Lorenz V, Armstrong R (1999) The geochronology and significance of ash-fall tuffs in the glaciogenic Carboniferous-Permian Dwyka Group of Namibia and South Africa. J Afr Earth Sci 29(1):33–49

Biggin AJ, de Wit MJ, Langereis CG, Zegers TE, Voûte S, Dekkers MJ, Drost K (2011) Palaeomagnetism of Archean rocks of the Onverwacht Group, Barberton Greenstone Belt (southern Africa):

evidence for a stable and potentially reversing geomagnetic field at ca. 3.5 Ga. Earth Planet Sci Lett 302(3–4):314–328

Bingen B, Jacobs J, Viola G, Henderson IHC, Skår Ø, Boyd R, Thomas RJ, Solli A, Key RM, Daudi EXF (2009) Geochronology of the Precambrian crust in the Mozambique belt in NE Mozambique, and implications for Gondwana assembly. Precambrian Res 170(3–4): 231–255

Bose MN, Kar RK (1976) Palaeozoic sporae dispersae from Zaïre (Congo). XI: Assises glaciaires et periglaciaire from the Lukuga Valley. Annales du Musée Royal de l'Afrique centrale, Tervuren (Belgique), Série in 8. Sciences Geologiques 77:1–19

Bose MN, Kar RK (1978) Biostratigraphy of the Lukuga Group in Zaire. Annales du Musée Royal de l'Afrique centrale, Tervuren (Belgique), Série in 8. Sciences Geologiques 82:97–114

Boulouard C, Calandra F (1963) Etude palynologique de quelques sondages de la République du Congo (Congo ex-Belge). Unpublished report R/ST-no.7376, SNPA Direction exploration et production Pau, France

Boutakoff N (1948) Les formations glaciaires et post-glaciaires fossilifères, d'âge permo-carbonifère (Karoo inférieur) de la region de Walikale (Kivu, Congo belge). Mémoire de l'Institut géologique. Université de Louvain, IX (II), 214pp

Caen-Vachette M, Vialette Y, Bassot J-P, Vidal P (1988) Apport de la géochronologie isotopique à la connaissance de la géologie gabonaise. Chroniques de la recherche minière 491:35–54

Cahen L (1954) Géologie du Congo Belge. Vaillant-Carmanne, Liège, p 577p

Cahen L (1981) Précisions sur la stratigraphie et les corrélations du Goupe de la Haute-Lueki et des formations comparables (Triasique a Liasique? d'Afrique Centrale). Rapport annuel du Musée Royal d'Afrique centrale, Tervuren (Belgique), Départment de Géologie et de Minéralogie, pp 81–96

Cahen L, Lepersonne J (1978) Synthèse des connaissances relatives au Groupe (anciennement Série) de la Lukuga (Permien du Zaïre). Annales du Musée Royal du Congo belge, Tervuren (Belgique), Série in-8, Sciences géologiques, 82, pp 115–152

Cahen L, Ferrand JJ, Haarsma MJF, Lepersonne J, Vebeek T (1959) Description du Sondage de Samba. Annales du Musée Royal du Congo belge, Tervuren (Belgique), Série in-8, Sciences géologiques, 29, 210p

Cahen L, Ferrand JJ, Haarsma MJF, Lepersonne J, Vebeek T (1960) Description du Sondage de Dekese. Annales du Musée Royal du Congo belge, Tervuren (Belgique), Série in-8, Sciences géologiques, 34, 115p

Cahen L, Snelling NJ, Delhal J, Vail J (1984) The geochronology and evolution of Africa. Clarendon, Oxford, p 512

Catuneanu O, Wopfner H, Eriksson PG, Cairncross B, Rubidge BS, Smith RMH, Hancox PJ (2005) The Karoo basins of south-central Africa. J Afr Earth Sci 43(1–3):211–253

Colin J-P (1981) Paleontological study of the Esso/Texaco well Gilson-1, Zaire. Unpublished report EPR-E.WA19.81

Colin J-P, Jan du Chêne J (1981) Paleontological study of the Esso/Texaco well Mbandaka-1, Zaire. Unpublished report EPR-E.WA15.81

Crosby AG, Fishwick S, White N (2010) Structure and evolution of the intracratonic Congo Basin. Geochem Geophys Geosyst 11(6):1–20

Dabo M, Gueye M, Ngom PM, Diagne M (2008) Orogen-parallel tectonic transport: transpression and strain partitioning in the Mauritanides of NE Senegal. In: Ennih N, Liégeois J-P (eds) The boundaries of the West African Craton, vol 297. Geological Society, London, pp 483–497, Special Publications

Daly MC, Lawrence SR, Kimun'a D, Binga M (1991) Late Paleozoic deformation in central Africa: a result of distant collision. Nature 350:605–607

Daly MC, Lawrence SR, Diemu-Tshiband K, Matouana B (1992) Tectonic evolution of the Cuvette Centrale, Zaire. J Geol Soc 149 (4):539–546

De Carvalho H (1981) Geologia de Angola. 4 folhas, escala 1: 1 000 000. Laboratório Nacional de investigação Cientifica Tropical, Junta de Investigaçãoes Cientifica do Ultramar

De Carvalho H, Tassinari C, Alves PM, Guimarães F, Simões MC (2000) Geochronological review of the Precambrian in western Angola: links with Brazil. J Afr Earth Sci 31(2):383–402

De Wit MJ, Jeffery M, Berg H, Nicolayson LO (1988) Geological Map of sectors of Gondwana reconstructed to their position ~150 Ma (with explanatory notes), scale 1: 1.000.000. American Association of Petroleum Geologist, Tulsa

De Wit MJ, De Brito Neves BB, Trouw RAJ, Pankhurst RJ (2008) Pre-Cenozoic correlations across the South Atlantic region: "the ties that bind". In: Pankurst RJ, Trouw RAJ, Brito Neves BB, de Wit MJ (eds) West Gondwana: Pre-Cenozoic correlations across the South Atlantic Region, 294. Geological Society of London, London, pp 1–8, Special Publications

De Wit MJ, Bowring S, Buchwaldt R, Dudas F, MacPhee D, Tagne-Kamga G, Dunn N, Salet AM, Nambatingar D (2014) Proterozoic crust of the Central Sahara Shield (in press)

Defretin-Lefranc S (1967) Étude sur les phyllopodes du Bassin du Congo. Annales du Musée Royal de l'Afrique centrale, Tervuren (Belgique), Série in-8, Sciences géologiques, 56, 122p

Delhal J (1991) Situation géochronologique 1990 du Precambrian du Sud-Kasai et de l'Ouest-Shaba. Rapport annuel du Musee Royal de l'Afrique centrale, Tervuren (Belgique), Département de Géologie et de Minéralogie, pp 119–125

Desthieux F (1995) Carte Géologique de la République du Congo au 1:1.000.000. Geological Survey of South Africa

Dos Santos EJ, Schmus WRV, Kozuch M, Neves BBDB (2010) The Cariris Velhos tectonic event in Northeast Brazil. J S Am Earth Sci 29(1):61–76

Esso-Zaire SARL (1981a) Geological completion report: Gilson-1. Unpublished report

Esso-Zaire SARL (1981b) Geological completion report: Mbandaka-1. Unpublished report

Evrard P (1960) Sismique. Annales du Musée Royal du Congo belge, Tervuren (Belgique), Série in-8, Sciences géologiques, 33, 87p

Eyles N, Eyles CH (1992) Glacial Depositional Systems. In: Walker RG, James NP (eds) Facies models: response to sea level change. Ontario, Geological Association of Canada, pp 73–100

Fedo CM, Sircombe KN, Rainbird RH (2003) Detrital zircon analysis of the sedimentary record. Rev Mineral Geochem 53(1): 277–303

Fourmarier P (1914) Le basin charbonnier d'âge permo-triassique de la Lukuga. Etude géologique de la region de la Lukuga et de la Lubumba au voisinage du lac Tanganyika. Ann Soc Geol Belg 41(C):77–227

Frimmel H, Tack L, Basei M, Nutman A, Boven A (2006) Provenance and chemostratigraphy of the Neoproterozoic West Congolian Group in the Democratic Republic of Congo. J Afr Earth Sci 46(3):221–239

Gonzalez F, Moreno C, Playford G (2011) The Gondwana-Laurussia convergence process: evidence from the Middle Mississipian (Visean) palynostratigraphic record. Geol Mag 148(2):317–328

Grekoff N (1960) Ostracodes du Bassin du Congo. II. Crétacé. Annales du Musée Royal de l'Afrique centrale, Tervuren (Belgique), Série in-8, Sciences géologiques, 35, 70p

Isbell JL, Cole DI, Catuneanu O (2008) Carboniferous-Permian glaciation in the main Karoo Basin, South Africa: Stratigraphy, depositional controls, and glacial dynamics. In: Fielding CR, Frank TD, Isbell JL (eds) Resolving the late paleozoic ice age in time and space. Geological Society of America Special Paper, vol 441, pp 71–82

Jamal DL, Zartman RE, De Wit MJ (1999) U – Th – Pb single zircon dates from the Lurio Belt, Northern Mozambique: Kibaran and Pan-African orogenic events highlighted. J Afr Earth Sci 111:32

Jamotte A (1932) Contribution à l'étude géologique du basin charbonnier de la Lukuga. Annales du Service des Mines. Comité Spécial du Katanga 2:1–75

Jelsma HA, Perrit SH, Amstrong R, Ferreira HF (2011) Shrimp U-Pb zircon geochronology of basement rocks of the Angolan Shield, western Angola. Abstract, 23rd CAG, Johannesburg, 8th–14th Jan

Johnson MR, Van Vuuren CJ, Hegenberger WF, Key R, Shoko U (1996) The Stratigraphy of the Karoo Supergroup in southern Africa: an overview. J Afr Earth Sci 23(1):3–15

Johnson MR, Van Vuuren CJ, Visser JNJ, Cole DI, De Villiers Wickens H, Christie ADM, Roberts DL, Brandl G (2006) Sedimentary rocks of the Karoo Supergroup. In: Johnson MR, Anhaeusser CR, Thomas RJ (eds) The geology of South Africa. Council for Geoscience, Pretoria, South Africa, pp 461–499

Jones L, Mathieu PL, Strenger H (1959) Magnétisme. Annales du Musée Royal de l'Afrique central, Tervuren (Belgique), Série in-8, Sciences géologiques, 27, 30p

Kadima E, Delvaux D, Sebagenzi SN, Tack L, Kabeya SM (2011) Structure and geological history of the Congo Basin: an integrated interpretation of gravity, magnetic and reflection seismic data. Basin Res 23(5):499–527

Kasanzu C (2014) Dating the unroofing and cooling histories of the Archean Tanzanian Craton, Eastern Africa: using a combination of apatite fission track and (U-Th)/He thermochronometric techniques. Unpublished PhD thesis, University of Cape Town

Kokonyangi J, Armstrong R, Kampunzu A, Yoshida M, Okudaira T (2004) U-Pb zircon geochronology and petrology of granitoids from Mitwaba (Katanga, Congo): implications for the evolution of the Mesoproterozoic Kibaran belt. Precambrian Res 132(1–2):79–106

Kreuser T (1984) Karroo Basins in Tanzania. In: Klerkx J, Michot J (eds) African geology. Musée royal de l'Afrique centrale, Tervuren, pp 231–246

Lawrence SR, Makazu MM (1988) Zaire's central basin: prospectivity outlook. Oil Gas J 86(38):105–108

Lenoir JL, Liegeois J-P, Theunissen K, Klerkx J (1995) The Palaeoproterozoic Ubendian shear belt in Tanzania: geochronology and structure. J Afr Earth Sci 19(3):169–184

Lepersonne J (1951) Les subdivisions du système du Karoo au Kwango (Congo belge). Annales de la Société Géologique de Belgique, LXXIV, pp 123–138

Lepersonne J (1974) Carte géologique du Zaïre au 1: 2.000.000 et notice explicative. Kinshasa, République du Zaïre: Direction de la Géologie/Musée Royal de l'Afrique centrale, Tervuren (Belgique)

Lepersonne J (1978) Structure géologique du bassin interieur du Zaïre. Académie royale des Sciences d'outre-Mer. Classe des Sciences Naturelles et Médicales, Bruxelles (Belgique), XX-2, 27p

Lerouge C, Cocherie A, Toteu SF, Penaye J, Milési J-P, Tchameni R, Nsifa EN, Mark Fanning C, Deloule E (2006) Shrimp U-Pb zircon age evidence for Paleoproterozoic sedimentation and 2.05Ga syntectonic plutonism in the Nyong Group, South-Western Cameroon: consequences for the Eburnean-Transamazonian belt of NE Brazil and Central Africa. J Afr Earth Sci 44(4–5):413–427

Link K, Koehn D, Barth MG, Aanyu K, Foley SF (2010) Continuous cratonic crust between the Congo and Tanzania blocks in western Uganda. Int J Earth Sci (Geologische Rundschau) 99:1559–1573

Linol B (2013) Sedimentology and sequence stratigraphy of the Congo and Kalahari Basins of south-central Africa and their evolution during the formation and break-up of West Gondwana. PhD thesis, Nelson Mandela Metropolitan University, 375p

Linol B, de Wit MJ, Barton E, de Wit MJC, Guillocheau (2014) U-Pb detrital zircons dates and source provenance analysis of Phanerozoic sequences of the Congo Basin, Central Gondwana (accepted at Gondwana Research)

Lombard AL (1961) La série de la Haute Lueki (partie orientale de la cuvette congolaise). Bulletin de la Société belge de Géologie, de Paléontologie et d'Hydrologie 70:65–72

Ludwig KR (2003) Isoplot/version 3.0: A geochronological toolkit for Microsoft Excel. Berkeley Geochronology Center Special Publication, 4, 71p

Maheshwari H, Bose MN, Kumaran KP (1977) Mesozoic sporae dispersae from Zaire. II: The Loia and Bokungu Groups in the Samba borehole. III: Some miospores from the Stanleyville Group. Annales du Musée Royal de l'Afrique centrale, Tervuren (Belgique), Série in-8, Sciences géologiques, 80, 60p

Master S, Rainaud C, Armstrong R, Phillips D, Robb L (2005) Provenance ages of the Neoproterozoic Katanga Supergroup (Central African Copperbelt), with implications for basin evolution. J Afr Earth Sci 42(1–5):41–60

Milani EJ, de Wit MJ (2008) Correlations between the classic Paraná and Cape Karoo sequences of South America and southern Africa and their basin infills flanking the Gondwanides: du Toit revisited. In: Pankurst RJ, Trouw RAJ, Brito Neves BB, de Wit MJ (eds) West Gondwana: Pre-Cenozoic correlations across the South Atlantic Region, vol 294. Geological Society of London, London, pp 319–342, Special Publications

Modie BN, Le Hérissé A (2009) Late Palaeozoic palynomorph assemblages from the Karoo Supergroup and their potential for biostratigraphic correlation, Kalahari Karoo Basin, Botswana. Czech Repub Geol Surv Bull 84(2):337–358

Montañez IP, Poulsen CJ (2013) The Late Paleozoic Ice Age: an evolving paradigm. Annu Rev Earth Planet Sci 41:629–656

Moore TL, Scotese CR (2013) Ancient Earth: assembly of Pangea, Vers. 1.0, iOS Mobile Application

Mounguengui MM, Lang J, Guiraud M, Jocktane O (2002) Sedimentary dynamics and structural geology of pre-rift deposits of the interior basin of Gabon. J Afr Earth Sci 35(2):315–329

Nagudi B, Koeberl C, Kurat G (2003) Petrography and geochemistry of the Singo granite, Uganda, and implications for its origin. J Afr Earth Sci 36(1–2):73–87

Newton AR, Shone RW, Booth PWK (2006) The Cape Fold Belt. In: Johnson MR, Anhaeusser CR, Thomas RJ (eds) The geology of South Africa. Pretoria, South Africa, Council for Geoscience, pp 521–530

Opdyke ND, Mushayandebvu M, de Wit MJ (2001) A new palaeomagnetic pole for the Dwyka System and correlative sediments in sub-Saharan Africa. J Afr Earth Sci 33(1):143–153

Pedrosa-Soares AC, Alkmim FF, Tack L, Noce CM, Babinski M, Silva LC, Martins-Neto MA (2008) Similarities and differences between the Brazilian and African counterparts of the Neoproterozoic Araçuaí-West Congo orogen. Geol Soc Lond, Spec Publ 294(1): 153–172

Poidevin JL (1985) Le Proterozoic supérieur de la Républic Centrafricaine. Annales du Musée Royal de l'Afrique centrale, Tervuren (Belgique), Serie in-8, Sciences Geologiques, 91, 75p

Rakotosolofo NA, Torsvik TH, Ashwal LD, Eide EA, de Wit MJ (1999) The Karoo Supergroup revisited and Madagascar-Africa fits. J Afr Earth Sci 29(1):135–151

Rocha-Campos AC (1976) Direction of movement of late Paleozoic glaciers in Angola (western Africa). Instituto de Geociêcias, USP, BoletimIG 7:39–44

Rolin P (1995) Carte tectonique de la République Centrafricaine au 1:1.500.000. BRGM

Sachse VF, Delvaux D, Littke R (2012) Petrological and geochemical investigations of potential source rocks of the central Congo Basin, DRC. Am Assoc Pet Geol Bull 96(2):245–275

Scotese CR (2014) PaleoAtlas for ArcGIS, vol 4. Late Permian, PALEOMAP Project, Evanston, IL

Streel MT, Theron JN (1999) The Devonian-Carboniferous boundary in South Africa and the age of the earliest episode of the Dwyka glaciation: New palynological result. Episodes 22(1):41–44

Tack L, Wingate MTD, Lie J (2001) Early Neoproterozoic magmatism (1000-910 Ma) of the Zadinian and Mayumbian Groups (Bas-Congo): onset of Rodinia rifting at the western edge of the Congo craton. Precambrian Res 110:277–306

Tack L, Wingate MTD, De Waele B, Meert J, Belousova E, Griffin B, Tahon a, Fernandez-Alonso M (2010) The 1375 Ma "Kibaran event" in Central Africa: Prominent emplacement of bimodal magmatism under extensional regime. Precambrian Res 180(1–2):63–84

Tait J, Delpomdor F, Preat A, Tack L, Straathof G, Nkula VK (2011) Neoproterozoic sequences of the West Congo and Lindi/Ubangi Supergroups in the Congo Craton, Central Africa. In: Arnaud E, Halverson GP, Shields-Zhou G (eds) The Geological Record of Neoproterozoic Glaciations. Geological Society of London, Memoirs, 36, pp 185–194

Toteu SF, Van Schmus WR, Penaye J, Michard A (2001) New U-Pb and Sm-Nd data from north-central Cameroon and its bearing on the pre-Pan African history of central Africa. Precambrian Res 108(1–2):45–73

Toteu SF, Fouateu RY, Penaye J, Tchakounte J, Mouangue ACS, Van Schmus WR, Deloule E, Stendal H (2006) U-Pb dating of plutonic rocks involved in the nappe tectonic in southern Cameroon: consequence for the Pan-African orogenic evolution of the central African fold belt. J Afr Earth Sci 44(4–5):479–493

Trouw RA, de Wit MJ (1999) Relation between the Gondwanide Orogen and contemporaneous intracratonic deformation. J Afr Earth Sci 28(1):203–213

Veevers JJ, Cole DI, Cohan EJ (1994) Southern Africa: Karoo Basin and Cape Fold Belt. In: Veevers JJ, Powell C.McA (eds) Permian-Triassic Pangean Basins and fold belts along the Panthalassan margin of Gondwanaland. Geological Society of America, Colorado, Memoir, 184, pp 223–280

Visser JNJ (1995) Post-glacial Permian stratigraphy and geography of southern and central Africa: boundary conditions for climatic modelling. Palaeogeogr Palaeoclimatol Palaeoecol 118(3–4): 213–243

Wopfner H, Diekmann B (1996) The Late Palaeozoic Idusi Formation of southwest Tanzania: a record of change from glacial to post-glacial conditions. J Afr Earth Sci 22(4):575–595

Facies Analyses, Chronostratigraphy and Paleo-Environmental Reconstructions of Jurassic to Cretaceous Sequences of the Congo Basin

8

Bastien Linol, Maarten J. de Wit, Erika Barton, Francois Guillocheau, Michiel C.J. de Wit, and Jean-Paul Colin†

8.1 Introduction

The Congo Basin (CB) covers about 1.8 million km^2 (Fig. 8.1), centered on the Democratic Republic of Congo (DRC, formerly Zaire), and extends to the west into Angola, Republic of Congo and Gabon, and to the north into the Central African Republic (CAR). The structure and stratigraphy of this vast continental basin was essentially described from two exploration programs in the 1950s and 1970s, including extensive field mapping, geophysical surveys (e.g. Evrard 1960) and drilling of four deep boreholes, each between 2 km and 4.5 km deep. The first two, Samba-1 and Dekese-1, were cored with excellent core recovery of ca. 78 % and 84 %, respectively (Cahen et al. 1959, 1960), representing unique stratigraphic sections through the center of the basin. The complete core-collections are archived at the Royal Museum for Central Africa (RMCA) in Tervuren, Belgium. The two younger exploration wells, Gilson-1 and Mbandaka-1, were logged by geophysics (Gamma Ray, Resistivity, Sonic) and cored occasionally to check significant lithological changes (Esso-Zaire 1981a, b), but most of these core-samples are now lost.

In this study, we present new facies analysis and regional correlations of the Jurassic-Cretaceous (J–K) sequences, based on field mapping in the Kwango Valley along the southwestern margin of the CB, and by re-logging the four wells/cores in the center of the basin (Fig. 8.1 for locations). This sequence-stratigraphic basin analysis enables more detailed paleogeographic interpretations of the CB during the period between the initial break-up between East and West Gondwana (ca. 160–180 Ma) and the final separation of Africa from South America (ca. 80–130 Ma). In addition, U-Pb geochronology of detrital zircons from these J–K sequences, collected both from the cores and in outcrops, characterizes the maximum ages and source provenances for the sediments.

8.2 Current Litho- and Bio-Stratigraphy (the Congo Supergroup)

The complete J–K succession is about 1,000 m thick in the center of the CB, and includes: the Stanleyville, Loia, Bokungu and Kwango Groups (Cahen 1954; Lepersonne 1974; Cahen 1983a, b). It overlies a major unconformity that truncates Precambrian basement and Carboniferous to Triassic sequences (Chap. 7, this Book). Along the margins of the basin, the lateral equivalents of these J–K sequences correlate with the Calonda Formation in Angola (de Carvalho 1981) and the Carnot and Moukka-Oudda Formations in the CAR (sourced from the southeast; Censier and Lang 1999), and all of which contain alluvial diamond paleo-placers suggesting concurrent kimberlite intrusions

†Author was deceased at the time of publication.

B. Linol (✉) • E. Barton
AEON-ESSRI (African Earth Observatory Network – Earth Stewardship Science Research Institute), Nelson Mandela Metropolitan University, Port Elizabeth, South Africa

Geological Sciences Nelson Mandela Metropolitan University, Port Elizabeth, South Africa
e-mail: bastien.aeon@gmail.com; erikabarton@yahoo.com

M.J. de Wit
AEON-ESSRI (African Earth Observatory Network – Earth Stewardship Science Research Institute), Nelson Mandela Metropolitan University, Port Elizabeth, South Africa
e-mail: maarten.dewit@nmmu.ac.za

F. Guillocheau
Géosciences-Rennes, UMR 6118 Université de Rennes 1 – CNRS, OSUR, Université de Rennes 1, Campus de Beaulieu, 35042 Rennes cedex, France
e-mail: francois.guillocheau@univ-rennes1.fr

M.C.J. de Wit
Delrand Resources Pty Ltd, Toronto, Ontario, Canada
e-mail: mdewit@delrand.com

M.J. de Wit et al. (eds.), *Geology and Resource Potential of the Congo Basin*, Regional Geology Reviews,
DOI 10.1007/978-3-642-29482-2_8,

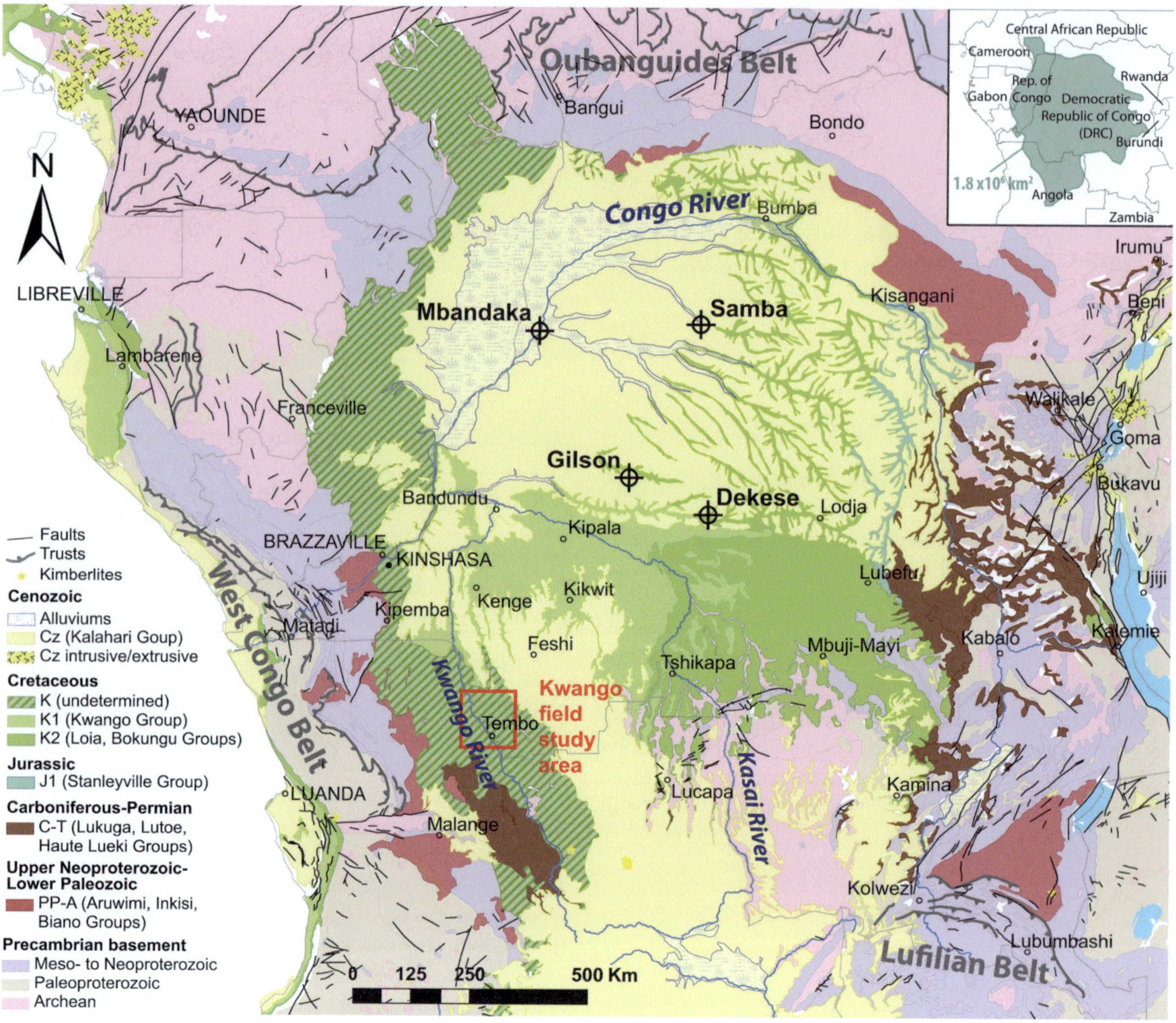

Fig. 8.1 Simplified geological map of the CB highlighting the J–K sequences, with locations of the studied four deep boreholes and the field area in the Kwango Valley (*red box*). This map was compiled from the national geological maps of DRC (Lepersonne 1974), Angola (de Carvalho 1981), Republic of Congo (Desthieux 1995), and CAR (Rolin 1995). Precambrian basement geology is from de Wit et al. (1988); kimberlites are from Jelsma et al. (2009)

and significant erosion (de Wit et al. 2011). All these J–K sequences of the CB are here referred to as the Congo Supergroup, as it represents a unique stratigraphic record in sub-Saharan Africa and which can possibly be correlated with other Mesozoic continental sections, such as in South America (Linol 2013; see Chap. 13, this Book).

8.2.1 The Stanleyville Group

The lowermost, Stanleyville Group is a 400 m thick section of fossiliferous sandstones and mudstones, mainly outcropping along the eastern margin of the CB (Cahen 1983b). Its type-section, at Kisangani (ex-Stanleyville) in northeast DRC (Fig. 8.1 for location), is generally subdivided into two (Fig. 8.2a):

1. A lower subgroup (equivalent to 'Complexes' 1 to 8 of Cahen 1983b), 50 m thick, comprises grey-green carbonated sandstones and mudstones with black shales and limestone intercalations.
2. An upper subgroup (equivalent to 'Complexes' 8 to 14 of Cahen 1983b), about 320 m thick, comprises red–brown mudstones alternating with sandstones. Its base is defined by a thin and relatively continuous pebble-bed (Fig. 8.2a).

Paleontological evidence from the Stanleyville Group, essentially from non-marine ostracods, such as in particular *Praecypridea* cf. *acuticyatha* and *Theriosynoecum trinodasa* (e.g. Fig. 8.2b; Grekoff 1957; Colin 1994) suggests a Late

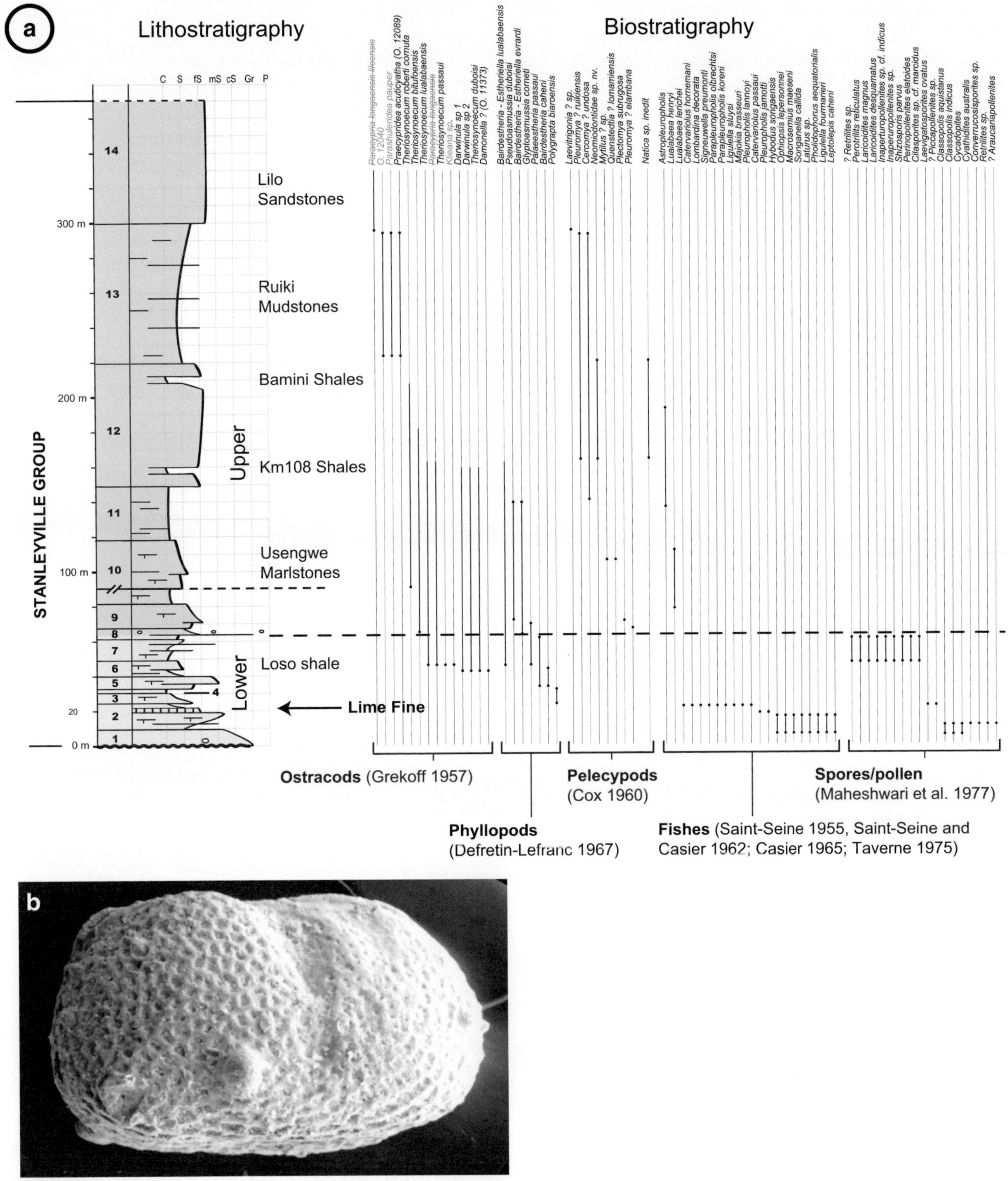

Fig. 8.2 (**a**) Litho- and bio-stratigraphy of the Stanleyville Group near Kisangani (Fig. 8.1 for location). This type-section was reconstructed from borehole-cores and field descriptions summarized by Cahen (1983b). The paleontological and palynological data indicate a Late Jurassic age by comparison with other records from the USA and Brazil. (**b**) *Theriosynoecum trinodosa*, a characteristic ostracod from the Upper Jurassic Stanleyville Group. The specimen is 0.85 mm long, from the collection archived at the RMCA museum in Tervuren

Jurassic age by comparison with equivalent biostratigraphic data from the USA (the Morrison Formation; Sames et al. 2010) and Brazil (the Alianca Formation; Krömmelbein and Weber 1971). Along the northeastern margin of the CB, the Stanleyville Group intercalates near its base with a 40 cm thin marker horizon of limestones, known as 'Lime Fine' (Cahen 1983b), which includes a well-preserved Kimmeridgian fish fauna (e.g. *Ophiopsis lepersonnei*; Saint Seine 1953, 1955; Saint Seine and Casier 1962; Taverne 1975). Furthermore, the large quantity of small and juvenile fish fossils within this horizon suggests proximity to marine spawning grounds (i.e. laguna; Taverne 1975).

Biostratigraphically, the Stanleyville Group is also identified in the center of the basin in the Samba-1 and Gilson-1 boreholes (Cahen et al. 1959; Colin 1981), as well as from borehole cores at Kinshasa (Defretin-Lefranc 1967). It is best preserved in the northern part of the central CB (ca. 322 m at Samba) and thins westward across Gabon with more continental deposits (e.g. the M'vone Series; Mounguengui et al. 2008).

8.2.2 The Loia, Bokungu and Kwango Groups

The overlying sequences are assigned to the Loia, Bokungu and Kwango Groups, which outcrop extensively within the CB (Cahen 1954; Fig. 8.1). Their biostratigraphy is less detailed than that for the Upper Jurassic Stanleyville Group, and is based on a number of isolated fossiliferous localities (Cahen 1983a). Paleontology and palynology indicate a middle Cretaceous age (Albian-Cenomanian) for the Loia and Bokungu Groups in the center of the basin, and a Late Cretaceous age for the Kwango Group in the southwestern part of the central CB (Maheshwari et al. 1977; Colin 1994). These three groups correlate biostratigraphically with post-rift sequences outcropping along the Atlantic margin, in Gabon (e.g. the Madiela Series; Mounguengui et al. 2008).

The Loia and Bokungu Groups are best described in subsurface from the Units S4 and S3 of the Samba section, respectively (equivalents of 'Couches' [Beds] 4 and 3 of Cahen et al. 1959), overlying a distinct erosion surface etched across ferruginous sandstones of the Stanleyville Group.

1. The lower, Loia Group (Unit S4) is 280 m thick, and comprises green and analcime-rich, poorly sorted, sandstones and mudstones with black shale intercalations (Cahen 1983a). The top of this sequence is a distinct silicified layer (see Sect. 5.2).
2. The overlying Bokungu Group (Unit S3) is 363 m thick, and comprises alternating red sandstones with purplish carbonated mudstones and marlstones that are commonly highly fractured ('clotted texture' described by Cahen et al. 1959).

   Both the Loia and Bokungu Groups are fossiliferous, containing numerous non-marine phyllopods (*Bairdestheria kasaiensis*, *Pseudoasmussia banduensis*, *P. dekeseensis*, *Asmussia dekeseensis*, *A. ubangiensis* and *Euestheria sambaensis*; Marlière 1950; Defretin-Lefranc 1967) and ostracods (*Cypridea elisabethaensis*, *C. maringaensis*, *C. yakokoensis*, *Ilyocypris lomamiensis*, *I. minor*, *Metacypris polita*, *M. pustulosa* and *Pattersoncypris rotunda*; Grekoff 1957; Colin 1994), but the complex biozonation of these (endemic) species does not allow differentiating the two groups chronostratigraphically. However, based on this fossil record, the two groups were also clearly identified in all the other three boreholes (Cahen et al. 1960; Boulouard and Calandra 1963; Colin 1981; Colin and Jan du Chêne 1981; Chap. 7, this Book). New palynostratigraphy in the Samba section (Unit S4), between −622 m and −676 m depth, indicates a late-Lower to Middle Albian age by comparison with West African marine sections (Palynozone CXIIa of Doyle et al. 1982; Moron, personal communication, 2011).
3. The uppermost Kwango Group is about 300–400 m thick, and defined from outcrops in the Kwango region of southwest DRC (Lepersonne 1945, 1949, 1951). It overlies a major unconformity across Precambrian basement (e.g. the Kasai Craton) and, in some places, relics of the Carboniferous to Triassic successions (the Lukuga and Haute Lueki Groups) and the middle Cretaceous Loia Group (Cahen 1951). It is commonly divided into two 'formations' (c.f. Lepersonne 1951):
   - A lower, Inzia Formation comprises 50–110 m thick red sandstones intercalated with red–brown mudstones. At Tshikapa, in the Kasai region of southern DRC (Fig. 8.1 for location), its basal conglomerates are diamondiferous (Cahen 1951).
   - An upper, Nsele Formation comprises conglomerates and coarse red sandstones, between about 100–200 m thick.

The age of the Kwango Group is essentially based on paleontology of another distinctive fish fauna (e.g. *Rhipis moorseli*, Decertids and *Diplomystus*; Saint Seine 1953; Casier 1965), and freshwater phyllopods (*Bairdestheria kitariensis*, *Pseudoestheria lepersonnei* and *Estheria lerichei*; Marlière 1950; Defretin-Lefranc 1967) and ostracods (*Afrocythere?* 536, *Cypridea kitariensis*, *Darwinula kwangoensis*, *Ilyocypris compressa*, *I. luzubiensis*, *Metacypris* K3099, *Dolerocypris kinkoensis* and *Paracypria makawaensis*; Grekoff 1960; Colin 1994), collected at Kipala and Kitari-Kimbau-Schwetz in the southwestern part of the central CB (Fig. 8.1 for location). Paleo-ichthyological studies have suggested a Cenomanian-Turonian age for the lower part of the group (Casier 1965; Taverne 1976). In the upper part

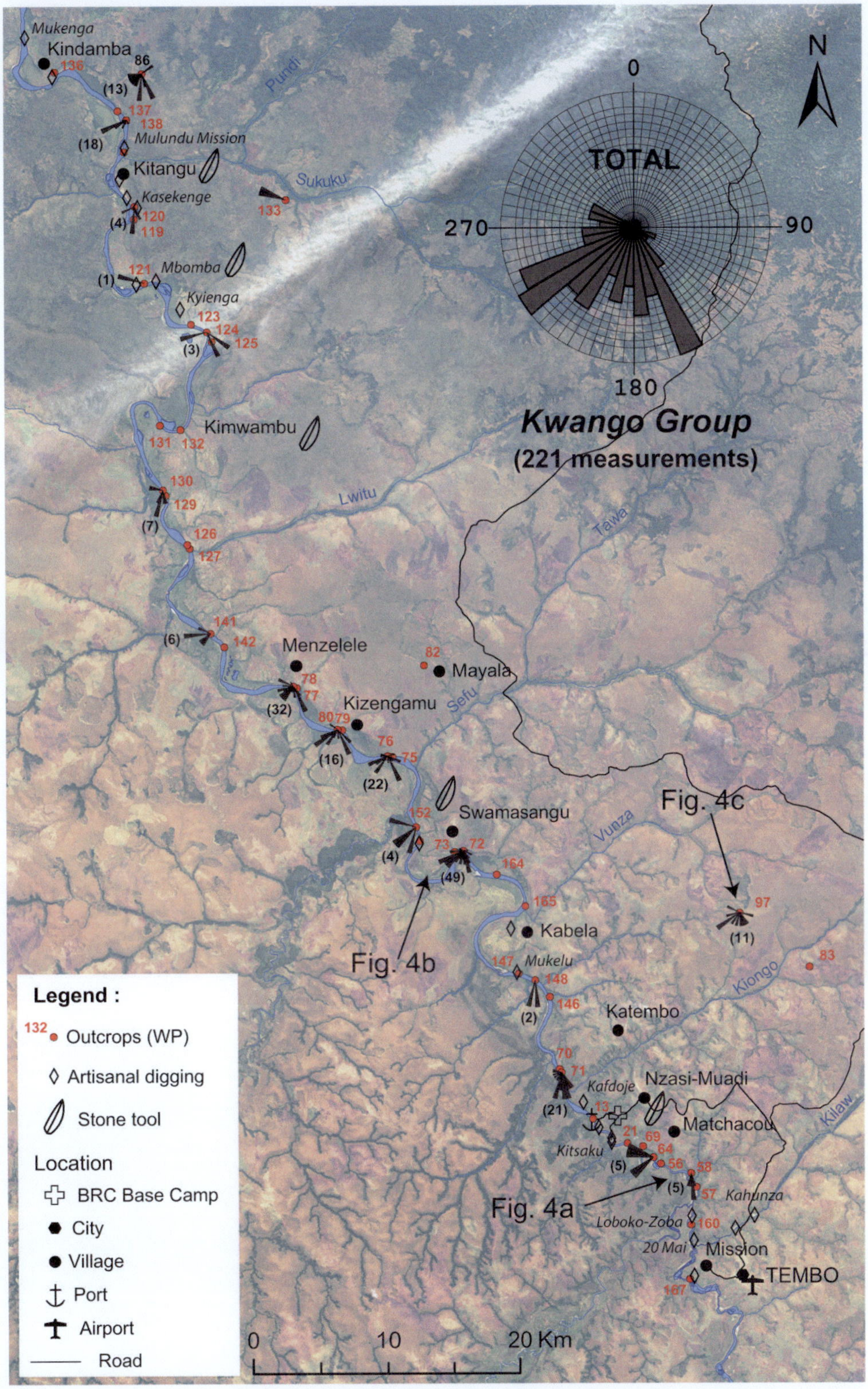

Fig. 8.3 Field study area along the Kwango River, southwest DRC (Fig. 8.1 for location), with locations of investigated outcrops (*WP* Waypoints; numbers in *red*) and paleocurrent rose-diagrams (number of measurements in *brackets*)

(e.g. the Nsele Formation) more recent biostratigraphy (Gobbo-Rodrigues et al. 2003) indicates a Campanian-Maastrichian age based on new correlations with ostracods from Brazil (the Bauru Group). These dates provide a maximum Late Cretaceous age for the regional peneplanation surface that cuts across the Kwango Group, and which defines the base of the overlying (Cenozoic) Kalahari Group (Cahen and Lepersonne; see Chaps. 10 and 14, this Book).

Fig. 8.4 Examples of studied outcrops in the Kwango Valley (Fig. 8.3 for locations): (**a**) near Matchacou (WP69), showing low-angle cross-stratifications that indicate predominant paleo-winds to the south; (**b**) at Swamasangu (WP72-73) along the Kwango River; and (**c**) at inland donga (WP97) at the Vunza stream water head

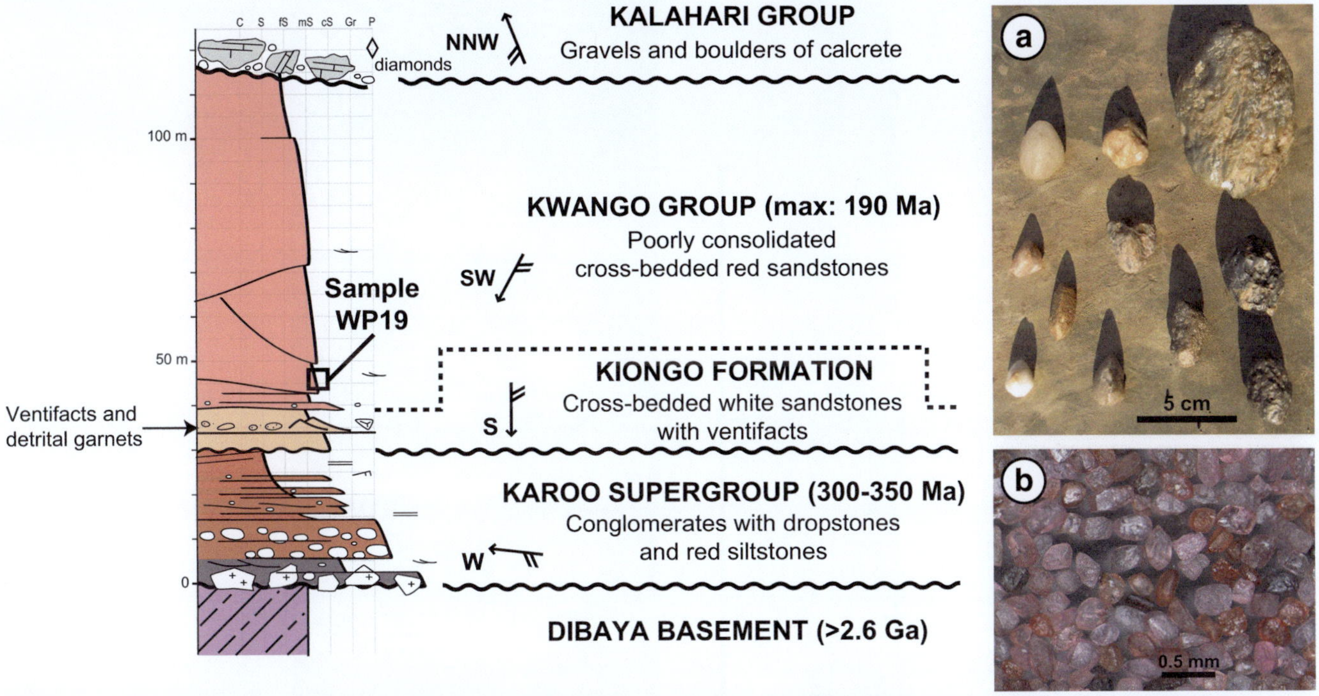

Fig. 8.5 Stratigraphic section across the Kwango nick-point. (**a**) Ventifacts and (**b**) heavy mineral assemblage dominated by sub-rounded pink garnets with frosted-glass appearance indicate aeolian deflations (hiatuses) near the base of the Kwango Group (e.g. the Kiongo Formation; Linol 2013)

8.3 Mapping in the Kwango Valley Area, Southwest Congo Basin

During fieldwork in the Kwango Valley (Fig. 8.3), an extensive succession of red sandstones of the Kwango Group was mapped, unconformably overlying Precambrian basement and Carboniferous-Permian ('Karoo-like') fluvial-glacial deposits (Linol 2013; Chap. 7, this Book). Field observations show this succession to be monotonous, comprising almost exclusively large-scale cross-bedded, red or white, fine to medium sandstones (e.g. Fig. 8.4a). Neither the Inzia and Nsele Formations (Lepersonne 1951), nor the basal diamondiferous conglomerates as identified at Tshikapa (Cahen 1951) were recognized in the studied area, implying there are regional facies changes from south to north, and from west to east within the Kwango Group.

The best sections of the (Lower) Kwango Group in the Kwango Valley are exposed along the banks of the Kwango River (e.g. Fig. 8.4b) downstream from a major, 150 m high, nick-point of Archean granites and migmatites (the Dibaya Complex; Cahen et al. 1984), and within large dongas (erosional gullies) away from the river (e.g. Fig. 8.4c).

The base of the succession is marked by a distinct (indurate), 5–10 m thick unit of coarser white cross-bedded sandstones, at the confluence between the Kwango and Kiongo Rivers (Fig. 8.3 for location). This formation (e.g. the Kiongo Formation) contains large reworked pebbles of sandstone (intraclasts 10–20 cm in diameter), disseminated wind-faceted pebbles and cobbles of quartz and black silica, and relatively abundant detrital garnets (Fig. 8.5). These observations suggest that there are major aeolian deflations (hiatuses) near the base of the Kwango Group.

Overlying red sandstones, exposed at several cliffs downstream along the Kwango River (e.g. Swamasangu, Kisengamu and Menzelele; Fig. 8.6), and inland, at the stream water-heads (e.g. Vunza; Fig. 8.7), all show large-scale trough and planar cross-stratifications, with cross-bed sets of up to 10 m in thickness that are characteristic of aeolian dune deposits (Kocurek 1996). These outcrops of red sandstones are often truncated by flat erosion surfaces, and covered by blocks and boulders of silcrete and calcrete, and river terraces that are exploited for alluvial diamonds (see Chap. 16, this Book).

Paleocurrents directions measurements from the Kwango Group indicate predominant paleo-winds blowing toward the south and the southwest (see Fig. 8.3). Also, the vast extent of this aeolian succession suggests a paleo-desert environment (Kocurek 1996). This is very similar to time-equivalent J–K aeolian sequences in Namibia (the Etjo and Twyfelfontein Formations; Jerram et al. 1999; Miller 2008) and eastern Brazil (the Sergi and Botucatu Formations; Scherer and Goldberg 2007, 2010), and suggest a vast expansion of arid conditions across central West Gondwana before the opening of the South Atlantic Ocean (Sellwood and Valdes 2008; Myers et al. 2011; see also Chap. 13, this Book).

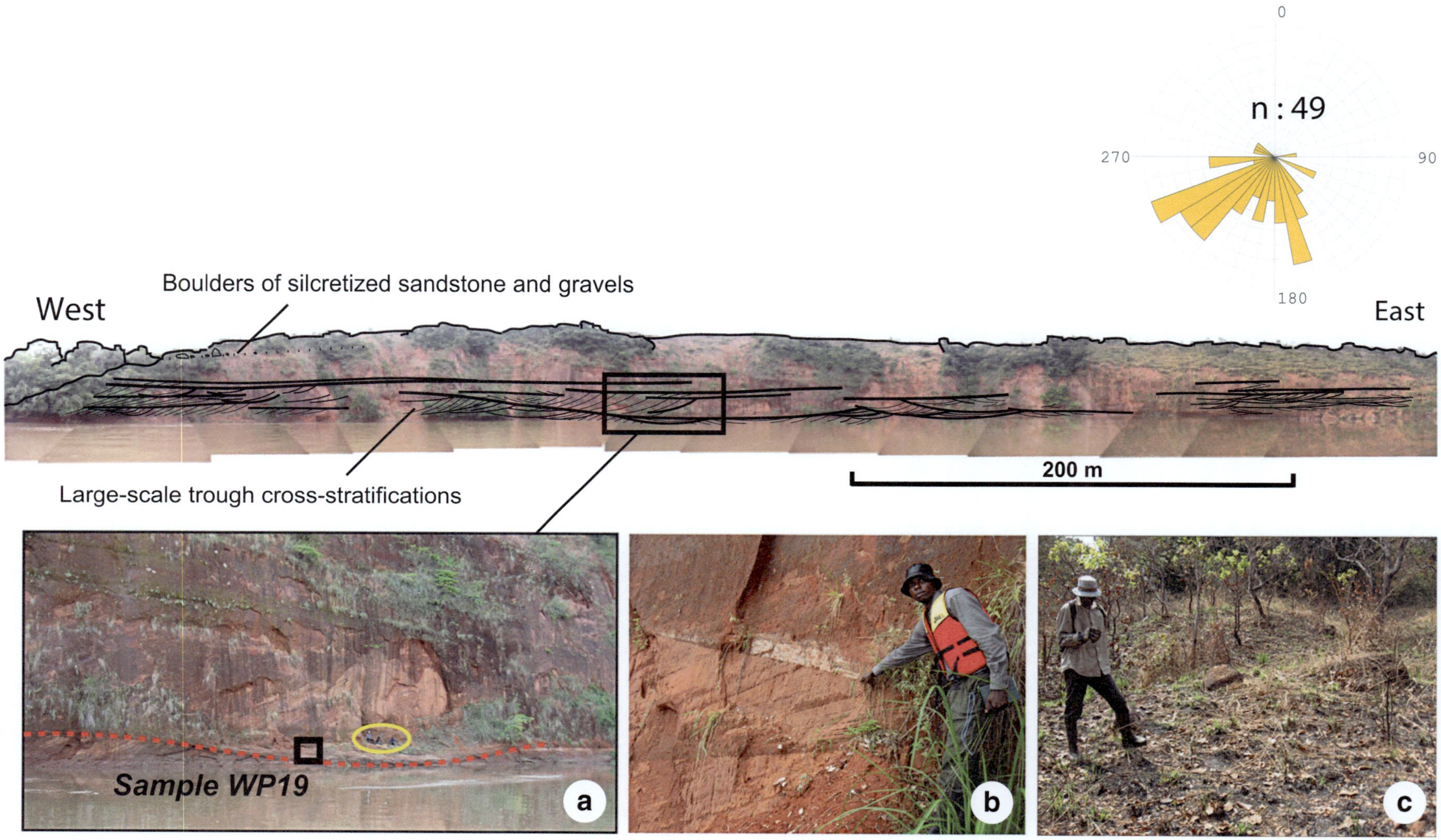

Fig. 8.6 Section of large-scale cross-bedded red sandstones at Swamasangu, along the Kwango River (WP72-73; Fig. 8.3). (**a**) 10 m thick cross-bed sets characteristic of aeolian dunes; note persons for scale (*yellow* ellipse). (**b**) Reactivation erosional surface. (**c**) Large boulders of silcretized sandstones

Fig. 8.7 Inland donga of red sandstones at the Vunza stream water head (WP97; Fig. 8.3). (**a**) Aeolian stratification. (**b**) Meniscate burrow (*arrow*) of type Crane fly larvae (Pemberton et al. 1992). (**c**) In-situ calcrete concretions

8.4 Facies Analysis and Depositional Modes Derived from the Samba and Dekese Cores

The J–K sequences in the center of the CB are primarily described from re-logging the Samba and Dekese cores at the RMCA museum in Belgium (Fig. 8.8a).

From the two examined core-sections, 19 facies and subfacies are documented (Table 8.1). These include: (a) conglomerates; (b) sandstones; (c) fine-grained facies and shales; and (d) paleosols. Further descriptions and photos of the cores are in Linol (2013).

The different facies successions are integrated into four depositional models (Figs. 8.9, 8.10, 8.11, 8.12 and 8.13): (1) shallow marine, (2) aeolian, (3) fluvial-deltaic, and (4) lacustrine, which serve as a basis for the following regional correlations and paleo-environmental interpretations.

8.5 Sequence Stratigraphic Correlations and Paleo-Environmental Interpretations

The vertical changes of facies and depositional environment variations observed within the Samba and Dekese cores are used to define shallowing and deepening upward cycles of different scale, or depositional sequences (e.g. Galloway and Hobday 1996). Based on this sequence-stratigraphic analysis (Fig. 8.8b), correlations are then further extended to the

Fig. 8.8 (continued)

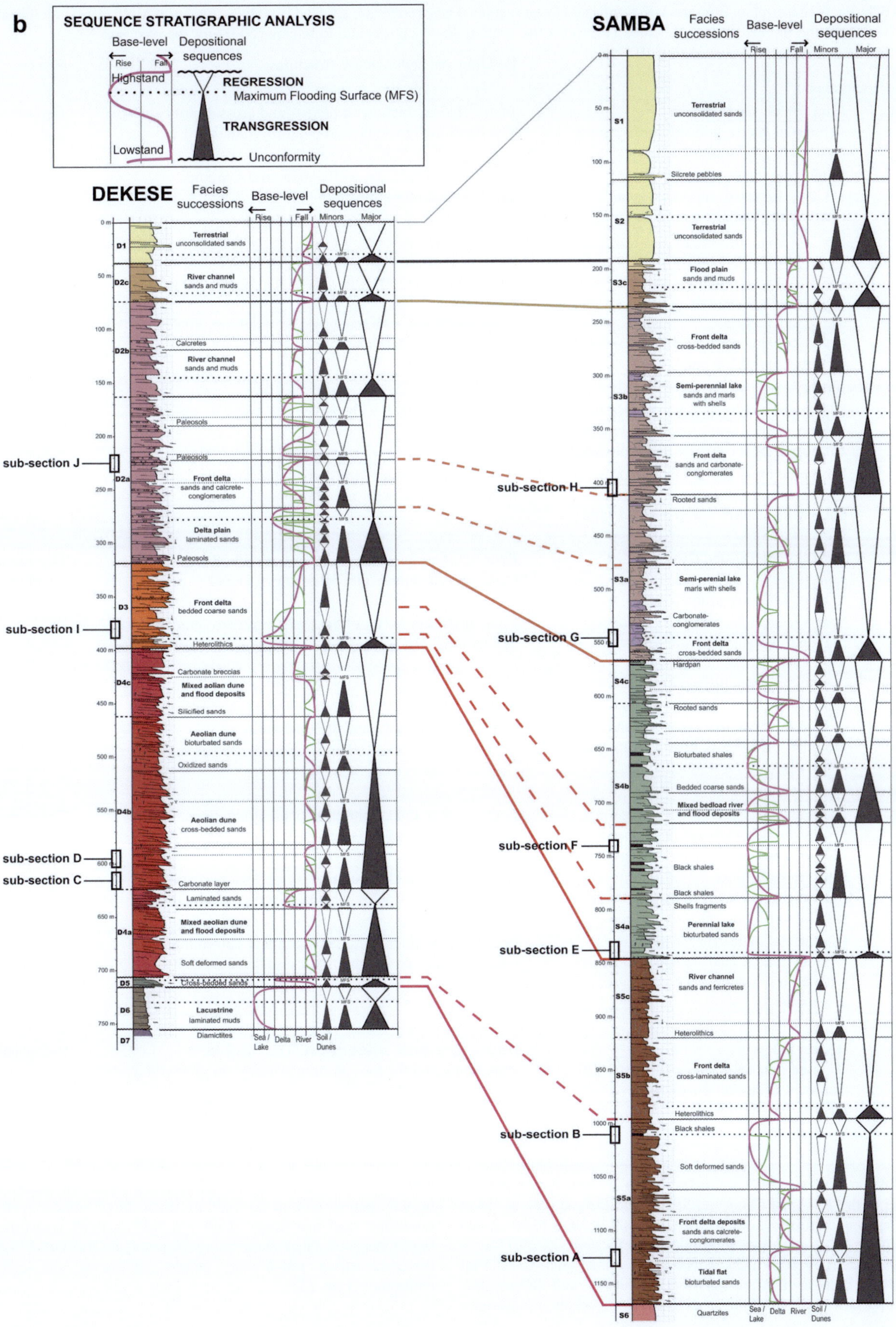

Fig. 8.8 (**a**) Photo-mosaic of core boxes containing the J–K (and Cenozoic) Samba and Dekese sections, with location of core-samples used for U-Pb detrital zircon geochronology. Each box represents ca. 5 m of core; all are stored at the RMCA museum in Tervuren (see Linol

Table 8.1 Facies description from the J–K Samba and Dekese cores. Facies codes are adapted from the schemes of Mutti (1992) for the turbidites, Miall (2006) for the fluvial deposits, Kocurek (1996) for the aeolian dune deposits and Retallack (2001) for the paleosols

Lithofacies		Bedding and sedimentary structures	Process
A. Conglomerates			
G-Stm	Mud pebble sandstones: Fine- to coarse-grained sands, rip-up mud clasts (1–7 cm)	Beds 0.1–3 m thick, with planar- and cross-stratifications. The bases are erosional and the tops gradational	Deposition from mainly bed-load
Gmc	Muddy carbonate conglomerates: Clast- to matrix-supported small rounded pebbles of carbonated mud (0.5–3 cm)	Beds 0.1–1.5 m thick, with planar-stratifications. The bases and tops are sharp	
C-Gt	Calcretized conglomerates: Clast-supported small pebbles (1–5 cm), cemented by carbonate	Beds 5–25 cm thick, with cross-stratifications. The bases are erosional and the tops gradational to sandstones	
Gc	Carbonate breccias: Clast- to matrix-supported fragments of carbonate (0.5–5 cm)	Beds 10–50 cm, structureless. The bases are erosional and the tops gradational	
B. Sandstone facies			
Sr	Ripple cross-laminated fine sandstones: Fine-grained sands to silts and muds with occasional calcretes and root traces	Beds 0.2–3 m thick with ripples and undulate (wavy) laminations	Deposition from high and low regimes
Stf-b	Bioturbed cross-bedded fine sandstones: Fine- to medium-grained sands and muds with shallow conic burrows type *Monocraterion*	Beds 0.5–1.5 m thick, with cross-stratifications and mud drapes. The bases are erosional	
Sm, Sl, and St	Massive, Planar- and Cross-bedded sandstones: Fine- to coarse-grained sands	Beds 0.5–5 m thick, structureless, or with planar- or cross- stratifications. The bases are erosional	Deposition from mainly bed-load
Slt and Slt-b	Large-scale cross-bedded sandstones: Fine to very coarse-grained and well-sorted sands, ferruginous or silicified, with diffuse bioturbations	Cross-beds sets 1–10 m thick, with inversely-graded laminations. The bases are erosional	
Smc and Smc-b	Massive and carbonated heterolithics: Fine to very coarse-grained sands with granules, diffuse bioturbations, burrows and shells	Beds 0.2–5 m thick, with soft-sediment deformation structures. The bases and tops are sharp	Debris-flows
C. Fine-grained facies and shales			
Fb and Fb-b	Black shales: Organic-rich muds and fine-grained sands Horizontal burrows type *Planolites*	Beds 0.2–2 m thick, with horizontal laminations. The bases and tops are sharp	Deposition from mainly suspended-load
Fm and Fl	Massive or laminated mudstones: Fine-grained sands to silts and muds with bioturbations, calcretes and root traces	Beds 0.1–5 m thick, structureless, or with horizontal laminations and ripples. The bases are gradational	
Fmc	Carbonated mudstones to marlstones: Fine-grained sands to silts and muds with diffuse bioturbations, burrows and shells	Beds 0.2–5 m thick, with soft-sediment deformation structures. The bases and tops are sharp	
D. Paleosols			
Pr	Rooted paleosols: Small pebbles, sands and muds with calcretes, roots and rhizoliths	Beds 1–3 m thick, structureless. The bases are gradational and the tops are truncated by conglomerates	Pedogenesis

Fig. 8.8 (continued) 2013 for details). (**b**) Sequence-stratigraphic analysis of the J–K (and Cenozoic) Samba and Dekese core-sections, with locations of sub-sections A to J (Figs. 8.9, 8.10, 8.11, 8.12 and 8.13). In the Dekese section, 40 m thick red mudstones (Unit D6) are sharply overlain by 10 m thick fluvial green coarse sandstones (Unit D5), 310 m thick aeolian red sandstones (Unit D4) and 360 m thick orange, purplish red delta front sandstones and flood plain mudstones with abundant paleosols (Units D2 and D3). These sequences correlate in the Samba section to 320 m thick fluvial-marine red–brown sandstones and mudstones (Unit S5), 280 m thick lacustrine green sandstones with black shale intercalations (Unit S4) and 370 m thick fluvial-lacustrine red–brown, purple sandstones and carbonated mudstones (Unit S3)

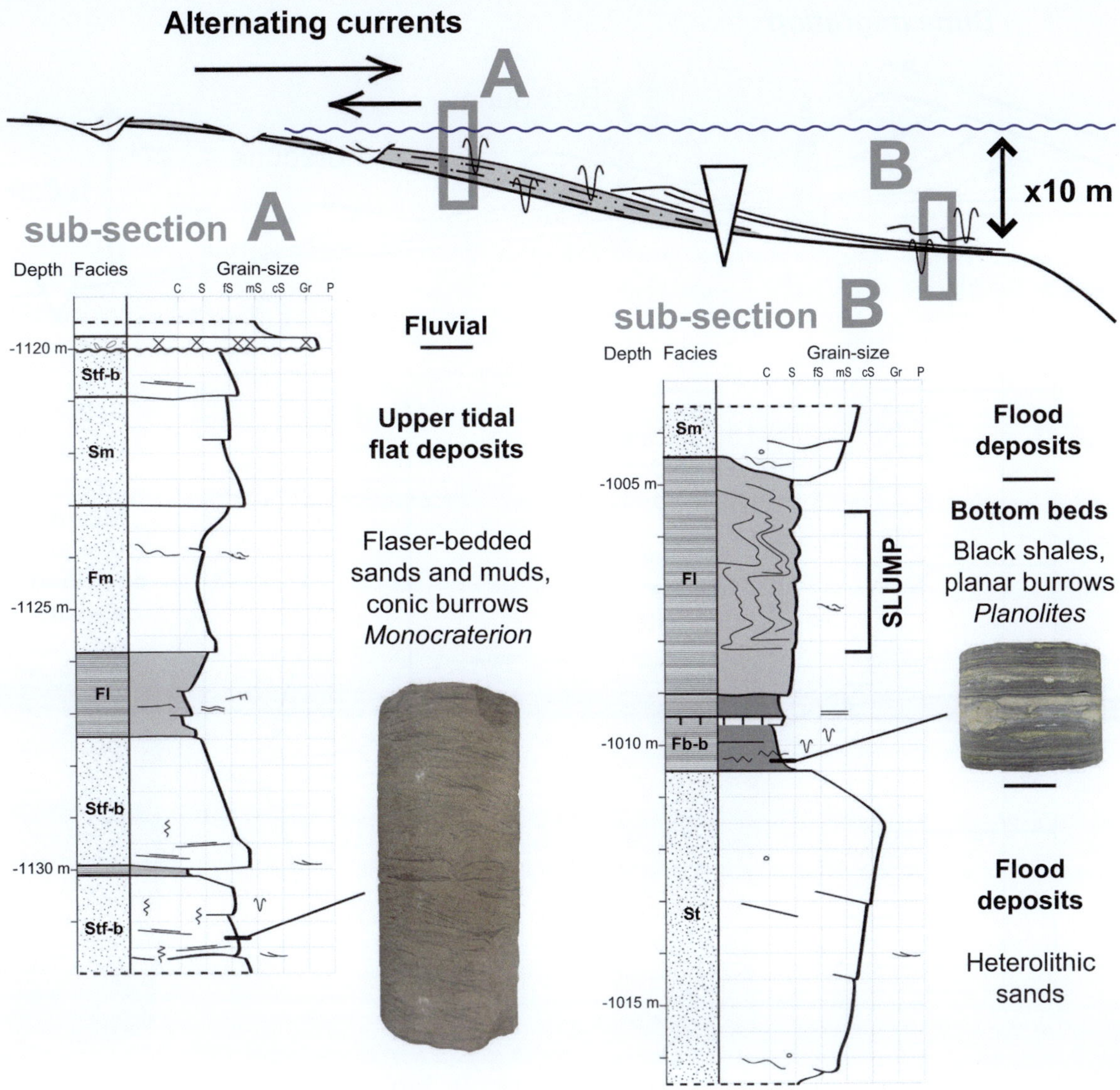

Fig. 8.9 Shallow marine facies successions from the lower part of the Samba core-section (Unit S5; Fig. 8.8b). Sub-section A: bioturbated and flaser-bedded fine sandstones (Stf-b facies), characteristic of tidal flat deposits (e.g. Dalrymple 1992). Sub-section B: heterolithic sandstones with a thin bed of bioturbated black shales (Fb-b facies) and soft-deformed rhythmites (e.g. slump 4 m thick)

Gilson and Mbandaka sections, re-interpreted from the well-logs. Along this correlation profile (Fig. 8.14), 880 km long across the central CB, the J–K sequences correspond to the Stanleyville, Loia, Bokungu and Kwango Groups, as described below.

8.5.1 The Stanleyville Group

The Unit S5 (322 m thick) of the Samba section is correlated with the Upper Jurassic Stanleyville Group, based on litho- and bio-stratigraphy (Cahen 1983b). It overlies a marked erosion surface across red quartzitic sandstones (the Haute Lueki Group; Chap. 7, this Book), at depth −1,167 m (Fig. 8.8b), and includes two main depositional sequences:

- A lower sequence (sub-Unit S5a) is 157 m thick and mainly transgressive. It comprises fine sandstones with cross-laminations and mud drapes, and abundant shallow conic burrows, interpreted as tidal (shallow marine) and flood deposits (Fig. 8.9). In the upper part, a marker bed of black shale records the first main maximum transgression, or Maximum Flooding Surface (MFS). Overlying thinly laminated shales and sandstones are truncated by an erosion surface at depth −995 m and, in turn, overlain by ferruginous and fractured red sandstones (Fig. 8.8b).

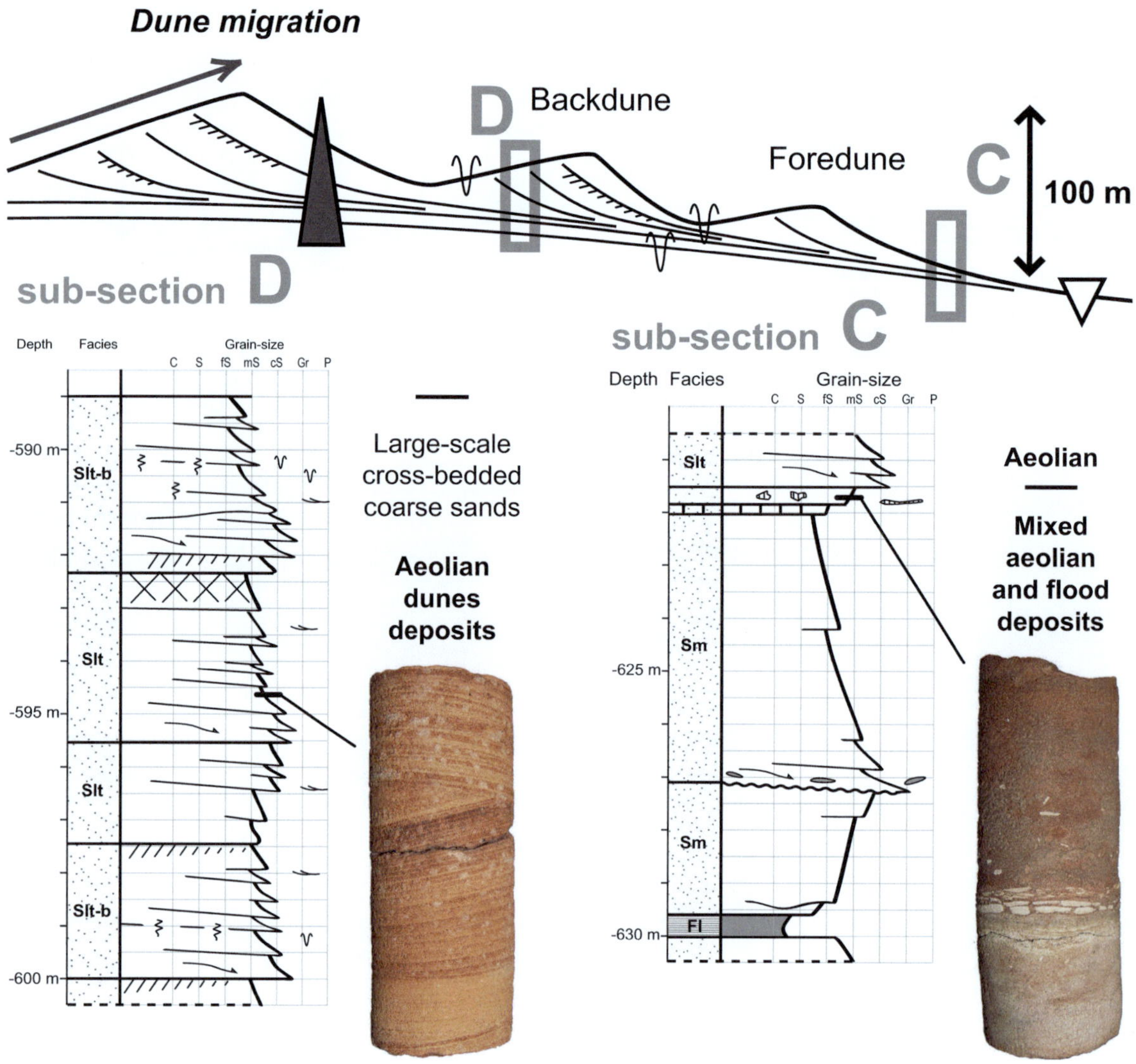

Fig. 8.10 Aeolian facies successions from the middle part of the Dekese core-section (Unit D4; Fig. 8.8b). Sub-section C: massive sandstones (Sm facies) with a thin cracked layer of carbonate. Sub-section D: large-scale cross-bedded ferruginous sandstones with inversely graded laminations (Slt facies), characteristic of aeolian dunes (e.g. Kocurek 1996)

- An upper sequence is 165 m thick and largely regressive, comprising delta front bars of cross-laminated muddy red sandstones (sub-Unit S5b), overlain by fluvial coarser sandstones with ferricretes (sub-Unit S5c) that indicate more frequent emersions. This thick upward-coarsening sequence is terminated with an erosion surface at depth −845 m (Fig. 8.8b), covered by thin carbonate-conglomerates and thick green shales of the Loia Group.

These two main depositional sequences in the Samba section, first shallow marine (e.g. Fig. 8.9), and then more fluvial and arid, likely correspond to the lower and upper Stanleyville subgroups defined in the Kisangani type-section (see Fig. 8.2a). In the lower part of the Samba section, the marker bed (MFS) of black shales can possibly be correlated with the horizon 'Lime Fine' at Kisangani, which marks a short Kimmeridgian marine incursion (Taverne 1975).

8.5.2 The Loia, Bokungu and Kwango Groups

In the Dekese section, diamictites with black shales and red mudstones of the Carboniferous-Permian and Triassic successions (the Lukuga and Haute Lueki Groups; Chap. 7, this Book) are truncated by a well-marked erosion surface at depth −715 m (Fig. 8.8b), overlain by a 10 m thick transgressive sequence of fluvial green heterolithic sandstones (Unit D5), and then 307 m thick almost exclusively large-scale cross-bedded red sandstones (Unit D4). The latter

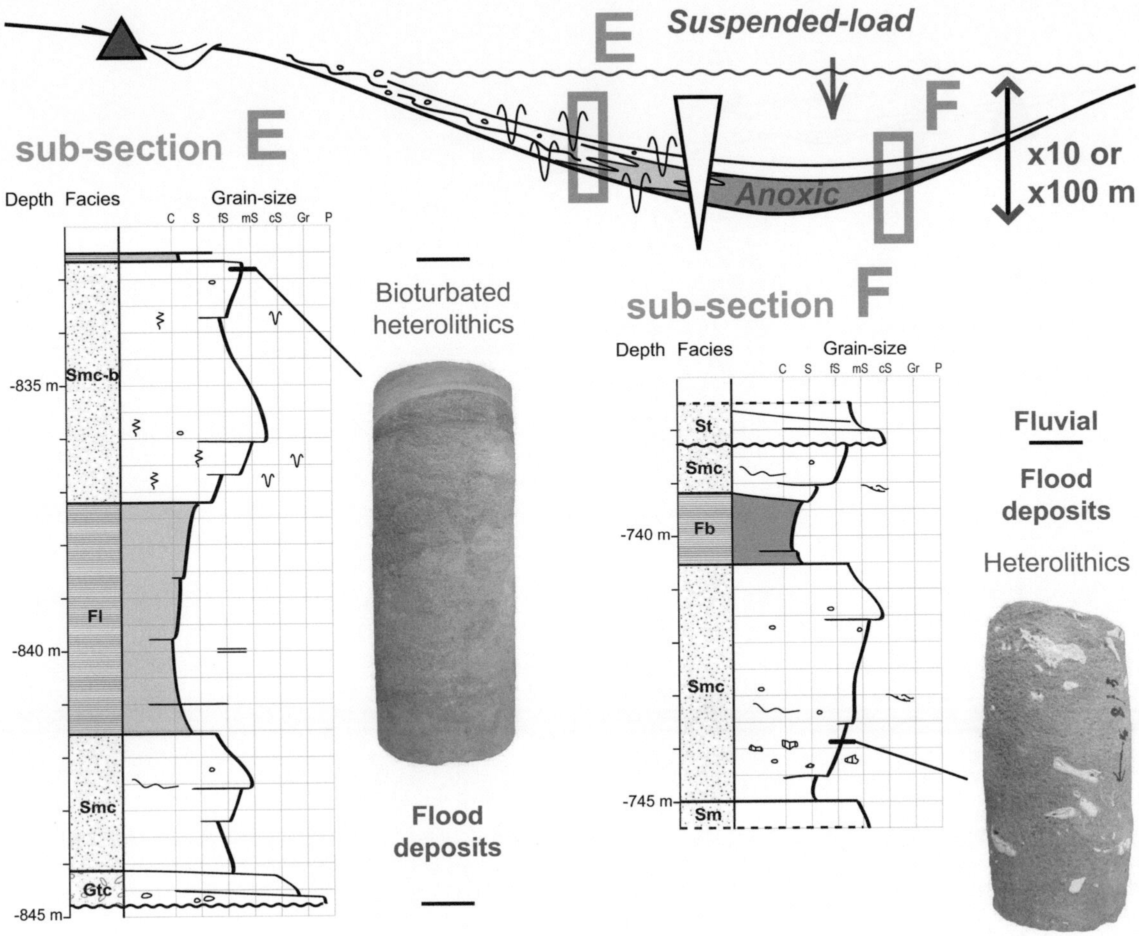

Fig. 8.11 Perennial lake facies succession from the middle part of the Samba core-section (Unit S4; Fig. 8.8b). Sub-section E: bioturbated heterolithic sandstones (Smc-b facies) alternating with green shales (Fl facies). Sub-section F: heterolithic sandstones with clasts of carbonate and bed intercalations of black shales (Fb facies) and cross-bedded sandstones (St facies), indicating a persistent and episodically anoxic lake (e.g. Talbot and Allen 1996)

aeolian succession comprises two main depositional sequences:

- In the lower part, mixed aeolian and flood deposits (sub-Unit D4a) record the main maximum transgression (MFS). This sequence terminates with a thin (1 cm thick) carbonate layer at depth −622 m that can represent a desiccation surface and a hiatus (Fig. 8.10).
- An overlying sequence includes numerous superimposed aeolian dunes (sub-Unit D4b) and intercalates near the top with two carbonate breccias (sub-Unit D4c), which could also attest of prolonged episodes of desiccation and hiatus.

This aeolian succession is also recognized within the Unit G4 of the Gilson section (Fig. 8.14), where it corresponds to 163 m thick, well-sorted, and ferruginous coarse sandstones, and can correlates with distinctive (very coarse), 36 m thick arkoses forming the Unit M4 of the Mbandaka section. We propose to name this aeolian succession: the Dekese Formation, as it is best described from the Dekese cores ('Couches' [Beds] Ca, Cb and Ba of Cahen et al. 1960). This formation is largely equivalent to the aeolian red sandstones of the Lower Kwango Group mapped from outcrops along the southwestern margin of the CB (see Fig. 8.4). Across the correlation profile in the center of the basin (Fig. 8.14), it is constrained in age (at Gilson) by biostratigraphy between the Upper Jurassic (Kimmeridgian) Stanleyville Group and the middle Cretaceous (Albian) Loia Group.

In the Samba section, the Units S4 and S3 are the stratotypes of the Loia and Bokungu Groups, respectively (Cahen 1983a), both dated biostratigraphically to the Albian-Cenomanian (Colin 1994). These two distinct lacustrine successions, first periodically anoxic with several beds of black shales, and then shallower with more carbonate-rich and fluvial sediments are separated by a fractured horizon of

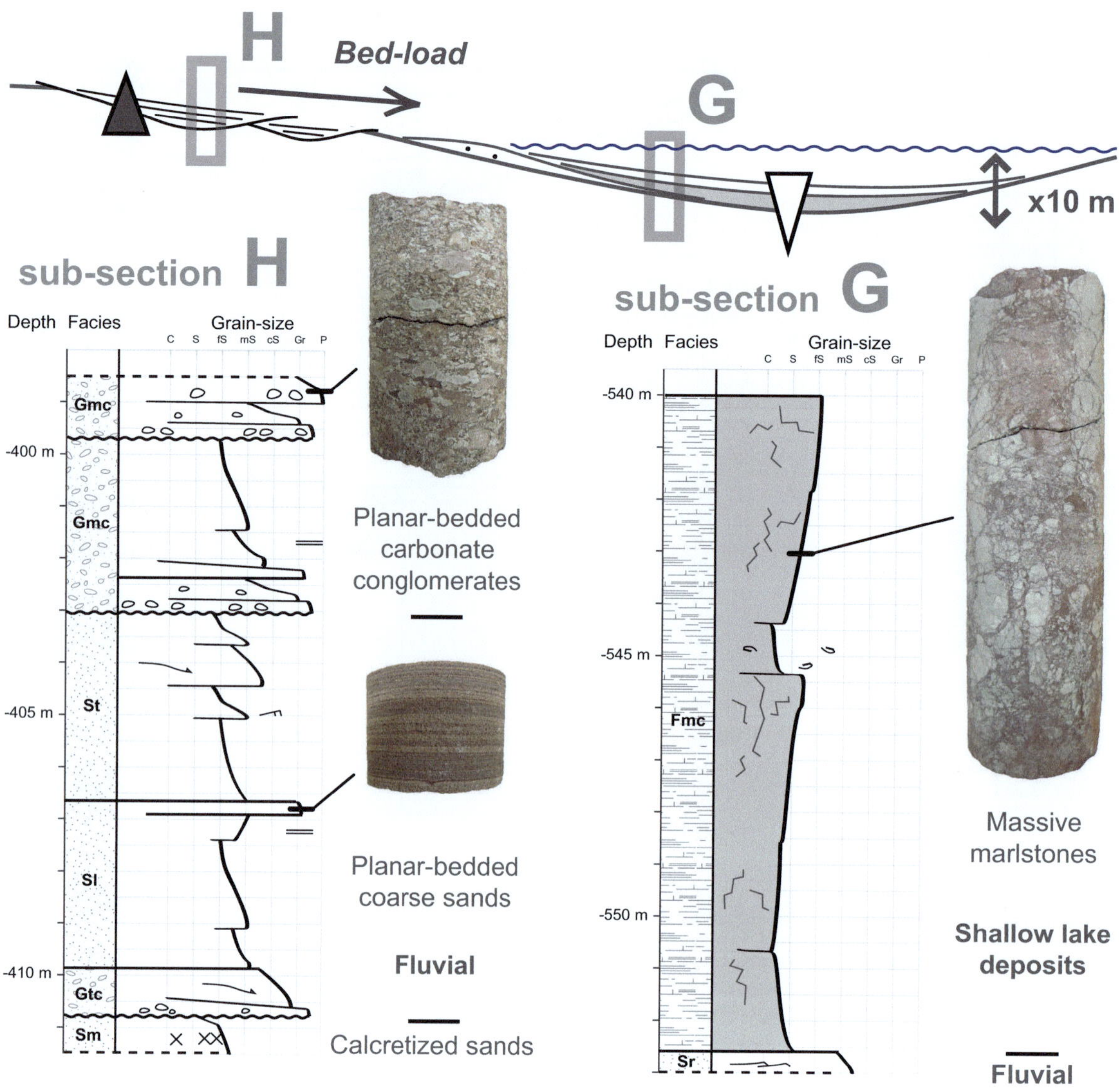

Fig. 8.12 Semi-perennial lake facies succession from the upper part of the Samba core-section (Unit S3; Fig. 8.8b). Sub-section G: purplish carbonated mudstones and marlstones (Fmc facies) with fossil remains. Sub-section H: planar-bedded sandstones (Sl facies) interbedded with intraformational (fluvial) carbonate-conglomerates (Gmc facies), indicating alternating shallow lacustrine and fluvial conditions

silicified mudstone (hardpan) at depth –565 m that marks a hiatus of sedimentation (Fig. 8.8b).

The Loia Group (Unit S4) is 280 m thick and includes two main depositional sequences:

- A lower sequence starts with relatively thick (8 m), thinly laminated green shales and upward-coarsening flood deposits (sub-Unit S4a) that record the first main maximum transgression (MFS). Overlying alternations of sandstones with mudstones intercalate with several beds of black shales (sub-Unit S4b) that indicate alternating episodes of lacustrine stagnation and fluvial activity (e.g. Fig. 8.11). At the top of this sequence, the most pronounced relative lake-level fall corresponds to two successive thin beds of horizontally laminated very coarse sandstones, at depths −719 m and −690 m (Fig. 8.8b).
- An upper sequence also comprises mixed fluvial and lacustrine deposits with beds of black shales that record another main maximum transgression (MFS). This sequence shallows upward with rooted sandstones overlain by alternating carbonated mudstones and sandstones (sub-Unit S4d).

These two sequences are also detected within the Gilson and Mbandaka sections (Fig. 8.14). In the Mbandaka section, the Unit M3 starts with 44 m thick green shales (sub-Unit M3a) that signal the first main maximum transgression (MFS). These are sharply overlain by 81 m thick, near

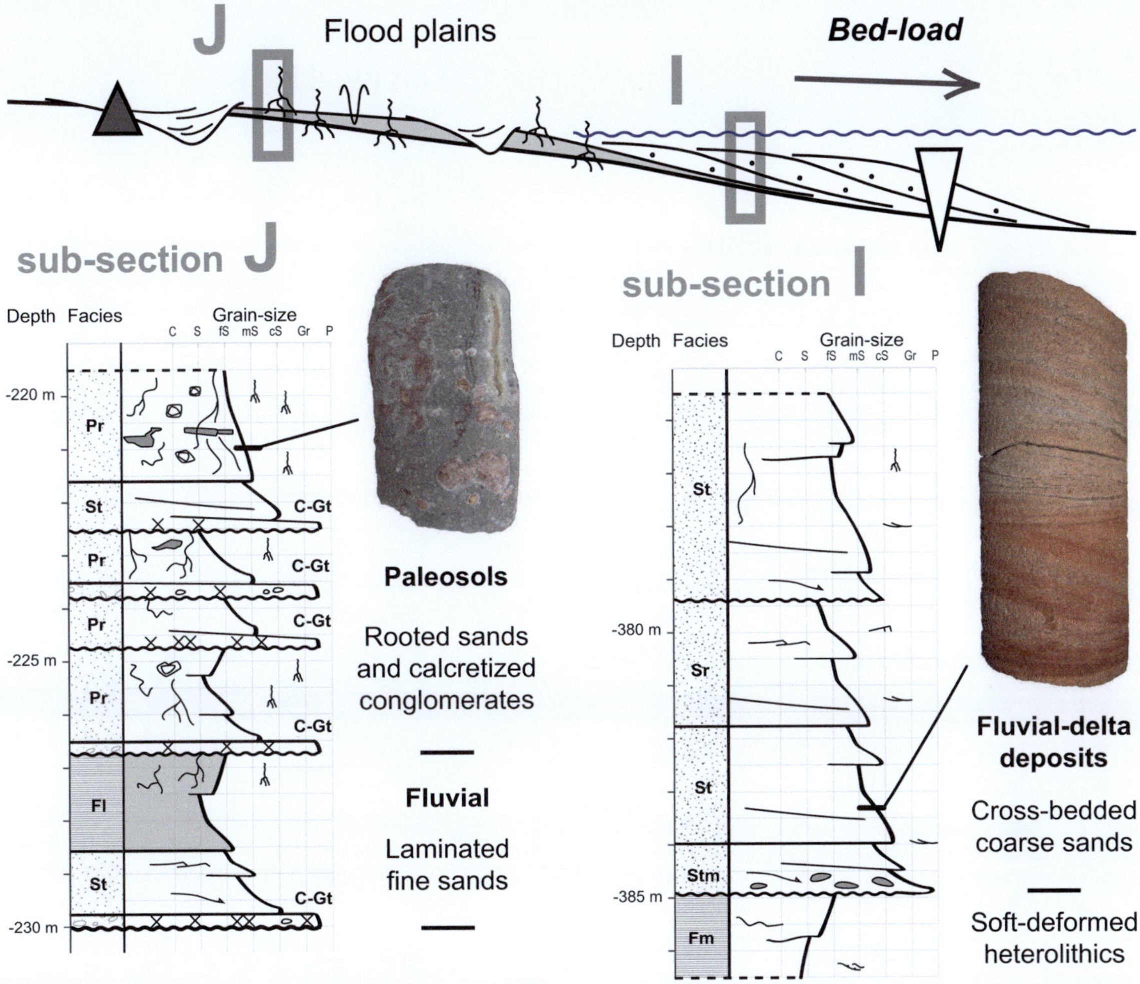

Fig. 8.13 Fluvial-delta facies succession from the upper part of the Dekese core-section (Units D2 and D3; Fig. 8.8b). Sub-section I: soft-deformed mudstones (Fm facies) and coarse sandstones with compound cross-stratifications (Stm, St and Sr facies). Sub-section J: mudstones and paleosols (Pr facies) interbedded with calcretized conglomerates (C-Gt facies), indicating ephemeral fluvial and arid conditions

regular, alternations of coarse sandstones with shales (sub-Unit M3b), and then 76 m thick claystones (sub-Unit M3a). In a similar manner, in the Gilson section, the Unit G3 includes in the lower part relatively thick (44 m) shales with intercalated beds of sandstones (sub-Unit G3b), overlain by 117 m thick alternations of sandstones with mudstones (sub-Unit G3a). All these sequences correlate with a (northward) prograding delta front in the Dekese section (Unit D3; Fig. 8.14). Here, an 80 m thick regressive sequence includes low-stand heterolithics and soft-sediment deformed red mudstones sharply overlain by upward-coarsening front bars, and terminates at the top with a 5 m thick unit of paleosols, which correlates with the hardpan separating the Loia and Bokungu Groups in the Samba section (Fig. 8.8b).

The Bokungu Group (Unit S3) in the Samba section is 373 m thick, more carbonate-rich, and comprises three main depositional sequences:

- A lower sequence (sub-Unit S3a) starts with a low-stand front delta overlain by relatively thick (22 m) fossiliferous carbonated mudstones and marlstones that record the first main maximum transgression (MFS). This sequence shallows upward with progressively more mudstones and intraformational conglomerates (Fig. 8.12). It ends at depth −412 m with a surface of emersion represented by calcretized sandstones (Fig. 8.8b).
- A middle sequence comprises fining upward planar-bedded sandstones with carbonate conglomerates, overlain by carbonated mudstones and marlstones with successive front bars (sub-Unit S3b). This sequence is

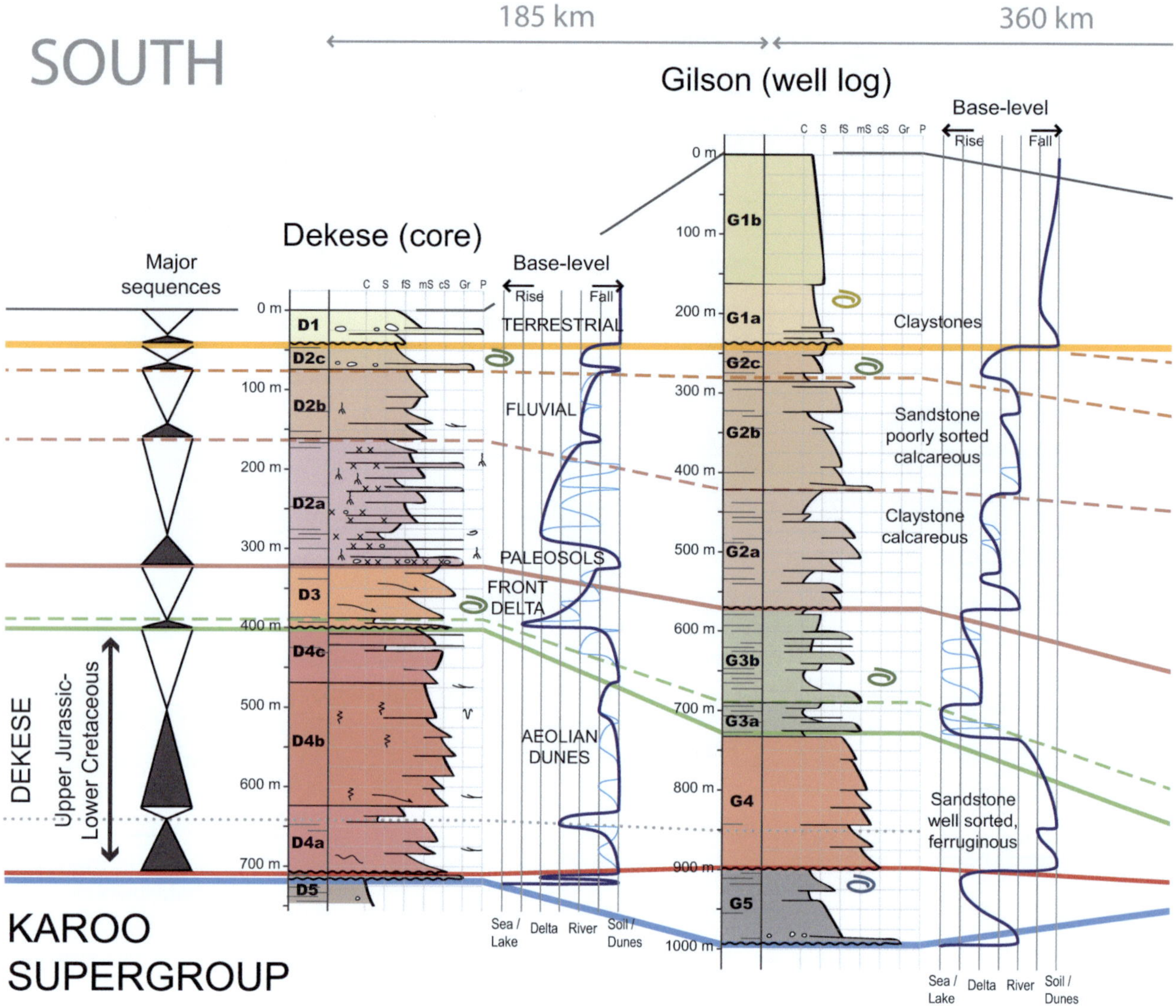

Fig. 8.14 (continued)

truncated by an erosion surface at depth –297 m (Fig. 8.8b), and covered by fluvial conglomerates.

- An upper sequence is mainly transgressive, comprising cross-bedded red sandstones fining upward into flood plain mudstones (sub-Unit S3c).

These three main depositional sequences are also identified within the Mbandaka, the Gilson, and the Dekese sections (Fig. 8.14). In the Mbandaka section, the Unit M2 (445 m thick) comprises in the lower part 212 m thick upward-shallowing carbonated mudstones (sub-Unit M2a), followed by two distinct sandstone members (sub-Units M2b and M2c, in total 110 m thick), and in turn overlain by 122 m thick siltstones coarsening upward into sandstones (sub-Unit M2d). In the Gilson section, the Unit G2 (332 m thick) comprises 151 m thick carbonated sandstones and mudstones (sub-Unit G2a), sharply overlain by 138 m thick poorly sorted sandstones and mudstones (sub-Unit G2b), and then 43 m thick siltstones (sub-Unit G2a). In the Dekese section, the Unit D2 (288 m thick) is mainly regressive, comprising superimposed sequences of delta plain and swamp deposits with numerous paleosols and calcretes (e. g. Fig. 8.13).

In the lower part (sub-Unit D2a), fossiliferous mudstones can record the first main transgression (MFS). This sequence shallows upward with alternations of sandstones and mudstones with calcretized conglomerates. It is sharply overlain by two successive fining upward intervals of channel infillings (sub-Units D2b and D2c). The top of these fluvial-lacustrine sequences, in all four boreholes, is an erosion surface overlain by unconsolidated conglomeratic sandstones and mudstones attributed to the (Cenozoic) Kalahari Group, in total between 37 m and 242 m thick (Fig. 8.14).

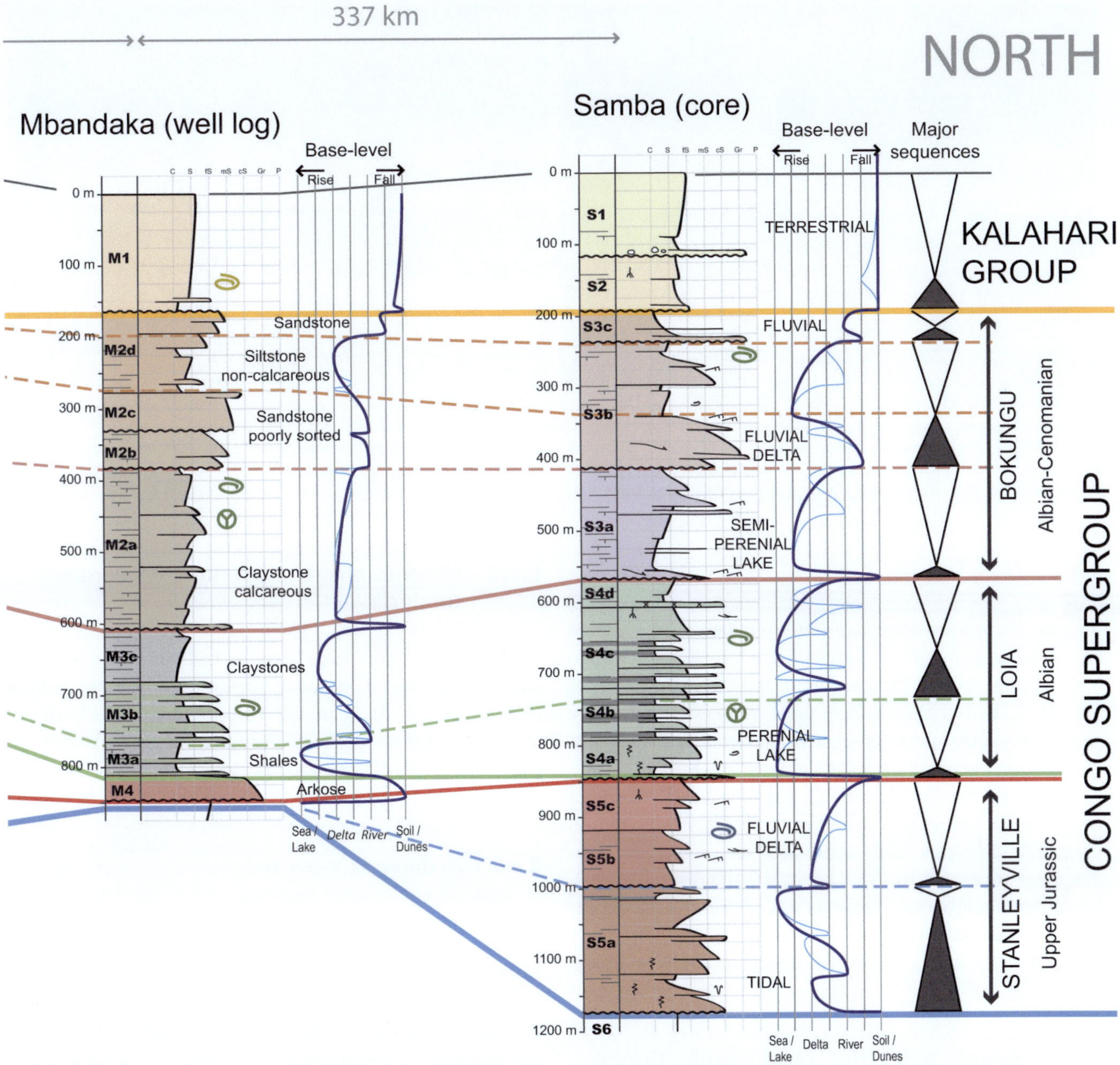

Fig. 8.14 Regional correlation profile of the J–K Stanleyville, Loia, Bokungu and Kwango sequences (and the Cenozoic Kalahari Group) across the central CB (Fig. 8.1 for borehole locations). The Upper Jurassic Stanleyville Group in the Samba section comprises a main transgressive sequence of tidal and flood deposits (sub-Unit S5a) overlain by regressive front bar sandstones (sub-Units S5b and S5c). The overlying middle Cretaceous Loia and Bokungu Groups in the Samba (Units S3 and S4), Gilson (Units G2 and G3) and Mbandaka (Units M2 and M3) sections comprise several regressive sequences of lacustrine deposits that correlate laterally to (northward) prograding fluvial delta sandstones with paleosols in the Dekese section (Units D2 and D3)

8.6 U-Pb Detrital Zircons Dates and Source Provenances Analysis

Detrital zircons collected from the (aeolian) J–K succession in the CB were U-Pb dated by Laser Ablation Multi-Collector Inductively Coupled Plasma Mass Spectrometer (LA-MC-ICP-MS; e.g. Cocherie and Robert 2008) at the AEON EarthLAB. The method for samples preparation and the experimental and analytical procedures are described in detail in Linol (2013); see also Chap. 7, this Book.

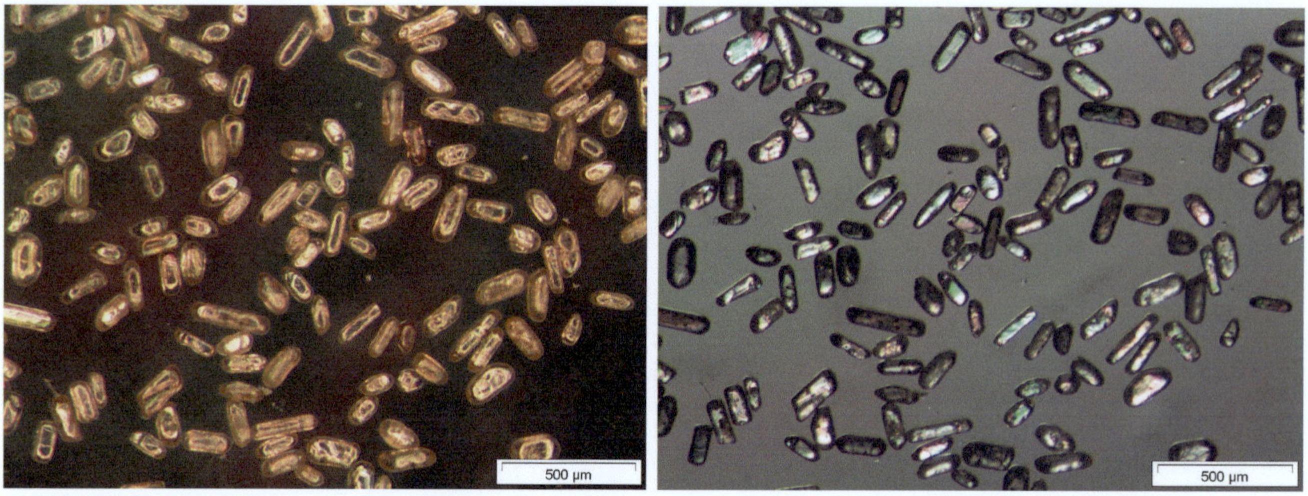

Fig. 8.15 Microphotographs (*left*: reflected light; *right*: polarized light) of selected detrital zircons of core-sample D600 from the Dekese Formation (Fig. 8.8a for location). Zircons from this aeolian formation (Unit D4) are small to medium in size (200–400 µm), elongated, well rounded and generally transparent

8.6.1 U-Pb Results

In total, 201 zircons were analyzed from the Lower Kwango Group outcropping in the southwestern CB (in the Kwango Valley), and from the Dekese cores (Unit D4) in the center of the basin. Detrital zircons from these samples are relatively small (100–300 µm), elongated, sub-rounded to well-rounded, and generally transparent (e.g. Fig. 8.15).

The U-Pb results (Tables 8.2 and 8.3; details of which are given in Linol 2013; Linol et al. 2014) are plotted on Concordia-diagrams and compared on frequency age-diagrams in Fig. 8.16. The youngest grains (<500 Ma), which commonly have low uranium and lead concentrations, were analyzed repeatedly to improve the lower intercept dates and determine the maximum age of sedimentation (Fig. 8.17).

8.6.1.1 Sample WP19 (the Lower Kwango Group)

63 zircons (65 analyses) were dated from sample WP19 (the Lower Kwango Group), collected in outcrop near Swamasangu (Fig. 8.6 for sample location). Results are summarized in Table 8.2.

8.6.1.2 Samples D470 and D600 (The Dekese Formation)

Two core-samples: D470 and D600 were taken from the Dekese Formation at depths of –470 m and –600 m (Unit D4), respectively, in the middle part of the Dekese section (Fig. 8.8a for samples location). 76 analyses were performed on 70 grains in sample D470, and 72 analyses on 68 grains in sample D600. Results are summarized in Table 8.3.

8.6.2 Maximum Age and Source Provenance Analysis

U-Pb detrital zircon dates from the aeolian, Kwango Group and the Dekese Formation shows five age-populations (Fig. 8.16): (1) Archean, (2) Paleoproterozoic (Eburnian), (3) early Neoproterozoic to Mesoproterozoic (Kibaran), (4) Cambrian to late Neoproterozoic (Pan African), and (5) mid-Phanerozoic.

8.6.2.1 Archean (2.5–3.1 Ga)

In total, 30 zircons date between 2.5 Ga and 3.1 Ga in the samples, indicating some contributions from Archean aged sources to the CB, such as for example the Kasai and NE Angola Cratons along the southern margin of the basin (Cahen et al. 1984; Chap. 2, this Book). These zircons can also have been derived from recycling older sediments, as suggested by the small and rounded shape of the grains (e.g. Fig. 8.15).

8.6.2.2 Paleoproterozoic (1.8–2.2 Ga)

28 zircons date between 1.8 Ga and 2.2 Ga, overlapping in age with the conventional Eburnian period of west and central Africa (Cahen et al. 1984). Such dates are relatively common within basement rocks of the West Congo, Lufilian and Ubendian Belts along the western, southern and eastern margins of the CB, respectively (Lenoir et al. 1995; Tack et al. 2001; Rainaud et al. 2005), as well as from their overlying Neoproterozoic sequences (Master et al. 2005; Frimmel et al. 2006).

Table 8.2 Summary of U-Pb detrital zircon dates from sample WP19 (the Lower Kwango Group)

Age-population	U-Pb dates ($^{206}Pb/^{207}Pb$)	Notes
1. Mesoarchean	One (oldest) zircon dates at 3133 ± 29 Ma	
2. Neoarchean	5 zircons date between 2596 ± 27 Ma and 2657 ± 27 Ma	Indicate some contributions from Neoarchean aged-terrains. These dates are similar to dated basement rocks in southern DRC and northern Angola (Delhal 1991; Jelsma et al. 2011), western Tanzania and Uganda (Link et al. 2010), and Gabon (Caen-Vachette et al. 1988)
3. Paleoproterozoic	10 zircons date between 1.7 Ga and 2.2 Ga, with 6 comprised between 1979 ± 29 Ma and 2114 ± 29 Ma	Overlap the conventional Eburnian period (e.g. Cahen et al. 1984). Such dates are common within the Ruwenzori and Ubendian Belts in Uganda and western Tanzania (Lenoir et al. 1995; Nagudi et al. 2003), and within the basements of the West Congo Belt in western DRC and the Lufilian Arc in Zambia (Tack et al. 2001; Rainaud et al. 2005), as well as from their overlying upper Neoproterozoic to lower Paleozoic sequences (Master et al. 2005; Frimmel et al. 2006)
4. Early Neoproterozoic to late Mesoproterozoic (*Second most common age-population*)	20 zircons date between 850 Ma and 1200 Ma, with 17 bracketed between 922 ± 34 Ma and 1044 ± 33 Ma	This major peak at 1000 Ma correspond to the general Kibaran age in Africa (e.g. Cahen et al. 1984), although its type area in eastern DRC, Rwanda and Burundi has been shown to be older, ca. 1375 Ma (Tack et al. 2010)
5. Late Neoproterozoic (*Most common age-population*)	24 zircons date between 544 ± 37 Ma and 854 ± 38 Ma	Indicate large contributions from Pan African aged-terrains, such as the West Congo, Oubanguides, Mozambique and Lufilian Belts
6. Early Paleozoic	One zircon dates at 426 ± 8 Ma ($^{206}Pb/^{238}U$ date)	No Silurian sources have been identified across central Africa. Such dates are common within the Andes and proto-Andes in Patagonia (Pankhurst et al. 1998; Vaughan and Pankhurst, 2008)
7. Early Jurassic	Two (youngest) zircons date at 204 ± 4 Ma and 192 ± 4 Ma	Overlap with the CAMP and Karoo LIPs, ca. 200 Ma and 182 Ma respectively (Marzoli et al. 1999; Svensen et al. 2012; Blackburn et al. 2013), and Carboniferous to Jurassic felsic volcanism and magmatism of the Choiyoi and Chon Aike igneous provinces in southern South America (Kay et al. 1989)

Table 8.3 Summary of U-Pb detrital zircon dates from samples D470 and D600 (the Dekese Formation)

Age-population	U-Pb dates ($^{206}Pb/^{207}Pb$)	Notes
1. Mesoarchean	One (oldest) zircon dates at 2942 ± 17 Ma	
2. Early Paleoproterozoic to Neoarchean	15 zircons date between 2.3 Ga and 2.7 Ga, with 11 concentrated between 2553 ± 11 Ma and 2665 ± 16 Ma	These dates can correspond to the Kasai, NE Angola, Ntem, Mboumou and Tanzanian Cratons
3. Paleoproterozoic	15 zircons date between 1.8 Ga and 2.3 Ga, with 4 restricted between 1964 ± 13 Ma and 1974 ± 17 Ma, and 2 at 2075 ± 12 Ma and 2080 ± 17 Ma	Indicate some contributions from Eburnian aged-terrains
4. Mesoproterozoic	6 zircons date between 1.2 Ga and 1.5 Ga, with only 2 concordant at 1272 ± 23 Ma and 1548 ± 13 Ma	This suggests that the Kibaran Belt, dated at ca. 1375 Ma (Tack et al. 2010), was not major source
5. Early Neoproterozoic to late Mesoproterozoic (*Second most common age-population*)	35 zircons date between 844 ± 19 Ma and 1187 ± 48 Ma	Indicate significant contributions from late Mesoproterozoic aged-terrains (distinctively younger than the Kibaran type-area)
6. Early Paleozoic to late Neoproterozoic (*Most common age-population*)	64 zircons date between 513 ± 21 Ma and 837 ± 21 Ma	Coincide with Pan African tectono-metamorphic events in Africa (e.g. Cahen et al. 1984). These dates are abundant within the West Congo, Oubanguides, Mozambique and Lufilian Belts
7. Permian-Triassic	Two (youngest) zircons date at 240 ± 6 Ma and 286 ± 7 Ma ($^{206}Pb/^{238}U$ dates)	Coincide with Permian-Triassic kimberlites discovered in northern Angola (Jelsma et al. 2011), or more likely to felsic volcanism and magmatism of the Choiyoi Province in western Argentina and eastern Chile (e.g. Munizaga et al. 2008)

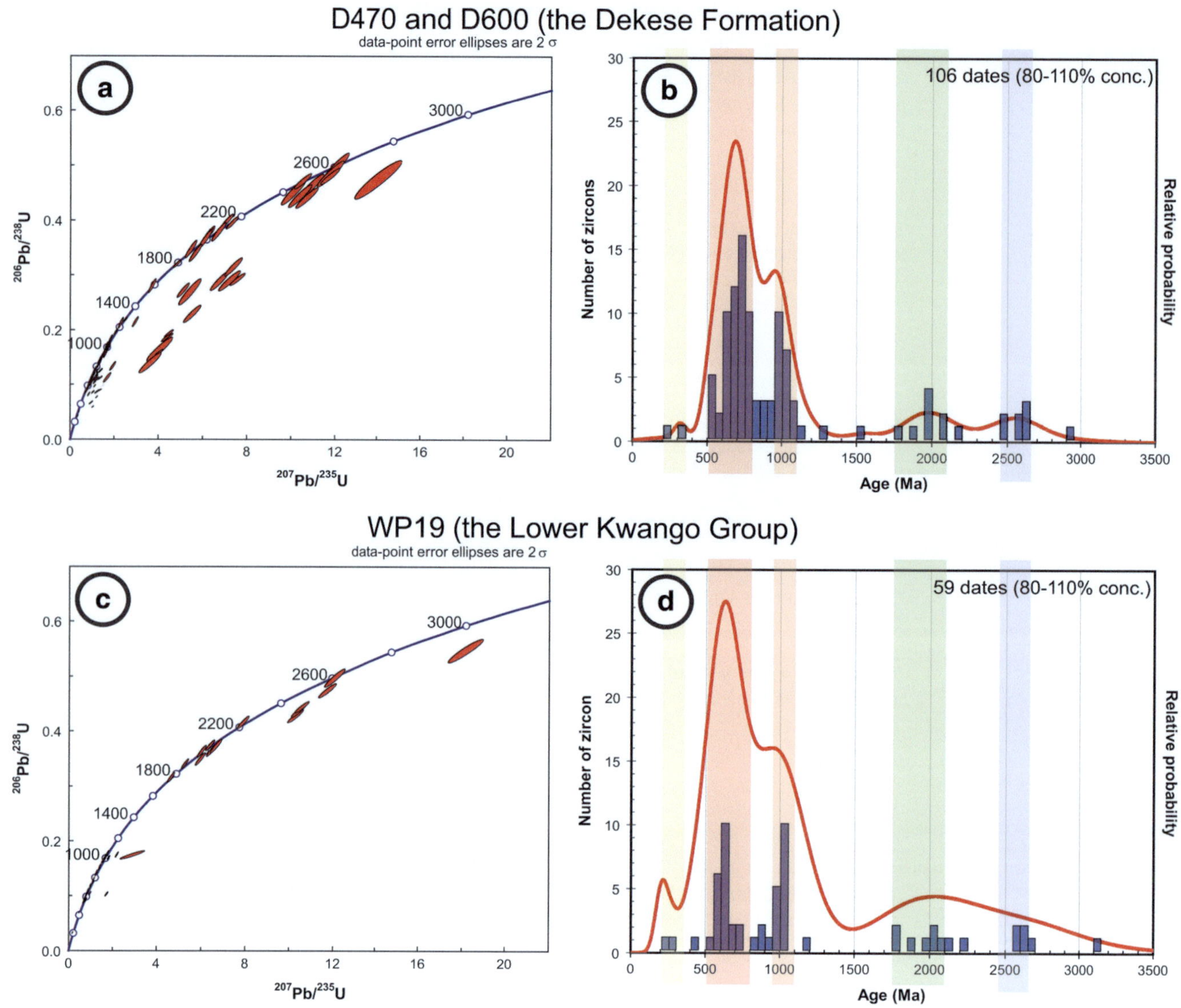

Fig. 8.16 U-Pb Concordia- and frequency age-diagrams of detrital zircons from (**a** and **b**) sample WP19 of the Lower Kwango Group in the southwest CB, and (**c** and **d**) samples D470 and D600 of the Dekese Formation in the middle part of the Dekese cores, plotted using Isoplot 3.0 (Ludwig 2003)

8.6.2.3 Early Neoproterozoic to late Mesoproterozoic (900–1100 Ma)

In total, 56 zircons date between 900 Ma and 1100 Ma, forming a major peak centered at 1000 Ma (Fig. 8.16). This 1 Ga age-population is identical to other analyzed samples from the underlying Carboniferous-Permian and Triassic successions (Linol 2013; Chap. 7, this Book), and corresponds to the general 'Kibaran' period in Africa (e.g. Cahen et al. 1984), although its type-area along the eastern margin of the CB, in eastern DRC, Rwanda and Burundi has been shown recently to be older and restricted in age to ca. 1375 Ma (Tack et al. 2010). This may requires a new definition of the Kibaran, but this is beyond the purpose of this study. Here, the general absence of ca. 1.4 Ga detrital zircons in the J–K sequences of the CB suggests that the type-Kibaran Belt was not a major source, possibly because it was covered by sediments or it was below sea-level during the Mesozoic.

8.6.2.4 Cambrian to late Neoproterozoic (500–850 Ma)

97 zircons date between 500 and 850 Ma (Fig. 8.16), corresponding to the widespread Pan African tectono-metamorphic events in Africa (Cahen et al. 1984). Such dates are abundant within the West Congo, Oubanguides, Mozambique and Lufilian Belts, and their overlying foreland sequences that completely surround the CB (Chaps. 2 and 3).

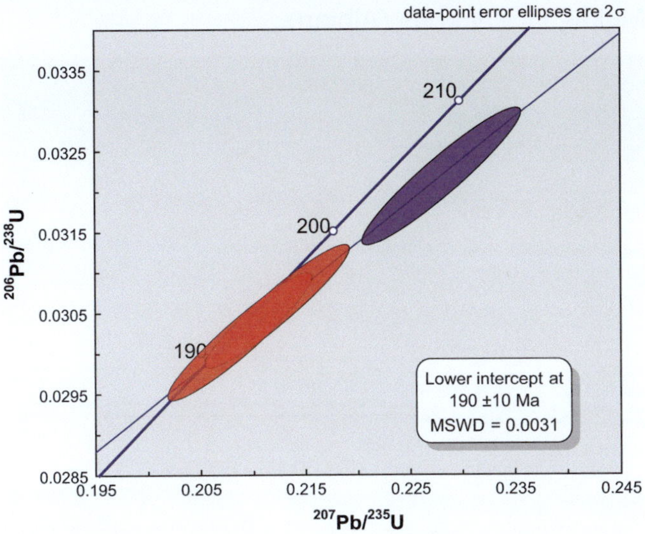

Fig. 8.17 U-Pb Concordia diagram for the two youngest detrital zircons of sample WP19. The lower intercept age is 190 ± 10 Ma, which provides a maximum age for this lower Kwango sequence

8.6.2.5 Mid-Phanerozoic (ca. 430 Ma, 240–280 Ma and 190 Ma)

One zircon grain has a $^{206}Pb/^{238}U$ date of 426 ± 8 Ma (mid-Silurian), although no sources of this age have been reported in central Africa (Cahen et al. 1984). There are, however, large potential source rocks within the Andes in South America (e.g. Patagonia; Pankhurst et al. 1998; Vaughan and Pankhurst, 2008), as well as detrital zircons with this age-range within lower Paleozoic sequences in southern Africa, such as the Cape Supergroup (Fourie et al. 2011).

Two zircons in sample D470 have concordant dates of 240 ± 6 Ma and 286 ± 7 Ma, and two youngest zircons in sample WP19 date at 204 ± 4 Ma and 192 ± 4 Ma ($^{206}Pb/^{238}U$ dates), which provide a maximum Early Jurassic age for the Kwango Group (Fig. 8.17). These Mesozoic dates may possibly correspond to kimberlite activity, as recently discovered in northern Angola (Jelsma et al. 2011), and/or to volcanism of the Central Atlantic (CAMP) and Karoo Large Igneous Provinces (LIPs), ca. 200 Ma and 182 Ma, respectively (Marzoli et al. 1999; Svensen et al. 2012; Blackburn et al. 2013). However, these rock sequences (kimberlites and basalts) are unlikely to yield abundant zircons. A more likely source for these zircons, therefore, is the extensive late Paleozoic to Jurassic silicic volcanic belt in southern South America (the Choiyoi and Chon Aike Provinces; Kay et al. 1989), from which zircons in abundant air fall tuffs are also recorded in the Permo-Triassic rocks of Paraná Basin in Brazil (e.g. Rocha-Campos et al. 2011) and the main Karoo Basin in southern Africa (Stollhofen et al. 2000; Fildani et al. 2007, 2009; Rubidge et al. 2013).

8.7 Regional Synthesis

The J–K succession of the CB (the Congo Supergroup) attains a maximum thickness of 1000 m in the center of the basin (Lepersonne 1974; Linol 2013). It comprises shallow marine, aeolian and fluvial-lacustrine sequences (the Stanleyville, Loia, Bokungu and Kwango Groups), deposited following the period of initial break-up between East and West Gondwana (ca. 160–180 Ma) and during the subsequent opening of the South Atlantic Ocean, ca. 80–130 Ma (Fig. 8.18).

8.7.1 Upper Jurassic to Lower Cretaceous Sequences

The lowermost, Upper Jurassic Stanleyville Group (370 m thick) comprises in the lower part black shales with one limestone intercalation ('Lime Fine'), which contains a distinct juvenile fish fauna from marine spawning sites during a short Kimmeridgian transgression of the proto-Indian Ocean in the northern CB (Taverne 1975). This MFS is correlated to a marker bed of black shales overlying tidal and flood deposits in the lower part of the Samba section (Unit S5), and is in turn overlain by upward-coarsening fluvial-deltaic red sandstones and mudstones with ferricretes that record a major regression and an emersion of the CB (Fig. 8.14). Geochemical analyses of the upper part of the Stanleyville Group in this section (Myers et al. 2011) have further suggested that hot and arid conditions, between 25° and 40 °C, prevailed during this period.

The Stanleyville Group is overlain to the south by aeolian dunes in the Gilson (Unit G4) and Dekese sections (Unit D4), and here named the Dekese Formation (Fig. 8.14). This new aeolian unit also correlates with widespread aeolian red sandstones of the Lower Kwango Group mapped in the southwestern CB. U-Pb dates of detrital zircons from this sequence indicate a maximum age of aeolian deposition younger than Early Jurassic, with two dates at 190 Ma (Fig. 8.17). With this new chronostratigraphy, the Lower Kwango Group (and the Dekese Formation) can now be correlated confidently with the J–K aeolian sequences in Namibia and eastern Brazil (Miller 2008; Scherer and Goldberg 2010), both of which are covered by Lower Cretaceous flood basalts of the Paraná-Etendeka LIP emplaced at ca. 132 Ma (Renne et al. 1992). These correlations attest to the presence of a giant, 'Sahara-like' paleo-desert across central West Gondwana, just before the separation of Africa from South America (Fig. 8.18; see Chap. 13, this Book). Additional zircons dates of ca. 430 Ma (mid-Silurian) and 240–290 Ma (Permian-Triassic) from these aeolian sediments of the CB also further support a far-field influence

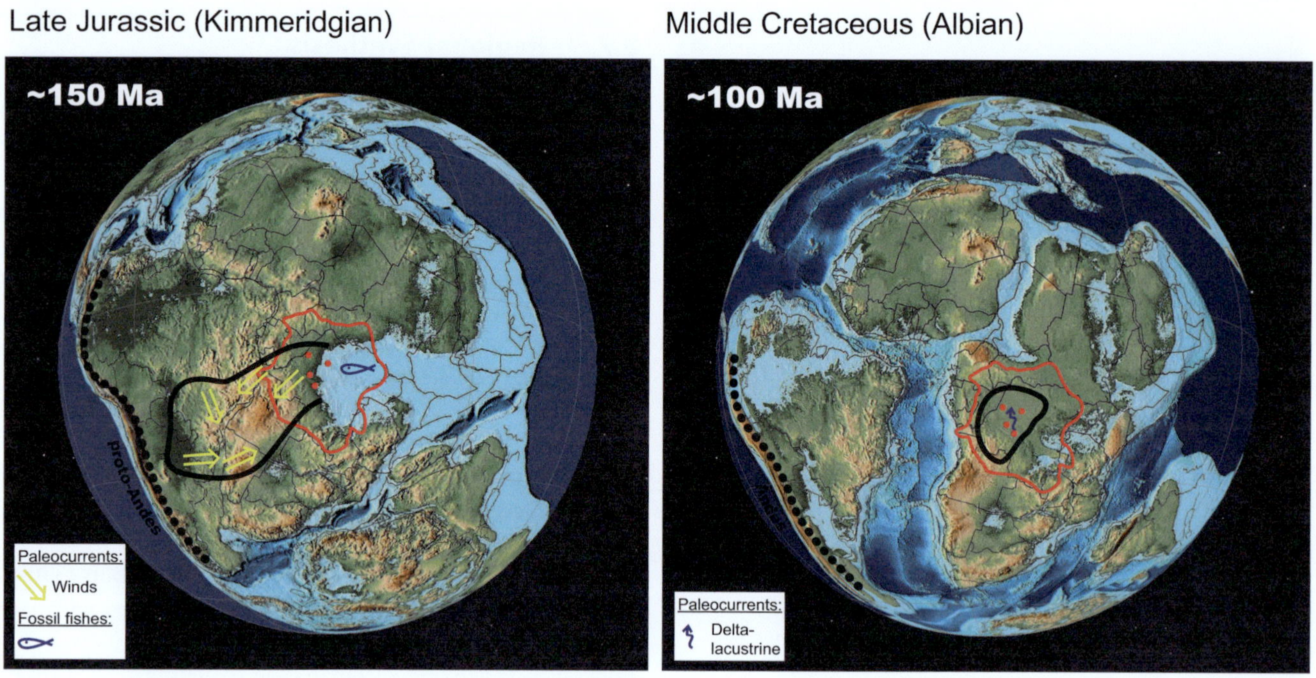

Fig. 8.18 Schematic paleogeographic reconstructions (*left*: Late Jurassic; *right*: middle Cretaceous) showing the extension of depositional sequences (*thick black line*) and paleocurrent directions (*arrows*) in central West Gondwana, superimposed on paleoglobes from Moore and Scotese (2012). Red dots and outline are the borehole locations and modern Congo drainage basin, respectively

of the active proto-Andes along the southwestern margin of Gondwana (e.g. Ramos and Aleman 2000).

8.7.2 Middle Cretaceous Sequences

The two succeeding middle Cretaceous (Albian-Cenomanian) Loia and Bokungu Groups represent two superimposed cycles of lacustrine sedimentation: first episodically anoxic, and then (ephemeral) more carbonate-rich. Depocenters in the Samba (Units S3 and S4) and Mbandaka sections (Units M2 and M3) pass laterally southwards into a prograding delta in the Dekese section (Fig. 8.14).

The first, Loia Group (Unit S4) is a green and analcime-rich section containing several (seven) black shale intercalations that may overlap with global anoxic events during the Albian, ca. 100–110 Ma (e.g. Ogg et al. 2008). Also, the presence of analcime suggests volcanic contributions during sedimentation. This could possibly characterize the period following the eruption of the Paraná-Etendeka LIP during the onset of opening of the South Atlantic (Fig. 8.18).

The second, Bokungu Group (Unit S3) is more carbonate-rich, with fractured carbonated mudstones and purplish marlstones, intraformational conglomerates, and devoid of black shales. The increase of carbonated sediments between the Loia and Bokungu Groups is clearly discernible in all the four borehole-sections (Fig. 8.14), and thus possibly reflect a regional, or even global, climatic warming during the mid-Cretaceous, as during this period average global ocean temperatures have been estimated to have risen up to 20–30 °C (e.g. Huber et al. 2002; Pucéat et al. 2003).

8.7.3 Upper Cretaceous Sequences

The uppermost Kwango Group was not recognized in the Kwango Valley area or in any of the four boreholes. This Upper Cretaceous sequence appears to be limited to the southwestern part of the central CB, where the youngest Cretaceous fossil suite was found, containing another unique fish fauna, also possibly of marine affinities (Taverne 1976). However, the abundant ostracods and phyllopods are all non-marine, which rather suggests an inland lake or a closed lagoonal environment during this final episode of sedimentation of the CB.

Conclusion

The J–K sequences of the Congo Supergroup in the CB overlie a major unconformity across Carboniferous to Triassic sequences, linked to regional uplift and erosion during the initial period of Gondwana break-up. Near the base, this 300 m to 1,000 m thick succession includes an Upper Jurassic, shallow marine to fluvial sequence (the Stanleyville Group), in turn overlain by an Upper Jurassic to Lower Cretaceous aeolian sequence (the Dekese Formation), recording a first cycle of basin expansion, emersion and desertification. U-Pb geochronology using detrital zircons collected from this aeolian sequence reveals abundant late Mesoproterozoic (1 Ga) and late Neoproterozoic (550–850 Ma) source areas of Pan African terrains to the CB. In addition, the occurrence of Paleozoic (ca. 430 Ma and 290 Ma) and Mesozoic zircons (ca. 240 Ma and 190 Ma) attests to the emergence of new different sources, most likely derived from the proto-Andes, located ca. 5,000 km to the south of the CB in a Gondwana setting (Fig. 8.18). In the CB, the three overlying middle and Upper Cretaceous, lacustrine to progressively more fluvial sequences (the Loia, Bokungu and Upper Kwango Groups) record a second major cycle of basin expansion and contraction, terminated by a regional peneplanation surface overlain by calcretized and silcretized sands and alluviums of the Kalahari Group.

Acknowledgments We acknowledge funding through the Inkaba yeAfrica and !Khure Africa programs, supported by the DST/NRF of South Africa. B. Linol particularly thanks Edmond Thorose from BRC DiamondCore Ltd. for his guidance in the field in southwestern DRC, and Ariel Boven at the RMCA—Royal Museum of Central Africa in Belgium for his help in accessing the archived cores. We also thank Anthony Tankard and an anonymous referee for reviews that improved this chapter. This AEON contribution number 128 and Inkaba yeAfrica number 98.

References

Blackburn TJ, Olsen PE, Bowring SA, McLean NM, Kent DV, Puffer J, McHone G, Rasbury ET, Et-Touhami M (2013) Zircon U-Pb Geochronology links the end-Triassic extinction with the Central Atlantic Magmatic Province. Science 340(6135):941–945

Boulouard C, Calandra F (1963) Etude palynologique de quelques sondages de la République du Congo (Congo ex-Belge). Unpublished report R/ST-no.7376, SNPA Direction exploration et production Pau, France

Caen-Vachette M, Vialette Y, Bassot J-P, Vidal P (1988) Apport de la géochronologie isotopique à la connaissance de la géologie gabonaise. Chron Rech Min 491:35–54

Cahen L (1951) Données nouvelles concernant la géologie at la géomorphologie du Kasai oriental et l'origine du diamant. Annales de la Société Géologique de Belgique, Tervuren, (Belgique), LXXIV, pp 105–122

Cahen L (1954) Géologie du Congo Belge. Vaillant-Carmanne, Liège, p 577

Cahen L (1983a) Brèves précisions sur l'âge des groupes crétaciques post-Wealdien (Loia, Bokungu, Kwango) du Bassin interieur du Congo (Republique du Zaïre). Rapport annuel du Musée Royal de l'Afrique centrale, Tervuren (Belgique), Département de Géologie et de Minéralogie, pp 61–72

Cahen L (1983b) Le Groupe de Stanleyville (Jurassic superieur et Wealdien de l'intérieur de la République du Zaïre): Révision des connaissances. Rapport annuel du Musée Royal de l'Afrique centrale, Tervuren (Belgique), Département de Géologie et de Minéralogie, pp 73–91

Cahen L, Lepersonne J (1952) Équivalence entre le système du Kalahari du Congo belge et les Kalahari beds d'Afrique austral. Mémoires de la Société Belge Géologique, Paléontologique, et Hydrologique, Tervuren (Belgique), Série in-8, Sciences Géologiques, 4, 63p

Cahen L, Ferrand JJ, Haarsma MJF, Lepersonne J, Vebeek T (1959) Description du Sondage de Samba. Annales du Musée Royal du Congo belge, Tervuren (Belgique), Série in-8, Sciences géologiques, 29, 210p

Cahen L, Ferrand JJ, Haarsma MJF, Lepersonne J, Vebeek T (1960) Description du Sondage de Dekese. Annales du Musée Royal du Congo belge, Tervuren (Belgique), Série in-8, Sciences géologiques, 34, 115p

Cahen L, Snelling NJ, Delhal J, Vail J (1984) The geochronology and evolution of Africa. Clarendon Press, Oxford, p 512pp

Casier E (1965) Poissons fossiles de la Série du Kwango (Congo). Annales du Musée royal de l'Afrique Centrale, Tervuren (Belgique), Série in-8, Sciences géologiques, 50, 69p

Censier C, Lang J (1999) Sedimentary processes in the Carnot Formation (Central African Republic) related to the palaeogeographic framework of Central Africa. Sediment Geol 127(1–2):47–64

Cocherie A, Robert M (2008) Laser ablation coupled with ICP-MS applied to U-Pb zircon geochronology: A review of recent advances. Gondwana Res 14:597–608

Colin J-P (1981) Paleontological study of the Esso/Texaco well Gilson-1, Zaire. Unpublished report EPR-E.WA19.81

Colin J-P (1994) Mesozoic-Cenozoic lacustrine sediments in the Zaïre Interior Basin. In: Gierlowski-Kordesch E, Kelts K (eds) Global geological record of Lake Basins, vol 4. Cambridge University Press, Cambridge, pp 31–36

Colin J-P, Jan du Chêne J (1981) Paleontological study of the Esso/Texaco well Mbandaka-1, Zaire. Unpublished report EPR-E.WA15.81

Cox LR (1960) Further mollusca from the Lualaba beds of the Belgian Congo. Annales du Musée Royal du Congo belge, Tervuren (Belgique) 37, pp 1–15

Dalrymple RW (1992) Tidal depositional systems. In: Walker RG, James NP (eds) Facies models: response to sea level change. Ontario, Geological Association of Canada, pp 195–218

De Carvalho H (1981) Geologia de Angola. 4 folhas, escala 1: 1 000 000. Laboratório Nacional de investigação Cientifica Tropical, Junta de Investigaçãoes Cientifica do Ultramar

De Wit MJ, Jeffery M, Berg H, Nicolayson LO (1988) Geological Map of sectors of Gondwana reconstructed to their position ~150 Ma (with explanatory notes), scale 1: 1.000.000. American Association of Petroleum Geologist, Tulsa

De Wit MJC, Köstlin EO, Liddle RS (2011) Prospecting in Africa: narratives by early De Beers explorers in the search for diamonds. De Beers Consolidated Mines Limited, Johannesburg, South Africa, p 374p

Defretin-Lefranc S (1967) Étude sur les phyllopodes du Bassin du Congo. Annales du Musée Royal de l'Afrique centrale, Tervuren (Belgique), Série in-8, Sciences géologiques, 56, 122p

Delhal J (1991) Situation géochronologique 1990 du Precambrian du Sud-Kasai et de l'Ouest-Shaba. Rapport annuel du Musée Royal de l'Afrique centrale, Tervuren (Belgique), Département de Géologie et de Minéralogie, pp 119–125

Desthieux F (1995) Carte Géologique de la République du Congo au 1:1.000.000. Geological Survey of South Africa, South Africa

Doyle JA, Jardiné S, Doerenkamp A (1982) Afropollis, a new genus of early angiosperm pollen, with note on the Cretaceous

palynostratigraphy and paleoenvironments of northern Gondwana. Bull Cent Rech Explor-Prod Elf-Aquitaine 6(1):39–117
Esso-Zaire SARL (1981a) Geological completion report: Gilson-1. Unpublished report
Esso-Zaire SARL (1981b). Geological completion report: Mbandaka-1. Unpublished report
Evrard P (1960) Sismique. Annales du Musée Royal du Congo belge, Tervuren (Belgique), Série in-8, Sciences géologiques, 33, 87p
Fildani A, Drinkwater NJ, Weislogel A, McHargue T, Hodgson DM, Flint SS (2007) Age controls on the Tanqua and Laingsburg deep-water systems: new insights on the evolution and sedimentary fill of the Karoo Basin, South Africa. J Sediment Res 77(11): 901–908
Fildani A, Weislogel A, Drinkwater NJ, McHargue T, Tankard A, Wooden J, Hodgson D, Flint S (2009) U-Pb zircon ages from the southwestern Karoo Basin, South Africa: Implications for the Permian-Triassic boundary. Geology 37(8):719–722
Fourie PH, Zimmermann U, Beukes NJ, Naidoo T (2011) Provenance and reconnaissance study of detrital zircons of the Palaeozoic Cape Supergroup in South Africa: revealing the interaction of the Kalahari and Rio de la Plata Cratons. Geol Rundsch 100:527–541
Frimmel H, Tack L, Basei M, Nutman A, Boven A (2006) Provenance and chemostratigraphy of the Neoproterozoic West Congolian Group in the Democratic Republic of Congo. J Afr Earth Sci 46 (3):221–239
Galloway WE, Hobday DK (1996) Terrigenous clastic depositional systems: applications to fossil fuel and groundwater resources, 2nd edn. Springer-Verlag, Berlin, p 489
Gobbo-Rodrigues SR, Coimbra JC, Petri SRJB (2003) Kwango Series (Congo), Bauru Group (Brazil) and Neuquen Basin (Argentina) ages, based on ostracods and vertebrates. XVIII Congresso Brasileiro de Paleontologia, Brasilia, Brazil, pp 152–153
Grekoff N (1957) Ostracodes du Bassin du Congo. I. Jurassique supérieur et Crétace inférieur du nord du basin. Annales du Musée Royal de l'Afrique centrale, Tervuren (Belgique), Série in-8, Sciences géologiques, 19, 97p
Grekoff N (1960) Ostracodes du Bassin du Congo. II. Crétacé. Annales du Musée Royal de l'Afrique centrale, Tervuren (Belgique), Série in-8, Sciences géologiques, 22, 36p
Jelsma HA, Barnett W, Richards S, Lister G (2009) Tectonic setting of kimberlites. Lithos 112S:155–165
Jelsma HA, Perrit SH, Amstrong R, Ferreira HF (2011) Shrimp U-Pb zircon geochronology of basement rocks of the Angolan Shield, Western Angola. Abstract, 23rd CAG, Johannesburg, 8th–14th January
Jerram D, Mountney N, Holzförster F, Stollhofen H (1999) Internal stratigraphic relationships in the Etendeka Group in the Huab Basin, NW Namibia: understanding the onset of flood volcanism. J Geodyn 28:393–418
Kay SM, Ramos VA, Mpodozis C, Sruoga P (1989) Late Paleozoic to Jurassic silicic magmatism at the Gondwana margin: Analogy to the Middle proterozoic in North America? Geology 17:324–328
Kocurek G (1996) Desert aeolian systems. In: Reading HG (ed) Sedimentary environments processes facies and stratigraphy, 3rd edn. Blackwell Science, Oxford, pp 125–153
Krömmelbein K, Weber R (1971) Ostracoden des 'Nordost-Brasilianischen Weaken'. Beihefte zum Geologischen Jahrbuch 115:1–93
Lenoir JL, Liegeois J-P, Theunissen K, Klerkx J (1995) The Palaeoproterozoic Ubendian shear belt in Tanzania: geochronology and structure. J Afr Earth Sci 19(3):169–184
Lepersonne J (1945) La Stratigraphie du Système du Kalahari et du Système du Karoo au Congo occidental. Bulletin du Service Géologique du Congo belge et Ruanda-Urundi 1:27–50
Lepersonne J (1949) Les grands traits de la géologie du Kasaï occidental et l'origine secondaire du diamant. Bulletin de la Société Belge de Géologie, de Paléontologie et d'Hydrologie 58(2):284–289
Lepersonne J (1951) Les subdivisions du système du Karoo au Kwango (Congo belge). Annales de la Société Géologique de Belgique, LXXIV, pp 123–138
Lepersonne J (1974) Carte géologique du Zaïre au 1: 2.000.000 + Notice explicative. Kinshasa, République du Zaïre: Direction de la Géologie/Musée Royal de l'Afrique centrale, Tervuren (Belgique)
Link K, Koehn D, Barth MG, Aanyu K, Foley SF (2010) Continuous cratonic crust between the Congo and Tanzania blocks in western Uganda. Int J Earth Sci (Geol Rundsch) 99:1559–1573
Linol B (2013) Sedimentology and sequence stratigraphy of the Congo and Kalahari Basins of south-central Africa and their evolution during the formation and break-up of West Gondwana. PhD thesis, Nelson Mandela Metropolitan University, 375p
Linol B, de Wit MJ, Barton E, de Wit MJC, Guillocheau (2014) U-Pb detrital zircons dates and source provenance analysis of Phanerozoic sequences of the Congo Basin, Central Gondwana (accepted at Gondwana Research)
Ludwig KR (2003) Isoplot/version 3.0: a geochronological toolkit for Microsoft Excel, vol 4. Berkeley Geochronology Center Special Publication, 71p
Maheshwari H, Bose MN, Kumaran KP (1977) Mesozoic sporae dispersae from Zaire. II: The Loia and Bokungu Groups in the Samba borehole. III: Some miospores from the Stanleyville Group. Annales du Musée Royal de l'Afrique centrale, Tervuren (Belgique), Série in-8, Sciences géologiques, 80, 60p
Marlière R (1950) Ostracodes et phyllopodes du système du Karoo au Congo Belge. Annales du Musée du Congo belge 6:1–43
Marzoli A, Renne PR, Piccirillo EM, Ernesto M, Bellieni G, Min AD (1999) Extensive 200-Million-Year-Old Continental Flood Basalts of the Central Atlantic Magmatic Province. Science 284:616–618
Master S, Rainaud C, Armstrong R, Phillips D, Robb L (2005) Provenance ages of the Neoproterozoic Katanga Supergroup (Central African Copperbelt), with implications for basin evolution. J Afr Earth Sci 42(1–5):41–60
Miall AD (2006) The geology of fluvial deposits: sedimentary facies, basin analysis, and petroleum geology. Springer, Berlin, p 582
Miller R (2008) The geology of Namibia. In: Upper Palaeozoic-Cenozoic, vol 3. Geological Survey, Windhoek, Namibia
Moore TL, Scotese CR (2012) Ancient Earth: breakup of Pangea, Vers. 1.0, iOS Mobile Application
Mounguengui MM, Lang J, Guiraud M (2008) Sedimentary dynamics and extensional structuring related to early Cretaceous rifting of Neocomian and Barremian deposits of the interior basin of Gabon. J Afr Earth Sci 51(5):239–256
Munizaga F, Maksaev V, Fanning CM, Giglio S, Yaxley G, Tassinari CCG (2008) Late Paleozoic–Early Triassic magmatism on the western margin of Gondwana: Collahuasi area, Northern Chile. Gondwana Res 13(3):407–427
Mutti E (1992) Turbidite Sandstones. AGIP, Istituto de geologia, Università di Parma, Special Publication, 275p
Myers TS, Tabor NJ, Jacobs LL (2011) Late Jurassic paleoclimate of Central Africa. Palaeogeogr Palaeoclimatol Palaeoecol 311(1–2): 111–125
Nagudi B, Koeberl C, Kurat G (2003) Petrography and geochemistry of the Singo granite, Uganda, and implications for its origin. J Afr Earth Sci 36(1–2):73–87
Ogg JG, Ogg G, Gradstein FM (2008) The concise geologic time scale. Cambridge University Press, Cambridge, p 177
Pankhurst RJ, Leat PT, Sruoga P, Rapela CW, Márquez M, Storey BC, Riley TR (1998) The Chon Aike province of Patagonia and related rocks in West Antarctica: a silicic large igneous province. J Volcanol Geotherm Res 81:113–136
Pemberton SG, MacEachern JA, Frey RW (1992) Trace fossil facies model: environmental and allostratigraphic significance. In: Walker RG, James NP (eds) Facies models: response to sea level change. Ontario, Geological Association of Canada, pp 47–72

Pucéat E, Lecuyer C, Sheppard SMF, Dromart G, Reboulet S, Grandjean P (2003) Thermal evolution of Cretaceous Tethyan marine waters inferred from oxygen isotope composition of fish tooth enamels. Paleoceanography 18(2): 1029, 12

Rainaud C, Master S, Armstrong RA, Robb LJ (2005) Geochronology and nature of the Palaeoproterozoic basement in the Central African Copperbelt (Zambia and the Democratic Republic of Congo), with regional implications. J Afr Earth Sci 42(1–5):1–31

Ramos VA, Aleman A (2000) Tectonic evolution of the Andes. In: Cordani UG, Milani EJ, Thomaz Filho A, Campos DA (eds) Tectonic evolution of South America, 31st International Geological Congress. Rio de Janeiro, Brazil, pp 635–685

Renne PR, Ernesto M, Pacca IG, Coe RS, Glen JM, Prévot M, Perrin M (1992) The age of Paraná flood volcanism, rifting of gondwanaland, and the Jurassic-Cretaceous boundary. Science 258(5084): 975–979

Retallack G (2001) Soils of the past: an introduction to paleopedology, 2nd edn. Blackwell Science, Oxford, p 404

Rocha-Campos AC, Basei MA, Nutman AP, Kleiman LE, Varela R, Llambias E, Canile FM, da Rosa ODCR (2011) 30million years of Permian volcanism recorded in the Choiyoi igneous province (W Argentina) and their source for younger ash fall deposits in the Paraná Basin: SHRIMP U-Pb zircon geochronology evidence. Gondwana Res 19(2):509–523

Rolin P (1995) Carte tectonique de la République centrafricaine, au 1:1.500.000. BRGM

Rubidge BS, Erwin DH, Ramezani J, Bowring SA, de Klerk W (2013) High-precision temporal calibration of Late Permian vertebrate biostratigraphy: U-Pb zircon constraints from the Karoo Supergroup, South Africa. Geology 41(3):363–366

Saint Seine P (1953) Poissons de la cuvette congolaise. Compte rendus de la Société géologique de France 16:343–345

Saint Seine P (1955) Poissons fossiles de l'étage de Stanleyville (Congo belge). 1ère partie: La faune des argilites et schistes bitumineux. Annales du Musée Royal de l'Afrique centrale, Tervuren (Belgique), Série in 8, Sciences Géologiques, 14, 126p

Saint Seine P, Casier E (1962) Poissons fossiles des couches de Stanleyville (Congo). 2ème partie: La faune marine des calcaires de Songa. Annales du Musée Royal de l'Afrique centrale, Tervuren (Belgique), Série in 8, Sciences Géologiques, 44, 52p

Sames B, Whatley R, Schudack ME (2010) Praecypridea: a new non marine ostracod genus from the Jurassic and Early Cretaceous of Europe, North and South America. J Micropalaeontol 29(2): 163–176

Scherer CMS, Goldberg K (2007) Palaeowind patterns during the latest Jurassic - earliest Cretaceous in Gondwana: Evidence from aeolian cross-strata of the Botucatu Formation, Brazil. Palaeogeogr Palaeoclimatol Palaeoecol 250(1–4):89–100

Scherer CMS, Goldberg K (2010) Cyclic cross-bedding in the aeolian dunes of the Sergi Formation (Upper Jurassic), Recôncavo Basin: Inferences about the wind regime. Palaeogeogr Palaeoclimatol Palaeoecol 296(1–2):103–110

Sellwood BW, Valdes PJ (2008) Jurassic climates. Proc Geol Assoc 119(1):5–17

Stollhofen H, Stanistreet IG, Bangert B, Grill H (2000) Tuffs, tectonism and glacially related sea-level changes, Carboniferous-Permian, southern Namibia. Palaeogeogr Palaeoclimatol Palaeoecol 161: 127–150

Svensen H, Corfu F, Polteau S, Hammer Ø, Planke S (2012) Rapid magma emplacement in the Karoo Large Igneous Province. Earth Planet Sci Lett 325–326:1–9

Tack L, Wingate MTD, Lie J (2001) Early Neoproterozoic magmatism (1000-910 Ma) of the Zadinian and Mayumbian Groups (Bas-Congo): onset of Rodinia rifting at the western edge of the Congo craton. Precambrian Res 110:277–306

Tack L, Wingate MTD, De Waele B, Meert J, Belousova E, Griffin B, Tahon A, Fernandez-Alonso M (2010) The 1375 Ma "Kibaran event" in Central Africa: Prominent emplacement of bimodal magmatism under extensional regime. Precambrian Res 180(1–2): 63–84

Talbot MR, Allen PA (1996) Lakes. In: Reading HG (ed) Sedimentary environments processes facies and stratigraphy, 3rd edn. Blackwell Science, Oxford, pp 83–124

Taverne L (1975) Etude ostéologique de Leptolepis caheni, Téléostéen fossile du Jurassique supérieur (Kimmeridgien) de Kisangani (ex-Stanleyville, Zaïre) précédemment décrit dans le genre Paraclupavus. Revue Zoologique Africaine 89:821–853

Taverne L (1976) Les téléostéens fossiles du crétacé moyen de Kipala (Kwango, Zaïre). Annales du Musée Royal de l'Afrique centrale, Tervuren (Belgique), Série in-8, Sciences géologiques, 79, 50p

Vaughan APM, Pankhurst RJ (2008) Tectonic overview of the West Gondwana margin. Gondwana Res (Focus) 13:150–162

Mesozoic Sedimentary Cover Sequences of the Congo Basin in the Kasai Region, Democratic Republic of Congo

9

Eric Roberts, Hielke A. Jelsma, and Thomas Hegna

9.1 Introduction

The Congo Basin of Central Africa is one of the largest continental sedimentary basins in the world (Daly et al. 1992), with a surface area of over 1,200,000 km^2 (Fig. 9.1). Despite its size, its gold, tin and copper resources, the potential for hydrocarbons (Clifford 1986), and its extensive alluvial and primary diamond deposits along the southern margin (Fieremans 1955, 1961, 1996; Bluck et al. 2005), it is remarkable that so little is actually known about the Congo Basin.

The term Congo Basin (also known as the Zaire Basin and Cuvette Congolaise) is widely used by geologists and geographers alike to define the area drained by the modern Congo River and the sedimentary cover sequences preserved within this enormous circular depression (Fig. 9.1) (e.g., Cahen 1954, 1983; Gouldie 2004; Giresse 2005). Intra-basinal correlation of cover sequences is problematic due to inconsistent nomenclature between countries and between portions of the basin, general inaccessibility with decades of conflict, minimal outcrop exposure across vast areas, limited seismic, core and well data (often of proprietary nature), a general lack of datable volcanics (with the exception of kimberlites), and the sporadic distribution of age and paleoenvironmentally diagnostic fossils (see Giresse 2005).

The Congo Basin has traditionally been described as a classic, saucer-shaped intracratonic basin, thick in the center and thinning uniformly towards the margins (e.g., see Cahen et al. 1960; Daly et al. 1992; Giresse 2005), and interpreted as a passive downwarp, likely cored by a failed Proterozoic rift (Daly et al. 1992; Kadima et al. 2011a, b) and rimmed by topographic highs including Proterozoic orogenic belts (Kibaran and Ubangi belts), the uplifted Atlantic margin and East African Rift shoulder, and basement highs of the Central African Shield (Fig. 9.2). Although much of the basin (i.e., Cuvette Centrale; Figs. 9.1, and 9.2) is indeed within a fully intracratonic setting, the sedimentary cover sequences extend well beyond this region.

The Congo Basin has been summarized as one of relative complexity during the Late Proterozoic-Early Paleozoic (Daly et al. 1992), whereas the Late Paleozoic-Cenozoic basin history has typically been interpreted as a period of extended regional tectonic stability, without noticeable subsidence and with passive sediment infilling (Sahagian 1993; Giresse 2005). Recent geophysical investigations by Downey and Gurnis (2009) indicate that a large negative free-air gravity anomaly is centered over the Congo Basin and they interpreted the basin as currently active and being depressed due to generation of dynamic subsidence by downward flow of an anomalously dense region in the mantle lithosphere (or asthenosphere, Moucha 2011). Although Downey and Gurnis (2009) cannot constrain how or when this density anomaly developed, they hypothesize that it is most likely coincident with deposition of Mesozoic-Quaternary cover sequences across the basin. In contrast, Kadima et al. (2011b) interpret a residual gravity anomaly as the remaining crustal thinning associated with a Neoproterozoic failed rift proposed by Daly et al. (1992). These studies indicate a more complex basin model than previously envisioned, which emphasizes the need for improved, field-based structural and stratigraphic investigations to validate geophysics-based basin models.

A synthesis of the major published investigations in the basin was presented by Giresse (2005), and can be briefly

E. Roberts
School of Earth and Environmental Sciences, James Cook University, Townsville, QLD 4811, Australia
e-mail: eric.roberts@jcu.edu.au

H.A. Jelsma (✉)
De Beers Group Services, Exploration, Private Bag X01, Southdale 2135, South Africa
e-mail: Hielke.Jelsma@debeersgroup.com

T. Hegna
Department of Geology, Western Illinois University, Tillman Hall 113, 1 University Circle, Macomb, IL 61455, USA
e-mail: TA-Hegna@wiu.edu

M.J. de Wit et al. (eds.), *Geology and Resource Potential of the Congo Basin*, Regional Geology Reviews,
DOI 10.1007/978-3-642-29482-2_9, © Springer-Verlag Berlin Heidelberg 2015

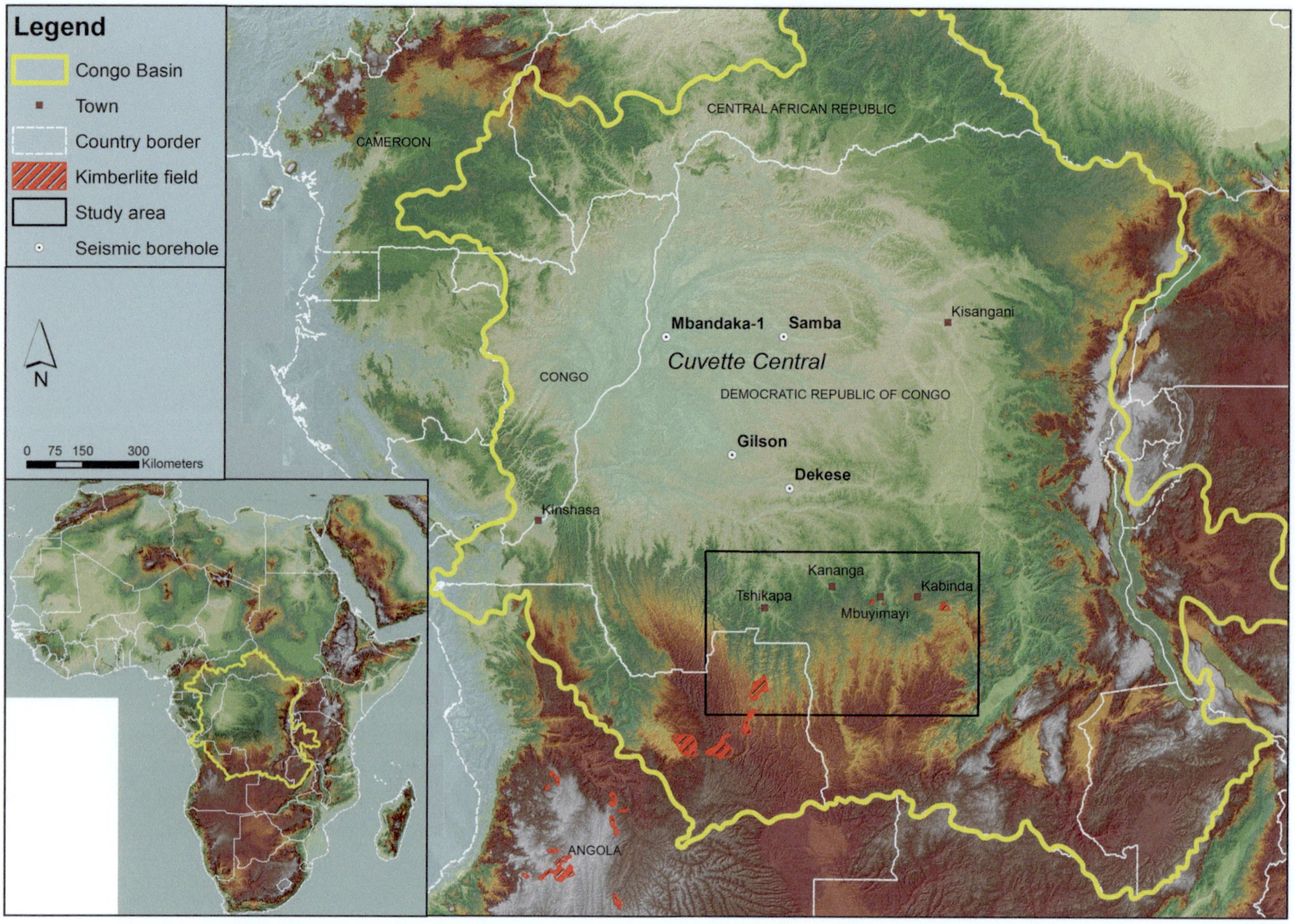

Fig. 9.1 SRTM model showing location of Congo Basin, occupying much of the Democratic Republic of Congo (DRC), along with portions of the Republic of Congo (Congo-Brazzaville, RCB), Central African Republic (CAR) and Angola. Note in the close up view the position of deep cores in the Cuvette Centrale and *black box* showing the study area in the Kasai portion of the basin

summarized as preliminary mapping during and shortly after the colonial Belgian Congo years (Cahen and Lepersonne 1951, 1954, 1978; Cahen 1954; Raucq and Lepersonne 1966; Raucq 1979), coupled with early geophysical explorations (Évrard 1957; Jones et al. 1960) that culminated with the drilling and detailed analysis of two deep boreholes in the central portion of the basin (Samba and Dekese wells, Figs. 9.1 and 9.2; Cahen et al. 1959, 1960). During and following this interval, most of the work in the basin focused on paleontological investigations of modest microfossil and fish faunas recovered from the wells and elsewhere in the basin (de Saint-Seine 1952, 1953, 1955; Grekoff 1957, 1958, 1960; Cox 1960; de Saint-Seine and Casier 1962; Defretin-Lefranc 1967; Taverne 1976, 1984; Cahen and Lepersonne 1978; Cahen 1983; Colin 1994). After the colonial mapping and exploration, the only significant new investigations in the basin have been the oil and gas exploration programs by Esso and Texaco in the Cuvette Centrale during the mid-1970s to early-1980s. This work included acquisition of a modest amount of magnetic and 2-D reflection seismic data, along with the drilling of two deeper boreholes (Gilson and Mbandaka wells; see Figs. 9.1 and 9.2) into the deepest parts of the basin (Lawrence and Makazu 1988). Our best insight into the tectonics and deep structure of the basin comes from a seminal basin synthesis and interpretation of some of this data by Daly et al. (1992).

In an attempt to better comprehend the sedimentary and tectonic evolution of the greater Congo Basin, the distribution, depositional environments, and structural controls on Mesozoic and Cenozoic cover sequences (particularly outside of the Cuvette Centrale), a study was conducted of borehole cores and logs made available from an extensive drilling program in the Kasai region of the Congo Basin in south-central DRC and northern Angola by De Beers Exploration (Fig. 9.3). Sedimentological data from 22 complete exploration boreholes (cores) and investigation of surface exposures in the area are presented herein (Fig. 9.3). The purpose of this paper is to apply detailed facies analysis, coupled with detrital zircon geochronology and paleontological investigation of the Mesozoic cover sequences along the

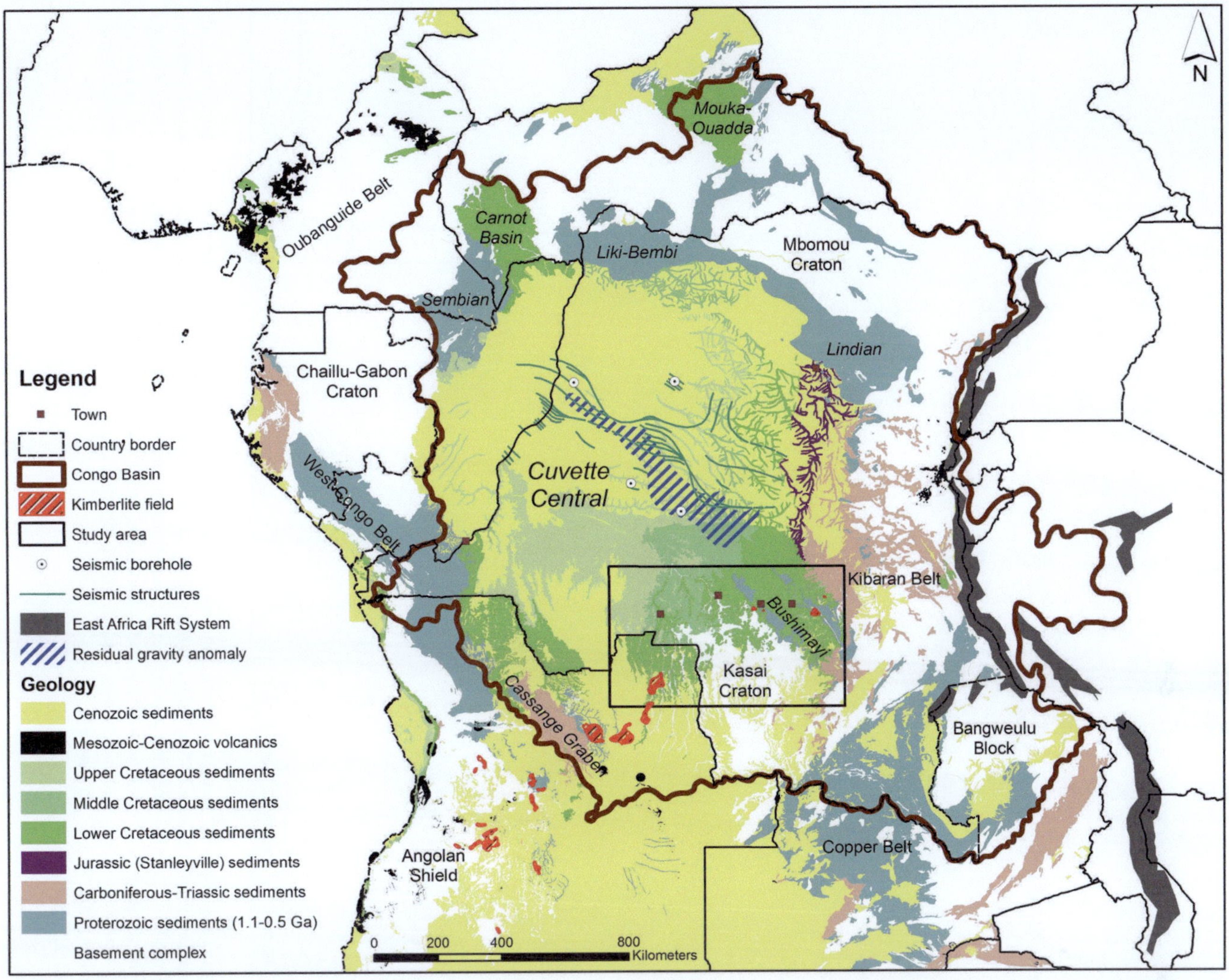

Fig. 9.2 Simplified geological map of the greater Congo Basin and surrounding areas of Central Africa with *black box* around Kasai field area. The margins of the basin are comprised of topographic highs, including the uplifted Atlantic margin, East African Rift shoulder, and basement highs of the Central African Shield

southern margin of the Congo Basin to reconstruct the age, depositional environments, and stratigraphy. This investigation highlights some hereto unrecognized structural and stratigraphic complexities, elucidating new aspects of the sedimentary and tectonic evolution of this part of the basin.

9.2 Background and Geologic Setting of the Congo Basin

9.2.1 Tectonic Setting

The Congo Basin occupies the core of the Central African Shield (Fig. 9.2), which Dirks et al. (2003) define as an assembly of Archean fragments, including the Kasai, Mbomou, and the Chaillu-Gabon cratons, which were once joined with the São Francisco Craton, to form the amalgamated Central African Landmass at the time of Gondwana assembly (~550 Ma; De Waele et al. 2008).

The nuclei of the Central African Shield are thought to have stabilized during the Palaeoproterozoic (following the Eburnean event). The eastern shield margin underwent convergent Mesoproterozoic tectonism during the formation of the Kibaran and Irumide Belts (Pedreira and DeWaele 2008; De Waele et al. 2008), and subsequent Neoproterozoic rifting led to the deposition of an epeiric platform sequence along the southern margin (Johnson et al. 2005). The continent-wide Late Neoproterozoic-Early Paleozoic Pan-African Orogeny significantly affected the shield, leading to the formation of the West Congo, Lufilian, Damaran and Kaoko, Oubanguide, and Zambezi belts along the margins of the shield (Pedreira and DeWaele 2008). Final assembly of the Gondwanan supercontinent during the Early Paleozoic (Late Pan-African Event) resulted in continued deformation

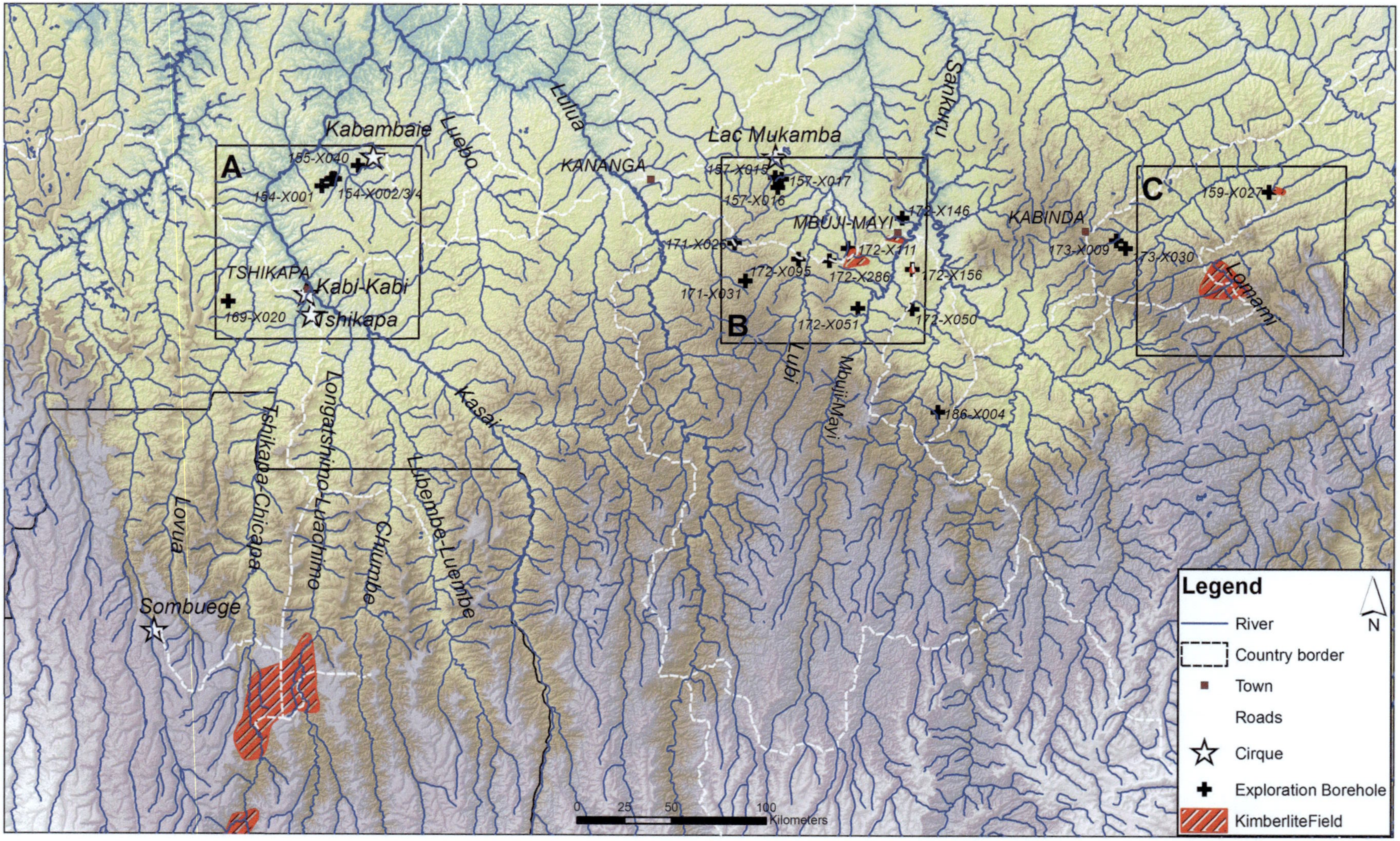

Fig. 9.3 Shaded relief map of the Kasai study area and the location of the De Beers exploration cores investigated in this project. The locations of known kimberlite fields, in relationship to the study localities, are also shown. Note the highly bisected topography and modern fluvial drainage network. The Kasai margin of the Congo Basin is characterized by broad savanna-like interfluve plateaus, and (in places deep) river valleys, with the major rivers progressively incising through cover sediments into underlying basement lithologies (Figs. 9.1, 9.2, and 9.3). Major rivers in the area are from west to east the Lovua, Tshikapa (alt. Chicapa), Longatshimo (alt. Luachimo), Tshiumbe (alt. Chiumbe), Lubembe (alt. Luembe), Kasai, Luebo, Lulua, Lubi and Mbuji-Mayi-Sankuru flowing northward and draining into the trunk Kasai River. Different drainage patterns reflect variations in the underlying geological terrain, with linear drainage patterns marking bedrock incision; broad, flat river valleys developed on Grès Polymorphs and Kalahari sands, and dendritic drainage patterns with cirque development on Cretaceous lithologies

and uplift in the West Congo and Lufilian Belts (Daly et al. 1992). Break-up of Gondwana and the opening of the South Atlantic Ocean resulted in an elevated rift shoulder forming the present western boundary of the Congo Basin. Although separation of South America from southern Africa began during the earliest Cretaceous and continued well into the middle Cretaceous and was followed by a major Santonian compressional event across central African (sensu Bosworth 1992) and the development of the Cenozoic East African Rift System, Daly et al. (1992) observed no evidence of the influence of these events in the Cuvette Centrale. Sahagian (1993) and other workers suggest that following separation in the Early Cretaceous, the Congo Basin remained largely stable, dominated by thermal relaxation and passive infilling, at or near sea-level for the last 100 Ma, remarkably with no noticeable effects (e.g., deformation; regional hiatuses; tectonic subsidence) associated with any of the major tectonic events affecting the continent during this interval (i.e., the Santonian compressional event; Super Plume and East African Rift).

9.2.2 Structure

Within the Cuvette Centrale, seismic and gravity data have permitted the recognition of a suite of major structures, basement highs and sub-basins. The presence of an angular unconformity between the Permian Lukuga Group and overlying beds in the Dekese core and in the Lukuga Valley, on the eastern margin of the basin, provides an indication of Late Paleozoic faulting and deformation in the basin. This interpretation was confirmed by Daly et al. (1992) in their seismic analysis of the Cuvette Centrale. They identified a series of dominantly southwest verging thrust faults that formed two major NW–SE trending subsurface basement highs (residual gravity anomalies shown on Fig. 9.2), which they called the Kiri and Lonkonia highs. Daly et al. (1992) interpret two phases of deformation, first during the Early Paleozoic, followed by reactivation of these structures during the Late Paleozoic.

Outside of the Cuvette Centrale, little is known about the structural geology of the basin. Most of the structures and deformation are interpreted to be largely constrained within the Paleozoic sequence, whereas the Mesozoic and Cenozoic stratigraphy are grossly characterized as uniform and undisturbed across the basin. Although geological structures are not indicated on the geological map of the Belgian Congo by Cahen and Lepersonne (1951), the distribution of pre-Mesozoic sequences and interpretation of satellite imagery provides some new insights into the structural geology. For instance, the southern part of the basin is characterized by dominant fault orientations that are inferred Mesozoic sedimentary depocenters, trending ENE–WSW and NW–SE following the trends of the Luana, Lulua, Bushimay (Fig. 9.2) and Cassange depocenters. Similar orientations are shown by the modern drainages such as the linear stretches of the Lubembe, Kasai, Luebo, Lulua and Lomami rivers (Fig. 9.3).

9.2.3 Stratigraphy

9.2.3.1 Pre-Mesozoic Sequences (Sequences 1–4)

The greater Congo Basin (Cuvette Centrale plus margins) preserves up to 9 km of sedimentary fill spanning the Late Proterozoic to Quaternary (Cahen 1954). Based on seismic data and cores, Daly et al. (1992) distinguished five major stratigraphic sequences in the Cuvette Centrale that range in age from Late Neoproterozoic to late Cenozoic. The age and compositional make-up of basement flooring the basin varies quite significantly across the basin, ranging from Archean crystalline rocks (Kasai, Mbomou, Chaillu-Gabon cratons) to Meso- to Neoproterozoic deformed sedimentary sequences (Cahen 1954; Cahen and Lepersonne 1978; Daly et al. 1992). Beyond the Cuvette Centrale on the southern margin of the basin, the ancestry of the Kasai Craton is well indicated from analyses of zircons recovered from Quaternary sediments of the Luebo River, which indicate a provenance area that included Archean-Paleoproterozoic, 3.2–2.1 Ga crustal components (Batumike et al. 2009). The basement comprises gneisses, migmatites and granulites and intrusive granitoids that have been assigned to the Sandoa-Kapanga, Kasai-Lomami and Dibaya Complexes (Cahen 1954; Delhal 1977; Delhal and Liégeois 1982; Walraven and Rumvegeri 1993), with remnant supracrustal rocks including greenstone belts and the Lulua and Luana fault-bound grabens (Delhal et al. 1989; Andre 1993; Pereira et al. 2003; Perritt et al. 2011).

The first cover sequence in the Cuvette Centrale (Sequence 1) recognized by Daly et al. (1992) comprises Late Neoproterozoic (Vendian) shallow marine stromatolitic limestones with subordinate evaporites that overlie truncated basement reflectors, correlated with the (~645 Ma) Ituri Group in northeastern DRC (Verbeek 1970). Deposition predates, or is at least partially syndepositional with, deformation interpreted during the Pan-African event. Stromatolitic limestones are also known from the Mbuji-Mayi Supergroup, which is an up to 4 km thick, siliciclastic-carbonate intracratonic failed-rift basin, but these sediments are older, constrained in age between >1.1 Ga and 0.9 Ga (Raucq 1970; Cahen et al. 1974; Kadima et al. 2011a, b; Delpomdor et al. 2014, in press). The Mbuji-Mayi basin is fault-bound and trends NW–SE, and projects, based on a residual gravity anomaly (Kadima et al. 2011a,b), below cover sediments towards the Dekese, Gilson and Mbandaka wells. Sequence 1 is terminated by deformation associated with the early Pan-African event.

Subsequent deposition of marine clastics (Sequence 2) over a regional unconformity was interpreted by Daly et al. (1992) as a marine transgression that flooded the craton. Acritarchs recovered from the Mbandaka well were correlated with the Lokomo Group of northeastern DRC and attributed to the Early Cambrian. Continued (late) Pan-African syn- and post-depositional compression resulted in the development of a regionally extensive subaerial, erosional unconformity (Daly et al. 1992).

Sequence 3 is referred to by Daly et al. (1992) as the Aruwimi Group, and represents a basin-wide Ordovician-Devonian transgressive-regressive marine cyclothem. The Aruwimi Group is characterized by basal transgressive sandstones, overlain by shales, and capped by progradational sandstones (Banalia Arkose). The latter have correlatives in the northern DRC (Bembe Group) and Angola (Bembe Formation, Inkisi Group), and are of Ordovician age (<500 Ma: Frimmel et al. 2006; Jelsma et al. 2012). Following this cyclothem, a long depositional hiatus is recorded across the basin.

Sequence 4 comprises Upper Carboniferous-Lower Permian glacial, peri-glacial and interglacial lacustrine deposits of the Lukuga Group, correlated to the Dwyka-Ecca interval of the Karoo Supergroup of southern Africa (Cahen and Lepersonne 1978). In the Congo Basin, deposits of this sequence are exposed across the eastern margin, as well as within the Dekese borehole, where it is >450 m thick (Cahen et al. 1960). Interestingly, the Lukuga Group was not identified by Cahen et al. (1959) in the Samba Well. A thin veneer of correlative late Paleozoic tillites outcrops in river drainages along the southern portion of the basin in Angola (Lutôe Series; Rocha-Campos and dos Santos 1981). In the Dekese Well, these units have high dip angles (>30–90°; Cahen et al. 1960), indicating a major episode of contractional deformation in the basin. The top of this sequence is characterized by a marked angular unconformity.

The final depositional sequence (Sequence 5) represents a regionally extensive, generally flat-lying, mostly continental series of Triassic-Cenozoic clastic units. This sequence comprises five major units, identified as the Triassic Haute Lueki Group, the Middle-Late Jurassic Stanleyville Group, the Lower Cretaceous Loia and Bokungu Groups and capped by the Kwango Group, which Daly et al. (1992) inexplicably refer to as Tertiary, contrary to most published reports (e.g., see Giresse 2005).

9.2.3.2 Mesozoic and Cenozoic Cover Sequences (Sequence 5)

The Mesozoic and Cenozoic cover sequences are the best exposed units across the basin in outcrop, and they have been the subject of most previous investigations conducted in the Congo Basin. They also represent the most debated and poorly understood sequences across the basin, perhaps only because they have been subject to more work (Fig. 9.4). Giresse (2005) in a comprehensive literature review of the Mesozoic and Cenozoic history of the basin admirably attempted to synthesize the stratigraphic findings from many of the key regions across the basin (see Table 1 in Giresse 2005). Although considerable research by numerous authors on the paleontology of the Congo Basin has greatly improved the age assessments for certain units, uncertainty remains largely due to the limited nature of fossil-bearing strata, the great size of the basin and difficulty in making long distance correlations, and also because some aspects of the biostratigraphy are still debated and remain somewhat ambiguous.

The base of the Mesozoic succession is dated as Upper Triassic, and referred to as the Haute Lueki Group. It is exposed along the eastern margin of the basin and has been identified in the Cuvette Centrale by Daly et al. (1992) in seismic sections, but has not been unambiguously identified in either the Samba or Dekese cores (Cahen et al. 1959, 1960). The Haute Lueki Group is separated from overlying units by a regional hiatus and unconformity (Lepersonne 1977; Daly et al. 1992), spanning the Early Jurassic. Correlatives of the Haute Lueki Group are exposed in the Cassange Graben in Angola (Atunes et al. 1990). The Haute Lueki and Cassange Groups have been correlated with the Beaufort Group of southern Africa, on the basis of fish fossils (Atunes et al. 1990).

The Stanleyville Group comprises a succession of calcareous sandstones and shales, some of which are bituminous, that were deposited in a lacustrine basin south of Kisangani. Paleontological work by De Saint-Seine (1955) and Grekoff (1957) and palynological work by Colin (1994) indicates that the Stanleyville Formation is Middle to Late Jurassic in age. A recent paleoclimate investigation of the Stanleyville Formation was conducted on samples from the Samba Core, which suggest quite a hot and arid depositional environment during this time (Myers et al. 2012).

The overlying continental Loia, Bokungu and Kwango groups are of Cretaceous age. Ostracodes (Grekoff 1957), fossil fish (Casier 1961; Lepersonne 1974) and pollen (Boulouard and Calandara 1963) date the Loia Group to the Neocomian-Aptian (145–112 Ma), the Bokungu Group to the Albian (112–100 Ma), and the Kwango Group to the Cenomanian-Turonian (100–89 Ma).

9.2.3.3 Kalahari and Younger Sediments

Kalahari remnants are preserved on isolated topographic highs in the southern part of the basin, such as at Mt. Bunzo in the Kasai region of the DRC (Giresse 2005) and

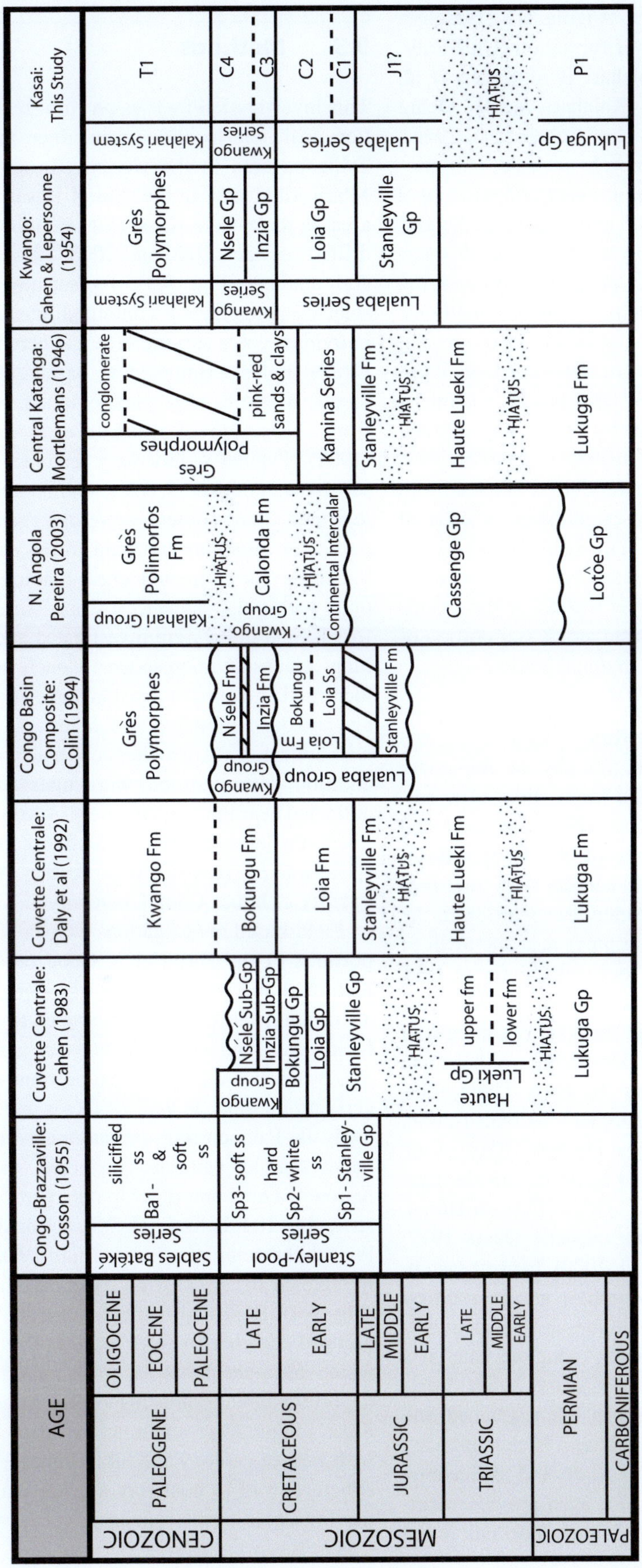

Fig. 9.4 Stratigraphic correlation chart showing the various nomenclatures and stratigraphic correlations proposed for the Mesozoic cover sequences of the Congo Basin. The last column represents the interpreted stratigraphy for the Kasai region from this study

on interfluve plateaus in northern Angola. The Kalahari Group has been divided into a lower formation (Grès Polymorphs or Grès Polymorfico; Kalahari B of Cahen et al. 1946) and an upper formation (Kalahari Sands, Ochre Sands or Sables Ochres, Kalahari C of Cahen et al. 1946). The <20 m thick Grès Polymorphs reflect the end-Cretaceous to mid-Tertiary planation event and associated aeolian deposition (see Giresse 2005 and references therein) and comprise calcretised and locally silcretised sands overlying deeply kaolinised weathering profiles. The overlying Sables Ochres Sands reflect a return to fluvial conditions. The last planation event was formed in the Pliocene and characterized by gravels with clasts of reworked Grès Polymorphs as well as laterites and phosphatic sediments (Giresse 2005). These are overlain by Quaternary alluvial deposits.

Renewed interest in the sedimentary cover sequences in the DRC has recently led to the re-examination of original cores. In this volume, Linol et al. (Chap. 8, this Book) review and update the Mesozoic and Cenozoic cover sequences of the Cuvette Centrale, whilst this paper focuses on the examination and characterization of the southern Kasai portion of the Congo Basin, and is discussed in detail below.

9.2.3.4 Diamonds and Kimberlites

Direct age constraints on the stratigraphy of the cover sequences are scarce and limited to kimberlite volcanoes. Alluvial and eluvial diamonds have been recovered and mined within the southern margin of the Congo Basin since the beginning of the twentieth century with total production to date of >600 million carats. Recognition in the 1940s of diamondiferous kimberlite pipes near the town of Mbuji-Mayi first demonstrated a local source for at least some of the diamonds in the region. To date, >500 occurrences have been discovered within the southern margin of the basin and two major episodes of kimberlite emplacement have been recognized, first during the Early Cretaceous, documented in northeastern Angola (Albian-Aptian, 145–113 Ma, Luxinga and Catoca clusters: Eley et al. 2008; Robles-Cruz et al. 2012), and again during the Late Cretaceous, documented in the DRC (Maastrichtian, ~70 Ma, Mbuji-Mayi and Tshibwe clusters: Davis 1977; Schärer et al. 1997; Kerschhofer et al. 2000). In northeastern Angola, the Early Cretaceous kimberlites are overlain by Calonda Formation sediments, which are an important collector of diamonds. The basal Calonda "conglomerate" is a 30 cm–2 m thick matrix-supported gravel with intraformational and extra-basinal clasts in a fine-grained and poorly sorted muddy sandstone matrix and is characterized by a clast assemblage that includes agate and chalcedony pebbles. The kimberlite emplacement ages provide maximum constraints on the age of the Calonda Formation and correlative Loia Group in the DRC.

9.3 Methods

This investigation focuses on cores recovered from exploration drill hole localities from three primary areas: (1) the Tshikapa area in the western Kasai; (2) the Kananga to Mbuji-Mayi area in the central Kasai; and (3) the Kabinda area, in the eastern Kasai (Fig. 9.3). In total, 196 holes were drilled between 2006 and 2008, 22 of which were selected and studied (Fig. 9.3), representing 16 localities. The selected cores have a combined intersect of 2,816 m (with individual cores varying in length between 59 m and 282 m). Fifteen holes terminated in basement and seven in sediments. All cores were photographed, and detailed core logs were constructed following the methods outlined by Johnson (1992). Typical sedimentary features, including grain size, sorting, bed thickness, color, sedimentary structures, mineralogy and fossil content, were observed and recorded while particular attention was paid to the distribution, thickness, thickness variation and clast lithologies of conglomerate units within the sequences. Cores were sampled for microfossils, and these were investigated and imaged using scanning electron microscope. Conchostracan fossils were identified and used to assist in correlations and help interpret age and depositional environments. A suite of other samples was collected for geochemical and provenance studies. In addition, detrital zircons were analyzed from a core sample collected near the base of core 173-X009 in the Kabinda area via U-Pb laser ablation inductively coupled plasma mass spectrometry (LA-ICPMS) at the Advanced Analytical Facility at James Cook University.

Field-based investigations were conducted across various portions of the Kasai area. Good outcrop exposures were identified and studied, primarily along river drainages, but also within large erosional cirques to supplement the core facies data and constrain alluvial architecture (Fig. 9.3). Photo mosaics and measured sections were constructed for certain outcrops, as well as facies and alluvial architecture analysis. Paleocurrent measurements and sandstone provenance samples were also collected where possible. Examination of cores and outcrop data permits the subdivision of strata into 10 lithofacies (LF), based on distinctive sedimentological characteristics (sensu Miall 1978). Each lithofacies is interpreted in terms of its basic depositional characteristics (Fig. 9.4). Repeated combinations of lithofacies have been grouped together to form six lithofacies associations (FA), which form the basis for interpreting the architecture and depositional environments of the cover sequences in the Kasai region. Core and outcrop data was synthesized, along with surface geology and SRTM imagery to identify regional structures and to construct a fence diagram for correlating the Mesozoic cover sequences in the Kasai portion of the Congo Basin.

9.4 Sedimentology and Depositional Environments of the Kasai Cover Sequences

9.4.1 Lithofacies

LF1A and 1B comprise pebble to cobble breccias and subordinate conglomerates, with clasts typically ranging in size from 2–7 cm. Lithofacies 1A and 1B are identical, except that 1A is clast-supported (Fig. 9.5a), whereas 1B is matrix-supported (Fig. 9.5b). Matrix is typically characterized by brick red sandy mudstone to poorly-sorted muddy sandstone (Fig. 9.5b). This facies tends to be oligomictic, but may be polymictic. Four primary basement-derived clast types are typically observed across the Kasai region, which include Proterozoic stromatolitic limestones and basalts (east: Mbuji-Mayi and Kabinda) and, less common, granitic and metamorphic clasts (west: Tshikapa). A diagnostic feature common with many examples of this lithofacies, are calcrete coatings around the clasts (Fig. 9.5c). Calcrete rims range from 3–8 mm in thickness and may coat the entire clast, or in many cases, the coatings are thicker on the upper surfaces of the clasts, and commonly absent on the basal surfaces. Most clasts associated with this LF are locally derived from nearby basement blocks; in particular Proterozoic limestone and basalt clasts.

LF2A and 2B comprise small- to medium pebble conglomerates, with most clasts falling within the 0.5 to 2.5 cm range, with isolated clasts reaching up to 15 cm in diameter. LF2A is clast-supported, well-sorted and generally upward-fining, whereas LF2B is matrix-supported, poorly-sorted and lacks obvious grading (Fig. 9.5d—LF2B). Clasts are generally moderately to well-rounded and represent a much broader spectrum of lithologies than LF1A and 1B clasts, containing metamorphics, vein quartz, agate, chalcedony, black chert, limestone, basalt, granitic pebbles, and rare to abundant intraformational mudstone rip-up clasts. Matrix is dominantly sand-sized, although there are examples of mudstone matrix. This lithofacies is typically brick red colored.

LF3A and 3B comprise granulestones to coarse-grained sandstones. As with the previous two lithofacies, LF3A and 3B are distinguished by clast versus matrix support, respectively. In the case of matrix-supported LF3B, the matrix is typically deep red claystone/mudstone. The color of this facies ranges from almost white, in the case of quartzose clast-supported granulestones (LF3A), to deep red, as in the case of mud-supported granulestones (LF3B). Granule lithology is most typically dominated by quartz, metamorphic lithics, and intraformational mudstone rip-ups. Clasts are typically subangular to subrounded and show moderate to good sorting. Beds show weak normal grading and are thin, a few cm to a few meters thick at most. Well-preserved trough to planar cross-bedding is diagnostic of this lithofacies and basal erosional scouring into underlying beds is common. This lithofacies represents high-energy, traction dominated depositional environments.

LF4A and 4B comprise ungraded to normally graded, massive granulestones to coarse-grained sandstones (LF4A and 4B, respectively) that are identical in character to LF3A and 3B, but are completely massive and lack evidence of cross-bedding (Fig. 9.5e—LF4A).

LF5A and 5B are two of the most common lithofacies observed in the study. They comprise fine- to medium-grained sandstones with well-defined small- to large-scale trough or tabular cross-bedding. Plane bedding is commonly observed between cross-bedded intervals. Color ranges from dominantly orange in LF5A, to dominantly red and purple colors with common heavy mineral laminations in LF5B.

LF5A is diagnosed by having larger scale cross-bedding with higher angle foresets, and in some cases, slight inverse grading (i.e., pinstripe bedding) along foresets. In other locations, smaller-scale current ripple cross-lamination is common. LF5A is typified by clean, extremely well-sorted sandstone with no mud content, and well-rounded quartz grains. Another important feature is the presence of frosted grains.

LF5B differs in having a modest to high mud content and poorer sorting (Fig. 9.6), with sub-rounded to rounded grains with isolated granule to pebble sized clasts, and is less compositionally mature. In addition, LF5B lacks the large, high angle cross-bedding and pinstripe bedding. Burrows or root traces are moderately common, particularly in LF5B, although no macro- or micro-fossils have been observed in this facies.

LF6A and 6B are similar to LF5A and 5B, except that they are completely massive with no indication of bedding. Lithofacies 6A has no mud matrix, whereas LF6B has minor to abundant mud matrix. In places, bleaching and small iron-oxide concretions are present. Isolated pebbles are rare to common, particularly in LF6B. Burrow or root structures are also common in some samples, particularly in LF6B. In one core (155-X040), well preserved, robust bivalves were identified in LF6B. Their massive character is most likely due to bioturbation, bleaching or other post-depositional process that has obfuscated whatever original bedding may have existed.

LF7 comprises massive mudstones and siltstones (Fig. 9.7). Rare examples of remnant planar laminations are present. Moderate to intense bioturbation is commonly present, along with less common desiccation cracks. Small to quite large, and sometimes amalgamated, carbonate nodules are locally present (Fig. 9.7). Color is typically red

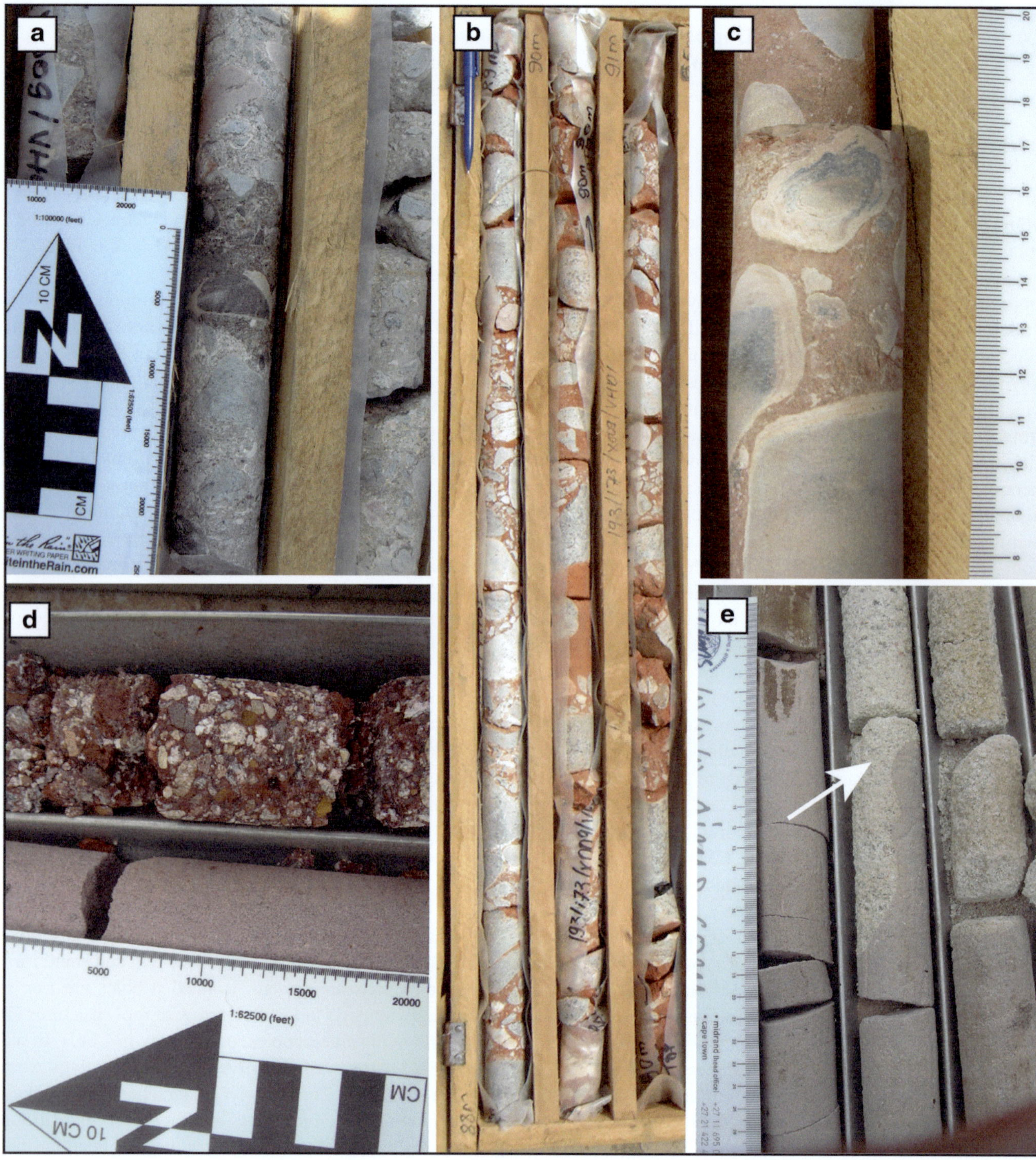

Fig. 9.5 Conglomeratic lithofacies. A LF1A, unit C3, core 173-X009; B. LF1B, unit C3, core 173-X009; C. Basalt clast (LF1B) with calcrete coating, unit C3, core 173-X030; D. LF2B, unit C2, core 154-X002; E, LF4A, unit C4, core 172-X146

to purple, with minor gray units. Beds range from a few centimeters to several meters thick, but are typically in the range of 1–2 m. Poor core recovery is typical of this lithofacies and the thickness of many units may be under appreciated.

LF8A and 8B comprise laminated siltstones and mudstones characterized by 1–3 mm thick horizontal laminae that are often wavy and deformed (Fig. 9.7). The only difference between LF8A and 8B is color, with LF8A typically gray in color, ranging from grayish red purple (5RP 4/2) to

Fig. 9.6 Sandstone Lithofacies. A. LF5b, unit C2, core 154-X001; B. LF5b, with small lens of LF7 (center), unit C2, core 154-X002; C. LF5a, core 157-X017, interpreted as aeolian dune foresets

grayish red (10R 4/3) to light brownish gray (5YR 6/2), whereas LF8B is more oxidized in appearance, ranging from pale reddish brown (10R 5/4) to light red (5R 6/6) to pale yellow (5Y 8/3) (Fig. 9.7). Minor to moderate bioturbation is present. Dewatering structures, including pipes and dish structures are present in some localities and desiccation cracks are locally abundant, particularly in LF8B (Fig. 9.7d). White bedded chert bands are present locally and are commonly wavy and deformed. Thin wave and current ripple-laminated siltstone beds are very commonly interbedded with the horizontally laminated siltstones and mudstones (Fig. 9.7e). Small-scale horizontally oriented stylolites are observed locally (Fig 9.7a). Ostracodes and conchostracans are present locally and are quite common in certain beds. Small, disarticulated fish bones are moderately common, along with rare articulated small fish skeletons.

LF9 comprises laminated siltstones and claystones containing rare to abundant isolated granule to cobble sized clasts (Fig. 9.8). Clasts are dominantly pebble-sized, angular to subrounded and represented by a wide range of lithologies, including red sandstone, limestone, basalt, quartzite, granite and foliated metamorphic clasts. Several clasts were observed with striations. Thin, fine-grained sandstone lenses are also present. A very distinctive chocolate

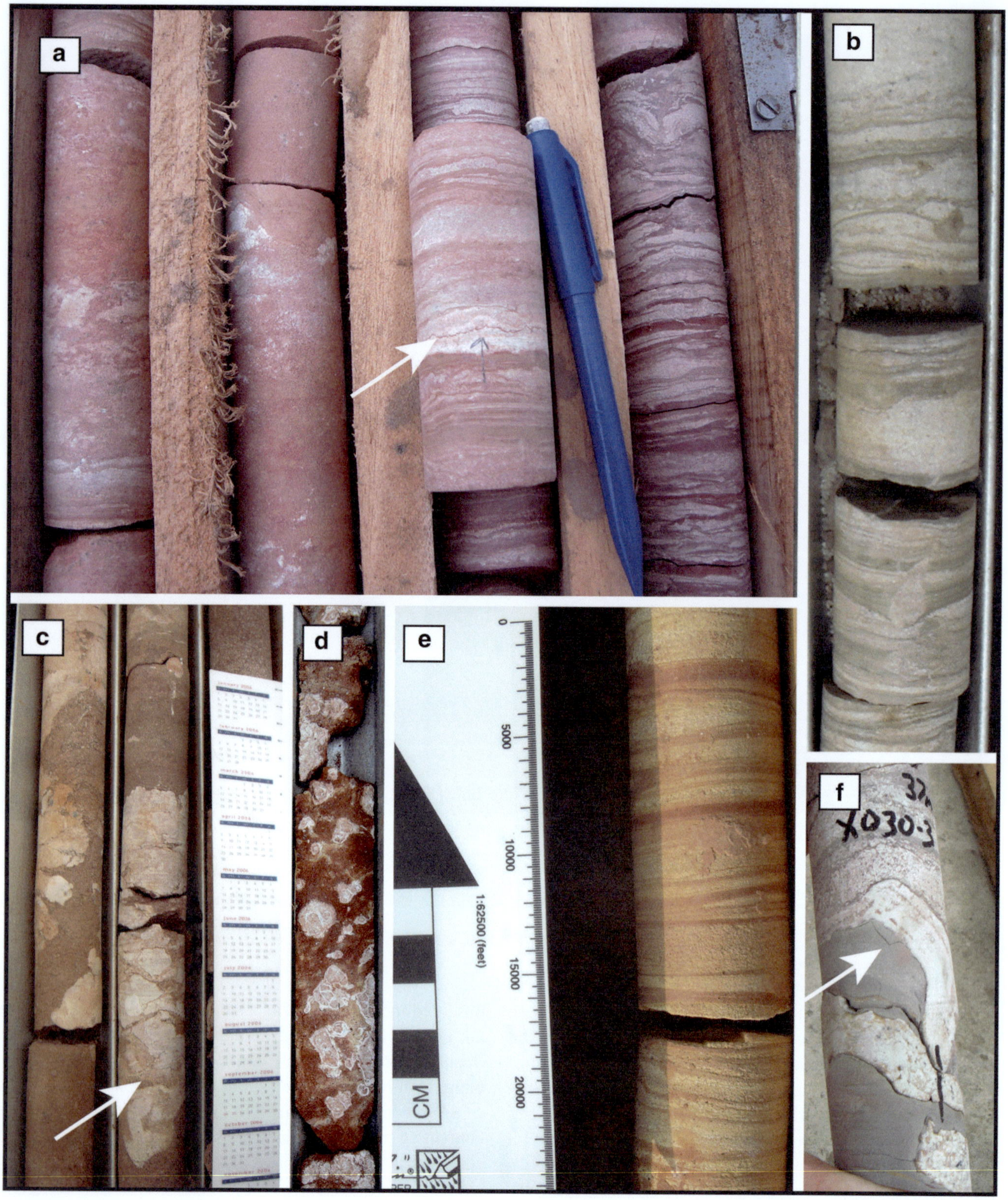

Fig. 9.7 Fine-grained Lithofacies. A. LF8A, unit C4, core 159-X027 with stylolites (*arrow*); B. LF8B, unit C4, core 172-X146; C. Massive mudstones of LF7 with large calcium carbonate concretions (*arrow*) from Unit C2, core 172-X050; D. Massive mudstones of LF7 with small calcium carbonate nodules in Unit C4, core 172-X111; E. Finely interlaminated siltstones and claystones of LF8A, unit C4, core 173-X030; F. Deformed bedded silicified layers of probable evaporative origin in LF8A, unit C4, core 173-X030

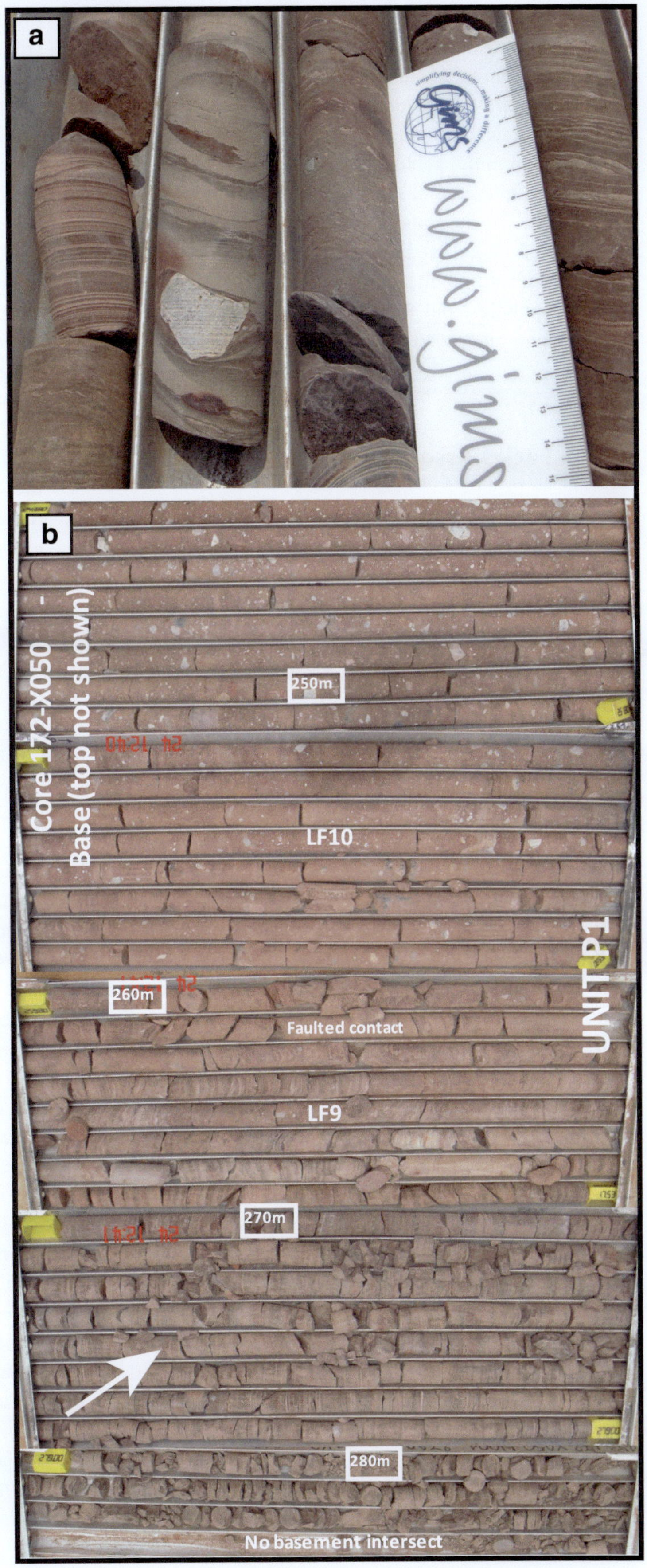

Fig. 9.8 Glacial lithofacies. A. Dropstone in laminated shales of LF9, core 172-X50; B. Lower 60 m of core 172-X050 dominated by glaciolacustrine laminated shales with dropstones basally (LF9) and grading upwards into glacial diamictites (LF10) This section of the core is the only area in the study area where Unit P1 was identified. White arrow shows location of Photo A

brown color to dark reddish brown color is diagnostic of this facies. Beds have steep dip angles and abundant micro-faults. Interlaminated siltstone and claystone beds are regular in places are less regular in other areas.

LF10 is a dark brown diamictite composed of randomly distributed granule, pebble and cobble size clasts floating in a weakly laminated to massive sandy mudstone matrix (Fig. 9.8B). Clasts range from angular to moderately well rounded and include red sandstone, limestone, basalt, dolerite, quartzite and granite clasts. There appear to be striations on some pebbles, but this is inconclusive. Most beds are massive and ungraded, but with zones of weak parallel bedding.

9.4.2 Facies Associations

Five different facies associations (FA) were identified and described below based on the repeated associations of different lithofacies, sedimentary structures and architectural elements. These five FAs were interpreted in terms of major depositional environments and form the basis for our understanding and interpretation of the paleoenvironments and depositional systems of the Mesozoic cover sequences across the Kasai study area.

9.4.2.1 FA1: Glaciolacustrine Environments

The glaciolacustrine facies association is composed of an important succession of lithofacies including basal laminated shales and siltstones (LF8A) grading upwards into LF9 and LF10 (Figs.9.8B), which are unconformably overlain by poorly sorted conglomerates (LF2A and 2B) with thin sandstone (LF5A) partings. LF9 is interpreted as lacustrine shale with dropstones, deposited under glaciogenic conditions and LF10 as a glacial diamictite deposited as till in a shallow subaqueous to subaerial setting. This lithofacies association is only found at a single locality in the basal 60 m of core 172-X050; with a clearly unconformable fault contact with overlying FA2 strata. The overall upward shallowing and coarsening succession appears to represent a deepwater lake setting with glacial rafting and progressive shallowing and infilling. The transition from LF9 to LF10 marks a transition from lacustrine to subaerial glacial tillites, followed by reworking and re-deposition by glacial outwash processes (LF2A, 2B, 5A).

9.4.2.2 FA2: Alluvial Fan Environments

Alluvial fan environments are common within the Kasai cover sequences and commonly include any combination of the following lithofacies: LF1A, 1B, 2A, 2B, 3A, 3B, 4A and 4B (with less common examples of LF6A, 6B, and 7). LF1A, 1B, 2A and 2B are dominant and characterized by poorly sorted, poorly rounded and randomly oriented clasts (Fig. 9.5). Matrix support is common in many units, suggesting deposition by debris flow processes was common, most likely in mid-to-outer fan settings. LF3A, 3B, 4A and 4B represents deposition associated with high energy, traction dominated depocenters. The high mud matrix contents in LF3B and 4B suggest deposition during flooding events associated with rapid sedimentation in mud-choked rivers. In some units, pedogenic calcrete covers many of the clasts indicating prolonged surface exposure under arid conditions, as would be expected in arid alluvial fan settings (Fig. 9.5). Thickened rinds on the top surfaces suggest prolonged surficial exposure and paleosol development during extended periods of surface stabilization.

The abundance of alluvial fan depocenters is associated with the tectonic complexity of the Kasai area, specifically the diverse, uneven and faulted basement topography. The majority of FA2 deposits are observed from core located along inferred fault zones. The basal portions of Cores 172-X286, -X111, and -X146, 173-X009, 157-X015, -X016, and –X017 all have substantial FA2 deposits (Fig. 9.9), which appear to be associated with synorogenic deposition around basement uplifts in the eastern part of the study area. FA2 deposits commonly transition to FA3 deposits and the contacts between the two facies associations ranges from erosional (unconformable) to gradational.

9.4.2.3 FA3: Fluvial Environments

The most common depositional environments identified across the basin are relatively high-gradient, low-accommodation (probably braided) fluvial environments of FA3. Lithofacies are typically dominated by sandstones of LF5B and 6B (with less common examples of LF5A and 6A) and interpreted to represent classic bedload-style fluvial channel deposits deposited under alternating lower-upper flow regime conditions.

Trough and tabular cross-bedding is a common feature of this facies association, commonly showing evidence of soft sediment deformation and slump features in outcrop exposures. Deformed and overturned crossbedding is common, and are typically simple, unfaulted, flat-lying structures with weakly to moderately curved hinges that face up-current. The sediment was likely water-saturated at the time of deformation, and possibly also liquidized as suggested by the smooth and unfaulted fold shape and extended upper limb of the fold. Deformation may be attributed to fluid drag on the upper surface of the cross-bedded sand body as indicated by the upstream facing and recumbent nature of the folds, the confinement to a single cross-bed set and by truncation upwards by the erosion surface beneath the set above. The origin of these structures is unclear, but likely a result of rapid sedimentation or seismic activity. No paleocurrent information

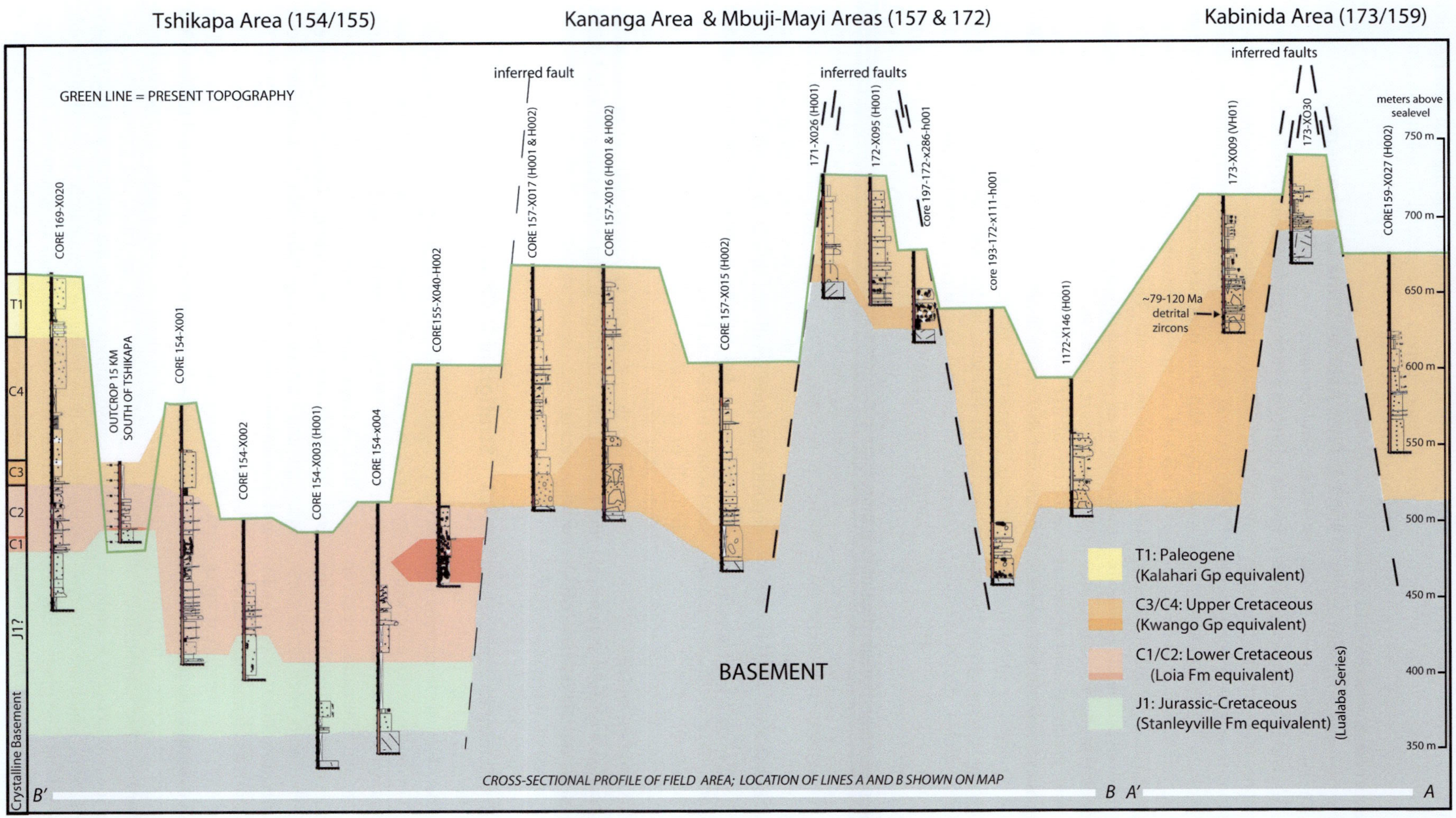

Fig. 9.9 Fence diagram for seventeen core logs and one outcrop section from across the Kasai study area. Depth to basement is based on basement intercept depths and the regional surface topography is used as the baseline for hanging all sections and core logs. The Kasai study area shows a complex stratigraphic framework with irregular basement topography and seven distinct stratigraphic units (note only six shown on fence diagram) that can be correlated across the study area

was obtained on cross bedding in cores and too few palaeocurrent indicators were measured in outcrop to present a clear indication of paleoflow.

The bases of many fluvial sequences across the field area are sharp and characterized by thin conglomeratic units. Conglomerate beds are quite variable within this FA, and may include beds of LF2A, 2B, 3A, and 3B, with 4A, and 4B (e.g., Fig 9.5D). These conglomeratic units are interpreted as either basal channel thalweg deposits or gravel bars within large channels. It represents deposition associated with high energy, traction dominated depocenters, similar to LF3.

Fine-grained deposits, interpreted as overbank floodplain facies, as well as in ponded depressions within channels, are limited to LF7 and are quite rare within this FA, suggesting the presence of high topographic gradients and minimal subsidence of the basin during these depositional phases. What rare fine-grained facies are observed, tend to preserve features indicative of moderate to intense pedogenesis. These features include carbonate nodules, deep oxidation, slickensides and desiccation cracks (Fig. 9.7C, D), all of which suggest that semi-arid conditions prevailed, and that ephemeral flow conditions existed in many of these channel systems.

A distinctive characteristic associated with some FA3 deposits are heavy mineral laminations, particularly along foresets in cross-bedded sandstones (LF5A, B) (Fig. 9.6A, B). Many FA3 deposits also have conglomeratic lenses composed of intraformational mudstone rip-up clasts, suggesting local derivation, due to channel incision.

9.4.2.4 FA4: Ephemeral Lake Environments

The most diagnostic facies associated with FA4 ephemeral lakes are finely laminated siltstones and claystones (LF8A and 8B) and less commonly, massive mudstones (LF7) (Fig. 9.7). Other lithofacies present may include LF6A, 6B, 4A, 5A, 5B, 2A, 2B, 1A, and 1B. Sedimentary structures are relatively common, dominated by small-scale ripple cross-laminations and small mudcracks. Fine-scale current and wave ripples suggest shallow conditions with alternating tidal and current processes operating. Although not varved, many examples show a very cyclical alternation between fine claystone and siltstone laminae suggesting alternating energy conditions. Some lacustrine units such as in cores 173-X009 and 173-X030 also preserve thin (1-5 cm thick), irregular zones of a white, partially laminated mineral identified as chert in XRD that appears to be curled up or deflected upwards (Fig. 9.7F). We interpret these features as likely expansion cracks, or teepee structures, in primary bedded evaporitic chert deposits. The abundance of mudcracks in FA4 supports this idea, as does the abundance of conchostracan fossils in these beds, which are also typically characteristic of ephemeral lake settings (Eugster 1986). Other fossils, such as ostracodes, rare bivalves and fish bones are also an important component of this FA and provide further support of a general lacustrine origin and arid to semi-arid climate.

9.4.2.5 FA5: Aeolian Dune Environments Aeolian Environments

FA5 is a fairly localized, but important depositional environment in the Kasai cover sequences. It is characterized by LF5A, 4A and 6A beds, which are interpreted as aeolian wind ripple and dune deposits, along with less common beds of LF8B and 2A that are possibly associated with fringing interdune environments. The key features of LF5A and 6A that suggest an aeolian origin for these deposits are the fine-grained, mature nature of these sandstones, most of which represent quartz arenites (Fig. 9.6C). Evidence of frosted grains and distinctive inverse grading along cross-bedding foresets ("pinstripe" laminations) are also present. Large-scale tabular foresets, in the order of several meters, are observed at several localities in outcrop and certain core samples are suggestive of this and appear to have high-angle foresets, relative to those observed in typical fluvial facies.

Additionally, we commonly observe an interbedded relationship between FA4 and 5 with sharp to gradational contacts. Overall, the thickness of aeolian sand dune deposits are quite small and this is suggestive of relatively arid conditions, but without a large or consistent upwind sand source to produce larger dunes or dune fields. Although difficult to reconstruct from core, these deposits give the sense that they represent more isolated barchan style dunes. They do not appear to be regionally extensive across the Kasai at any one interval, but rather fairly isolated dunes or dune fields.

9.4.3 Stratigraphic Correlation

We investigated drill core from along a ~500 km long (E-W) by 100 km wide (N-S) transect along ~6° South latitude, from west of Tshikapa, through Kananga, to east of Kabinda (Fig. 9.3). A fence diagram was constructed from 17 drafted core logs and one outcrop section, along with basement intercept and surface topography data to correlate facies and depositional units across the study area (Fig. 9.9). The stratigraphy is complicated and more variable than expected, making stratigraphic correlations with units described in other parts of the basin extremely difficult and tenuous. We identified and subdivided the stratigraphy into six different lithostratigraphic units that are based on gross patterns of lithofacies associations within the study area (Figs. 9.4 and 9.9). Rather than trying to prematurely force stratigraphic correlations with beds hundreds of kilometers away (e.g., in the Cuvette Centrale), we have assigned these units to informal stratigraphic names—P1, J1, C1, C3, C4, and T1 and made tentative correlations to the major Permian-Tertiary stratigraphy within the basin (Fig. 9.4).

Significantly, the base of the Kasai stratigraphy is highly variable across the field area and in no single locality did we observe all six stratigraphic units together in the same section or core (Fig. 9.9). The lowest stratigraphic unit, P1, is characterized by organic-rich lacustrine siltstones and shales with dropstones and glacial diamictites and outwash. A glacial origin is indicated by the presence of dropstones and striated pebbles, and the unit resembles the Permian Lukuga Formation described elsewhere in the basin (Cahen 1983). P1 beds were observed only in a single core (172-X050) in the Kasai area (Fig. 9.8).

In the Tshikapa area, the lowest unit consistently observed at the base of many cores and in outcrop overlying basement is termed J1. This unit is characterized by mature, cross-bedded quartz arenites, with minor thinly laminated lacustrine shales. It is likely correlative with the Middle to Late Jurassic Stanleyville Formation at the base of the Lualaba Series, although this is considered a loose correlation until supporting biostratigraphic or chronostratigraphic data can be established.

Above this unit, and also restricted to the Tshikapa area, is a distinctive succession of maroon to purple colored, heavy mineral-rich, sandstones and rare mudstones, designated as C2. Locally, there are thin basal conglomerates present at the base of C2, which we refer to as C1. This unit is very common in the nearby Lundas region of Angola, where it is referred to as the basal Calonda Formation (e.g., Pereira et al. 2003). The Calonda Formation is an important collector of diamonds and overlies Early Cretaceous kimberlite pipes that have been dated between 145 Ma and 113 Ma (Eley et al. 2008; Robles-Cruz et al. 2012), giving a maximum age constraint to the age of deposition. C2 appears to be conformable on top of C1 deposits. On the basis of definition and age, we tentatively correlate the C1 and C2 deposits to the Early Cretaceous Loia Formation in the upper part of the Lualaba Series in other parts of the basin.

Above C2, there is a fundamental shift in sedimentology, characterized by two distinctive units, termed C3 and C4, which make up most of the upper part of the stratigraphy in Kasai. On the west side of the basin, the shift is subtle, marked by a change to finer-grained, orange sandstones of C4 that lack the distinctive heavy mineral laminae of C2. However, on the eastern side of the basin, we typically observe conglomerates of alluvial fan origin that sit on bedrock below the C4 beds, which we tentatively refer to as C3. These conglomerates are dominated by reworked Proterozoic limestones and basalts that show evidence of pedogenesis and reworking. They fundamentally differ from C2 conglomerates, but their stratigraphic assignment is considered provisional.

Conformably overlying C3, or in some places bedrock in the Kananga, Mbuji-Mayi and Kabinda area, is unit C4, which is dominated by fine-grained, crossbedded, orange sandstones, massive orange sandstones, and laminated maroon, orange and gray shales and siltstones. C4 represents a variety of semi-arid depositional environments, from small aeolian dunes, to flashy fluvial channels, to shallow ephemeral lakes. C3 and C4 are interpreted to correlate to the Kwango Series. In general, C4 is similar in character and depositional environments to that of J1, complicating regional correlations across the Kasai.

The final stratigraphic unit observed in the Kasai Region, T1, has only been observed in one core, west of Tshikapa, at the top of the section. It represents a variable succession of beds, including ferruginous sandstones and conglomerates, phosphates, and unconsolidated sands of probable aeolian origins.

9.4.4 Core Description

9.4.4.1 Tshikapa Area

Six primary cores and several outcrops were examined and logged in detail to the north and west of Tshikapa. From west to east, these include: (1) core 169- X020; (2) Tshikapa outcrop 1, (3) cores 154-X001 and -X002 (Fig. 9.10); (4) cores 154-X003 and -X004 (Figs. 9.11); (5) core 155-X040 and (6) Kabambaie outcrop (Fig. 9.12). In general, cores and outcrops from the Tshikapa area were drilled at some of the lowest elevations in the Kasai study area and reach the greatest stratigraphic depths. Not surprisingly, the stratigraphy observed in this area is distinctly different and generally older than what is observed in either the Kananga/Mbuji-Mayi or Kabinda areas. Some of the best outcrop exposures in the Kasai region are found in this area (Fig. 9.12). J1 is observed in all but one of the Tshikapa cores, as well as in outcrop, and is primarily characterized as very well-sorted and rounded quartz sandstones and thin laminated claystones of mixed fluvial, aeolian and ephemeral lacustrine origins.

C1 conglomerates are thin, or missing in most cores and outcrops. At the base of core 155-X040 is bleached white sandstone with concretions and vertical calcium carbonate veins overlain by a matrix supported pebble to cobble conglomerate that has large angular granitic clasts within fine-medium grained, well-sorted sand with bivalves at the top. Other evidence of bleaching and hydrothermal alteration is present in other areas of C2.

The highest elevation Tshikapa cores (169-X020 and 154-X001) both preserve what is interpreted as C4 (possibly Kwango equivalent) strata. These strata differ from the underlying C2 beds in having slightly greater sorting and finer grain size, along with a having a whitish to dull orange color. They are typically fluvial with minor aeolian

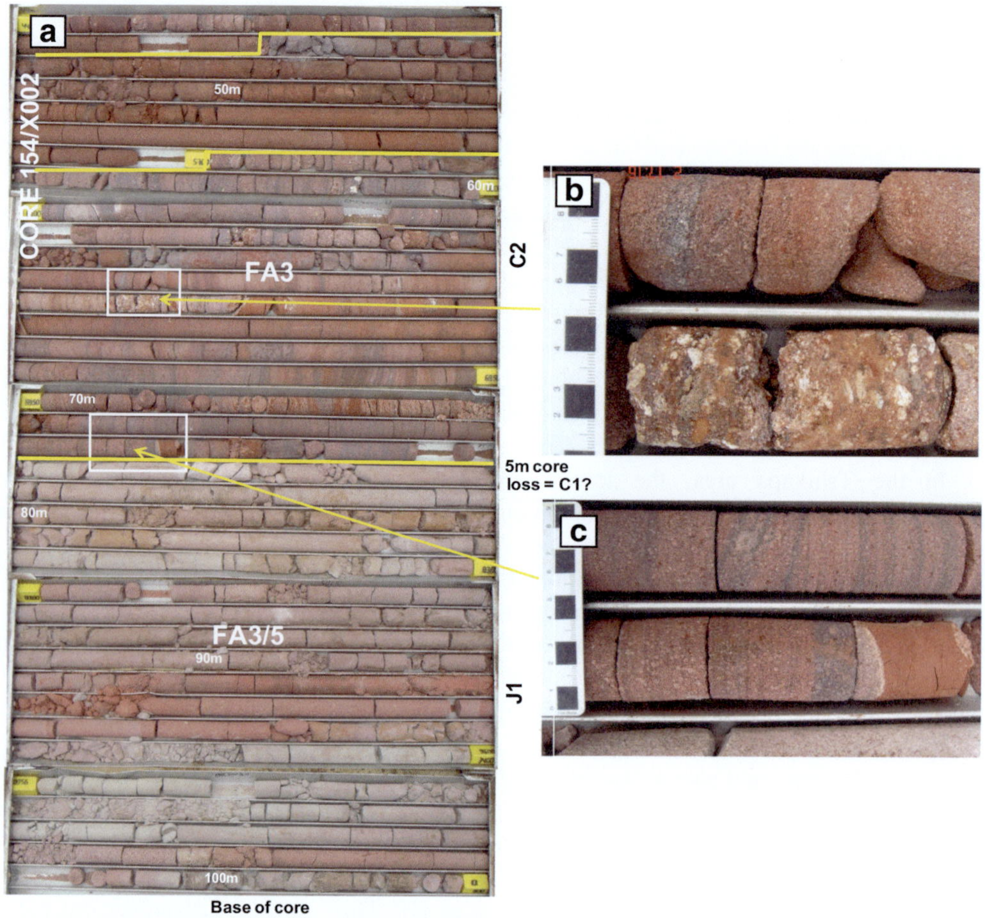

Fig. 9.10 Photo of core 154-X002

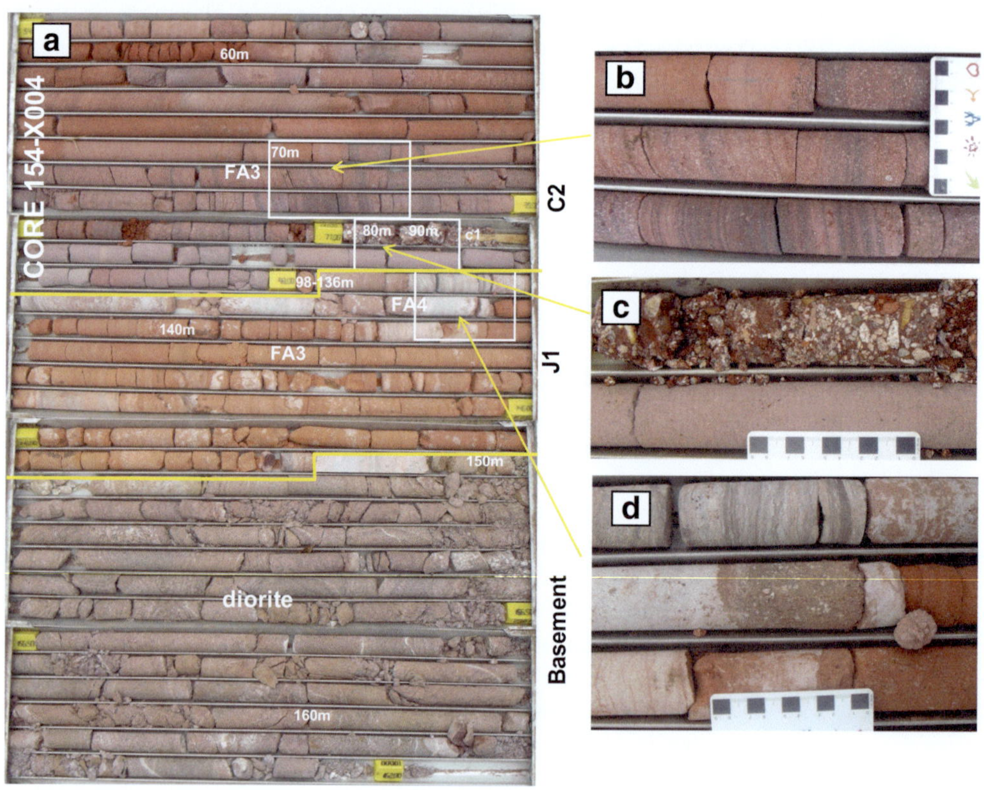

Fig. 9.11 Photo of Core 154-X004

Fig. 9.12 Photograph of large cirque within Kabambaie catchment (coordinates 21.09089°E, -5.77704°S, and elevation 650–550 m) in the Tshikapa area. Unit C2 is observed at the base, and this is overlain by a thin band of C3 and a much thicker C4 succession

influences. T1 is observed above this level only in core 169-X020, and is characterized by mostly loose, well-rounded, well-sorted sands, thin ferruginous sandstone and conglomerate beds and unusual thin bedded phosphates. Overall, the sedimentology of the Tshikapa area is internally consistent, but very different from the stratigraphy observed in the Kananga, Mbuji-Maya or Kabinda areas.

9.4.4.2 Kananga and Mbuji-Mayi Areas

Between Kananga and Mbuji-Mayi, eight cores were examined in detail (Fig. 9.9), including: cores 157-X017 and -X016 (Fig. 9.13); cores 157-X015 and 171-X026 ; cores 172-X095, -X286, -X111, -X146 and –X050 (Fig. 9.8B, but not included in fence diagram). Although this region is characterized by basement highs and highly variable topography, there appears to be a reasonably consistent stratigraphy across the region. In particular, coarse alluvial fans (FA3) characterize the base of the stratigraphy in most cores, which frequently overlie crystalline basement. Alluvial fan conglomerates within these cores range from coarse, immature proximal fan facies to better-sorted, distal fan facies. In many cases alluvial fan dominated strata conformably grade into fluvial channel and paleosol deposits. Up section across much of the region, small-scale wind-blown aeolian dunes interfinger with fluvial channels. On the western margin, near Kananga, aeolian facies dominate the middle-upper portion of the C4 stratigraphy. The middle and particularly upper portions of the stratigraphy across the area is characterized by numerous small and spatially restricted, ephemeral lakes (FA4) that commonly interfinger with the aeolian dunes or fluvial channels. The age and stratigraphic assignment of these beds is somewhat enigmatic due to a general discordance between biostratigraphic indicator fossils that suggest correlation to J1 or C2, and the presence of much younger detrital zircons in the Kabinda Area that have led us to tentatively place the basal alluvial fan conglomerates of

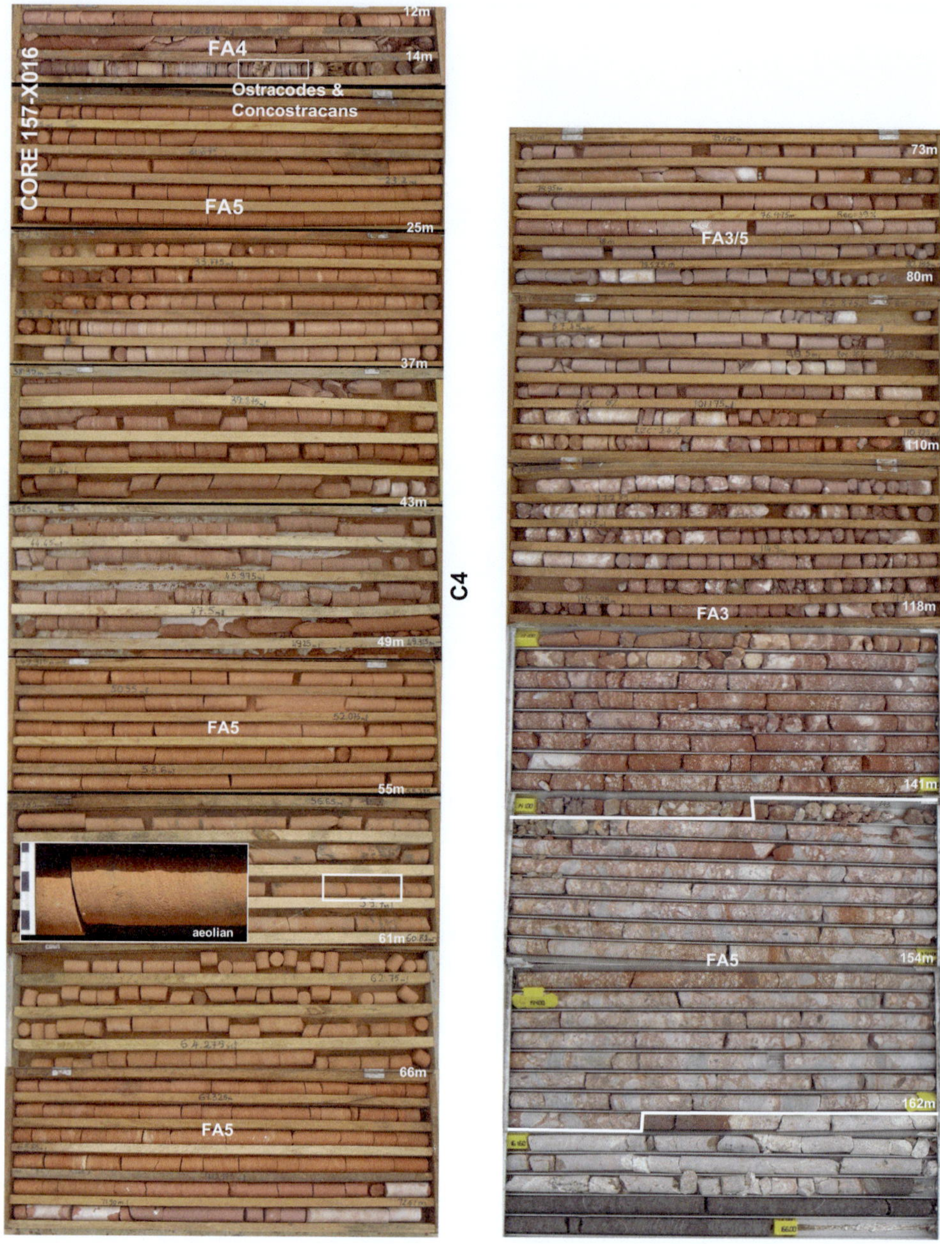

Fig. 9.13 Photo of core 154-X016

the Kabinda area within C3 and the overlying aeolian, fluvial and lacustrine deposits within C4.

9.4.4.3 Kabinda Area

Only three cores were studied and logged in the Kabinda region, viz: 173-X009 and -X030 (Fig. 9.14), and 159-X027. The cores in this area, particularly 173-X009 and 173-X030, were drilled at the highest-elevation areas in the study area. The base of the section here is variable, with a major alluvial fan succession at the base of 173-X009 on the western side, but sparse fan deposits exist at the base of 173-X030, which is perched above the highest elevation basement intercepts observed in the core records. To the far east of the study area, basement was not reached in core 159-X027 and no evidence of the basal C3 conglomerate was observed. Here, fluvial or mixed fluvial-aeolian facies are present at the base of the section, which are similar to both J1 near Tshikapa and C4 near Kabinda. However, based on detrital zircon evidence from a nearby core, (173-X009) we favor a C4 assignment for these deposits.

Most of unit C4 in cores 173-X030 and 173-X009 is characterized by a relatively thick succession of ephemeral lake beds, which are interspersed with small fluvial channels and aeolian dunes. Deposition of C3 and C4 strata across the Kabinda area appears to have transpired during the Late Cretaceous (post 80 Ma) based on detrital zircon data constraining maximum age, and in a fairly arid to semi-arid climate. This finding is in contrast to the geologic map of

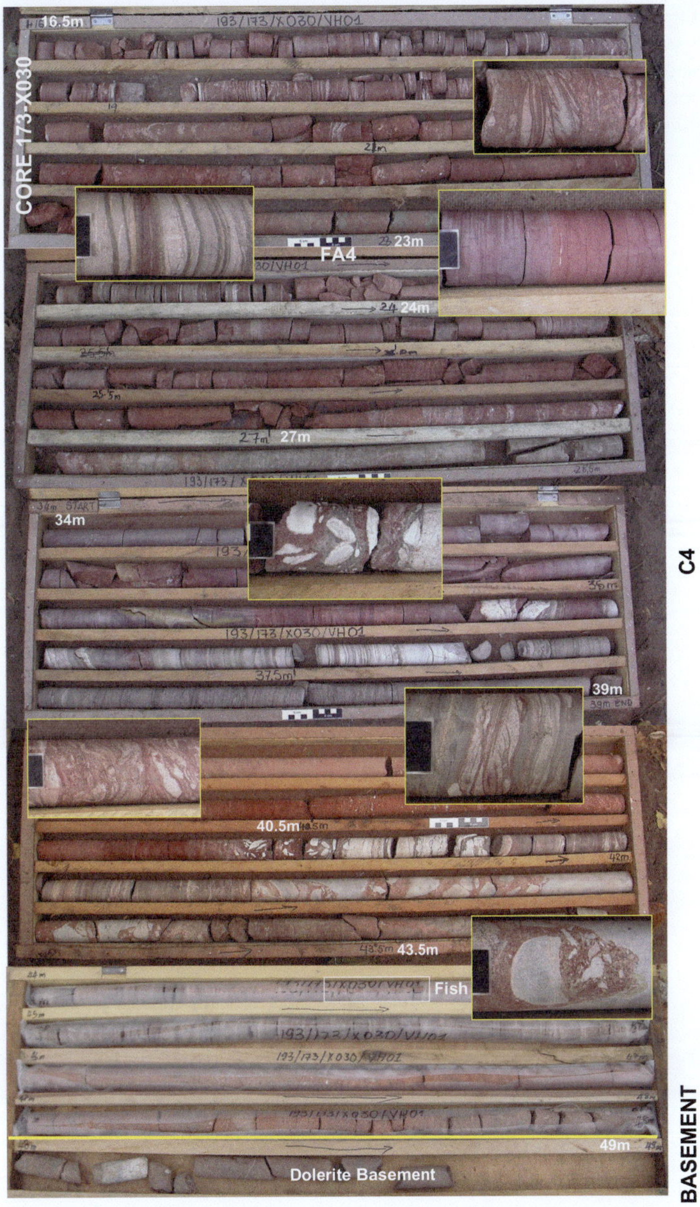

Fig. 9.14 Photo of the base of core 173-X030

DRC, which has the sedimentary cover sequences in this area mapped as Early Cretaceous.

9.5 Detrital Zircon Geochronology

9.5.1 Methods

A single pilot sample for detrital zircon geochronology was acquired from core 173-X009. Sample 173-X009-3B was collected at a depth of 65 m below surface (near the base of the section) from lithofacies LF1A (Figs. 9.9 and 9.15). The sample was crushed and the heavy minerals were then separated using lithium polytungstate and a Frantz magnetic separator, and then the non-magnetic grains were handpicked. Zircons were mounted in a 25 mm epoxy resin puck, polished to expose their mid-sections, and imaged using a Jeol JSM5410LV scanning electron microscope with attached cathodoluminescence detector for identification of zoning, cracks and impurities. All work was conducted at the Advanced Analytical Centre at James Cook University, using a Coherent GeolasPro 193 nm ArF Excimer laser ablation system connected to a Bruker 820-MS (formerly Varian 820-MS), following the methodology described in Tucker et al. (2013). Fifty-eight zircon grains were analyzed using a 24 μm spot size. The primary standard used was GJ-1 (609 Ma, Jackson et al. 2004) and the secondary standard used was an in-house zircon standard (Temora-2 [TEM-2] 416.8 Ma, Black et al. 2003). NIST 612 was analyzed at the beginning and end of each session,

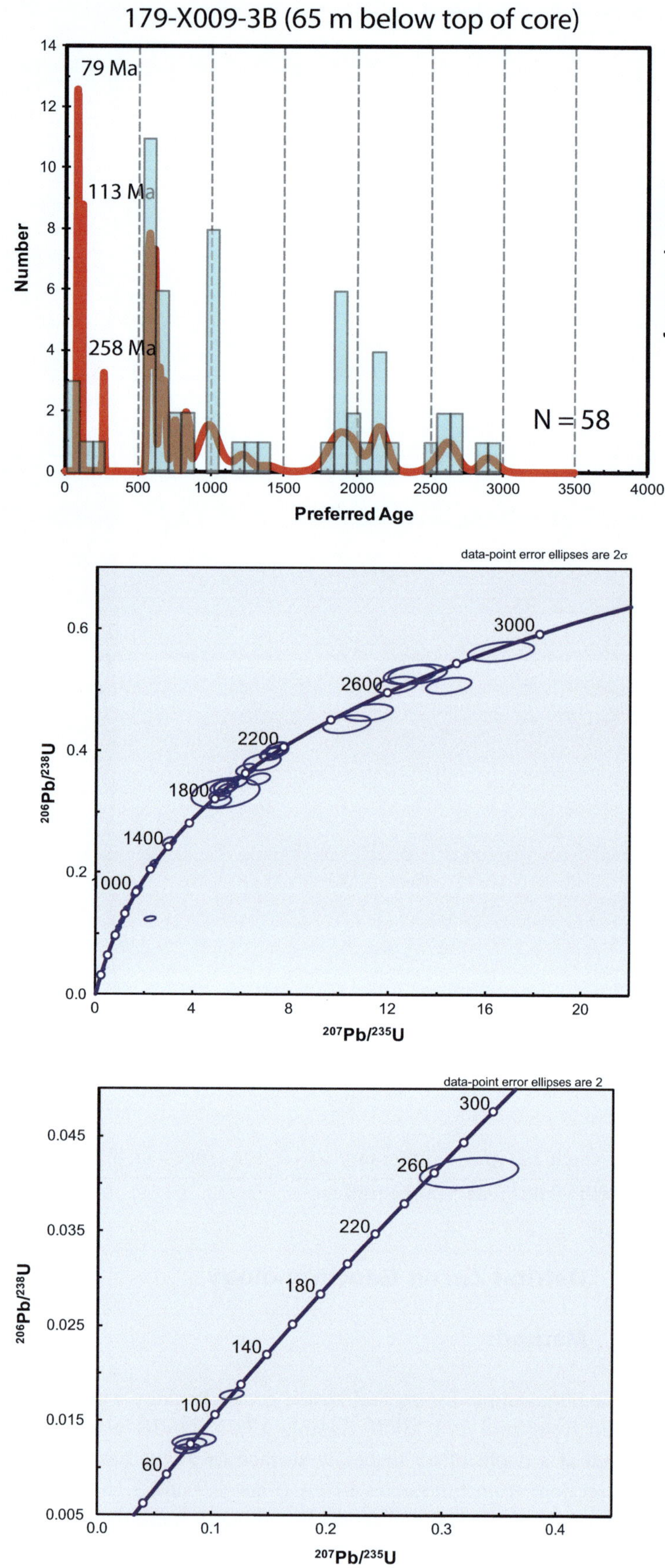

Fig. 9.15 Age-probability spectrum and histogram (*above*) and U-Pb Concordia plots (*below*) for detrital zircon sample 179-X009-3B

and at least once in between, for the purpose of calibrating Th and U concentrations. Data were processed using GLITTER^{TM} software package and plotted in ISOPLOT Ex (Ludwig 2003). A common lead correction was applied to all grains <800 Ma using the ISOPLOT function Age7Corr.

9.5.2 Results

The majority of the detrital zircons analyzed in this sample were dominantly of late Paleozoic and Precambrian age, however four Cretaceous-age grains were identified that provide important maximum depositional age constraint for this sample. The youngest zircon population includes a Late Cretaceous population of three concordant grains that yield a mean age of ~79.0 ± 6.8 Ma (Fig. 9.15) and is consistent with the age of kimberlite magmatism in the Mbuji-Mayi region (~70 Ma). In addition, a single Mid-Cretaceous grain (~113 Ma) that is statistically insignificant, but consistent in age with some of the younger kimberlites (~120 Ma) in the Lundas region in Angola was identified. The former population provides a maximum depositional age constraint for this unit and a tie point for correlating this portion of the stratigraphy. This finding also provides context for evaluating the biostratigraphic ranges of conchostracan fossils from this part of the stratigraphy, suggesting that several of the forms may have younger upper age limits than previously thought. Moreover, these data suggest that conchostracan biostratigraphy in particular should be used with caution in the basin and should not form the sole basis for stratigraphic correlation or age assignments (see expanded discussion of biostratigraphy below). Given the small number of grains analyzed, a more intensive detrital zircon investigation of the Kasai cover sequences is warranted to build a more robust detrital zircon provenance dataset that can be used to test the stratigraphic correlations presented in this study, and elsewhere in the Congo Basin.

9.6 Paleontology

9.6.1 Ostracodes

Small, podocopid ostracodes were identified in three cores: one from the Kananga area (157-X016) and two from the Kabinda area (173-X009 and 173-X030). They are present in multiple horizons in the two Kabinda cores, even abundant enough to form local ostracode pavements. Podocopids are typically freshwater ostracodes, and their presence here is consistent with the interpretation of fully continental depositional environments. However, ostracodes occur in many different types of freshwater environments (Carbonel et al. 1988), so without further study, they cannot constrain the environment more precisely.

9.6.2 Conchostracans

Overall, conchostracans (Branchiopoda) are very poorly studied in Africa. Equatorial Africa is an exception to this: several authors have produced important monographs detailing the conchostracan faunas (Defretin 1953; Defretin-Lefranc 1967; Marlière 1948, 1950; Tasch 1987). However, those materials studied are often insufficiently illustrated by modern standards, making them difficult to compare to other specimens.

Conchostracan remains were identified in three cores: one from the Mbuji-Mayi area (172-X146) and two from the Kabinda area (173-X009 and 173-X030). Core 173-X009 had conchostracans present in only one horizon (32 m level); core 173-X030 yielded conchostracans from two horizons between the 31 to 45 m levels; and core 172-X146 yielded conchostrans at two levels, between 35–39 m and at 67 m. All identifiable remains seem to represent new species of spinicaudatan conchostracans. The species present show a high endemicity—although one genus occurs in multiple horizons/cores, species were only identified in restricted horizons and did not reoccur in other horizons/cores.

Several cores (172-X146 (2 horizons), 173-X009, and 173-X030) contained species that likely belong to the genus *Pseudestherites*. Much of the material was too fragmentary to determine species-level affinities easily; however, they all possess punctae (minute pits) that are as much as an order of magnitude smaller (1–6 μm in diameter versus as much as 70 μm) than other known species of *Pseudestherites*—suggesting an identity as a new species (Fig. 9.16). There are, at present, no described species of *Pseudestherites* from Africa, but there is one undescribed species found during field work in Zimbabwe (EMR & HJ) that is presently being studied (Hegna et al. in preparation). *Pseudestherites* is known from the Late Jurassic to Early Cretaceous of China and Early Cretaceous of Argentina. Specimens from cores 173-X009 and 173-X030 could represent the youngest occurrence of *Pseudestherites* (Late Cretaceous). The detrital zircon sample from near the base of core 173-X030 indicates that these new specimens are not older than ~79 Ma, however further work is required to confirm whether specimens from core 172-X146 are indeed the same and of similar age.

Several specimens of a new species belonging to the family Loxomegaglyptidae were found in core 173-X030 (45 m from top, Kabinda area, Kasai-Oriental province) (Fig. 9.16). The specimens are likely Late Cretaceous in age, as the core is thought to correlate closely with core

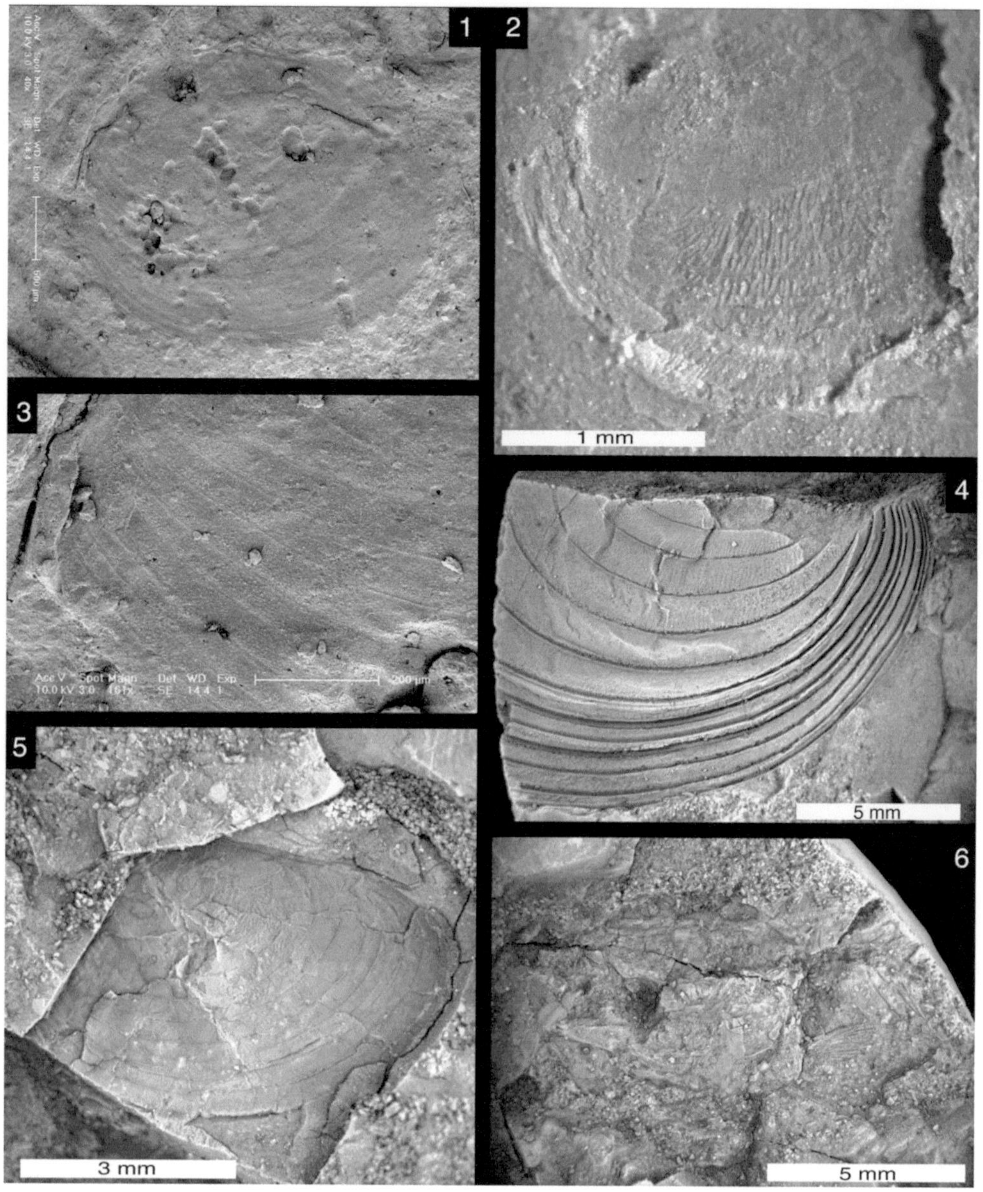

Fig. 9.16 A selection of fossil specimens from studied cores. 1. SEM image of *Pseudestherites* sp. (YPM 227690) from core 172-X146 at 36–39 m from *top*, Mbuji-Mayi area. 2. Light microscope image of *Congestheriella* sp. (YPM 519557) 173-X009 at 32 m from *top*, Kabinda area. 3. Close-up SEM image of the ornamentation on the *left side* of the *Pseudestherites* sp. (YPM 227690) specimen illustrated in Fig. 32A. 4. Light microscope image of *Stanleyviella* sp. (YPM 227693) from core 172-X146 at 35.8–35.9 m from top, Mbuji-Mayi area. 5. Light microscope image of *Paleoleptestheria* sp. (YPM 227707) from core 173-X030 at 45 m from top, Kabinda Area. 6. Light microscope image of an unidentified actinopterygian fish, 173-X030 at 45 m from top, Kabinda Area. Note: specimens in (4–6) were whitened with ammonium chloride prior to photography. All specimens are housed at the Yale Peabody Museum (YPM)

173-X009, which bears Late Cretaceous detrital zircons. They closely resemble several loxomegaglyptid genera: *Paraleptestheria* (known from the Tertiary of China) and *Paleoleptestheria* (which has a Triassic to Tertiary range and is known from China, Kazakhstan, Mongolia, Russia (Transbaikal and Siberian regions), and Thailand). The similarity of the two genera and the relationship of their geologic ranges suggest that they may represent ancestor-descendant lineages. At face value, this would make the new species represented in the core sample the first representative of this lineage on the African continent. However, Defretin-Lefranc (1967) described a species called *Euestheria sambaensis* found near Kinshasa and Samba (on the Maringa River) which, despite being poorly illustrated, may related to the lineage discussed above.

A species of *Stanleyviella* was identified from a single horizon in core 172-X146. This is an exclusively African genus, with the genus based on material collected from a core near Kisangani (see Defretin-Lefranc 1967 and Chen and Shen 1985). (Defretin-Lefrac's 1967) material is Late Jurassic (Stanleyville Series), so the material from this core in the Mbuji-Mayi area (Kasai-Oriental province) would be the youngest know occurrence of *Stanleyviella*. Alternatively, core 172-X146 may actually correlate to J1 rather than C4 as we have tentatively suggested.

Congestheriella is known from a single horizon in the 173-X009 core (Kabinida area, Kasai-Oriental province). The first representative of the genus was identified by Leriche (1913) in material from near Lualaba, DRC. It has since been identified from elsewhere in Africa, as well as

South America (Gallego et al. 2010). Previously known African species have been exclusively Late Jurassic, while South American species of *Congestheriella* have ranged from Late Jurassic to Early Cretaceous. The specimens from the 173-X009 core are thought to be Late Cretaceous based on zircon dates, making them the youngest representatives of the genus.

The presence of conchostracans is very suggestive about the paleoenvironment. Living conchostracans are almost exclusively ephemeral, freshwater animals (Webb 1979; Frank 1988; Dumont and Negrea 2002). Some living conchostracans can tolerate slightly elevated salinities (Webb 1979), but there are no living marine representatives (this may not have always been so, there are a very few fossil conchostracan faunas that co-occur with marine fossils, see Hegna 2012). In fact, they produce dormant resting cysts to withstand dry periods (Dumont and Negrea 2002). In the presence of predaceous fish, conchostracans are often preyed upon to the point of local extinction (Frank 1988). Thus, the occurrence of conchostracans suggests that the paleoenvironments were likely small, ephemeral bodies of freshwater (i.e. poorly drained) in a region with seasonal wet/dry periods.

9.6.3 Mollusks

Bivalve shell fragments were observed in a number of localities. However, only a single invertebrate macrofossil locality with identifiable material was found. This site represents a bivalve shell accumulation in core 155-X040 near the base of the core, at ~115 m from surface. Multiple partial specimens were found and several of these were collected for further taxonomic identification. At least two of the specimens appear to be articulated. Although the shells are incomplete and of limited utility, L. Tapanila (Personal Communication, 2008) was able to identify enough characters, including the dentition, a thick, robust shell and an oblate shape, to tentatively assign the bivalves as members of the Unionoidea, a family of a freshwater (rarely brackishwater) clams (mussels) that commonly live in fluvial and lacustrine environments. Although they are of little value for biostratigraphy, the identification of these shells as Unionoids is significant because they are dominantly a freshwater group and the thick shell morphotype of these specimens is consistent with a higher-energy fluvial mode of life (L. Tapanila, Personal Communication, 2008). This is important because a marine bivalve identified as *Pteria* sp. was reported from an outcrop in the Tshikapa/Luebo region from the Kwango Series by Cahen (1954). The taxonomic assignment of *Pteria* sp. in the Kwango Series is questioned here based on the sedimentological investigation of Cretaceous strata in this region and the overwhelmingly continental character of the Kwango strata in the Kasai region. It is possible, albeit unlikely, that marine deposits exist within C3/C4 (Kwango) strata; however it seems more likely that the specimen was either misidentified or collected from the Lualaba Group, rather than the Kwango Group.

9.6.4 Vertebrates

Two cores, 157-X016 and 173-X050, contained vertebrate fossil remains in the study area. In the upper part of this core (157-X016), several fish bones were observed. The bones appear to be mostly spines and isolated cranial bones. A single, as yet unidentified, partially articulated actinopterygian fish skeleton (1.5 cm long) was also identified from this core (157-X016; Fig. 9.16); however most of the other material is too fragmentary and isolated to permit detailed taxonomic assessment. The other core (173-X030) also preserves rare fish scales. Several other fish have been reported from the western margin of the Kasai area, including fragments of teleosts and a specimen of *Rhipis moorseli* (Cahen 1954).

Two other notable reports of vertebrates from the Kasai region include large bones of sauropod dinosaurs in a conglomerate facies below or at the base of the Kwango Series near Luebo. Cahen (1954) reported the conchostracan *Bairdestheria* at the same locality and through personal communication with W.E. Swinton about the dinosaur bones, suggested a Late Jurassic to Early Cretaceous age for these basal conglomerates. Pterosaur bones, identified as the Cretaceous taxon *Ornithocheirus*, were recovered from the upper Lualaba Series along the Lubilash River (Cahen 1954).

9.7 Discussion

This study represents the first detailed investigation of the Mesozoic cover sequences in the Kasai portion of the Congo Basin since the seminal exploratory work of Cahen (1954) and early geologic mapping of the region. Over the last 50+ years, most inferences about the Kasai stratigraphy have been drawn from the original work of Cahen (1954) and various reports, primarily focused on diamond exploration associated with the Mbuji-Mayi kimberlite fields. The extensive De Beers Exploration drilling program in the Kasai region has provided an important window and research opportunity, which, for the first time, permits detailed observation and insight into the Mesozoic cover sequences of the Kasai portion of the Congo Basin.

Klein (1995) highlighted the complicated and controversial origin of intracratonic basins, particularly owing to their long-lived histories that typically involve various combinations and successions of basin forming processes.

In light of this, attempting to correlate isolated strata and cores across the Congo Basin using a regional layer cake style stratigraphy is challenging and possibly misleading, particularly in the Kasai region where a complex network of faulting appears to have played a role in topographic development across the region, minimally from the late Mesozoic onwards. We constructed a simple 2-D fence from 17 core logs and one measured section, from Tshikapa to Kabinda, which clearly demonstrates this point and helps explain the complex stratigraphy (Fig. 9.9). Two prominent sets of faults are inferred from the basement topography and correspond with lineaments observed in the DEM images (Fig. 9.3). A combination of NW–SE and ENE–WSW oriented sets of faults appears to control stratigraphy on the eastern side of the basin, whereas there appears to be a less dominant structural control on the stratigraphy in the central and western portion of the basin. Most faults are interpreted as extensional normal faults, creating a series of basement highs and associated grabens. These structures explain the distribution of Proterozoic and rare Permian strata in the region, as well as the highly uneven intercepts with crystalline basement rocks that prominently control topography across the region. It is apparent from this study that prolonged and sporadic movement appears to have taken place on multiple faults, leading to heavily bisected and uneven thickness and depth distributions of J1-C4 strata in the region.

One of the most challenging aspects of this study revolves around dating and correlation of the cover sequences. Datable volcanic beds are rare and restricted to kimberlites that are often isolated and difficult to access. However, a fair abundance of both conchostracan and ostracode specimens were identified from multiple cores and various stratigraphic levels across the field area that provide relative dating opportunities. Ostracode and particularly, conchostracan biostratigraphy forms the framework for much of the Mesozoic continental stratigraphy of the Congo Basin, and hence we studied the conchostracan specimens from cores, although not yet the ostracodes, in an attempt to correlate the Kasai strata with other parts of the basin. Surprisingly, one of the most significant findings from this study is the contrasting results that we observed between several biostratigraphic age assessments, which are consistent with records from other parts of the basin, and the much younger-aged detrital zircon grains recorded from one of two samples we analyzed from the C4 beds in the Mbuji-Mayi/Kabinda area. Whereas the biostratigraphic data suggests a Late Jurassic to Early Cretaceous (Lualaba Group) age assignment for unit C4, consistent with the geologic map assignment of these strata, our detrital zircons from the lower part of C3 in the same core yielded a cluster of three ~79 Ma (mean = 79.0 ± 6.0 Ma) zircons and a fourth mid-Cretaceous (113 Ma) grain. These zircon ages closely resemble the ages of local kimberlites (~70 Ma; Mbuji-Mayi) and others further afield (~120 Ma; N. Angola). Although it is possible that the grains are anomalous or erroneous ages, but they do correspond quite well to known kimberlites in the region.

This is confounding on a number of levels. First, it calls into question the usefulness of conchostracans in the basin as index taxa for age assessment and correlation. Second, it suggests that if units C3 and C4 do indeed correlate with the Kwango Group in other parts of the basin, then the Cenomanian-Turonian age typically assigned to the Kwango Group (based on conchostracan and ostracode biostratigraphy) may be inaccurate, or that deposition of these strata is much longer-lived. However, as we have observed even within the Kasai region, a simpler explanation may be that there is significant local stratigraphic variation across the basin and that usage of terms like Stanleyville, Loia and Kwango is at the root of the problem. These stratigraphic terms seem to have limited utility outside of their type areas and better local stratigraphies and nomenclature would probably help clarify this issue. Although this would add to the confusion of having more nomenclature to deal with, it would ultimately lead to less confusion and more precise correlations.

This study provides important new data on the stratigraphy of the Kasai cover sequences; however it also highlights the previously underestimated complexity of the basin and importance of further work. In particular, expanded biostratigraphic and detrital zircon studies are necessary before a formalized local stratigraphy can be developed for the Kasai Region. The informal subdivisions, P1-T1, used in this study to subdivide the local stratigraphy are only placeholders until more data are available to constrain the age of these units and their stratigraphic relationship with units outside of the Kasai region.

Conclusions

Detailed investigation of 22 cores and various outcrop exposures from across the Kasai region provides a basis for reinterpreting the regional stratigraphy and depositional environments of these extensive cover sequences. Analysis of the sedimentary facies of the Kasai cover sequences reveals numerous discrete lithofacies that combine in various combinations to form five distinct facies associations, FA1-5, which are interpreted as glaciolacustrine environments, alluvial fans, fluvial channels, ephemeral lakes, and aeolian dune environments, respectively. These environments dominate the stratigraphy and suggest prolonged arid to semi-arid conditions throughout the Late Jurassic to Late Cretaceous in the region. These data, coupled with detrital zircon provenance, biostratigraphy, and facies stacking patterns, provide context for subdividing the Kasai cover sequences into a temporary, informal stratigraphy until additional work on the age of these units is available. The informal stratigraphy includes a basal glacio-lacustrine dominated sequence of Permian age, termed P1. This unit has limited spatial extent and probably represents an erosional remnant of a potentially much more widely distributed sequence of Permian-age rocks in the Kasai region. Overlying this, but only in the western part of the Kasai near Tshikapa, is a package of mature fluvial-aeolian sandstones and thin lacustrine shales of probable Late Jurassic age, termed J1. Also exposed exclusively in the western part of the basin, is a distinctive succession of maroon-purple, heavy mineral-rich, cross-bedded quartzose sandstones of predominantly fluvial origin, termed C2. Rare basal conglomerates, termed unit C1, sit conformably beneath C2 sandstones in isolated cores. In the central and eastern side of the basin from Kananga to Kabinda, immature, coarse-grained conglomerates of alluvial fan origin rest on a highly uneven basement topography. These synorogenic conglomerates are conformably overlain by a heterogeneous sequence of sandstones and laminated shales and siltstones of fluvial, aeolian and ephemeral lake origins. They are collectively termed C4 and represent the thickest and most widespread unit in the Kasai region. A pilot detrital zircon sample collected from the base of C3 yields a ~79 Ma population of detrital grains, which sharply contrasts the Late Jurassic to Early Cretaceous age suggested by conchostracan biostratigraphy. These results suggest a Late Cretaceous age (or younger) for units C3 and C4. The unexpected local variability and stratigraphic complexity observed in the cover sequences from the Kasai region demonstrate a more tectonically active Mesozoic-Cenozoic basin history then has previously been assumed, with synorogenic movements on major fault systems apparently playing a large role in the local stratigraphy. These findings necessitate even greater care and study in the Kasai and other portions of the basin before basin-scale correlations can be fully assessed.

Acknowledgements We would like to acknowledge the exploration manager of De Beers Exploration for support in the research of the Congo Basin geology and for permission to present this work. Samantha Perritt is thanked for her contributions to our understanding of the geology of the region and Mike de Wit, Anthony Revitt, Mike Roberts, Mike Lynn, John Ward and Renato Spaggiari for valuable discussions over the years. Mike de Wit and Maarten J. de Wit are thanked for their valuable review contributions. Former joint venture partners MIBA, BUGECO and African Diamonds are gratefully acknowledged for permission to carry out research on the drill cores.

References

André L (1993) Age Rb-Sr Proterozoique inferieur du magmatisme continental du Groupe de la Lulua (Kasai, Zaire): ses implications géodynamiques. Ann Soc Géol Belg 116:1–12

Atunes MT, Maisey JG, Marques MM, Schaeffer B, Thomson KS (1990) Triassic fishes from the Cassenge Depression (R.P. De Angola). Ciencieas da Terra (UNL) Lisboa, Numero Especial

Batumike JM, Griffin WL, O'Reilly SY, Belousova EA, Pawlitschek M (2009) Crustal evolution in the central Congo-Kasai Craton, Luebo, D.R. Congo: insights from zircon U-Pb ages, Hf isotope and trace-element data. Precambrian Res 170:107–115

Black LP, Kamo SL, Allen CM, Aleinikoff JN, Davis DW, Korsch RJ, Foudoulis C (2003) TEMORA 1: a new zircon standard for Phanerozoic U–Pb geochronology. Chem Geol 200:155–170

Bluck BJ, Ward JD, De Wit MCJ (2005) Diamond mega-placers: southern Africa and the Kaapvaal craton in a global context. In: McDonald I, Boyce AJ, Butler IB, Herrington RJ, Polya DA (eds) Mineral deposits and earth evolution, vol 248, Geol Soc Pub., pp 213–245

Bosworth W (1992) Mesozoic and early Tertiary rift tectonics in East Africa. Tectonophysics 209:115–137

Boulouard C, Calandara F (1963) Etude palynologique de quelques sondages de la Republique du Congo (Congo ex-belge). S.N.P.A., Direction Exploration and Production, Unpublished report, 6 p

Cahen L (1954) Geologie du Congo Belge. H. Vaillant-Carmanne, Liege, 577 pp

Cahen L (1983) Brèves précision sur l'âge des Groupes Crétaciques post-Wealden (Loia, Bokungu, Kwango) du bassin intérieur du Congo (Républic du Zaire). Mus. Roy. Afri. Centr., Tervuren (Belg.), Dep. Geol. Min. Rapp. Ann., 1981-1982, pp 61–72

Cahen L, Lepersonne J (1951) Carte géologique du Congo belge et du Ruanda-Urundi avec notice explicative. Int. roy. col. belg., Atlas général du Congo, 27 p

Cahen L, Lepersonne J (1954) État actuel des connaissances relatives aux séries mésozoïques de lintérieur du Congo. Bulletin de la Société Belge de Géologie 77:20–37

Cahen L, Lepersonne J (1978) Synthèse des connaissance relatives au group (anciennement sèrie) de la Lukuga (Permien du Zaïre)

Cahen L, Jamotte A, Lepersonne J, Mortelmans J (1946) État actuel des connaissances relatives à la stratigraphie des Systemes du Kalahari et du Karroo au Congo belge. Bull Serv Gàol, C.B. R.U. 2(II): 236–289

Cahen L, Ferrand JJ, Haarsma MJF, Lepersonne J, Verbeek T (1959) Description du sondage de Samba, Ann Mus R Congo Belge, Sci Géol, 29

Cahen L, Ferrand JJ, Haarsma MJF, Lepersonne J, Verbeek T (1960) Description du sondage de Dekese. Ann Mus Roy Congo Belge, Sci Géol, 34

Cahen L, Ledent D, Snelling NJ (1974) Données géochronologiques dans le Katangien inférieur du Kasai oriental et du Shaba nord-oriental (République du Zaïre). Mus Roy Afr Centr, Tervuren, Rapp Ann Dépat Géol Min:59–70

Carbonel P, Colin JP, Danielopol DL, Neustrueva I (1988) Paleoecology of limnic ostracodes: a review of some major topics. Palaeogeog Palaeoclim Palaeoecol 62:413–461

Casier E (1961) Matèriaux pour la faune ichtyologique èo crètacique du Congo. Ann Mus Roy Afr Central, in 8, Sc. Géol, 39, 96 pp

Chen PJ, Shen YB (1985) An introduction of fossil conchostracans. Science Press, Beijing, p 241

Clifford AC (1986) Africa oil—past, present, and future. In: Halbouty MT (ed) Future petroleum provinces of the world, vol 40, Am. Assoc. Pet. Geol. Memoir., pp 339–372

Colin JP (1994) Mesozoic-Cenozoic lacustrine sediments of the Zaire Interior Basin. In: Gierlowski-Kordesch E, Kelts K (eds) Global geological record of lake basins, vol 1. Cambridge University Press, Cambridge, pp 31–36

Cox LR (1960) Further Mollusca from the Lualaba beds of the Belgian Congo. Annales de Musée royal de l'Afrique centrale, Tervuren (Belgique). Série in 8. Sciences géologiques 37:1–15

Daly MC, Lawrence SR, Diemu-Tshiband K, Matouana B (1992) Tectonic evolution of the Cuvette Centrale, Zaire. J Geol Soc Lond 149:539–546

Davis GL (1977) Ages and uranium contents of zircons from kimberlites and associated rocks, vol 76. Carnegie Institution of Washington Yearbook, Washington, DC, pp 631–635

de Saint-Seine P (1952) Sur quelques vertébrés fossiles de la cuvette congolaise. CR Acad Sci Paris 234:1995–1997

de Saint-Seine P (1953) Poissons fossiles de la cuvette congolaise. CR Som Soc Géol Fr 16:343–345

de Saint-Seine P (1955) Poissons fossiles de l'étage de Stanleyville (Congo Belge). Ann Mus Roy Congo Belge, Ser. in 8, Sci Géologiques 14, 126 p

de Saint-Seine P, Casier E (1962) Poissons fossiles des couches de Stanleyville (Congo). 2e'me partie: La faune marine des calcaires de Songa. Ann Mus Roy Afrique centrale, Sér., in -8, Sci Géol 14, 126 pp

De Waele B, Johnson SP, Pisarevsky SA (2008) Paleoproterozoic to Neoproterozoic growth and evolution of the eastern Congo Craton. Its role in the Rodenia puzzle. Precambrian Res 160:127–141

Defretin S (1953) Quelques conchostraces Nord-Cameroun. Bulletin du Service des Mines, Territoire du Cameroun 1:111–120

Defretin-Lefranc S (1967) Étude sur les phyllopodes du Bassin du Congo. Annales, Musée Royal de l'Afrique Centrale, Tervuren, Belgique 56:1–121

Delhal J (1977). Le complexe tonalitique de Kanda-Kanda et données géochimiques et géochronologiques comparées des unités archéennes du Kasai. Mus. roy. Afr. centr., Tervuren (Belg.), Dépt. Géol. Min. Rapp. ann. 1976:65–82

Delhal J, Liégeois JP (1982) Le socle granito-gneissique du Shaba occidental (Zaïre): Pétrographie et géochronologie. Ann Soc Géol Belg 105:295–301

Delhal J, Deutsch S, Snelling NJ (1989) Datation du Complexe sédimentaire et volcanique de la Lulua (Protérozoïc Inférieur, Kasai, Zaïre). Mus Roy Afr Centr, Tervuren (Belg), Dépt Géol Min Rapp Ann 1987–1988:93–99

Delpomdor F, Linnemann U, Boven A, Gärtner A, Travin A, Blanpied C, Virgone A, Jelsma H, Préat A (2013) Depositional age, provenance, and tectonic and paleoclimatic settings of the late mesoproterozoic–middle neoproterozoic Mbuji-Mayi supergroup, democratic Republic of Congo. Palaeogeogr Palaeoclimatol Palaeoecol 389:4–34

Dirks PHGM, Blenkinsop TG, Jelsma HA (2003) The geological evolution of Africa. In: Stuwe K, Grasemann de Vivo (eds) Encyclopedia of life support systems. UNESCO Publishing, EOLSS Publishers Co Ltd. Oxford, UK. http://www.eolss.net

Downey NJ, Gurnis M (2009) Instantaneous dynamics of the cratonic Congo basin. J Geophys Res 114, B06401. doi:10.1029/2008JB006066

Dumont HJ, Negrea SV (2002) Branchiopoda. Backhuys Publishers, Leiden, p 398

Eley R, Grütter H, Louw A, Tunguno C, Twidale J (2008) Exploration geology of the Luxinga kimberlite cluster (Angola) with evidence supporting the presence of kimberlite lava. In: 9th international kimberlite conference, extended abstract, No. 9IKC-A-00166, 3 pp

Eugster HP (1986) Lake Magadi, Kenya: a model for rift valley hydrochemistry and sedimentation? Geological Society, London Special Publications 25: Siliclastic, chemical, Pedogenic and Organic Sediments in Contemporary Rift Enviornments:177–189

Évrard P (1957). Les recherches géophysiques et géologiques et les travaux de sondage dans la Cuvette Congolaise. Acad Roy Sci Coloniale Sci Tech Brux 7(1), 62 pp

Fieremans CL (1955) Etude géologique preliminaire des conglomérats diamantifères d'âge mésozoique au Kasaï (Congo belge). Univ Louvain Inst Géol Mém 19:223–294

Fieremans CL (1961) Origine et répartition de la minéralisation diamantifière au Kasai occidental (Congo) et dans le N E de la Lunda (Angola). Bull Soc Belge Geol 70:89–95

Fieremans CL (1996). A brief review of the occurrences of diamonds in the Republic of Zaïre. Spec. ed. Africa Geoscience Rev.: 91–94

Frank PW (1988) Conchostraca. Palaeogeog Palaeoclim Palaeoecol 62: 399–403

Frimmel HE, Tack L, Basei MS, Nutman AP (2006) Provenance and chemostratigraphy of the Neoproterozoic West Congolian Group in the Democratic Republic of Congo. J Afr Earth Sci 46:221–239

Gallego OF, ShenYB CN, Hernández M (2010) The genus Congestheriella Kobayashi, 1954 (Conchostraca, Afrograptioidea): redescription and new combination to Isaura olsoni Bock from Venezuela and a new species from Argentina (Upper Jurassic). Alavesia 3:11–24

Giresse P (2005) Mesozoic-Cenozoic history of the Congo Basin. J Afr Earth Sci 43:301–315

Gouldie AS (2004) The drainage of Africa since the Cretaceous. Geomorphology 67:437–456

Grekoff N (1957) Ostracodes du Bassin du Congo. I: Jurassique supérieur et Crétacé inférieur du Nord du bassin. Ann. Mus. Roy. Congo belge, in -8 Sc. Géol., 19, 97 pp

Grekoff N (1958) Ostracodes du Bassin du Congo. III: Tertiaire. Ann. Mus. Roy. Congo belge, in -8_ Sc. Ge´ol., 22, 36 pp

Grekoff N (1960) Corrélations et âge relatif de quelques séries continentals mésozoıques. In: Int. Geol. Congr., Rept 21st Sess., Norden, Part VI, Pre-Quaternary micro-paleontology:18–26.

Hegna TA (2012) The phylogeny and fossil record of branchiopod crustaceans: an integrative approach. Dissertation, Yale University

Jackson SE, Pearson NJ, Griffin WL, Belousova EA (2004) The application of laser ablation-inductively coupled plasma-mass spectrometry to in situ U-Pb zircon geochronology. Chem Geol 211:47–69

Jelsma H, Krishnan U, Perritt S, Preston R, Winter F, Lemotlo L, van der Linde G, Armstrong R, Phillips D, Joy S, Costa J, Facatino M. Posser A, Kumar M, Wallace C, Chinn I, Henning A (2012) Kimberlites from central Angola: a case study of exploration findings. In: Proc. 10th International Kimberlite Conference

Johnson MR (1992) A proposed format for general-purpose comprehensive graphic logs. Sed Geol 81:289–298

Johnson SP, Rivers T, DeWaele B (2005) A review of the Mesoproterozoic to early Palaeozoic magmatic and tectonothermal history of south-central Africa: implications for Rodinia and Gondwana. J Geol Soc Lond 162:433–450

Jones L, Mathieu PL, Strenger H (1960). Résultats scientifiques des missions du syndicat pour l étude géologique et miniére de la Cuvette Congolaise et travaux connexes. Géodèsie et géophysique. Magnetisme. Ann. Musée royal du Congo belge, Tervuren, 27, 30 pp

Kadima EK, Sebagenzi SMN, Lucazeau F (2011a) A Proterozoic-rift origin for the structure and the evolution of the cratonic Congo basin. Earth Planet Sci Lett 304:240–250

Kadima K, Delvaux D, Sebagenzi MN, Tack L, Kabeya M (2011b) Structure and geological history of the Congo Basin: an integrated interpretation of gravity, magnetic and reflection seismic data. Basin Res. doi:10.1111/j.1365-2117.2011.00500

Kerschhofer L, Schärer U, Deutsch A (2000) Evidence for crystals from the lower mantle: baddeleyite megacrysts of the Mbuji-Mayi kimberlite. Earth Planet Sci Lett 179:219–225

Klein GD (1995) Intracratonic basins. In: Busby CJ, Ingersoll RV (eds) Tectonics of sedimentary basins. Blackwell, Oxford, pp 459–478

Lawrence S, Makazu MM (1988) Zaire's Central basin: prospectivity outlook. Oil Gas J 86:105–108

Lepersonne J (1974) Notice expliclative de la carte geologique du Zaïre, au 1/2,000,000. Rep Zaire Serv. Geol. Kinshasa, Mus. Royal Africa Centrale, Tervuren, 67 p

Lepersonne J (1977) Structure geologique du basin interieur du Zaire. Bull. Acad. Roy. Belgique, Cl. Sci, 5e Ser. 63:941–965

Leriche M (1913) Les entomostracés des couches du Lualaba (Congo Belge). Revue Zoologique Africaine 3:1–11

Ludwig KR (2003) User's Manual for Isoplot 3.00. A geochronological toolkit for Microsoft Excel. Berkley Geochronology Centre Special Publication No.4.

Marlière R (1948) Ostracodes et Phyllopodes du système du Karroo au Congo Belge. Annales du Musée du Congo Belge Tervuren, Série in 8. Sciences Géologiques 2:1–61

Marlière R (1950) Ostracodes et Phyllopodes du systèm du Karroo au Congo belge et les régions avoisinantes. Annales du Musée du Congo Belge Tervuren, Série in 8. Sciences Géologiques 6: 1–43

Miall AD (1978) Lithofacies types and vertical profile models in braided river deposits: a summary. In: Miall AD (ed) Fluvial sedimentology, vol 5. Canadian Society of Petroleum Geologists Memoir, pp 597–604

Moucha R, Forte AM (2011) Changes in African topography driven by mantle convection. Nat Geosci 4:707–712

Myers TS, Tabor NJ, Jacobs LL (2012) Late Jurassic paleoclimate of Central Africa. Palaegeog Palaeoclim Palaeoecol 311:111–125

Pedreira AJ, DeWaele B (2008) Contemporaneous evolution of the Palaeoproterozoic-Mesoproterozoic sedimentary basins of the São Francisco-Congo Craton. Geol Soc Lond Spec Publ 294: 33–48

Pereira E, Rodrigues J, Reis B (2003) Synopsis of Lunda geology, NE Angola: Implications for diamond exploration. Comun Instituto Geológico e Mineiro 90:189–212

Perritt SH, Jelsma HA, Armstrong R (2011) Paleoproterozoic sedimentation and tectonism along the Lucapa Fracture Zone. In: 23rd Colloquium of African Geology, Johannesburg, 8–13 Jan 2011, abstract

Raucq P (1970) Nouvelles acquisitions sur le Système de la Bushimay. Mus Roy Afr Centr, Tervuren (Belg.), Annales, Serie In-8, Science Géologiques, 69, 156 p

Raucq P (1979) Notice explicative de la feuille Mbuji-Mayi (S7/E23) (carte géologique au 1/200,000, 1977). Dep Mines Serv Geol Républ Zaïre, Kinshasa, 50 pp

Raucq P, Lepersonne J (1966) Carte géologique au 1/200,000, de la feuille Dibaya (S7/E22). Musée royal de l'Afrique centrale Tervuren, Belgique

Robles-Cruz SE, Escayola M, Jackson S, Galí S, Pervov V, Watangua M, Gonçalves A, Melgarejo JC (2012) U–Pb SHRIMP geochronology of zircon from the Catoca kimberlite, Angola: implications for diamond exploration. Chem Geol 310–311:137–147

Rocha-Campos AC, Dos Santos PR (1981) The Itarare Subgroup, Aquidauana Group and San Gregorio Formation, Parana Basin, Southeastern South America. In: Hambrey MJ, Harland B, Chumakov NM et al (eds) Earth's Pre-Pleistocene glacial record. Cambridge Press, Cambridge, pp 842–852

Sahagian DL (1993) Structural evolution of African basins: stratigraphic synthesis. Basin Res 5:41–54

Schärer U, Corfu F, Demaiffe D (1997) U-Pb and Lu-Hf isotopes in baddeleyite and zircon megacrysts from the Mbuji-Mayi kimberlite: constraints on the subcontinental mantle. Chem Geol 143:R1–16

Tasch P (1987) Fossil Conchostraca of the Southern Hemisphere and continental drift. Paleontology, biostratigraphy, and dispersal. Geol Soc Am Memoir 165:1–290

Taverne L (1976) Les téléostéens fossiles du Crétacé moyen de Kipala (Kwango, Zaire). Annales du Museé Royal de l'Afrique Centrale, Tervuren, Série in 88. Sciences Géologiques 79:1–50

Taverne L (1984) À propos de Chanopsis lombardi du Crétacé inférieur du Zaïre. Rev Zool Afr 98:578–590

Tucker R, Roberts E, Hu Y, Kemp A, Salisbury S (2013) Detrital zircon age constraints for the Winton formation, Queensland: contextualizing Australia's late cretaceous dinosaur faunas. Gondwana Res 24:767–779

Verbeek T (1970) Geologie et Lithologie du Lindien. Annals du Musee Royale Afrique Centrale, Tervuren. Serie 8, No. 66

Walraven F, Rumvegeri BT (1993) Implications of whole-rock Pb-Pb and zircon evaporation dates for the early metamorphic history of the Kasai Craton, Southern Zaire. J Afr Earth Sci 16: 395–404

Webb JA (1979) A reappraisal of the palaeecology of conchostracans (Crustacea: Branchiopoda). Neues Jahrbuch für Geologie und Paläontologie. Abhandlungen 158:259–275

10 Formation and Collapse of the Kalahari Duricrust ['African Surface'] Across the Congo Basin, with Implications for Changes in Rates of Cenozoic Off-Shore Sedimentation

Bastien Linol, Maarten J. de Wit, Francois Guillocheau, Michiel C.J. de Wit, Zahie Anka, and Jean-Paul Colin†

10.1 Introduction

The Congo Basin (CB) of central Africa lies at about 400 m above mean sea level (amsl), and is linked to the south, across a central African drainage divide, to the high interior Kalahari Plateau (KP) at ca. 1,100 m amsl (Fig. 10.1). The CB and KP are flanked by distinct marginal escarpments along the South Atlantic and southwest Indian Ocean margins, and to the east by the East African Rift System (EARS). Their relatively flat interior is covered by an extensive Upper Cretaceous-Cenozoic succession of sand dunes, pan-lacustrine sediments and alluviums with hard-caps (duricrusts) of calcrete, silcrete and ferricrete, collectively named the Kalahari Group (SACS, 1980). This succession reaches a maximum thickness of about 500 m, but across southern and central Africa is generally less than 100 m thick, representing one of the world's most extensive, long-lived condensed stratigraphic sequences.

The Kalahari Group directly overlies Precambrian basement of the Kalahari and Central African Shields (Fig. 10.1b), late Paleozoic to mid-Mesozoic sequences of the Karoo Supergroup including Lower Jurassic flood basalts in southern Africa, dated at 178–183 Ma (the Stormberg Group; Jourdan et al. 2007), and Cretaceous volcanics and dykes in Namibia, dated at 127–132 Ma (the Etendeka Group; Miller 2008). By contrast, across the CB, the Kalahari Group overlies Upper Jurassic to Upper Cretaceous red sandstones (e.g. the Kwango Group) ranging in thickness between about 300 m and 1,000 m (Linol 2013; Chap. 8, this Book). The unconformity at the base of the Kalahari Group marks a subcontinental scale peneplanation surface that is commonly referred to as the 'African Surface' (e.g. King 1963; Partridge and Maud 2000; Haddon and McCarthy 2005; Decker et al. 2013), developed as a result of extensive denudation and uplift following the break-out of Africa from Gondwana during the opening of the Indian and South Atlantic Oceans, and the onset of the Kalahari epeirogeny (de Wit 2007). However, the precise age-range of this elevated, mega-denudation surface remains uncertain because the stratigraphy of the overlying Kalahari Group is poorly characterized, including many internal and local unconformities and abundant silcretes, calcretes and ferricretes that are difficult to date and correlate regionally. Moreover, because large parts of this Kalahari succession

†Author was deceased at the time of publication.

B. Linol (✉)
AEON-ESSRI (African Earth Observatory Network – Earth Stewardship Science Research Institute), Nelson Mandela Metropolitan University, Port Elizabeth, South Africa

Geological Sciences, Nelson Mandela Metropolitan University, Port Elizabeth, South Africa
e-mail: bastien.aeon@gmail.com

M.J. de Wit
AEON-ESSRI (African Earth Observatory Network – Earth Stewardship Science Research Institute), Nelson Mandela Metropolitan University, Port Elizabeth, South Africa
e-mail: maarten.dewit@nmmu.ac.za

F. Guillocheau
Géosciences-Rennes, UMR 6118 Université de Rennes 1 – CNRS, OSUR, Université de Rennes 1, Campus de Beaulieu, 35042 Rennes cedex, France
e-mail: francois.guillocheau@univ-rennes1.fr

M.C.J. de Wit
Delrand Resources Pty Ltd., Toronto, Ontario, Canada
e-mail: mdewit@delrand.com

Z. Anka
Helmholtz Centre Potsdam, GFZ German Research Centre for Geosciences, Telegrafenberg, 14473 Potsdam, Germany

TOTAL Exploration/New Venture, 2 place Jean Millier, La Defence, 92078 Paris, France
e-mail: zahie@gfz-potsdam.de

M.J. de Wit et al. (eds.), *Geology and Resource Potential of the Congo Basin*, Regional Geology Reviews,
DOI 10.1007/978-3-642-29482-2_10, © Springer-Verlag Berlin Heidelberg 2015

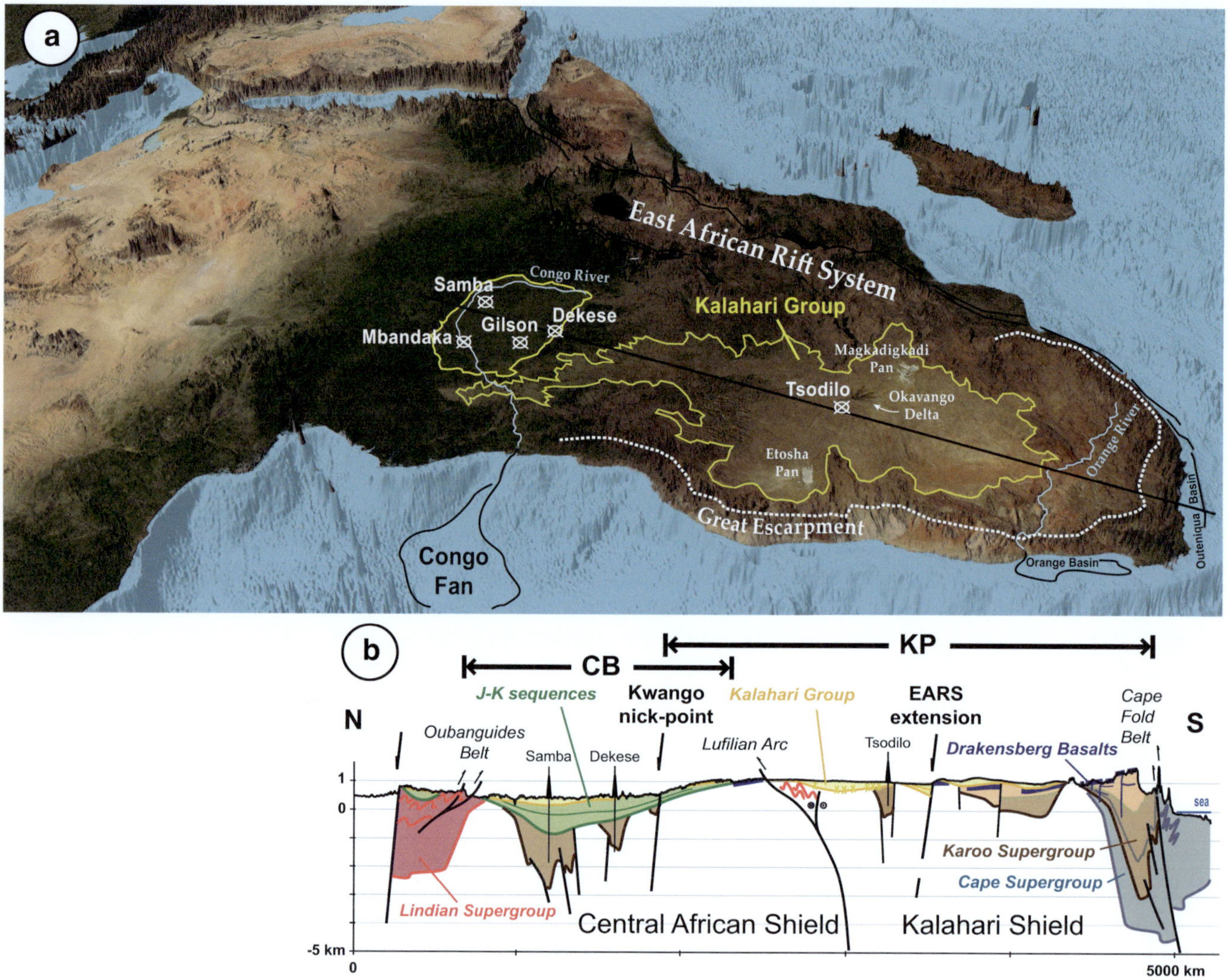

Fig. 10.1 (**a**) Digital elevation model of sub-Saharan Africa and (**b**) N–S cross-section of the CB and KP, highlighting the vast extension of Cenozoic sediments and duricrusts of the Kalahari Group (in *yellow*), and with location of studied boreholes. Note that the transition between the KP and CB is not related to the boundary between the Kalahari and Central African Shields

are unconsolidated, there is relatively limited data available from drill-cores (Haddon 2000; Miller 2008).

Here, we present new sedimentological and stratigraphic data from drilling into the Kalahari Group on top of the KP, in the Ngamiland region of northwest Botswana, and from field investigations in the Kwango Valley of the southwest CB, flanking the transition to the KP in the southern Democratic Republic of Congo (DRC). On the basis of well/core and seismic data, we also extend this Kalahari sequence to the center of the CB (Fig. 10.2). Here, very little of the Kalahari duricrusts cover is preserved, leaving a denuded landscape ('Bad-Lands') of Cretaceous red-beds sometimes covered by residual blocks and large boulders of silcrete and calcrete that suggest relatively recent collapse of the Kalahari duricrusts and accelerated erosion across the CB of its underlying poorly consolidated red-beds. Because the off-shore sedimentation history of the Congo Fan along the Atlantic margin reveals a sudden episode of rapid sedimentation in the Oligocene (e.g. Anka and Séranne 2004), we propose a new model of rapid disintegration of the Kalahari duricrusts carapace and preferential erosion ('flushing out') of the soft underlying red-beds across the CB, driven by increased fluvial activity in response to global cooling in the mid- to late Cenozoic (e.g. Zachos et al. 2001).

10.2 The Kalahari Group

The Kalahari Group covers most of southern and central Africa (albeit poorly exposed), extending continuously from the Orange River in South Africa, through Namibia, Botswana, western Zimbabwe, Zambia, Angola, to the Congo River in DRC and the Republic of Congo, covering some 2.7 million km^2 (Figs. 10.1a and 10.2).

A thickness map of the Kalahari Group in southern Africa, compiled mainly from borehole data of water wells

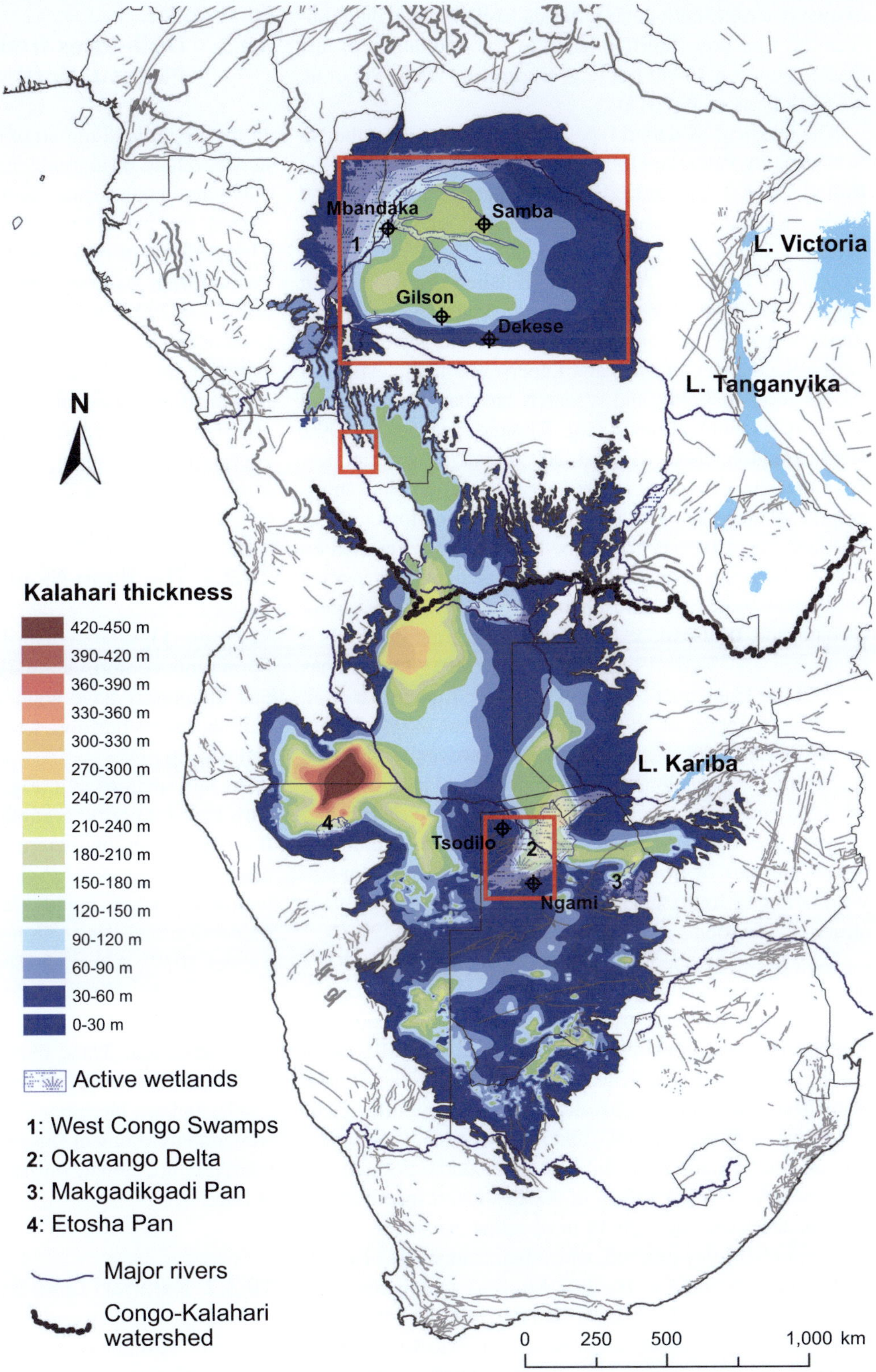

Fig. 10.2 Isopach map of the Kalahari Group, extended from Haddon (1999) by Linol (2013). Three study areas referred to in the text are shown by *red rectangles*

by Haddon (1999), has now been extended to include central Africa (Fig. 10.2), based on new field observations in the Kwango Valley along the northern flank of the KP, and re-examination of the seismic and well/core data from the 1950s and 1970s in the center of the CB (Linol 2013; see also Chaps. 7 and 8, this Book). This isopach map emphasizes that the thickest sediment accumulations are preserved along the western margin of the KP: in northern Namibia (ca. 200–450 m thick), in central Angola (ca. 200–300 m thick), in northern Botswana (ca. 150–250 m thick), and in the western part of the central CB (ca. 100–200 m thick). The distinct decrease in thicknesses eastward across south-central Africa likely reflects the influence of Cenozoic uplift related to the formation of the EARS, between 20–40 Ma (e.g. Chorowicz 2005; Pik et al. 2008; Roberts et al. 2012).

The classical 'Kalahari type-section', as first described in Botswana by Passarge (1904), consists of basal gravels (the Botletle Beds), calcretized or silcretized sandstones and marls ('Kalahari Limestones') and aeolian sands ('Kalahari Sands'). Lithostratigraphic equivalents were later also recognized in South Africa (the Kalahari Beds; du Toit 1954), Zimbabwe ('Pipe Sandstones'; Maufe 1936), Zambia (the Barotse Formation; Money 1972) and DRC ('Polymorph Sandstones'; Cahen and Lepersonne 1952). Much of this historic mapping and research has been summarized since in regional reviews (e.g. Thomas and Shaw 1990, 1993; Haddon 2000; Giresse 2005), but very little new field data have been generated to provide detailed stratigraphic sections and test these regional correlations.

Most recent studies across the Kalahari have focused on near-surface pan sediments, silcretes/calcretes and aeolian sand dunes to provide information about Quaternary paleoclimate and landscape evolution (e.g. Summerfield 1983; Holmgren and Shaw 1996; Ringrose et al. 1999; Thomas et al. 2003; Huntsman-Mapila et al. 2006; Kampunzu et al. 2007; Burrough et al. 2009; Hürkamp et al. 2011; Moore et al. 2012; Eckardt et al. 2013). However, significant reworking, bioturbation and post-depositional modifications of these sediments often hamper reliable age-dating and stratigraphic correlations. It has also been argued that a significant part of the Kalahari Group results from in-situ weathering of its basement rocks, further hampering construction of a robust stratigraphy (McFarlane et al. 2010).

Only in the Etosha Pan of northern Namibia (Figs. 10.1a and 10.2 for location) recent borehole and paleontological investigations have described in detail a more complete Kalahari succession (Miller 2008; Miller et al. 2010). Here, the Kalahari Group includes basal gravels of the Beiseb Formation (probably equivalent to the Botletle Beds described by Passarge 1904), and two thick sand-dominated fan sequences, named the Olukonda and Andoni Formations, maximum 150 m and 550 m in thicknesses, respectively. It intercalates at the top with 50 m of saline clays that constitute the present-day pan bed, and which contain fossil vertebrate suites dated from the late Miocene to Pliocene, ca. 4–6 Ma (Miller et al. 2010). This relatively thick Kalahari succession in turn passes laterally into an 80 to 120 m thick, massive 'calcrete' named the Etosha Calcrete Formation, considered to represent a gigantic groundwater deposit that developed <4 Ma under arid conditions (Miller 2008). These terrestrial carbonates, however, are complex because of multiple phases of deposition and dissolution, and are thus difficult to interpret chronologically (e.g. Nash and McLaren 2003; Wanke and Wanke 2007; Linol et al. 2009).

10.3 Drill-Cores from the Central Kalahari Plateau, Northwest Botswana

In the Ngamiland region of northwest Botswana (Fig. 10.3), new borehole stratigraphy has also characterized widespread calcretes in subsurface, between 10 m and 60 m thick, and named the Nxau-Nxau Calcrete Formation (Linol 2013). These carbonate rocks overly a regional unconformity across Precambrian basement (the Damara Supergroup; Haddon and Roos 2001) and diamictites with black shales and redbeds of the Karoo Supergroup that are deformed and locally intruded by kimberlites in the Nxau-Nxau area, dated at 83 Ma (Batumike et al. 2007; de Wit 2013a). Calcretization also frequently penetrates into the Karoo Supergroup and Precambrian basement, and xenoliths of calcrete occur in the kimberlites (de Wit 2013b).

10.3.1 The Nxau-Nxau Calcrete Formation

In cores (Figs. 10.4 and 10.5), the Nxau-Nxau Calcrete Formation comprises sandy and muddy carbonate rocks that commonly have abundant micro-karst structures, root moulds and/or burrows, and cracks filled with coarser sands and gravels.

In thin sections (e.g. Fig. 10.5), the matrix is generally a dark brown micritic groundmass supporting detrital quartz. It is partially or totally replaced by sparite due to subsequent diagenesis. Fenestral fabric and circum-granular cracks within the more muddy facies also indicate episodes of immersion and desiccation (e.g. Wright and Tucker 1991; Tanner 2010). Although biogenic features, such as alveolar structures, rhyzoliths and burrows are abundant, indicating intense biological activity (e.g. Goudie 1983), no fossils were observed. Thus, these facies have characteristics of both lacustrine and pedogenic deposits (c.f. Alonso-Zarza 2003), and are interpreted to be deposited in a shallow lake wetland environment (Fig. 10.5). This suggests a humid and relatively hot climate during early deposition of the Kalahari Group.

10.3.2 Kalahari Lake Sediments

Previous sedimentological studies of Kalahari sediments have demonstrated wet conditions during the Pleistocene and suggested the occurrence of extensive paleo-lakes (e.g. Du Plessis and Le Roux 1995; Holmgren and Shaw 1996; Thomas et al. 2003; Huntsman-Mapila et al. 2006; Eckardt et al. 2013). However, these studies often lack deep borehole data, particularly into the Kalahari lakes and wetlands (e.g. Okavango Delta, Makgadikgadi Pan and Lake Ngami;

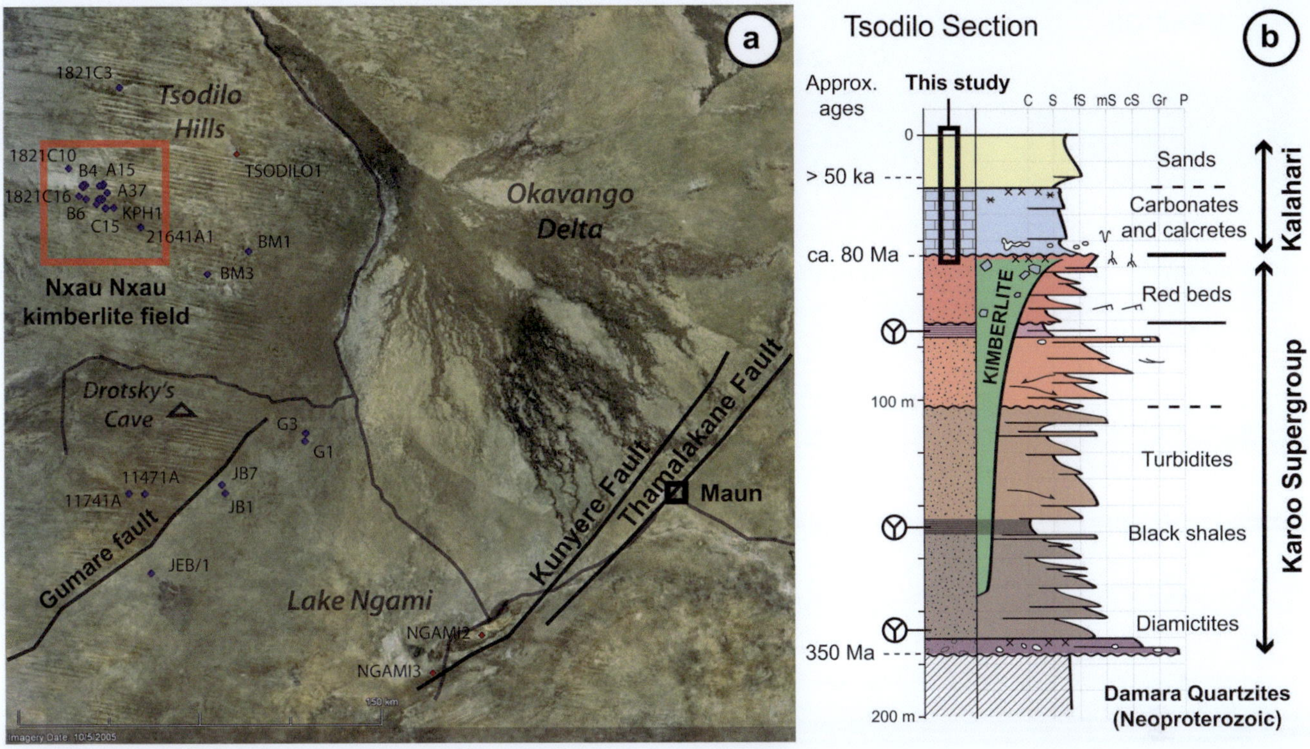

Fig. 10.3 (**a**) Location map of the Kalahari drilling project in the Ngamiland region of northwest Botswana, showing the studied boreholes (*blue* and *red dots*), and (**b**) simplified stratigraphic section (for more details see Linol 2013)

Figs. 10.1a and 10.2) that have great potential to preserve long Cenozoic sedimentary records. Below, we describe new sequence analysis of the Kalahari Group from drilling up to ca. 100–125 m depth at the edge of the Tsodilo Hills and into Lake Ngami (Fig. 10.6). Details of the drilling technology and core recovery are provided online: http://www.aeon.org.za/content/pdf/kalahar_drilling_project_v7_10308.pdf.

10.3.2.1 Tsodilo Hills

At the foot of the Tsodilo Hills, the Tsodilo-1 borehole (18°46'58"S; 21°43'11"E) intercepted a ca. 42 m succession of sands and calcretes of the Kalahari Group. This section overlies pre-Kalahari (e.g. Karoo-age) heterolithic sandstones and green mudstones that are increasingly silicified upward to form a 4 m thick silcrete. This distinct hardpan, likely formed by deep weathering and groundwater activity at the discontinuity that defines the base of the Kalahari Group, possibly correlates with the Nxau-Nxau (and Etosha) Calcrete Formation. The overlying Kalahari succession at Tsodilo Hills comprises a lower, 32 m thick sand-dominated fluvial sequence, overlain by 8 m thick sandy carbonates and a 2 m thick fossiliferous calcrete that cap the flat present day topography (Fig. 10.6a). Previous radiocarbon dates (Thomas et al. 2003), between 32 ka and 36 ka, obtained from molluscs and shells collected at surface, date this uppermost sequence of pan-lacustrine and calcrete deposits to the late Pleistocene (Fig. 10.6c).

10.3.2.2 Lake Ngami

Near the center of Lake Ngami (Fig. 10.6b), the Ngami-2 borehole (20°29′39″S; 22°43′44″E) is 125 m deep, without reaching the base of the lake sediments. The recovered Kalahari succession comprises calcareous fine sands with at depth, below −60 m, seven intervals of darker brown, organic-rich muds that record successive episodes of lake expansion (i.e. flooding). These mud-rich horizons are fossiliferous with ostracods, charophytes and charophyte gyrogonites (Fig. 10.7) that suggest an age younger than Miocene and confirm the existence of a deep paleo-lake. The ostracod shell fragments, from depth—106–107 m, were dated using AMS ^{14}C technique (e.g. Rethemeyer et al. 2012) and resulted in an age greater than 50 ka, beyond the precise radiocarbon detection (Fig. 10.6c). Near the surface, at a depth of −3.2 m, a horizon of organic-rich sediments was previously ^{14}C dated at 18 ka (Huntsman-Mapila et al. 2006).

In summary, the Kalahari Group in the Ngamiland region of northwest Botswana comprises basal clastic-carbonates and calcretes (the Nxau-Nxau Calcrete Formation) that pass laterally and upward into a fluvial sequence of Kalahari sands, which in turn intercalates near the top with Pleistocene pan-lacustrine and younger calcrete deposits (e.g. Thomas et al. 2003; Huntsman-Mapila et al. 2006). Correlations of these sequences (at the Tsodilo Hills and Lake Ngami; Fig. 10.6c) with the Kalahari stratigraphy established at the

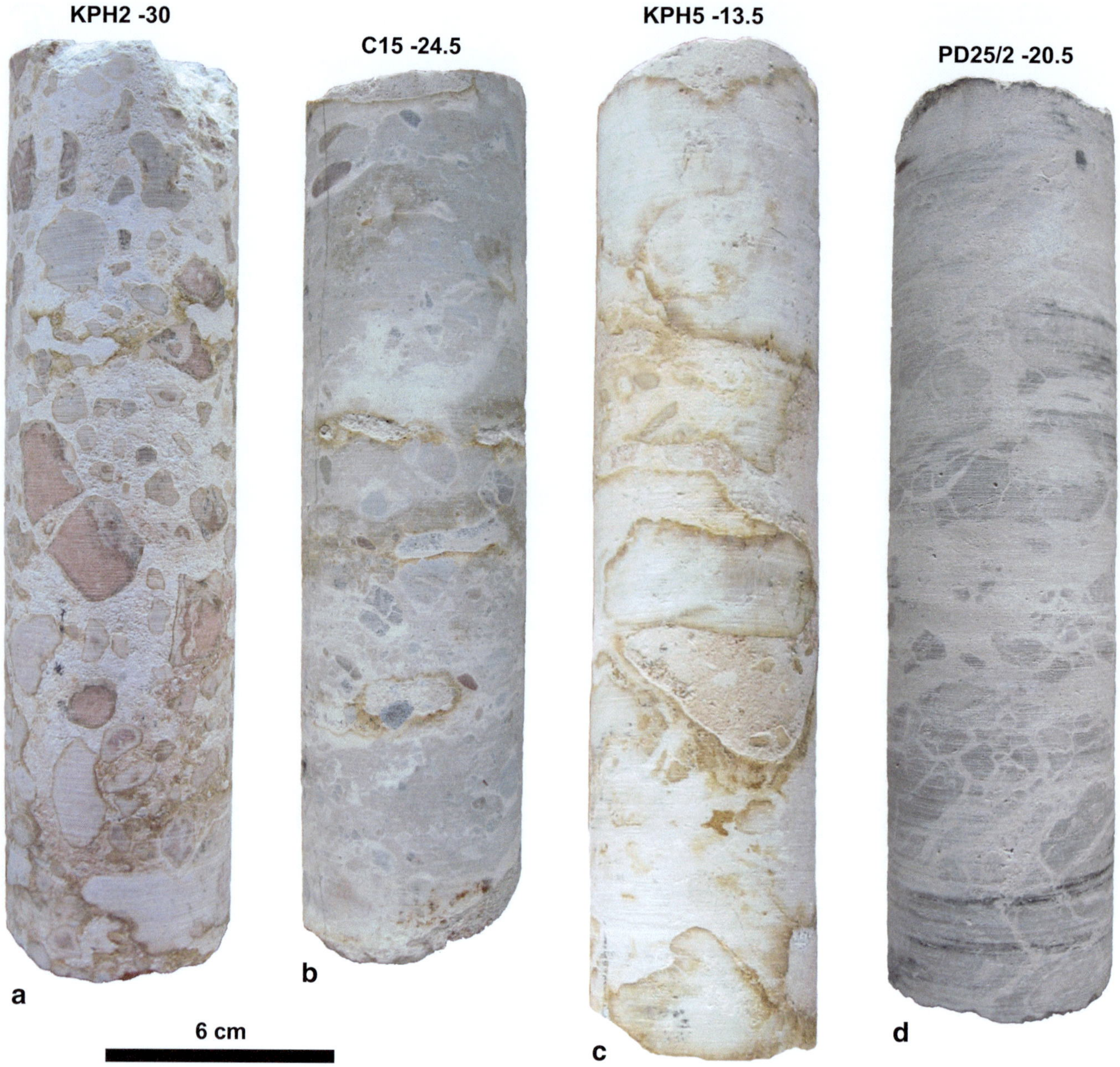

Fig. 10.4 Calcrete cores: (**a**) clastic-carbonates with large intraclasts (CCi facies); (**b**) and (**c**) clastic-carbonates with small pebbles and granules (CCg facies) and micro-karst structures; and (**d**) muddy-carbonates with cracks (CCm facies)

Etosha Pan in Namibia (Miller 2008) indicate increasingly arid conditions during deposition of the uppermost Kalahari Group regionally across southern Africa.

10.4 Denudation Surfaces and Kalahari Sediments of the Congo Basin

The KP extends northward into the CB, terminating at a scarp (Fig. 10.8), between 400 m and 600 m high that extends from the Kwango and Kasai regions in southern DRC and northern Angola, to the Batéké region in the Republic of Congo and eastern Gabon (Fig. 10.2).

Along the upper part of the KP at the southwestern margin of the CB, the Kalahari Group overlies a regional erosion surface across red sandstones of the Upper Jurassic to Upper Cretaceous Kwango Group (Linol 2013; Chap. 8, this Book). Here, the type section of the Kalahari Group ranges between about 100 m and 300 m thick and is generally subdivided into two (Cahen and Lepersonne 1952):

1. A lower, massive sequence comprises 60–80 m thick extensively silicified 'limestones', calcareous sandstones and silcretes with chalcedony, known as the 'Polymorph

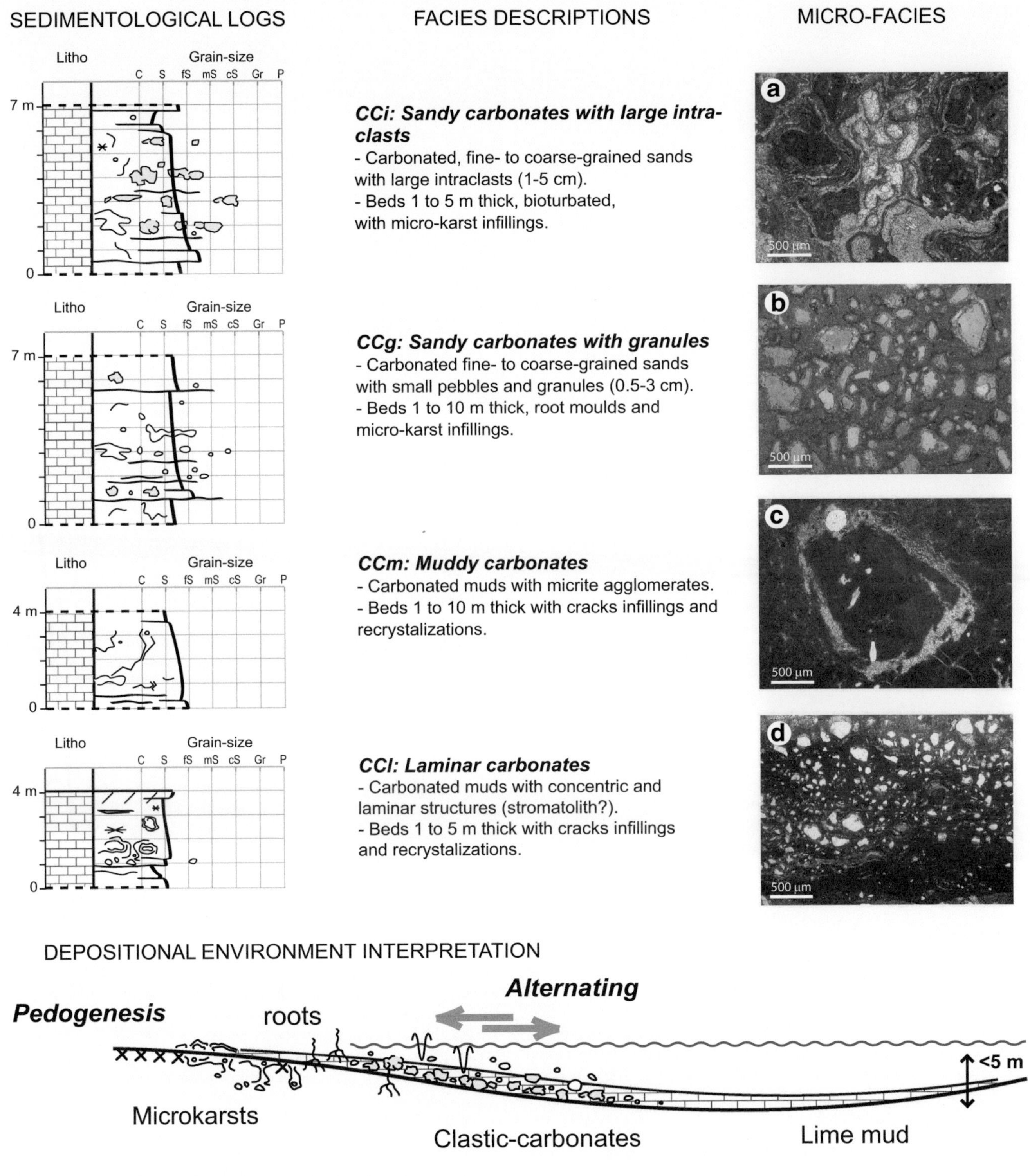

Fig. 10.5 Facies descriptions of the Nxau-Nxau Calcrete Formation, and shallow lake wetland depositional model. Microfacies: (**a**) Alveolar structure filled with sparite (*white*); (**b**) 'floating' quartz grains (*white*) coated with sparite; (**c**) circum-granular crack; and (**d**) laminar micrite (*dark brown*)

Sandstones' (c.f. Lepersonne 1945; see Chap. 14, this Book). These rocks are often cavernous, bioturbated and fossiliferous with shells that display early silicification (De Ploey 1968). Ostracods (*Cypris farnhami*, *C. lerichei*, *Erpetocypris* sp., *Gomphocythere?* sp., *Oncocypria* sp. and *Stenocypris bunzaensis*; Leriche 1927; Polinard 1932; Grekoff 1958; Colin 1994) and other lacustrine fossils (charophytes and gastropods) suggest an Eocene

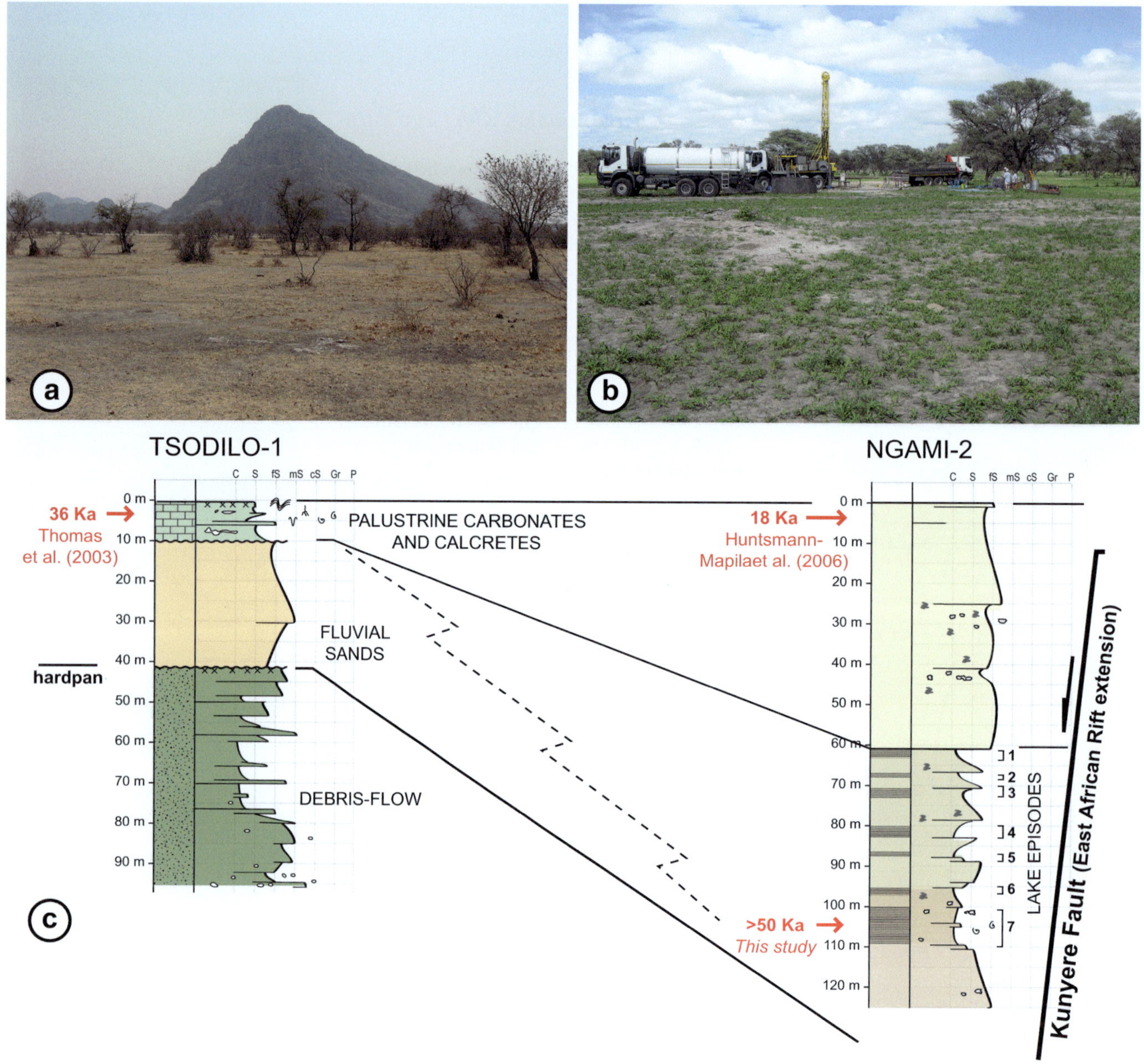

Fig. 10.6 AEON drill sites (**a**) at the Tsodilo Hills and (**b**) at Lake Ngami, and (**c**) sequence correlations of the Kalahari Group between these two sites (Figs. 10.2 and 10.3a for locations), with location of ^{14}C dated samples (in *red*). In the Tsodilo-1 borehole, the Kalahari Group comprises a lower, 30 m thick fluvial sequence overlain by an upper sequence of palustrine carbonates and calcretes, in total 10 m thick. In the Ngami-2 section, the equivalent lacustrine succession is more than 125 m thick, including at depth, below −60 m, seven horizons of fossiliferous, dark organic-rich mud. These correlations indicate wetter conditions during deposition of the lower Kalahari Group

age for this lower Kalahari sequence. Lithostratigraphically, this sequence of duricrusts correlates well with the similar formations of calcrete and silcrete described at the base of the Kalahari Group in northern Namibia and Botswana (the Nxau Nxau and Etosha Calcrete Formations). Differences in lithology between these deposits (silcretes versus calcretes) likely reflect regionally variable weathering conditions related to more humid and hotter climate in central Africa.

2. An upper sequence of unconsolidated sands, up to 120 m thick, overlies another weathering surface, marked by ferricrete deposits at the top of the 'Polymorph Sandstones' (Cahen and Lepersonne 1952). This sequence comprises fine ochreous sands, silts and kaolinitic muds devoid of stratification, possibly due to intense bioturbation, and which are interpreted as sheet wash deposits (De Ploey 1968; see also Chap. 14, this Book).

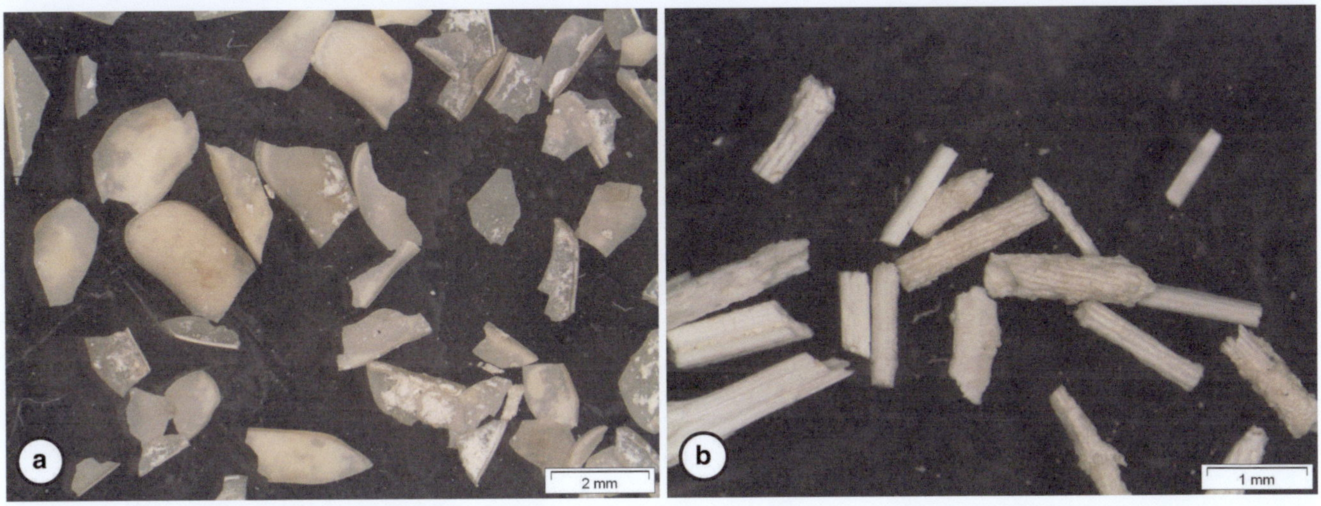

Fig. 10.7 Sieved-residues of sample—106–107 m from the Ngami-2 borehole (Fig. 10.6c). (**a**) Ostracode shell fragments dated at ca. >50 ka by ^{14}C method, and (**b**) charophyte stems

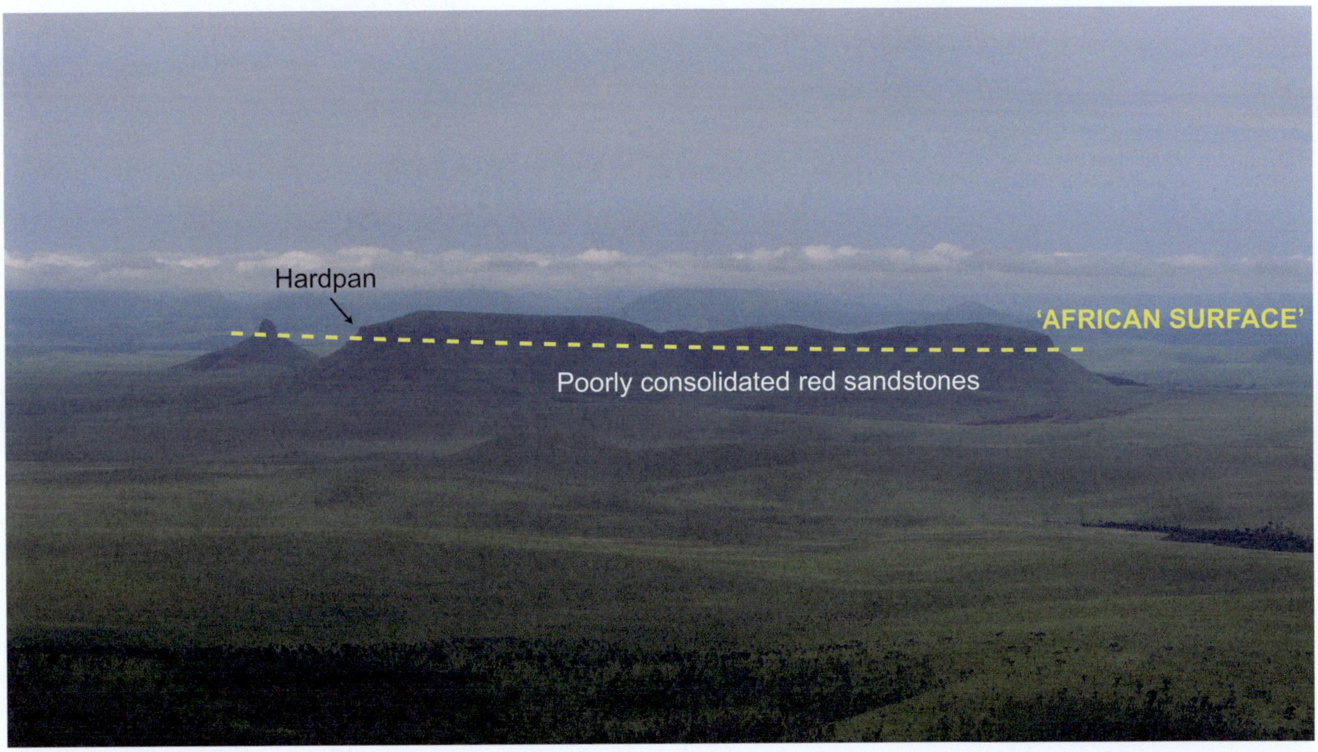

Fig. 10.8 Northernmost extension of the KP (ca. 1,100 amsl) in the southwest CB (ca. 400 amsl). A hardpan of silcrete and calcrete ('Polymorph Sandstones') overlies a regional peneplanation surface [the'African Surface'] across red sandstones of the Kwango Group. This distinct pedogenic unit at the base of the Kalahari Group can be correlated with similar calcretes in Botswana and Namibia

10.4.1 Silcrete and Calcrete Boulders on Flats and River Terraces of the Kwango Valley

The Kwango Valley is deeply incised into the Kwango Group red sandstones along the northern flank of the KP in the southwest CB (Fig. 10.2 for location). Along the scarp at the upper part of the KP, large blocks of silcretized sandstones (5–15 m in diameter) have broken-away from the lower sequence of the Kalahari Group (the 'Polymorph Sandstones'), and are displaced lower down along the slopes (Fig. 10.9a). In the lower-lying Kwango Valley, at elevations between 400 m and 600 m amsl, remnant outcrops of red sandstones form flat erosional surfaces devoid of silcrete or calcrete caps (Fig. 10.9b). These are commonly covered with residual blocks of silcrete and calcrete (0.5–5 m in diameter), and often best preserved within the

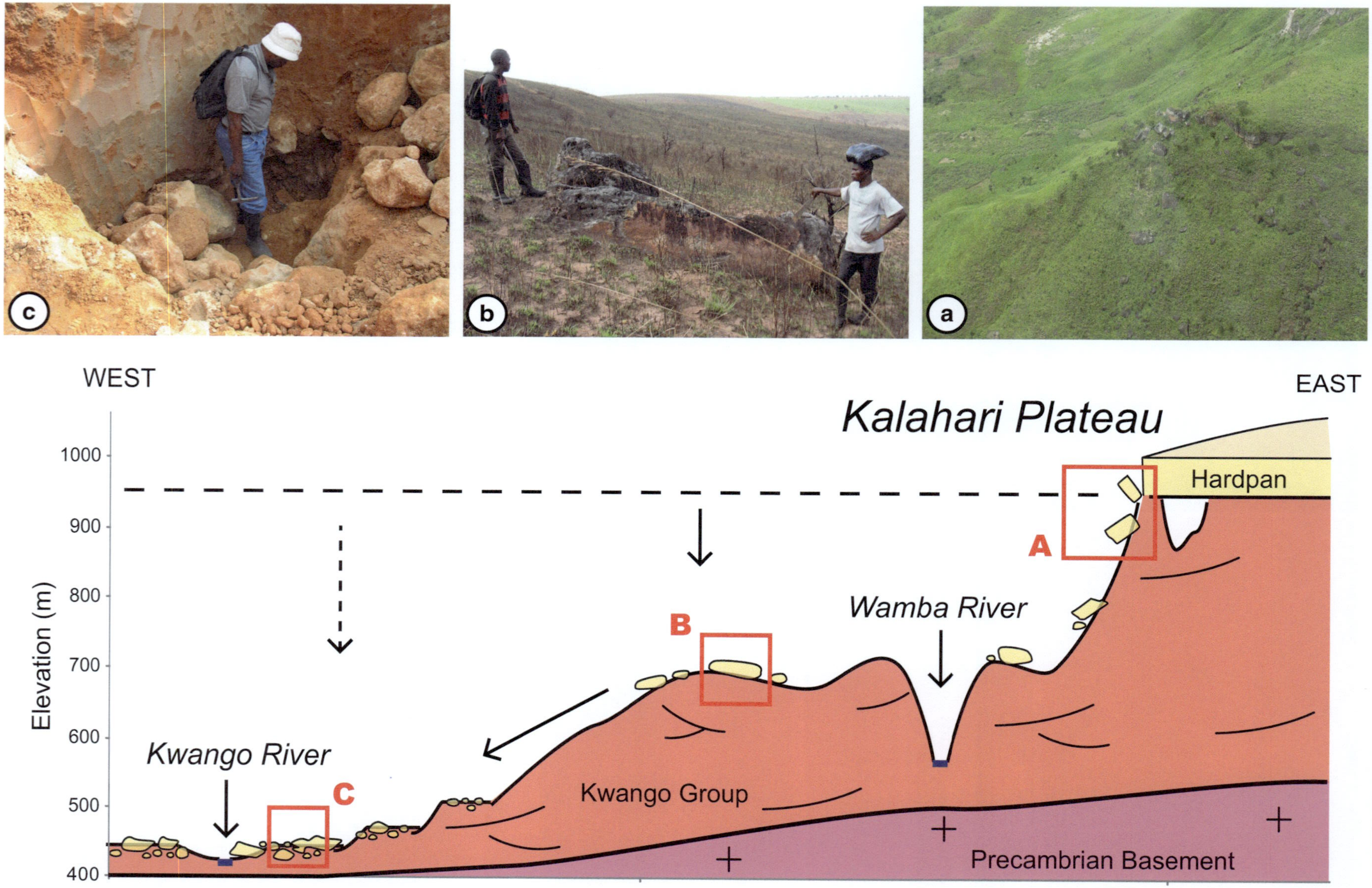

Fig. 10.9 Simplified E–W cross-section across the Kwango Valley in southwestern DRC. (**a**) Large blocks of silcrete and calcrete ('Polymorph Sandstones') fragmented from the upper part of the adjacent KP, (**b**) preserved on slopes, and (**c**) re-deposited within the Kwango river terraces

Fig. 10.10 Lupemban stone tools typically made out of silcrete rock fragments ('Polymorph Sandstones'), found at artisanal diggings within the Kwango river terrace deposits

erosional gullies and ravines, indicating fragmentation and erosion of an overlying unit of Kalahari duricrusts, like the 'Polymorph Sandstones'.

Lower down, along the large meandering Kwango River, river terraces comprise basal conglomerates (1 to 5 m thick) supporting blocks and boulders of silcrete and calcrete, ranging between 1 to 10 m in length (Fig. 10.9c). This type of deposit also appears to have originated from the fragmentation of the Kalahari 'polymorph' duricrusts and re-deposited by groundwater along the river terraces (e.g. the 'detrital model' of Goudie 1983). These giant boulders of silcrete and calcrete form good trap for alluvial diamonds (Chap. 16, this Book), and also contain some human artefacts (e.g. Fig. 10.10). The latter are stone tools mainly with elongated bifacial points constructed out of 'Polymorph Sandstones', and characteristic of the Lupemban industry in central Africa (e.g. Clark and Brown 2001), which is as old as middle Pleistocene in age (300 Ka). These indicate that deposition of river sands (and underlying calcrete boulders?) onto the Kwango terraces is a relatively recent event.

10.4.2 The Kalahari Group Across the Central Congo Basin

Based on seismic, litho- and bio-stratigraphy, the Kalahari Group has also been identified across the CB (Fig. 10.11).

Old seismic refraction surveys (Evrard 1960) mapped the sediment thicknesses at 111 stations across the entire central CB (Fig. 10.11a). On this refraction data the Kalahari sequence is imaged as a thin distinctive upper unit with a low velocity range of ca. 1,800–2,200 m/s (Linol 2013; see also Chap. 7, this Book). The reconstructed surface elevation of this uppermost main reflector shows the thickest accumulations (ca. 180–200 m) to be located beneath the Middle Congo River and Lake Ndombe, in the western part of the central CB (Fig. 10.2). It is bounded to the south by an E-W unconformity that coincides at surface with the Lukenie River (Fig. 10.11a).

In the four deep boreholes drilled in the center of the CB (Fig. 10.11b), this Kalahari succession corresponds to unconsolidated sandstones with pebbles, mainly of quartz, flint (e.g. silcretes) and sandstones, and varicolored claystones (Cahen et al. 1959, 1960; Esso-Zaire 1981a, b). Its thickness ranges from 37 m at Dekese to 242 m thick in the Gilson-1 well. In the Samba section, only the uppermost 80–110 m ('Couche' [Bed] 1 of Cahen et al. 1959) was originally attributed to the Kalahari Group. However, new logging of the cores (Linol 2013) has identified a marked erosional surface etched across red mudstones at a depth of -192 m (Fig. 10.11c), which more likely corresponds to the base of the Kalahari Group, consistent with the seismic data and the depth of the first Cretaceous fossils found in this section, at -195 m (Cahen et al. 1959). Biostratigraphically, the Kalahari Group is dated from the Eocene to the Oligo-Miocene in the lower parts (Units G1 and M1) of the

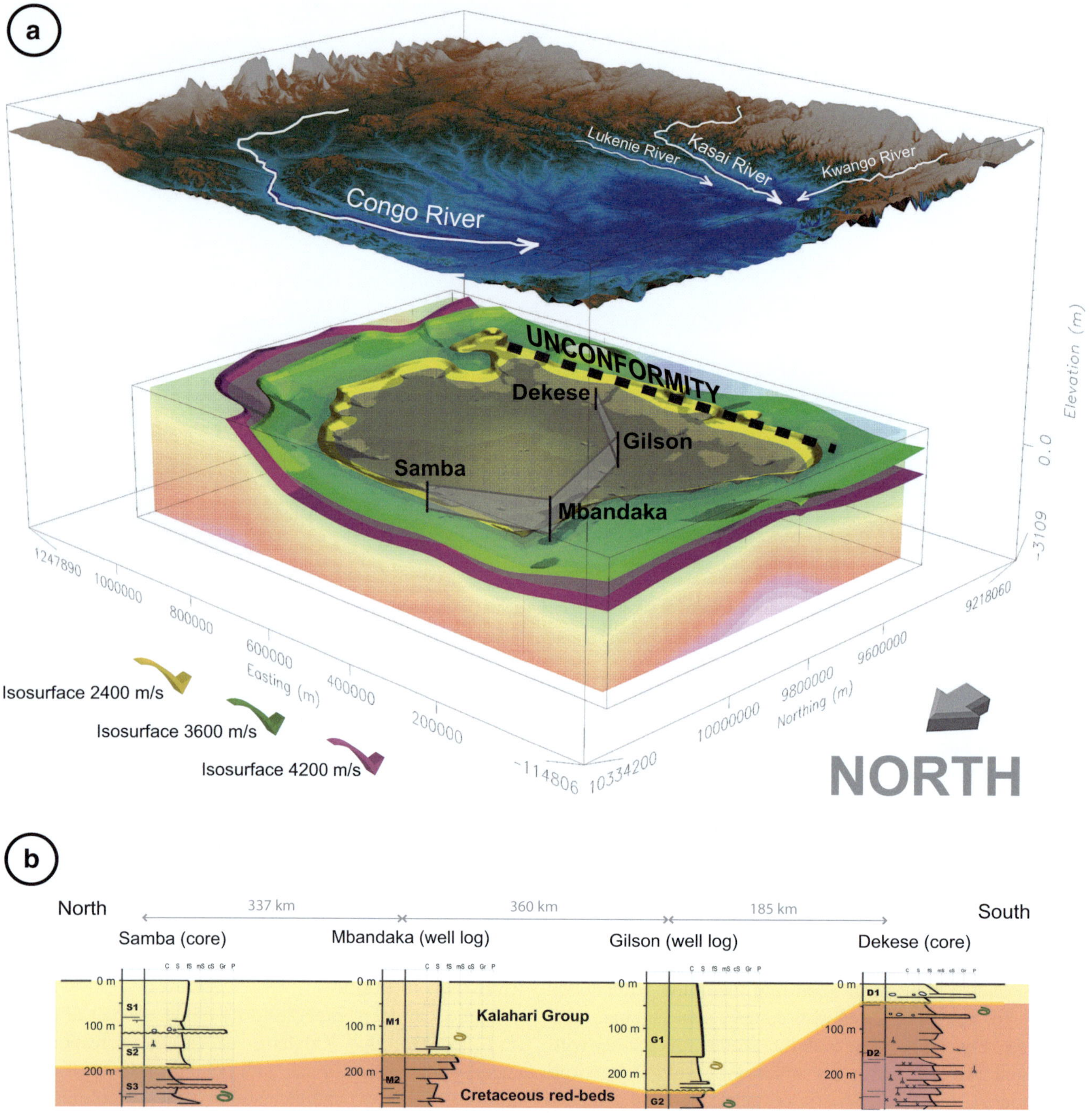

Fig. 10.11 (continued)

Gilson-1 and Mbandaka-1 wells (Fig. 10.11b), based on freshwater ostracods (*Potamocypris?* sp., *Eucypris* sp., *Cyprinotus* sp. and *Cypridopsis* sp.) and charophytes (*Grambastichara*) that can be tentatively correlated to the 'Polymorph Sandstones' (Colin 1981; Colin and Jan du Chêne 1981).

In summary, seismic and borehole correlations identify a vast extent of fluvial sediments of the Kalahari Group in the center of the CB. This succession contains relatively abundant pebbles of silcrete and calcrete, similar to those observed in the Kwango Valley, and which we interpret to have been derived also from the calcretized Kalahari carapace ('Polymorph Sandstones').

10.5 Regional Correlation and Synthesis

Across the central Kalahari (in Botswana and Namibia), a distinct, condensed sequence (10–100 m thick) of calcretes (the Nxau-Nxau and Etosha Calcrete Formations) cap a regional unconformity across Precambrian basement, the (Carboniferous-Triassic) Karoo Supergroup, and Upper

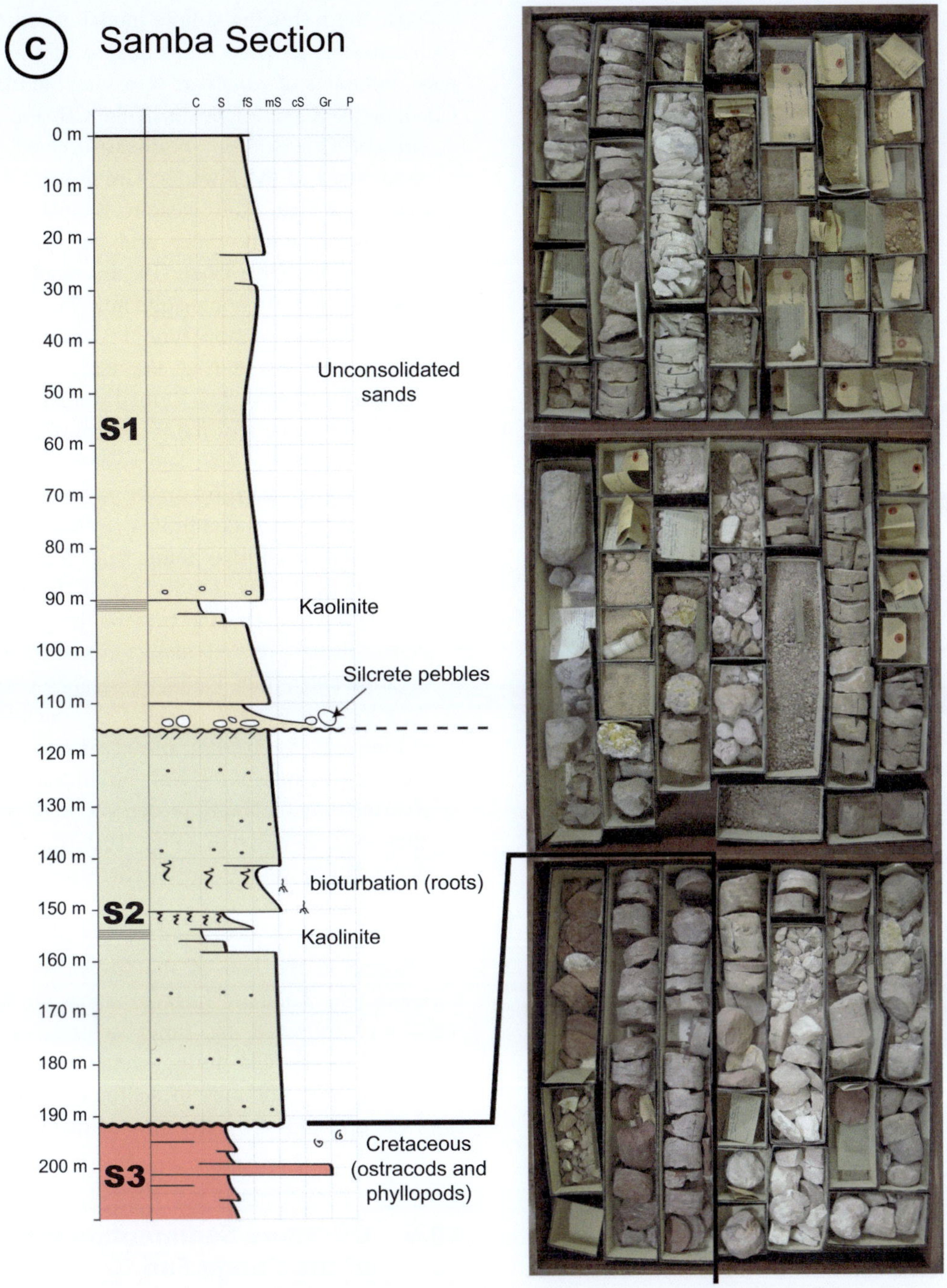

Fig. 10.11 (**a**) 3-D seismic block diagram showing a relatively flat uppermost main reflector at 2,400 m/s (surface in *yellow*), and (**b**) borehole stratigraphic correlations of the Kalahari Group across the central CB (see also Chaps. 7 and 8, this Book). (**c**) New Kalahari stratigraphic column of Samba cores (photos on the *right*) archived at the RMCA museum in Tervuren, Belgium. The Kalahari succession comprises two sequences separated by a horizon with pebbles of silcrete, similar to those observed in the Kwango Valley along the southwest margin of the CB

Cretaceous (83 Ma) kimberlites that corresponds to the 'African Surface' (e.g. du Toit 1954; King 1963; Summerfield 1983; Partridge and Maud 2000; Decker et al. 2013; Linol 2013). This basal sequence of the Kalahari Group extends continuously to the northern flank of the KP, overlapping the southwestern margin of the CB (Fig. 10.8), where it merges with the characteristic calcrete and silcrete deposits of the 'Polymorph Sandstones' (Cahen and Lepersonne 1952). Below the KP, in the Kwango Valley and across the central CB, abundant boulders and pebbles of calcrete and silcrete appear to be derived from the disintegration of this polymorph unit of duricrusts. We thus

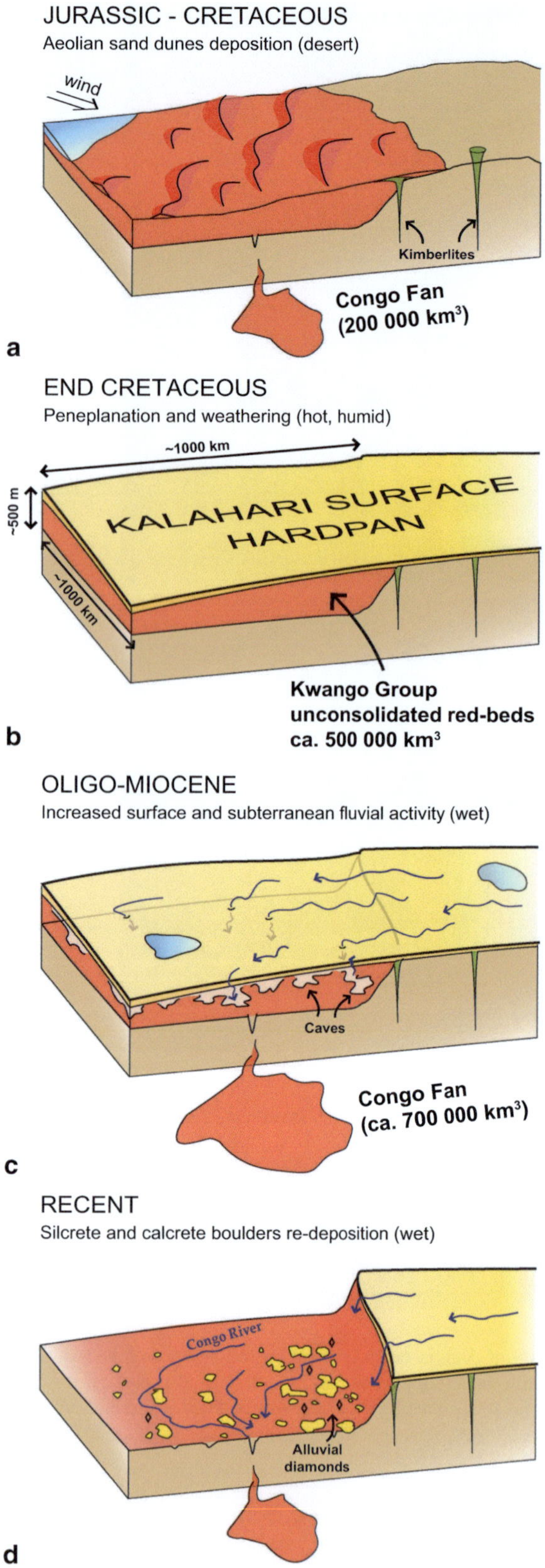

Fig. 10.12 Regional model of formation and collapse of the Kalahari hard-cap (the 'Kalahari Surface') across the CB, and re-deposition of the Kwango red-beds into the off-shore Congo Fan. A calculated volume of 0.5 million km^3 sediments removed on-shore closely matches the estimated off-shore accumulation of 0.7 million km^3 of Oligocene to Recent sediments

propose the following simple model of regional excavation (subterranean erosion) and collapse of this Kalahari carapace, remnants of which are now found in many lower-lying Cenozoic to Recent deposits of the CB (Fig. 10.12):

1. Jurassic-Cretaceous northeast-derived aeolian red sandstones of the Kwango Group were deposited across central Africa (and eastern Brazil), forming a giant Gondwana desert under arid conditions (Myers et al. 2011; Linol 2013; Chap. 13, this Book). The sandstones remained largely unconsolidated (Fig. 10.12a).
2. With the onset of more humid conditions at the end of the Cretaceous and during the early Cenozoic, Kalahari silcretes and calcretes formed a thick layer of hard-cap rocks across most of the basement of the KP and the red sandstones of the CB, and is here named the 'Kalahari Surface' (Fig. 10.12b).
3. Increasingly wet conditions, following the Paleocene-Eocene thermal maximum (e.g. Zachos et al. 2001), resulted in fluvial activity across the Kalahari Surface, excavating and eroding preferentially underlying poorly consolidated (soft red) sandstones of the Kwango Group, creating subsurface karsts topography (Fig. 10.12c).
4. Accelerated erosion in the CB removed an estimated volume of 0.5 x 10^6 km^3 of red sandstones that ultimately led to the demise of the Kalahari Surface during sustained riverine erosion and to re-deposition of giant boulders of silcrete and calcrete (Fig. 10.12d). The most recent deposits trap Stone Age artefacts and alluvial diamonds concentrates.

In summary, a prolonged period of peneplanation and weathering at the end of the Cretaceous formed a thick carapace (the Kalahari Surface) across the KP and the CB. This was followed by rapid subterranean erosion and 'flushing' of the CB during wetter times that must have generated large volumes of sediment to an off-shore sink after the opening of the South Atlantic Ocean.

10.6 Off-Shore Sedimentation History of the Congo Fan

The Congo deep-sea fan is one of the largest submarine fan systems in the world (ca. 300,000 km^2; Fig. 10.13), formed on the continental margin of west-equatorial Africa

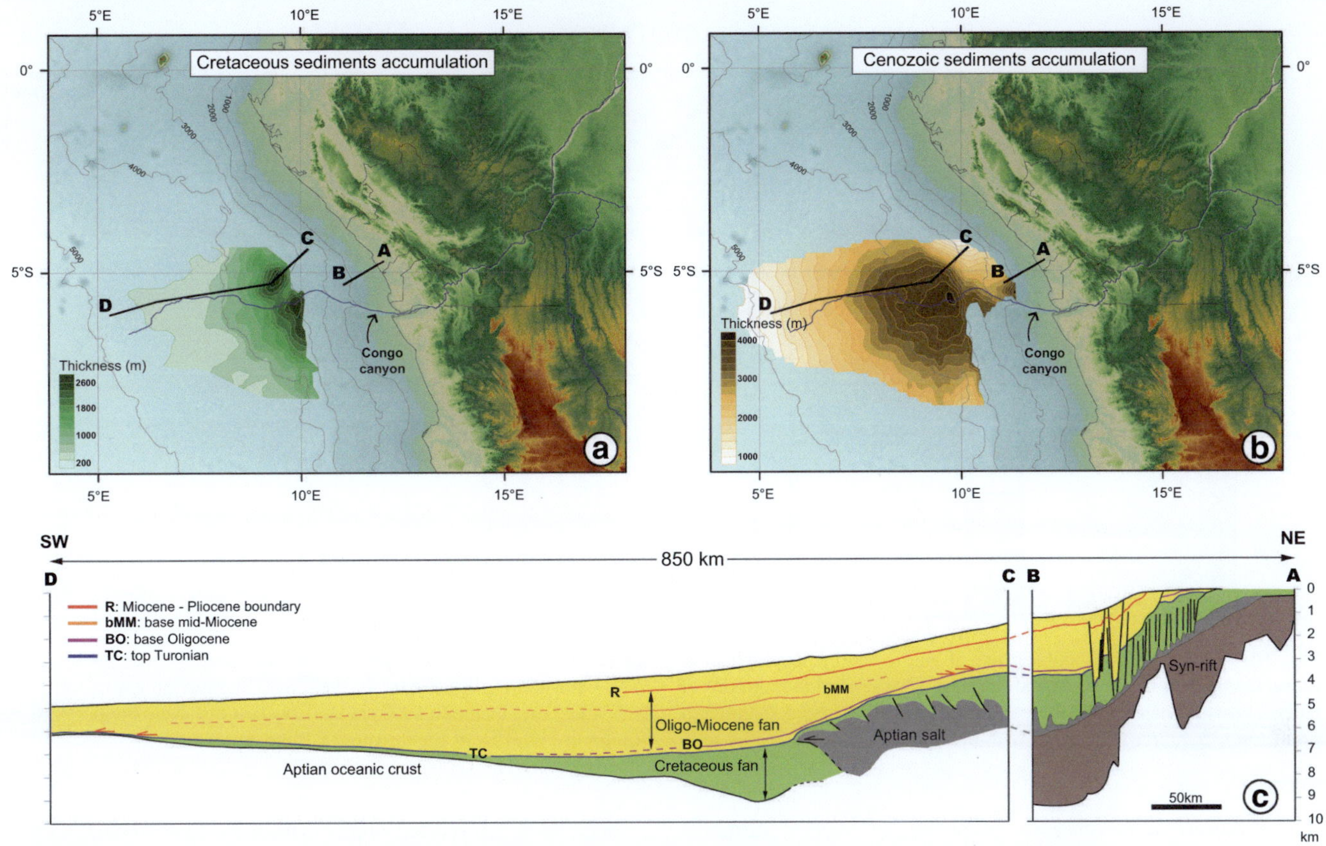

Fig. 10.13 Off-shore isopach maps of (**a**) Cretaceous and (**b**) Cenozoic sediment accumulations of the Congo Fan, and (**c**) interpreted reflection seismic profiles (from Anka and Séranne 2004). Note the relative thickness between the Cretaceous and the Oligo-Miocene deep-sea fans

following the Early Cretaceous rifting of the South Atlantic (e.g. Moulin et al. 2010; Heine et al. 2013). It is currently sourced by the Congo River (the world's second largest river by drainage area; Runge 2007), and extends over 1,000 km offshore from the shelf as far as to the abyssal plain (Savoye et al. 2000; Droz et al. 2003). The existence of a direct connection between the Congo River outlet and the submarine fan through an impressive submarine canyon is one of the most important characteristics of this system (Barbonneau et al. 2002), linking directly the marine stratigraphic record of the Congo Fan to the on-shore erosion and sediment transport dynamics of the Congo River (Anka et al. 2009).

Detailed analysis of deep-offshore seismic reflection data (Anka and Séranne 2004) has revealed that the volume of the Cenozoic fan (Oligocene–Recent) is at least 0.7×10^6 km^3. In addition, an Albian-Turonian (mid-Cretaceous) wedge that extends across the Continent-Ocean boundary beneath the Cenozoic fan (Fig. 10.13c), and whose depocenter is centered near the present-day Congo canyon, indicates that an older Cretaceous fan was also sourced by a paleo-Congo located nearby the present-day river (Anka et al. 2010). These Cretaceous deposits, together with their proximal age-equivalent sequences on the shelf, contain a minimum of 0.2 x10^6 km^3 sediments, representing almost one-third of the volume of the Cenozoic fan (Anka and Séranne 2004; Anka et al. 2009). These observations suggest that the Congo River has been one of the major off-shore sediment suppliers since the opening of the South Atlantic, and that the location of its outlet has remained relatively stable ever since. Thus, the observed variations in the off-shore sedimentation history of the Congo Fan (Fig. 10.14) are most likely primarily controlled by changes in the on-shore CB.

The two episodes of high-sediment supply, represented by the deposition of both (Cretaceous and Oligo-Miocene) submarine fan systems, are separated by a long period (Coniacian-Eocene; 89–34 Ma) of condensed sedimentation and basin starvation (Fig. 10.14). On-shore, however, apatite fission track results generally suggest accelerated erosion and denudation at that time (e.g. Brown et al. 1990; Spiegel et al. 2007; Tinker et al 2008; Turner et al. 2008). We propose that off-shore resumption of high sedimentation following this event was driven by enhanced continental erosion of the CB due to onset of karst topography beneath

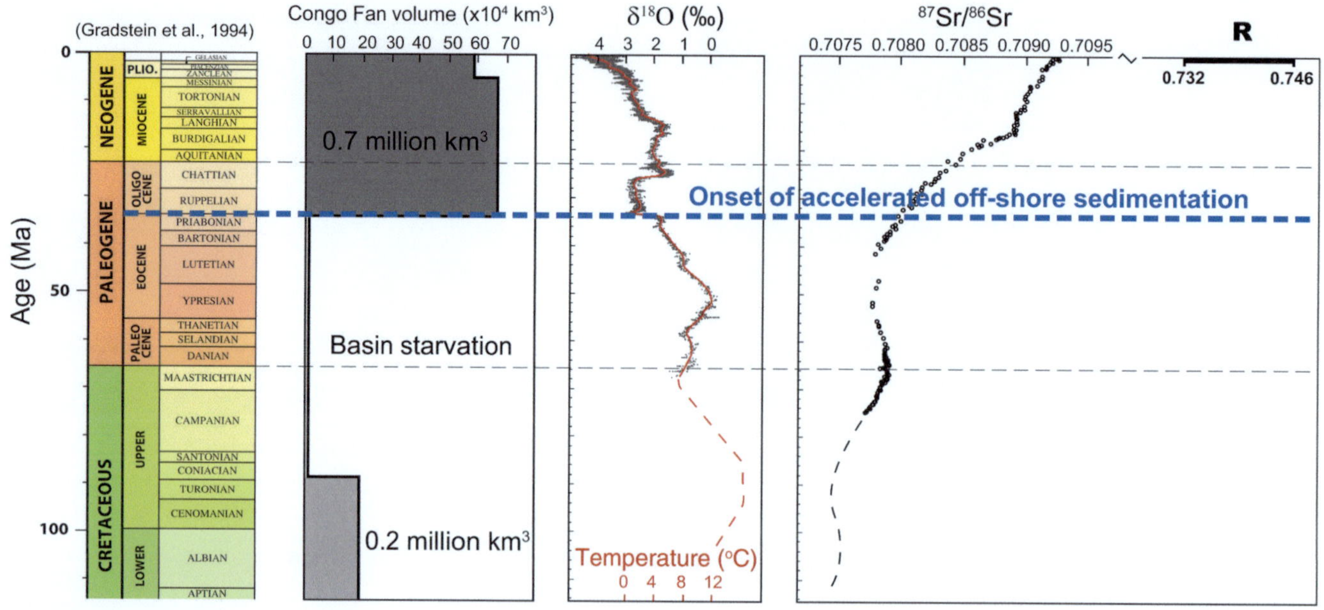

Fig. 10.14 Comparison between the off-shore sedimentation history of the Congo Fan (from Anka and Séranne 2004) and global proxies of paleo-climate. The onset of Cenozoic accelerated off-shore sedimentation (*dotted blue line*) correlates with the transition from greenhouse to ice conditions inferred from oxygen isotope records (from Zachos et al. 2001) and the rapid increase in the $^{87}Sr/^{86}Sr$ ratio of sea-water (from McCauley and DePaolo 1997) at the Eocene-Oligocene boundary. R = Range in $^{87}Sr/^{86}Sr$ of the on-shore CB red sandstones

the Kalahari duricrusts in response to wetter climatic conditions following major global cooling during the Eocene-Oligocene transition (e.g. Zachos et al. 2001), and doing so 'flushing-out' the unconsolidated red-beds (Fig. 10.12). This may also have been further enhanced by uplifts of the Angolan Highland (Walford and White 2005; Pritchard et al. 2009; Roberts and White 2010) and the EARS between 20 Ma and 40 Ma (Lavier et al. 2001; Pik et al. 2008; Roberts et al. 2012).

Conclusion

The Kalahari Group overlies a subcontinental scale peneplanation surface [the 'African Surface'] across the KP and the CB that resulted from significant exhumation and erosion-weathering following the separation of Africa from South America, during the opening of the South Atlantic Ocean. On the top of the KP this major denudation surface is capped (protected) regionally by a thin sequence of terrestrial carbonates and calcretes (e.g. the Etosha and Nxau-Nxau Calcrete Formations) covering Paleozoic and Precambrian basement rocks. This lowermost Kalahari carapace unit extends northward across the CB where it merges with the 'Polymorph Sandstones' that, in contrast, overlies a thick sequence of poorly consolidated Jurassic-Cretaceous red-beds. At the scarp marking the northernmost extent of the KP, the Kalahari carapace is fragmented and large boulders and pebbles of silcrete and calcrete have been re-deposited regionally within the lower-lying, Cenozoic to Recent alluviums of the central CB. Rapid preferential erosion of the red-beds of the CB, underneath the cap rock Kalahari Surface, led to the rapid disintegration (caving) and collapse of this carapace across an area $>1 \times 10^6$ km² (Fig. 10.12). Initiation of the erosion of the CB can be matched to the marine records of accelerated sediments accumulation associated to the onset of the Congo Fan during the Eocene-Oligocene transition (Fig. 10.14). This sudden regional 'transfer' of sediment from on-shore to off-shore may be linked to increasingly humid conditions due to rapid global cooling at that time; since our initial results of $^{87}Sr/^{86}Sr$ of the red sandstones reveal very high initial ratio and is consistent with the onset of rapid increase in the $^{87}Sr/^{86}Sr$ ratio of sea-water (e.g. McCauley and DePaolo 1997), which may not be solely related therefore to the onset of the Alpine-Himalaya uplift/erosion as previously suggested (e.g. Raymo et al. 1988).

Acknowledgments We acknowledge funding through the Inkaba yeAfrica and !Khure Africa programs, supported by the DST/NRF of South Africa. B. Linol is particularly grateful to James Bruchs from his invitations to the core shed of Tsodilo Resources Ltd. in Maun. Frank Eckardt and Tyrel Flugel are thanks for their help with Figures 10.1A and 10.2. We also thank A. Tankard and G.A. Botha for valuable reviews that improved the chapter. This is AEON contribution number 129 and Inkaba yeAfrica number 99.

References

Alonso-Zarza AM (2003) Palaeoenvironmental significance of palustrine carbonates and calcretes in the geological record. Earth Sci Rev 60:261–298

Anka Z, Séranne M (2004) Reconnaissance study of the Ancient Zaire (Congo) deep-sea fan (ZaiAngo Project). Mar Geol 209:223–244

Anka Z, Séranne M, Lopez M, Scheck-Wenderoth M, Savoye B (2009) The long-term evolution of the Congo deep-sea fan: a basin-wide view of the interaction between a giant submarine fan and a mature passive margin (ZaiAngo project). Tectonophysics 470(1–2):42–56

Anka Z, Séranne M, di Primio R (2010) Evidence of a large upper-Cretaceous depocentre across the Continent-Ocean boundary of the Congo-Angola basin. Implications for palaeo-drainage and potential ultra-deep source rocks. Mar Pet Geol 27(3):601–611

Barbonneau N, Savoye B, Cremer M, Klein B (2002) Morphology and architecture of the present canyon and channel system of the Zaire deep-sea fan. Mar Pet Geol 19(4):445–467

Batumike J, Belousova EA, Griffin WL (2007) U-Pb Dating of Perovskite: Nxau Nxau Kimberlites, Botswana. Confidential report to Tsodilo resources. GEMOC Macquarie University, Australia, 13p

Brown RW, Rust DJ, Summerfield MA, Gleadow AJW, de Wit MJC (1990) An early cretaceous phase of accelerated erosion on the south-western margin of Africa: Evidence from apatite fission track analysis and the offshore sedimentary record. Int J Radiat Appl Instr D Nucl Tracks Radiat Measur 17(3):339–350

Burrough SL, Thomas DSG, Bailey RM (2009) Mega-Lake in the Kalahari: a late pleistocene record of the Palaeolake Makgadikgadi system. Quat Sci Rev 28:1392–1411

Cahen L, Lepersonne J (1952) Équivalence entre le système du Kalahari du Congo belge et les Kalahari beds d'Afrique austral. Mémoires de la Société Belge Géologique, Paléontologique, et Hydrologique, Tervuren (Belgique), Série in-8, Sciences Géologiques 4, 63 p

Cahen L, Ferrand JJ, Haarsma MJF, Lepersonne J, Vebeek T (1959) Description du Sondage de Samba. Annales du Musée Royal du Congo belge, Tervuren (Belgique), Série in-8, Sciences géologiques 29, 210 p

Cahen L, Ferrand JJ, Haarsma MJF, Lepersonne J, Vebeek T (1960) Description du Sondage de Dekese. Annales du Musée Royal du Congo belge, Tervuren (Belgique), Série in-8, Sciences géologiques 34, 115 p

Chorowicz J (2005) The East African rift system. J Afr Earth Sci 43(1–3):379–410

Clark JD, Brown KS (2001) The Twin Rivers Kopje, Zambia: stratigraphy, fauna, and artefact assemblages from the 1954 and 1956 excavations. J Archaeol Sci 28:305–330

Colin J-P (1981) Paleontological study of the Esso/Texaco well Gilson-1, Zaire. Unpublished report EPR-E.WA19.81

Colin J-P (1994) Mesozoic-Cenozoic lacustrine sediments in the Zaïre Interior Basin. In: Gierlowski-Kordesch E, Kelts K (eds) Global geological record of lake basins, vol 4. Cambridge University Press, Cambridge, pp 31–36

Colin J-P, Jan du Chêne J (1981) Paleontological study of the Esso/Texaco well Mbandaka-1, Zaire. Unpublished report EPR-E.WA15.81

De Ploey J (1968) Sédimentologie et origine des sables de la série des sables ocre et de la série des 'grès polymorphes' (Système du Kalahari) au Congo Occidental. Annales du Musée Royal de l'Afrique centrale, Tavuren (Belgique), Série in-8, Sciences géologiques 61, 72 p

De Wit MJ (2007) The Kalahari Epeirogeny and climate change: differentiating cause and effect from core to space. S Afr J Geol 110(2–3):367–392

De Wit MJC (2013a) The Xaudum Kimberlite province straddling the southern margin of the Angolan Craton. Abstract, 24th Colloquium of African Geology, Addis Ababa, Ethiopia, 8–14 Jan

De Wit MJC (2013b) The Xaudum kimberlite province in the southern part of the Congo Craton and the age of the basal Kalahari Group sediments. Abstract TopoAfrica Meeting, Saasveld, South Africa, 21–25 Jan

Decker JE, Niedermann S, de Wit MJ (2013) Climatically influenced denudation rates of the southern African plateau: clues to solving a geomorphic paradox. Geomorphology 190:48–60

Droz L, Marsset T, Ondréas Lopez M, Savoye B, Spy-Anderson F-L (2003) Architecture of an active mud-rich turbidite system: the Zaire Fan (Congo-Angola margin southeast Atlantic): results from ZaïAngo 1 and 2 cruises. AAPG Bull 87(7):1145–1168

Du Plessis PI, Le Roux JP (1995) Late Cretaceous alkaline saline lake complexes of the Kalahari Group in northern Botswana. J Afr Earth Sci 20(1):7–15

Du Toit A (1954) The geology of South Africa, 3rd edn. Oliver and Boyd, Edinburgh, p 611 pp

Eckardt FD, Barker C, Meadows M (2013) Landscape inventories and remote sensing of southern Africa. In: Holmes P, Meadows M (eds) Southern African geomorphology: recent trends and new directions. Sun Press, Bloemfontein, pp 365–391

Esso-Zaire SARL (1981a) Geological completion report: Gilson-1. Unpublished report

Esso-Zaire SARL (1981b) Geological completion report: Mbandaka-1. Unpublished report

Evrard P (1960) Sismique. Annales du Musée Royal du Congo belge, Tervuren (Belgique), Série in-8, Sciences géologiques 33, 87 p

Giresse P (2005) Mesozoic-Cenozoic history of the Congo Basin. J Afr Earth Sci 43:301–315

Goudie AS (1983) Calcretes. In: Goudies AS, Pye K (eds) Chemical sediments and geomorphology. Academic, London, pp 93–131

Grekoff N (1958) Ostracodes du Bassin du Congo. III. Tertiaire. Annales du Musée Royal de l'Afrique centrale, Tervuren (Belgique), Série in-8, Sciences géologiques 22, 36 p

Haddon I (1999) Isopach map of the Kalahari Group, at a scale of 1: 2.500.000. Council for Geoscience, Pretoria, South Africa

Haddon I (2000) Kalahari group sediments. In: Partridge TC, Maud RR (eds) The Cenozoic of Southern Africa, vol 40, Oxford monographs on geology and geophysics. Oxford University Press, New York, pp 173–181

Haddon IG, McCarthy TS (2005) The Mesozoic-Cenozoic interior sag basins of Central Africa: the Late-Cretaceous - Cenozoic Kalahari and Okavango basins. J Afr Earth Sci 43(1–3):316–333

Haddon IG, Roos HM (2001) Sub-Kalahari geological map, at a scale of 1: 2.500.000. Council for Geoscience, Pretoria, South Africa

Heine C, Zoethout J, Müller RD (2013) Kinematics of the South Atlantic rift. Solid Earth 4:215–253

Holmgren K, Shaw P (1996) Palaeoenvironmental reconstruction from near-surface pan sediments: an example from Lebatse Pan, southeast Kalahari, Botswana. Geogr Ann 79(1–2):83–93

Huntsman-Mapila P, Ringrose S, Mackay aW, Downey WS, Modisi M, Coetzee SH, Tiercelin J-J, Kampunzu aB, Vanderpost C (2006) Use of the geochemical and biological sedimentary record in establishing palaeo-environments and climate change in the Lake Ngami basin, NW Botswana. Quat Int 148(1):51–64

Hürkamp K, Völkel J, Heine K, Bens O, Leopold M, Winkelbauer J (2011) Late Quaternary environmental changes from aeolian and fluvial geoarchives in the southwestern Kalahari, South Africa: Implications for past African climate dynamics. South Afr J Geol, Special Inkaba yeAfrica 114:459–474

Jourdan F, Féraud G, Bertrand H, Kampunzu AB, Tshoso G, Watkeys MK, Le Gall B (2007) Karoo large igneous province: Brevity, origin, and relation to mass extinction questioned by new 40Ar/39Ar age data. Geology 33(9):745–748

Kampunzu AB, Ringrose S, Huntsman-Mapila P, Harris C, Vink BW, Matheson W (2007) Origins and palaeo-environments of Kalahari duricrusts in the Moshaweng dry valleys (Botswana) as detected by major and trace element composition. J Afr Earth Sci 48:199–221

King LC (1963) South African scenery: a textbook of geomorphology, 3rd edn. Olivier and Boyd, Edinburg, p 308p

Lavier L, Steckler M, Brigaud F (2001) Climatic and tectonic control on the Cenozoic evolution of the West African margin. Mar Geol 178:63–80

Lepersonne J (1945) La Stratigraphie du Système du Kalahari et du Système du Karoo au Congo occidental. Bulletin du Service Géologique du Congo belge et Ruanda-Urundi 1:27–50

Leriche M (1927) Les fossils des grès polymorphes (couches de Lubilash) aux confins du Congo et de l'Angola. Annales de la Société Géologique de Belgique 50:44–51

Linol B (2013) Sedimentology and sequence stratigraphy of the Congo and Kalahari Basins of south-central Africa and their evolution during the formation and break-up of West Gondwana. PhD thesis, Nelson Mandela Metropolitan University, 375 p

Linol B, de Wit MJ, de Wit MCJ, Thorose E, Bruchs J, Eckardt F, Guillocheau F (2009) Stratigraphy of calcretes and related climate proxies from the Kalahari Plateau: New clues to 100 Ma of climate fluctuations across south-central Africa. Abstract, 5th EGU, Cape Town, 12–16 Jan

Maufe HB (1936) New sections in the Kalahari beds at the Victoria Falls, Rhodesia. J R Anthropol Inst Great Britain Ireland 66: 348–368

McCauley SE, DePaolo DJ (1997) The Marine 87Sr/86Sr and δ18O Records, Himalayan Alkalinity Fluxes, and Cenozoic climate models. In: Ruddiman WF (ed) Tectonic uplift and climate change. Plenum Press, New York, pp 427–467

McFarlane MJ, Eckardt FD, Coetzee SH, Ringrose S (2010) An African surface weathering profile in the Kalahari of North West Ngamiland, Botswana: processes and products. Zeitschrift für Geomorphologie 54(3):273–303

Miller R (2008) The geology of Namibia, in three volumes. Geological Survey, Windhoek, Namibia

Miller R, Pickford M, Senut B (2010) The geology, palaeontology and evolution of the Etosha Pan, Namibia: implications for terminal Kalahari deposition. S Afr J Geol 113:307–334

Money NJ (1972) An outline of the geology of western Zambia. Record Geol Surv Zambia 12:103–123

Moore AE, Cotterill FPD, Eckardt FD (2012) The evolution and ages of Makgadikgadi palaeo-lakes: consilient evidence from Kalahari drainage evolution south-central Africa. S Afr J Geol 115:385–413

Moulin M, Aslanian D, Unternehr P (2010) A new starting point for the South and Equatorial Atlantic Ocean. Earth-Sci Rev 98(1–2):1–37

Myers TS, Tabor NJ, Jacobs LL (2011) Late Jurassic paleoclimate of Central Africa. Palaeogeogr Palaeoclimatol Palaeoecol 311 (1–2):111–125

Nash DJ, McLaren SJ (2003) Kalahari valley calcretes: their nature, origins, and environmental significance. Quat Int 111:3–22

Partridge TC, Maud RR (2000) Macro-scale geomorphic evolution of southern Africa. In: Partridge TC, Maud RR (eds) The Cenozoic of Southern Africa. Oxford University Press, Oxford, pp 3–18

Passarge S (1904) Die Kalahari. Dietrich Reimer, Berlin, p 823 p

Pik R, Marty B, Carignan J, Yirgu G, Ayalew T (2008) Timing of East African Rift development in southern Ethiopia: implication for mantle plume activity and evolution of topography. Geology 36: 167–170

Polinard E (1932) Découverte des gisements fossilifères d'eau douce sur les versants de la Lubudi au Katanga méridional. Annales de la Société Géologique de Belgique 55:63–81

Pritchard D, Roberts GG, White NJ, Richardson CN (2009) Uplift histories from river profiles. Geophys Res Lett 36:L24301. doi:10.1029/2009GL040928

Raymo ME, Ruddiman WF, Froelich PN (1988) Influence of late Cenozoic mountain building on ocean geochemical cycles. Geology 16:649–653

Rethemeyer J, Fülöp R-H, Höfle S, Wacker L, Heinze S, Hajdas I, Patt U, König S, Stapper B, Dewald A (2012) Status report on sample preparation facilities for C-14 analysis at the new CologneAMS center. Nucl Instr Meth Phys Res, B 294, p.168–172. doi:10.1016/j.nimb.2012.02.012

Ringrose S, Downey B, Genecke D, Sefe F, Vink B (1999) Nature of sedimentary deposits in the western Makgadikgadi basin, Botswana. J Arid Environ 43:375–397

Roberts G, White NJ (2010) Estimating uplift rates histories from river profiles using African examples. J Geophys Res-Solid Earth 115, B02406

Roberts EM, Stevens NJ, O'Connor PM, Dirks PHGM, Gottfried MD, Clyde WC, Amstrong RA, Kemp AIS, Hemming S (2012) Initiation of the western branch of the East African Rift coeval with the eastern branch. Nat Geosci 5:289–294

Runge J (2007) The Congo river, Central Africa. In: Gupta A (ed) Large river: geomorphology and management. Wiley, New York, pp 293–309

SACS (South African Committee for Stratigraphy) (1980) Stratigraphy of South Africa. Part 1(comp. L.E. Kent). Lithostratigraphy of the Republic of South Africa, South West Africa/Namibia, and the Republics of Bophuthatswana, Transkei and Venda. Handbook geological survey of South Africa, vol 8. 690 p

Savoye B et al (2000) Structure et évolution récent de l'éventail turbiditique du Zaire: premiers résultats scientifiques des missions d'exploration Zaiango 1&2 (marge Congo-Angola). Comptes rendus de l'Académie des Sciences de Paris 331:221–220

Spiegel C, Kohn BP, Belton DX, Gleadow AJW (2007) Morphotectonic evolution of the central Kenya rift flanks: implications for late Cenozoic environmental change in East Africa. Geology 35(5): 427–430

Summerfield MA (1983) Silcrete as a palaeoclimatic indicator: evidence from southern Africa. Palaeogeogr Palaeoclimatol Palaeoecol 41:65–79

Tanner LH (2010) Continental carbonates as indicators of Paleoclimate. In: Alonzo-Zarza AM, Tanner LH (eds) Carbonates in continental settings: geochemistry, diagenesis and applications. Developments in sedimentology, vol 62. Elsevier, Amsterdam, pp 179–214

Thomas DSG, Shaw PA (1990) The deposition and development of the Kalahari Group sediments, Central southern Africa. J Afr Earth Sci 10(1–2):187–197

Thomas DSG, Shaw PA (1993) The evolution and characteristics of the Kalahari, southern Africa. J Arid Environ 25:97–108

Thomas DS, Brook G, Shaw P, Bateman M, Haberyan K, Appleton C, Nash D, McLaren S, Davies F (2003) Late Pleistocene wetting and drying in the NW Kalahari: an integrated study from the Tsodilo Hills, Botswana. Quat Int 104(1):53–67

Tinker J, de Wit M, Brown R (2008) Mesozoic exhumation of the southern Cape, South Africa, quantified using apatite fission track thermochronology. Tectonophysics 455(1–4):77–93

Turner JP, Green PF, Holford SP, Lawrence SR (2008) Thermal history of the Rio Muni (West Africa)-NE Brazil margins during continental breakup. Earth Planet Sci Lett 270:354–367

Walford HL, White NJ (2005) Constraining uplift and denudation of West African continental margin by inversion of stacking velocity data. J Geophys Res-Solid Earth 110, B04403

Wanke H, Wanke A (2007) Lithostratigraphy of the Kalahari Group in northeastern Namibia. J Afr Earth Sci 48:314–328

Wright VP, Tucker ME (1991) Calcretes. Blackwell Scientific Publications, New York, p 360p

Zachos J, Pagani M, Sloan L, Thomas E, Billups K (2001) Trends, rhythms, and aberrations in global climate 65 Ma to present. Science 29:686–693

Part III

Geodynamics, Tectonic Evolution and Basin Models

11 Multiphase Phanerozoic Subsidence and Uplift History Recorded in the Congo Basin: A Complex Successor Basin

Bastien Linol, Maarten J. de Wit, Francois Guillocheau, Cecile Robin, and Olivier Dauteuil

11.1 Introduction

The stratigraphic record of the Congo Basin (CB) represents a long (ca. 550 million years) history of sedimentation, erosion and preservation in the heartland of Gondwana and, following Gondwana break-up, in central Africa. This long-lived Phanerozoic basin features large in many tectonic models of Africa, yet paradoxically it remains very poorly understood because of limited recent field and geophysical data acquired in this politically unstable region, especially the war-torn Democratic Republic of Congo (DRC), host to the greater part of the basin. In particular, there is considerable controversy about the origin and influence of its associated long wavelength (ca. 1,500 km) negative gravity anomaly (Fig. 11.1). This anomaly is commonly linked to a long-lived dynamic 'cold spot' in the underlying mantle and related to convective down-welling, resulting in a long history of slow and continuous subsidence of the CB since the late Precambrian (Hartley and Allen 1994; Downey and Gurnis 2009; Crosby et al. 2010; Kadima et al. 2011; Buiter et al. 2012). But, these studies are based on meagre information of the basin fill and rely on the stratigraphy from the 1950s and 1970s that, as will be shown here, is frequently inadequate and often inaccurate (Linol 2013; see Chaps. 7, 8, 9 and 10, this Book).

In this contribution, we use revised stratigraphic and sedimentological data of the four historic deep boreholes (each ca. 2 to 4.5 km in depth) drilled near the center of the basin to calculate a new first-order model for the subsidence and uplift history across this vast region of central Africa (Fig. 11.1 for location). To achieve this, the four borehole-sections are backstripped using a procedure especially adapted to this terrestrial (non-marine) basin, as described below. The results reveal a more complex (multiphase) geodynamic evolution of the CB during the Phanerozoic, including distinctive phases of rapid subsidence separated by episodic uplifts. We then propose a complex successor basin model for the CB during and following the period between the final amalgamation of Gondwana (and Pangea) supercontinent and its subsequent break-up resulting in the successive openings of the Indian and South Atlantic Oceans.

B. Linol (✉)
AEON-ESSRI (African Earth Observatory Network – Earth Stewardship Science Research Institute), Nelson Mandela Metropolitan University, Port Elizabeth, South Africa

Geological Sciences, Nelson Mandela Metropolitan University, Port Elizabeth, South Africa
e-mail: bastien.aeon@gmail.com

M.J. de Wit
AEON-ESSRI (African Earth Observatory Network – Earth Stewardship Science Research Institute), Nelson Mandela Metropolitan University, Port Elizabeth, South Africa
e-mail: maarten.dewit@nmmu.ac.za

F. Guillocheau • C. Robin • O. Dauteuil
Géociences-Rennes, UMR 6118 Université de Rennes 1 – CNRS, OSUR, Université de Rennes 1, Campus de Beaulieu, 35042 Rennes cedex, France
e-mail: francois.guillocheau@univ-rennes1.fr; cecile.robin@univ-rennes1.fr; olivier.dauteuil@univ-rennes1.fr

11.2 Stratigraphic Record and Subsidence Analysis

Based on stratigraphic data, the evolution of the CB is analyzed by modeling its subsidence and uplift history. The vertical displacements of the basin basement are defined as subsidence, which increases when the basin sinks or decreases when it uplifts, resulting from different processes that can generate vertical movement: thermal cooling, sediment (and water) loading and tectonics (e.g. McKenzie 1978; Cochran 1983; Allen and Allen 1990; Vail et al. 1991).

The total subsidence of the basin is reflected in its sedimentary infill (sequences) that becomes progressively more compacted as it is buried, and interrupted by phases of erosion during uplift that create regional unconformities.

M.J. de Wit et al. (eds.), *Geology and Resource Potential of the Congo Basin*, Regional Geology Reviews,
DOI 10.1007/978-3-642-29482-2_11,

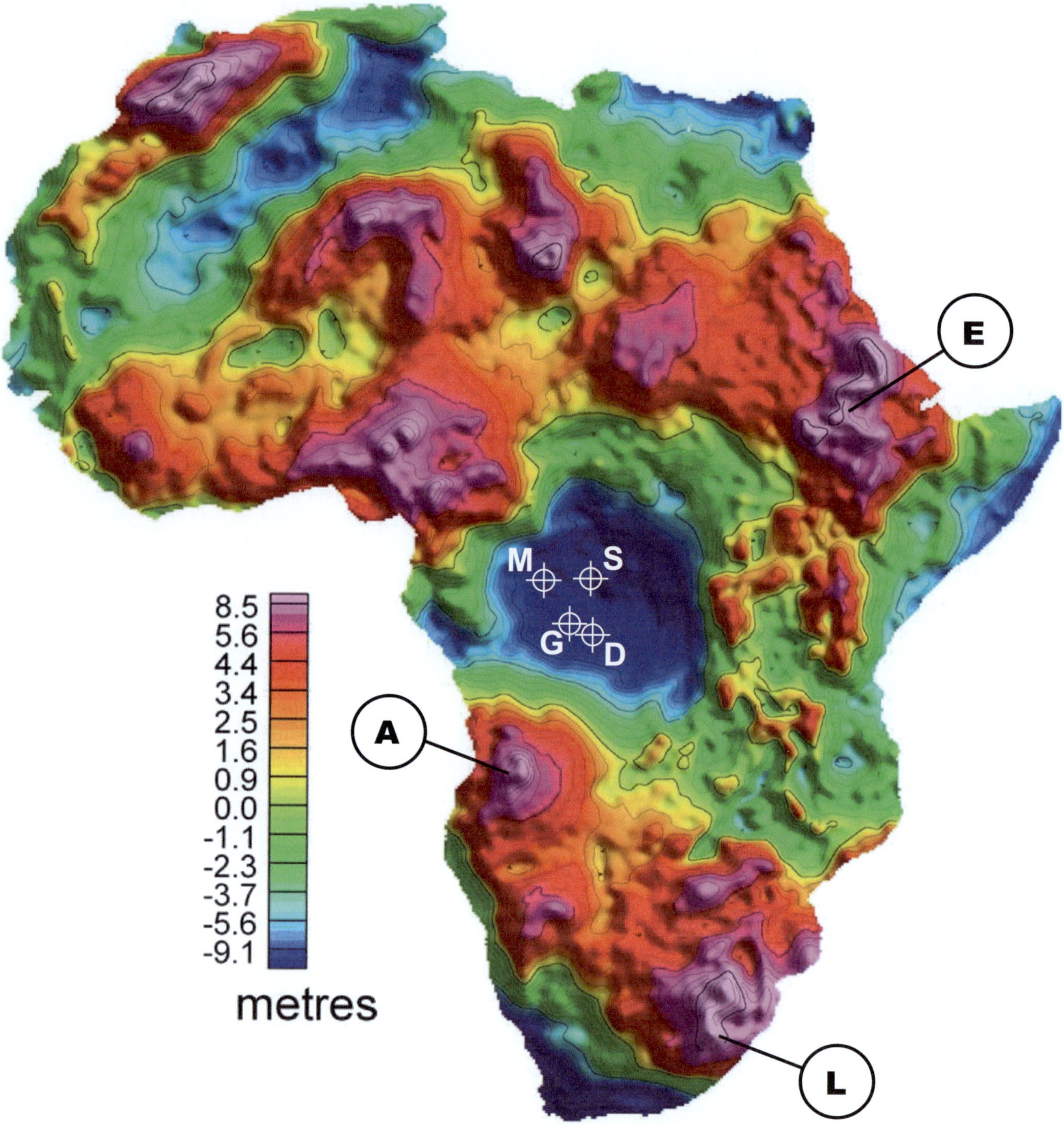

Fig. 11.1 Geoid gravity anomaly of Africa showing the large depression of the CB (from Doucouré and de Wit 2003). Also shown are the locations of the four deep boreholes: (S) Samba, (D) Dekese, (G) Gilson and (M) Mbandaka, drilled in the 1950s and 1970s near the center of the basin. A, Angolan Highland; E, Ethiopian Highland; L, Lesotho Highland

During burial, the expulsion of fluid and the increase in temperature and pressure reduce the porosity, and consequently the thickness of the stratigraphic sequences by up to 50 % to 70 % (Allen and Allen 1990). This history can be best-reconstructed using backstripping techniques (Allen and Allen 2005). We extend this method to additionally take into account episodes of hiatus (no deposition) and erosion (material removal) in continental domain.

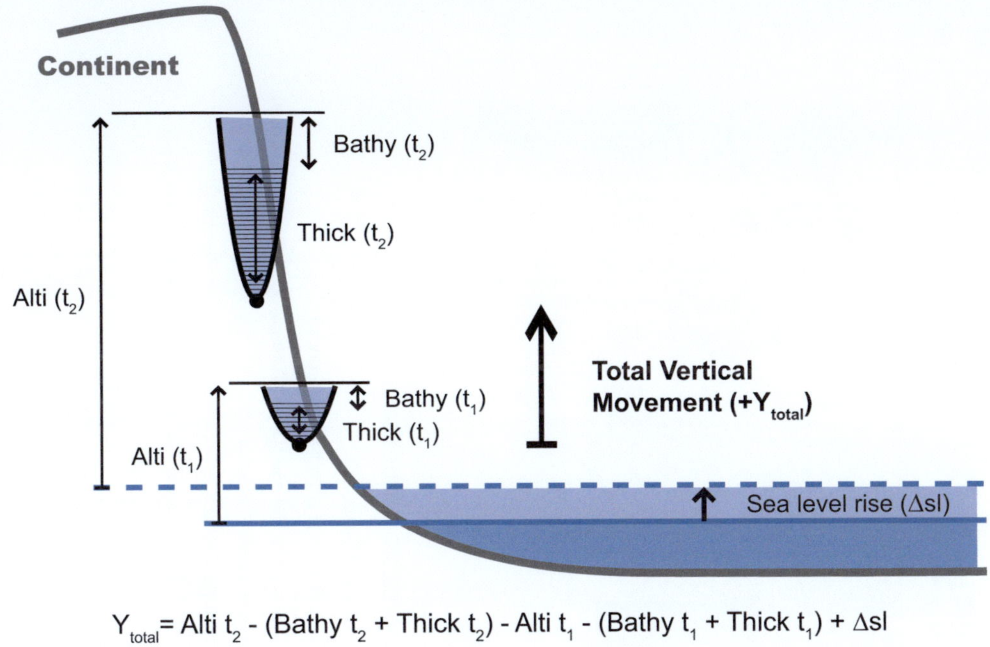

Fig. 11.2 Method of calculating the total vertical movement of a terrestrial basin (from Linol 2013). Subsidence or uplift (Y_{total}) equals the variation of altitude of the basin's surface (Alti), sediment thickness (Thick) and bathymetry (Bathy), corrected to known eustatic sea level fluctuations (Δsl)

11.2.1 Method of Backstripping

Backstripping corrects for sediment compaction at the respective depths and times of deposition of each stratigraphic unit, and is then adjusted to sea level fluctuations and paleo-bathymetry to quantify a total subsidence of the basin basement. By removing the sediment and water loadings, backstripping also calculates the tectonic subsidence that corresponds to the vertical displacement of the basin induced by tectonic and thermal processes only (e.g. Vail et al. 1991; Allen and Allen 2005). Although this method is well-known for marine settings, no such procedure exists in the literature for terrestrial sequences where overburden removal or erosion occurred. Here, we first describe a new method of backstripping especially adapted for the continental CB (Linol 2013).

In continental domain, the total subsidence of a basin integrates variations of accommodation (i.e. the space available for sediments accumulation) and the paleo-altitudes of the basin surface, i.e. base level above sea level (Fig. 11.2).

The accommodation simply corresponds to the thickness of the deposited (decompacted) sediments plus the paleo-bathymetry. The latter is evaluated from the sedimentological data. For example, tidal bundles form at shallow depth, less than 10–20 m (e.g. Dalrymple 1992). At greater depths, such as for turbidites, the paleo-bathymetry is more difficult to assess, but in a lacustrine context it normally ranges from 0 m to about 2,000 m (e.g. the Black Sea) and, in general, anoxic layers start to form around 100–200 m depth (e.g. Murray et al. 2007). By contrast, paleo-bathymery is nil if the basin emerges and records terrestrial sedimentation (e.g. aeolian dunes), fluvial sequences and paleosols.

11.2.1.1 Decompaction

The amount of decompaction is calculated using empirical porosity-depth relationships, linked to the lithological properties of each stratigraphic unit (c.f. Allen and Allen 2005). In this case, we use field estimation of the proportion of clay (and silt) compared to sand (Fig. 11.3) to determine the value of these physical parameters: porosity-depth coefficient, surface porosity and dry sediment density (see Tables 11.1–11.8). The decompacted sedimentary thicknesses are then calculated from the general decompaction equation:

$$\begin{aligned} \mathrm{EpS} &= \mathrm{z2 - z1} - (\Phi_0/\mathrm{c}) * (\exp(\text{-c.z}_1) - \exp(\text{-c.z}_2)) \\ &\quad + (\Phi_0/\mathrm{c}) * (\exp(\text{-c.z}'_1) - \exp(\text{-c.z}'_2)) \end{aligned} \tag{11.1}$$

where EpS is the thickness of the stratigraphic unit before compaction; z_1 and z_2 the present base and top of the unit (after compaction), and z'_1 and z'_2 before compaction, respectively; c is the porosity-depth coefficient; Φ_0 is the surface porosity, and Φ is the porosity at depth z. The porosity exponentially decreases with depth as:

$$\Phi = \Phi_0 \exp(\text{-c.z}) \tag{11.2}$$

11.2.1.2 Water and Sediment Loadings

The water and sediment loadings are calculated in using a standard 1D Airy-compensated model. The density contrast of water and sediment reflects the density of the mantle:

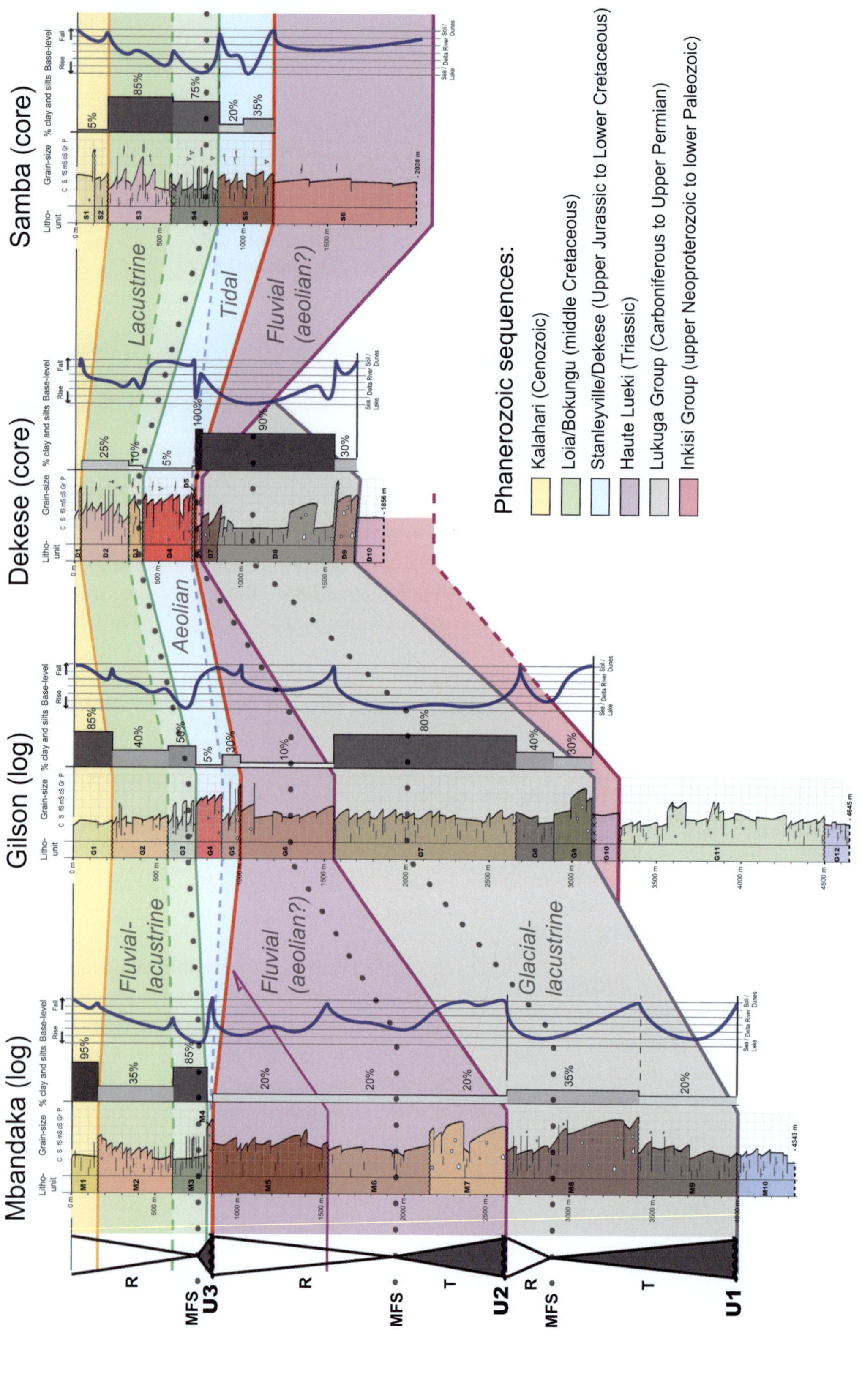

Fig. 11.3 Sequence stratigraphic correlations (*T* Transgression, *R* Regression, *MFS* Maximum flooding surface) and paleoenvironmental data of the four deep borehole-sections in the center of the CB (Fig. 11.1 for locations). U1, U2 and U3 mark major unconformities determined from analysis of seismic (Linol 2013; see Chap. 7, this Book)

Table 11.1a Stratigraphic data of the Samba section

Unit	Age (Ma)	Base depth (km)	Top depth (km)	Porosity-depth coeff. (km^{-1})	Surface porosity (%)	Dry sediment density (km/m^3)	Bathymetry (km)	Eustasy (km)	Density of deposit fluid (km/m^3)	Elevation (km)
Hiatus	158	1.167	1.167	0.270	0.490	2,650	0.000	0.130	1	−0.330
S5 lower	153	1.167	0.995	0.354	0.539	2,685	0.020	0.160	1,030	0.000
S5 upper	149	0.995	0.845	0.318	0.518	2,670	0.010	0.160	1,000	0.000
Overburden	143	0.845	0.845	0.282	0.497	2,655	0.000	0.120	1	−0.040
S4	100	0.845	0.565	0.450	0.595	2,725	0.100	0.260	1,000	−0.040
S3	93.6	0.565	0.192	0.474	0.609	2,735	0.020	0.280	1,000	−0.020
Overburden	65.5	0.192	0.192	0.342	0.532	2,680	0.010	0.230	1,000	−0.070
S1 + S2	33.9	0.192	0.001	0.282	0.497	2,655	0.000	0.220	1	−0.150
Hiatus	0	0.001	0.001	0.270	0.490	2,650	0.000	0.000	1	−0.370

Table 11.2 Eroded units in the Samba section

Overburden unit age (Ma)	Erosion age (Ma)	Eroded thickness (km)	Eustasy (km)
143	112	0.300	−0.040
65.5	55.8	0.300	−0.070

$$\text{IsosSed} = \text{EpS} * (\rho\text{M- }\rho\text{PS}) / (\rho\text{M- }\rho\text{Mil}) \quad (11.3)$$

$$\text{IsosWat} = \text{PaleoBat} * (\rho\text{Mil}) / (\rho\text{M- }\rho\text{Mil}) \quad (11.4)$$

where IsosSed and IsosWat are the sediment and water loadings, respectively, and PaleoBat is the paleo-bathymetry; ρMil: the density of the deposit fluid (1,000 kg/m^3 for fresh water, 1,030 kg/m^3 for marine water, 1 kg/m^3 for air); ρM: the density of the mantle (3,300 kg/m^3); and ρPS: the density of the considered sedimentary section.

11.2.1.3 Overburden

Because not all the stratigraphic units are preserved in all of the four studied borehole-sections, overburden units with estimated thickness and petrological properties are incorporated in this modeling. This assumes continuous deposition of the modeled units across the entire central CB. These units are then removed (eroded) resulting in a decrease of the thickness of the considered section. The erosion and the associated uplift do not change the compaction of the underlying units because the compaction is an irreversible process. Only once the stratigraphic sequences are buried to a greater depth than before the erosion, are the units compacted further.

11.2.2 Input Data

The input data required for this computation is summarized and listed in Figs. 11.3 and 11.4 and Tables 11.1–11.8, respectively. This includes: (1) sediment-type—percentage of clay and silt compared to sand; (2) age— in millions of years; (3) thickness—in kilometers; and (4) petro-physical characteristics—porosity-depth coefficient, surface porosity and dry sediment density—for each stratigraphic unit; (5) paleo-bathymetry—>0 if marine and lacustrine or =0 if fluvial and aeolian; (6) eustasy—deduced from the long-term sea-level curves of Haq et al. (1987); Haq and Schutter (2008); (7) density of depositional fluid (fresh vs. salt water or air); and (8) paleo-elevation—in kilometers above paleo-eustatic sea-level.

11.2.2.1 Paleo-Elevations of the Congo Basin

The paleo-elevations of the CB are estimated from paleo-environmental interpretations, as described below (Fig. 11.4), and calibrated to the eustatic sea-level curve of Haq et al. (1987) and Haq and Schutter (2008). Although there is debate about the global amplitude of this curve (e.g. Müller et al. 2008, 2011; Rowley 2013), it best describes the relative sea level variations observed in and around Africa over the last 500 million years.

At the base of the CB, the well-recognized, thick Carboniferous-Permian sequences of diamictites and black shales (the Lukuga Group) are considered to have been sourced through large west-facing paleo-glacial valleys and fjords flanking the eastern and southern margins of the basin (Cahen and Lepersonne 1978; Linol 2013; Chap. 7, this Book), and possibly intermittently connected with the Paraná Basin of southeastern Brazil, where there is evidence for marine turbidites (the Itararé Group; Vesely 2007; see Chap. 13, this Book). Thus, during the late Paleozoic, the low-lying paleo-surface of the CB is estimated to have been at approximately 100 m above the paleo-eustatic sea level (Fig. 11.4). This paleo-elevation is then elevated by about 200 m during deposition of the overlying, more proximal, thick Triassic sequences of conglomerates, red sandstones and siltstones of the Haute Lueki Group (Fig. 11.4). An emergence of the CB and increased regional relief of south-central Africa at the end of the Paleozoic is also consistent with the stratigraphic record of the Karoo Basins of southern Africa, where an abrupt change from turbidites (the Ecca Group) to fluvial sediments (the Beaufort Group)

Table 11.3 Stratigraphic data of the Dekese section

Unit	Age (Ma)	Base depth (km)	Top depth (km)	Porosity-depth coeff. (km^{-1})	Surface porosity (%)	Dry sediment density (km/m^3)	Bathymetry (km)	Eustasy (km)	Density of deposit fluid (km/m^3)	Elevation (km)
Hiatus	318.1	1.677	1.677	0.270	0.490	2,650	0.000	0.000	1	−0.200
D9	298	1.677	1.554	0.342	0.532	2,680	0.020	0.080	1,000	−0.120
D7 + D8	251	1.554	0.755	0.486	0.616	2,740	0.100	−0.040	1,000	−0.240
D6	199.6	0.755	0.715	0.510	0.630	2,750	0.020	0.000	1,000	−0.460
Overburden	183	0.715	0.715	0.318	0.518	2,670	0.020	0.060	1,000	−0.400
D5	149	0.715	0.705	0.294	0.504	2,660	0.010	0.160	1,000	0.000
D4	143	0.705	0.398	0.282	0.497	2,655	0.000	0.120	1	−0.040
Hiatus	112	0.398	0.398	0.270	0.490	2,650	0.000	0.180	1	−0.120
D3	100	0.398	0.318	0.294	0.504	2,660	0.020	0.260	1,000	−0.040
D2	93.6	0.318	0.037	0.330	0.525	2,675	0.010	0.280	1,000	−0.020
Overburden	65.5	0.037	0.037	0.342	0.532	2,680	0.010	0.230	1,000	−0.070
D1	33.9	0.037	0.001	0.270	0.490	2,650	0.000	0.220	1	−0.150
Hiatus	0	0.001	0.001	0.270	0.490	2,650	0.000	0.000	1	−0.370

Table 11.4 Eroded units in the Dekese section

Overburden unit age (Ma)	Erosion age (Ma)	Eroded thickness (km)	Eustasy (km)
183	158	0.700	−0.330
65.5	55.8	0.300	−0.070

Table 11.5 Stratigraphic data of the Gilson section

Unit	Age (Ma)	Base depth (km)	Top depth (km)	Porosity-depth coeff. (km^{-1})	Surface porosity (%)	Dry sediment density (km/m^3)	Bathymetry (km)	Eustasy (km)	Density of deposit fluid (km/m^3)	Elevation (km)
Hiatus	350	3.107	3.107	0.270	0.490	2,650	0.000	0.120	1	−0.080
G9	318.1	3.107	2.889	0.342	0.532	2,680	0.020	0.000	1,000	−0.200
G8	298	2.889	2.646	0.366	0.546	2,690	0.050	0.080	1,000	−0.120
G7	251	2.646	1.536	0.462	0.602	2,730	0.080	−0.040	1,000	−0.240
G6	199.6	1.536	0.998	0.294	0.504	2,660	0.020	0.000	1,000	−0.460
Overburden	183	0.998	0.998	0.318	0.518	2,670	0.020	0.060	1,000	−0.400
G5	149	0.998	0.890	0.342	0.532	2,680	0.010	0.160	1,000	0.000
G4	143	0.890	0.728	0.282	0.497	2,655	0.000	0.120	1	−0.040
Hiatus	112	0.728	0.728	0.270	0.490	2,650	0.000	0.180	1	−0.120
G3	100	0.728	0.567	0.390	0.560	2,700	0.100	0.260	1,000	−0.040
G2	93.6	0.567	0.234	0.366	0.546	2,690	0.020	0.280	1,000	−0.020
Overburden	65.5	0.234	0.234	0.342	0.532	2,680	0.010	0.230	1,000	−0.070
G1	33.9	0.234	0.001	0.474	0.609	2,735	0.000	0.220	1	−0.150
Hiatus	0	0.001	0.001	0.270	0.490	2,650	0.000	0.000	1	−0.370

Table 11.6 Eroded units in the Gilson section

Overburden unit age (Ma)	Erosion age (Ma)	Eroded thickness (km)	Eustasy (km)
183	158	0.7	−0.33
65.5	55.8	0.3	−0.07

marks a regional unconformity near the Permian-Triassic boundary (e.g. Turner 1999; Catuneanu et al. 2005; Johnson et al. 2006).

At the base of the overlying Jurassic-Cretaceous sequences of the CB (the Stanleyville, Loia, Bokungu and Kwango Groups) there is strong biostratigraphic and sedimentological evidence for a short Kimmeridgian marine incursion (Taverne 1975; Chap. 8, this Book). Thus, at that time, the basin surface is considered to have remained just below this local paleo-eustatic maximum (i.e. 160 m above present day sea-level; Fig. 11.4) during deposition of the Upper Jurassic to Lower Cretaceous shallow marine to continental Stanleyville Group and Dekese Formation. The paleo-elevation of the CB is then again increased by about 100 m to account for the overlying middle and Upper Cretaceous fluvial-lacustrine sequences (the Loia, Bokungu and Upper Kwango Groups) and to remain just above the high

Table 11.7 Stratigraphic data of the Mbandaka section

Unit	Age (Ma)	Base depth (km)	Top depth (km)	Porosity-depth coeff. (km^{-1})	Surface porosity (%)	Dry sediment density (km/m^3)	Bathymetry (km)	Eustasy (km)	Density of deposit fluid (km/m^3)	Elevation (km)
Hiatus	350	3.991	3.991	0.270	0.490	2,650	0.000	0.120	1	−0.080
M9	318.1	3.991	3.408	0.318	0.518	2,670	0.020	0.000	1,000	−0.200
M8	298	3.408	2.622	0.354	0.539	2,685	0.050	0.080	1,000	−0.120
Hiatus	251	2.622	2.622	0.270	0.490	2,650	0.000	−0.040	1	−0.240
M6 + M7	199.6	2.622	1.536	0.318	0.518	2,670	0.020	0.000	1,000	−0.460
M5	183	1.536	0.844	0.318	0.518	2,670	0.020	0.060	1,000	−0.400
Hiatus	158	0.844	0.844	0.270	0.490	2,650	0.000	0.130	1	−0.330
Hiatus	149	0.844	0.844	0.270	0.490	2,650	0.000	0.160	1	0.000
M4	143	0.844	0.808	0.270	0.490	2,650	0.000	0.120	1	−0.040
Hiatus	112	0.808	0.808	0.270	0.490	2,650	0.000	0.180	1	−0.120
M3	100	0.808	0.607	0.474	0.609	2,735	0.100	0.260	1,000	−0.040
M2	93.6	0.607	0.161	0.354	0.539	2,685	0.020	0.280	1,000	−0.020
Overburden	65.5	0.161	0.161	0.342	0.532	2,680	0.010	0.230	1,000	−0.070
M1	33.9	0.161	0.001	0.498	0.623	2,745	0.000	0.220	1	−0.110
Hiatus	0	0.001	0.001	0.270	0.490	2,650	0.000	0.000	1	−0.330

Table 11.8 Eroded unit in the Mbandaka section

Overburden unit age (Ma)	Erosion age (Ma)	Eroded thickness (km)	Eustasy (km)
65.5	55.8	0.300	−0.070

Cenomanian-Turonian paleo-eustatic maximum, i.e. 280 m above the present-day sea-level (Fig. 11.4). This low-lying paleo-surface of the CB is compatible with a potential Late Cretaceous marine connection to explain paleo-spawning sites of Kwango fish fossils found at Kipala near the center of the basin (Casier 1965; Taverne 1976; Giresse 2005; Chap. 8, this Book). The overlying terrestrial sequences of the Cenozoic Kalahari Group are relatively poorly constrained by sedimentological and biostratigraphic data and here simply considered to be deposited at the same elevation as the present-day basin surface, ca. at the elevation of borehole heads (e.g. 370 m above sea-level; Fig. 11.4).

In summary, the paleo-elevation of the CB is relatively well-constrained by sedimentological and stratigraphic data for the Jurassic-Cretaceous, but is less precisely known for the Carboniferous to Triassic sequences. However, in their geomorphic study of the CB, Stankiewicz and de Wit (2006) identified two recurrent surface elevations: at 457 m and 610 m above the present-day sea level. The lower (457 m) peneplain was interpreted as the base level of the basin during the Cretaceous. The higher (610 m) peneplain was proposed to represent an older, late Paleozoic erosion surface. These two paleo-surfaces reflect a relative depositional base level drop of 150 m between about 300 Ma and 100 Ma, which is independently consistent with the paleo-topographic model of the CB described above (Fig. 11.4).

11.2.3 Results

The total subsidence curves for the Samba, Dekese, Gilson and Mbandaka sections are analyzed in Fig. 11.5. Variations of both the total (Subtot) and tectonic (Subtect) subsidences for each section are shown in Fig. 11.6.

The curves of total subsidence for the CB demonstrate the existence of two main episodes of rapid subsidence (Fig. 11.5):

Episode A

A first and most pronounced episode of subsidence starts at ca. 350 Ma with the onset of the main Carboniferous-Permian (Dwyka) glaciation of Gondwana (e.g. Milani and de Wit 2008) and terminates at about 180 Ma, and which corresponds to the deposition of the Lukuga and Haute Lueki Groups in the CB. This late Paleozoic to early Mesozoic single phase of subsidence (Phase 1; Fig. 11.5) has a magnitude of between 1,500 m at Dekese and 3,500 m at Mbandaka, and a calculated rate of subsidence of 10–20 m/Ma. In the Dekese section, an episode of uplift occurs at around 200 Ma (Fig. 11.6), resulting in the attenuation of the Triassic Haute Lueki Group here to a thickness of only 40 m (see Fig. 11.3).

Episode B

The second main episode of subsidence starts at ca. 160 Ma in the Samba and Mbandaka sections in the northern part of the basin, and about 10 Ma later (ca. at 150 Ma) in the Gilson and Dekese sections in the south-central CB (Fig. 11.5). This late Mesozoic subsidence corresponds to the deposition of the Stanleyville, Loia, Bokungu and Kwango Groups, and has a magnitude of between 800 m at Dekese and 1,500 m at Samba, and a mean calculated rate of subsidence of 5–10 m/

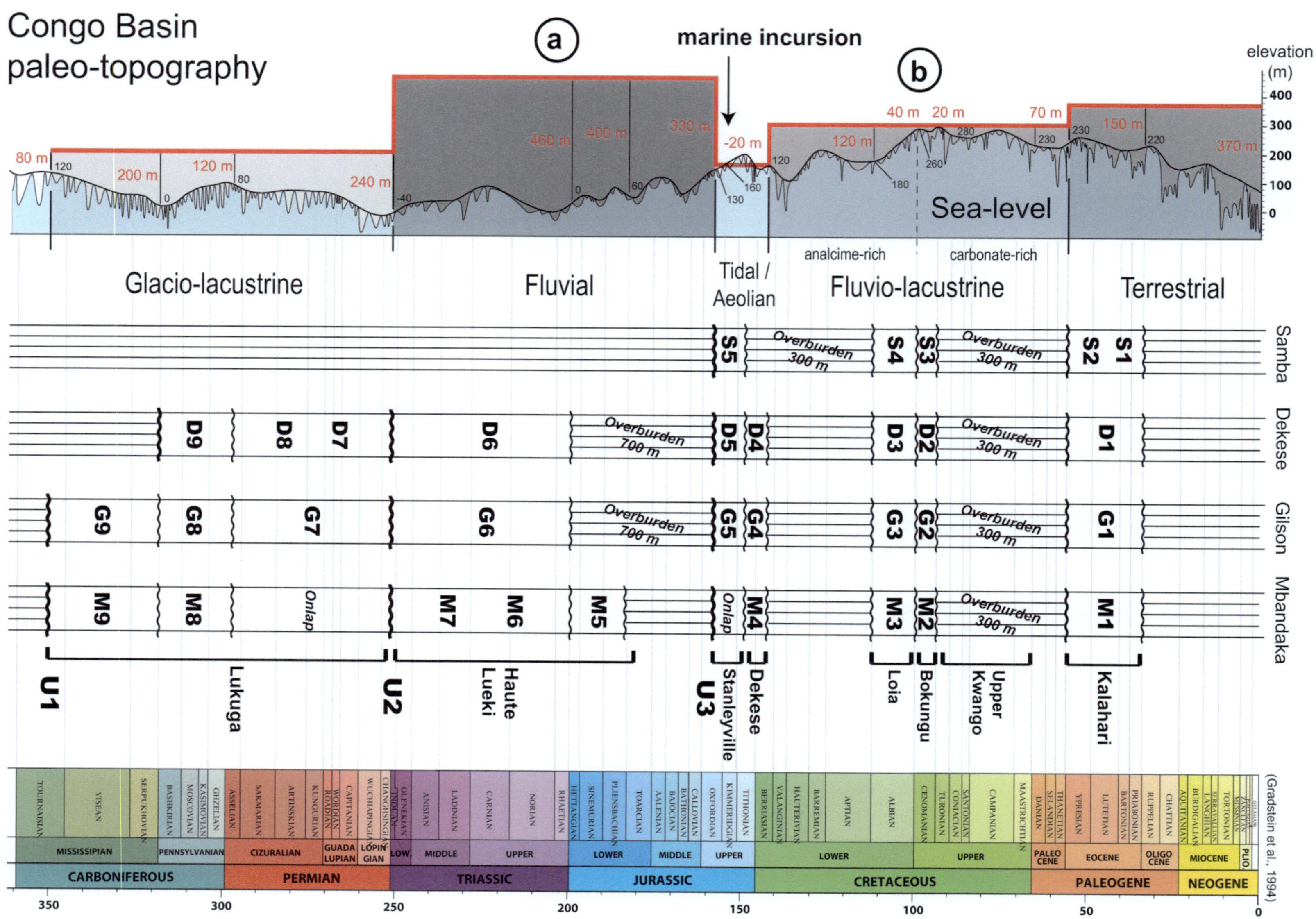

Fig. 11.4 Stratigraphic records (center section) and paleo-topographic model of the CB (*thick red line*). The paleo-elevations (values in *red*, *top* section) are estimated from sedimentological data and are adjusted to the long-term sea level curves of Haq et al. (1987) and Haq and Schutter (2008). When corrected to the eustatic sea level maximum for the Late Jurassic marine transgression in east-central Africa (ca. 160 m above present-day sea level), the Triassic and Cretaceous paleo-topographic highs correspond to two peneplains (**a** and **b**) traced across the CB by Stankiewicz and de Wit (2006)

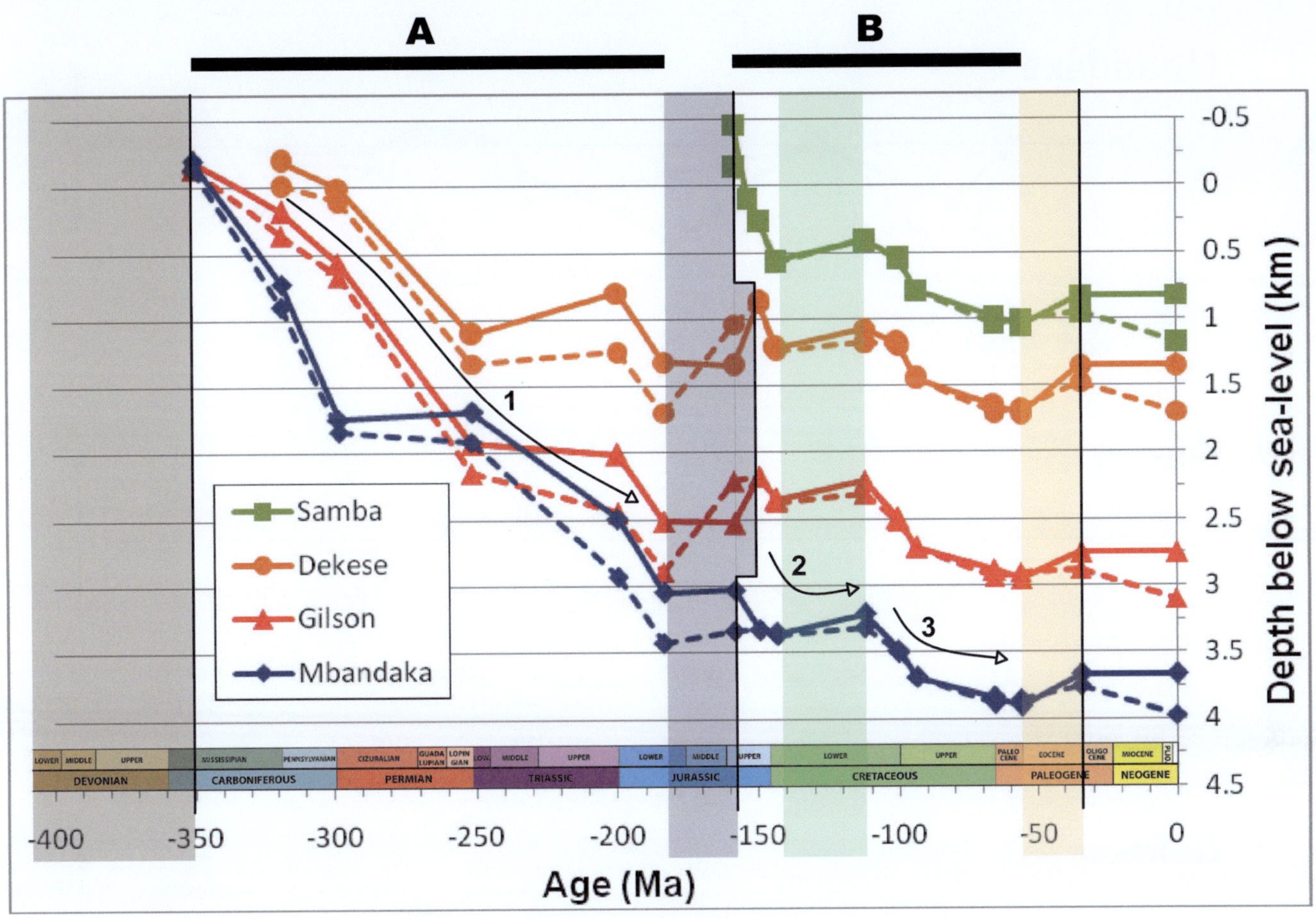

Fig. 11.5 Total subsidence curves for the Samba, Dekese, Gilson and Mbandaka well/core data in the central CB (Fig. 11.1 for location). The curves show two main episodes of subsidence (**A** and **B**), including three phases of rapid subsidence (1–3, *arrowed*) separated by episodes of slower subsidence or uplift (*highlighted columns*). *Dotted lines* correspond to calculated subsidence without paleo-topography input (paleo-elevation = 0)

Ma. It contains two distinct phases of rapid subsidence [ca. 10–50 m/Ma]: from 150–160 Ma to 140 Ma (Phase 2), and from 110 Ma to 50 Ma (Phase 3). These two phases coincide with the main period of break-up of Gondwana, and West Gondwana, that respectively resulted in the opening of the Indian and South Atlantic Oceans (e.g. Jokat et al. 2003). These two phases of subsidence are remarkably similar in all the four sections (Fig. 11.5), and all terminated with a major uplift at about 30–50 Ma (Fig. 11.6), recorded by the condensed Kalahari Group (see Fig. 11.3).

11.3 Tectono-Stratigraphic Synthesis

In summary, the four curves of total subsidence in the CB show two main episodes of basin subsidence [ca. 5–20 m/Ma]: during (a) the Carboniferous to Early Jurassic, and during (b) the Late Jurassic to early Paleogene (Fig. 11.5). These results differ significantly from the very slow [ca. 2–4 m/Ma] and continuous subsidence since the late Precambrian (Neoproterozoic) proposed in previous studies (e.g. Crosby et al. 2010; Kadima et al. 2011). The most important differences between these subsidence models arise from the stratigraphic databases used. Below, we describe some of the recent improvements to the stratigraphic framework and regional correlations of the CB (Linol 2013), as a basis from which to evaluate further its subsidence and tectonic evolution during the Phanerozoic (Fig. 11.7).

Following the Pan African orogens (ca. 530–650 Ma; de Wit et al. 2008), widespread upper Neoproterozoic to lower Paleozoic continental red-bed sediments (the Inkisi, Aruwimi and Biano Groups) were deposited across peneplained Precambrian basement of central Africa. These red-beds are about 1,000–1,500 m thick flanking the peripheral Pan African fold-and-thrust belts of the CB (Lepersonne 1974; Alvarez et al. 1995; Master et al. 2005; Tait et al. 2011; Chap. 6, this Book). Yet their stratigraphic equivalents

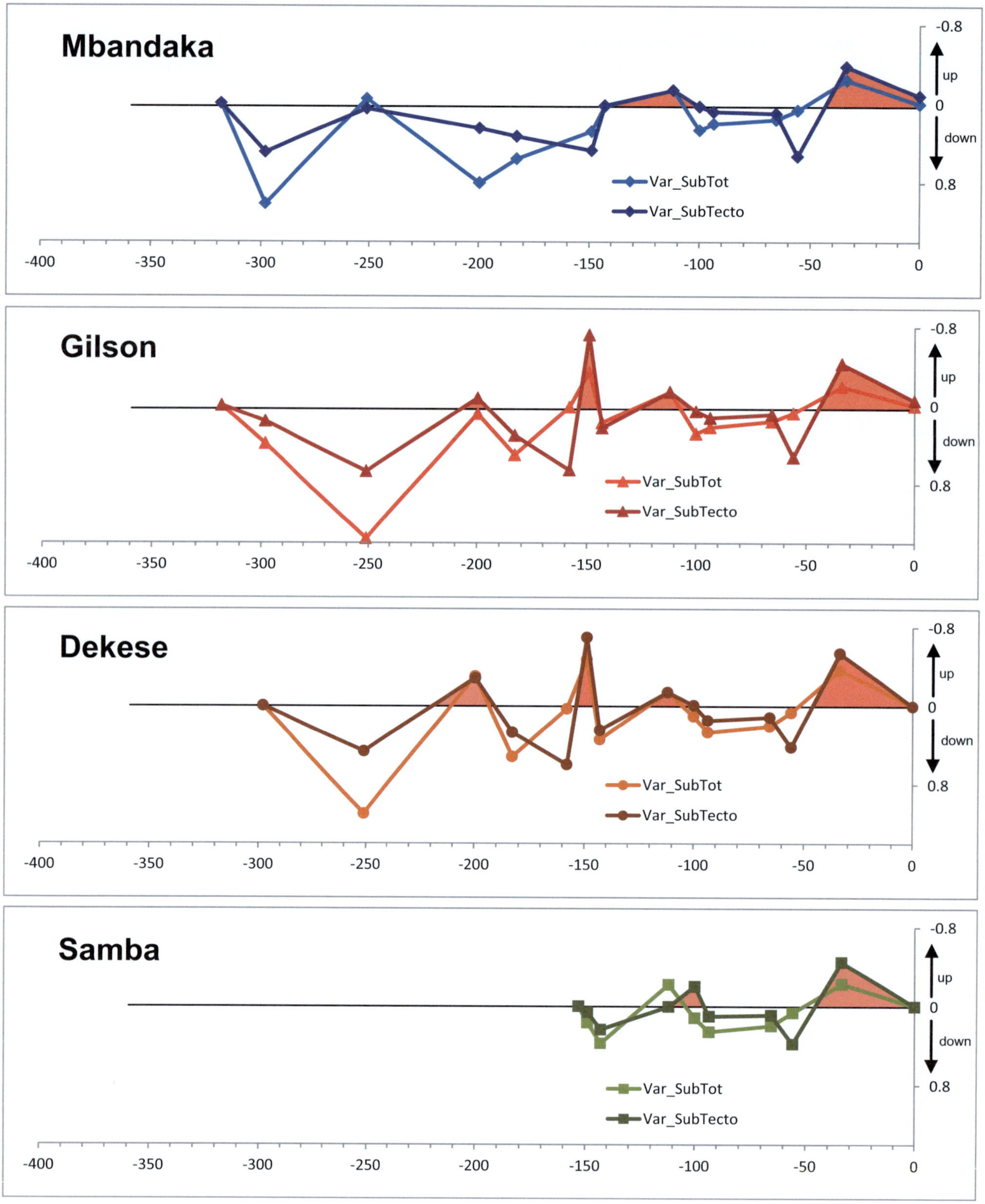

Fig. 11.6 Tectonic (SubTecto) and total (SubTot) subsidence variations for the Samba, Dekese, Gilson and Mbandaka well/core data (Fig. 11.1 for location). Negative variations (highlighted in *red*) indicate upward motions of the basement, corresponding to episodes of uplift of the CB

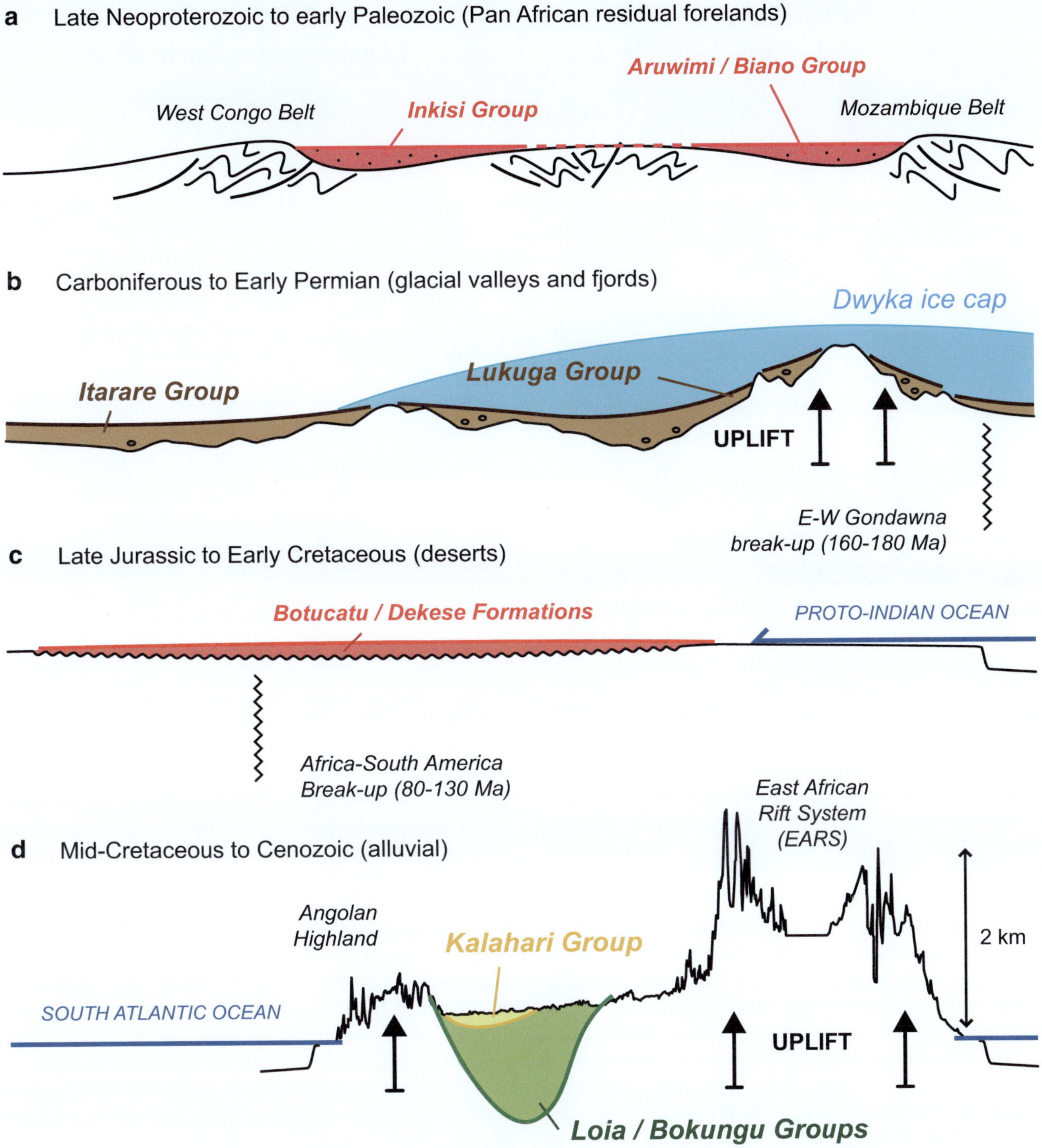

Fig. 11.7 Sketch cross-sections of the Phanerozoic evolution of the CB. (**a**) Upper Neoproterozoic to lower Paleozoic red-beds deposited along surrounding Pan African fold-and-thrust belts. (**b**) Carboniferous-Permian glacial sequences derived from a main highland in east Africa. (**c**) Upper Jurassic to Lower Cretaceous aeolian dunes extended across southern Congo to eastern Brazil. (**d**) Middle to Upper Cretaceous and Cenozoic fluvial-lacustrine sequences related to marginal uplifts and the Kalahari epeirogeny

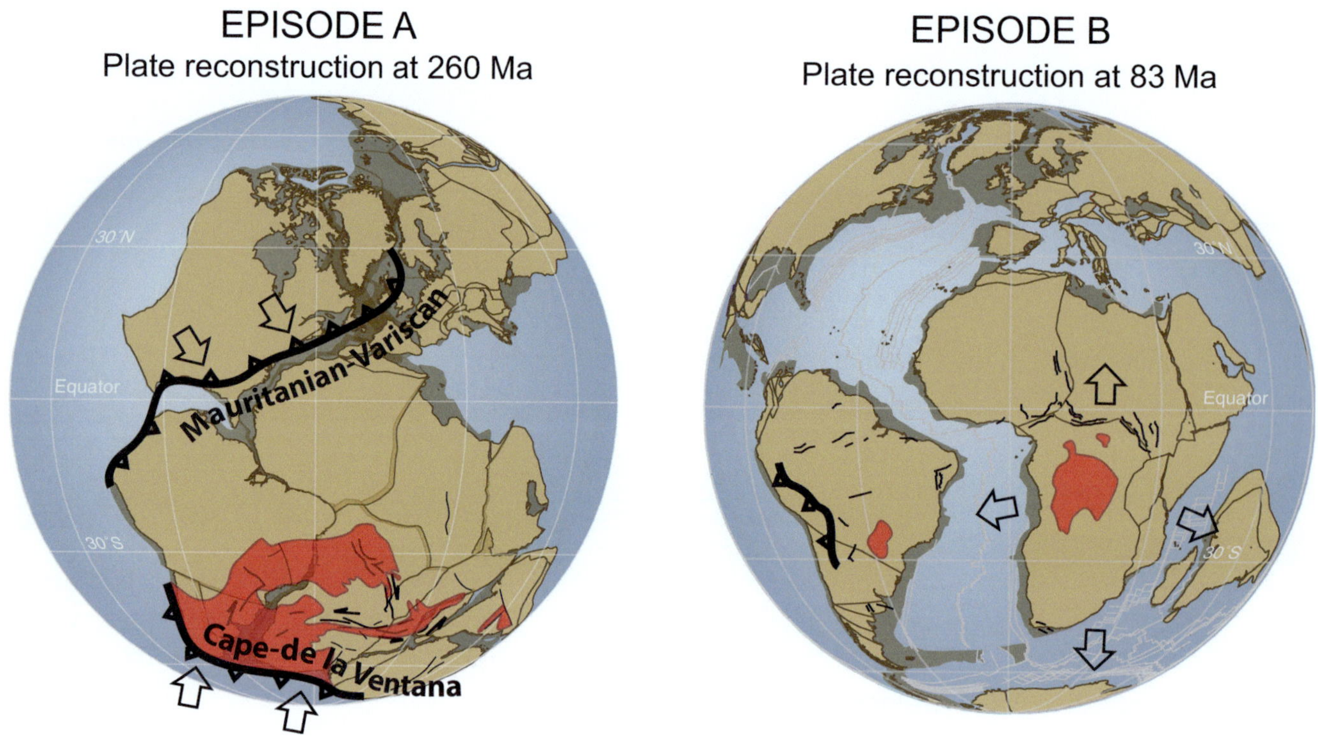

Fig. 11.8 Gondwana and Pangea plate-reconstructions (http://www.gplates.org), illustrating the two different geodynamic contexts during the two main episodes of subsidence of the CB (**A** and **B** in Fig. 11.5)

are largely attenuated or absent in all the four borehole-sections (Fig. 11.3). This suggests a long episode, up to 150 million years, of non-deposition and/or erosion in the center of the CB during the early Paleozoic (Fig. 11.7a).

During and following the main Dwyka glaciation (Fig. 11.7b), and the amalgamation of Gondwana into Pangea (ca. 300 Ma; Domeier et al. 2012; Kroner and Romer 2013), the Carboniferous to Triassic Lukuga and Haute Lueki Groups were deposited during the first and longest (~150 million years) Episode A of accelerated subsidence of the CB (Figs. 11.5 and 11.8). The lowermost glacial and deglaciation sequences (the Lukuga Group) are best preserved in the Mbandaka (1,400 m thick), Gilson (1,300 m thick) and Dekese (900 m thick) sections of the western and southern parts of the basin (Fig. 11.3). These sequences are interlayered with relatively thick, dark (organic-rich) siltstones and black shales that herald the initial phase of basin deepening (i.e. transgression). This transgression in the CB is largely equivalent to that of the Dwyka and Lower Ecca Groups in the Karoo Basins of southern Africa, and most probably links to the final rapid deglaciation of Gondwana and sea level rise followed by isostatic rebound during the Early Permian (e.g. Visser 1995; Scheffler et al. 2003; Frank et al. 2008; Montañez and Poulsen 2013). The overlying fluvial (and aeolian?) sequences of the Triassic Haute Lueki Group of the CB are remarkably thick in the Mbandaka (1,750 m), Samba (900 m) and Gilson (550 m) sections (Fig. 11.3), and represent a more rapid phase of basin infilling (i.e. regression). Although the ages of these units are relatively poorly constrained (Linol 2013; see Chap. 7, this Book), these are here considered to be equivalent to the Beaufort Group of the Karoo of southern Africa, ca. 180–260 Ma (Catuneanu et al. 2005; Johnson et al. 2006). This marked Permian-Triassic regression of the CB is therefore roughly contemporaneous across central and southern Africa and overlaps in time with the Cape-de la Ventana orogen (ca. 250 Ma) flanking the southern margin of Gondwana (e.g. the Gondwanides; du Toit 1937; Daly et al. 1992; Trouw and de Wit 1999; Newton et al. 2006; Pángaro and Ramos 2012; see also Chap. 13, this Book).

The upper, most pronounced and laterally extensive unconformity across the Carboniferous to Triassic sequences in the CB coincides with the onset of break-up between East and West Gondwana and the resulting opening of the Indian Ocean (ca. 160–180 Ma; Reeves 1999; Jokat et al. 2003; Moulin et al. 2010; Torsvik et al. 2012; Heine et al. 2013). Thereafter, the Stanleyville, Loia, Bokungu and Kwango

Groups were deposited during a second, slower main Episode B of subsidence that includes two distinctive phases of subsidence (Fig. 11.5): during the Late Jurassic to Early Cretaceous (10–50 m/Ma), and during the middle to Late Cretaceous (10–15 m/Ma), separated by episodes of uplift at 160–180 Ma, 110–140 Ma and 30–50 Ma (Fig. 11.6). The lower, shallow marine to fluvial Stanleyville Group, 320 m thick at Samba in the northern part of the basin, and the overlying aeolian Dekese Formation, maximum 300 m thick at Dekese in the south-central CB, record the first phase of regression and desertification during the Late Jurassic to Early Cretaceous (Fig. 11.7c). Succeeding a hiatus and/or erosion of about 25 million years, the overlying middle and Upper Cretaceous fluvial-lacustrine Loia, Bokungu and Upper Kwango Groups, maximum 650 m thick in the Samba and Mbandaka sections (Fig. 11.3), record the second phase of subsidence and regression of the CB (Fig. 11.5). It is terminated by a regional uplift during the early Cenozoic that corresponds to limited preservation of the Kalahari Group (Fig. 11.7d), possibly linked to the Kalahari epeirogeny (de Wit 2007) and the onset of doming of the Angolan Highland and along the East African Rift System (EARS), ca. 30–40 Ma (e.g. Walford and White 2005; Pik et al. 2008).

Discussion and Conclusion

A large number of driving mechanisms have been proposed for the formation and development of the CB: (1) long-lived tectonic uplift along its margins, (2) long-lived thermal subsidence following an initial phase of late Precambrian rifting, (3) long-lived dynamic subsidence related to the effect of a high-density lithospheric mantle root, (4) long-lived lithospheric delamination, and (5) long-lived dynamic topography due to downward flow within the mantle (Sahagian 1988; Hartley and Allen 1994; Downey and Gurnis 2009; Crosby et al. 2010; Kadima et al. 2011; Buiter et al. 2012). All these models assume a single long (~700 million years) and extremely slow (ca. 2–4 m/Ma) subsidence history of the CB since the Neoproterozoic. However, our new subsidence analysis, based on revised sedimentological and stratigraphic data from the four historic deep boreholes drilled in the center of the basin (Fig. 11.3), reveal that the geodynamic evolution of the CB is much more complex, including at least two main episodes of rapid subsidence (ca. 5–20 m/Ma), punctuated by multiple uplifts (Figs. 11.5 and 11.6):

1. Late Paleozoic to early Mesozoic accelerated subsidence (180–350 Ma), with a calculated rate of 10–20 m/Ma, follows a prolonged erosional hiatus in the central CB. This long (~150 million years) erosion time-interval was possibly associated with the successive, Ordovician and Carboniferous Gondwana ice-sheets that covered much of central Africa (e.g. Visser 1995; Ghienne 2003). Thereafter, the first and most pronounced episode of subsidence of the CB possibly linked to long wavelength flexural loading of the Gondwana continental lithosphere related to the Mauritanian-Variscan orogen (ca. 325–275 Ma; Dabo et al. 2008; Van Staal et al. 2009; Kroner and Romer 2013; Chap. 13, this Book) during amalgamation of Gondwana into Pangea, and the Cape-de la Ventana orogen (ca. 245–278 Ma; Newton et al. 2006), as it has been proposed for the main Karoo and Paraná Basins of south-western Gondwana (e.g. Daly et al. 1992; Pysklywec and Quintas 2000; Milani and de Wit 2008; Fig. 11.8).
2. Two late Mesozoic phases of subsidence of the CB (at 140–160 Ma and 60–110 Ma), with calculated rates ranging between 10 and 50 m/Ma, are separated by three regional episodes of uplift at 160–180 Ma (e.g. 'Karoo'), 120–140 Ma (e.g. 'Paraná-Etendeka'), and 30–50 Ma (e.g. 'Ethiopian'). These episodic uplifts are coeval with continental rifting and the successive outpouring of Large Igneous Provinces (LIPs, or hotspot plumes) associated with the opening of the Indian and South Atlantic Oceans during the main period of Gondwana break-up (e.g. Jokat et al. 2003; Torsvik et al. 2012; Heine et al. 2013), as well as the Late Cretaceous Kalahari epeirogeny (de Wit 2007), and the Cenozoic onset of doming of the Angolan Highland and EARS (e.g. Walford and White 2005; Pik et al. 2008).

In conclusion, this multiphase Phanerozoic history of the CB suggests that initial basin subsidence was linked to the Carboniferous-Permian glaciation/deglaciation of Gondwana and its accretion to Laurentia (Europe and North America) to form Pangea during the Mauritanian-Variscan orogen (280–330 Ma; Dabo et al. 2008; Kroner and Romer 2013), and further amplify by following Permian-Triassic deformation during the Cape-de la Ventana orogen (ca. 250 Ma;) along the southern margin of Gondwana (Daly et al. 1992; Trouw and de Wit 1999; Newton et al. 2006; Milani and de Wit 2008). Subsequent (late Mesozoic) intermittent phases of subsidence recorded in the CB most likely related to extensional tectonic, marginal uplifts and the successive cooling of LIPs around Africa (Fig. 11.8). Although it appears difficult to relate these phases of subsidence to specific events, it is clear from this study that the initiation and development of the CB related to both geodynamic and global climate processes during the formation and break-up of the Pangea supercontinent.

Acknowledgments We acknowledge funding through the Inkaba yeAfrica and !Khure Africa programs, supported by the DST/NRF of South Africa. We thank Anthony Tankard and an anonymous referee for critical reviews that improved the chapter. This is AEON contribution number 130 and Inkaba yeAfrica contribution number 100.

References

Allen PA, Allen JR (1990) Basin analysis. Principles and applications. Blackwell Scientific Publications, Oxford, 459 p

Allen PA, Allen JR (2005) Basin analysis: principles and applications. 2nd edn. Wiley-Blackwell, 560 p

Alvarez P, Maurin JC, Vicat J-P (1995) La Formation de l'Inkisi (Supergroupe Ouest-congolien) en Afrique centrale (Congo et Bas-Zaire): un delta d'âge Paléozoique comblant un bassin en extension. J Afr Earth Sci 20(2):119–131

Buiter SJH, Steinberger B, Medvedev S, Tetreault JL (2012) Could the mantle have caused subsidence of the Congo Basin? Tectonophysics 514–517:62–80

Cahen L, Lepersonne J (1978) Synthèse des connaissances relatives au Groupe (anciennement Série) de la Lukuga (Permien du Zaïre). Annales du Musée Royal du Congo belge, Tervuren (Belgique), Série in-8, Sciences géologiques, 82, pp 115–152

Casier E (1965) Poissons fossiles de la Série du Kwango (Congo). Annales du Musée royal de l'Afrique Centrale, Tervuren (Belgique), Série in-8, Sciences géologiques, 50, 69 p

Catuneanu O, Wopfner H, Eriksson PG, Cairncross B, Rubidge BS, Smith RMH, Hancox PJ (2005) The Karoo basins of south-central Africa. J Afr Earth Sci 43(1–3):211–253

Cochran JR (1983) Effects of finite rifting times on the development of sedimentary basins. Earth Planet Sci Lett 66:289–302

Crosby AG, Fishwick S, White N (2010) Structure and evolution of the intracratonic Congo Basin. Geochem Geophys Geosyst 11(6):1–20

Dabo M, Gueye M, Ngom PM, Diagne M (2008) Orogen-parallel tectonic transport: transpression and strain partitioning in the Mauritanides of NE Senegal. In: Ennih N, Liégeois J-P (eds) The Boundaries of the West African Craton. Geological Society, London, Special Publications, 297, pp 483–497

Dalrymple RW (1992) Tidal depositional systems. In: Walker RG, James NP (eds) Facies models: response to sea level change. Ontario, Geological Association of Canada, pp 195–218

Daly MC, Lawrence SR, Diemu-Tshiband K, Matouana B (1992) Tectonic evolution of the Cuvette Centrale, Zaire. J Geol Soc 149(4):539–546

De Wit MJ (2007) The Kalahari epeirogeny and climate change: differentiating cause and effect from core to space. South Afr J Geol 110(2–3):367–392

De Wit MJ, Stankiewicz J, Reeves C (2008) Restoring Pan-African-Brasiliano connections: more Gondwana control, less Trans-Atlantic corruption. In: Pankurst RJ, Trouw RAJ, Brito Neves BB, de Wit MJ (eds) West Gondwana: pre-cenozoic correlations across the South Atlantic Region. Geological Society of London, Special Publications, 294, pp 399–412

Domeier M, Van der Voo R, Torsvik TH (2012) Paleomagnetism and Pangea: the road to reconciliation. Tectonophysics 514–517:14–43

Doucouré CM, de Wit MJ (2003) Old inherited origin for the present near-bimodal topography of Africa. J Afr Earth Sci 36:371–388

Downey NJ, Gurnis M (2009) Instantaneous dynamics of the cratonic Congo basin. J Geophys Res 114(B6):1–29

Du Toit AL (1937) Our wandering continents: an hypothesis of continental drifting. Oliver and Boyd, Edinburgh, p 366 p

Frank TD, Birgenheier LP, Montañez IP, Fielding CR, Rygel MC (2008) Late Paleozoic climate dynamics revealed by comparison of ice-proximal stratigraphic and icedistal isotopic records. In: Fielding CR, Frank TD, Isbell JL (eds) Resolving the late Paleozoic Ice age in time and space. Geological Society of America, Special Paper, 441, pp 331–342

Ghienne JF (2003) Late Ordovician sedimentary environments, glacial cycles, and post-glacial transgression in the Taoudeni Basin, West Africa. Palaeogeogr Palaeoclimatol Palaeoecol 189:117–145

Giresse P (2005) Mesozoic-Cenozoic history of the Congo Basin. J Afr Earth Sci 43:301–315

Haq BU, Schutter SR (2008) A chronology of Paleozoic. Science 322: 64–68

Haq BU, Hardenbol J, Vail PR (1987) Chronology of fluctuating sea levels since the Triassic. Science 235(4793):1156–67

Hartley RW, Allen PA (1994) Interior cratonic basins of Africa: relation to continental break-up and role of mantle convection. Basin Res 6(2–3):95–113

Heine C, Zoethout J, Müller RD (2013) Kinematics of the South Atlantic rift. Solid Earth 4:215–253

Johnson MR, Van Vuuren CJ, Visser JNJ, Cole DI, de V. Wickens H, de V. Wickens H, Christie ADM, Roberts DL, Brandl G (2006) Sedimentary rocks of the Karoo Supergroup. In: Johnson MR, Anhaeusser CR, Thomas RJ (eds) The geology of South Africa. Council for Geoscience, Pretoria, South Africa, pp 461–499

Jokat W, Boebel T, König M, Meyer U (2003) Timing and geometry of early Gondwana break-up. J Geophys Res: Solid Earth 108(B9)

Kadima E, Ntabwoba SSM, Lucazeau F (2011) A Proterozoic-rift origin for the structure and the evolution of the cratonic Congo basin. Earth Planet Sci Lett 304(1–2):240–250

Kroner U, Romer RL (2013) Two plates – Many subduction zones: the Variscan orogeny reconsidered. Gondwana Res 24:298–329

Lepersonne (1974) Carte géologique du Zaïre au 1: 2.000.000 et notice explicative. Kinshasa, République du Zaïre: direction de la Géologie/Musée Royal de l'Afrique centrale, Tervuren (Belgique).

Linol B (2013) Sedimentology and sequence stratigraphy of the Congo and Kalahari Basins of south-central Africa and their evolution during the formation and break-up of West Gondwana. PhD thesis, Nelson Mandela Metropolitan University, 375 p

Master S, Rainaud C, Armstrong R, Phillips D, Robb L (2005) Provenance ages of the Neoproterozoic Katanga Supergroup (Central African Copperbelt), with implications for basin evolution. J Afr Earth Sci 42(1–5):41–60

McKenzie D (1978) Some remarks on the development of sedimentary basins. Earth Planet Sci Lett 40:25–32

Milani EJ, de Wit MJ (2008) Correlations between the classic Paraná and Cape Karoo sequences of South America and southern Africa and their basin infills flanking the Gondwanides: du Toit revisited. In: Pankurst RJ, Trouw RAJ, Brito Neves BB, de Wit MJ (eds) West Gondwana: Pre-Cenozoic correlations across the South Atlantic Region. Geological Society of London, Special Publications, 294, pp 319–342

Miller KG, Mountain GS, Wright JD, Browning JV (2011) A 180-million-year record of sea level and ice volume variations from continental margin and deep-sea isotopic records. Oceanography 24(2):40–53

Montañez IP, Poulsen CJ (2013) The Late Paleozoic ice age: an evolving paradigm. Annu Rev Earth Planet Sci 41:629–656

Moulin M, Aslanian D, Unternehr P (2010) A new starting point for the South and Equatorial Atlantic Ocean. Earth-Sci Rev 98(1–2):1–37

Müller RD, Sdrolias M, Gaina C, Steinberger B, Christian Heine C (2008) Long-term sea-level fluctuations driven by ocean basin dynamics. Science 319:1357–1362

Murray JW, Stewart K, Kassakian S, Krynytzky M, DiJulio D (2007) Oxic, suboxic and anoxic conditions in the Black Sea. In: Gilbert A, Yanko-Hombach V, Panin N (eds) Climate change and coastline migration as factors in human adaptation to the Circum-Pontic region: from past to forecast. Kluwer, pp 437–452

Newton AR, Shone RW, Booth PWK (2006) The Cape Fold Belt. In: Johnson MR, Anhaeusser CR, Thomas RJ (eds) The geology of South Africa. Council for Geoscience, Pretoria, South Africa, pp 521–530

Pángaro F, Ramos VA (2012) Paleozoic crustal blocks of onshore and offshore central Argentina: new pieces of southwestern Gondwana collage and their role in the accretion of Patagonia and the evolution of Mesozoic south Atlantic sedimentary basins. Mar Pet Geol 37:162–183

Pik R, Marty B, Carignan J, Yirgu G, Ayalew T (2008) Timing of East African Rift development in southern Ethiopia: implication for mantle plume activity and evolution of topography. Geology 36: 167–170

Pysklywec RN, Quintas MCL (2000) A mantle flow mechanism for the late Palaeozoic subsidence of the Paranà basin. J Geophys Res 105 (16):359–370

Reeves CV (1999) Aeromagnetic and gravity features of Gondwana and their relation to continental break-up: more pieces, less puzzle. J Afr Earth Sci 28:263–277

Rowley DB (2013) Sea level: Earth's dominant elevation—implications for duration and magnitudes of sea level variations. J Geol 121(5):445–454

Sahagian D (1988) Epeirogenic motions of Africa as inferred from Cretaceous shoreline deposits. Tectonics 7(1):125–138

Scheffler K, Hoernes S, Schwark L (2003) Global changes during Carboniferous-Permian glaciation of Gondwana: linking polar and equatorial climate evolution by geochemical proxies. Geology 31(7):605–608

Stankiewicz J, de Wit MJ (2006) A proposed drainage evolution model for Central Africa - did the Congo flow east? J Afr Earth Sci 44(1): 75–84

Tait J, Delpomdor F, Preat A, Tack L, Straathof G, Nkula VK (2011) Neoproterozoic sequences of the West Congo and Lindi/Ubangi Supergroups in the Congo Craton, Central Africa. In: Arnaud E, Halverson GP, Shields-Zhou G (eds) The geological record of neoproterozoic glaciations. Geological Society of London, Memoirs, 36, pp 185–194

Taverne L (1975) Etude ostéologique de Leptolepis caheni, Teleosteen fossile du Jurassique superieur (Kimmeridgien) de Kisangani (ex-Stanleyville, Zaïre) précedemment décrit dans le genre Paraclupavus. Revue Zoologique Africaine 89:821–853

Taverne L (1976) Les téléostéens fossiles du crétacé moyen de Kipala (Kwango, Zaïre). Annales du Musée Royal de l'Afrique centrale, Tervuren (Belgique), Série in-8, Sciences géologiques, 79, 50p

Torsvik TH, Van der Voo R, Preeden U, Mac Niocaill C, Steinberger B, Doubrovine PV, van Hinsbergen DJJ, Domeier M, Gaina C, Tohver E, Meert JG, McCausland PJA, Cocks LR (2012) Phanerozoic polar wander, palaeogeography and dynamics. Earth Sci Rev 114(3–4): 325–368

Trouw RA, de Wit MJ (1999) Relation between the Gondwanide Orogen and contemporaneous intracratonic deformation. J Afr Earth Sci 28(1):203–213

Turner BR (1999) Tectonostratigraphical development of the Upper Karoo foreland basin: orogenic unloading versus thermally-induced Gondwana rifting. J Afr Earth Sci 28(1):215–238

Vail PR, Audemard F, Bowman SA, Eisner PN, Perez-Cruz C (1991) The stratigraphic signatures of tectonics, eustasy and sedimentology: an overview. In: Einsele G, Ricken W, Seilacher A (eds) Cycles and events in stratigraphy. Springer, Berlin, pp 617–659

Van Staal CR, Whalen JB, Valverde-Vaquero P, Zagorevski A, Rogers N (2009) Pre-Carboniferous, episodic accretion-related, orogenesis along the Laurentian margin of the northern Appalachians. Geological Society of London, Special Publications, vol 327, pp 271–316

Vesely FF (2007) Sistemas subaquosos alimentados por fluxos hiperpicnais glaciogenicos: modelo deposicional para arenitos do Grupo Itararé, Permocarbonífero da Bacia do Paraná. Boletim de Geociências da Petrobras 15(1):7–25

Visser JNJ (1995) Post-glacial permian stratigraphy and geography of southern and central Africa: boundary conditions for climatic modelling. Palaeogeogr Palaeoclimatol Palaeoecol 118(3–4): 213–243

Walford HL, White NJ (2005) Constraining uplift and denudation of West African continental margin by inversion of stacking velocity data. J Geophys Res-Solid Earth 110, B04403

Thermal Regime and Evolution of the Congo Basin as an Intracratonic Basin

12

Francis Lucazeau, John Armitage, and Étienne Kadima Kabongo

12.1 Introduction

The interior of the continents is generally considered as the most stable part of the Earth surface that survive through a succession of tectonic episodes. The heat-flow and the lithospheric thickness of these stable pieces of continents ("cratons" or "shields") are usually much lower and thicker than the younger neighbouring lithosphere (Artemieva 2006). Within these stable continents, cratonic basins can subside progressively over longer periods of time than other rifted continental basins, because the thicker lithosphere has a longer thermal relaxation (Xie and Heller 2009) and extension proceeds at lower strain rates (Armitage and Allen 2010). The Congo Basin (CB) located in the central part of central Africa is a classical example of an intracratonic basin (Fig. 12.1). It is a sub-circular basin filled in its upper part by Mesozoic and Cenozoic sediments, and surrounded by different Archean and Paleo-Proterozoic units. The origin of this basin can be related to a late Proterozoic rift (Daly et al. 1992; Crosby et al. 2010; Kadima Kabongo et al. 2011b) that has been inferred from a lower crustal gravity anomaly when the effect of sediment is removed (Fig. 12.1). From Early to Late Proterozoic, several extensional basins (such as the West Congolese Supergroup, Sembe-Ouesso basin, Sangha Aulacogen, Mbuji-Mayi Supergroup, Lufilian Copperbelt in Katanga and Northern Zambia basin) were formed on the Congo Craton basement during the Rodinia breakup (see Chap. 6, this Book). A recent stratigraphic study (Linol et al. 2013c, d) suggests a possible reactivation of the subsidence during two episodes, the first one in the late Paleozoic (350–300 Ma) and the second one in the late Jurassic (155–145 Ma) (Fig. 12.2).

The subsidence of the Congo basin is to a large extent poorly constrained, as the sedimentation is not continuous and the stratigraphy is not always well established. As the deepest parts of the basin have not been drilled, there is a large uncertainty about the thickness and the age of the oldest (non metamorphic) sedimentary sequences, which can reach to depths of 5–6 km in the central part and locally 8–9 km in the Dekese and Gilson areas (Kadima Kabongo et al. 2011b). These sediments cover an Archean and Late-to-Meso-proterozoic crystalline basement, and are probably late Neo-Proterozoic in age (Kadima Kabongo et al. 2011a). Here, we discuss the past and the present-day thermal regime of the CB, from published data mainly and unpublished oil exploration reports. The stratigraphic record, the "tectonic" subsidence and the maturation of the organic matter are used to reconstruct the paleo-geotherms, but the Cretaceous and Neogene kimberlites also provide constraints on the P-T conditions at the time they took place. As some of the subsidence variations are difficult to explain by the lithospheric thermal regime, we also introduce the possibility that instabilities may have developed at the base of the continental lithosphere with subsequent dynamic subsidence or uplift contributions to the topography.

12.2 The Present Thermal Structure of the Congo Basin

The thermal regime of continents is determined by the distribution of radiogenic heat sources (mostly concentrated in the upper crust) and by the mantle heat-flow at the base of the lithosphere. In stable continents, variations of heat-production in the specific geological units can explain most of the variations of the surface heat-flow: in the North American Craton for instance, the mantle heat-flow is uniformly between 11 and 15 mW m^{-2}, while larger

F. Lucazeau (✉) • J. Armitage
Dynamique des Fluides Géologiques, Institut de Physique du Globe de Paris/Sorbonne Paris Cité, UMR CNRS 7154, 1 rue Jussieu, 75005 Paris, France
e-mail: lucazeau@ipgp.fr; armitage@ipgp.fr

É.K. Kabongo
UNILU, Service de Géophysique Appliquée, B.P. 1825 Lubumbashi, DR Congo
e-mail: Kadima.Kabongo@unilu.ac.cd

M.J. de Wit et al. (eds.), *Geology and Resource Potential of the Congo Basin*, Regional Geology Reviews,
DOI 10.1007/978-3-642-29482-2_12, © Springer-Verlag Berlin Heidelberg 2015

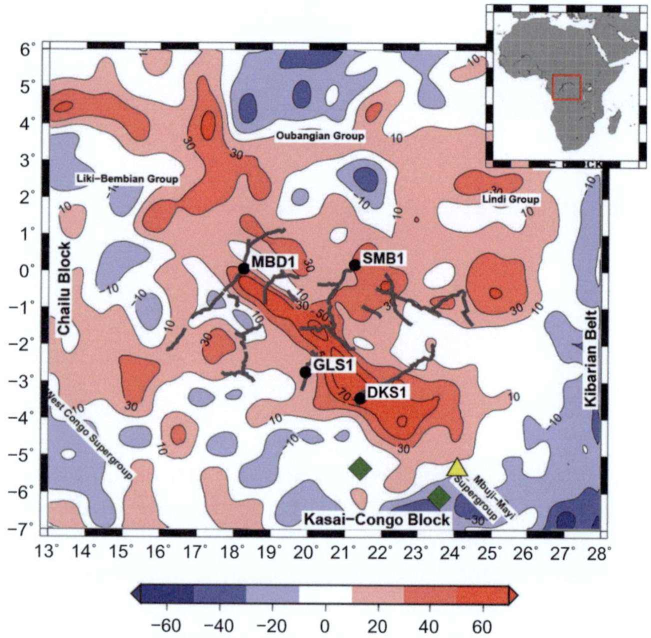

Fig. 12.1 Residual gravity map obtained by removing the attraction of sediment (Kadima Kabongo et al. 2011b). The map shows a NW–SE gravity high interpreted as the remnant effect of crustal thinning during the Proterozoic rifting and eventually reactivated during the break-up of Gondwana. The location of the four deep boreholes is shown by *black dots*, the seismic by *grey lines*, the heat-flow at Mbuji-Mahi (44 mW m^{-2}) by a *yellow triangle* and the xenoliths by *green* diamonds. The inset shows the location of the Intracratonic Congo Basin (CB)

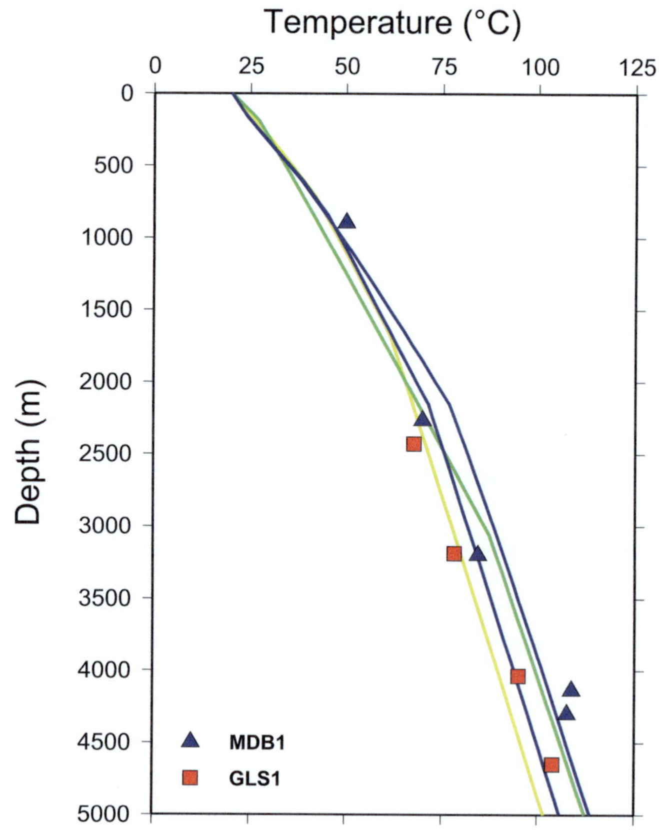

Fig. 12.2 Bottom Hole Temperatures in Gilson 1 and M'bandaka 1 wells and the present-day geotherm calculated for different erosion hypothesis (see Fig. 12.10)

variations of the surface heat-flow (~16 mW m^{-2} at the Grenville-Appalachian transition) are caused by the crustal radiogenic heat-production (Mareschal and Jaupart 2004). A similar pattern is observed in south and east Africa (Nyblade et al. 1990), with variations at the surface that are also related to the crustal heat-production (Jaupart and Mareschal 1999). The vertical distribution of heat-producing elements in the crust is well-constrained along the overturned Vredefort crustal section exposed in the Kaapvaal craton, which provides an estimate for the mantle heat-flow of 17 mW m^{-2} (Jones 1988), and more likely 20 mW m^{-2} if one accounts for paleoclimatic effects (Jaupart and Mareschal 1999). The higher mantle heat-flow can be explained by a thinner lithosphere in South Africa (~180 km according to Fishwick 2010) than in Canada (250–275 km in eastern Canada reported by Jaupart and Mareschal 1999).

In the Congo craton and Congo basin, there is conversely no published data except at one site in the Mbuji-Mayi region (Fig. 12.1) where the heat-flow is 44 $\pm$ 20 mW m^{-2} (Sebagenzi et al. 1993). However, several previous studies that addressed the thermal regime of the CB have either used empirical estimates based on measurements in similar geological contexts (e.g. Artemieva 2006) or the average of measurements in the Katangan belt published by Sebagenzi et al. (1993) rather than data from Mbuji-Mayi (Sachse et al. 2012). Artemieva (2006) considered that both heat-flow and thickness of the lithosphere are not known in the CB, and therefore speculated that they are similar to other reworked Archean cratons during late Proterozoic or Paleozoic (e.g. the Sino-Korean craton). In that case, the surface heat-flow would be $\sim$ 50–55 mW m^{-2} and the lithosphere thickness $\sim$ 150 km, but several other empirical studies (Shapiro and Ritzwoller 2004; Goutorbe et al. 2011) suggest significantly lower values (35–50 mW m^{-2}).

However, there exist several bottom hole temperatures (BHT) measurements in the Gilson 1 and M'bandaka 1 oil wells (Fig. 12.1) that can be processed together with geophysical logs to derive estimates of surface heat-flow. Although large uncertainties due to mud circulations in the wells are associated with these temperature measurements, they are deep enough to provide reliable estimates. On the Congo continental margin for instance (Lucazeau et al. 2004), the differences between conventional surface measurements and oil estimates are generally less than 10 %. The processing requires however careful estimates of the thermal conductivity to account for the lithology and porosity variations. Here, we combined lithology and geophysical logs, and core samples porosity measurements: the inferred bulk thermal conductivity ranges between 1.8 and 2.2 W m^{-1} K^{-1} . We estimated the temperature gradient to 20 m km^{-1} (Kadima Kabongo et al. 2011b), which leads to a surface heat-flow in the CB of 35–45 mW m^{-2} in agreement with the previous measurements (40–48 mW m^{-2}) at Mbuji-Mayi (Sebagenzi et al. 1993). This value ranges between the average for the Kaapvaal craton granites domes in South Africa (33 ± 2 mW m^{-2}) and the Witwatersrand basin (51 ± 6 mW m^{-2}) within this craton, but here we have no information on the crustal heat-production. In the exposed Vredefort section in South Africa, the radiogenic component is high, between 1.0 and 1.8 μW m^{-3}, and we assumed therefore that the enriched crust in the CB (fixed arbitrarily to 10 km) varies within this range: three models are considered with respective heat production of 1.25, 1.50 to 1.75 μW m^{-3} . The remaining lower crust has a standard 0.4 μW m^{-3} value (Mareschal and Jaupart 2004). Then, we can determine the geotherms assuming a similar thermal conductivity in the crust and in the mantle (including composition and temperature dependence) as assumed by Mareschal and Jaupart (2004). For a surface heat-flow of 40 mW m^{-2}, the geotherm intercepts the 1,250 °C isentropic profile between 200 and 300 km depending on the crustal heat-production (Fig. 12.3).

The base of the thermal lithosphere can be compared to the base of the seismic lithosphere, which varies between 200 and 250 km below the Congo basin according to recent tomography (Fishwick 2010). Most of other tomographic studies provide similar conclusions (see review in Buiter et al. 2012). The heat-flow estimate of 40 mW m^{-2} is therefore consistent with the seismic observations, providing that the total radiogenic contribution to surface heat-flow is 24–27 mW m^{-2} and the mantle heat-flow 13–16 mW m^{-2}. There is also evidence from kimberlites xenoliths and xenocrysts at the south and southeast of the CB that the base of the lithosphere is ~ 210 km (Batumike et al. 2009), corresponding to a 35 mW m^{-2} surface heat-flow according to the authors. Our calculations (Fig. 12.3) show that the xenoliths data can be also explained by a surface heat-flow of 40 mW m^{-2} and a total radiogenic contribution of 27 mW m^{-2} (Fig. 12.3). The xenoliths temperature/depth conditions from the Kaapvaal craton (James et al. 2004) show a significantly warmer geotherm (Fig. 12.3) in agreement with the shallower lithosphere-asthenosphere boundary. This good agreement between the surface heat-flow, the seismic tomography and the equilibration of xenoliths in these two different regions suggests that the present-day thermal regime of the CB is reasonably well constrained. Kimberlites from the Mbuji Mayi region intrude Cretaceous terranes up to 120 Ma and two megacrystic zircons are dated at ~ 70 Ma (Batumike et al. 2009). Therefore, the thermal regime of the CB remained stable at least since mid-Cretaceous.

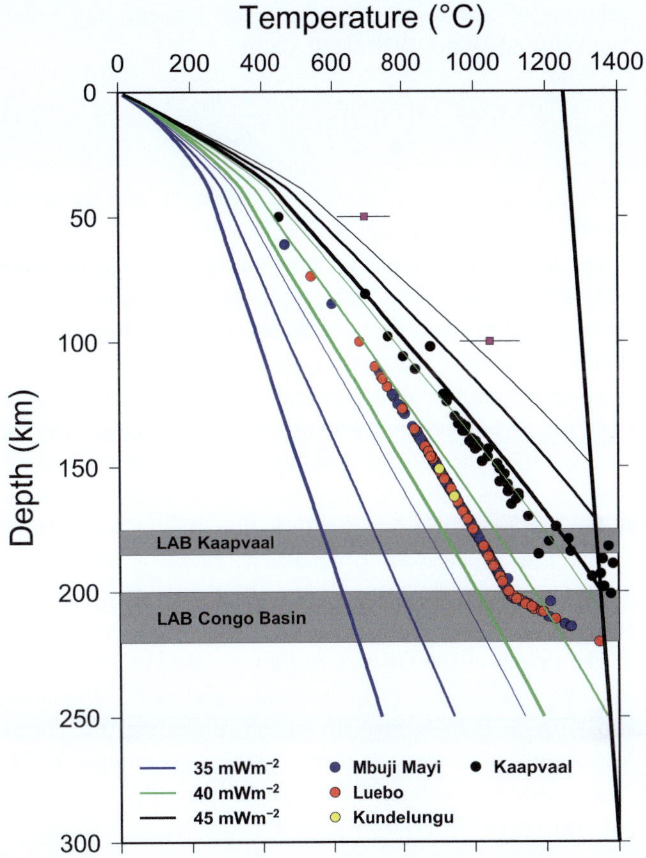

Fig. 12.3 Temperature-depth profiles for different surface heat-flows and radiogenic heat-productions: the *thick lines* correspond to higher heat-production (1.75 μW m^{-3}) than the *thin lines* (1.50 and 1.25 μW m^{-3}) in a 10 km enriched upper crust. The heat-production in the lower crust (10–40 km) corresponds to a standard value (0.4 μW m^{-3}). The geotherms are calculated for stationary conditions. The *coloured dots* correspond to the xenocryst geotherms given by garnet- olivine equilibration temperature (T_{Ni}) and the Cr-saturated garnet pressure (P_{Cr}) for three locations in the Intracratonic Congo Basin area (Batumike et al. 2009). The *black dots* are xenoliths T/depth from Kaapvaal craton (James et al. 2004). The *purple squares* correspond to temperatures at 50 and 100 km provided by Artemieva (2006). The *grey area* corresponds to the LAB depths given by seismic studies (Fishwick 2010)

12.3 Subsidence of the Congo Basin

The sediments accumulated in sedimentary basins infill a preexisting depression known as the "tectonic subsidence" related to the actual mechanism of basin initiation. The tectonic subsidence can be reconstructed back in time substituting the sediment load by an equivalent water (or air) load. The equation giving the subsidence Y from the sediment thickness S, the paleobathymetry W_d and the variation of sea level Δ_{SL} above the present day value (Steckler and Watts 1978) is given by:

$$Y = S\frac{\rho_m - \rho_s}{\rho_m - \rho_w} + W_d - \Delta_{SL}\frac{\rho_m}{\rho_m - \rho_w} \quad (12.1)$$

where ρ_m, ρ_s, ρ_w are the mantle, the sediment and the water (air) densities respectively.

The determination of the "tectonic subsidence" requires the knowledge of the sediment thickness, the paleobathymetry and the eustatic level at each step back in time. Therefore, the stratigraphic and sedimentologic constraints are the most critical information. The ideas about the formation and evolution of the CB have changed since the time the seismic lines were acquired and the deep boreholes were drilled in the seventies, but recently several studies attempted to relate the formation of the CB to a large gravity low superposed to the basin (Downey and Gurnis 2009; Crosby et al. 2010; Kadima Kabongo et al. 2011b; Buiter et al. 2012).

The origin of the CB was initially related to the deposition of a Paleozoic to Neogene "platform" sequence overlying a Precambrian basement including the Neo-Proterozoic sediments (Cahen 1954; Lepersonne 1977; Cahen and Lepersonne 1978). More recently, Lawrence and Makazu (1988), Daly et al. (1992) and Kadima Kabongo et al. (2011a) used the seismic reflection lines correlated with four deep wells and outcrop data to subdivide the sedimentary succession into three main seismo-stratigraphic units (A, B, C) separated by two regional unconformities (U2 and U3). The basal Unit A corresponds to the "Schisto-Calcaire" sequence defined in the M'bandaka and Gilson wells (Esso Zaïre 1981a, b), and correlated to the outcropping "Ituri" and "Lokoma" groups that belong to the Cryogenian/Ediacaran Lindi Supergroup (700–630 Myr). The top of Unit A is limited by a Pan-African unconformity (U2). The middle Unit B corresponds to discontinuous Paleozoic deposits including the "Red beds" sequence observed at the bottom of Dekese and Samba wells (Cahen et al. 1959, 1960) and the Lukuga Formations. The "Red beds" sequence is overlain in turn by the "Lukuga" Formation in the Dekese, Gilson and M'bandaka wells or is directly capped by the Upper-Jurassic/lower-Cretaceous unconformity as in the Samba well (Cahen et al. 1959). The upper Unit C corresponds to the Mesozoic ("Haute Lueki", "Stanleyville", "Loia", "Bokungu" and "Kwango") and Cenozoic ("Grès Polymorphes" and "Sables") deposits. Recently, this stratigraphy was revised (Linol 2013, unpublished; Linol et al. 2013c, d) using modern sequence stratigraphy analysis and new biostratigraphy and chronology of detrital zircons from the Dekese and Samba wells. This study focuses mainly on the upper Paleozoic and Mesozoic deposits, and shows that there are successive episodes (~20–40 Myr) of uplift and subsidence periods (Figs. 12.4 and 12.5). The main difference between this study and previous studies is related to the interpretation of the "Red-beds" sequence. This sequence is attributed to the Karoo Supergroup in this new study, and to the Neo-Proterozoic (Cahen et al. 1959, 1960) or the Early Paleozoic (Daly et al. 1992; Kadima Kabongo et al. 2011b) in the previous works. This debate is probably not finalised, because of the absence of the typical characteristics of the Gondwana series (diamictites and varves of the Gondwana glaciation and fossils like Gangamopteris and Glosseopteris) in the "Red beds" sequences, but it is beyond the scope of this chapter (see discussion in other chapters of this book, e.g. Linol et al. 2013a, b). The important aspect is that the subsidence of the Congo basin starts within the upper Neo-Proterozoic (Cryogenian), i.e. prior the "Red beds" deposition.

The long term accumulation of sediments in the CB has been related to the formation of a Neo-Proterozoic rift (Daly et al. 1992; Crosby et al. 2010; Kadima Kabongo et al. 2011b), but several authors (Heine et al. 2008; Downey and Gurnis 2009; Moucha and Forte 2011) related it to dynamic subsidence. However, dynamic subsidence cannot be the primary cause of basin formation and evolution over several hundred million years. Subsidence related to mantle flow is likely an additional effect superposed on the long-term evolution. The alternation of periods of subsidence and

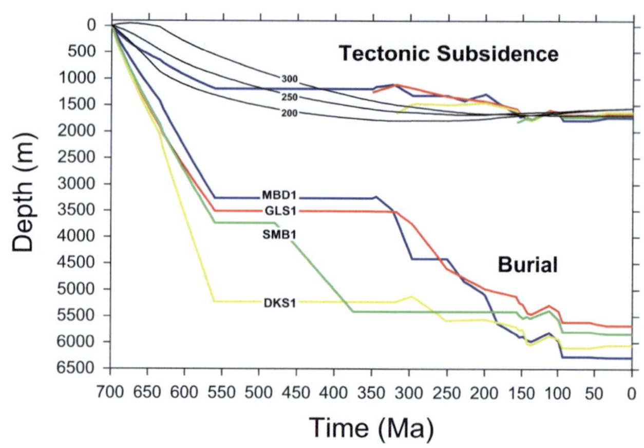

Fig. 12.4 Tectonic Subsidence and burial of the basement for the four deep wells drilled in the Intracratonic Congo Basin. The *curves* have been constructed with the recent stratigraphy of Linol et al. (2013c, d) for the upper part and the stratigraphic interpretation of Kadima Kabongo et al. (2011a) for the lower part (before PanAFrican Uncomformity). The tectonic subsidence is calculated with a method similar to that proposed by Sclater and Christie (1980) including the effects of sediment compaction, paleobathymetry or paleoelevation taken from the study of Linol et al. (2013c, d) and the first-order variations of the sea-level. We show also three tectonic subsidence curves calculated with a numerical 1D model (Lucazeau and Le Douaran 1985) including a finite duration rifting and the blanketting effect of sediments, for three different lithosphere thickness (200, 250 and 350 km)

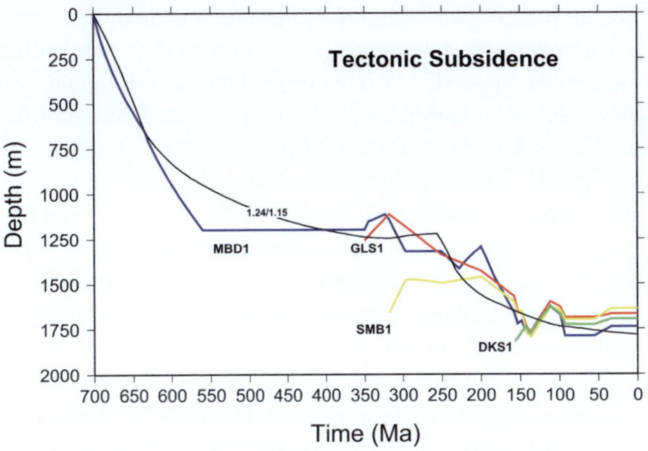

Fig. 12.5 Tectonic Subsidence of the basement for the four deep wells drilled in the Intracratonic Congo Basin. A two-stages rifting model is also shown with a thinning factor $\beta = 1.24$ during the Neo-Proterozoic and 1.15 during reactivation in Lower Triasic

uplift (Linol 2013, unpublished; Linol et al. 2013c, d), as we discuss in Sect. 12.4, is potentially is a good proxy for the contributions of such mantle perturbations to the subsidence evolution. The long-term subsidence is more likely caused by the thermal relaxation of a thick lithosphere (Armitage and Allen 2010; Crosby et al. 2010; Kadima Kabongo et al. 2011b). In Atlantic-type margins or rifted-basin like the North-Sea where the lithosphere is thin, the initial subsidence is high and the thermal subsidence is short-lived. The CB however formed within a region of thick lithosphere. The duration of the subsidence is therefore related to the diffusion time constant τ of the lithosphere:

$$\tau = \frac{L^2}{\pi^2 \kappa} \tag{12.2}$$

where L is the thickness of the lithosphere and κ the thermal diffusivity. The time constant of the subsidence for a 200 km thick lithosphere is therefore four times that of a 100 km thick lithosphere, which would explain the long-lived subsidence.

Two recent works analysed the tectonic subsidence of the CB. Crosby et al. (2010) showed that the tectonic subsidence inferred from the M'bandaka 1 and Gilson 1 oil wells (Fig. 12.1) can be well explained by the Cambrian rifting (560–490 Ma) of a 200 km thick lithosphere and a lithospheric thinning factor $\beta \simeq 1.2$. They used a finite difference numerical model that includes the cooling of the lithosphere when the lithosphere is thinning. Kadima Kabongo et al. (2011b) estimated a slightly higher $\beta \simeq 1.4$, in agreement with the crustal thinning inferred from their gravity study. However, they used a different stratigraphic interpretation based on Kadima Kabongo et al. (2011a), a simplified backstripping of the sediment load (constant densities, no variation of elevation and eustatic sea-level changes) and the original tectonic subsidence model of McKenzie (1978) where the rifting is "instantaneous". Here, we use a modified stratigraphy to take into account the recent improvements as well as the paleoelevation and first-order eustatisc history proposed by Linol et al. (2013c, d) a more elaborated technique of backstripping that takes into account the decompaction of sediment and its effects on density and thermal conductivity, and a numerical model of thermal evolution that accounts for a finite duration rifting and the effects of the sediment blanketing on this evolution (Lucazeau and Le Douaran 1985). This model is described in Appendix 1. As the stratigraphy given by Linol et al. (2013c, d) is limited to the Paleozoic, we extrapolated the stratigraphy of the Proterozoic and Lower Phanerozoic from the work of Kadima Kabongo et al. (2011a). This new procedure leads to a final value of the tectonic subsidence between 1,650 and 1,750 m (Fig. 12.4), which can be interpreted with the rifting of the lithosphere between 700 and 630 Ma by the same β factor of 1.4 that we used in the previous study. As the subsidence in the oldest period is only constrained by extrapolation of the stratigraphy on seismic lines, it is difficult to promote one of the two models with lithosphere thickness of 200 or 250 km that both produce the expected tectonic subsidence (Fig. 12.4). On the other hand, a 300 km thick lithosphere results in a time delay of about 100 Ma before the onset of subsidence that would not support the existence of a pre-PanAfrican sequence (Fig. 12.4).

The initial basin forming extensional event can be related to rifting (Fig. 12.1) and the formation of aulocogens in the late Neo-Proterozoic, such as the NE-trending Sangha aulocogen (Kadima Kabongo et al. 2011a). The relationship between subsidence initiation and landward dipping extensional structures observed here is similar to intracratonic basins of North America (Illinois and Michigan Basins), South America (Parana Basin) and the West Siberian Basin (Armitage and Allen 2010; Allen and Armitage 2012).

12.4 Subsidence Anomalies

The subsidence of the CB shows several anomalies that cannot be obviously related to a Neo-Proterozoic rifting (Fig. 12.4). A significant increase of the subsidence during the Permian to Triassic can be related to a second period of extension that would be coeval with the formation of early Permian rift basins along the northern African margin (Guiraud et al. 2005). This rifting event may have led to a general tensional regime across Gondwana (Guiraud et al. 2005), and hence altering subsidence within a suite of intracratonic basins including the CB and Berkine Basin (Algeria) to the north (Boote et al. 1998; Yahi et al. 2001). In this case, the subsidence of the CB can be decomposed in two rifting stages, a Neo-Proterozoic stage associated with $\beta = 1.24$ and a late Paleozoic stage with $\beta = 1.15$ (Fig. 12.5).

However, several short period anomalies (20–40 Ma) superposed to the long term subsidence of the CB (Fig. 12.5) cannot be explained by successions of tectonic episodes. Alternatively, in-plane stresses could change subsidence patterns of actively stretching or folding regions (Karner 1986), allowing for distal continental collisions to affect subsidence, but the continual change is suggestive of an internal mechanism rather than far-field events. Dynamic subsidence related to sub-lithospheric mantle flow can explain such anomalies, as for instance the mechanism proposed by Crosby et al. (2010) to explain the increase of subsidence during Cenozoic. Here we suggest that the 20–40 Ma subsidence perturbations are related to the gravitational instabilities caused by the changes in lithospheric thickness (and hence contrasts of density, temperature and viscosity) at the edges of the CB. The present-day thickness of lithosphere (Fishwick 2010) shows large variations around the CB from roughly 100 to 200 km that will cause movement of the lithosphere and asthenosphere. The form of mantle flow most often considered in relation to a change in lithosphere thickness is a "corner flow" (King and Anderson 1998; Sleep 2007; van Wijk et al. 2010; Hardebol et al. 2012), where the change in thickness drives a downward drip to form at the point of change in lithosphere thickness. What has received little attention is the lateral movement of the lithosphere and asthenosphere where there are lateral changes in density (e.g. Huismans and Beaumont 2011; Armitage et al. 2013).

To understand how such a change of lithospheric thickness can drive the mantle flow and affect the subsidence with the proper periods, we used an idealised viscous flow model described in Appendix 2. The initial thermal structure is of a linear geotherm from 0 °C to a temperature of 1,315 °C at the base of the lithosphere at either 100 or 200 km (Fig. 12.6). The melt depletion within the lithosphere is modelled using a parameter, X, that represents a completely compatible tracer (see Appendix 2). This tracer has an initial condition of a linear reduction from $X = 2$ to $X = 1$ at the base of the thermally defined lithosphere (Fig. 12.6). The 2-D model is set up with a 100 km wide transition zone in melt depletion and temperature such that the base of the compositional and thermally defined lithosphere increases from 100 to 200 km (Fig. 12.7a). This initial condition is then allowed to evolve through time within a Cartesian domain of fixed boundaries of dimensions 2,800 km long by 700 km deep (257 × 257 nodes) with boundary conditions of free-slip. Temperature is held fixed at 0 °C at the top and 1,315 °C at the base. The sides have a zero temperature gradient. The equations of viscous mantle flow are solved using Citcom (Moresi et al. 1996; Nielsen and Hopper 2004), and the topography is calculated from the normal stress acting on the fixed upper surface of the model.

The upper mantle evolves through time, and for the first 50 Ma there is a single large "drip-like" instability that pulls the surface down within the transition zone of the thicker region of lithosphere (Fig. 12.7b and 12.8 red line). The return flow causes a topographic high above the edge of the thinner region of lithosphere (Fig. 12.8). This return flow however does not generate melt due to decompression as it is too week and deep (as also predicted in Nielsen and Hopper 2004). As the system continues to evolve, Rayleigh Taylor instabilities form with a characteristic wavelength that is 3–4 times the thickness of the lithosphere (Jaupart et al. 2007, Fig. 12.7c, d).

Crucially, the lateral change in density set up by the change in lithosphere thickness drives migration of the transition in lithosphere domains, changing the shape of the transition in lithosphere thickness and hence altering topography. The result is a dynamically changing topography through time (Fig. 12.8), with an episodic rise and fall as convection cells form and move. By plotting the change in

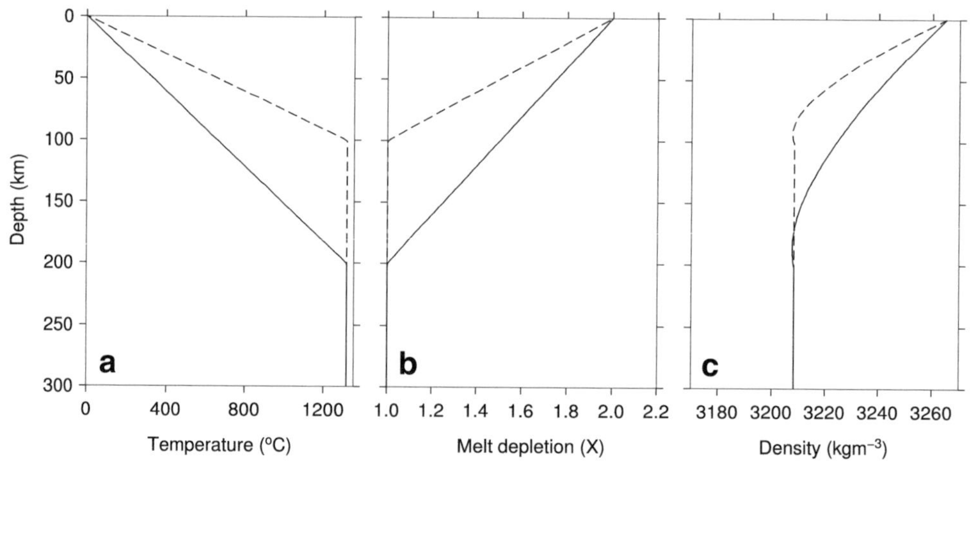

Fig. 12.6 Initial condition for the upper mantle. (**a**) Temperature of the upper mantle for the 100 km thick lithosphere, *dashed line*, and the 200 km thick lithosphere, *solid line*. (**b**) Melt depletion, the concentration of a hypothetical completely compatible trace element that remains in the solid as melt is generated. Dashed line for 100 km thick lithosphere and *solid line* for 200 km thick lithosphere. (**c**) The resulting density structure for the 100 km thick (*dashed line*) and 200 km thick lithopshere (*solid line*)

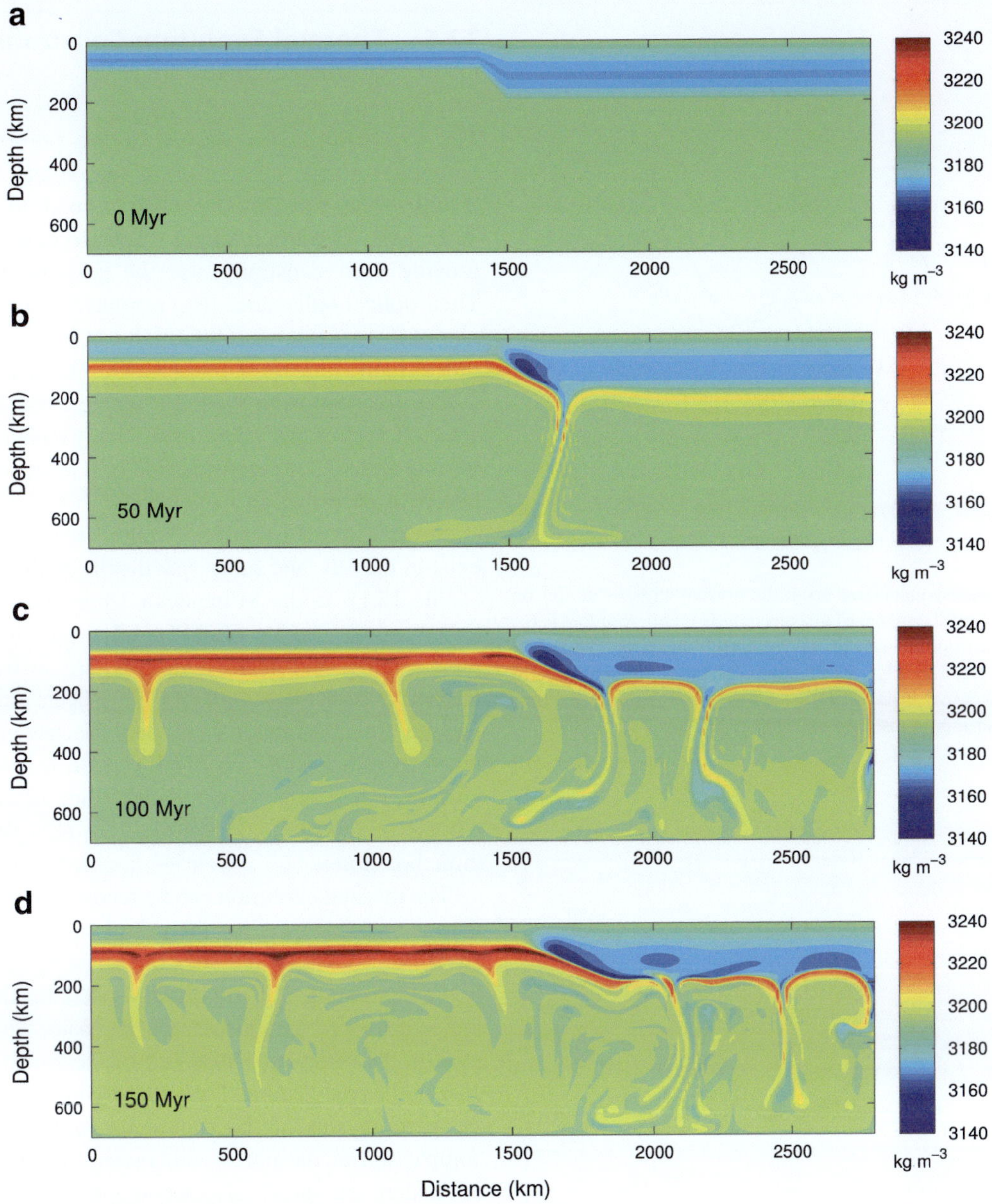

Fig. 12.7 Density of the model domain as convective instabilities form and the lateral change in density drives horizontal flow of the transition region. (**a**) Density at 0 Myr, which is the model initial condition with a 100 km wide transition zone between lithosphere thickness of 100 km on the *left* and 200 km on the *right*. (**b**) 50 Myr: where a single drip-like instability has formed due to the corner flow set up by the transition in lithosphere thickness. (**c**) 100 Myr: where convective instabilities have formed beneath both the thick and thin lithosphere with a characteristic spacing of roughly 400–600 km. (**d**) 150 Myr: where the interaction between the lateral change in density and convective instabilities has caused a change in the pattern of instabilities, the cause of change in surface topography

elevation at two locations, within the thinner lithosphere (1,250 km, Fig. 12.7) and within the thicker region of lithosphere (1,750 km), we find that locally the episodic change in elevation has a magnitude of roughly 200 m, with changes occurring over periods of 20 to 40 Ma (Fig. 12.9). This prediction is assuming that the upper mantle deforms by a very simple diffusion creep. It is more probable that the upper mantle deforms principally by dislocation creep that is pressure and temperature dependent (Asaadi et al. 2011; Lévy and Jaupart 2011). This would increase the rate of change of topography, as instabilities would form at supra-exponential rates rather than simply exponential (Conrad and Molnar 1997; Armitage et al. 2013). Edge-driven convection is clearly not a basin forming mechanism, but the changes in topography due to the natural movement of a evolving continental lithosphere will cause regional changes

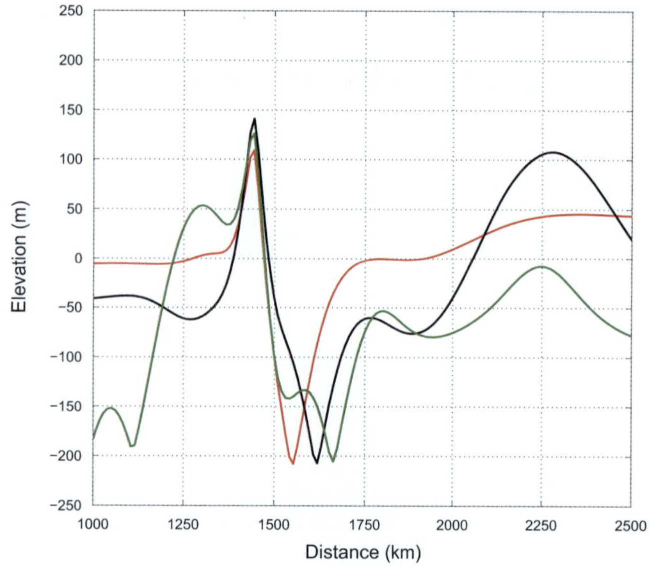

Fig. 12.8 Topography calculated from the normal stresses acting on the upper boundary of the model of viscous mantle creep. *Red line* is for 50 Ma of model evolution (see Fig. 12.7b), *green line* is at 75 Ma and the *black line* is at 125 Myr. Topography changes through lateral migration of convective instabilities set up by the transition in lithosphere thickness at model distance of 1,400–1,500 km

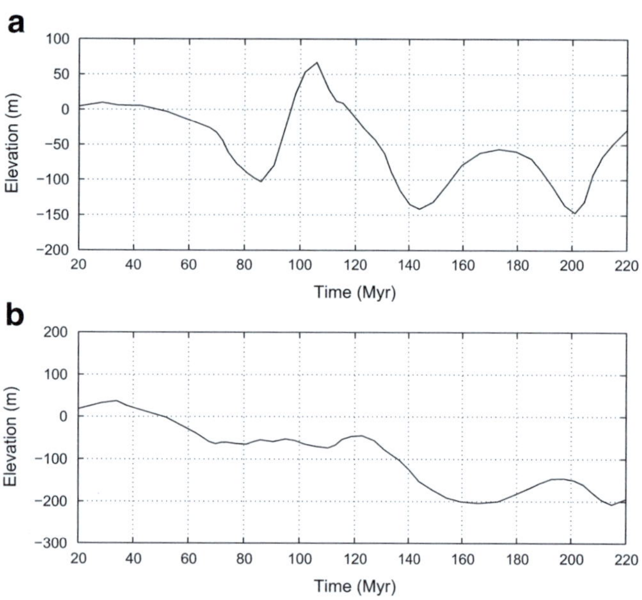

Fig. 12.9 Profiles of topography taken through time at a model distance of (**a**) 1,250 and (**b**) 1,750 km (see Fig. 12.8). The elevation of the surface alternates by a magnitude of 100–200 m over periods of 20–40 Ma depending on location

in subsidence as cratonic basins evolve. It should have also a small effect on the thermal regime compared to the thinning process associated with rifting.

12.5 Thermal Evolution Constrained by Vitrinite Data

The low heat-flow observed at the present time is not favourable to the maturation of the organic matter, except if higher heat-flow or deeper burial conditions have existed in the past. Vitrinite particles dispersed in the sediments can provide such constraints on the past thermal conditions. Their optical reflectance (R_o) is related to the transformation rate of the organic matter, which can be modelled by a first order kinetic process (Ungerer 1990; Burnham and Sweeney 1989). In addition to time and temperature, R_o can be also affected by pressure (Carr 1999), but we have not accounted for this effect. The determination of R_o from the temperature history is described in Appendix 3.

Several unpublished and published R_o measurements exist in the CB, and more specifically in the deep boreholes (Table 12.1). In the M'bandaka 1 borehole, several unpublished values can be found in oil exploration reports, but their quality and their reliability are difficult to assess. Recently, new measurements have been published on the Dekese 1 and Samba 1 well cores (Sachse et al. 2012). All these measurements are displayed on a same plot of the vitrinite reflectance versus depth (Fig. 12.10) and several models of the thermal evolution are also shown for these three boreholes.

The thermal evolution can be reconstructed with the same thermal models used for the calculation of subsidence (see Appendix 1), which includes the effect of sediment deposition and erosion. The important aspects for the maturity are the heat-flow history at base of the basin, which depends in turn on the interaction between the sediment and the lithosphere thermal regime ("thermal blanketing"), but also on the petrophysics of the sediments, which depends on the compaction of the pores with increasing depth. In the case of erosion, one should also account that the compaction state is "frozen" to the conditions of the maximum lithostatic pressure. The porosity distribution is therefore a key parameter, as it determines the density, the thermal conductivity or the radiogenic heat-production of sediments. For instance, higher porosity leads to lower thermal conductivity, but for a given heat-flow value, to higher thermal gradients than for more compacted rocks. Here we use a 1D finite difference model (Lucazeau and Le Douaran 1985) modified to account for one or several erosion stages (Pagel et al. 1997). The background heat-flow is obtained by considering one or several thinning stages of the lithosphere; therefore, it is not prescribed as in many other models, but fully constrained by the lithospheric history. As in the previous section, the modelling follows the most recent stratigraphic interpretation (Linol et al. 2013c, d) and the previous extrapolation to

Table 12.1 Value of R_o obtained in the Congo Basin deep boreholes

Well	Depth (m)	Ro (%)	Number	Stratigraphic age	Formation	References
Samba 1	657–825	0.64	3	Mid-Cretaceous	Loia	Sachse et al. (2012)
Dekese 1	1,050–1,530	0.69	28	Permian	Lukuga	Sachse et al. (2012)
M'bandaka 1	1,985	0.61	10	Trias	Haute-Lueki	Unpublished
M'bandaka 1	2,725	1.0	1	Paleozoic		Unpublished
M'bandaka 1	4,030	2.56	1	Neo-Proterozoic		Unpublished

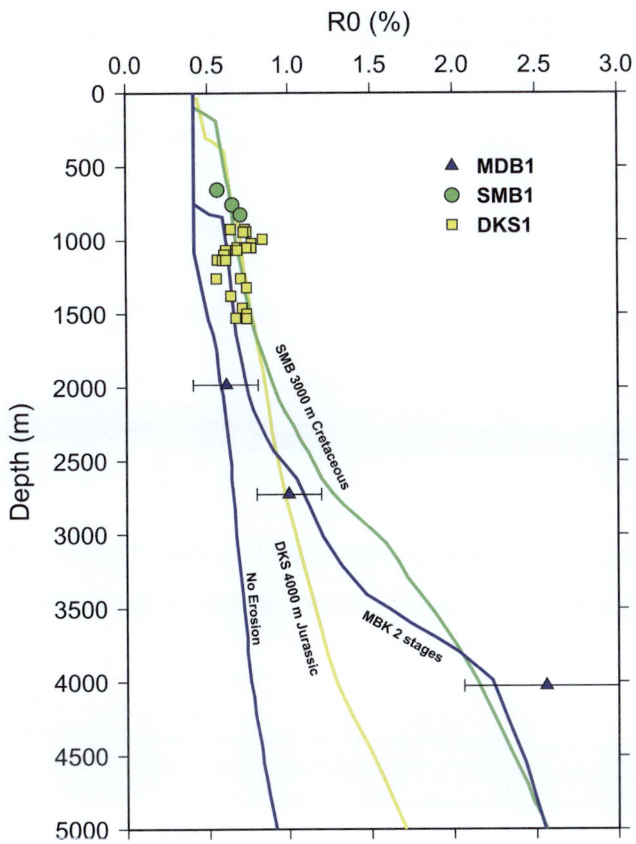

Fig. 12.10 Vitrinite reflectance data and predicting models with various thermal and burial histories. One of the *blue lines* corresponds to the hypothesis with no erosion (same profile for all boreholes): the burial and temperature history is shown in Fig. 12.11a for well MBD1. The other *blue line* corresponds to the two stages erosion (8,000 m in Paleozoic and 4,000 m in Jurassic with an additional heat-source during Neo-Proterozoic): the burial and temperature history is shown in Fig. 12.11b for well MBD1. The yellow line corresponds to one 4,000 m stage of erosion during Jurassic: the burial and temperature history is shown in Fig. 12.11c for well DKS1. The green line corresponds to one stage of 3,000 m erosion during Cretaceous with an additional heat-source: the burial and temperature history is shown in Fig. 12.11d for well SMB1

the Neo-Proterozoic (Kadima Kabongo et al. 2011a). We do not intend to present here an exhaustive analysis, but few different scenarios (Fig. 12.10) that can explain the maturation data and satisfy the thermal and geological constraints described previously.

First, we reconstruct the thermal history of sediments as described previously in the subsidence analysis (no erosion). This is illustrated for the evolution at M'bandaka 1. This "no erosion" scenario shows almost no variation of the surface heat-flow (Fig. 12.11e), although the isotherms become progressively shallower, because of the lower thermal conductivity of the uppermost sediments. In this scenario, only the Neo-Proterozoic can reach temperatures greater than 120 °C, with a maximum of 133 °C at the bottom of the basin. The present-day calculated Ro profile (blue line with label "no erosion" in Fig. 12.10) is much lower than all observed values, which clearly shows that the organic matter has recorded higher temperatures in the past. This is also a regional effect as the three boreholes where Ro has been determined are affected. The possible causes of this discrepancy are a lower conductivity, a higher heat-flow or a deeper burial. The lower conductivity hypothesis is very unlikely: it requires higher porosity for long periods of time. The existence of higher heat-flow is possible, but conflicts with the apparently stable geotherm. The most likely interpretation is that these rocks have been buried at deeper levels than their present depth and then have been exhumed during some erosional period.

If we assume that all R_o values in M'bandaka 1 are reliable, one should add at least two burial/erosion periods. The large difference between the two deepest R_o values (1.0 and 2.56 %) is difficult to explain with low temperature gradients: we need both to add 8,000 m burial (Fig. 12.11b) and to increase the heat-flow during the Neo-Proterozoic rifting by a greater thinning of the mantle ($\delta = 10$ assuming the two layers stretching model of Royden and Keen 1980). The first erosion stage would be related to the inversion of the basin during the tail end of the Pan-African orogeny and subsequent extension between Laurentia and Gondwana during break-up (Daly et al. 1992; Kadima Kabongo et al. 2011a), and therefore the in excess 8,000 m of sediment should have been deposited (and/or thickened) and eroded during the lower Paleozoic stratigraphic gap. A second 4,000 m burial/erosion stage (Fig. 12.11b) is needed to explain the intermediate value ($R_o = 1$ %) at 2,725 m. We can relate this stage to the fragmentation of the Gondwana land, which corresponds to the lower-middle Jurassic sedimentation gap in the stratigraphic record. Because of the initial thermal anomaly, the

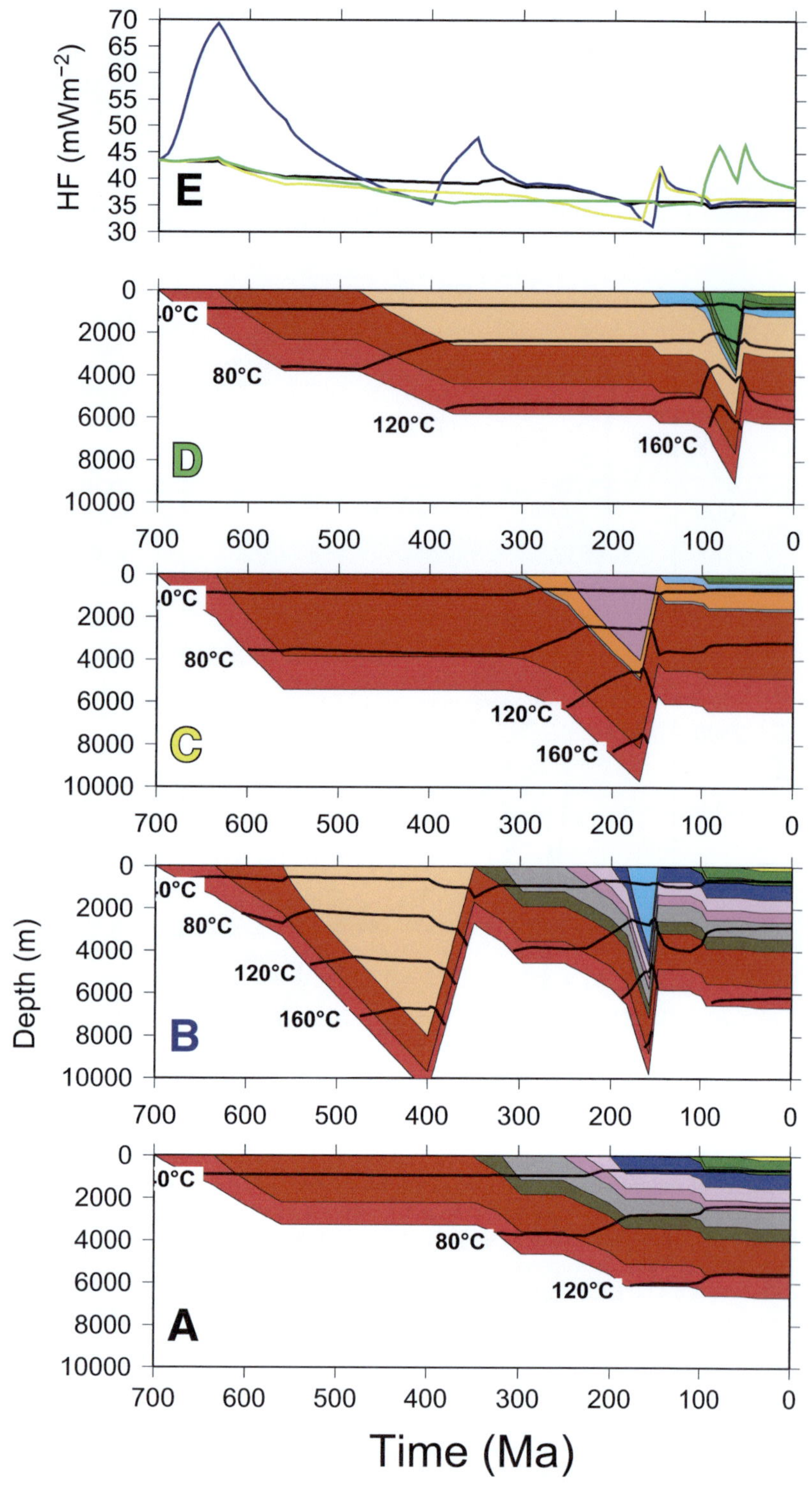

Fig. 12.11 Burial, temperature history and surface heat-flow at three boreholes where vitrinite reflectance data are available. (**A**) At M'bandaka with no erosion. (**B**) At M'bandaka for the 2 erosion stages, 8,000 m during Paleozoic and 4,000 m during Jurassic. (**C**) At Dekese with 4,000 m erosion during Jurassic. (**D**) At Samba with 3,000 m erosion and an additional heat-source during Cretaceous. (**E**) The heat-flow evolution with geological time. The vitrinite reflectance data are shown with the modelled profiles in Fig. 12.10

surface heat-flow increases to a maximum of 70 mW m^{-2}, which affects the lithospheric thermal regime for about 200 Ma (Fig. 12.11e) during the lower Paleozoic: therefore, it is possible to obtain higher temperature gradient during the Paleozoic burial stage and increase the maturation of the Neo-Proterozoic with respect to the Paleozoic (blue curve labelled MBK 2 stages in Fig. 12.10). The surface heat-flow displays two other peaks related to the erosion stages, which do not affect the lithospheric thermal regime and the maturation. The present-day temperature gradient is different for the two scenarios at M'bandaka 1 (Fig. 12.2): the "no erosion" scenario is characterized by a higher porosity and therefore a lower thermal conductivity than the scenario with high erosion for a same present-day heat-flow. This scenario obviously requires too much burial and erosion with no other geological evidence, which suggests that the maturation data at M'bandaka are either unreliable or are not representative of regional processes (local fluid circulations?).

The other R_o data have been obtained at shallower depths (Sachse et al. 2012), between 925 and 1,530 m at Dekese and between 657 and 735 m at Samba. Sachse et al. (2012) use the easy R_o method (Burnham and Sweeney 1989) implemented in the PetroMod commercial software to estimate the burial in these two boreholes. However, they used a much higher value of the present-day heat-flow (62 mW m^{-2}) based on the average of measurements in the Katagan belt (Sebagenzi et al. 1993) near the Zambia territory where Chapman and Pollack (1977) obtained similar high values that they attributed to an incipient arm of the East African rift system. Moreover, Sachse et al. (2012) progressively increased the heat-flow before the rifting to 70–80 mW m^{-2} . Consequently, they proposed a minimum erosion of 900–1,500 m, which is much less from our estimate for M'bandaka 1. This is not a methodology problem related to the calculation of R_o, as we can reproduce their results with a similar hypothesis, but rather it is related to the heat-flow assumptions. Therefore, we examine alternative interpretations based on a more realistic thermal regime.

The first scenario corresponds to burial/erosion during the lower-middle Jurassic period as assumed in the scenario for M'bandaka. This scenario can be applied to the Dekese 1 borehole where the maturation data come from the Permian "Lukuga" formation. It requires $\sim$ 4,000 m of burial/erosion (Fig. 12.11c). The second scenario is suggested by the maturation of the mid-Cretaceous "Loia" formation at Samba 1. In order to record high temperatures, the "Loia" formation should have been buried and exhumed later, probably during the late Cretacous/Paleocene gap. In this case, we need again 4,000 m of burial/erosion to raise R_o to the observed values. In order to reduce the magnitude of the burial, we tried to introduce a thermal anomaly that does not perturb the mantle lithosphere (otherwise we cannot end up with the present-day geotherm) by imposing a high temperatures at the bottom of the crust, just as if some magma issued from the sub-lithospheric mantle were trapped at that time. This heat-source produces a significant increase of the surface heat-flow and an uplift of the isotherms during the burial (Fig. 12.11d). It still requires a significant amount of burial and erosion (3,000 m), but it can match the maturation data from the three different boreholes (green curve in Fig. 12.10).

All these scenarios show however that it is very difficult (or even impossible) to explain the maturation data in the CB with realistic hypothesis for both the thermal regime and the burial/erosion history. It is difficult to change significantly thermal parameters: the lithospheric geotherm is constrained by surface heat-flow, xenoliths P/T equilibrium and seismic lithosphere thickness, and given the large thickness of the lithosphere, it takes too much time for any lithospheric perturbation to relax that it cannot be detected at the present time. The surface geothermal gradient is constrained by deep temperatures in oil wells, and unless the sediment series were such undercompacted in the past that they decreased thermal conductivity, there is no reason that it varied significantly. A further problem is that datasets of the three different boreholes lead to different interpretations of the burial/erosion history. All these interpretations assume that heat transfers in the basin are limited by the inefficiency of conductive heat transfers in the lithosphere. Alternatively, Makhous and Galushkin (2003) have suggested that the maturation jumps in the Illizi basin (Algeria) can be attributed to heat advection by volcanism and/or hydrothermal fluids. Such processes in the CB are possible but need to be documented: they would help to decrease significantly the amount of burial/erosion when only conductive heat transfers are considered.

Conclusions

Several pieces of information were used to estimate the past and present thermal regime of the Congo Basin. The present-day heat-flow was estimated from the deep temperatures and geophysical logs in two oil wells drilled in the central part of the basin. The resulting heat-flow value is low and consistent with one published value in the south of the basin (Mbuji-Mayi region). A stable geotherm was inferred from the present-day surface heat-flow and the seismic lithosphere thickness: it agrees well with the xenoliths pressure/temperature conditions. On the other hand, the magnitude and the duration of the tectonic subsidence agree well with the thermal subsidence of a thick lithosphere rifted during the NeoProterozoic by a thinning factor $\beta = 1.4$, and the residual gravity anomaly also agree with a crustal thinning of the same magnitude. Overall, these results suggest that the lithosphere slowly relaxed to the present thermal regime. Several anomalies with respect to this long-term evolution are observed during the subsidence history: some of them can be related to the far field tectonic stresses causing reactivation or inversion of the subsidence evolution, others with shorter periods (20–40 Ma) are more likely related to the instabilities caused by thickness variations at the base of lithosphere. None have such significant effect that it could affect the high maturation observed in the basin. A possible cause of this maturation is that organic matter was buried at sufficient depths in the past and then exhumed during erosion. With the appropriate thermal conditions, it requires however a minimum of 4 km of burial and erosion, and taking into account some of the data, it could exceed 10 km! In addition, the burial/erosion inferred at one borehole is not always consistent with that at other boreholes. Therefore, as heat-flow cannot vary significantly in the past and as burial/erosion is also questionable, alternative explanations to increase sediment temperatures in the past are needed. Hydrothermal fluids or volcanism are obviously efficient mechanisms that could contribute to anomalous maturation (Makhous and Galushkin 2003), but they should be long enough that the kinetic transformation are possible (few Ma)! In this paper, we tested an intermediate scenario including advection of heat at the base of the crust (magma trapping) and diffusion of this heat within the crust and basin. This hypothetical scenario produce a maturation profile that can match all boreholes data without violating thermal observations (present day heat-flow, xenoliths P/T and time, lithosphere thickness). Obviously, additional geological constraints are needed to infer scenarios for the maturation of organic matter in a cold cratonic lithosphere.

Acknowledgements Marteen De Witt, Robert Harris, Jean Braun and Nicky White are thanked for their constructive remarks and reviews. This is IPGP contribution # 3394.

Appendix 1: Thermal Model

The thermal evolution is described by the 1D heat equation:

$$\partial_z[\lambda(z,t)\partial_z T] + A(z,t) = \rho c(z,t) d_t T \qquad (12.3)$$

where $\lambda(z, t)$ is the thermal conductivity at depth z and time t, $\rho c(z, t)$ is the specific heat per volume unit and $A(z, t)$ is the heat production. This equation is solved numerically by a finite differences method initially developed by Lucazeau and Le Douaran (Lucazeau and Le Douaran 1985). We use a Lagrangian frame, which allows to describe more easily the sedimentation (by adding nodes), the erosion (by removing nodes) or compaction and extension (by changing the thickness of a cell). The initial conditions assume thermal equilibrium within a 200 km thick lithosphere, with a 40 km thick crust and a 10 km enriched radiogenic upper crust. The upper and and lower boundary conditions are fixed temperatures (20 and 1,350 °C). This corresponds to a surface heat-flow of 43 mW m^{-2} and a mantle heat-flow of 16 mW m^{-2} (i.e. the upper limit assumed for the north American craton).

The time-dependant equation is solved implicitly, which gives less precision but more stability. In a first stage, we consider the rifting of the lithosphere. Modelled strain rates are typically low and therefore temperatures do not increase significantly due to the upward advection of mantle material, as the mantle cools at the same time as it deforms. This reduces the degree of subsidence during the stretching phase. The strain rate $\dot{\varepsilon}$ is related to the thinning factor β by:

$$\beta = exp(\dot{\varepsilon}\Delta t) \qquad (12.4)$$

The subsidence of the CB constrains the thinning of the crust to a maximum of about 1.4. If we assume that extension ranges from 700 to 630 Ma, the strain rate is of the order of $10^{-16}\ s^{-1}$.

Lithosphere Physical Properties

The thermal conductivity of the mantle includes a lattice component λ_l, which depends on temperature and pressure, and a radiative component λ_r:

$$\lambda_l = 4.13\sqrt{\frac{298}{T_{abs}}\left(1 + 0.032\left(3.2410^{-5}(z - z_{crust}) + 1.14\right)\right)} \qquad (12.5)$$

$$\lambda_r = 3.6810^{-10} T_{abs}^3 \qquad (12.6)$$

These expressions are derived from a compilation of literature data (Jaupart and Mareschal 1999).

The thermal conductivity of the crust also depends on the temperature and the thermal conductivity λ_0 measured in laboratory conditions (Durham et al. 1987):

$$\lambda = 2.264 - \frac{618.2}{T_{abs}} + \lambda_0\left(\frac{355.6}{T_{abs}} - 0.3025\right) \qquad (12.7)$$

Isostasy

We assume Airy-type isostasy, with a constant weight per surface unit at a reference level Z_{ref} in the asthenosphere. The initial state provides the reference weight M_0 assuming that elevation corresponds to the sea-level:

$$M_0 = \rho_c \int_0^c (1 - \alpha T(z))dz + \rho_m \int_c^L (1 - \alpha T(z))dz + (Z_{ref} - L)\rho_{asth} \qquad (12.8)$$

where α is the coefficient of thermal expansion, ρ_c and ρ_m the density of the crust and mantle in laboratory conditions, T_L the temperature at the base of the lithosphere and $\rho_{asth} = \rho_m(1 - \alpha T_L)$ the density of asthenosphere.

At each time, the weight M1 is calculated to estimate the subsidence S:

$$S = \frac{M_1 - M_0}{\rho_{asth} - \rho_w} \tag{12.9}$$

Sediment physical properties

The porosity varies with depth z:

$$\phi(z) = \phi_0 exp\left(\frac{-z}{z_c}\right) \tag{12.10}$$

where ϕ_0 is the porosity at the surface and z_c the compaction length. The thickness of a cell varies accordingly:

$$\Delta z = \Delta z_0 \frac{(1 - \phi_0)}{(1 - \phi(z))} \tag{12.11}$$

where Δz and Δz_0 are the compacted and decompacted thicknesses of a cell. The thermal conductivity follows a geometric mean in the model:

$$\lambda_{bulk} = \lambda_{water}^{\phi(z)} \lambda_{matrix}^{1-\phi(z)} \tag{12.12}$$

where λ_{bulk} is the bulk conductivity, $\lambda_{water} = 0.6\ \mathrm{W\ m^{-1}\ K^{-1}}$ is the conductivity of water, and λ_{matrix} is the conductivity of the rock matrix.

The density (and the specific volumetric heat) follow harmonic means:

$$\rho_{bulk} = \rho_{water}\phi(z) + \rho_{matrix}(1 - \phi(z)) \tag{12.13}$$

where λ_{bulk} is the bulk density, λ_{water} is the density of water, and λ_{matrix} is the density of the rock matrix.

Appendix 2: Model of Lithospheric Instabilities

To understand how a change in lithosphere thickness at the margins of the Congo Basin will affect subsidence we form an idealised model of viscous flow within a 2-D Cartesian domain. We assume that deformation of the lithosphere and asthenosphere can be effectively captured as a Stokes flow with the Boussinesq approximation. The numerical model is as outlined in Armitage et al. (2008, 2013). The key equation is the momentum balance,

$$-\frac{\partial \tau_{ij}}{\partial x_j} + \frac{\partial p}{\partial x_i} = \Delta\rho g \lambda_i \tag{12.14}$$

where τ_{ij} is the deviatoric stress, p is pressure, g is gravity and λ_i is a unit vector in the vertical direction. The change in density, $\Delta\rho$ is a function of temperature and melt depletion, F,

$$\frac{\Delta\rho}{\rho_0} = \Delta\rho_T + \Delta\rho_F = \alpha T + \beta F \tag{12.15}$$

where ρ_0 is the mantle density at the surface, $\alpha = 3.3 \times 10^{-5}\ \mathrm{K^{-1}}$ is the thermal expansion coefficient and β is the coefficient for change in density due to melt depletion (see below). Melt depletion F is tracked by a parameter X, which is a hypothetical completely compatible trace element. The value of X increases from 1 as melting progresses. The trace element can be related to melt depletion, F, assuming that the melt and solid remain in equilibrium with each other, by a simple mass balance, $F = (X - 1)/X$. We can then base the initial density variation with depth due to melt depletion on density changes predicted from melting experiments of natural and model rocks. We assume that change in density is a linear function of the removal of melt (Equation 12.15). This allows for β to be calculated from a reference state of melt depletion (Scott 1992),

$$\beta = \frac{\Delta\rho_{ref} X_{ref}}{\rho_0 (X_{ref} - 1)} \tag{12.16}$$

where $X_{ref} = 1.3$, which is equivalent to F = 23 %. From the melting of model fertile lherzolite compositions and estimates from natural rocks, complete clynopy-roxene removal occurs at melt depletion of 23 %. This is our reference state of melt depletion, and using $\Delta\rho_{ref} = 34\ \mathrm{kg\ m^{-3}}$ as estimated for Proterozoic lithosphere (Poudjom Djomani et al. 2001), $\beta = 0.04$.

We consider a simple temperature dependent diffusion creep that is a reasonable approximation for the deformation of the Earth based on geological observations (Watts and Zhong 2000). Viscosity is given by,

$$\eta = \eta_0 \eta_{ref} exp\left(\frac{E}{RT}\right) \tag{12.17}$$

where $E = 120\ \mathrm{kJmol^{-1}}$ (Watts and Zhong 2000). The scaling viscosity, η_0 to dimensionalise dimensionless viscosity, is set from the thermal Rayleigh number,

$$Ra = \frac{\alpha g \rho_m \Delta T d^3}{\kappa \eta_0} = 4.877 \times 10^6 \tag{12.18}$$

where d = 700 km is the depth to the base of the model; $\kappa = 10^{-6}\ \mathrm{m^2\ s^{-1}}$ is the thermal diffusivity and $\Delta T = 1{,}315$ K is the temperature difference between the surface and depth d. Using this Rayleigh number, the scaling mantle viscosity is $\eta_0 = 10_{20}$ Pa s. Finally $\eta_{ref} = 1.129$, which is calculated from a reference state ($T_{ref} = 1{,}588$ K) such that viscosity $\eta = 10_{20}$ Pa s at 200 km depth. The relatively small activation energy is justified by the observation within numerical experiments that a larger activation energy would predict too large a flexural rigidity or too thick an elastic plate near sea-mounts (Watts and Zhong 2000). This rheology is an obvious simplification, but it allows us to explore the general behaviour of the upper mantle with a geologically reasonable viscosity structure.

Appendix 3: Calculation of R_o

The calculation of R_o is similar to previous studies based on the transformation rate of vitrinite (Burnham and Sweeney 1989; Ungerer 1990); it is assumed that the vitrinite transformation can be described by a set of N independent Arrhenius first-order kinetic reactions:

$$\frac{dx_i}{dt} = A_i exp\left(\frac{E_i}{R\,T_{abs}}\right) x_i(t) \tag{12.19}$$

where $\frac{dx_i}{dt}$ is the rate of the *i*th reaction, A_i the Arrhenius frequency factor (s^{-1}), E_i the activation energy (kcal/mole), R the gas constant (0.001987 kcal/mole K), T_{abs} the absolute temperature and x_i the initial hydrocarbon potential for the *it* h reaction (mg hydrocarbon/TOC). x_i (*t*) can be determined by integration of the previous equation:

$$x_i(t) = x_i(t_0)\left(1 - exp\left(-A_i R \frac{T^2 exp\left(\frac{-E_i}{RT_{abs}}\right) + T_0^2 exp\left(\frac{-E_i}{RT_{0abs}}\right)}{E_i \frac{T-T_0}{t-t_0}}\right)\right) \tag{12.20}$$

We have used a spectrum of 15 reactions (46–74 kJ) to determine the transformation rate of vitrinite and then an empirical relationship for the conversion to R_o. The effect of pressure, which tends to decrease the R_o value for a given heating rate (Carr 1999), has not been included.

References

Allen P, Armitage J (2012) Cratonic basins. In: Tectonics of sedimentary basins. In: Busby C, Azor A (eds) Recent advances. Blackwell Publishing, pp 602–620

Armitage JJ, Allen PA (2010) Cratonic basins and the long-term subsidence his- tory of continental interiors. J Geol Soc 167:61–70. doi:10.1144/0016-76492009-108

Armitage JJ, Henstock TJ, Minshull TA, Hopper JR (2008) Modelling the composition of melts formed during continental breakup of the Southeast Greenland margin. Earth Planet Sci Lett 269:248–258. doi:10.1016/j.epsl.2008.02.024

Armitage JJ, Jaupart C, Fourel L, Allen PA (2013) The instability of continental passive margins and its effect on continental topography and heat flow. J Geophys Res 101:11503–11518. doi:10.1002/jgrb.50097

Artemieva IM (2006) Global 1x1 thermal model TC1 for the continental lithosphere: Im- plications for lithosphere secular evolution. Tectonophysics 416:245–277. doi:10.1016/j.tecto.2005.11.022

Asaadi N, Ribe NM, Sobouti F (2011) Inferring nonlinear mantle rheology from the shape of the Hawaiian swell. Nature 473:501–504. doi:10.1038/nature09993

Batumike J, Griffin W, O'Reilly S (2009) Lithospheric mantle structure and the diamond potential of kimberlites in southern D.R. Congo. Lithos 112S:166–176. doi:10.1016/j.lithos.2009.04.020

Boote DRD, Clark-Lowes DD, Traut MW (1998) Palaeozoic petroleum sys- tems of North Africa. Geol Soc Lond Spec Publ 132:7–68. doi:10.1144/GSL.SP.1998.132.01.02

Buiter SJ, Steinberger B, Medvedev S, Tetreault JL (2012) Could the mantle have caused subsidence of the Congo Basin? Tectonophysics 514–517:62–80. doi:10.1016/j.tecto.2011.09.024

Burnham AK, Sweeney JJ (1989) A chemical kinetic model of vitrinite matu- ration and reflectance. Geochim Cosmochim Acta 53:2649–2657. doi:10.1016/0016-7037(89)90136-1

Cahen L (1954) Géologie du Congo Belge. H Vaillant-Carmanne, Liège

Cahen L, Lepersonne J (1978) Synthèse des connaissances relatives au groupe (an- ciennement série) de la Lukuga (permien du Zaïre), Ann. Mus. Roy. Afr. Cent., Tervuren (Belgique), série in-8, Sci. Géol., 82, 115–152

Cahen L, Ferrand JJ, Haarsma MJ, Lepersonne J, Verbeek T (1959) Description du Sondage de Samba, Ann. Mus. Roy. Congo belge, Tervuren (Belgique), série in-8, Sci. Géol., 29, 210–210

Cahen L, Ferrand JJ, Haarsma MJ, Lepersonne J, Verbeek T (1960) Description du Sondage de Dekese, Ann. Mus. Roy. Congo belge, Tervuren (Belgique), série in-8, Sci. Géol., 34, p 115

Carr AD (1999) A vitrinite reflectance kinetic model incorporating overpressure re- tardation. Mar Petrol Geol. 16: 355–377. doi:10.1016/S0264-8172(98)00075-0

Chapman DS, Pollack HN (1977) Heat Flow And Heat Production In Zambia: Evidence for lithospheric thinning in central Africa. Tectonophysics 41:79–100. doi:10.1016/0040-1951(77)90181-0

Conrad CP, Molnar P (1997) The growth of Rayleigh–Taylor-type instabilities in the lithosphere for various rheological and density structures. Geophys J Int 129:95–112. doi:10.1111/j.1365-246X.1997.tb00939.x

Crosby AG, Fishwick S, White N (2010) Structure and evolution of the in- tracratonic Congo Basin. Geochem Geophys Geosyst. 11: Q06010–Q06 010. doi:10.1029/2009GC003014

Daly MC, Lawrence SR, Diemu-Tshiband K, Matouana B (1992) Tectonic evolution of the Cuvette Centrale, Zaire. J Geol Soc Lond 149:539–546. doi:10.1144/gsjgs.149.4.0539

Downey NJ, Gurnis M (2009) Instantaneous dynamics of the cratonic Congo basin. J Geophys Res Solid Earth. 114: B06401. doi: 10.1029/2008JB006066

Durham WB, Mirkovich VV, Heard HC (1987) Thermal diffusivity of igneous rocks at elevated pressure and temperature. J Geophys Res 92:11615–11634. doi:10.1029/JB092iB11p11615

Esso Zäire SARL (1981a) Geological completion report. Mbandaka 1, Tech. Rep. un- published report, Esso Zäire SARL

Esso Zäire SARL (1981b) Geological completion report. Gilson 1, Tech. Rep. unpub- lished report, Esso Zäire SARL

Fishwick S (2010) Surface wave tomography: Imaging of the lithosphere-asthenosphere boundary beneath central and southern Africa? Lithos 120:63–73. doi:10.1016/j.lithos.2010.05.011

Goutorbe B, Poort J, Lucazeau F, Raillard S (2011) Global heat flow trends re- solved from multiple geological and geophysical proxies. Geophys J Int 187:1405–1419. doi:10.1111/j.1365-246X.2011.05228.x

Guiraud R, Bosworth W, Thierry J, Delplanque A (2005) Phanerozoic geological evolution of Northern and Central Africa: An overview. J Afr Earth Sci 43:83–143. doi:10.1016/j.jafrearsci.2005.07.017

Hardebol NJ, Pysklywec RN, Stephenson R (2012) Small-scale convection at a continental back-arc to craton transition: application to the southern Canadian Cordillera. J Geophys Res. 117: B01 408. doi: 10.1029/2011JB008431

Heine C, Muller RD, Steinberger B, Torsvik TH (2008) Subsidence in intra- continental basins due to dynamic topography. Phys Earth Planet Inter 171:252–264. doi:10.1016/j.pepi.2008.05.008

Huismans R, Beaumont C (2011) Depth-dependent extension, two-stage breakup and cratonic underplating at rifted margins. Nature 473:74–78. doi:10.1038/nature09988

James DE, Boyd FR, Schutt D, Bell DR, Carlson RW (2004) Xenolith con- straints on seismic velocities in the upper mantle beneath southern Africa. Geochem Geophys Geosyst. 5: Q01 002. doi: 10. 1029/2003GC000551

Jaupart C, Mareschal JC (1999) The thermal structure and thickness of continental roots. Lithos 48:93–114. doi:10.1016/S0024-4937(99) 00023-7

Jaupart C, Molnar P, Cottrell E (2007) Instability of a chemically dense layer heated from below and overlain by a deep less viscous fluid. J Fluid Mech 572:433–469. doi:10.1017/S0022112006003521

Jones MQW (1988) Heat flow in the Witwatersrand basin and environs and its signif- icance for the South African shield geotherm and lithosphere thickness. J Geophys Res 93:3234–3260

Kadima Kabongo E, Delvaux D, Sebagenzi Ntabwoba Mwene S, Talk L, Kabeya SM (2011a) Structure and geological history of the Congo Basin: an integrated interpretation of gravity, magnetic and reflection seismic data. Basin Res. 23: 499–527. doi: 10.1111/j. 1365-2117.2011.00500.x

Kadima Kabongo E, Sebagenzi Ntabwoba Mwene S, Lucazeau F (2011b) A Proterozoic-rift origin for the structure and the evolution of the cra- tonic Congo basin. Earth Planet Sci Lett. 304: 240–250. doi: 10.1016/j.epsl.2011.01.037

Karner GD (1986) Effects of lithospheric in-plane stress on sedimentary basin stratigra- phy. Tectonics 5:573–588. doi:10.1029/ TC005i004p00573

King SD, Anderson DL (1998) Edge-driven convection. Earth Planet Sci Lett 160:289–296

Lawrence SR, Makazu MM (1988) Zaire's Central basin: prospectivity outlook. Oil Gas J 86:105–108

Lévy F, Jaupart C (2011) Temperature and rheological properties of the mantle beneath the North American craton from an analysis of heat flux and seismic data. J Geophys Res 116(B01):408. doi:10.1029/ 2010JB007726

Lepersonne J (1977) Structure gélogique du bassin intérieur du Zaïre, Bull. Aca. Roy. Bel., Classe des Sciences, Ve série, LXIII, 941–965

Linol B (2013) Sedimentology and sequence stratigraphy of the Congo and Kalahari Basins of South-central Africa and their evolution during the formation and Break-up of West Gondwana, Ph.D. thesis. Nelson Mandela Metropolitan University, unpublished

Linol B, de Wit M, Barton E, Guillocheau F, de Wit M, Colin J (2013a) Facies analysis, chronostratigraphy and paleo-environmental reconstructions of the Jurassic-Cretaceous sequences of the CB. In: de Wit M, Guillocheau F, Fernandez-Alonso M, Kanda N, Wit MD (eds) The geology and resource potential of the Congo Basin, Chap. 4a, unpublished. Springer, Heidelberg

Linol B, de Wit M, Barton E, Guillocheau F, de Wit M, Colin JP (2013b) Carboniferous-Triassic stratigraphy and paleogeography of Karoo-like sequences of the CB. In: de Wit M, Guillocheau F, Fernandez-Alonso M, Kanda N, Wit MD (eds) The geology and resource potential of the Congo Basin, Chap. 3b, unpublished. Springer, Heidelberg

Linol B, de Wit MJ, Guillocheau F, Robin C, Dauteuil O (2013c) Multiphase phanerozoic subsidence and uplift history recorded in the CB – A complex successor basin. In: de Wit M, Guillocheau F, Fernandez-Alonso M, Kanda N, Wit MD (eds) The geology and resource potential of the Congo basin, Chap. 6b, unpublished. Springer, Heidelberg

Linol B, de Wit MJ, Milani E, Guillocheau F (2013d) New regional correlations between the CB and the Paraná Basin of central West Gondwana. In: de Wit M, Guillocheau F, Fernandez-Alonso M, Kanda N, Wit MD (eds) The geology and resource potential of the Congo Basin, Chap. 7b, unpublished. Springer, Heidelberg

Lucazeau F, Le Douaran S (1985) The blanketting effect of sediments in basins formed by extension : a numerical model. Application to the gulf of Lion and the Viking graben. Earth Planet Sci Lett 74:92–102. doi:10.1016/0012-821X(85)90169-4

Lucazeau F, Brigaud F, Bouroullec JL (2004) High resolution Heat Flow Density in lower Congo basin from probe measurements, oil exploration data and BSR. Geochem Geophys Geosyst 5:1–24. doi:10.1029/2003GC000644

Makhous M, Galushkin YI (2003) Burial history and thermal evolution of the northern and eastern Saharan basins. AAPG Bull 87:1623–1651. doi:10.1306/04300301122

Mareschal JC, Jaupart C (2004) Variations of surface heat flow and lithospheric thermal structure beneath the North American craton. Earth Planet Sci Lett 223:65–77. doi:10.1016/j.epsl.2004.04.002

McKenzie DP (1978) Some remarks on the development of sedimentary basins. Earth Planet Sci Lett 40:25–32. doi:10.1016/0012-821X (78)90071-7

Moresi L, Zhong S, Gurnis M (1996) The accuracy of finite element solutions of Stokes's flow with strongly varying viscosity. Phys Earth Planet Inter 97:83–94. doi:10.1016/0031-9201(96) 03163-9

Moucha R, Forte AM (2011) Changes in African topography driven by mantle convection. Nat Geosci 4:707–712. doi:10.1038/ngeo1235

Nielsen TK, Hopper JR (2004) From rift to drift: mantle melting during continental breakup, Geochem Geophys Geosyst. 5: Q07 003. doi: 10.1029/2003GC000662

Nyblade AA, Pollack HN, Jones DL, Podmore F, Mushayandebvu M (1990) Terrestrial heat flow in East and Southern Africa. J Geophys Res 95:17–371

Pagel M, Braun J-J, Disnar JR, Martinez L, Renac C, Vasseur G (1997) Thermal history constraints from studies of organic matter, clay minerals, fluid inclusions, and apatite fission tracks at the Ardeche paleo-margin (BA1 drill hole, GPF Program), France. J Sediment Res 67:235–245. doi:10.1306/D4268540-2B26-11D7-8648000102C1865D

Poudjom Djomani YH, O'Reilly SY, Griffin W, Morgan P (2001) The density structure of subcontinental lithosphere through time. Earth Planet Sci Lett 184:605–621. doi:10.1016/S0012-821X(00)00362-9

Royden LH, Keen CE (1980) Rifting Process and Thermal Evolution of the Conti- nental Margin of Eastern Canada determined from Subsidence curves. Earth Planet Sci Lett 51:343–361

Sachse VF, Delvaux D, Littke R (2012) Petrological and geochemical investiga- tions of potential source rocks of the central Congo Basin, Democratic Republic of Congo. AAPG Bull 96:245–275. doi:10.1306/07121111028

Sclater JG, Christie PAF (1980) Continental stretching: an explanation of the post-mid-cretaceous subsidence of the central North Sea basin. J Geophys Res 85:3711–3730. doi:10.1029/JB085iB07p03711

Scott DR (1992) Small-scale convection and mantle melting beneath mid-ocean ridges. In: Geophys. Monogr. Ser., vol 71. AGU, Washington, DC, pp 327–352. doi: 10.1029/GM071p0327

Sebagenzi MN, Vasseur G, Louis P (1993) First heat flow density determinations from Southeastern Zaire (Central Africa). J Afr Earth Sci 16:413–423. doi:10.1016/0899-5362(93)90100-5

Shapiro NM, Ritzwoller MH (2004) Inferring surface heat flux distributions guided by a global seismic model: particular application to Antarctica. Earth Planet Sci Lett 223:213–224. doi:10.1016/ j.epsl.2004.04.011

Sleep NH (2007) Edge-modulated stagnant-lid convection and volcanic passive margins. Geochem Geophys Geosyst. 8: Q12 004. doi: 10. 1029/2007GC001672

Steckler MS, Watts AB (1978) Subsidence of the Atlantic-type continental margin off New York. Earth Planet Sci Lett 41:1–13

Ungerer P (1990) State of art of research in kinetic modelling of oil formation. Org Geochem 16:1–25. doi:10.1016/0146-6380(90) 90022-R

van Wijk J, Baldridge W, van Hunen J, Goes S, Aster R, Coblentz D, Grand S, Ni J (2010) Small-scale convection at the edge of the Colorado Plateau: Im- plications for topography, magmatism, and evolution of Proterozoic lithosphere. Geology 38:611–614. doi:10. 1130/G31031.1

Watts AB, Zhong S (2000) Observations of flexure and the rheology of oceanic lithosphere. Geophys J Int 142:855–875. doi:10.1046/j. 1365-246x.2000.00189.x

Xie X, Heller PL (2009) Plate tectonics and basin subsidence history. Geol Soc Am Bull 121:55–64. doi:10.1130/B26398.1

Yahi N, Schaefer RG, Littke R (2001) Petroleum generation and accumulation in the Berkine basin, eastern Algeria. Am Assoc Petrol Geol Bull. 85: 1439–1467. http://techplace.datapages.com/data/ bulletns/2001/08aug/1439/1439.htm

New Regional Correlations Between the Congo, Paraná and Cape-Karoo Basins of Southwest Gondwana

13

Bastien Linol, Maarten J. de Wit, Edison J. Milani, Francois Guillocheau, and Claiton Scherer

13.1 Introduction

The recognition of the complementary shapes of Africa and South America as early as the sixteenth century generally suggested that the two continents were once united (e.g. Oreskes 2001). This idea was latter transformed into the hypothesis of the displacement of the continents, or 'Continental Drift', by Wegener (1912), and then supported by field observations by Keidel (1916) and Du Toit (1926, 1927, 1937) who, first correlated in detail the litho- and bio-stratigraphy of the Paleozoic Cape and Karoo sequences of southern Africa, with the Sierra de la Ventana and Paraná sequences of South America. Thereafter, pioneering Belgian geologists Robert (1946) and Cahen (1954) also established robust correlations between the Congo Basin (CB) of central Africa and the Karoo and Kalahari Basins of southern Africa, thus suggesting a unified stratigraphy across large parts of southwest Gondwana. Subsequently, during the two decades 1947–1967, advances in paleomagnetism and the discovery of sea-floor spreading finally allowed more accurate plate-reconstructions. Based on this new data, the first detailed Geological Map of Gondwana was produced by de Wit et al. (1988), clearly documenting similarities and differences in the stratigraphy of Africa, South America, Madagascar, India, Antarctica, Australia and New Zealand when viewed in this Gondwana framework (Fig. 13.1).

Today, paleogeographic reconstructions of Gondwana can be routinely computed and animated through geological times using modern software systems (e.g. http://www.gplates.com; www.reeves.nl/Gondwana). But these models need to be tested and continuously refined further through new field observations and correlations between the different continents (e.g. Iannuzzi and Boardman 2007: 'Problems in western Gondwana geology' workshop). Here, based on recent field-work in the Democratic Republic of Congo (DRC) and in Brazil, and modern basin analysis of the Congo, Paraná and Cape-Karoo Basins (Milani 1997; Milani and de Wit 2008; Linol 2013), we describe new regional stratigraphic correlations of their Phanerozoic sequences with the aim to improve paleogeographic and geodynamic models of south-west Gondwana.

13.2 General Sequence Stratigraphic Framework

The CB of central Africa and the Paraná Basin (PB) of eastern South America, now on the opposite margins of the South Atlantic, are two large and long-lived Phanerozoic basins that preserve comparatively similar stratigraphic records (Fig. 13.2).

The subcircular CB is the largest basin (ca. 1.8 million km^2), formed near the center of Gondwana, and now completely surrounded by Precambrian basement (Fig. 13.1). Its

B. Linol (✉)
AEON-ESSRI (Africa Earth Observatory Network – Earth Stewardship Science Research Institute), Nelson Mandela Metropolitan University, Port Elizabeth, South Africa.

Geological Sciences, Nelson Mandela Metropolitan University, Port Elizabeth, South Africa.
e-mail: bastien.aeon@gmail.com

M.J. de Wit
AEON-ESSRI (Africa Earth Observatory Network – Earth Stewardship Science Research Institute), Nelson Mandela Metropolitan University, Port Elizabeth, South Africa.
e-mail: maarten.dewit@nmmu.ac.za

E.J. Milani
Petrobras Research Center, 950 Horácio Macedo Av., 21941-915 Rio de Janeiro, Brazil
e-mail: ejmilani@petrobras.com.br

F. Guillocheau
Géociences-Rennes, UMR 6118 Université de Rennes 1 – CNRS, OSUR, Université de Rennes 1, Campus de Beaulieu, 35042 Rennes cedex, France
e-mail: francois.guillocheau@univ-rennes1.fr

C. Scherer
Instituto de Geociências, Universidade Federal do Rio Grande do Sul, P.O. Box 15001, CEP 91501-970, Porto Alegre, Brazil
e-mail: claiton.scherer@ufrgs.br

M.J. de Wit et al. (eds.), *Geology and Resource Potential of the Congo Basin*, Regional Geology Reviews,
DOI 10.1007/978-3-642-29482-2_13,

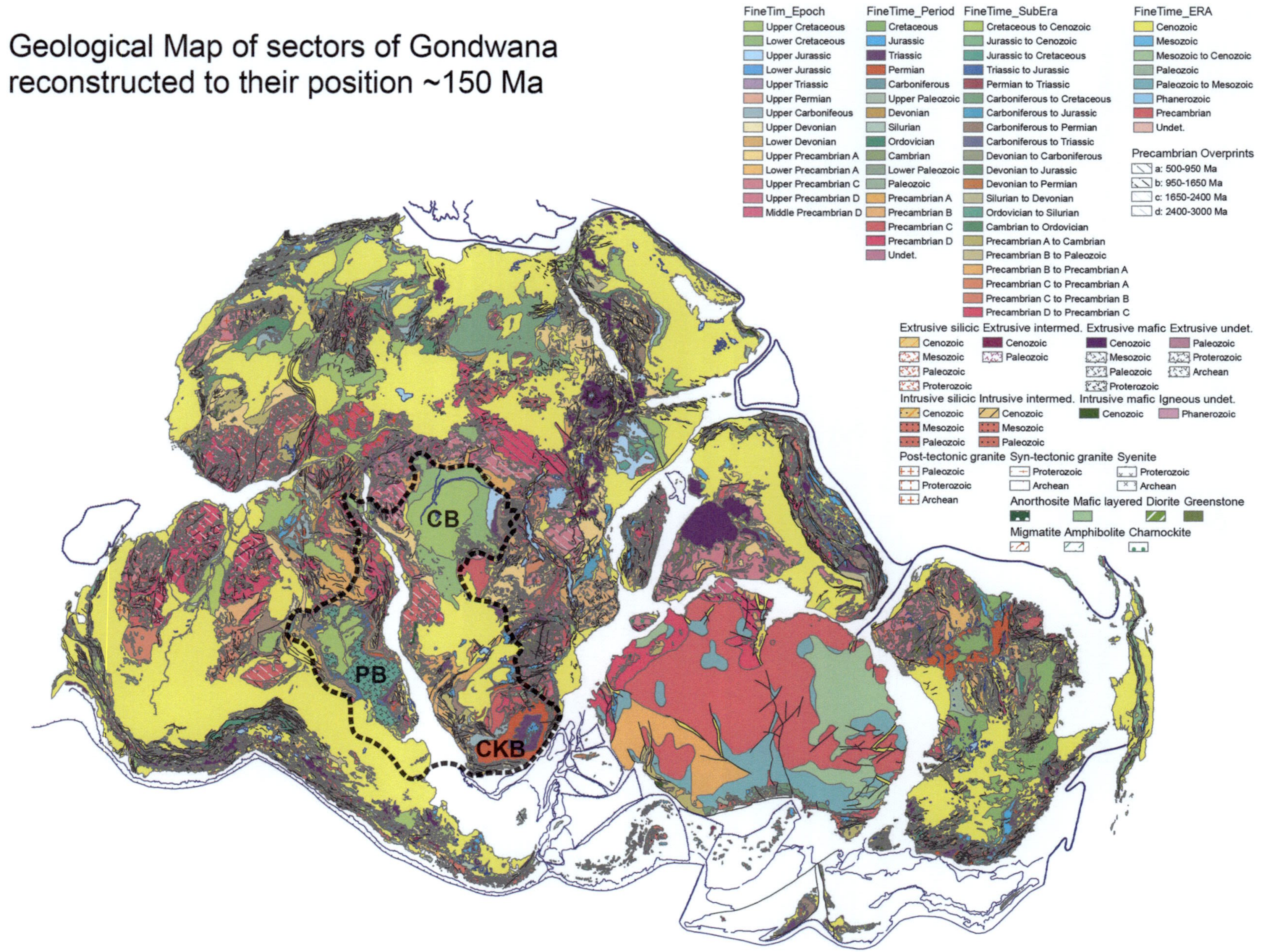

Fig. 13.1 Geological map of Gondwana (updated GIS from de Wit et al. 1988), highlighting three main Phanerozoic basins: the Congo, Paraná and Cape-Karoo Basins (CB, PB and CKB, respectively), and their possible connections in the framework of a larger Central West Gondwana Basin (CWGB) complex (*thick dotted line*)

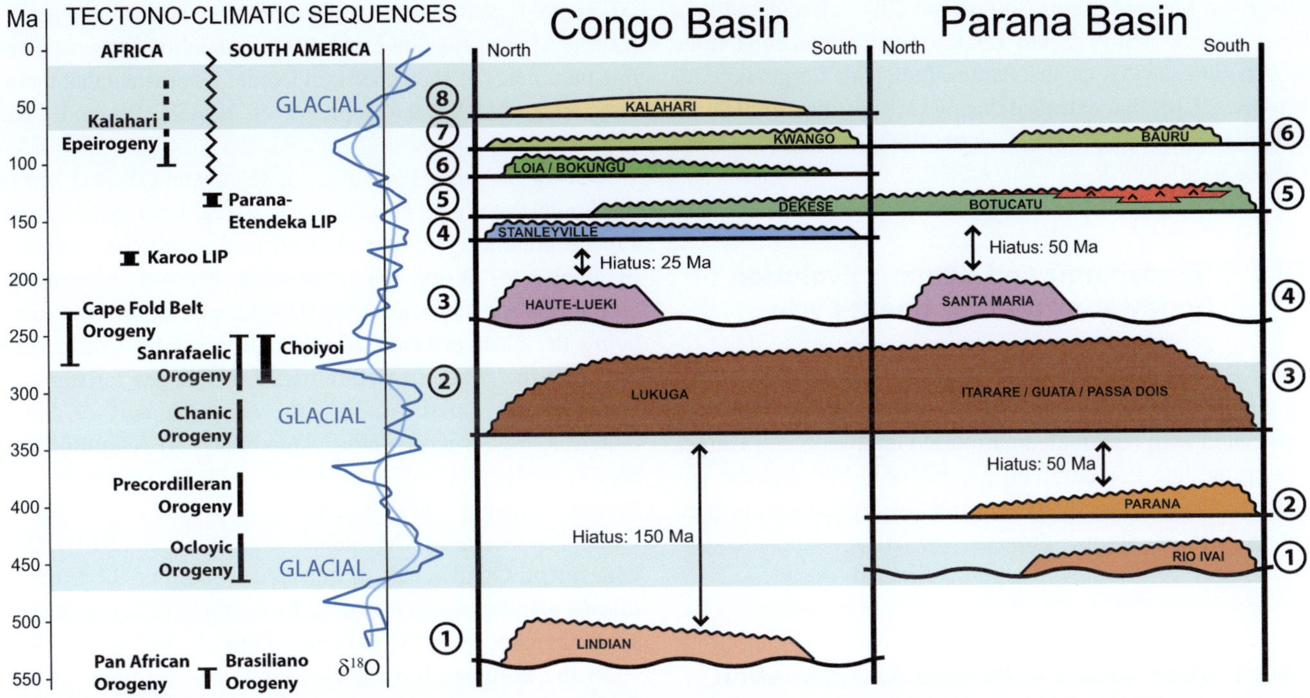

Fig. 13.2 Compared Phanerozoic stratigraphic sequences of the CB and PB (*circled numbers* on the *left* and *right*, respectively), and correlation with regional tectonic and magmatic events of southwest Gondwana (adapted from Milani and de Wit 2008), and $\delta^{18}O$ curves from Veizer et al. (2000), highlighting main episodes of glaciation (*blue shades*)

stratigraphic record comprises five major depositional successions (or supersequences) ranging in age from late Neoproterozoic to Cenozoic, with a total thickness of 4–6 km in the center of the basin (Cahen 1954; Lepersonne 1974; Daly et al. 1992; Linol 2013; Chaps. 6–10, this Book). The oldest succession (Sequence 1; Fig. 13.2), part of the upper Neoproterozoic-lower Paleozoic Lindian Supergroup, comprises extensive 'red-beds', between 1,000 and 1,500 m thick, covering deformed Neoproterozoic (Pan African) carbonate and siliclastic platforms (Lepersonne 1974; Poidevin 1985). Overlying a major unconformity that represents a hiatus of up to 150 million years, the two succeeding successions, Carboniferous-Permian, and Triassic (Sequences 2 and 3, respectively; Fig. 13.2), totaling 900 to 3,000 m in thickness, represent the northernmost extension of the classical Karoo Supergroup of southern Africa (Cahen and Lepersonne 1978; Cahen 1981). An overlying Mesozoic succession, defined as the Congo Supergroup (Chap. 8 , this Book), is between 300 and 1,000 m thick and comprises four sequences (Sequences 4 to 7; Fig. 13.2) from shallow marine (tidal), to aeolian, lacustrine, and then fluvial. This Upper Jurassic to Upper Cretaceous succession is in turn truncated by a regional peneplanation surface and covered by silcretes, calcretes and alluviums of the Cenozoic Kalahari Group (Sequence 8), with a maximum thickness of 250 m in the center of the basin (Chap. 10, this Book).

The PB is a 1.5 million km^2, N–S elongated basin of west Gondwana (Fig. 13.1). It is flanked to the north by Precambrian basement (e.g. the Sao Francisco Craton) and to the southwest by the Andes and proto-Andes, constructed above the long-lived (~600 Ma) paleo-Pacific margin of southwestern Gondwana (e.g. Ramos and Aleman 2000). The PB preserves six main depositional sequences (supersequences), ranging in age from Late Ordovician to Late Cretaceous, with a total thickness of about 7 km (Milani et al. 2007). The first two sequences comprise the Ordovician-Silurian Rio Ivaí and the Devonian Paraná Groups (Sequences 1 and 2; Fig. 13.2), in total 1,500 m thick. These lowermost successions correspond to two superimposed shallow marine platform sequences, similar (but thinner) to the lower Paleozoic section of the Cape Supergroup in South Africa (Milani and de Wit 2008). Overlying a major erosional surface that represents a hiatus of about 50 million years, the succeeding Carboniferous-Permian succession (Sequence 3; Fig. 13.2) comprises up to 2,500 m thick glacio-marine to progressively fluvial sedimentary rocks, attributed to the Itararé, Guatá and Passa Dois Groups (Schneider et al. 1974). These sequences are equivalent of large parts of the Karoo Supergroup of southern Africa (e.g. Veevers et al. 1994; Milani and de Wit 2008). Three overlying Mesozoic sequences: the Santa Maria, São Bento and Bauru Groups (Sequences 4 to 6; Fig. 13.2) correspond to 1,000 m thick fluvial-aeolian sediments (like in the CB), intercalated near the top with nearly 2 km thick Lower Cretaceous flood basalts and dacites of the Paraná-Etendeka Large Igneous Province (LIP) that extruded during early rifting of the South Atlantic at ca. 132 Ma (Peate et al. 1992; Renne et al. 1992; Gibson et al. 2006).

13.3 Geodynamic and Climatic Evolution of Gondwana Supercontinent Cycle

The CB, PB and Cape-Karoo Basin (CKB) evolved during continent-wide geodynamic and global climate changes associated with the amalgamation of Gondwana and Pangea during the late Precambrian (ca. 500–800 Ma) and the late Paleozoic (ca. 250–300 Ma), respectively, and during their subsequent break-up, through the Meso-Cenozoic (ca. 0–200 Ma), to a complete supercontinental 'cycle'.

13.3.1 Gondwana and Pangea Amalgamation

The formation of Gondwana occurred over a period of about 300 Ma during the late Precambrian (ca. 500–800 Ma), associated with extensive 'Himalayan-type' orogens and a succession of intense glaciations (e.g. 'Snowball Earth'; Hoffman 1999). Several Archean cratons amalgamated along Paleo- and Meso-Proterozoic belts, fragments of a previous (Rodinia) supercontinent (and here referred as 'Shields'), merged along a regional network of late Neoproterozoic to Early Cambrian fold-and-thrust belts to form Gondwana. These mobile belts are referred to as Pan African and Brasiliano in Africa and South America, respectively (Fig. 13.3).

In southern Africa, the (Archean) Azanian Craton flanked by the 1–2 Ga Namaqua-Natal Belt comprises the Kalahari Shield (Fig. 13.3), which in turn amalgamated to the north with the Central African Shield along the Damara-Lufilian Belts during the southern Pan African orogen (de Wit et al. 2008). The Central African Shield that includes the Kasai, NE Angolan, Ntem, Mboumou and Tanzanian Cratons surrounding the CB (see Chap. 2, this Book), is flanked to the north by the 0.5–1 Ga E–W striking Oubanguides Belt, along which it amalgamated with the North African Shield during the northern (Sahara) Pan African orogen. This vast Pan African mobile belt has also been linked to the (Brasiliano) Sergipano Belt of northeast Brazil (Toteu et al. 2006; de Wit et al. 2008, and references therein). In a similar manner, the N–S trending West Congo Belt of west-central Africa and its counterpart, the Ribeira-Araçuaí Belts in eastern Brazil, linked the Central African and Sao Francisco Shields (Pedrosa-Soares et al. 2008). In turn, the Sao Francisco and the 2.0–2.5 Ga Rio de la Plata Shields, that in part underlies the PB (Rapela et al. 2011), amalgamated to the west with the Amazonian Shield along the Araguaia-Brasilia Belts during the eastern Brasiliano orogen (e.g. Cordani et al. 2013). All together, these sutured South American and south-central African Shields formed the foundations (e.g. crystalline basement) of southwest continental Gondwana that thereafter supported the development of the CB, PB and CKB (Fig. 13.4).

During the early Paleozoic, as Gondwana drifted across the South Pole (e.g. Torsvik et al. 2012), large parts of this supercontinent again experienced two successive episodes of glaciation, during the Ordovician (Pakhuis glaciation), and during the Carboniferous (Dwyka glaciation), the latest being the most pronounced with the great Dwyka ice cap covering the southern two-thirds of Africa (as far north as Ethiopia and Saudi Arabia), the southern half of South America, Madagascar, India, Antarctica and Australia (e.g. Martin 1981; Veevers et al. 1994), and during which substantial erosion and extensive deposition of glacial sequences occurred (Montañez and Poulsen 2013). This widespread Gondwanan stratigraphic record of the Carboniferous glaciation was first used by du Toit (1927) to demonstrate the concept of 'Continental Drift'.

In the northern hemisphere, a number of other terrains (now known as Laurentia and Baltica) had also merged to form Laurasia (e.g. Cocks and Torsvik; 2006; Domeier et al. 2012, and references therein) that, in turn, collided with Gondwana to form Pangea at ca. 275–325 Ma along the Mauritanian-Variscan Belts that extends from eastern United States (Appalachian), through the Mauritanides in northwest Africa, to Brittany (Armorican) in western Europe (Lefort 1989; Dabo et al. 2008; Van Staal et al. 2009; Kroner and Romer 2013; Fig. 13.4). Shortly thereafter, orogenesis along the southwestern margin of Gondwana resulted in the formation of the Sierra de la Ventana-Cape Fold Belts (ca. 245–278 Ma; Newton et al. 2006), which extends from northern Argentina to southern South Africa (de Wit and Ransome 1992; Pángaro and Ramos 2012, and references therein), as part of the larger 'Samfrau Orogen' of du Toit (1937). It has since been postulated that this main period of plate convergence, between 240–330 Ma, resulted in large-scale intracontinental deformation with long wavelength flexures across southwest Gondwana (Daly et al. 1992; Trouw and de Wit 1999; Milani and de Wit 2008). This deformation was also associated with abundant volcanic and magmatic arcs activity along the proto-Andes (e.g. Kleiman and Japas 2009), which produced widespread ignimbrites and air-fall tuffs in the surrounding sedimentary basins (Stollhofen et al. 2000; López-Gamundí 2006; Fildani et al. 2009; Rocha-Campos et al. 2011).

13.3.2 Synopsis of Pangea and Gondwana Break-Up

Pangea, and subsequently Gondwana break-up started at about 200 and 180 Ma, respectively (e.g. Reeves 1999; Jokat et al. 2003; Moulin et al. 2010; Torsvik et al. 2012; Heine et al. 2013), and was accompanied by successive

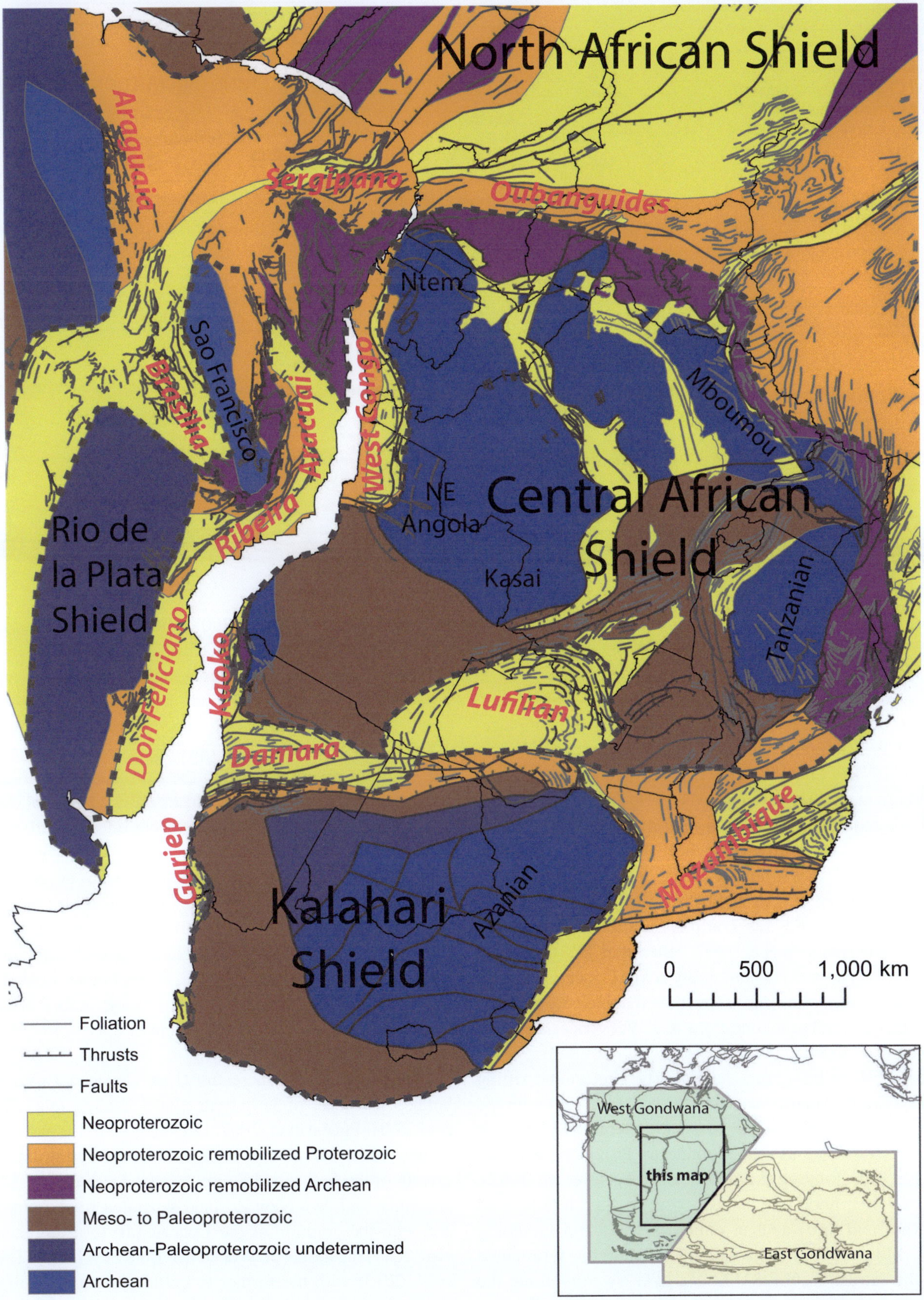

Fig. 13.3 Precambrian basement map of West Gondwana (updated GIS from de Wit et al. 2008), showing the Pan African-Brasiliano mobile belts (named in *red*) and Proterozoic Shields (*dotted outlines*)

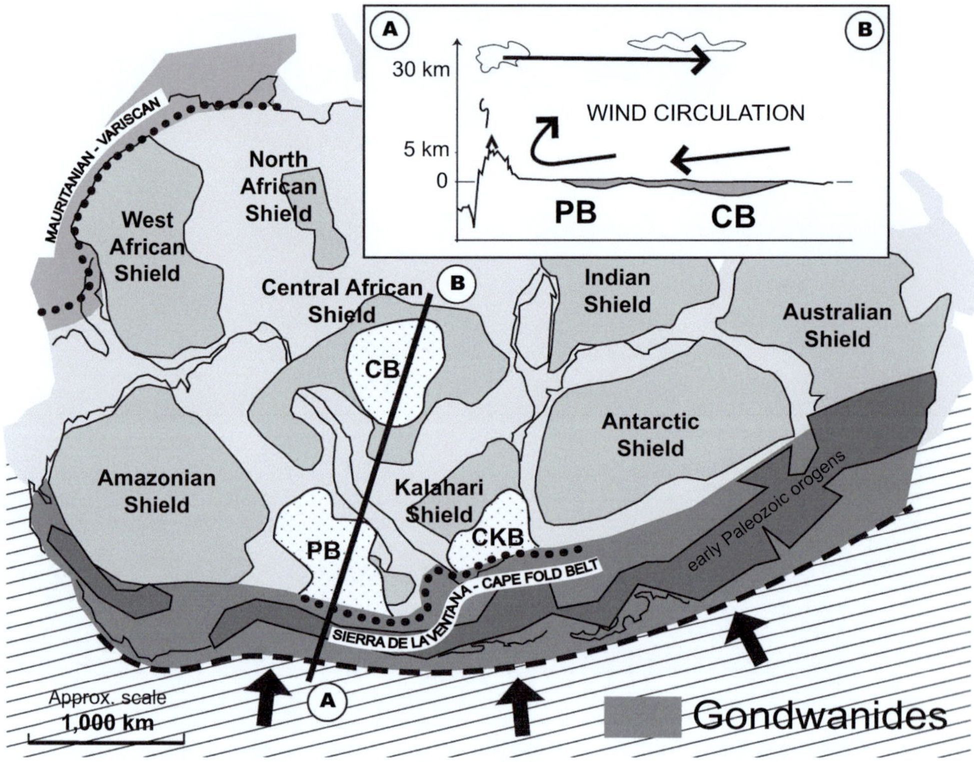

Fig. 13.4 Regional Phanerozoic map (modified from Pankurst and Vaughan 2009) and cross-section of Gondwana, showing the CB, PB and CKB, flanked to the south by an extensive, long-lived Phanerozoic orogenic belt, named the Gondwanides

outpouring of extensive continental flood basalts, as part of Large Igneous Provinces (LIPs, or 'hot-spot plumes'; Fig. 13.5).

Gondwana first separated from Laurentia (North America) in the Late Triassic with rifting marked by the emplacement of the Central Atlantic Magmatic Province (CAMP: 3–11 $\times$ 10^6 km^3) between 201–202 Ma (Marzoli et al. 1999; Blackburn et al. 2013). Then, in the Early Jurassic the Karoo LIP erupted in southeastern Africa and in southern Antarctica as the Ferrar Province (Fig. 13.5a), totaling 2.5 $\times$ 10^6 km^3 (Encarnación et al. 1996). Across South Africa, Botswana and in Zimbabwe, the Karoo lavas and abundant dykes and sill complexes are dated between 178–183 Ma (Duncan et al. 1997; Jones et al. 2001; Jourdan et al. 2007). These lavas commonly cover Upper Triassic to Lower Jurassic aeolian dune sequences that indicate contemporaneous hot and arid conditions (e.g. Veevers et al. 1994; Johnson et al. 2006).

Some 50 million years later, Early Cretaceous rifting between South America and Africa was associated with the eruption of the Paraná LIP in southeastern Brazil and in western Namibia as the Etendeka Province between 131–132 Ma (totaling 2.3 $\times$ 10^6 km^3; Peate et al. 1992), both of which flood basalts also cover and are interlayered with (hot and arid) aeolian dunes (Scherer and Goldberg 2007; Miller 2008). This magmatic activity terminated with the intrusion of 128–132 Ma dyke swarms along the Brazilian and African margins (Deckart et al. 1998; Kirstein et al. 2001; Trumbull et al. 2007) and the final opening of the South Atlantic Ocean (Fig. 13.5b).

Thereafter, Gondwana break-up continued during the mid-Cretaceous with the final separation of Antarctica from Africa (e.g. Fairhead 1988; Maurin and Guiraud 1993; Fig. 13.5c) with the opening the Weddell Sea and the southern oceans following the extrusion of the Agulhas LIP between 90–110 Ma (Gohl et al. 2011). At about the same time, continental rifting occurred in north-central Africa (Fig. 13.5d), and abundant kimberlites were emplaced across southern Africa (Jelsma et al. 2009), attesting of important intraplate reorganization and intermittent interactions with mantle processes. This time-interval (mid-Cretaceous) also coincides with a global climate maximum (e.g. Pucéat et al. 2003; Forster et al. 2007) and accelerated exhumation and erosion across southern Africa during the onset of the Kalahari epeirogeny (de Wit 2007; Tinker et al. 2008).

The period shortly thereafter (Late Cretaceous-Cenozoic) heralds the onset of long-term global cooling (e.g. Zachos et al. 2001; Pucéat et al. 2007; Decker et al. 2013) particularly as new ocean circulation regimes evolved during the rapid growth of the Indian and South Atlantic Oceans (Fig. 13.5e). Australia, and then New Zealand separated from Antarctica, and with the eruption of the Deccan LIP (66 Ma; Ganerød et al. 2011) the Indian plate detached from Seychelles (Torsvik et al. 2013) with the former to collide thereafter with Asia (Fig. 13.5f), forming the Himalaya and Tibet Plateau (e.g. Searle 2013). Following the formation of the Ethiopian LIP

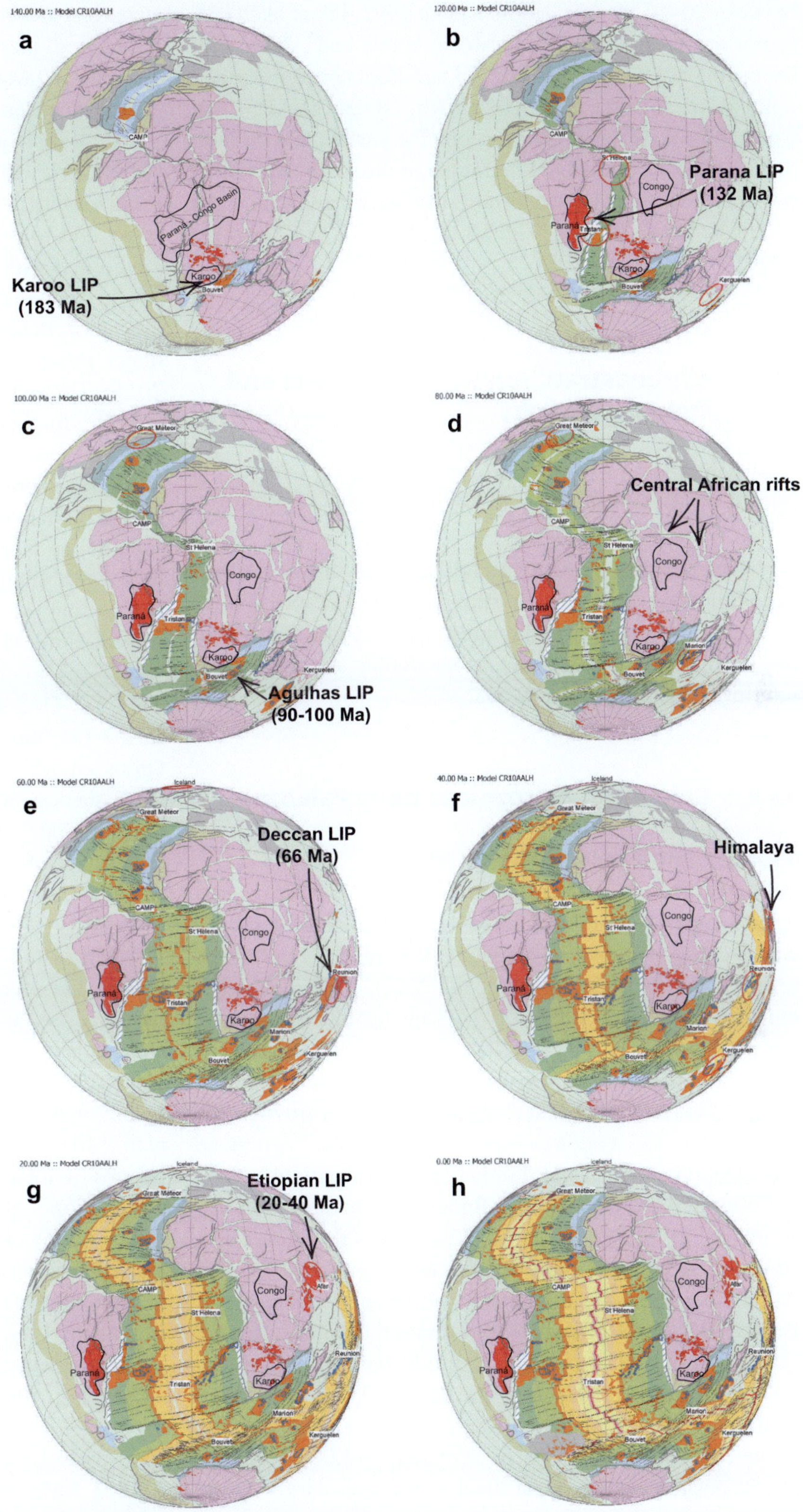

Fig. 13.5 Sequences of Gondwana break-up (http://www.reeves.nl/gondwana), highlighting successive extrusions of major Large Igneous Provinces (LIPs)

between 20–40 Ma (e.g. Hofman et al. 1997; Pik et al. 2008), Cenozoic rifting continued between Arabia and Africa with the opening of the Red Sea (Fig. 13.5g). Further extension and uplift resulted in the East African Rift System (EARS) that extends from Afar and Kenya, along the eastern and western flanks of the Tanzanian Craton farther to the south, and with a possible embryonic branch towards the Okavango Delta in the central Kalahari (Reeves 1972; Modisi et al., 2000; Chorowicz 2005).

13.4 Chronostratigraphic Correlations and Paleogeography

Based on modern basin analysis of the CB (Linol 2013) and the PB (Milani 1997), we derive new stratigraphic correlations between their Phanerozoic sequences (Fig. 13.6). Together with some references to the better-known CKB of southern Africa, as well as detrital zircon provenance studies from these basins, this regional synthesis aims to improve the paleogeography and paleo-topographic models of the interior of southwest Gondwana.

13.4.1 Late Neoproterozoic to Early Paleozoic

13.4.1.1 End Precambrian-Cambrian

Following the Pan African orogens (ca. 500–800 Ma), widespread upper Neoproterozoic to lower Paleozoic red-beds were deposited across the North and Central African Shields (e.g. Poidevin 1985; Morag et al. 2011). This succession outcrops abundantly south of the Oubanguides Belt along the northern margin of the CB, and was originally referred to as the Aruwimi Group (upper part of the Lindian Supergroup; Lepersonne 1974). It comprises massive quartzites (the Galamboge Formation), red schists with limestone intercalations (the Alolo Formation) and relatively thick (~1,000 m) arkoses of the Banalia Formation, totaling 1,500 m in thickness (Fig. 13.6). Only the uppermost (Banalia) sequence is now considered to post-date the Pan African deformation (Tait et al. 2011; see also Chap. 6, this Book). Based on this new lithostratigraphy, the revised 'Banalia Group' is correlated with southward prograding conglomeratic red sandstones of the Inkisi Group (~1,000 m thick) along the West Congo Belt (Alvarez et al. 1995), and rippled red siltstones of the Biano Group (~1,000 m thick), also deposited southward, at the southeastern margin of the CB (Master et al. 2005). Detrital zircon provenance analyses from these red-beds (Master et al. 2005; Frimmel et al. 2006; Jelsma et al. 2011) reveal two prominent age-peaks: at 1000 Ma and 650 Ma that characterize predominant late Mesoproterozoic to early Neoproterozoic (Kibaran), and late Neoproterozoic (Pan African) aged sources in north-central Africa (e.g. the Central Saharan Belt). The youngest zircons dated at 540 Ma provide a maximum early Cambrian age to these sequences (the Inkisi Group; Jelsma et al. 2011).

These three earliest Paleozoic sequences surrounding the CB are herewith correlated with the Upper Nama Group of southwestern Namibia, deposited with paleocurrents to the south and to the southeast from the Damara Belt (Germs 1995), and which unconformably overlies the Precambrian-Cambrian boundary (Grotzinger and Miller 2008). U-Pb dating of detrital zircons from this section indicate two major peaks at 1050 Ma and 540–600 Ma (Blanco et al. 2011). These dates are similar to those from the CB, and all indicate dominant Kibaran and Pan African sources for the red-beds across a vast area of south-central Africa. Equivalent (post-Pan African) sequences are apparently absent in the PB (Milani et al. 2007), except possibly within local rifts, such as the Camarinha Basin along the western Brasiliano fold belt (Fig. 13.7). Thus, at this stage, southwest Gondwana red-beds were preferentially preserved across proto-central Africa.

13.4.1.2 Ordovician-Silurian and Devonian

In the PB, the oldest preserved sediments are part of the Ordovician-Silurian Rio Ivaí Group (Fig. 13.6). This lowermost succession comprises fluvial-marine sandstones and conglomerates (the Alto Garcas Formation), covered by diamictites of the Iapo Formation and followed by deglaciation black shales of the Vila Maria Formation that record a first, Early Silurian (Llandovery) maximum marine transgression in the southern part of the basin (Mizusaki et al. 2002). The overlying Devonian Paraná Group is also a transgressive marine platform, and includes conglomerates and sandstones with trilobite tracks (the Furnas Formation) and black shales of the Ponta Grossa Formation that record a second maximum flooding of the PB during the Late Devonian (Milani et al. 2007). Both, the Rio Ivaí and the Paraná Groups (in total 1,400 m thick) have been correlated with the Upper Cape Supergroup of South Africa, including the Upper Ordovician Pakhuis glacial deposits (Milani and de Wit 2008). These correlations suggest the development of a giant shallow marine shelf along the southwestern margin of Gondwana during the Silurian-Devonian, more than 1,000 km wide, stretching from southern Brazil to South Africa (Fig. 13.8).

13.4.2 Late Paleozoic to Early Mesozoic

13.4.2.1 Carboniferous-Early Permian

In southeastern Brazil, the lower Paleozoic sequences of the PB (e.g. the Furnas Formation) are truncated by a regional unconformity marked with glacial pavements, such as the

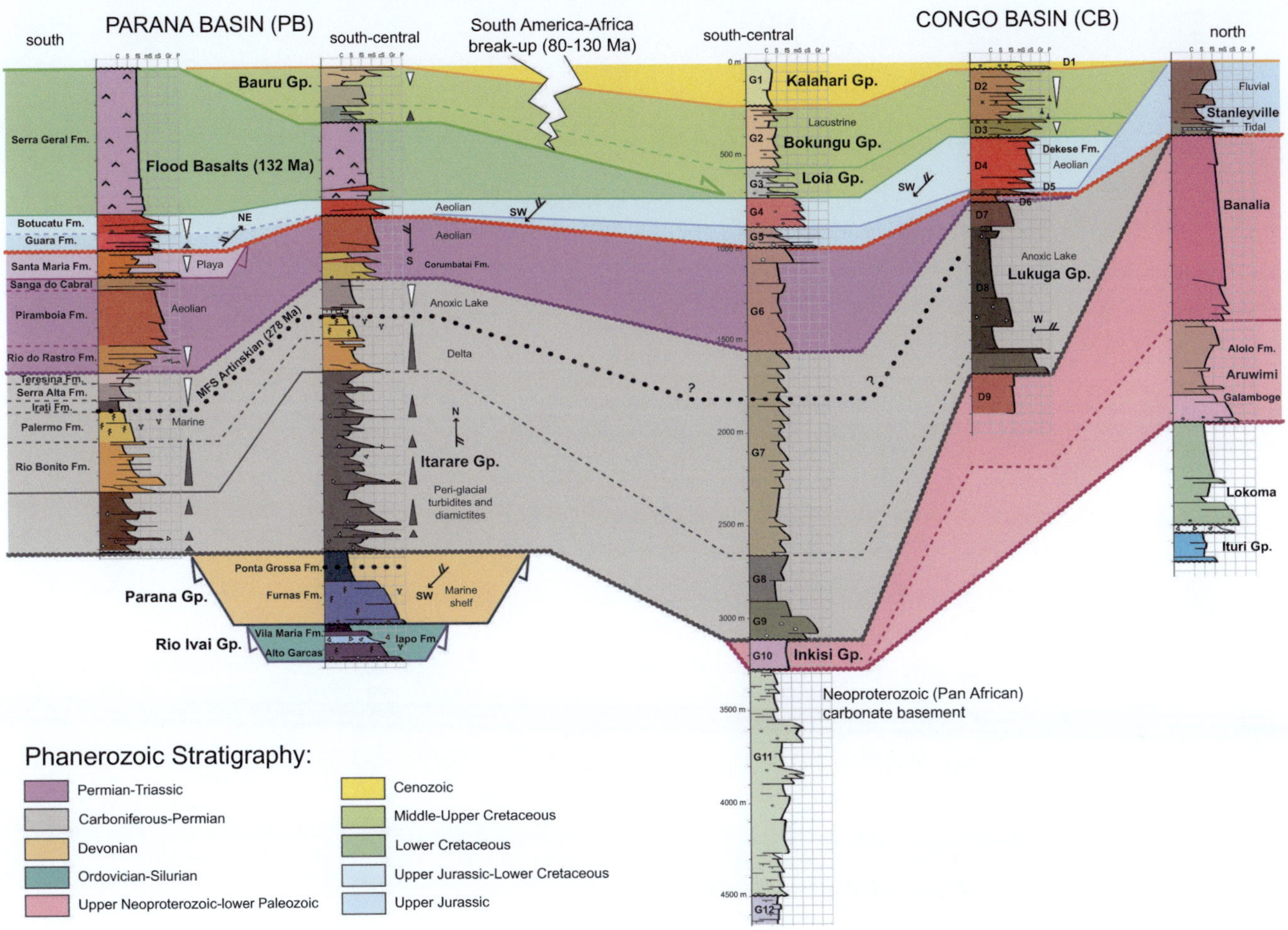

Fig. 13.6 New regional correlations between Phanerozoic sequences of the CB of central Africa and the PB of southeastern Brazil, based on core/well data and field investigations (e.g. 'Coluna White', La Serra do Rio Rastro, Brazil)

benchmark at Witmarsum (Fig. 13.9) that indicates N–S advances and retreats of the Gondwana ice caps. The overlying Carboniferous Itararé Group (maximum 1,500 m thick) includes glacio-marine turbidites, diamictites and rhythmites, forming five fining upward peri-glacial sequences (Vesely and Assine 2006; Vesely 2007) that can confidently be correlated with the Dwyka Group of southern Africa (Milani and de Wit 2008) and now, herewith, also correlated with the Lower Lukuga Group in the CB (Fig. 13.6). In detail, these widely separated sequences, however, show important variations in thickness, lithology and sediment provenances that likely reflect regional differences in paleo-environments.

In central Africa, the Carboniferous-Permian Lukuga Group is best preserved (300–600 m thick) in west-facing paleo-glacial valleys along the eastern margin of the CB (Boutakoff 1948; Cahen and Lepersonne 1978), and links to depot-centers up to 1,400 m thick in the center of the basin (Linol 2013; Chap. 7, this Book). Detrital zircons from the Lukuga diamictites date predominantly at 1.8–2.1 Ga, likely derived from large mid-Paleoproterozoic (Eburnian) sources in Uganda and Tanzania, thus suggesting a major paleo-relief in east Africa (e.g. the Tanzanian Highland; Fig. 13.10). In contrast, in southern Africa the Dwyka Group reaches a maximum thickness of 800 m in the southern CKB (Catuneanu et al. 2005; Johnson et al. 2006) and thins northward to about 250 m thick in the central Kalahari (Smith 1984). Ice-flow directions suggest a highland to the north and to the northeast, e.g. the 'Cargonian Highland' of du Toit (1954). U-Pb detrital zircon dates from the Dwyka tillites (Craddock and Thomas 2011) have two major peaks: at 1100 Ma and 550 Ma, sourced from late Mesoproterozoic (Kibaran) and late Neoproterozoic (Pan African) age-sequences of the Kalahari and Central African Shields. U-Pb detrital zircons results from tillites in southern South America (Craddock 2011) yield a dominant peak at 400–600 Ma that suggest important Brasiliano-Pan African contributions to the PB, such as for example from the Don

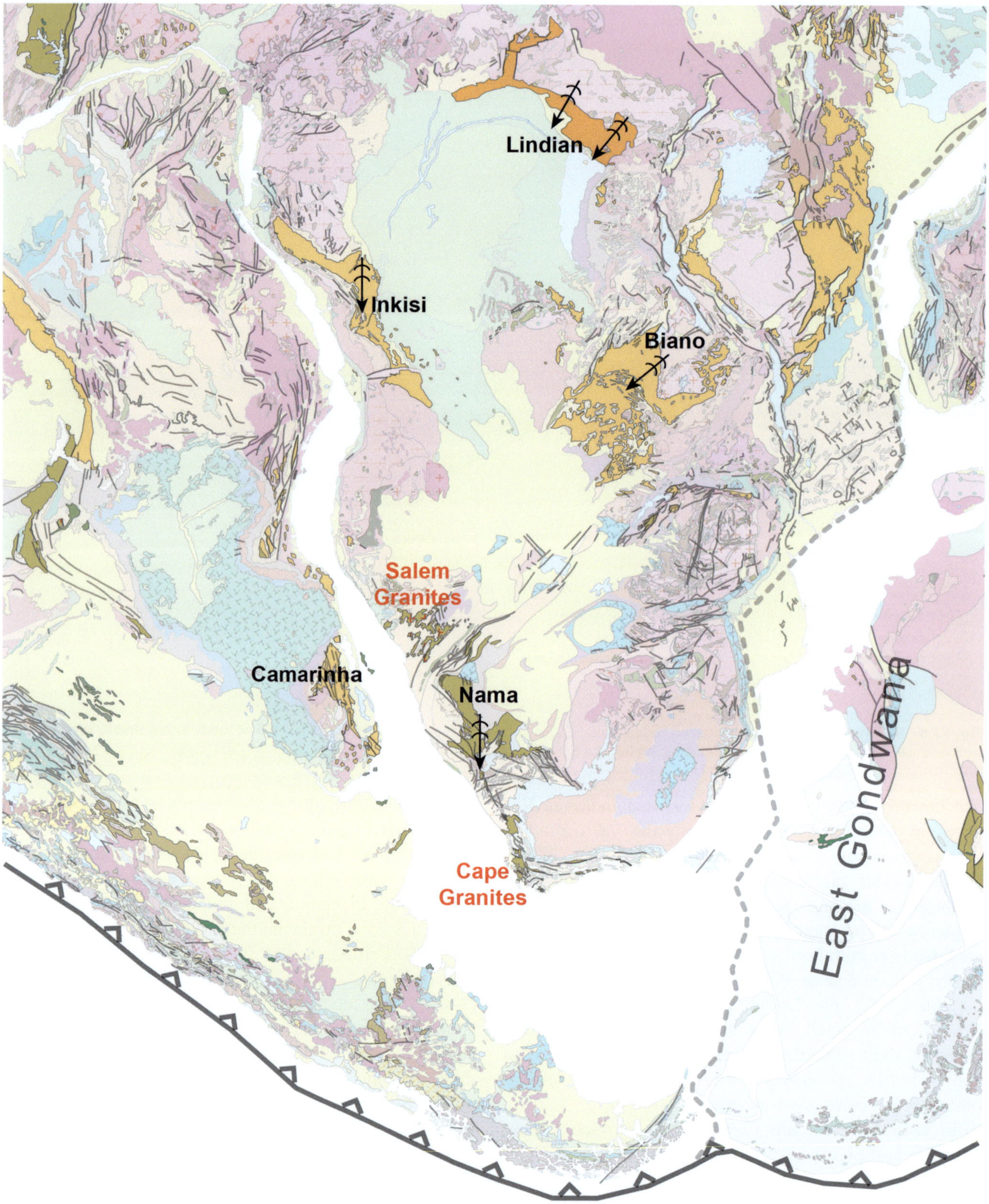

Fig. 13.7 Late Neoproterozoic-early Paleozoic paleo-map of southwest Gondwana, highlighting main stratigraphic sequences (Fig. 13.1 for legend). Paleocurrent directions (*arrows*) from red-bed sediments across south-central Africa are regionally to the south

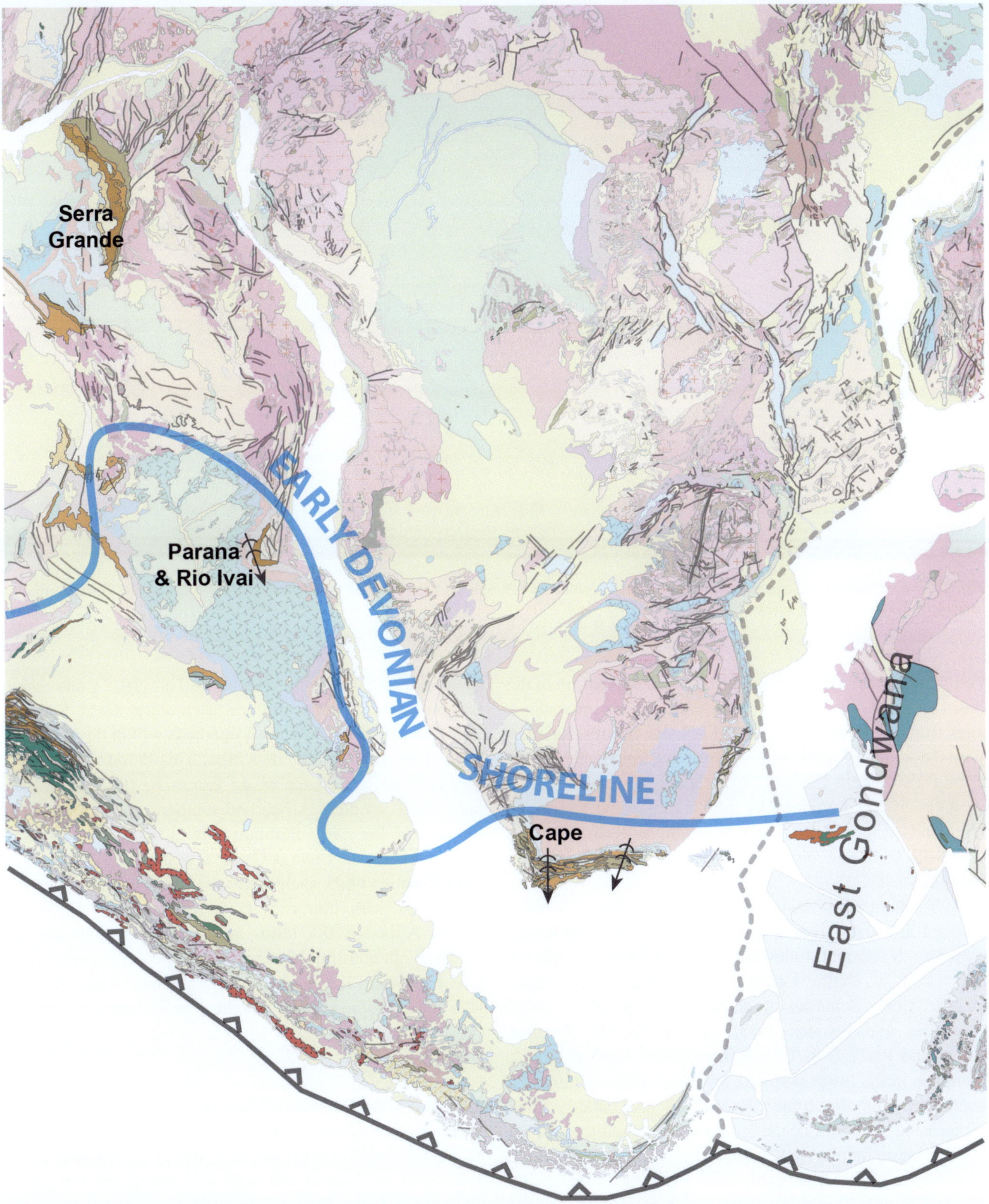

Fig. 13.8 Ordovician-Devonian paleo-map of southwest Gondwana, highlighting main stratigraphic sequences (Fig. 13.1 for legend). Southerly deposited shallow marine sediments in southern South Africa and southeastern Brazil suggest a vast marine platform (*thick blue line*). Reconstruction of Falkland Islands is from Marshall (1994)

Fig. 13.9 Carboniferous glacial pavement at Witmarsum, southwest Paraná. Large groove marks (*double sided arrows*) onto the Furna Formation indicate N–S advances and retreats of the ice cap

Feliciano-Damara Belts along the southeastern margin of the basin (see Fig. 13.3). Thus, the detrital zircons provenance ages from the Itararé, Dwyka and Lukuga Groups, together with the sediment and ice-flow dispersion directions indicate a long wavelength Carboniferous topography across southwest Gondwana dominated by the Cargonian and Tanzanian Highlands (Fig. 13.10). These paleo-highs generally coincide with central remnants of Braziliano and Pan African orogens.

Subsequent rapid deglaciation during the Early Permian resulted in a short-lived marine transgression across southwest Gondwana and, following isostatic rebound, extensive deposition of shallow water, organic-rich black shales (Visser 1995; Scheffler et al. 2003; Frank et al. 2008; Montañez and Poulsen 2013). In the PB, the Permian Guatá and Passa Dois Groups (maximum 2,600 m thick), comprise in the lower part vast low-stand deltas prograding southward (the Rio Bonito Formation) and notably thick bioturbated siltstones of the Palermo Formation that record a final maximum marine transgression during the Artinskian (e.g. Souza 2006; Milani et al. 2007; Fig. 13.6). It is sharply overlain by 40–70 m thick, organic-rich black shales of the Irati Formation, U-Pb dated with igneous zircons from interbedded ash beds (tungsten) at ca. 278 ± 3 Ma (Santos et al. 2006). This distinct (condensed) Irati sequence was correlated biostratigraphically to the Whitehill Formation of South Africa for the first time by du Toit (1927), and which is now confirmed by zircons accurately dated between 275–279 Ma from similar ash interbeds within the Whitehill Formation across southern Africa. This chronostratigraphy supports the occurrence of an extensive 'Black Sea-like' lake across southern Africa and southern South America at the end of the Early Permian (Fig. 13.10). In the CB of central Africa, the Lukuga Group also includes significantly thick Permian black shales (e.g. the Lukuga Formation) and sandstones with thin coal beds (Cahen and Lepersonne 1978). Although the litho- and bio-stratigraphy are not well-enough constrained to propose more detail correlations (see Fig. 13.6), it is likely that the gigantic Early Permian lake may, at times, have extended as far north as the CB (Fig. 13.10).

13.4.2.2 Late Permian-Triassic

Permian-Triassic sedimentation across southwest Gondwana became entirely continental and arid (e.g. Veevers et al. 1994) and was associated with widespread arc-volcanism and magmatism along the proto-Andes (e.g. Ramos and Aleman 2000). This extensive silicic magmatic belt (Fig. 13.11) comprises the Permian-Triassic Choiyoi Igneous Province in Chile and Argentina, and the Triassic to Middle Jurassic Chon Aike Province in Patagonia (Kay

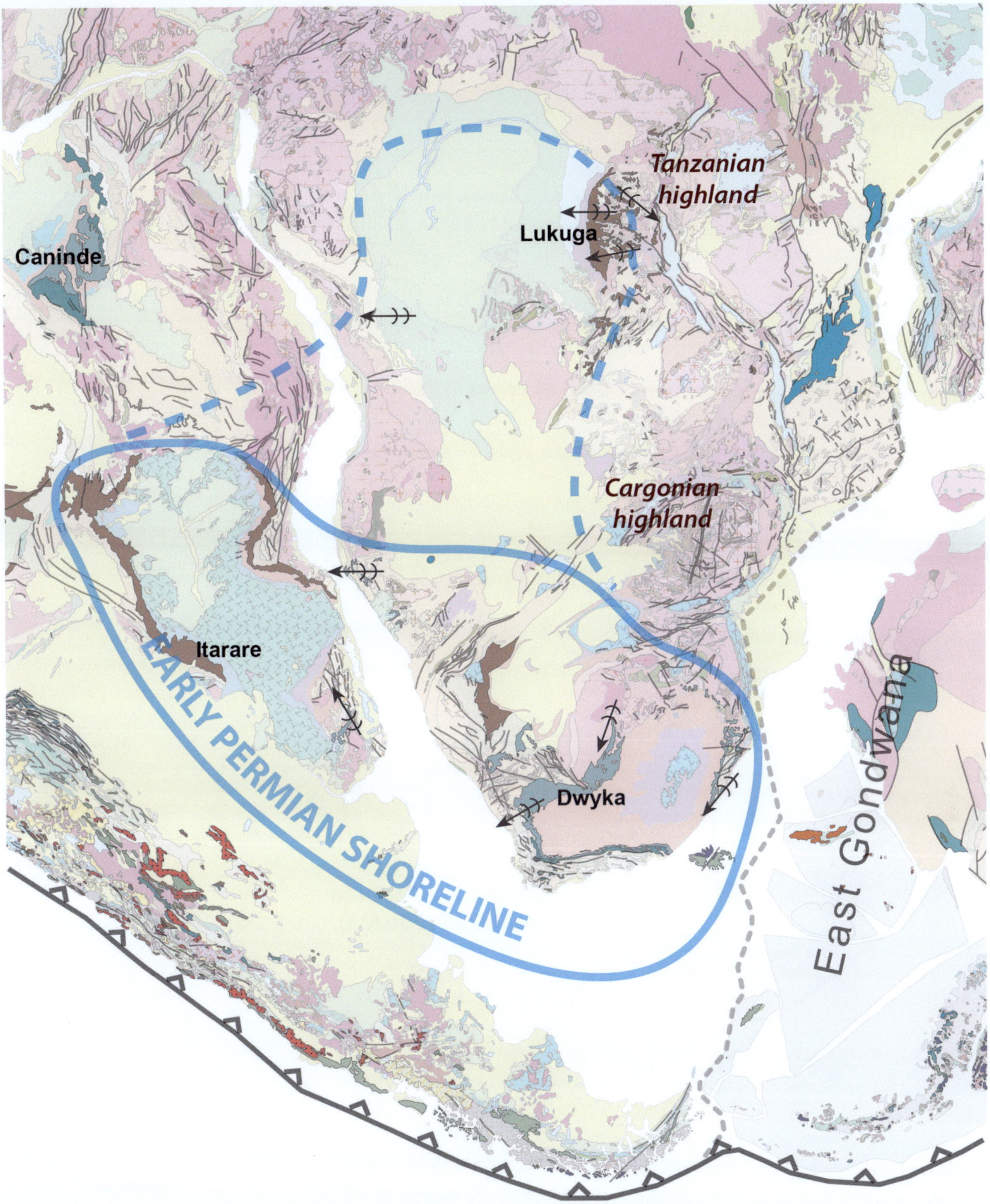

Fig. 13.10 Carboniferous-Early Permian paleo-map of southwest Gondwana, highlighting main stratigraphic sequences (Fig. 13.1 for legend). Sediment and ice-flow dispersion directions (*arrows*) point to major paleo-reliefs in south-central Africa (the Cargonian and Tanzanian Highlands). *Thick blue line* shows an Early Permian giant lake across southern Africa and southeastern Brazil, possibly extending as far north as in Congo (*dotted line*). Reconstruction of Falkland Islands is from Marshall (1994)

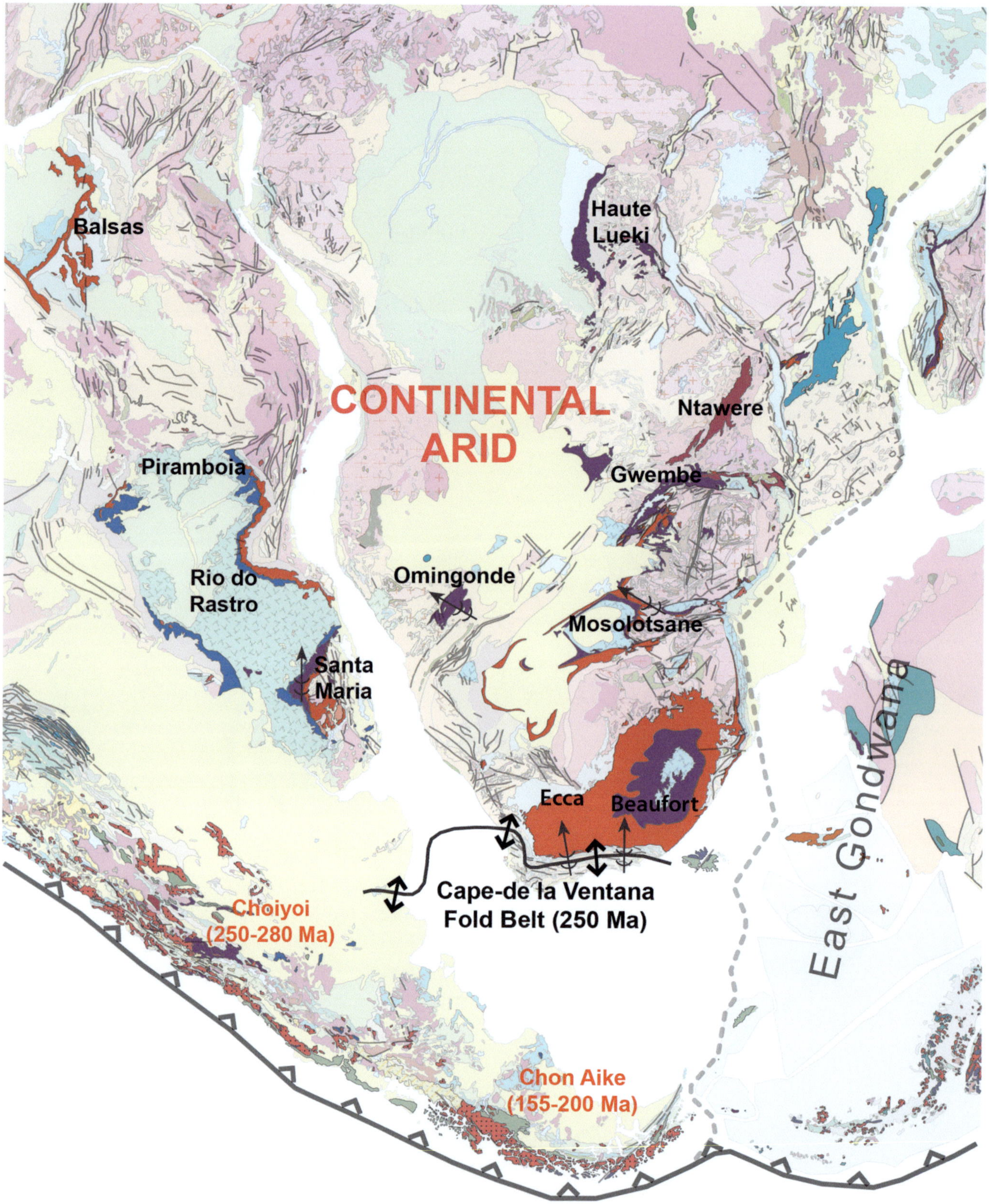

Fig. 13.11 Late Permian-Triassic paleo-map of southwest Gondwana, highlighting main stratigraphic sequences (Fig. 13.1 for legend). Paleocurrent directions (*arrows*) across southern Africa and southeastern Brazil are regionally to the north from the rising Sierra de la Ventana-Cape Fold Belt (from Pángaro and Ramos 2012)

et al. 1989; Kleiman and Japas 2009). In adjoining sub-Andean basins, well-preserved ignimbrites, recently dated between 250–280 Ma, characterize the main period of volcanic activity (Rocha-Campos et al. 2011) and, as a result, the large number of air-fall tuffs reported with this age-range in the PB and the CKB are generally attributed to this same source (e.g. López-Gamundí 2006).

In the PB, the Early Permian Irati Formation is overlain by transgressive black shales (the Serra Alta Formation), 850 m thick siltstones of the Teresina Formation and 650 m thick southward prograding delta fronts of the Rio do Rastro Formation (Milani et al. 2007; Fig. 13.6). This thick regressive section is similar to that of the Upper Ecca and Beaufort Groups in the CKB (Milani and de Wit 2008), deposited with paleocurrents largely to the north from the rising Cape Fold Belt, ca. 245–278 Ma (Cole 1992; Newton et al. 2006; Tankard et al. 2009). Late Permian-Triassic sedimentation then culminated in fluvial-aeolian and playa-lake (arid) red-bed deposits both in the PB, with the Pirambóia, Corumbataí, Sanga do Cabral Formations and the Santa Maria Group (Assine et al. 2004; Zerfass et al. 2004; Milani et al. 2007), and in the CKB, with the Molteno, Eliot and Clarens Formations (Johnson et al. 2006). These arid sequences are also equivalent to the Omingonde Formation in Namibia (Miller 2008) and the Mosolotsane Formation in Botswana (Bordy et al. 2010). By contrast, in central Africa the Triassic sequences (the Haute Lueki and Cassange Groups) are less precisely known, but have been shown to be relatively thick (500–1,400 m) in the center of the CB (Linol 2013; Chap. 7, this Book) and, most likely also represent arid, fluvial and aeolian equivalents of the Brazilian red-beds (Fig. 13.11).

13.4.3 Late Mesozoic to Cenozoic

13.4.3.1 Late Jurassic-Cretaceous

The Triassic sequences of the CB and PB (the Santa Maria and Haute Lueki Groups) are truncated by regional unconformities, and overlain by Jurassic-Cretaceous sequences (Fig. 13.6). In the CB, the Upper Jurassic Stanleyville Group comprises 400 m thick shallow marine to fluvial sediments (Cahen 1983b; Linol 2013), and intercalates in the lower part with a thin carbonate horizon ('Lime Fine') with fish fossils that record a short, Kimmeridgian marine transgression of the proto-Indian Ocean in the northeastern part of the basin (Taverne 1975). It is onlapped to the south by aeolian dunes, maximum 300 m thick in the center of the CB (the Dekese Formation; Chap. 8, this Book), which extend along the southwest margin of the basin as the Lower Kwango Group, into Namibia (the Etjo/Twyfelfontein Formation; Jerram et al. 1999; Miller 2008), and in eastern Brazil as the Sergi and Botucatu Formations (Scherer and Goldberg 2007, 2010). In the PB the latter is conformably overlain by Lower Cretaceous lavas of the Serra Geral Formation (Fig. 13.12).

These Upper Jurassic to Lower Cretaceous aeolian sequences (also 'red-beds') that pre-date the extrusion of the Paraná-Etendeka LIP imply an immense (>2.5 million km^2), 'Sahara-like' paleo-desert near the center of West Gondwana just prior to the opening of the South Atlantic (Fig. 13.13). U-Pb detrital zircons geochronology from this aeolian sequence in the CB (Linol 2013; see Chap. 8, this Book) dated some Permian-Triassic (ca. 200 Ma and 240–290 Ma) and Jurassic (ca. 190 Ma) zircons, most likely derived from the proto-Andes, as for the Permo-Carboniferous air-fall tuffs reported in the PB and CKB (e.g. Fildani et al. 2009; Rocha-Campos et al. 2011). However, the paleo-wind directions measured from aeolian dunes in the southwest CB and in the northern PB are northerly, opposite in direction of the source for the wind-blown zircons from the proto-Andes. We interpret this to reflect complex atmospheric circulation regimes where high-atmosphere winds responsible for the ash fall deposits blow to the north, from the proto-Andean margin of southwest Gondwana toward Tethys, whilst strong south-directed winds were blowing close to the surface (see Fig. 13.4).

In the CB, the Upper Jurassic to Lower Cretaceous aeolian sequence (the Dekese Formation; Fig. 13.6) is sharply overlain by two superimposed, Albian-Cenomanian (middle Cretaceous) successions of lacustrine deposits, corresponding to the Loia and Bokungu Groups (Cahen 1983a), in total 650 m thick. These two singular (more humid) and fossiliferous sequences can be correlated to syn-rift marine sequences covering the thick Aptian salt along the west-African Atlantic margin (e.g. Mounguengui et al. 2008), but have apparently no equivalents in the PB. In southeastern Brazil, the 132 Ma basalts of the Serra Geral Formation are directly overlain by 200–300 m thick fluvial sediments of the Upper Cretaceous Bauru Group (de Paula e Silva et al. 2009), which is now correlated by biostratigraphy to the Upper Kwango Group in the western CB (Gobbo-Rodrigues et al. 2003). This renewed Late Cretaceous fluvial sedimentation both in the CB and PB may related to marginal uplifts and erosion following rifting of the South Atlantic (Fig. 13.14).

13.4.3.2 Cenozoic

Cenozoic terrestrial sedimentation is generally condensed in south-central Africa (Linol 2013; Chap. 10, this Book). In the CB and on top of the Kalahari Plateau, the Jurassic-Cretaceous sequences (e.g. the Kwango Group) are truncated by a regional peneplanation surface well-marked by duricrusts (calcretes and silcretes), and in turn covered by alluvial sands and muds of the Kalahari Group, maximum 250–500 m thick. Similar pedogenic carbonates and calcrete deposits have also been described in the southern PB (e.g. the Queguay Formation; Tófalo and Pazos 2010), and thus

Fig. 13.12 Jurassic-Cretaceous aeolian dunes of the Botucatu Formation covered by 132 Ma flood basalts of the Serra Geral Formation, east of Porto Alegre. At the contact the lavas preserve the dune morphology

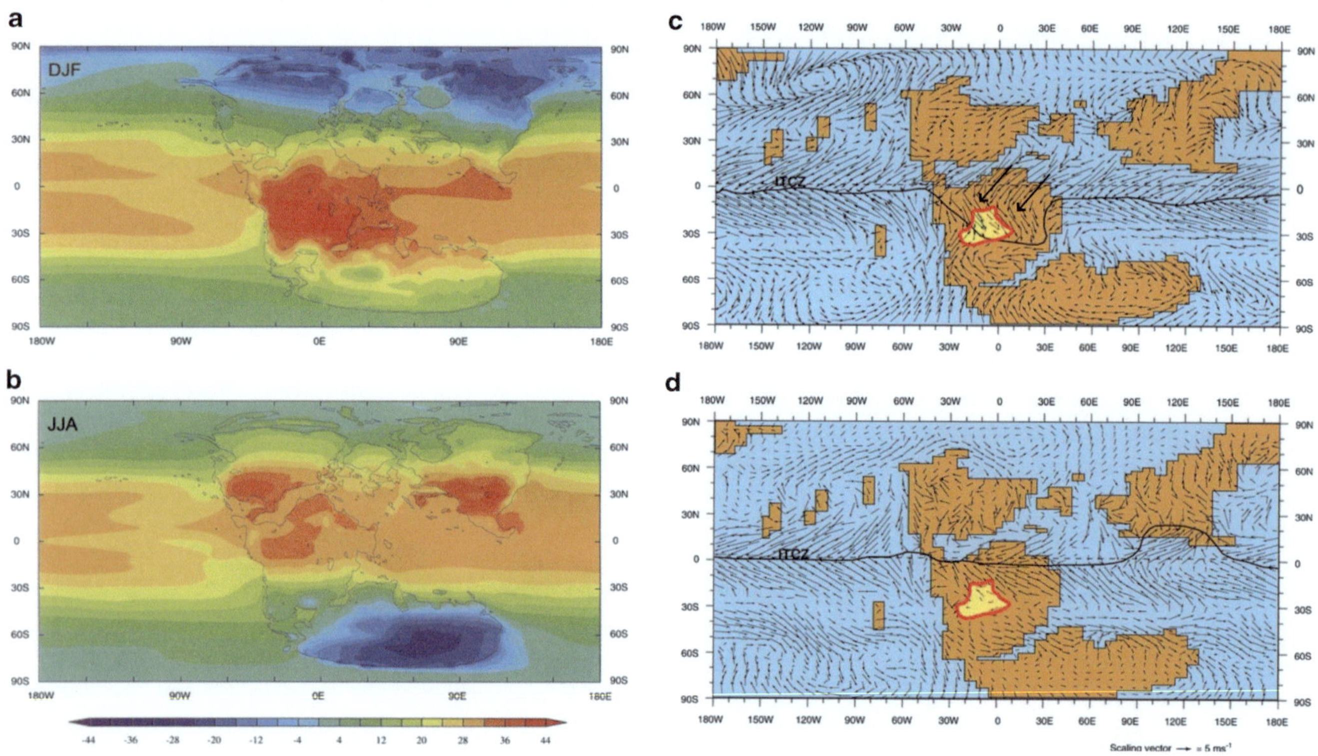

Fig. 13.13 Global paleo-climate models for the Late Jurassic showing a giant paleo-desert (*circled* in *red*) in Central West Gondwana (modified from Scherer and Goldberg 2010). (**a**, **b**) Paleo-temperatures (Sellwood and Valdes 2008), and (**c**, **d**) paleo-atmospheric circulations (Moore et al. 1992). These models predict hot and arid conditions (35–40 °C) and strong winds blowing to the southwest (*arrows*) at low latitudes across the Central West Gondwana surface

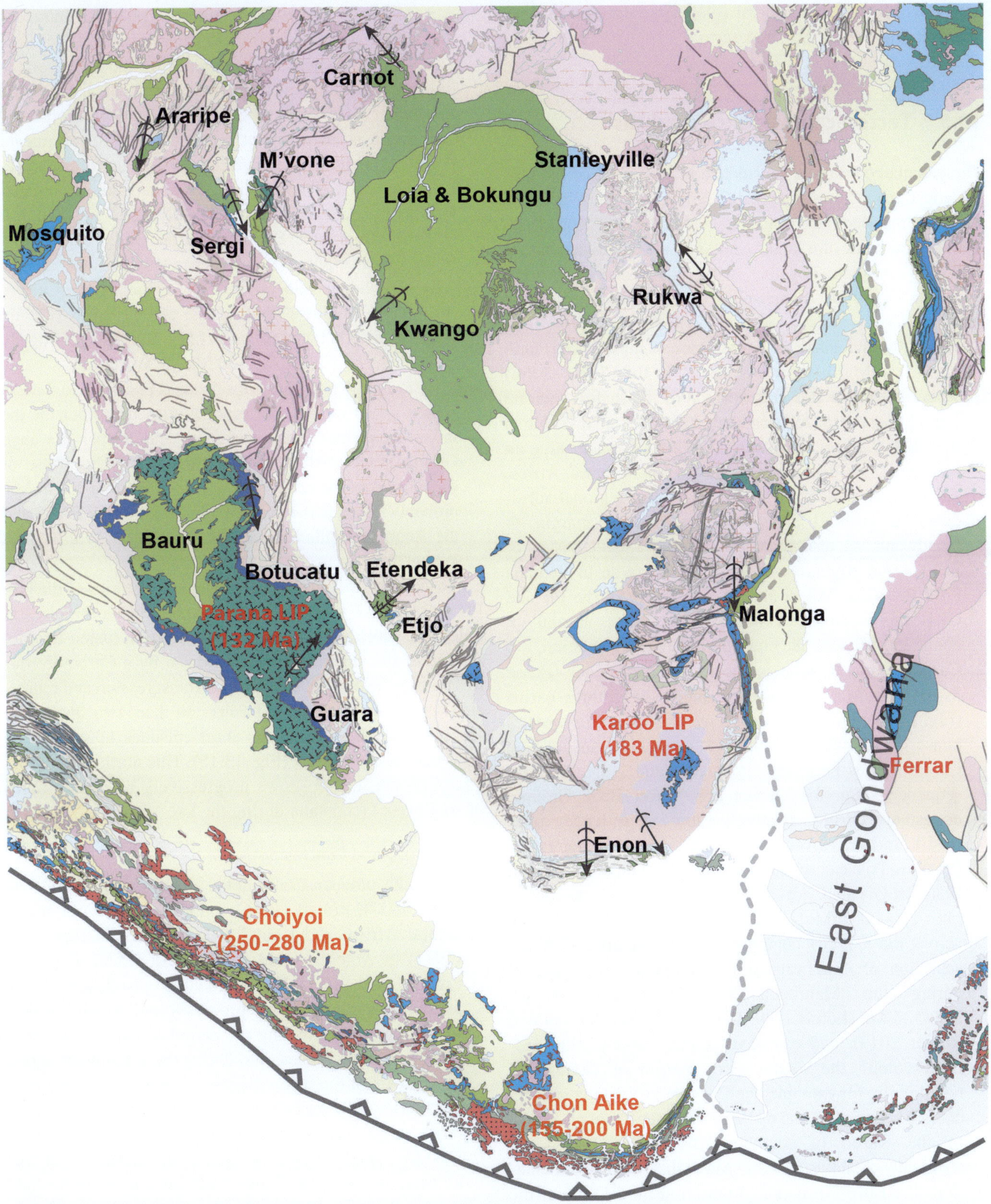

Fig. 13.14 Jurassic-Cretaceous paleo-map of southwest Gondwana, highlighting main stratigraphic sequences (Fig. 13.1 for legend). Sediment dispersion directions (*arrows*) indicate a complex topography across south-central Africa and South America

may possibly reflect a general (global) cooling and subsequent drying during the Cenozoic.

13.5 Tectonic Subsidence Analysis

The calculated tectonic subsidence of basement of the CB, PB and CKB (Zalan et al. 1990; Milani and de Wit 2008; Chap. 11, this Book) are compared to test their interdependence in the framework of a larger Central West Gondwana Basin (CWGB) complex. Together, these three curves show the existence of four main phases of rapid subsidence (Fig. 13.15):

1. A first, Ordovician to Devonian phase of subsidence [Phase 1: 5–20 m/Ma] is most pronounced in the CKB (~2 km) where it corresponds to deposition of the shallow marine Cape Supergroup (Johnson et al. 2006). It is also visible in the PB (~0.5 km), but is absent in the CB, suggesting an episode of non-deposition and/or erosion in central Africa. At this early Paleozoic time, the greatest regional subsidence is focused along the southwestern margin of Gondwana, as also recorded in the foreland basins of Bolivia and Argentina flanking the Sierra de la Ventana Fold Belt (e.g. Milani and de Wit 2008). In the PB and CKB, this subsidence terminates more or less abruptly at about 350 Ma with the onset of the Dwyka glaciation.
2. Carboniferous to Triassic accelerated subsidence [Phase 2: 10–15 m/Ma] is remarkably similar in all three basins (Fig. 13.15). This second and longest (~150 million years) phase of subsidence comprises, generally transgressive Carboniferous-Permian glacial and deglaciation sequences (e.g. the Dwyka and Lower Ecca Groups) deposited during relatively low subsidence rates [5–10 m/Ma], overlain by thicker and regressive Permian-Triassic sequences (e.g. the Upper Ecca and Beaufort Groups), deposited during increased subsidence rates [10–20 m/Ma]. This regional phase of subsidence across the CWGB complex overlaps in time with the accretion of Gondwana to Laurentia to form Pangea, during the Mauritanian-Variscan orogen at ca. 275–325 Ma (Lefort 1989; Dabo et al. 2008; Kroner and Romer 2013), and extensive subduction-accretion processes along the southwestern margin of Gondwana, resulting in, for example, the Cape-de la Ventana orogen, ca. 250 Ma (Trouw and de Wit 1999; Newton et al. 2006). In the Karoo of southern Africa, this subsidence terminates at about 180–200 Ma with aeolian sedimentation (the Eliot Formation; Johnson et al. 2006) and the outpouring of the Karoo LIP during the initial break-up between East and West Gondwana (e.g. Jokat et al. 2003).
3. A Late Jurassic to Early Cretaceous phase of intracontinental subsidence is most pronounced in the CB [Phase 3: 25 m/Ma], where it corresponds to deposition of the fluvial-marine Stanleyville Group and the aeolian Dekese Formation (~0.5 km thick). This relatively short phase of subsidence is also distinct in the PB with the Botucatu Formation (Milani et al. 2007), but is not uniformly widespread in the CKB, where instead sedimentation is confined to Jurassic-Cretaceous rift basins within the Cape Fold Belt (e.g. the intramontain Enon Formation; Dingle et al. 1983). In the CB, this phase of subsidence terminates by an episode of uplift at 120–140 Ma that coincides with active rifting and shoulders uplift in the central South Atlantic (Heine et al. 2013) and the onset of the Kalahari epeirogeny in southern Africa, ca. 80–120 Ma (de Wit 2007).
4. Mid- to Late Cretaceous subsidence is restricted to the CB [Phase 4: 10 m/Ma], corresponding to deposition of the fluvial-lacustrine Loia, Bokungu and Upper Kwango Groups (Linol 2013; Chap. 8, this book). This phase of subsidence terminates by an episode of uplift at 30–50 Ma, probably linked to the onset of doming along the EARS and the Angolan Highland (e.g. Pik et al. 2008; Walford and White 2005).

To summarize, the related subsidence histories of the CB, PB and CKB of southwest Gondwana indicate initial early Paleozoic subsidence along the southwestern (paleo-Pacific) margin of the supercontinent, followed by a Carboniferous to Triassic cycle of subsidence (i.e. transgression and regression) common to all three basins and then, two final, Late Jurassic and mid-Cretaceous short episodes of subsidence best recorded in the CB (Fig. 13.15). This northward propagation of subsidence shows progressively increasing uplift of southwest Gondwana during the late Mesozoic.

Summary and Conclusion

Since the pioneer stratigraphic work of Du Toit (1926, 1927, 1937), Robert (1946) and Cahen (1954) in south-central Africa, and Keidel (1916) and Schneider et al. (1974) in Argentina and southeastern Brazil, it is now established that the CB, PB and CKB episodically shared tectonic continental-scale processes and climatic stability, during and following the period of amalgamation of Gondwana and Pangea, and during the subsequent period of break-up of these two mega-continents. Comparison of the stratigraphic records and subsidence histories of these three large basins of Central West Gondwana (e.g. CWGB complex) is consistent with interpretation that all these basins were significantly influenced by far-field processes during the evolution of long-lived mountain systems along the southwestern margin of Gondwana—Gondwanides (Fig. 13.4).

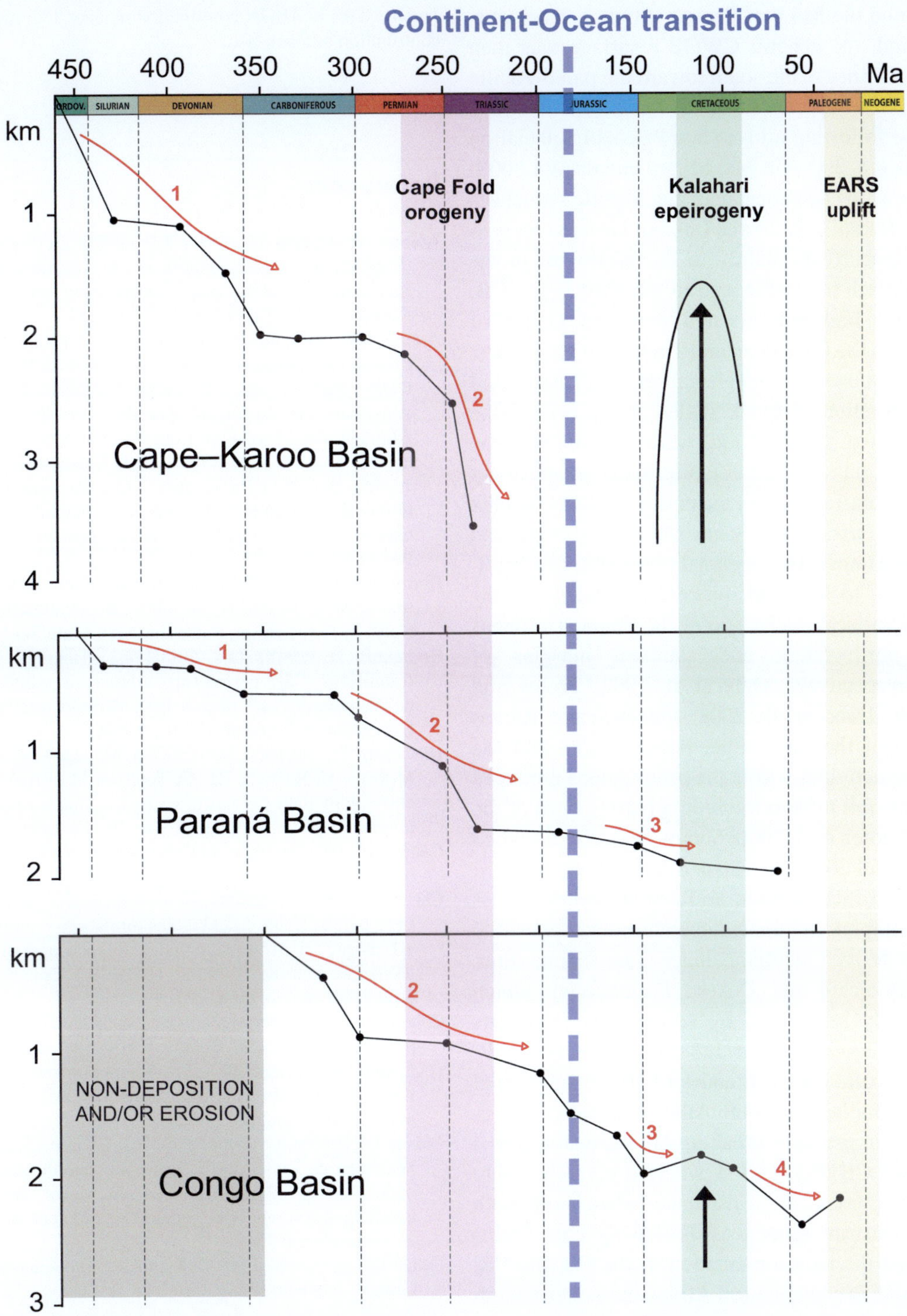

Fig. 13.15 Backstripped tectonic subsidences of basement of the CKB (Cloetingh et al. 1992), PB (Zalan et al. 1990; Milani and de Wit 2008) and CB (Linol 2013; Chap. 11, this Book). The three curves indicate: (1) initial early Paleozoic subsidence in the CKB and PB, and (2) generalized Carboniferous to Triassic subsidence in all the three basins, followed by (3) Late Jurassic and (4) mid-Cretaceous episodic subsidence events that are most pronounced in the CB. Black arrows pointing-up represent main uplifts related to the Kalahari epeirogeny

Following the Pan African-Brasiliano orogens (ca. 500–800 Ma), widespread northerly-derived upper Neoproterozoic-Cambrian red-beds (the upper Lindian Supergroup and the Upper Nama Group) were preferentially preserved across the Central African and Kalahari Shields (Fig. 13.7). These earliest Phanerozoic sequences were onlapped in the extreme south by a wide Ordovician-Devonian shallow marine platform (the Cape

Supergroup and the Rio Ivaí-Paraná Groups). Thus, during this period, the general CWGB's surface was then consistently inclined to the south toward the paleo-Pacific (Fig. 13.8).

During the following mid- to late Paleozoic glaciation of Gondwana (ca. 300–350 Ma; Milani and de Wit 2008) significant erosion and widespread glaciogenic sequences (the Dwyka, Itararé and Lower Lukuga Groups), deposited with paleocurrents mainly to the southwest in the CKB, to the north in the PB, and to the west in the CB, indicate major highlands located near the center of Gondwana (e.g. the Cargonian and Tanzanian Highlands; Fig. 13.10). Subsequent deglaciation and associated extensive deposition of black shales at the end of the Early Permian (ca. 270–280 Ma) were then followed by Late Permian-Triassic sedimentation under increasingly hot and arid conditions across the entire CWGB complex (the Beaufort, Passa Dois, Santa Maria, Cassange and Haute Lueki Groups). This regional regression can possibly be linked to far-field processes related to the Mauritanian-Variscan and Cape-de la Ventana orogens along the northwestern and southern margins of Gondwana, respectively (Daly et al. 1992; Trouw and de Wit 1999; Dabo et al. 2008; Pángaro and Ramos 2012). At about the same time, between 250–280 Ma large volcanic activities within the proto-Andes produced widespread air-fall tuffs in the sedimentary record of the CWGB (Stollhofen et al. 2000; Santos et al. 2006; Fildani et al. 2009; Linol 2013; see Chap. 8, this Book).

Triassic, and then Jurassic arid sedimentation across the CWGB complex culminated in widespread deposition of northerly-derived aeolian dunes (the Eliot, Etjo, Piramboia, Botucatu and Dekese Formations), which was episodically interrupted by voluminous extrusions of continental flood basalt at 183 Ma (the Karoo LIP) and at 132 Ma (the Paraná-Etendeka LIP), during initial opening of the Indian and South Atlantic Oceans, respectively. This main period of break-up of Gondwana is best recorded in the CB (Fig. 13.15), where it corresponds to two distinct phases of subsidence associated with shallowing lacustrine sequences. Following Late Cretaceous renewed fluvial sedimentation in the CB and PB, epeirogeny both in south-central Africa and eastern South America led to limited preservation of Cenozoic sequences and widespread development of calcretes, suggesting higher rates of evaporation during a more humid climate.

Acknowledgments We acknowledge funding through the Inkaba yeAfrica and !Khure Africa programs. We also acknowledge Petrobras for providing support in Brazil. B. Linol particularly thanks Renata da Silva Schmitt, convener of the Gondwana14 conference in Buzios and leader of the Gondwana Map Project for welcoming him in the new Gondwana mapping center at the Universidade Federal do Rio de Janeiro. This is AEON contribution number 131 and Inkaba yeAfrica contribution number 101.

References

Alvarez P, Maurin JC, Vicat J-P (1995) La Formation de l'Inkisi (Supergroupe Ouest-congolien) en Afrique centrale (Congo et Bas-Zaire): un delta d'âge Paléozoique comblant un bassin en extension. J Afr Earth Sci 20(2):119–131

Assine ML, Piranha JM, Carneiro CDR (2004) Os paleodesertos Pirambóia e Botucatu. In: Mantesso-Neto V, Bartorelli A, Carneiro CDR, Brito-Neves BB (eds) Geologia do Continente Sul-Americano: evolução da obra de Fernando Flávio Marques de Almeida. Beca Prod Cult Ltda, São Paulo, pp 77–92

Blanco G, Germs GJB, Rajesh HM, Chemale F Jr, Dussin IA Jr, Justino D (2011) Provenance and paleogeography of the Nama Group (Ediacaran to early Palaeozoic, Namibia): Petrography, geochemistry and U-Pb detrital zircon geochronology. Precambrian Res 187(1–2):15–32

Bordy EM, Segwabe T, Makuke B (2010) Sedimentology of the upper triassic-lower jurassic (?) mosolotsane formation (Karoo Supergroup), Kalahari Karoo Basin, Botswana. J Afr Earth Sci 58(1):127–140

Boutakoff N (1948) Les formations glaciaires et post-glaciaires fossilifères, d'âge permo-carbonifère (Karoo inférieur) de la region de Walikale (Kivu, Congo belge). Mémoire de l'Institut géologique. Université de Louvain, IX (II), 214 pp

Blackburn TJ, Olsen PE, Bowring SA, McLean NM, Kent DV, Puffer J, McHone G, Rasbury ET, Et-Touhami M (2013) Zircon U-Pb geochronology links the end-triassic extinction with the central Atlantic magmatic province. Science. doi: 10.1126/science.1234204

Cahen L (1954) Géologie du Congo Belge. Vaillant-Carmanne, Liège, p 577 p

Cahen L (1981) Précisions sur la stratigraphie et les corrélations du Goupe de la Haute-Lueki et des formations comparables (Triasique a Liasique? d'Afrique Centrale). Rapport annuel du Musée Royal d'Afrique centrale, Tervuren (Belgique), Départment de Géologie et de Minéralogie, pp 81–96

Cahen L (1983a) Brèves précisions sur l'âge des groupes crétaciques post-Wealdien (Loia, Bokungu, Kwango) du Bassin interieur du Congo (Republique du Zaïre). Rapport annuel du Musée Royal de l'Afrique centrale, Tervuren (Belgique), Département de Géologie et de Minéralogie, pp 61–72

Cahen L (1983b) Le Groupe de Stanleyville (Jurassic superieur et Wealdien de l'intérieur de la République du Zaïre): Révision des connaissances. Rapport annuel du Musée Royal de l'Afrique centrale, Tervuren (Belgique), Département de Géologie et de Minéralogie, pp 73–91

Cahen L, Lepersonne J (1978) Synthèse des connaissances relatives au Groupe (anciennement Série) de la Lukuga (Permien du Zaïre). Annales du Musée Royal du Congo belge, Tervuren (Belgique), Série in-8, Sciences géologiques, 82, pp 115–152

Catuneanu O, Wopfner H, Eriksson PG, Cairncross B, Rubidge BS, Smith RMH, Hancox PJ (2005) The Karoo basins of south-central Africa. J Afr Earth Sci 43(1–3):211–253

Chorowicz J (2005) The East African rift system. J Afr Earth Sci 43(1–3):379–410

Cloetingh S, Lankreijer A, de Wit MJ, Martinez I (1992) Subsidence history analysis and forward modelling of the Cape and Karoo Supergroups. In: de Wit MJ, Ransome I (eds) Inversion tectonics of the Cape Fold Belt Karoo and Cretaceous basins of southern Africa. Rotterdam, A.A. Balkema, pp 239–248

Cocks LRM, Torsvik TH (2006) European geography in a global context from the Vendian to the end of the Palaeozoic. In: Gee DG, Stephenson RA (eds) European lithosphere dynamics. Geological Society of London, pp 83–95

Cole DI (1992) Evolution and development of the Karoo Basin. In: de Wit MJ, Ransome I (eds) Inversion tectonics of the Cape Fold Belt Karoo and Cretaceous basins of southern Africa. A.A. Balkema, Rotterdam, pp 87–99

Cordani UG, Pimentel MM, de Araújo CEG, Basei MAS, Fuck RA, Girardi VAV (2013) Was there an Ediacaran Clymene Olcean in central South America? Am J Sci 313:517–539

Craddock J, Thomas R (2011) Detrital zircon provenance ages of the "Dwyka tillite" in South Africa and the Falkland Islands. Abstract, GeoSynthesis, Cape Town, 30 Aug–1 Sept

Craddock J (2011) Provenance of late Paleozoic tillites across Gondwana. Abstract, Gondwana14, Buzios, 25–30 Sept

Dabo M, Gueye M, Ngom PM, Diagne M (2008) Orogen-parallel tectonic transport: transpression and strain partitioning in the Mauritanides of NE Senegal. In: Ennih N, Liégeois J-P (eds) The boundaries of the West African Craton, Special Publications, vol 297. Geological Society, London, pp 483–497

Daly MC, Lawrence SR, Diemu-Tshiband K, Matouana B (1992) Tectonic evolution of the Cuvette Centrale, Zaire. J Geol Soc 149(4):539–546

Deckart K, Féraud G, Marques LS, Bertrand H (1998) New time constraints on dyke swarms related to the Paraná-Etendeka magmatic province, and subsequent South Atlantic opening, southeastern Brazil. J Volcanol Geotherm Res 80:67–83

Decker JE, Niedermann S, de Wit MJ (2013) Climatically influenced denudation rates of the southern African plateau: clues to solving a geomorphic paradox. Geomorphology 190:48–60

De Paula e Silva F, Kiang CH, Caetano-Chang MR (2009) Sedimentation of the Cretaceous Bauru Group in São Paulo, Paraná Basin, Brazil. J South Am Earth Sci 28(1):25–39

De Wit MJ, Stankiewicz J, Reeves C (2008) Restoring Pan-African-Brasiliano connections: more Gondwana control, less Trans-Atlantic corruption. In: Pankurst RJ, Trouw RAJ, Brito Neves BB, de Wit MJ (eds) West Gondwana: Pre-Cenozoic correlations across the South Atlantic Region, Special Publications, vol 294. Geological Society of London, pp 399–412

De Wit MJ (2007) The Kalahari Epeirogeny and climate change: differentiating cause and effect from core to space. South Afr J Geol 110(2–3):367–392

De Wit MJ, Ransome IGD (1992) Inversion tectonics of the Cape Fold Belt, Karoo and Cretaceous Basins of Southern Africa. A.A. Balkema, Rotterdam, p 269 p

De Wit MJ, Jeffery M, Berg H, Nicolayson LO (1988) Geological Map of sectors of Gondwana reconstructed to their position ~150 Ma (with explanatory notes), scale 1: 1.000.000. American Association of Petroleum Geologist, Tulsa

Dingle RV, Siesser WG, Newton AR (1983) Mesozoic and tertiary geology of southern Africa. A.A. Balkema, Rotterdam, p 375p

Domeier M, Van der Voo R, Torsvik TH (2012) Paleomagnetism and Pangea: the road to reconciliation. Tectonophysics 514–517: 14–43

Du Toit AL (1954) The geology of South Africa, 3rd edn. Oliver and Boyd, Edinburgh, p 611 p

Du Toit AL (1937) Our wandering continents: an hypothesis of continental drifting. Oliver and Boyd, Edinburgh, p 366 p

Du Toit AL (1927) A geological comparison of South America with South Africa. Publication of the Carnegie Institution of Washington. No. 381. Washington DC, Carnegie Institution of Washington

Du Toit AL (1926) The geology of South Africa. Olivier and Boyd, Edinburgh

Duncan RA, Hooper PR, Rehacek J, Marsh JS, Duncan AR (1997) The timing and duration of the Karoo igneous event, southern Gondwana. J Geophys Res 102:18127–18138

Encarnación J, Fleming TH, Elliot DH, Eales HV (1996) Synchronous emplacement of Ferrar and Karoo dolerites and the early break up of Gondwana. Geology 24(6):535–538

Fairhead JD (1988) Mesozoic plate tectonic reconstructions of the central South Atlantic Ocean: the role of the West and Central African rift system. Tectonophysics 155(1–4):181–191

Fildani A, Weislogel a, Drinkwater NJ, McHargue T, Tankard A, Wooden J, Hodgson D, Flint S (2009) U-Pb zircon ages from the southwestern Karoo Basin, South Africa: implications for the Permian-Triassic boundary. Geology 37(8):719–722

Forster A, Schouten S, Baas M, Sinninghe Damsté JS (2007) Mid-Cretaceous (Albian-Santonian) sea surface temperature record of the tropical Atlantic Ocean. Geology 35(10):919–922

Frank TD, Birgenheier LP, Montañez IP, Fielding CR, Rygel MC (2008) Late Paleozoic climate dynamics revealed by comparison of ice-proximal stratigraphic and ice distal isotopic records. In: Fielding CR, Frank TD, Isbell JL (eds) Resolving the late Paleozoic ice age in time and space. Geological Society of America, Special Paper, 441:331–342

Frimmel H, Tack L, Basei M, Nutman A, Boven A (2006) Provenance and chemostratigraphy of the Neoproterozoic West Congolian Group in the Democratic Republic of Congo. J Afr Earth Sci 46(3):221–239

Ganerød M, Torsvik TH, Van Hinsbergen DJJ, Gaina C, Corfu F, Werner S, Owen- Smith TM, Ashwal LD, Webb SJ, Hendriks BWH (2011) Palaeoposition of the Seychelles microcontinent in relation to the deccan traps and the Plume generation zone in late cretaceous-early palaeogene time. In: Van Hinsbergen DJJ, Buiter SJH, Torsvik TH, Gaina C, Webb SJ (eds) The formation and evolution of Africa: a synopsis of 3.8 Ga of Earth history. Geological Society of London, Special Publications, 357:229–252

Germs GJB (1995) The Neoproterozoic of southwestern Africa, with emphasis on platform stratigraphy and paleontology. Precambrian Res 73:137–151

Gibson SA, Thompson RN, Day JA (2006) Timescales and mechanisms of plume-lithosphere interactions: 40Ar/39Ar geochronology and geochemistry of alkaline igneous rocks from the Paraná-Etendeka large igneous province. Earth Planet Sci Lett 251:1–17

Gobbo-Rodrigues SR, Coimbra JC, Petri SRJB (2003) Kwango Series (Congo), Bauru Group (Brazil) and Neuquen Basin (Argentina) ages, based on ostracods and vertebrates. XVIII Congresso Brasileiro de Paleontologia, Brasilia, Brazil, pp 152–153

Gohl K, Uenzelmann-Neben G, Grobys N (2011) Growth and dispersal of a Southeast African large igneous province. South Afr J Geol 114(3–4):379–386

Grotzinger JP, Miller RM (2008) Nama Group. In: Miller RM (ed) The geology of Namibia. Windhoek, Geological Survey of Namibia, pp 229–272

Heine C, Zoethout J, Müller RD (2013) Kinematics of the South Atlantic rift. Solid Earth 4:215–253

Hoffman PF (1999) The break-up of Rodinia, birth of Gondwana, true polar wander and the snowball Earth. J Afr Earth Sci 28(1): 17–33

Hofman C, Courtillot V, Féraud G, Rochette P, Yirgu G, Ketefo E, Pik R (1997) Timing of the Ethiopian flood basalt event and implication for plume birth and global change. Nature 389:838–841

Iannuzzi R, Boardman DR (2007) Problems in western Gondwana Geology – 1st Workshop – South America-Africa correlations: du Toit revisited. Gramado-RS-Brazil, 27–29 Aug, extended abstracts, 197 p

Jelsma HA, Barnett W, Richards S, Lister G (2009) Tectonic setting of kimberlites. Lithos 112S:155–165

Jelsma HA, Perrit SH, Amstrong R, Ferreira HF (2011) Shrimp U-Pb zircon geochronology of basement rocks of the Angolan Shield, western Angola. Abstract, 23rd CAG, Johannesburg, 8–14 Jan

Jerram D, Mountney N, Holzfo F, Stollhofen H (1999) Internal stratigraphic relationships in the Etendeka Group in the Huab Basin, NW Namibia: understanding the onset of flood volcanism. J Geodyn 28:393–418

Johnson MR, Van Vuuren CJ, Visser JNJ, Cole DI, de V. Wickens H, Christie ADM, D.L. R, Brandl G (2006) Sedimentary rocks of the Karoo supergroup. In: Johnson MR, Anhaeusser CR, Thomas RJ (eds) The geology of South Africa. Pretoria, South Africa, Council for Geoscience, pp 461–499

Jokat W, Boebel T, König M, Meyer U (2003) Timing and geometry of early Gondwana breakup. J Geophys Res Solid Earth 108(B9)

Jones DL, Duncan RA, Briden JC, Randall DE, Macniocaill C (2001) Age of the Batoka basalts, northern Zimbabwe, and the duration of Karoo large igneous province magmatism. Geochem Geophys Geosyst 2: 14 p

Jourdan F, Féraud G, Bertrand H, Watkeys M, Renne PR (2007) Distinct brief major events in the Karoo large igneous province clarified by new 40Ar/39Ar ages on the Lesotho basalts. Lithos 98(1–4):195–209

Kay SM, Ramos VA, Mpodozis C, Sruoga P (1989) Late Paleozoic to Jurassic silicic magmatism at the Gondwana margin: analogy to the Middle proterozoic in North America? Geology 17:324–328

Keidel J (1916) La geología de las sierras de la Província de Buenos Aires y sus relaciones con las montañas de Sud Africa y los Andes. Anales del Ministerio de Agricultura de la Nación, Sección Geología, Mineralogía y Minería, Buenos Aires 3:1–78

Kirstein LA, Kelley S, Hawkesworth C, Turner S, Mantovani M, Wijbrans J (2001) Protracted felsic magmatic activity associated with the opening of the South Atlantic. J Geol Soc London 158: 583–592

Kleiman LE, Japas MS (2009) The Choiyoi volcanic province at 34°S–36°S (San Rafael, Mendoza, Argentina): implications for the late palaeozoic evolution of the southwestern margin of Gondwana. Tectonophysics 473:283–299

Kroner U, Romer RL (2013) Two plates – many subduction zones: the Variscan orogeny reconsidered. Gondwana Res 24:298–329

Lefort JP (1989) Basement correlation across the North Atlantic. Springer, Berlin, p 148 p

Lepersonne (1974) Carte géologique du Zaïre au 1: 2.000.000 et notice explicative. Kinshasa, République du Zaïre: Direction de la Géologie/Musée Royal de l'Afrique centrale, Tervuren (Belgique)

Linol B (2013) Sedimentology and sequence stratigraphy of the Congo and Kalahari Basins of south-central Africa and their evolution during the formation and break-up of West Gondwana. PhD thesis, Nelson Mandela Metropolitan University, 375 p

López-Gamundí O (2006) Permian plate margin volcanism and tuffs in adjacent basins of west Gondwana: Age constraints and common characteristics. J South Am Earth Sci 22(3–4):227–238

Maurin J-C, Guiraud R (1993) Basement control in the development of the early cretaceous West and Central African rift system. Tectonophysics 228(1–2):81–95

Marshall JEA (1994) The Falkland islands: a key element in Gondwana paleogeography. Tectonics 13:499–514

Martin H (1981) The late Palaeozoic Gondwana glaciation. Geologische Rundschau 70(2):480–496

Marzoli A, Renne PR, Piccirillo EM, Ernesto M, Bellieni G, Min AD (1999) Extensive 200-million-year-old continental flood Basalts of the Central Atlantic Magmatic province. Science 284:616–618

Master S, Rainaud C, Armstrong R, Phillips D, Robb L (2005) Provenance ages of the Neoproterozoic Katanga Supergroup (Central African Copperbelt), with implications for basin evolution. J Afr Earth Sci 42(1–5):41–60

Milani EJ, Gonçalves de Melo JH, de Souza PA, Fernandes LA, França AB (2007) Bacia do Paraná. Boletim de geociências da petrobras 15 (2):265–287

Milani EJ, de Wit MJ (2008) Correlations between the classic Paraná and Cape Karoo sequences of South America and southern Africa and their basin infills flanking the Gondwanides: du Toit revisited. In: Pankurst RJ, Trouw RAJ, Brito Neves BB, de Wit MJ (eds) West Gondwana: Pre-Cenozoic correlations across the South Atlantic region. Geological Society of London, Special Publications, 294, pp 319–342

Milani EJ (1997) Evolução tectono-estratigráfica da Bacia do Paraná e seu relacionamento com a geodinâmica fanerozóica do Gondwana sul-ocidental, in 2 volumes. Thesis, Universidade Federal do Rio Grande do Sul, Porto Alegre

Miller R (2008) The geology of Namibia, in three volumes. Geological Survey, Windhoek, Namibia

Mizusaki AMP, Melo JHG, Vignol-Lelarge ML, Steemans P (2002) Vila Maria Formation (Silurian, Paraná Basin, Brazil): integrated radiometric and palynological age determinations. Geol Mag 139(4):453–463

Montañez IP, Poulsen CJ (2013) The late Paleozoic ice age: an evolving paradigm. Annu Rev Earth Planet Sci 41:629–656

Moore GT, Hayashida DN, Ross CA, Jacobson SR (1992) Paleoclimate of the Kimmeridgian/Tithonian (Late Jurassic) world: I. Results using a general circulation model. Palaeogeogr Palaeoclimatol Palaeoecol 93(1–2):113–150

Morag N, Avigad D, Gerdes A, Belousova E, Harlavan Y (2011) Detrital zircon Hf isotopic composition indicates long-distance transport of North Gondwana Cambrian-Ordovician sandstones. Geology 39(10):955–958

Moulin M, Aslanian D, Unternehr P (2010) A new starting point for the South and Equatorial Atlantic Ocean. Earth-Sci Rev 98(1–2):1–37

Mounguengui MM, Lang J, Guiraud M (2008) Sedimentary dynamics and extensional structuring related to early Cretaceous rifting of Neocomian and Barremian deposits of the interior basin of Gabon. J Afr Earth Sci 51:239–256

Newton AR, Shone RW, Booth PWK (2006) The cape fold belt. In: Johnson MR, Anhaeusser CR, Thomas RJ (eds) The geology of South Africa. Pretoria, South Africa, Council for Geoscience, pp 521–530

Oreskes N (2001) Plate tectonics: an insider's history of the modern theory of the Earth. Westview Press, Colorado, p 424 p

Pángaro F, Ramos VA (2012) Paleozoic crustal blocks of onshore and offshore central Argentina: New pieces of southwestern Gondwana collage and their role in the accretion of Patagonia and the evolution of Mesozoic south Atlantic sedimentary basins. Mar Pet Geol 37: 162–183

Pankurst RJ, Vaughan APM (2009) The tectonic context of the Early Palaeozoic southern margin of Gondwana. In: Bassett MG (ed) Early Palaeozoic Peri-Gondwana terranes: new insights from tectonics and biogeography. London, Geological Society, Special Publications, 325, pp 171–176

Peate DW, Hawkesworth CJ, Mantovani MSM (1992) Chemical stratigraphy of the Paranà lavas (South America): classification of magma types and their spatial distribution. Bull Volcanol 55: 119–139

Pedrosa-Soares AC, Alkmim FF, Tack L, Noce CM, Babinski M, Silva LC, Martins-Neto MA (2008) Similarities and differences between the Brazilian and African counterparts of the Neoproterozoic Araçuaí-West Congo orogen. Geol Soc Lond Spec Publ 294 (1):153–172

Pik R, Marty B, Carignan J, Yirgu G, Ayalew T (2008) Timing of East African Rift development in southern Ethiopia: implication for mantle plume activity and evolution of topography. Geology 36:167–170

Poidevin JL (1985) Le Proterozoic supérieur de la République Centrafricaine. Annales du Musée Royal de l'Afrique centrale, Tervuren (Belgique), Serie in-8, Sciences Geologiques, 91, 75 p

Pucéat E, Lecuyer C, Donnadieu Y, Naveau P, Cappetta H, Ramstein G, Hubert BT, Kriwet J (2007) Fish tooth δ18O revising Late Cretaceous meridional upper ocean water temperature gradients. Geology 35(2):107–110

Pucéat E, Lecuyer C, Sheppard SMF, Dromart G, Reboulet S, Grandjean P (2003) Thermal evolution of Cretaceous Tethyan marine waters inferred from oxygen isotope composition of fish tooth enamels. Paleoceanography 18(2), 1029, 12 p

Ramos VA, Aleman A (2000) Tectonic evolution of the Andes. In: Cordani UG, Milani, EJ, Thomaz Filho A, Campos DA (eds) Tectonic evolution of South America. Rio de Janeiro, Brazil: 31st international geological congress, pp 635–685

Rapela CW, Fanning CM, Casquet C, Pankhurst R, Spalletti L, Poiré D, Baldo EG (2011) The Rio de la Plata craton and the adjoining Pan-African/brasiliano terranes: their origins and incorporation into south-west Gondwana. Gondwana Res 20:673–690

Reeves C (1972) Rifting in the Kalahari? Nature 237(5350):95–96

Reeves C (1999) Aeromagnetic and gravity features of Gondwana and their relation to continental break-up: more pieces, less puzzle. J Afr Earth Sci 28:263–277

Renne PR, Ernesto M, Pacca IG, Coe RS, Glen JM, Prévot M, Perrin M (1992) The age of Paraná flood volcanism, rifting of gondwanaland, and the Jurassic-Cretaceous boundary. Science 258(5084):975–979

Robert M (1946) Le Congo physique. Troisième edition. H. Vaillant-Carmanne, Liége, 449 pp

Rocha-Campos AC, Basei MA, Nutman AP, Kleiman LE, Varela R, Llambias E, Canile FM, da Rosa ODCR (2011) 30million years of Permian volcanism recorded in the Choiyoi igneous province (W Argentina) and their source for younger ash fall deposits in the Paraná Basin: SHRIMP U-Pb zircon geochronology evidence. Gondwana Res 19(2):509–523

Santos RV, Souza P a, de Alvarenga CJS, Dantas EL, Pimentel MM, de Oliveira CG, de Araújo LM (2006) Shrimp U-Pb zircon dating and palynology of bentonitic layers from the Permian Irati Formation, Paraná Basin, Brazil. Gondwana Res 9(4):456–463

Scheffler K, Hoernes S, Schwark L (2003) Global changes during Carboniferous-Permian glaciation of Gondwana: linking polar and equatorial climate evolution by geochemical proxies. Geology 31(7):605–608

Scherer CMS, Goldberg K (2007) Palaeowind patterns during the latest Jurassic - earliest Cretaceous in Gondwana: Evidence from aeolian cross-strata of the Botucatu Formation, Brazil. Palaeogeogr Palaeoclimatol Palaeoecol 250(1–4):89–100

Scherer CMS, Goldberg K (2010) Cyclic cross-bedding in the eolian dunes of the Sergi Formation (Upper Jurassic), Recôncavo Basin: inferences about the wind regime. Palaeogeogr Palaeoclimatol Palaeoecol 296(1–2):103–110

Schneider RL, Muhlmann H, Tommasi E, Medeiros RA, Daemon RF, Nogueira AA (1974) Revissao estratigrafica da Bacia do Paraná. Anais do XXVIII Congresso Brasileiro de Geologia. Sociedade Brasileira de Geologia. Porto Alegre 1:41–65

Searle M (2013) Colliding continents: a geological exploration of the Himalaya, Karakoram, and Tibet. Oxford University Press, USA, 368 p

Sellwood BW, Valdes PJ (2008) Jurassic climates. Proc Geol Assoc 119(1):5–17

Smith R (1984) The lithostratigraphy of the Karoo Supergroup in Botswana, vol 26. Geological Survey/Ministry of Mineral Resources and Water Affairs Republic of Botswana, Lobatse, 239p.

Souza PA (2006) Late Carboniferous palynostratigraphy of the Itararé Subgroup, northeastern Paraná Basin, Brazil. Rev Palaeobot Palynol 138:9–29

Stollhofen H, Stanistreet IG, Bangert B, Grill H (2000) Tuffs, tectonism and glacially related sea-level changes, Carboniferous-Permian, southern Namibia. Palaeogeogr Palaeoclimatol Palaeoecol 161:127–150

Tait J, Delpomdor F, Preat A, Tack L, Straathof G, Nkula VK (2011) Neoproterozoic sequences of the West Congo and Lindi/Ubangi Supergroups in the Congo Craton, Central Africa. In: Arnaud E, Halverson GP, Shields-Zhou G (eds) The geological record of Neoproterozoic glaciations. Geological Society of London, Memoirs, 36, pp 185–194

Tankard A, Welsink H, Aukes P, Newton R, Stettler E (2009) Tectonic evolution of the Cape and Karoo basins of South Africa. Mar Pet Geol 26(8):1379–1412

Taverne L (1975) Etude ostéologique de Leptolepis caheni, Teleosteen fossile du Jurassique superieur (Kimmeridgien) de Kisangani (ex-Stanleyville, Zaïre) précedemment décrit dans le genre Paraclupavus. Revue Zoologique Africaine 89:821–853

Tinker J, de Wit M, Brown R (2008) Mesozoic exhumation of the southern Cape, South Africa, quantified using apatite fission track Thermochronology. Tectonophysics 455(1–4):77–93

Torsvik TH, Van der Voo R, Preeden U, Mac Niocaill C, Steinberger B, Doubrovine PV, van Hinsbergen DJJ, Domeier M, Gaina C, Tohver E, Meert JG, McCausland PJA, Cocks LR (2012) Phanerozoic polar wander, palaeogeography and dynamics. Earth Sci Rev 114(3–4):325–368

Torsvik TH, Amundsen H, Hartz EH, Corfu F, Kusznir N, Gaina C, Doubrovine PV, Steinberger B, Ashwal LD, Jamtveit B (2013) A Precambrian microcontinent in the Indian Ocean. Nat Geosci 6: 223–227

Tófalo OR, Pazos PJ (2010) Paleoclimatic implications (Late Cretaceous - Paleogene) from micromorphology of calcretes, palustrine limestones and silcretes, southern Paraná Basin, Uruguay. J South Am Earth Sci 29(3):665–675

Toteu SF, Fouateu RY, Penaye J, Tchakounte J, Mouangue ACS, Van Schmus WR, Deloule E, Stendal H (2006) U-Pb dating of plutonic rocks involved in the nappe tectonic in southern Cameroon: consequence for the Pan-African orogenic evolution of the central African fold belt. J Afr Earth Sci 44(4–5):479–493

Trouw RA, de Wit MJ (1999) Relation between the Gondwanide Orogen and contemporaneous intracratonic deformation. J Afr Earth Sci 28(1):203–213

Trumbull RB, Reid DL, de Beer C, van Acken D, Romer RL (2007) Magmatism and continental breakup at the west margin of southern Africa; a geochemical comparison of dolerite dikes from northwestern Namibia and the Western Cape. South Afr J Geol 110: 477–502

Van Staal CR, Whalen JB, Valverde-Vaquero P, Zagorevski A, Rogers N (2009) Pre-Carboniferous, episodic accretion-related, orogenesis along the Laurentian margin of the northern Appalachians. Geol Soc Lond Spec Publ 327:271–316

Veizer J, Godderis Y, Francois LM (2000) Evidence for decoupling of atmospheric CO2 and global climate during the Phanerozoic eon. Nature 408:698–701

Veevers JJ, Cole DI, Cohan EJ (1994) Southern Africa: Karoo Basin and Cape Fold Belt. In: Veevers JJ, and Powell CMcA (eds) Permian-Triassic Pangean Basins and fold belts along the Panthalassan margin of Gondwanaland. Geological Society of America, Colorado, Memoir, 184, pp 223–280

Vesely FF, Assine ML (2006) Deglaciation sequences in the Permo-Carboniferous Itararé Group, Paraná Basin, southern Brazil. J South Am Earth Sci 22:156–168

Vesely FF (2007) Sistemas subaquosos alimentados por fluxos hiperpicnais glaciogenicos: modelo deposicional para arenitos do Grupo Itararé, Permocarbonífero da Bacia do Paraná. Boletim de Geociências da Petrobras 15(1):7–25

Visser JNJ (1995) Post-glacial Permian stratigraphy and geography of southern and central Africa: boundary conditions for climatic modelling. Palaeogeogr Palaeoclimatol Palaeoecol 118(3–4): 213–243

Walford HL, White NJ (2005) Constraining uplift and denudation of West African continental margin by inversion of stacking velocity data. J Geophys Res-Solid Earth 110, B04403

Wegener A (1912) Die Enstehung der Kontinente. Geologische Rundschau 3:276–292

Zachos J, Pagani M, Sloan L, Thomas E, Billups K (2001) Trends, rhythms, and aberrations in global climate 65 Ma to present. Science 29:686–693

Zalan PV, Wolff S, de J. Conceicao JC, Marques A, Astolfi MAM, Vieira IS, Appi VT, Zanotto OA (1990) Bacia do Paraná. In: De Raja Gabaglia GP, Milani EJ (eds) Origem e Evolucao de Bacias Sedimentares. Petrobras, Rio de Janeiro, Brazil, pp 135–168

Zerfass H, Chemale F, Schultz CL, Lavina E (2004) Tectonics and sedimentation in southern South America during Triassic. Sediment Geol 166(3–4):265–292

Part IV

Neotectonics, Geomorphology and River Systems

Cenozoic Landscape Evolution in and Around the Congo Basin: Constraints from Sediments and Planation Surfaces

14

Francois Guillocheau, Roman Chelalou, Bastien Linol, Oliver Dauteuil, Cecile Robin, Francois Mvondo, Yannick Callec, and Jean-Paul Colin†

14.1 Introduction

The Congo Cuvette (here referred to as Congo Basin, CB) corresponds (Fig. 14.1) to a large, ovoid, depression (E–W: 1,300–1,500 km, N–S: 1,000–1,200 km), bordered:

- To the east, by the Mitumba Mounts, forming the western flank of the East-African Dome (western branch of the East African Rift System with an elevation of 1,800–2,000 m to 2,500–3,000 m, and a maximum elevation of 4,507 m (at the Mount Karisimbi volcano of the Virunga volcanic province),
- To the north, by the Cameroon Volcanic Line (Western Highlands and Adamawa-Yadé Plateau, called hereafter the Cameroon Highlands) with an elevation of 900–1,800 m and a maximum elevation of 3,010 m at Mount Oku (Western Highlands) and 4,040 to 4,095 m at Mount Cameroon (near the Coastal Plain) and the Ubangian Rise with an elevation of 500–1,000 m, and a maximum elevation of 1,330 m at Mount Toussoro (Central African Republic, CAR),
- To the south, by the Kwango-Lunde-Kasai Plateau with an elevation of 1,000–1,100 m to 1,200–1,400 m, flanked by the Katanga Plateaus and Plains (east) and the Angola Plateau (west),
- To the west, by the Central African Atlantic Swell that separates the CB from the South Atlantic coast and which comprises two subswells, the Angola Swell, south of the Congo River and with a mean elevation of 1,800–2,200 m and the Congo-Gabon-South Cameroon Swell, north of the Congo River and with a mean elevation of 600–900 m and a maximum elevation of 1,575 m at Mount Iboundji (Gabon).

The elevation of the base of the CB, here defined at the river water level, ranges between 320 m in the west to 380 m in the east, flanked by local reliefs (hills) on the order of 200–300 m above the river water level.

Because of the large size of the CB (≥1,000 km wavelength) a dynamic topography control linked to mantle processes has been inferred for the Cuvette (Hartley and Allen 1994; Downey and Gurnis 2009; Crosby et al. 2010; Moucha and Forte 2011; see also Raveloson et al., Chap. 1 this Book) In contrast, other authors favour a long-term thermo-tectonic control on the evolution of the basin, following an early Neoproterozoic rift below the central part of the CB (e.g. Kadima et al. 2011; see also Kadima et al., Chap. 4, this Book) or by sediment loading on a thick lithosphere (e.g. Buiter et al. 2012). Little attention has been paid in these models to the details of the Cenozoic evolution of the CB. This may be because there are only few reviews (e.g. Lepersonne 1978—without figures; Giresse 2005) and which are mostly ignored. Two competing models should be considered for the Cenozoic landscape evolution in and around the CB: (1) significant subsidence of the Central Basin from a Mesozoic plateau, or (2) low uplift of the

†Author was deceased at the time of publication.

F. Guillocheau (✉) • R. Chelalou • O. Dauteuil • C. Robin
Géociences-Rennes, UMR 6118 Université de Rennes 1 – CNRS, OSUR, Université de Rennes 1, Campus de Beaulieu, 35042 Rennes cedex, France
e-mail: francois.guillocheau@univ-rennes1.fr; roman.chelalou@univ-rennes1.fr; olivier.dauteuil@univ-rennes1.fr; cecile.robin@univ-rennes1.fr

B. Linol
Africa Earth Observatory Network, Nelson Mandela Metropolitan University, PO Box 77000, Port Elizabeth 6031, South Africa
e-mail: bastien.aeon@gmail.com

F. Mvondo
Département de Géosciences, Université de Douala, 2701 Douala, Cameroon
e-mail: francois_mvondo@hotmail.com

Y. Callec
BRGM-CONGO, 256 Boulevard Loango, BP 1758 Pointe Noire, DR Congo
e-mail: y.callec@brgm.fr

M.J. de Wit et al. (eds.), *Geology and Resource Potential of the Congo Basin*, Regional Geology Reviews,
DOI 10.1007/978-3-642-29482-2_14, © Springer-Verlag Berlin Heidelberg 2015

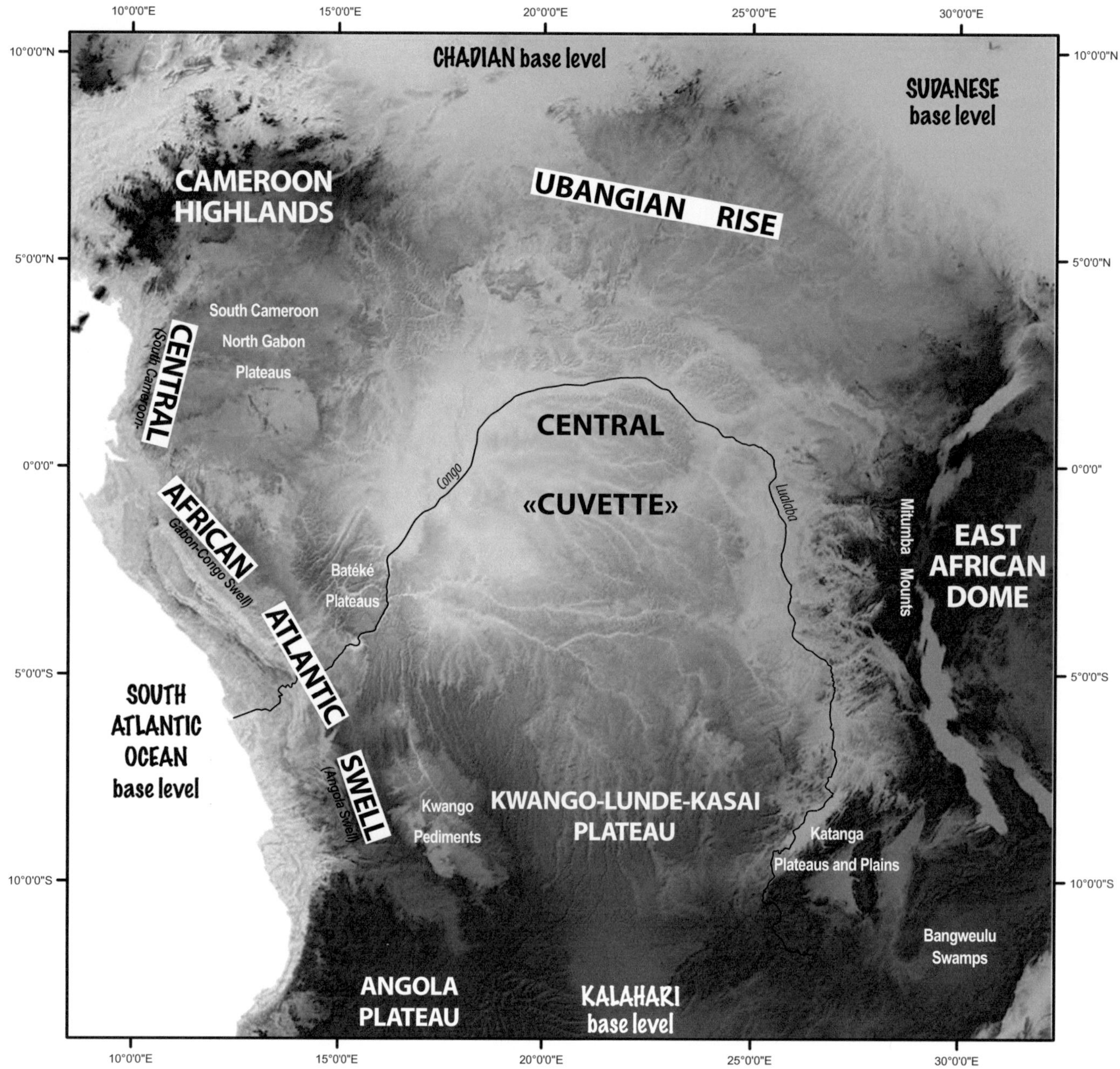

Fig. 14.1 Main geomorphological domains of the CB and surrounding highs

Central Basin from a flat domain close to sea level and substantial uplift of the regions surrounding the CB.

Our aim is to provide and integrate geomorphological and stratigraphical analyses of the CB and its flanking highlands to determine the timing of the Cenozoic evolution of the landscapes and associated vertical movements that have led to the present-day geomorphology of this central African region. Our analysis includes the topography of Cameroon and CAR, in the north, of northern Angola in the south, and of the Atlantic Ocean in the west.

This work is presented in two parts, (1) a stratigraphical and sedimentological analysis of the Central CB based on both outcrops and wells data and (2) a geomorphologic analyses focussed on the planation surfaces, the most characteristic landforms of the CB and surrounding regions, as already documented by the first geologists working on these landscapes (Veatch 1935; Jessen 1936; Cahen and Lepersonne 1948; Cahen 1954). Morphological analysis of the present-day river network is not included, for which the interested reader is referred to Flugel et al. (Chap. 15, this Book).

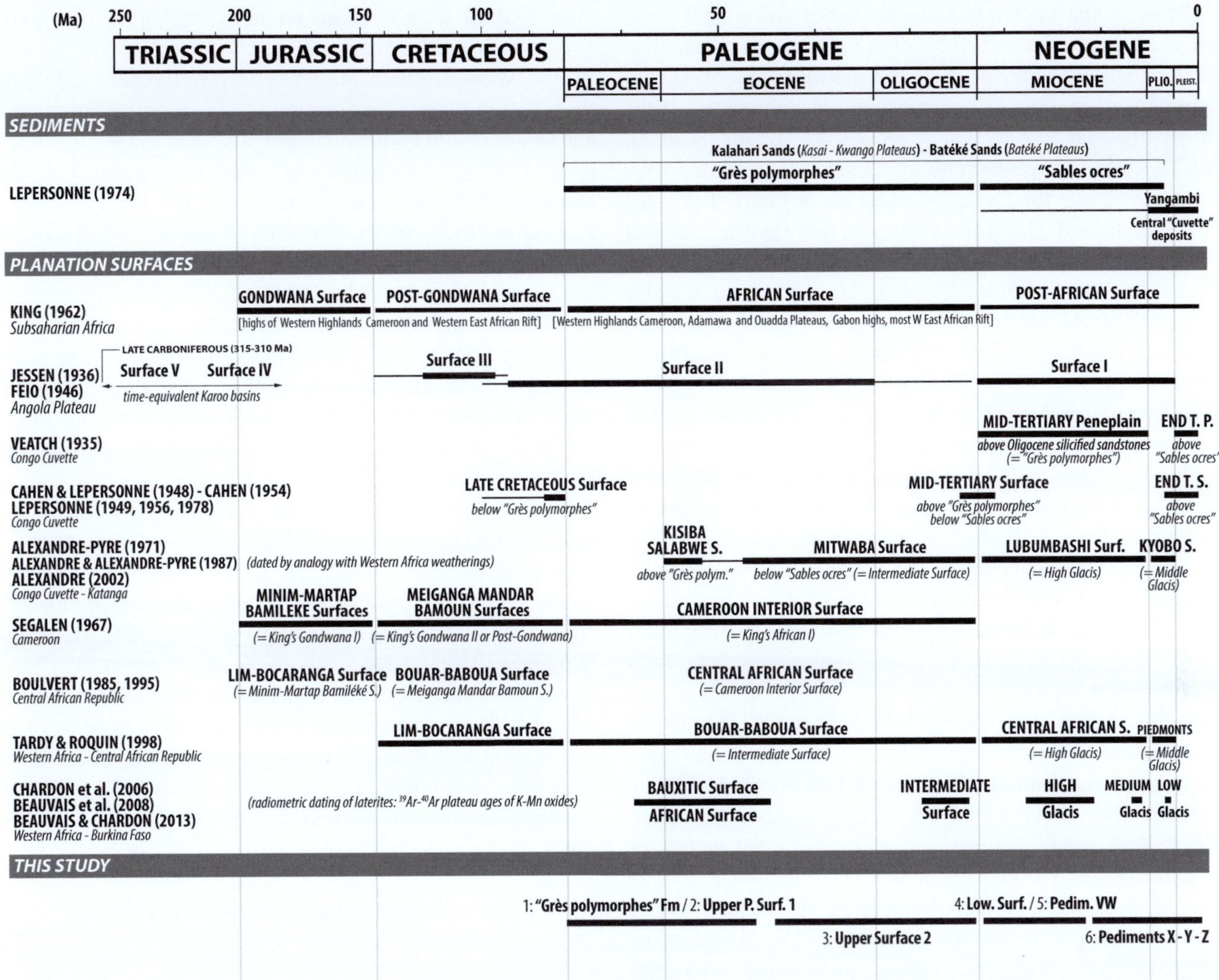

Fig. 14.2 Summary of the previous stratigraphic and geomorphologic studies carried out in the CB and surrounding highs—comparisons with the dated planation surfaces of West Africa and the results of this study

14.2 Previous Studies

Three successive Cenozoic sedimentary units are defined in the CB (e.g. Lepersonne 1974—Fig. 14.2): (1) "Grès polymorphes" (Paleogene), (2) "Sables ocres" (Neogene before Late Pliocene) and (3) Yangambi and Lodja-Salonga Beds (Plio-Pleistocene to Neogene). The first two, grouped as the Kalahari Sands (Maufe 1928–1929; Cahen and Lepersonne 1952; Polinard 1954) or Batéké Sandstones (Le Maréchal 1966; Giresse and K'Vadec 1971), crop out in the southern (Katanga, Kasai and Kwango Plateaus) and western (Batéké Plateaus) parts of the CB, while the third ones are restricted to the central part of the CB. These three lithological units are all incised by rivers during Pleistocene times (e.g. Lepersonne 1974).

The only dated sediments are the "Grès polymorphes" based on fossils (ostracodes, gastropods and characeans) found in silicified lacustrine carbonates (Katanga and Kasai). The ages provided by Leriche (1925 1926–1927, 1927—summary in Polinard, 1931–1932) were inconsistent: changing from Cenozoic (Late Pliocene to Quaternary—on gastropods in 1925), but then revised to Purbeckian (Jurassic–Cretaceous boundary—on characeans in 1926–1927). A subsequent biostratigraphic study (Grekoff 1958) on ostracodes, attributed an Eocene to Oligo-Miocene age to the "Grès polymorphes". Later still, Colin (1994) confirmed an Eocene age on the basis of characeans occurrence. In Angola, some isolated outcrops ascribed to the "Grès polymorphes" were dated as Pleistocene based on gastropods, ostracodes and characeans (Mouta and Dartevelle 1954).

Most of the geomorphological studies since 1935 have focussed on planation surfaces. The different interpretations of land-surfaces of the CB and surrounding areas are summarized in Fig. 14.2, against recent radiometric dates of weatherings surfaces in West Africa (Beauvais et al. 2008). Only few geomorphologic maps are available (e.g. King 1962 at the scale of Africa; Ségalen 1967 for Cameroon; and Boulvert 1996 for the Central African Republic). In the central CB, the chronology of land surfaces is based on their relationships with sediments (Veatch 1935; Cahen 1954; Alexandre 2002). Along the surrounding areas (Cameroon—Ségalen 1967; Central African Republic—Boulvert 1985, 1996; Katanga—Alexandre-Pyre 1971; Alexandre and Alexandre-Peyre 1987; Alexandre 2002), the chronology of land surfaces is based on comparisons with the sequence of weathering profiles and planation surfaces of West Africa (Ivory Coast, Burkina Faso, Mali) studied by the soil geologists and geomorphologists of the Strasbourg University and of the French overseas scientific research agency (O.R.S.T.O.M, now I.R.D.—"Institut de Recherche pour le Développement") (Michel 1973; Grandin 1976; Tardy and Roquin 1998; see Chardon et al. 2006 for a recent review).

This compilation (Fig. 14.2) shows numerous inconsistancies, including different ages for the same surfaces. This is likely due to different methods of age determinations. (1) Along the coast, surfaces are dated where they merge with the coastal basins (albeit with a lot of uncertainties abouth the accuracy of correlations). (2) The age of the surface is function of their elevation. Another cause of error is to confuse an exhumed basal sediments surface with a younger land-surface cross cutting and locally merging with this old exhumed surface.

Several other investigations were carried out in northwest Congo (Ituri), with a controversy about the importance of faulting along the western highs flanking Lake Albert Rift (Ruhe 1954, 1956, 1958, 1959; Lepersonne 1956, 1958a, b; Moeyersons 1975; Mbuluyo 1991), in Lower Congo (van Kerschaver 1975, 1983, 1984a, b; Kaseba et al. 1997) and in Gabon-Congo (Congo-Brazzaville or Congo Republic—Dresch 1946; Vogt 1962; Novikoff 1974; Petit 1990).

14.3 Methodology

14.3.1 Sedimentary Record

Most of the Cenozoic sediments are preserved in the central part of the CB and its southern extension into the Kalahari Basin. The only Cenozoic sediments preserved in the surrounding areas (and not reported on here) are paleovalleys dammed by lavas flows (e.g. on the Adamawa Plateau, Anloua area, Letterman 1984; Oustrière 1984) or thin (×10 m) accumulations on the Ubangian Rise (Bigotte and Bonifas 1968; Miauton 1980).

For *subsurface* data, only four deep wells are available in the CB (see Plate 2 for location): (1) two old ones (Samba and Dekese) with cores and cuttings, drilled by the "Syndicat pour l'Etude géologique et minière de la Cuvette congolaise" (Professional association for the geological and mining Study of the Congolese Cuvette) and described by Cahen et al. (1959, 1960) and now stored in the Royal Museum for Central Africa (RMCA) in Tervuren (Belgium) and (2) two petroleum wells (Mbandaka 1 and Gilson 1) drilled by ESSO ZAIRE (EXXON). We acquired access to the cuttings of Dekese and Samba wells. A petrological study including grain shape analysis using Scanning Electron Microprobe (exoscopy) was completed on cuttings. The study of Mbandaka 1 and Gilson 1 was based on well-logs (gamma-ray, resistivity and sonic) and cutting descriptions available on the masterlogs (Plate 2).

For *surface* data, published data of outcrops of "Grès polymorphes" and "Sables ocres" were studied (grain-size, grain shape and clay mineralogy analysis: Claeys 1947; Beugnies 1950; Le Maréchal 1966; de Ploey et al. 1968; Alexandre-Pyre 1971) supplemented by observations during a short field trip managed by the BRGM (French geological survey), north of Brazzaville (facies sedimentology). For the Kisangani outcrops (see Plate 2 for location), we used the excellent descriptions of de Heinzelin (1952).

14.3.2 Geomorphology: Planation Surfaces Analysis

This study is mainly focussed on the planation surfaces defined as nearly flat erosional surfaces cutting across the underlying heterogenous rocks (Migon 2004a). The two characteristic planation surfaces of the CB and the surrounding reliefs are etchplains and pediplains to pediments (see definitions in 14.5.1.1 and 14.5.1.2). Such planation surfaces can be deformed (tilted and/or domed) by long to very long wavelength deformations (×100 km and × 1,000 km), with possible major basement faults reactivations. These uplift induced a local base level fall and the growth of a new planation surface below the previous one. The consequences of such deformations are formation of stepped planation surfaces, with the oldest landform at higher elevation and the youngest one at lower elevation. The boundaries between such surfaces can be marked by sharp scarps, or may be gradational.

Other processes that can generate base level fall are eustasy with the relief connected to the sea (exoreic system) or climate changes for the endoreic systems through control by the watershed hydrological budget. The amplitude of the sea level variations since Eocene times continues to be controversial. Haq et al. (1987) suggest a sea level fall since Early Eocene of ca. 210–260 m. By contrast, Miller et al. (2005), and more recently Rowley (2013), suggest much lower values of between 40 and 120 m, and no more than 40 m since the Late Eocene (Miller et al. 2005), and no more than 20 m since 40 Ma (Rowley 2013). These recent data may indicate that the maximum amplitude of sea level fall since the end of the Eocene « Grès polymorphes » deposition is less than 50 m, and then of low importance.

The cornerstone of our study is a new map (Plate 2) of planation surfaces (and scarps) and incised rivers, in chronologic order. Dating of these landforms is based on the geometrical relationships between planation surfaces and dated rocks (e.g. magmatic rocks such as younger" granites and volcanics), sediments and weathering profiles. Two areas are critical for the accuracy of this analysis (Fig. 14.1): the Cameroon Volcanic Lines (including the Adamawa Plateau) and the Virunga-Bukavu (=Kivu) Volcanic Province of the East African Dome in Eastern Congo, Rwanda, and Burundi.

The analysis is based on a combination of Digital Elevation Model (SRTM 90 m processed with ArcGIS 10) and published geological maps, research papers and local reports. To characterize local base level falls and to evaluate possible different uplift at regional scale, a number of topographical sections are presented (Fig. 14.9) to illustrate (i) planation surface types and their extents, (ii) the amplitude of steps between surfaces and (iii) the change of slope gradients between surfaces.

14.4 Sedimentary Rocks

Following the international lithostratigraphic code, the "Grès polymorphes", "Sables ocres", "Yangambi", "Lodja" and "Salonga" units are hereafter ranked as formations (Fm) and the Batéké sands as a group (Gp). The "Grès polymorphes" and "Sables ocres" Fms are parts of the Kalahari Gp.

14.4.1 Outcrops of "Grès Polymorphes" and "Sables Ocres" Sediments

14.4.1.1 "Grès Polymorphes" Formation (Fm) (Figs. 14.3 and 14.4; Table 14.1)

The base of the "Grès polymorphes" Fm is a sharp contact, corresponding to the Late Cretaceous surface of Cahen (1954). In Congo-Brazzaville (Le Maréchal 1966; Giresse 1990), the "Grès polymorphes" Fm overlies kaolinite-rich aeolian sandstones of the Stanley-Pool Fm, dated as Late Jurassic to Early Cretaceous based on the fauna of the basal lacustine sediments (Egoroff and Lombard 1961–1962). This kaolinite results from a post-depositional weathering as indicated (Le Maréchal 1966) by (i) *in situ* weathered feldspars and micas and (ii) illuviation of kaolinite in the pores. In the Kwango region, the "Grès polymorphes" Fm overlie an iron-rich level (duricrust?) (Cahen 1954; Linol 2013). In Katanga (Alexandre 2002), this contact has not been observed. This suggests that a period of weathering in warm humid conditions preceded the deposition of the "Grés polymorphes" Fm.

Three main types of facies were defined by Le Maréchal (1966) and here validated further by field studies (Table 14.1, Fig. 14.4). The primary environment of sand deposition is a reg covered by aeolian dunes as suggested by the occurrence of (1) ventifacts (facetted aeolian pebbles), (2) round-frosted grains, (3) high oblique laminations (1 to 10 m) migration of large 2D to 3D dunes and (4) alternations of massive, ×1 cm-thick, medium-grained sands laminations and mm-thick fine-grained sands laminations, characteristic of grainfall on the lee-side of aeolian dunes (e.g. Brookfield and Silvestro 2010 for a review). These aeolian deposits are episodically reworked by probable alluvial processes as suggested by the flat truncations of the aeolian dunes (deflation surfaces) sometimes overlaid by sands with low preservation 2D to 3D current megaripples cross-beddings (bedload stream flows?). Based on the types of gastropods, the limestones are lacustrine in origin. These sediments were deposited in a hot desert interrupted by recurrent more humid periods (possibly driven by Milankovitch cycles) as witnessed by the interbedded alluvial and lacustrine sediments.

Subsequently, the upper quartz-rich sands and limestones of the top half of the formation were silicified. The silicification of the quartz-rich sands results from euhedral quartz overgrowths (Fig. 14.4.2), while the limestones are microquartzified (Fig. 14.4.4). Those mineralogic modifications are characteristic of groundwater silicifications (Thiry 1999; Thiry and Ribet 1999; Thiry and Maréchal 2001). Both the poorly consolidated and silicified sediments are dissolved. The desilicification of the silicified limestones and sandstones is recorded by vugs, filled by palissadic and euhedral silica. Desilicification can occur during groundwater circulation (Thiry and Maréchal 2001). But the occurrence of dissolved quartz and kaolinite precipitation in the poorly consolidated sands, the weathering of the rare feldspars, as well as illuviation structures (Mouyoungou 1990) suggest dissolution along a weathering profile superimposed on quartz-rich sands.

14.4.1.2 "Sables Ocres" Formation (Fm) (Figs. 14.3 and 14.4; Table 14.1)

The base of the "Sables ocres" Fm is a sharp contact with underlying white and brown colour sands. It corresponds to the Mid-Tertiary surface of Cahen (1954). In some places, this surface is aligned with iron concretions (the "limonitic" crusts of Cahen and Lepersonne), and iron crusted

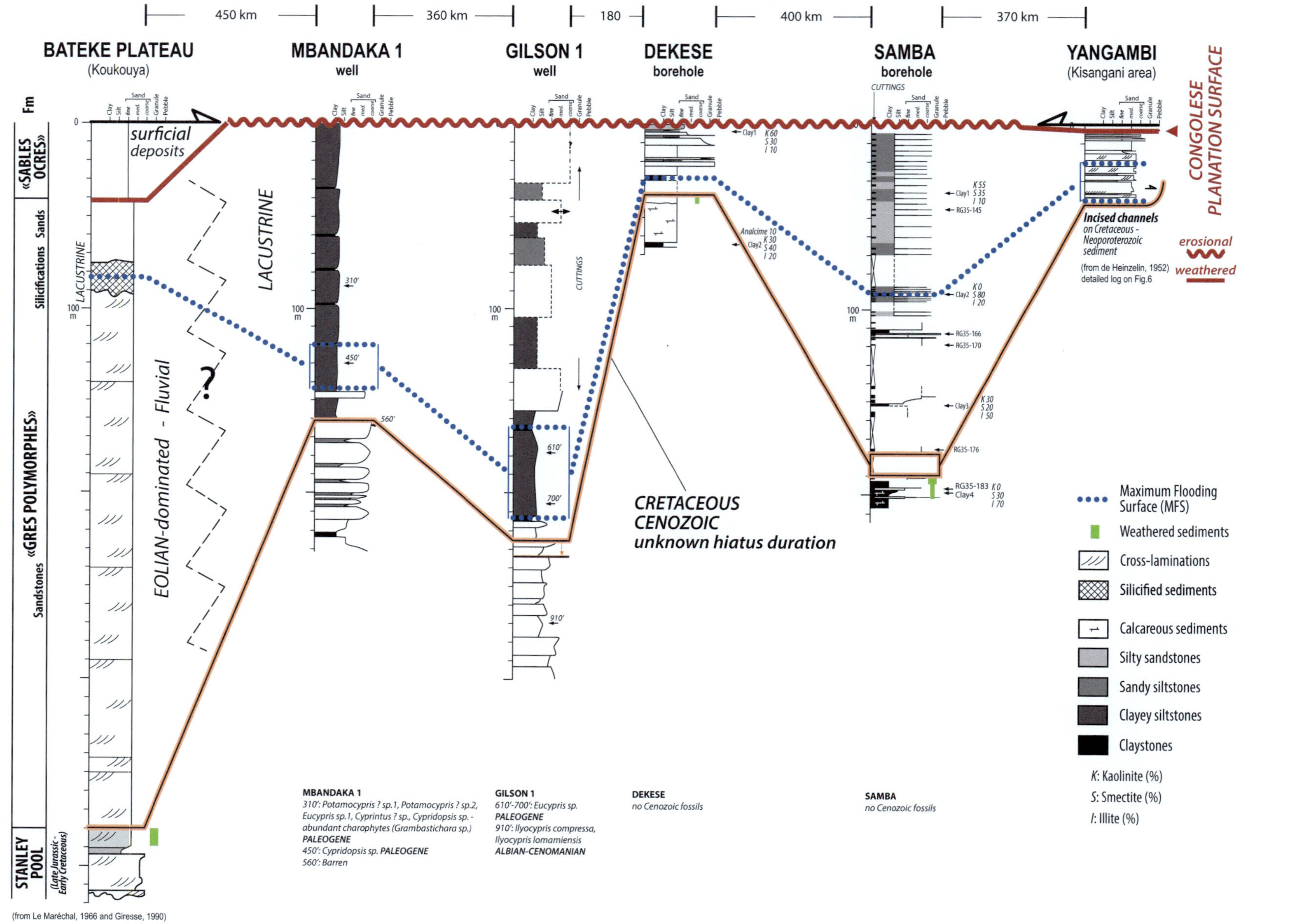

Fig. 14.3 Logs and correlations of the boreholes Dekese and Samba, the wells Gilson 1 and Mbandaka 1 and the surrounding outcropping domains, Batéké Plateaus (Koukouya Plateau) and Kisangani area (Yangambi cliffs). Clay mineralogy of Samba and Dekese boreholes from Vanderstappen and Verbeek (1964). Location maps on Plate 2 and Fig. 14.7

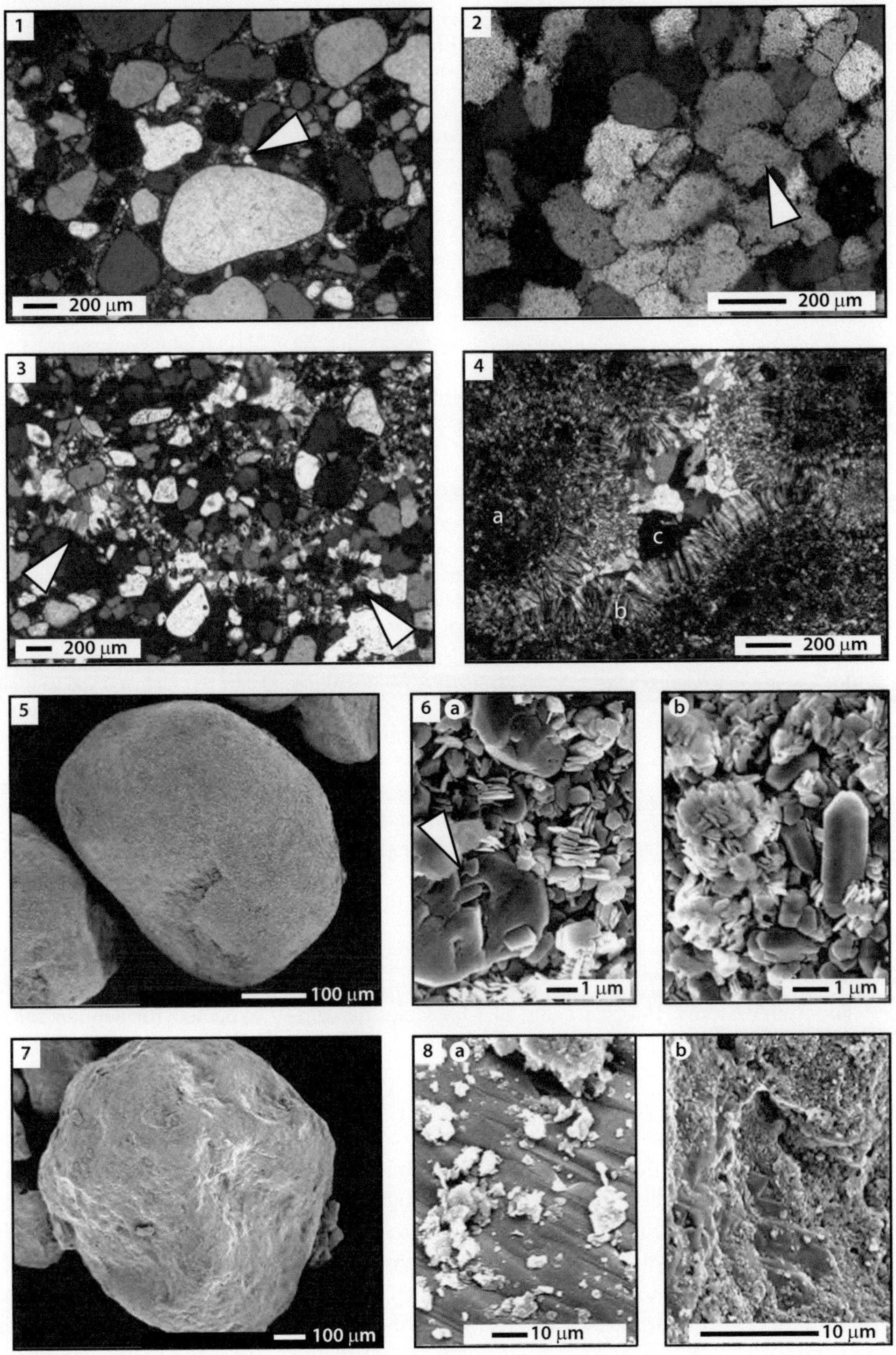

Fig. 14.4 "Grès polymorphes" (Gp) and "Sables ocres" (So) Formations: Thin-sections and Scanning Electron Microprobe pictures of Mount Mougengenge samples (10 km SE from the Kinshasa International airport, DRC). 1. Gp—thin-section of coarse-grained sandstones, subrounded to subangular (finest—*arrows*) grains. 2. Gp—thin-section of silicified sandstones (quartzites); the quartz grains are cemented by subhedral and sutured silica overgrowth; the original rounded shape of the grain is sometimes recognizable (*arrow*), underlined by dust rims. 3. Gp—thin-section of silicified sandstones with vugs filled by euhedral quartz (*arrows*). 4. Gp—thin-section of silicified limestones (microcrystalline quartz or microquartzitic matrix—(**a**) with vugs filled by (**b**) flamboyant palisadic silica (chalcedony) and (**c**) euhedral quartz. 5. Gp—rounded medium quartz grain coated by silica—probable aeolian origin. 6. Gp—detail of the grain surface: evidences of dissolution (**a**, *arrow*) and coating by automorphic quartz and kaolinite (**b**). 7: So—Subrounded irregular medium to coarse quartz grain. 8: So—detail of the grain surface: evidences of dissolution (**a**, **b**) and possible iron oxides/hydroxides coating (**a**)

pebbles or conglomerates (reworking of weathered "Grès polymorphes" in Kasai according to de Ploey et al. 1968). In Katanga (Alexandre 2002), the succession is as follows: (i) "Grès polymorphes", (ii) iron duricrust 1, (iii) gravels bed ("Cailloutis des plateaux"), (iv) iron duricrust 2, (v) "Sables ocres". Iron duricrust 1, sometimes directly overlain by the "Sables ocres" Fm, is a pisolitic crust superimposed on breccia of corroded quartz and weathered pebbles of Proterozoic sandstones and "Grès polymorphes". Iron duricrust 2 is a gravelly crust with coated red shinning iron nodules with some corroded coarse-grained sands. When directly located over the basement, the iron duricrust constitutes the uppermost part of a weathering profile of kaolin.

Table 14.1 Main sedimentological characteristics of the “Grès polymorphes” and “Sables ocres” Formations, based on published data and samples provided by the Royal Museum of Central Africa (Tervuren, Belgium) coming from the Mount Mougengenge (or Mense Peak), located 10 km far SE from the Kinshasa International Airport (DRC)

“GRÈS POLYMORPHES” FORMATION			
White fine to medium-grained $\pm$ ***well-sorted sandstones***	Mean grain-size Maximum Clay fnraction	150–350 µm 1 mm 3–8 %	Quartz (rare feldpars) and kaolinites subrounded dominant—well rounded to subangular (finest grains)—Fig. 4.1–4.5 round-frosted grains of the French-speaking sedimentary petrologists due to coating by a silicieous pellicule
More or less consolidated	Moderately to well-sorted **SEM:** Dissolution on all grains (Fig. 4.6a), later coated by automorphic quartz and kaolinite (Fig. 4.6b)		
Le Maréchal (1966), de Ploey et al., (1968)	**Sedimentary structures:** Large oblique laminasets ($\times 1$ m, up to 10 m-thick), simple or compound, resulting from the migration of large 2D–3D dunes with grainfall laminations, truncated by flat surfaces sometimes overlain by low preservation 2D to 3D current megaripple cross-beddings near the base, facetted pebbles (ventifacts) in Katanga (Mortelmans 1946)		
Silicified sandstones to quartzites (Tshidibi 1986; Mouyoungou 1990)	Same grain-size distribution and grain shape than above—two main types of silicification: 1. Cemented quartz grains by suhedral and sutured silica overgrowth (crystal $>$ 100 µm)—Fig. 4.2 2. More or less cemented quartz grains with small vugs (200–300 µm) with palissadic and euhedral quartz growth—Fig. 4.3 Mainly located in the second half of the formation		
Silicified limestones (Le Maréchal 1966; Tshidibi 1986	Marls to fine-grained limestones with continental subaqueous gastropods (*Limnea, Planorbis, Physia*) and some characeans later silicified by microcrystalline quartz with vugs filled, with different generations of euhedral quartz with different shapes (palisadic, radiolitic..) and different types of silica (chalcedony, lutecite, quartzine)—Fig. 4.4 Mainly located in the second half of the formation		
“SABLES OCRES” FORMATION			
Brown fine-grained homogenous sandstones	Mean grain-size Maximum Clay fraction Moderately sorted less than G.p. Fm	100–250 µm 500–700 µm 5–30 %	Quartz (no feldpars) and kaolinites - four types of quartz grains: 1. Subrounded (shinning—coarser grains—Fig. 4.7) to subangular (frosted-finer $<$300 µm) some with impact structures - few rounded (coarsest grains $>$500 µm) 2. Polyhedral non-worn shinning 3. Corroded and fractured 4. Automorphic (bipyramidal)
(Le Maréchal 1966) (de Ploey et al. 1968) (Alexandre-Pyre 1971)	**SEM:** Evidences of dissolution (Fig. 4.8a, b), grains later coated by different types of iron oxydes/hydroxydes **Sedimentary Structures:** Homogenous, with no evidence of sedimentary structures		

The “Sables ocres” Fm (see Table 14.1 for facies description) onlaps Proterozoic basement to the south (in Katanga; Cahen and Lepersonne 1952; Cahen 1954), and to the west (the Batéké Plateaus, Le Maréchal 1966).

The age and environmental setting of the “Sables ocres” Fm are both unresolved. In the absence of dating, there is no other data that indicates that this formation not diachronous is across the CB. In Congo-Brazzaville, the same facies extends across both over the “Grès polymorphes” Fm and the basement where these sands overlie “Stone-Lines” (beds of angular quartz pebbles). In Gabon, charcoals interbedded within these sandy facies were ^{14}C dated to the Holocene (c. 3000–2000 BP) and an aeolian origin has been proposed (Thiéblemont et al. 2013; Thiéblemont 2013), consistent with a very recent Pleistocene age for the “Sables ocres” Fm in this area.

Several observations question the generalization of this recent age to the “Sables ocres” Fm all around the Central Cuvette: (1) Based on SEM work, the “Sables ocres” grains appear more subrounded to subangular than rounded (as expected for aeolian sands), and grains are partly corroded by water dissolution; (2) No sedimentary structures indicative of aeolian sediments (e.g. horizontal sand sheets) are preserved.

The quartz grain dissolution is consistent with a model of intense weathering, as already mentioned above with respect to the occurrence of iron crusts and silicified levels in the “Sables ocres” Fm of the Batéké Plateau (Le Maréchal 1966; Giresse and K’Vadec 1971). The sand deposition may thus be a combined effect of aeolian and alluvial processes (e.g. de Ploey et al. 1968; interpreted as ephemeral rivers with sheet-wash by Alexandre-Pyre 1971), and subsequent weathering.

Because in Katanga (Alexandre 2002) the “Sables ocres” Fm is older than a major ‘cuirasse’ (iron duricrust 3) of possible Miocene age, we suggest that the “Sables ocres” is a convergent facies, shaped by the same processes, at different times during the Neogene. Thus, the “Sables ocres” Fm is *not* a time-marker horizon.

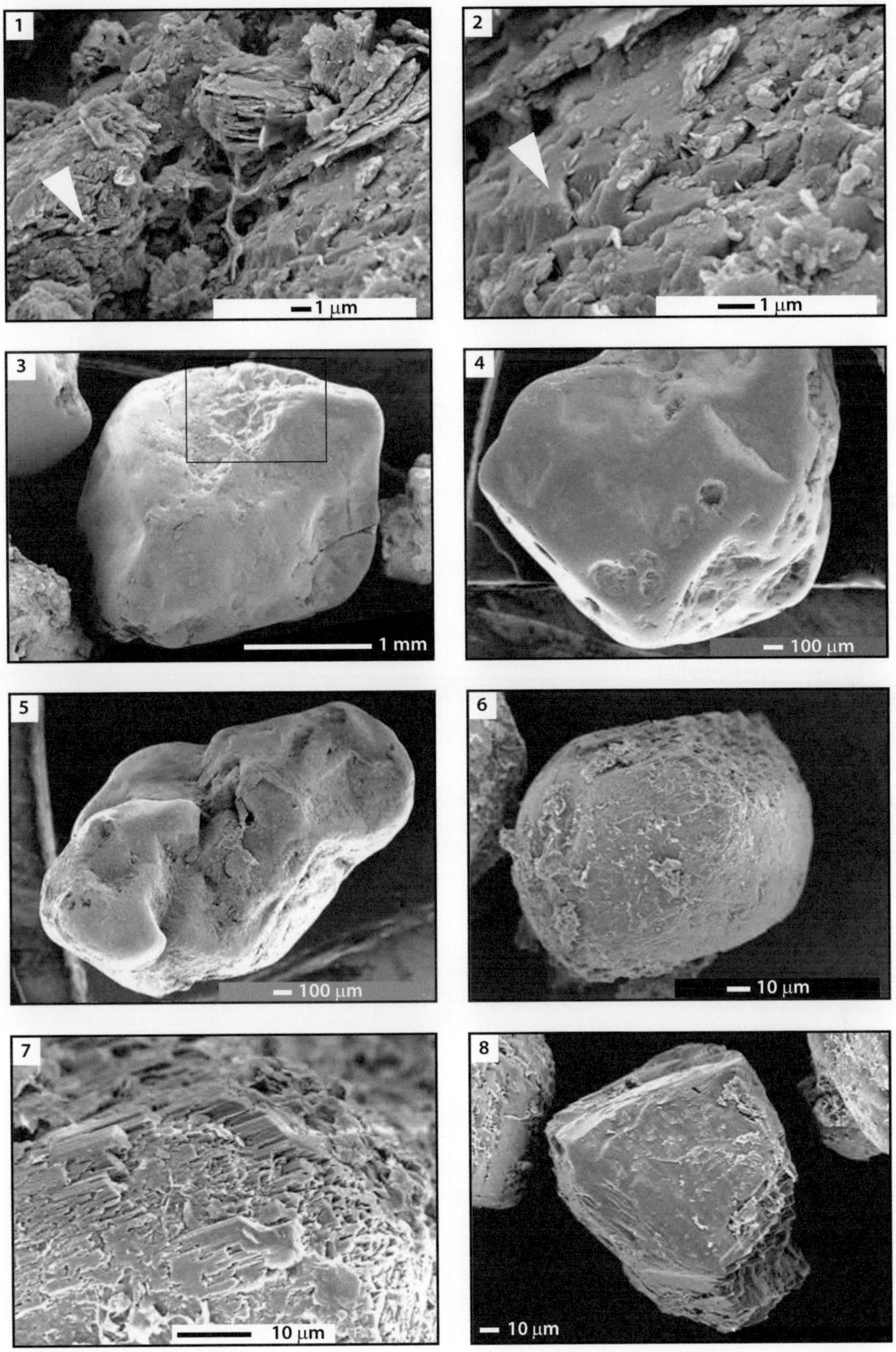

Fig. 14.5 Samba boreholes: Scanning Electron Microprobe pictures of the cuttings (samples location in Fig. 14.3). 1: RG35-145, stacked kaolinite sheets (*arrow*) on quartz grains. 2: RG35-145, dissolution structures (*arrow*) on quartz grain surface. 3: RG35-166, subangular very coarse quartz grain with dissolution structures (*rectangle*). 4: RG35-170, subangular coarse quartz grain. 5: RG35-176, blunt irregular (subangular) coarse quartz grain. 6: RG35-183, dissolved rounded aeolian fine quartz grain. 7: RG35-183, detail of the previous one showing the dissolution structures. 8: RG35-183, highly dissolved quartz grain leading to an angular shape

14.4.2 Boreholes and Wells: Dekese, Samba, Gilson 1, Mbandaka 1

14.4.2.1 Samba and Dekese Boreholes (See Plate 2 for Location)

The *Samba* borehole (190 m of Cenozoic deposits, Fig. 14.3), consists of three units:

- A *basal truncated weathering profile*, superimposed on Late Albian sediments (Linol et al., this volume), alternation of calcareous claystones and fine-grained (rounded aeolian grains—reworked?) sands (Fig. 14.5.6). These aeolian quartz sands are highly dissolved (Fig. 14.5.7) and its rounded grains can become angular (Fig. 14.5.8). The claystones transformation is low (no kaolinite,

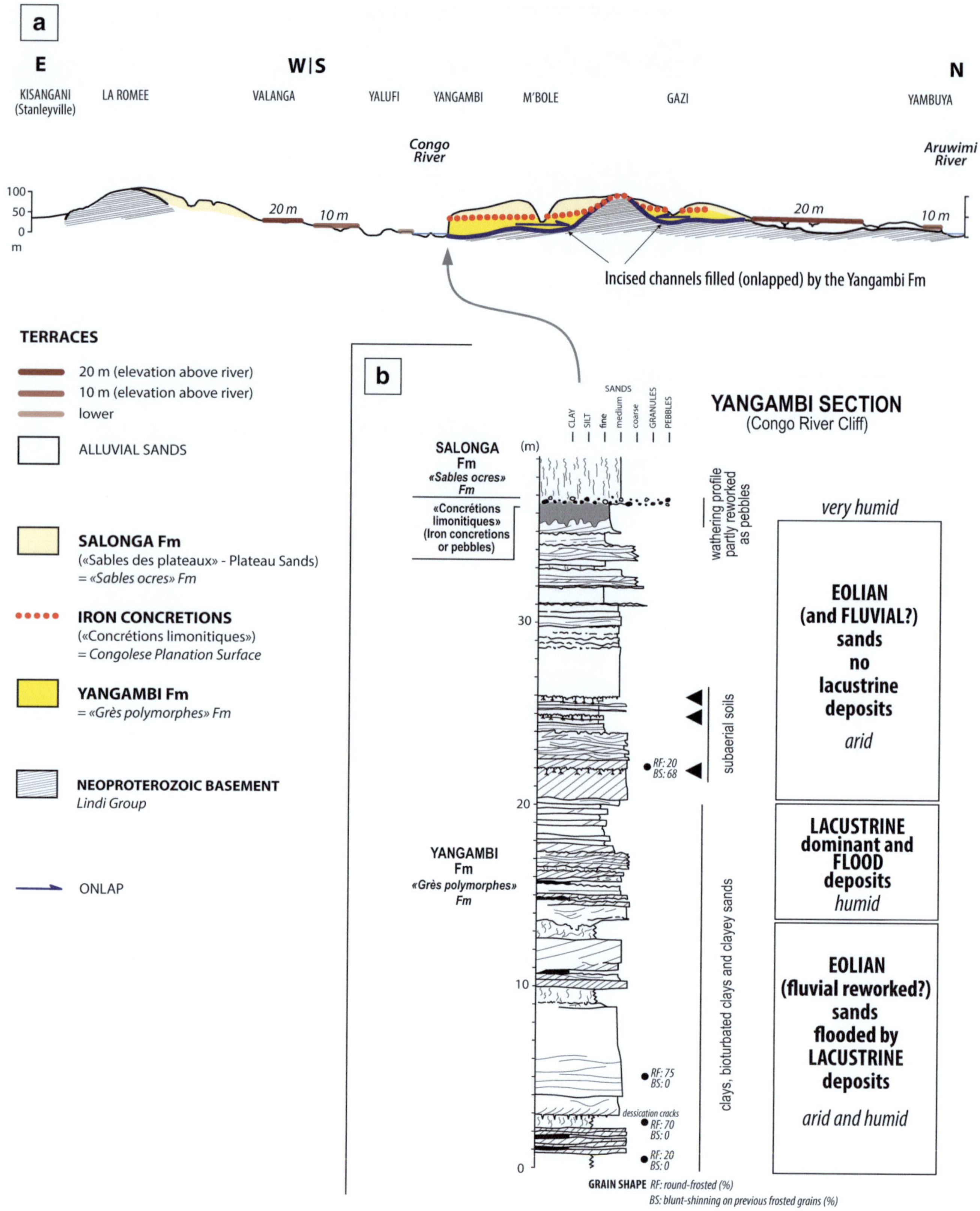

Fig. 14.6 Yangambi and Salonga Fm: Cenozoic sediments of the Yangambi (Kisangani) area cropping out along the Congo River (see Plate2 and Fig. 14.7 for location)—data from de Heinzelin (1952). (**a**) Synthetic 2D section of the Cenozoic sediments across the Congo River at the level of Yangambi—characterization of incised channels at the base of the Yangambi Fm (= "Grès polymorphes" Fm). b: Yangambi 1D section along the southern cliffs of the Congo River

Fig. 14.3) except near the contact with the overlying unit, where kaolinite occurs along horizontal fractures (level 2/29, Cahen et al. 1959). These features would be typical of the base of a laterite profile (fissured level). That suggests a truncation of most of the weathering profile before the deposition of the above coarse-grained unit.

- A *coarse-grained sandstones unit* (85 m), composed of subangular grains (Fig. 14.5.3 to 5.5) with local evidences of dissolution (Fig. 14.5.3). These grains are very irregular and only the crests and angles are blunted, indicating (limited?) fluvial transport. Two thin conglomeratic layers (0.5–1 cm angular to subangular pebbles of quartz, cherts) mark the top of this unit.
- A *sandy claystones unit* (105 m) comprising silty claystones with numerous coarse-grained sands similar to those of the underlying coarse-grained unit but with more evidences of dissolution at the surface of the grain (Fig. 14.5.2). The claystones are mainly made up of smectites and illites, kaolinite is only dominant toward the top (Fig. 14.3). These silty clays (by comparison with Mbandaka 1) could be of lacustrine origin (see discussion below), and were later weathered as suggested by the kaolinite increase toward the top.

  No fossils were found in any of the cuttings.

The *Dekese* borehole (35 m of Cenozoic deposits, Fig. 14.3) consists of three units:

- A *basal truncated weathering profile*, comprising faded claystones along cracks (level B/3, Cahen et al. 1960). This is interpreted as a truncation of the profile before the deposition of the above fine-grained unit.
- A *fine-grained sandstones unit* (≥10 m thick), composed of fine to very fine-grained sands with a clayey level. The occurrence of medium grained sandstones with very well rounded grains at the base suggests aeolian influences.
- The *coarse-grained unit* (≤25 m thick), bounded at its base by a conglomeratic layer (pebbles less than 10 cm in size) made up of quartzites, "Grès polymorphes" and cherts. This unit is composed of claystones, coarse-grained sandstones (subangular quartz grains) and fine-grained conglomerate (granules and pebbles less than 3 cm, quartz, chert). This suggests fluvial transport. The amount of smectite and illite is around 40 % (Fig. 14.3). No fossils were found in any of the cuttings.

14.4.2.2 Gilson 1 and Mbandaka 1 Petroleum Wells (See Plate 2 for Location)

The *Mbandaka 1* well (160 m, Fig. 14.3) is mainly made up of red-brown to grey silty claystones with one bed of medium to coarse grained-sandstones at the base (rounded quartz grains). Based on the occurrence of numerous ostracodes and characeans, these claystones are interpreted as lake deposits.

The *Gilson 1* well (225 m, Fig. 14.3) is made up of clayey silstones with very fine to fine-grained sandstones (10 m) overlaid by silty claystones similar to those of Mbandaka 1 (60 m) and then 10 to 30 m thick alternations of silty claystones and sandstones (coarsening upward). The sandstones are very fine to coarse-grained. Near the top, they are moderate- to well-sorted, with rounded-frosted quartz grains. This suggests an overall change from lacustrine at the base to aeolian environments at the top.

For both Mbandaka 1 and Gilson 1 (fossils lists in Fig. 14.3), the ostracodes and particularly the characeans (*Grambastichara* sp.) of the clayey lower part indicate (1) a similar age than the "Grès polymorphes" Fm and (2) a Paleogene (and probably Eocene) age (Colin and Jan du Chêne 1981).

14.4.3 Eastern Outcrops: Yangambi and Salonga Formations

In the eastern part of the Cuvette (Kisangani area—location on Plate 2), the Cenozoic sediments are subdivided into two units (Fig. 14.6): the "Yangambi" Fm overlaid by the "Salonga" Fm (de Heinzelin 1952). These unfossiliferous sediments were first considered as Plio-Pleistocene (de Heinzelin 1952) and later time equivalent of the "Grès polymorphes" and "Sables ocres" Fms (de Heinzelin 1962).

- These formations fill up a network of incised channels (Fig. 14.6a from de Heinzelin 1952) with around 80 m of incision.
- The *Yangambi Fm* (Fig. 14.6b) is a medium to coarse-grained poorly sorted sandstone with some granules layers and few claystones beds. The sandstone shows clear evidences of current megaripple cross-beddings. The study of the quartz grain shape and texture (Fig. 14.6b) by de Heinzelin (1952) shown the dominance of aeolian grains (round and frosted) on fluvial ones (dull and shinning). The description of the facies is precise enough to identify (i) bioturbated clayey sands that are characteristic, in continental environments, of lacustrine deposits and (ii) mud cracks and soils, records of emersion. In agreement with de Heinzelin (1952), we interpret this as an alternation between aeolian sediments reworked by rivers (floods deposits) and lake deposits (Fig. 14.6b). The Yangambi Fm evolves from alternating aeolian—lacustrine conditions to more lacustrine and then more aeolian.
- *Iron duricrust and weathering* at the top of the Yangambi Fm has preserved truncated laterite profile. The iron duricrust of the top of the laterite is preserved in only

restricted areas. Mostly the duricrust is reworked, leaving pebbles ("concretions") at the base of the Salonga Fm.

- The *Salonga Fm* (or "Sables des Plateaux"—Plateaus Sands—Fig. 14.6b), made up of medium to coarse-grained sands bearing aeolian grains (dominant—round-frosted: 10–55 %), mixed ones (blunt shinning on previous frosted grains: 4–48 %) and fluvial ones (blunt shinning: 4–40 %), shows a similar facies and then a similar depositional setting than the "Sables ocres" Fm.

These formations are eroded by at least three generations of alluvial terraces all with abundant ventifacts (de Heinzelin 1952).

14.4.4 Regional Correlations and CB Evolution Constructed from the Sedimentary Records

Regional correlations (Fig. 14.3) are based on the biostratigraphic ages as well as regional major lithology and facies changes. Two main groups of facies associations were characterized: (1) from the west (Batéké Plateau) to the southeast (Katanga), the outcropping "Grès polymorphes" Fm, sandstones dominated by aeolian facies, and (2) in the central part of the CB, subsurface deposits and outcrop of the Yangambi Fm, dominated by lacustrine facies which correspond to the Mio-Pliocene on the geological map of Zaire (Lepersonne 1974).

Our main new constraint is based on the biostratigraphy of ostracodes and characeans, which are of the same age (Paleogene and probably Eocene) as well in the silicified lacustrine deposits of the "Grès polymorphes" Fm as in the lacustrine silty claystones of the Gilson 1 and Mbandaka 1 wells. Although no fossils were found in the Yangambi Fm, a similar age than that of the subsurface Cenozoic sediments from the wells (Paleogene, Eocene?) is realistic because they have similar facies, lacustrine-dominated with aeolian facies and are both bounded by laterites.

To summarize, the sedimentary evolution of the CB can be consolidated into four steps:

1. An uppermost Cretaceous to base Paleogene base level fall, followed by lateritization and local river incision (Kisangani area). This first laterite profile (informally called Top Cretaceous sediment laterite) chemically eroded the underlying Late Jurassic to Cretaceous sediments before being partially removed ahead of the deposition of the "Grès polymorphes" Fm and its subsurface equivalent (Dekese, Gilson 1, Mbandaka 1, Samba).
2. Paleogene (Eocene?) lake to desert deposition comprising the "Grès polymorphes"—Yangambi Fms time interval which represents a retrogradational ("transgressive")—progradational ("regressive") stratigraphic cycle that records a lake base level rise and fall. A maximum lacustrine flooding surface can be defined and traced along the six studied sections (Fig. 14.3). The retrogradational trend preserves most of the aeolian facies of the "Gres polymorphes" Fm, while in the central Cuvette it corresponds to a low volume of lacustrine sediments except in the Samba borehole (alluvial facies aggradation). The progradational trend is a coarsening-upward evolution going from lacustrine conditions at the base to more fluvial and aeolian environments in both southern Dekese and Gilson 1 drillings. The occurrence of pebbles of "Grès polymorphes" Fm in the progradational trend of the Dekese borehole (Cahen et al. 1959) suggests either (1) intraformational deformation (exhumation of the Cretaceous rocks located between the Cuvette Cenozoic sediments and the Kasai Plateau) or (2) a younger age for these sediments.
3. An Eocene-Pliocene base level fall and a second episode of lateritization. The "Grès polymorphes" Fm is capped by a major weathering surface (informally called Top "Grès polymorphes" laterite, equivalent to Iron cuirasses 1 and 2 of Alexandre 2002, in Katanga).
4. Pleistocene to older: multiple episodes of deposition of "Sables ocres" facies. This faciès is only preserved in the southern outcrops, from the Batéké Plateaus (W) to Katanga (SE). The Salonga Fm, overlying the Yangambi Fm, could be a "Sables ocres" facies. The central Cuvette is free of this facies as shown by the four drillings. As mentioned before, in the absence of dating, this probably diachronic facies (Miocene to Holocene) cannot be used as a time marker of the CB evolution.

14.4.5 Eocene Paleogeography

The thickness of the "Grès polymorphes" and "Sables ocres" Fms is greatest in the western part of the CB (Batékés Plateaus, "Grès polymorphes" Fm: $\geq$ 300 m, "Sables ocres" Fm: $\leq$ 100 m, Fig. 14.3) decreasing toward the southeast (Katanga) where they pinch out (Mporokoso Plateau, Plate 1). Along the Kwango River, the thickness is less than 200 m (Cahen 1954, "Grès polymorphes" Fm: 60–80 m, "Sables ocres" Fm: maximum 120 m). In the northern Central Cuvette, the maximum thickness of the time equivalent lacustrine sediments is 230 m (Gilson 1) thinning out southward (Dekese) and eastward (Yangambi), but these thicknesses represent residuals because of an erosional surface at the top of these sediments (Congolese Surface, see below) (Plate 1).

A paleogeographic map compiled to depict the progradational trend of the "Grès polymorphes"—Yangambi Fms cycle is shown as Figure 14.7. Two distinct

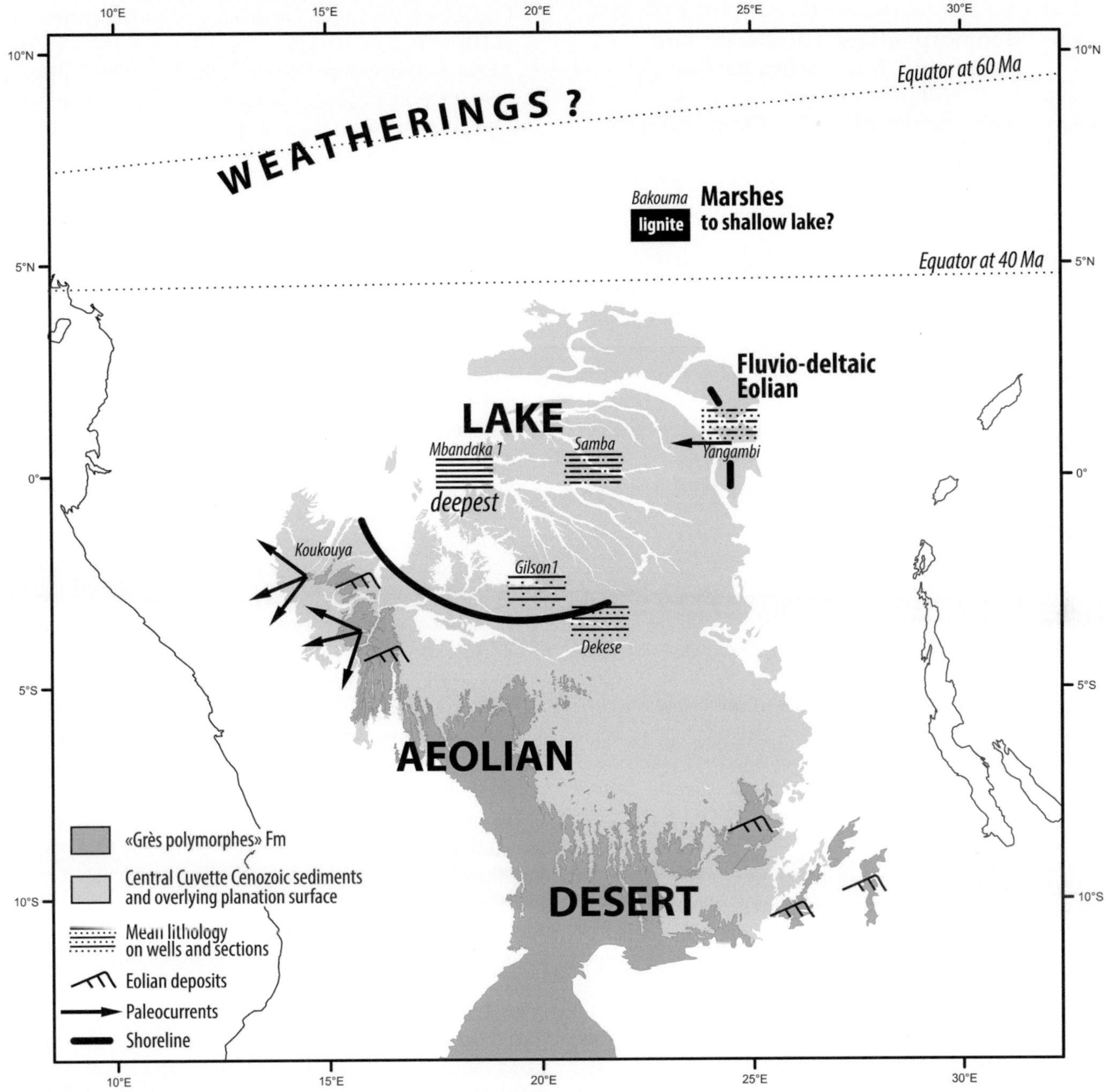

Fig. 14.7 Palaeogeographic map of the CB and surrounding highs during the progradational trend of the "Grès polymorphes"—Yangambi Fms cycle of Paleogene pre- to syn- Middle Eocene age, according to the proposed age model (5.2.7). Paleocurrents data from de Heinzelin (1952)—Yangambi area—Le Maréchal (1966)—Batéké Plateau and field studies north of Brazzaville. Centrafrican data (Bakouma) from Bigotte and Bonifas 1968 and Miauton 1980

environments are characterized: a hot arid desert with aeolian dunes and a lake with marshes. The lake paleoshoreline is traced using facies distribution. The lake is located in the northwestern part of the present-day CB. This implies that during the Paleogene (Eocene?) time interval a depression filled by a lake predefined the present-day CB. The facies distribution is consistent with the position of the equator during Paleocene-Eocene (Fig. 14.7): the humid environments (marshes, perennial lake) are closest to the equator and the hot arid desert environment occupied a transition between the lower and middle latitudes. The westward migration of aeolian dunes in Congo-Brazzaville is in good agreement with the direction of the winds operating at these latitudes, e.g. the westward trade winds.

The U-Pb single grain dating of "Grès polymorphes" zircons in Gabon (Séranne et al. 2008) suggests that these sands were reworked from older sediments across a flat topography at the scale of the Sub-saharan African Shields.

14.5 Geomorphology: Landforms and Associated Weathering Profiles

14.5.1 Main Landforms of the Congo Cuvette and Surrounding Areas

Three types of landforms are represented across the studied area. These are, from the smallest to the largest: (1) incised valleys and incised river channels; (2) small planation surfaces (10^3 to $\times 10^4$ km^2) of pediment types; (3) large to very large planation surfaces ($\times 10^4$ to 2×10^6 km^2) of pediplain and etchplain types. These three types of landforms can be associated in a single geomorphic system, hereafter called pediment system.

14.5.1.1 Incised Valleys and Channels

River incision produces two types of landforms based on a ratio of width to depth.

The *incised channels* represent wide structures (width: $\times 1$ to 25 km) with shallow incisions (depth: 3–40 m). These are predominantly located in the eastern part of the CB (Plate 1), viz. the Congo River (Lualaba) upstream of Lisala, the Lulonga (Maringa, Lupon), the Ruki (Monboyo, Busira) and the Kasai-Lukenie. These merge westward (except the Kasai-Lukenie) with the Likouala and Ubangi Swamps (Plate 1) that constitute a flat domain characterized by low sediment preservations (Desthieux 1993). These channels have associated river terraces frequently embedded with evidences of human artifacts indicating a Pleistocene age (Mortelmans 1947; Leakey 1949; Breuil and Janmart 1950; de Ploey 1967, 1969; Cahen 1978).

The *incised valleys* are 1 to 10 km wide with a depth of 40 to 200 m. The most incised valleys are located in Katanga. Along the Central African Atlantic Dome and the Cameroon Highlands similar deeply incised valleys are located along inactive basement faults.

Networks of erosional parallel rivers are characteristic of *piedmont* (e.g. Boulvert 1985, 1996) along the flanks of the CB and of the aborted rifts of Sudan and Chad. The base levels of these piedmonts are depositional flats (e.g. Likouala and Ubangi Swamps for the CB). The piedmonts are 120 and 150 km-long for a difference of elevation of 150–250 m (slopes of ca. 10^{-3} %).

14.5.1.2 Small Planation Surfaces: Pediments

Most of the landforms surrounding the CB are *pediments* as defined by Gilbert 1877 (see Tator 1952, 1953; Whitaker 1979; Dohrenwend 1994; White 2004; Dohrenwend and Parsons 2009) and characterized by:

- Gently inclined terrestrial erosional surfaces (with slopes of $\times 10^{-4}$ to $\times 1$ %) that sharply truncate underlying rocks without significant relief.
- Scarps or steepest slopes upstream.
- Lack of sediment covers or thin (ca. $\times 1$ m) discontinuous sediment layers.
- Absence of superimposed river channels or where they do exist, they are narrow ($\times 1$ m to 10 m large) and without significant incision ($\times 1$ m).
- Concave or rectilinear profiles.

In Africa, pediments have been described for a long time (e.g. Fair 1948; King 1949; Mabbutt 1955; Dresch 1950; Joly 1950). In the CB, the first pediments were described in Katanga by Mammerickx (1959, 1964). We further define two types of pediments (Fig. 14.8):

- *Piedmont pediments* range from a few tens of kilometres long for a width of a few hundreds of kilometres. They are bounded upstream by a scarp and merge downstream with a pediplain (see definition below 14.5.1.3) or depositional systems (marshes, lakes). The slope of the most recent piedmont pediments of the CB is between 2 and 8×10^{-4} %.
- *Pedi-valleys* are large erosional flat bottom valley-like systems, ranging from a few tens to one hundred kilometres in width and one hundred to few hundreds kilometres long. They are bounded laterally and upstream by steep slopes (mainly scarps). The longitudinal profile of the most recent pedi-valleys of the studied area have slopes from around 1 to 5×10^{-4} % downstream and to 4 to 8×10^{-4} % upstream. Scarps can enclose some pedi-valleys in which case their outlet is an incised valley.

Five successive generations of pediments are recognized in the CB, the older one being located above the younger one and degraded by the latter. We label successive pediments using the same rules as those of the fluvial terraces, viz. the youngest being labelled Z and the older labelled backward in the alphabetic order (e.g. Z to V). To distinguish a main geomorphic domain from an adjoining one where local base level may be different, we added indices (e.g. Z_c for pediment Z fitting to the local base level of the CB).

14.5.1.3 Large Planation Surfaces

Three large ($\times 10^4$ to $\times 10^5$ km^2) planation surfaces (upper planation surfaces 1 and 2—labelled *u1* and *u2*, lower planation surface—*l*) and one very large (1–2×10^6 km^2, Congolese Surface—*C*, see 14.5.1.4) were mapped and are of two types.

1. *Pediplain*, as defined by King (1953) in southern Africa, is a planation surface resulting from the parallel retreat (backwearing) of scarps (hillslopes) and composed of many coalescent pediments. All intermediate sizes exist between pediments and pediplains, and the upper limit is sometimes difficult to define (e.g. the Kwango large pedi-valleys or pediplains?).
2. *Etchplain* defines weathered top surfaces under hot very humid conditions that generate laterites (Millot 1980, 1981). They are also called *mantled* etchplains (Migon 2004b) and where the surface is eroded and stripped of

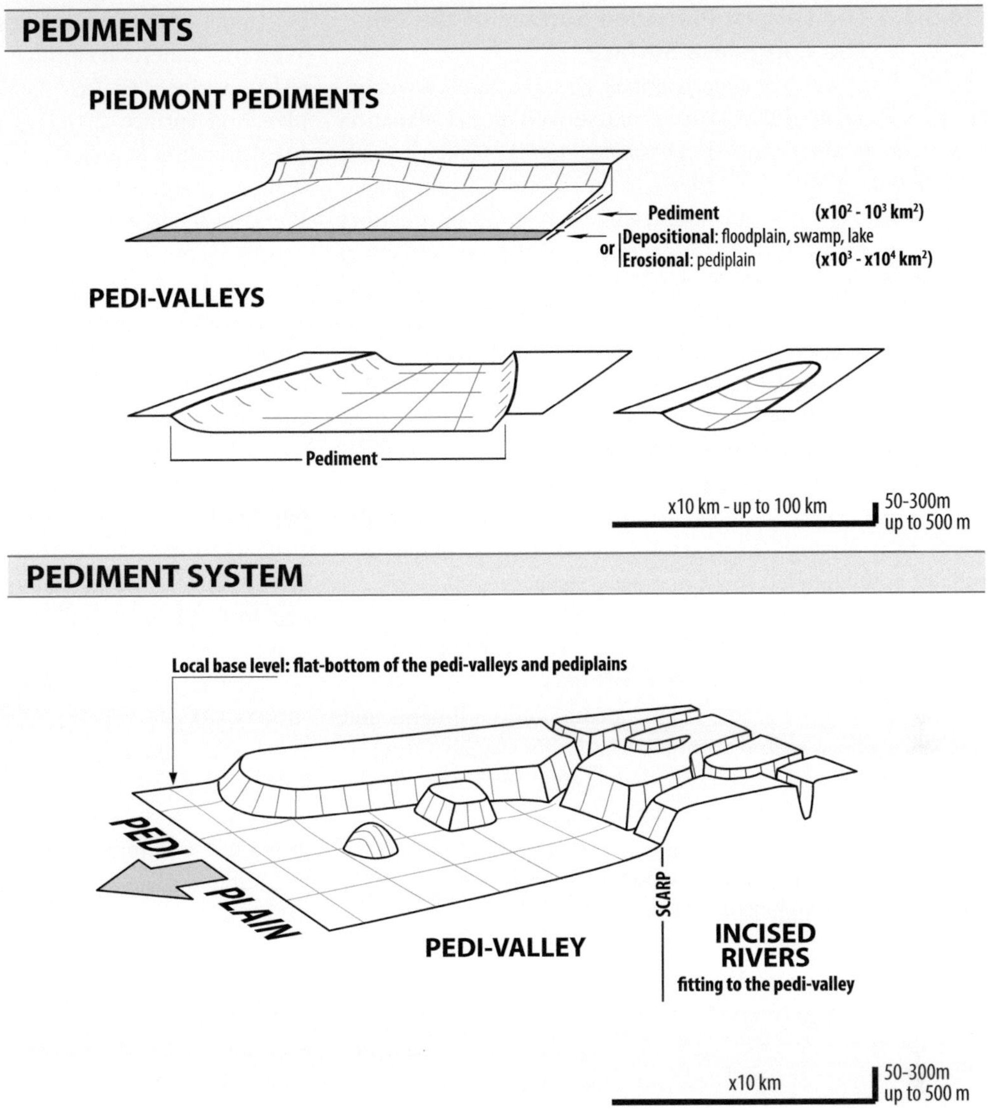

Fig. 14.8 The pediments of the CB and surrounding highs

most of the weathering products (saprolite), they are called *stripped* etchplains.

Stripped etchplains define either the base of the weathering profile (exhumed weathering front, Mabbutt 1961) or a surface with remnants of weathering profiles dissected by incised rivers or pediments (Twidale and Bourne 2013). Remnant relief called inselbergs, tors or bornhardts can have hillslopes hillslopes still weathered (Büdel 1957; Thomas 1989a,b) or are baren. Stripped etchplains cut by incised rivers and pediments may be the result (1) of simultaneous chemical weathering and physical stripping (Büdel 1957, 1982; Bremer 1993) or (2) of alternations of chemical and mechanical erosion periods due to climate changes (Thomas 1989a,b; Millot 1980; Beauvais et al. 2008; Beauvais and Chardon 2013).

All the large planation surfaces (*u1*, *u2*, *l*) and most of the pediments, are partly to fully duricrusted. Iron duricrusts form in two different settings, (1) capping a complete lateritic profile (with saprolite) as the end product of the weathering or (2) encrusting (cuirassing) thin (×1 m) conglomerates, veneering a poorly to unweathered basement.

Type (1) iron duricrusts are characteristic of mantled etchplains or stripped etchplains by pediments under relatively stable climatic conditions, while type (2) records alternative periods of mechanical erosion under more arid climatic settings (pediments incision) and chemical erosion (weathering) under very humid conditions. Upper planation surfaces (*u1* and *u2*) are etchplains. The youngest pediments (X, Y, Z) record contrasted climatic conditions, as suggested by the occurrence of iron duricrusted conglomerates (mechanical erosion) directly deposited on unweathered pediments (e.g. the Cabatuquila iron duricrust in the large pedi-valley of the Kwango in Angola; Marques 1992).

14.5.1.4 The Largest Planation Surface of the CB: The Congolese Surface

The CB and the Katanga, Kamina, Kasai, Lunde, Kwango, Batéké Plateaus (Plate 1) are characterized by flat planation surfaces (Cahen and Lepersonne 1948; Cahen 1954; Lepersonne 1949, 1956, 1978). In the Central CB, the planation surface is purely erosional, while across the surrounding plateaus, it corresponds to a remnant of the low erosional weathered surface across the top of the "Grès polymorphes" Fm. Mapping of these planation surfaces (Plate 2) indicates that they represent the same planation surface, the largest of which, here called Congolese Surface to avoid confusions with the previous surface names of King, Cahen or Lepersonne. The diachronic "Sables ocres" Fm are surficial deposits, covering this surface during Neogene times.

In the Central Cuvette, incised rivers connected to the local base level of the Likouala and Ubangi Swamps, dissected and degraded this Congolese Surface.

14.5.1.5 The Pediment System: An Assembly of the Previous Landforms

At the scale of $\times 10^3$ to 10^4 km^2, pediplains, pediments and incised valleys form a unique geomorphic system fitting to the same regional base level, here called *pediment system* (Fig. 14.8), comprising from downstream to upstream:

1. a pediplain,
2. piedmont pediments or numerous pedi-valleys, both upstream bounded by scarps and
3. a network of incised rivers connected to the flattest part of the downslope pediments that constitutes a local base level for these rivers.

This erosional river network degrades, or dissects, the oldest upstream stepped pediments, making them sometimes difficult to differentiate. Most of the incised valleys of the studied area, mainly located along the Cameroon Highlands, the Ubangian Rise and the western flank of the East African dome, are upstream parts of such pediment systems.

14.5.2 Relative Chronology—Relationships Between Landforms, Sediments and Magmatism (Plate 2)

The following analysis is based on the CB landforms map (Plate 2) and several geomorphologic cross-sections (Fig. 14.9). All the geographical names are provided on Plate 1.

14.5.2.1 Northwest: Cameroon Highlands (Cameroon Volcanic Line—E Nigeria, Cameroon, W Central African Republic)

Eight landforms are defined (Plate 2, Fig. 14.9 CH, see description on Table 14.2), from the oldest to the youngest, (1) Upper planation surfaces 1 (*u1u*), subdivided along the Western Cameroon Highlands into three surfaces, surface 1a (*u1a*), surface 1b (*u1b*) and surface 1p (*u1p*), (2) Upper planation surface 2 (*u2*), (3) Lower planation surface (*l*), (4) Pediments V and W (V_a, V_c, W_a, W_c) (5) Pediments X, Y (X_a, Y_a) and (6) Pediments Z (Z_a). Along the Ngaoundéré Plateau, planation surfaces *u1a*, *u1b* and *u1p* merge into a complex composite relief with numerous lava flows, filling old landforms (Humbel 1966). This composite irregular surface was mapped as undifferentiated Upper planation surface 1 (*u1u*). Pediments were labelled according to their base level, e.g. V_a for the Atlantic Ocean one and V_c for the CB local base level.

The chronological relationships between the magmatic rocks of the Cameroon Volcanic Line and the different landforms are summarized in Fig. 14.10. The Cameroon Volcanic Line (Déruelle et al. 1991, 2007; Njonfang et al. 2011) is composed of plutonic ("younger" granites, syenites, gabbros) and volcanic rocks (basalts, trachytes and rhyolites). Some magmatic rocks are highly weathered (first "old" basalts) or duricrusted by bauxites (trachytes).

The magmatic rocks provide a maximum age for the landforms that erode them. In contrast, basalt flows or aerial trachytic and rhyolithic domes covering planation surfaces provide a minimum age for these landforms. Only Upper planation surfaces 1 and Pediments V_a, X_a and Z-Y can be dated in this manner using the age of magmatic rocks (see age arguments on Table 14.2). Upper planation surfaces 1 (*u1*) were shaped between 67 ± 7 Ma and 45.5 ± 1.1 Ma and probably between 67 ± 7 Ma and 51,8 ± 1.2 Ma, according to the confidence granted to the datings of Moundi et al. (2007). Pediment V_a is younger than 46.7 ± 1.1 Ma. Pediment Y_a is younger than 9.9 ± 0.4 Ma and possibly as young as 5.6 ± 0.2 Ma (whole rock ^{39}K–^{40}Ar dates). Pediment Z–Y is younger than 9.3 Ma.

- The first volcanic rocks of the Cameroon Volcanic Lines—the widespread, highly weathered "old" basalts, extending from the southern Bamiléké Plateau (Mount Bangou, Fosso et al. 2005) to the Adamawa Plateau (Lasserre 1961)—sealed pre-existing landforms. On the Adamawa Plateau, west of Ngaoudéré (Humbel 1966), basalt flows filled flat bottom large valleys (5–10 km in width, 100 to 400 m deep—possible pediments).
- The South Cameroon Plateau has acted as a major water divide since the onset of incision of pediments V. Unlike pediments V and W, pediments X and Y are found only along the Atlantic side of the South Cameroon Plateau and north of the Cameroon Highlands, as a relatively "narrow" band (100–200 km in width).
- One of the famous occurence of bauxites (Fongo-Tongo—Bamiléké Plateau) can be dated to about Middle Miocene, because the bauxites rest on trachytes (flowing

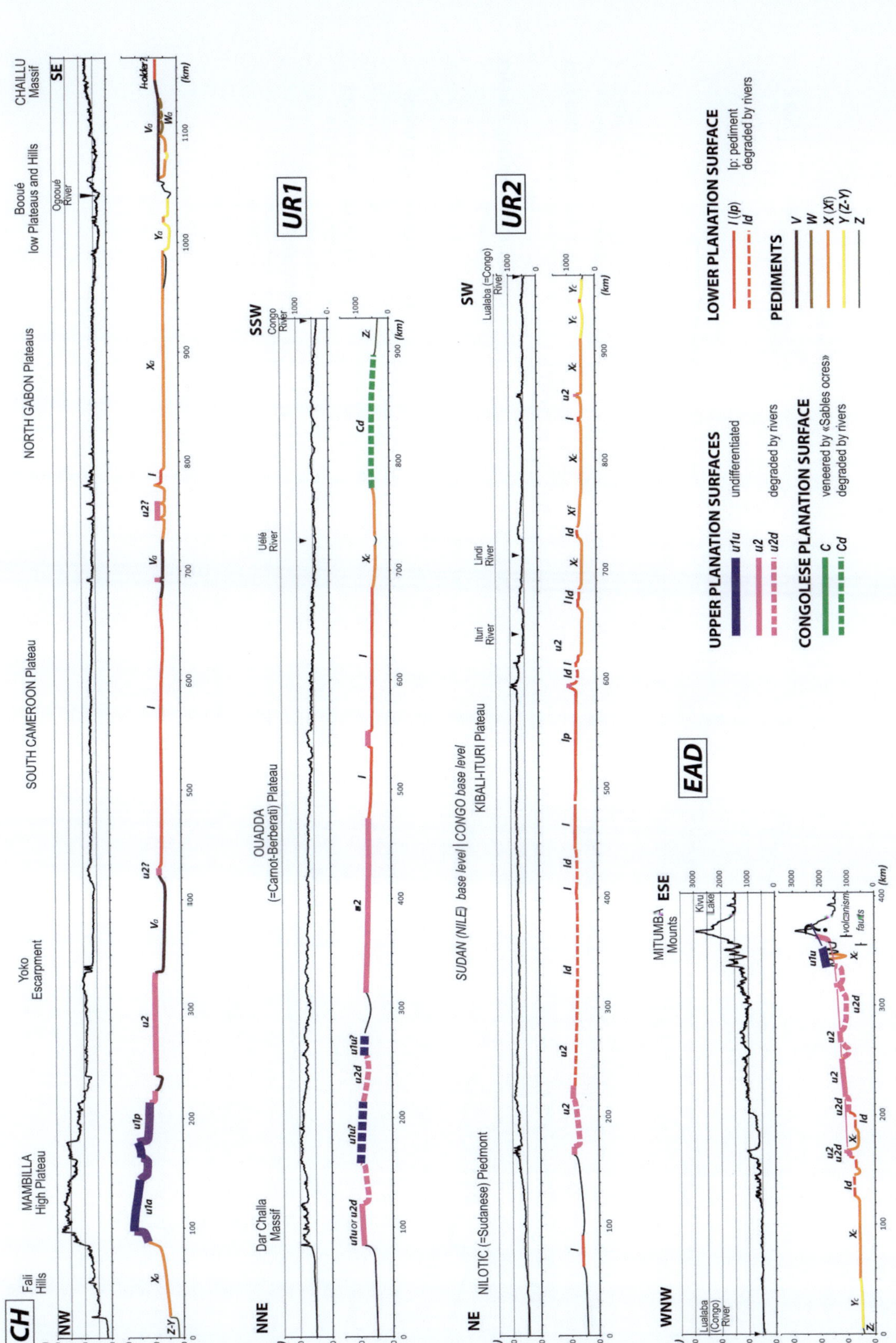

Fig. 14.9 (continued)

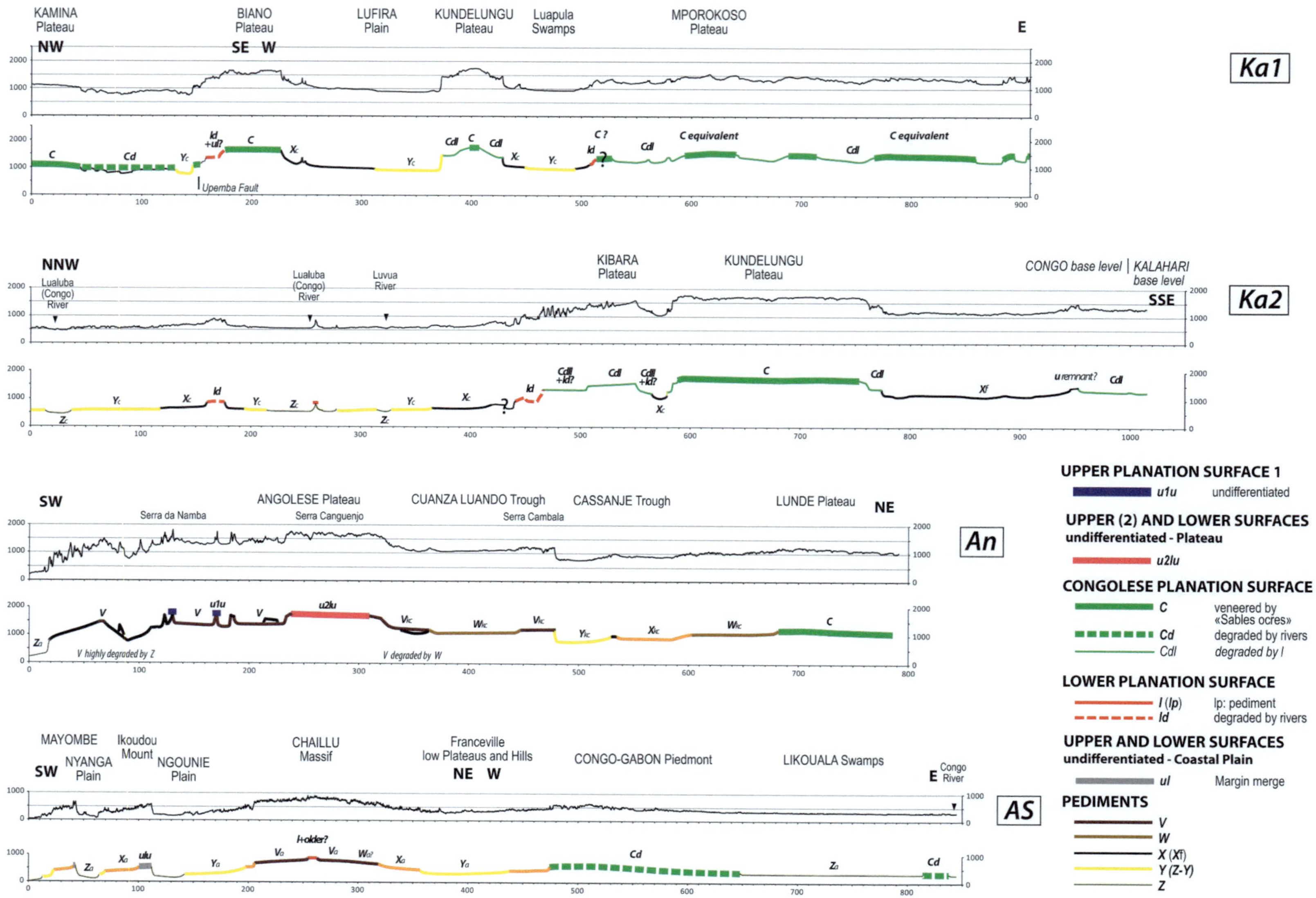

Fig. 14.9 Key topographic cross-sections with landforms of the CB and surrounding areas—cross-sections location on Plate 2

Table 14.2 Relative chronology, characteristics, synonymy and age of the landforms of the Cameroon Highlands

	ELEVATION	LANDFORM-WEATHERING (W.)	RELATIONSHIPS WITH THE OLDEST FORM	SYNONYMY	AGE
Upper Planation Surface 1a (*u1a*)	1,500–1,800 m	Complex geometry—several planation surfaces along the NE-SW basement faults: pediments or faults scarps? *Highest relief: Mambilla High Plateau, part of the Bamiléké Plateau* **w.**: bauxitic laterites on Proterozoic basement, «old» basalts or «recent» trachytes (Mambilla Pl., Schwarz 1997—Bamiléké Pl., Fongo-Tongo, Hiéronymus 1973, 1977, 1983)		Cameroon Minim—Martap—Bamiléké Surface (Ségalen 1967)	Mont Bangou: the top weathered surface of the « old » basalts (**u1a**) is overlain by basalts dated at **44.7 ± 1 Ma** (Fosso et al. 2005—whole rock $^{39}K–^{40}Ar$)
Upper P. Surface 1b (*u1b*)	1,200–1,300 m	Duricrusted planation surface - mantled etchplain? *Minim-Martap Plateau only* **w.** : bauxitic laterite on «old» basalts (Belinga 1966, 1972, 1973a,b)		Central African Republic Lim - Bocaranga Surface (Boulvert 1996)	Bamoun Plateau: pediment **u1p** cut through Mount Nkogam «younger» granites dated at **67 ± 7 Ma** (Caen-Vachette et al. 1987 - whole rock Rb/Sr age) and is overlain by basalts at **51.8 ± 2 Ma** (Moundi et al. 2007—whole rocks $^{39}K–^{40}Ar$—no data provided) and the Mbepit ryolithic dome at **45.5 ± 1.1 Ma** (Wandji et al. 2008—whole rocks $^{39}K–^{40}Ar$)
Upper P. Surface 1p (*u1p*)	1,100–1,500 m	pediments downslope *u1a* **w.**: no published descriptions	Scarps height: 100 to 200–300 m		
Upper Planatio Surface 2 (*u2*)	900–1,200 m	Pediplain ± degraded *southern part of the Adamawa Plateau* **w.**: no published descriptions	Transition to u1: – Gradational (Tchabbal Ngou, W Adamawa Plateau) – Scarps (100 m—Minim-Martap Plateau) (≤400 m—Bamiléké Plateau)	Meiganga—Mandara—Bamoun Surface (Ségalen 1967) Bouar—Baboua Surface (Boulvert 1996)	
Lower Planation Surface (*l*)	600–800 m	Nearly flat, weathered duricrusted pediplain *most of the South Cameroon Plateau* **w.**: laterite with iron cuirasses (Boyé and Fritsch 1973 ; Yongue-Fouateu et al. 2006)	Transition to u2: – Gradational (south Pangar Plat., east Tibati Plain) – Scarps (200 m—east Yoko Escarpment)	Interior Surface («Surface intérieure») (Ségalen 1967) Central-African Surface (Boulvert 1996)	
Pediments V–W (V_a, V_c, W_a, W_c)		Two generations of large pedi-valleys/ pediplains (V ,W) two directions of drainage, towards (1) the Atlantic Ocean (*Va, Wa*) and (2) the Congo Cuvette (*Vc*′, *Wc*) 50 to 200 km large, ≤400 km long	Scarps ≤ ×10 m (low pediment incision: 50–200 m) *detailed study in Yaoundé area* (Embrechts and de Dapper 1990)		Bamoun Plateau: Pediment $\mathbf{V_a}$ cut through basalts with a youngest age of **46.7 ± 1.1 Ma** (Wandji et al. 2008—whole rocks $^{39}K–^{40}Ar$)
Pediments X – Y (X_a, Y_a)		Two generations of pedi-valleys (V, W) one direction of drainage, towards the Atlantic Ocean (X_a and Y_a) × 10 km large, ≤100–150 km long	Scarps ≤500 m (Cameroon Highlands—Benue side) (more deeply incised than V–W)		Rumpi Hills: pediment **Ya** levels gabbros dated (Walgenwitz *in* Nkoumbou 1990) between 10.8 and 5.6 Ma (whole rocks $^{39}K-^{40}Ar$) and **10.6 ± 0.5 and 9.9 ± 0.4 Ma** (micas $^{39}K–^{40}Ar$)
Pediment Z (Z_a)		Large pedi-valleys to pediplains *Cameroon coastal plain—reentrant of Benue Trough* merge with pediment *Ya* (south Benue Trough)	Scarps ≤ 200–300 m		Adamawa Plateau: **Z-Y** erodes trachytes (Tchabal Nganha, Nono et al. 1994) with a youngest age of **9.28 Ma** (Marzoli et al. 1999—feldspars $^{39}Ar–^{40}Ar$)

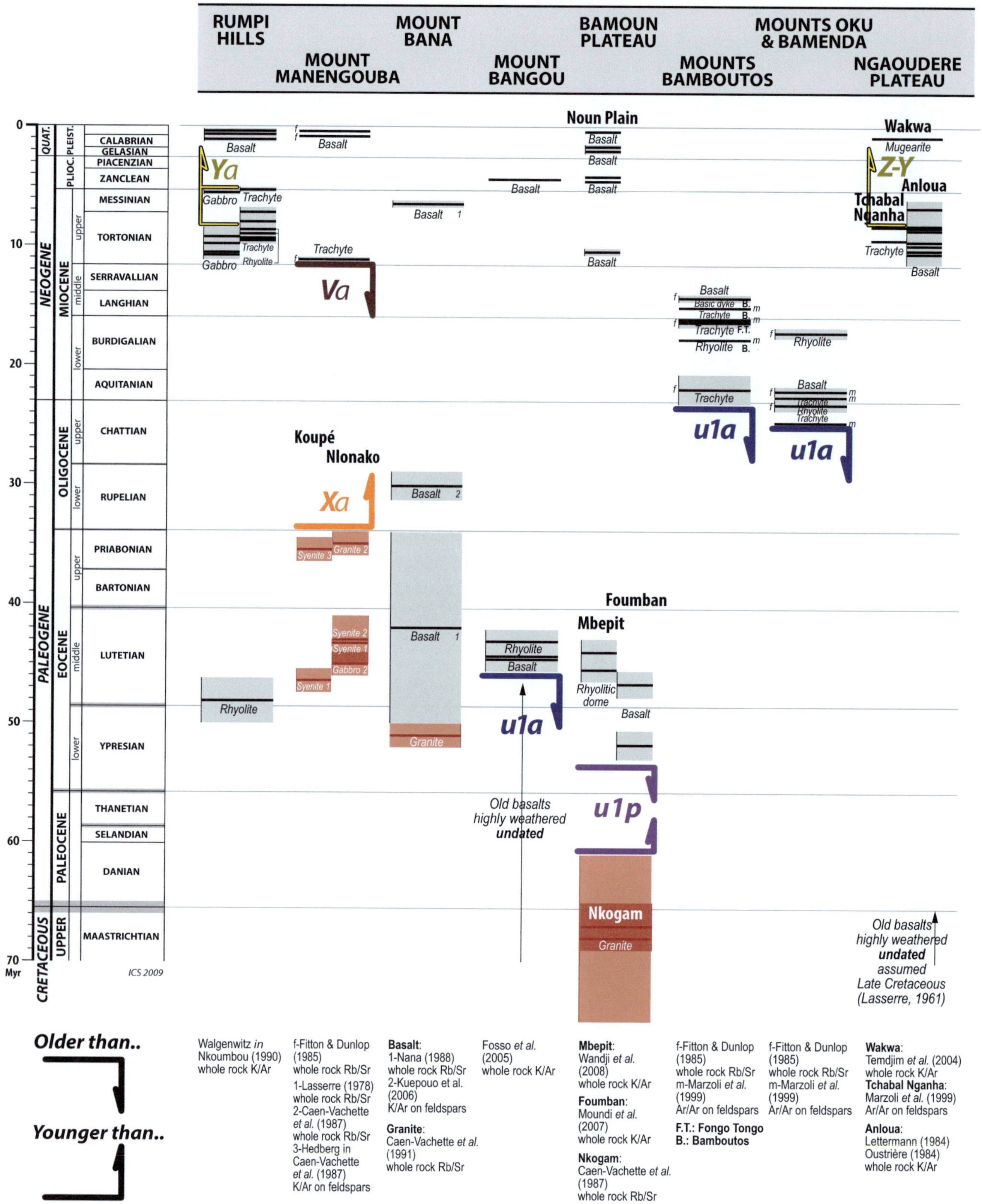

Fig. 14.10 Chronologic relationships between magmatism and planation surfaces for the Cameroon Highlands

Table 14.3 Relative chronology, characteristics and synonymy of the landforms of the Ubangian Rise

	ELEVATION	LANDFORM—WEATHERING (W.)	SYNONYMY
Upper Planation Surface 1 (*u1u*)	900–1,200 m	Planation surface ± degraded *Highest reliefs of the Bongo Mounts and Dar Challa Massif* **w.**: occurence of bauxites in the Bangali Massif (N8°39′–E20°07′, Boulvert 1996) *Tentatively correlated with u1 of the Cameroon Highlands on the basis of:* *– A difference of elevation between this surface and the underlying one (u2) and* *– The occurrence of bauxites*	Lim-Bocaranga Surface (Boulvert 1996)
Upper Planation Surface 2 (*u2, u2d*)	700–1,000 m	Pediplain ± degraded by incised rivers, then called **degraded Upper Surface 2** (*u2d*) *Ouadda(=Mouka) and Gadzi (=Carnot-Berberati) Plateau both made up of Lower Cretaceous sandstones southward tilted* **w.**: thin weathering mantle—some iron duricrusts (Mainguet 1975; Boulvert 1996)	Bouar-Baboua Surface (Boulvert 1996)
Lower Planation Surface (*l, lp, ld*)	500–800 m	3 types: – Nearly flat, weathered duricrusted pediplain (*l*) *most of the Ubangian Rise* – Pediplain with few duricrusts (*lp*) *south Uélé Plateau - NE Congo* – Degraded pediplain (*ld*) dissected by incised rivers and younger small pediments *between the gadzi and Ouadda Plateaus* **w.**: highly weathered and duricrusted by laterites (Boulvert 1996; Beauvais and Mazaltarim 1988; Beauvais and Roquin 1996; Beauvais 1999)—the bauxites described by Sluys (1946) along the Ituri River (Congo) are aluminium-rich kaolinites of « classical» laterite profile (Vanderstappen and Cornil 1955)	Central-African Surface (Boulvert 1996)
Pediments V - W (*V W_c*)		Large pedi-valleys/pediplains similar to the Cameroon Highland ones *eastern side of the Gadzi Plateau* 50 to 200 km large, ≤400 km long	
Pediments X (*X_c*)		Pedi-valleys—Congo Cuvette local base level 2 directions: WNW–ESE (dominant) and NNE–SSW (between the Bianga and Kembe Plateau) × 10 km large (20–80 km) controlled by the lithology: cutting through Neoproterozoic Lindian sediments and bounded upslope by Archean and Paleoproteroic crystalline rocks	
Pediment Z (*Z_c*)		Large pedi-valleys pediment Y - uncharacterized—merges with Z Z_c is the local basel level of the Central Congo Cuvette	

across surface *u1a*) dated at 16.3 ± 0.34 Ma (Marzoli et al. 1999—^{39}Ar–^{40}Ar on feldspars).

14.5.2.2 North: Ubangian Rise (Central African Republic and NW DRC)

This relatively low elevated rise (500–1,000 m) extends from the Cameroon Highlands to the Blue Mountains (NE Congo—northwestern limit of the East African Dome, Plate 1). To the west, this rise is the divide between the Congo (Ubangi) and the Logoné—Chari Rivers feeding the endoreic Chad Lake drainage, and to the east, the divide between the Congo and Nile Rivers. The highest relief of the Ubangi Rise (Plate 1), the Ouadda Plateau and the Bongo Mounts (1,300 m), are located at the intersection of these two divides. On both the Chadian and Sudanese areas, the northern side of the Ubangian Rise corresponds to piedmonts, the Ouham-Bamingi Piedmont on the Chadian side and the Nilotic Piedmont on the Sudanese side, with the Ouangjia-Vakaga Piedmont in between. The area was extensively studied by Boulvert (1985, 1996 and numerous reports available on the IRD website (http://horizon.documentation.ird.fr).

Six landforms are defined (Plate 2, Fig. 14.9 UR, see description on Table 14.3), from the oldest to the youngest, (1) Upper planation surface 1 (*u1u*), (2) Upper planation surface 2 (*u2, u2d*), (3) Lower planation surface (*l, lp, ld*), (4) Pediments V and W (*VW_c*) (5) Pediments X (X_c) and (6) Pediments Z (Z_c). The stepping between those different landforms is quite low. Planation surfaces can (1) merge upslope or (2) be bounded by steepest slopes rather than by scarps.

Along the central CB, the Upper planation surface 2 (*u2*) merges with the erosional Congolese Surface (*C*), whereas the Lower planation surfaces (*l, lp, ld*) are located below, with a marked scarp between these two planation surfaces. No age constraints are available.

14.5.2.3 East: Western Limb of the East African Dome (E DRC, Uganda, Rwanda, Burundi)

This domain (Plate 1) corresponds to the western flank of the East African Dome, and mainly the western border of the western branch of the East African Rift. This domain extends from the border between the DRC, Uganda and South Sudan (Blue Mountains) to northeastern Zambia (Mporokoso Plateau), crossing through Kivu and northeastern Katanga (former Shaba).

Six landforms are defined (Plate 2, Fig. 14.9 EAD, see description on Table 14.4), from the oldest to the youngest, (1) Upper planation surface 1 (*u1u*), (2) Upper planation surface 2 (*u2, u2d*), (3) Lower planation surface (*l, lp, ld*), (4) Pediments X (X_c), (5) Pediments Y (Y_c) and (6) incised channels.

Table 14.4 Relative chronology, characteristics and age of the landforms of the western flank of the East African Dome

	ELEVATION	LANDFORM–WEATHERING (W.)	RELATIONSHIPS WITH THE OLDEST FORM	AGE
Upper Planation Surface 1 (*u1u*)	1,400–2,700 m	Complex geometry - several planation surfaces along the rift: pediments or faults scarps? not studied in details *Highest relief: Blue Mountains, Mitumba Mounts, Itombwe Plateau - with doubt: Masseba and Marungu Plateaus* at regional-scale draw a half dome: maximum elevation in the Mitumba Mounts, minimum in the Blue Moutains (north) and the Masseba Plateau (south) **w.**: no published descriptions		
Upper Planation Surface 2 (*u2, u2d*)	900–1,800 m	Pediplain (*u2*) ± degraded (u2d) by incised rivers connected to younger pediments *Blue Mountains (Kibali-Ituri Plateau)—900 m—to Itombwe Plateau—1,800 m* merge to the west (Central Cuvette) with the Congolese Surface (*Cd*) **w.**: only studied in the Blue Moutains (Mbuluyo 1991) – No preserved weathering on *u2* (stripped etchplain?) – Iron-duricrusts on pediments degrading *u2*	scarps: ≥500 m—Itombwe Plateau 0–50 m—Kibali-Ituri Plateau	*Named by analogy with Cameroon Highlands (same stepping in reference to the Lower Planation Surface)*
Lower Planation Surface (*l, lp, ld*)	600–800 m	Lateral extension of the duricrusted pediplain of the Ubangian Rise—three types: – Duricrusted pediplain (*l*) – Upstream pediments of the pediplain with few duricrusts (*lp*) – Degraded pediments (*ld*) dissected by valleys connected to pediment *X* difficult to map south of the Bukavu Volcanic Province: (1) highly degraded and (2) several stepped pediments (equivalent Pediments V and W ?) **w.**: no published descriptions	Scarps 100–300 m (maximum height in the Mitumba Mounts)	
Pediment X (*Xc , Xf*)		N–S trending EW asymmetrical pediment systems (connected incised rivers: X_f) – West: pediplain (200–500 m below the Congolese surface (*Cd*) – East: deeply incised pedi-valleys (E–W to NW-SE trending) *Transition between the Central Cuvette and the East African Dome from Upemba Lake (south) to Itimbiri Pedi-valley (north)* Present-day Congo River outlet toward the local base level of the Likuala Swamps **w.**: no published descriptions	Scarps (pedi-valleys) 200–500 m	<u>**Lugulu basaltic flow**</u> **(W Kahuzi-Biega)**: - incised by the river network of the Lubimbi Rivers that fit with the local base level of pediment $\mathbf{X_c}$ – Lugulu basalts (undated) belongs to the initial episode of the Bukavu volcanic province – *oldest age*: debated, between 17.6 ± 0.5 Ma (Baubron, *in* Kampunzu et al. 1998) et 10.0 ± 2.2 Ma (Pasteels et al. 1989) – *Youngest age*: **5.9 ± 0.3 Ma** (whole rock ^{39}K–^{40}Ar Kampunzu et al. 1998—Kahuzi-Biega)
Pediment Y (*Yc*)		Pediplain *Southern Lualuba (=Congo) at the confluence with the Lukuga* Merge northward with pediplain X_c	Elevation between X_c and Y_c <100 m	
Incised channels		Cut across pediplains *Xc* and Y_c *Lualuba (=Congo) River and tributaries (Luvua, Lukuga and Luama Rivers)*	Greater incision 100-200 m, in the south (Lukuga and Luama Rivers)	

Table 14.5 Relative chronology, characteristics, synonymy and age of the landforms of the Katanga Plateaus and Plains

	ELEVATION	RELIEF FORM-WEATHERING (W.)	RELATIONSHIPS WITH THE OLDEST FORM	SYNONYMY	AGE
Congolese Surface (*C, Cd*)	1,700–1,400 m (Katanga) 1,200–1,000 m (Kasai-Lunde-Kwango)	Top « Grès polymorphes» Fm planation surface, covered (veneered) by the «Sables ocres» Fm *Katanga (Kundelungu, Kibara, Biano Plateaus)—Kamina, Kasai (=Cassai in Angola), Lunde and Kwango Plateaus* slightly domed—present-day Congo-Zambezi divide, from the Muchinga Mountains (Zambia) to the Angolese Plateau Detailed evolution in Katanga (Alexandre-Pyre 1971; Alexandre and Alexandre-Peyre 1987; Alexandre 2002—see Sect. 14.4.1.2) 1. Low erosion (×10 m?) of the "Grès polymorphes" Fm and weathering by iron duricrust 1 (pisolitic)—degraded Kisiba Salagwe Surf 2. Low erosion, removal of most of duricrust 1 and deposition of a fluvial gravels bed ("cailloutis des plateaux") 3. Weathering by iron duricrust 2 (red shinning coating), both on the gravels bed and surrounding basement rocks Mitwaba Surf. 4. Draping by the "Sables ocres" Fm			<u>**Katanga to Kwango Plateaus**</u>: The Congolese Surface, slightly erosional on the top of the «Gès polymorphes» Fm is younger than those aeolian deposits dated Paleogene and probably Eocene (see text)
Congolese Surface degraded by the Lower P. Surface (*Cdl*)	1,500–1,200 m	Same polygenic planation surface than *C* (based on maps of Alexandre and Alexandre-Peyre 1987; Alexandre 2002) – Not veneered by «Sables ocres» Fm – Degraded by pedi-valleys and rivers of the Lower planation surface (*l*) and Pediment Xc *Biano, Kibara, Kundelungu, Mporokoso Plateaus - north of the Congo-Zambezi divide* **w.**: relicts of Iron duricrusts 1 and 2			
Lower degraded Planation Surface (*ld*)		Highly degraded pediments by incised rivers of X_c Possible equivalent of *ld* (difficult mapping from the NE Mitumba Mounts)—Pediments V and W not impossible **w.**: no published descriptions	Progressive or sharp—scarps 200–300 m		
Pediments X (X_c,X_f)		Pediment systems: three large pedi-valleys (width: 50–200 km) passing upslope to smaller pedi-valleys (w.: 10–20 km) and then to connected incised valleys (X_f) degrading the Lower planation surface (*l*) *Kamalondo (Upemba) Depression, Lu_ra Plain, Mweru-Luapula Plain* **w.**: iron duricrust 3 (Alexandre 2002), dark brown shinning coating	Deply incised scarps ≥ 500 m X_f network extends 100 to 200 km far upstream of the pedi-valleys	Lubumbashi Surface (Alexandre 2002)	<u>**Kolwezi - Lubumbashi area**</u> (Kawama, Kasompi, Luishiwi): On the Congolese Surface (*Cdl*) degraded by *l* and X_f, oxidized manganese ores ("cobalt cap"), dated by ^{39}K-^{40}Ar (Decrée et al. 2010), range from 10.3 ± 0.1 Ma to 2.3 ± 0.17 Ma (five ages: 10.3, 5.3, 3.66, 3.54, 2.30) confirmed by an U-Pb age on supergene heterogenites at 3.1 ± 4 Ma. That implies: – A major uplift, before 10.3 Ma – Minor uplifts, between 10.3 and 3.66 Ma or between 10.3 and 5.3 Ma and 5.3 and 3.66 Ma **Our interpretation**: Major uplift = deep incision of pediment X_c Minor uplifts = incision of Y_c (and may be Z_c)
Pediments Y (Y_c)		Pedi-valleys superimposed on previous Pediments X **w.**: iron duricrust 4 (Alexandre 2002), ochre dull coating	Low amplitude incision – Highest (SE Kamalondo): 50–100 m – Lowest (Lu_ra and Mweru Plains)	Kyobo Surface (Alexandre 2002)	
Incised channels and rivers time-equivalent of X_c Y_c Z_c		– Outlets between two pedi-valleys (X_c – Y_c) *Luvua River (between Kibara and Masseba Plateau) Lu_ra River (between the Biano and Kibara Plateaus)* – Southern part of the pediment system Z_c	Incision: 200-500 m		

Few radiometric dates are available from this area (see age arguments on Table 14.4). The only dated rocks intersecting the planation surfaces are the volcanic rocks of the Bukavu (or South Kivu) Volcanic Province (Kampunzu et al. 1998). Only Pediment Xc can be dated by cross cutting of its upstream incised network of rivers on the Lugulu basaltic flow that belongs to the initial episode of the Bukavu Volcanic Province (Lubala et al. 1984; Kampunzu et al. 1998), made up of pre-rift tholeiitic to transitional basalts. The upstream part of the incised river network fitting to pediment X_c is younger than 5.9 ± 0.3 Ma.

14.5.2.4 South-Central: Katanga, Kamina/Kasai/Lunde/Kwango Plateaus and Central CB (S DRC, Zambia, Angola)

This domain of plateaus (Plate 1) stretches from the southern side of the CB (i.e. the northern limit of the Kalahari Plateau) to the transitions with the East African Dome (Mporokoso Plateau, northern-eastern Zambia) and with the Angolese Plateau (Bié Plateau). This is the Congo-Zambezi divide that define the boundary between two local base levels, the Ubangi-Likouala Swamps to the north and the Chobe—Okavango Swamps (Okavango and Machili Rifts), to the south (see also Flugel et al., Chap. 15, this Book).

These plateaus and the CB correspond to the Congolese Surface (Plate 2 and 14.5.1.3), buckled along an E-W trending long wavelength (×100 km) antiform. In the Central CB, the Congolese Surface cuts across from the southern plateaus to the Lukenie River (Plate 1), Archean rocks (Kasai Craton) and Late Cretaceous red sandstones (Kwango Gp), and from the Lukenie River to the Uele, Itimbiri and Congo Rivers, the lacustrine Paleogene (Eocene?) sediments (see 14.4.2 and 14.4.4).

The Congolese Surface is dissected by incised valleys (Kamina-Kasai-Lunde-Kwango Plateaus and Central CB) and/or pediments (Katanga, Mbuji-Mayo area—not mapped here; but see Roberts et al., Chap. 9, this Book, for more information).

Along the *southern flank of the CB*, the plateaus are deeply incised by northward facing valleys. All these incised rivers and channels fit to the local base level corresponding to the Ubangi-Likouala Swamps. Two arguments indicate that this incised drainage was initiated at time of formation of pediment X_c: (1) The present-day Congo River between Kisangani and Mbandaka is located on the outlet of pediment X_c (see 5.2.3.); (2) Along the eastern side of the CB, following the N-S trending X_c pediplain, the dissection of the Kamina Plateau, incised by several rivers, starts with pediments X_c (Plate 2).

The *Kalahari flank of the plateaus* is deeply incised by valleys, mainly on the Angolese Plateau. The rivers fit to the local base level of the Chobe Swamps (Plate 1) filled by the recent Kalahari Sands (younger than the "Sables ocres" Fm, here upslope incised).

In the *Katanga* area, six landforms are defined (Plate 2, Fig. 14.9 Ka, see description on Table 14.5), from the oldest to the youngest, (1) Congolese Surface (*C, Cd*), (2) Congolese Surface degraded by the Lower planation surface (*Cdl*), (3) degraded Lower planation surface (*ld*), (4) Pediments X (X_c, X_f), (5) Pediment Y (Y_c) and (6) incised channels.

The only available age constraints are from relative relationships with (1) dated sediments of the "Grès polymorphes" Fm (Alexandre 2002) and (2) dated supergene "cobalt caps" (Decrée et al. 2010—see arguments on Table 14.5). From this we deduce that all the studied landforms described above are younger than Paleogene (and probably Eocene). Pediment X_c is older than 10.3 ± 0.1 Ma and Y_c (and may be Z_c) were shaped between 10.3 and 3.66 Ma.

The three large depressions or plains of the Upemba/Kamalondo, Lufira and Lake Mweru (=Moero)/Luapala Swamps (Plates 1 and 2) were interpreted as rifts (e.g. Mortelmans 1953). Our study suggests that only the Mweru Lake is an asymmetrical rift (Chorowicz 2005), bordered by the NW Lake Mweru Fault, present-day sharp escarpment. The other ones, the Luapala Swamps, the Lufira Plain and the Upemba/Kamalondo Depression are not rifts as indicated by the absence of significative sediments, but pediments (pediments X_c). In this area, the only active fault is the Upemba Fault, old geological structure (Mortelmans 1953), reactivated after the formation of the Congolese Surface (maximum vertical displacement: 600–700 m) and prior to X_c. The Marungu, Kibara, Biano, Kundelungu and Mporokoso Plateaus (Fig. 14.9 Ka1) form a large block bounded to the NW by the Upemba Fault and titled toward the SSE (long wavelength deformation: ×100 km).

14.5.2.5 Southwest: Angola Plateau and the Central African Atlantic Swell, South of the Congo River (Angola, SW DRC)

This domain (Plate 1) extends from the Angolese Plateau to the Congo River and includes, from west to east (1) the Atlantic Ocean coastal plain, (2) the Angolese coastal Escarpment and (3) the Angolese Plateau passing northward into the large stepped pedi-valleys to pediplains of the Cuanza (=Kwanza) and Kwango present-day Rivers with two troughs, the Cuanza-Luando (Cuanza large pedi-valley) and Cassange (large pedi-valley) Troughs.

Here we focus on the Kwango and Cuanza large pedi-valleys to pediplains. The successive planation surfaces are labelled with the subscript kc for Kwango-Cuanza.

Six landforms are defined (Plate 2, Fig. 14.9 An, see description on Table 14.6), from the oldest to the youngest, (1) Upper planation surface 1 (*u1u*), (2) undifferentiated Upper planation surface 2 and Lower planation surface (*u2lu*), (3) Pediments V (V_{kc}), (4) Pediments W (W_{kc}) (5) Pediments X (Xk_c) and (6) Pediment Y (Yk_c).

The N-S trending Kwango large pedi-valley to pediplain (600 to 700 km long) that feeds into the Congo River south of the Batéké Plateau, is a type-case of pediment system (14.5.1.5, Fig. 14.8).

Table 14.6 Relative chronology, characteristics, synonymy and age of the landforms of the Central African Atlantic Swell—Angola Swell

	ELEVATION	LANDFORM-WEATHERING (W.)	RELATIONSHIPS WITH THE OLDEST FORM	SYNONYMY	AGE
Upper Planation Surface 1 (*u1u*)	2,500 m	Highly degraded planation surface—stripped etchplain *? isolated mounts and hills along the Atlantic side of the Bié Plateau* **w.**: bauxites mentioned by by Carvalho (1961)—unknown exact location		± Surface V (Jessen 1936)	*Named by analogy with Cameroon Highlands*
Upper P. Surface 2 Lower P. Surface (*u2lu*)	2,200–2,400 m	Weathered pediments at time of *u1u* stripping? *most of the Bié Plateau and Serra Canguejo* **w.**: similar than *u1u*	Difference of elevation with *u1u* 100 m	± Surface IV (Jessen 1936)	
Pediments V (V_{kc})		large pedi-valleys to pediplains - asymmetrical: narrower on the Biè Plateau flank (lithologic effect?) *well preserved east of Bié Plateau - remnants upstream of Cassange Escarpment (Serra Biringa to S. Cambala)* **w.**: 10 to 15 m-thick iron duricrusted conglomerate (well-rounded quartz pebbles, < 6 cm) upstream Cassange Esc. (Cabatuquila duricrust, Marques 1992)	Greater incision 500 m (north of Serra Canguenjo)		
Pediments W (W_{kc})		large pedi-valleys to pediplains - two drainages: - to the west (Atlantic Ocean), Cuenza - to the north (Central Cuvette), Kwango	Greater incision150 m decrease northward	Surface II–III (Jessen 1936) with different contours	<u>**Cassange Trough:**</u> pediment **Ykc** cut across pediment **Xkc** capped by silicified lacustrine deposits with gastropods, ostracods, characeans of (middle?) Pleistocene age (based on gastropods, Mouta and Dartevelle 1954) *Bangalas (Bango Mount) and Cassanza River, close from the Cassange Escarpment (Serra Bango Cassala)* related to the "Grès polymorphes" Fm ***comment***: similar facies but different age than the «Grès polym.» of the Kasai and Kwango Plateaus as indicated by their geomorphological setting
Pediments X (X_{kc})		large pedi-valleys to pediplains—asymmetrical with a sharp SW scarp (Cassange Escarpment) *Kwango pedi-valley only: southeast of Cassange Trough*	Incision 200-300 m decrease northward		
Pediment Y (Y_{kc})		Large pedi-valleys to pediplains *Kwango pedi-valley only: Cassange Trough* connection to its local base level (Congo River) trough a dense network of incised rivers	Re-incision of the Cassange Escarpment		

- The Kwango present-day catchment is a stepping of four large pedi-valleys to pediplains (V_{kc}, W_{kc}, Xk_c, Yk_c), the youngest one being nested and smaller than the older ones.
- Downstream the pediment system is intermediate between pediplains and large pedi-valleys bounded on both sides by scarps (e.g. Cassange) or steepest slopes (width between ×10 km to 100 km).
- Upstream the system corresponds to highly dissected slopes by a dense drainage of incised rivers connected to the pediplain/pedi-valley.

Pediments W_{kc} record the initiation of a new drainage of large pedi-valleys, the Cuanza on the west side, flowing toward the Atlantic Ocean and the Cuango (=Kwango) to the north, flowing toward the CB. At time of pediments X_{kc} (only recorded in the Kwango), the Cuanza large pedi-valley is inactive, fossilized; erosion is localized along the Atlantic Escarpment.

The only available date is of low resolution (Table 14.6). Pediment Y_{kc} is young, Plio-Pleistocene but possibly older, because the gastropods fossils used by Mouta and Dartevelle 1954 have poor stratigraphic resolution.

14.5.2.6 West: Central African Atlantic Swell, North of the Congo River (R of Congo—DRC, Gabon, Equatorial Guinea, South Cameroon)

This domain (Plate 1) is subdivided into two parts, separated by the Ogooué River (Gabon): the Chaillu Massif in the

Table 14.7 Relative chronology and characteristics of the landforms of the Central African Atlantic Swell–South Cameroon, Gabon, Congo Swelly

	ELEVATION	LANDFORM-WEATHERING (W.)	RELATIONSHIPS WITH THE OLDEST FORM
Lower Planation Surface and older? (*l*)	600–900 m	Relict weathered planation surface *South Cameroon Plateau (600–700 m), Central Equatorial Guinea Plateau, summit of the Chaillu Massif (800–900 m)* **w.**: Laterites with iron duricrusts (Cameroon: Yongue-Fouateu et al. 2006; Gabon: Novikoff 1974)—some bauxites (Makongonio area): relicts of Upper planation surfaces?	
Pediments V (V_a) and **Pediments W** (W_a)		Large pedi-valleys to pediplain 100 km-width (*South Cameroon Plateau*) –Piedmont pediments (*Chaillu Massif*) **w.**: no published descriptions	Difference of elevation (l, $V_{a'}$, W_a): low (scarps $\leq$ 100 m) - can merge in the same planation surface
Pediments X (X_a)		pedi-valleys—well-organized network - located on the Atlantic flank of the swell (except Oogoué) narrower than Pediments V–W (×10 km-width) –*South: Chaillu Massif—Mayombe* pedi-valleys parallel to the coast (20–60 km-width) controlled by the NW-SE Neoproterozoic structures –*Middle: present-day Ogooué River* large pediplain cutiing-across di_erent lithologies of the swell –*North: Equatorial Guinea, South Cameroon* pedi-valleys oriented along a NE-SW trend (Neoproterozoic Ubangides structures) **w.**: no published descriptions	– Scarps or steepest slopes mean height: 100 m–up to 300 m (Ngounié Plain and Niari Valley) –Scarps $\leq$100 m along the North Gabon/ Ivindo Plateau –Scarps $\leq$100 m
Pediments Y (Y_a) and **Pediments Z** (Z_a)		Pedi-valleys superimposed on Pediments X - enhance the previous network narrower than Pediments X – *South: Ngounié Plain/Niari Valley and Nyanga Plain* controlled by the dolomitic limestones of the Neoproterozoic "Schisto-calcaires" Gp –*Middle: Booué and Franceville low plateaus and hills* only pediment Y_a - shifted to the south of X_a	
Incised rivers		Three main rivers: Congo, Ogouoé and Sanaga Congo incision: Ubangi Flat to coastal plain Ogooué incision: along Pediment Y_a Sanaga incision: Lower p. surf. (*l*) to the coastal plain	Incision: 50–150 m

south and the South Cameroon–North Gabon Plateau, in the north.

Six landforms are defined (Plate 2, Fig. 14.9 AS, see description on Table 14.7), from the oldest to the youngest, (1) Lower planation surface (*l*) and older?, (2) Pediment V (V_a) and Pediment W (W_a), (3) Pediment X (X_a), (4) Pediment Y (Y_a) and Pediment Z (Z_a) and (5) incised rivers.

The Atlantic slope of the swell comprises planation surfaces that merge with marine sediments of the Atlantic passive margin. An Upper and Lower planation surfaces undifferentiated (*ulu*) is defined along the Atlantic Ocean (mainly Mayombe and Sierra de Cussianzumbo), with very low stepping amplitudes and sometimes a merge of the Upper and Lower planation surfaces. In the Lower Congo area, Kaseba et al. (1997) defined three planation surfaces with a difference of elevation of 150 m. The two younger ones are bauxitized and the older one is stripped from weathered materials.

No dates are available from any of these surfaces.

14.5.2.7 Regional Chronology and Dating of Landforms

The relative chronology of the landforms of the CB and surrounding highs (Fig. 14.11) is summarized as follows (from the oldest to the youngest).

- *Upper planation surfaces 1* (*u1*) are the oldest landforms preserved in the studied area. They are identified: (1) in the Cameroon Highlands by successive stepped planation surfaces (*u1a*, *u1b*, *u1p*); (2) on top of the Ubangian Rise (Bongo Mounts); (3) along the East African dome, from the Blue Mountains (flanking the western margin of Lake Albert) to the Marungu Plateau (Tanganyika Lake) and (4) on the Angola Plateau. The absence of spatial continuity between these relict forms, hampers demonstration of their synchronicity. In Katanga, these surfaces may be the equivalent of the lowermost iron duricrust (iron duricrust 1) located above the "Grès Polymorphes" Fm.
- *Upper planation surface 2* (*u2*) is more widely distributed, with numerous relicts, large enough to be

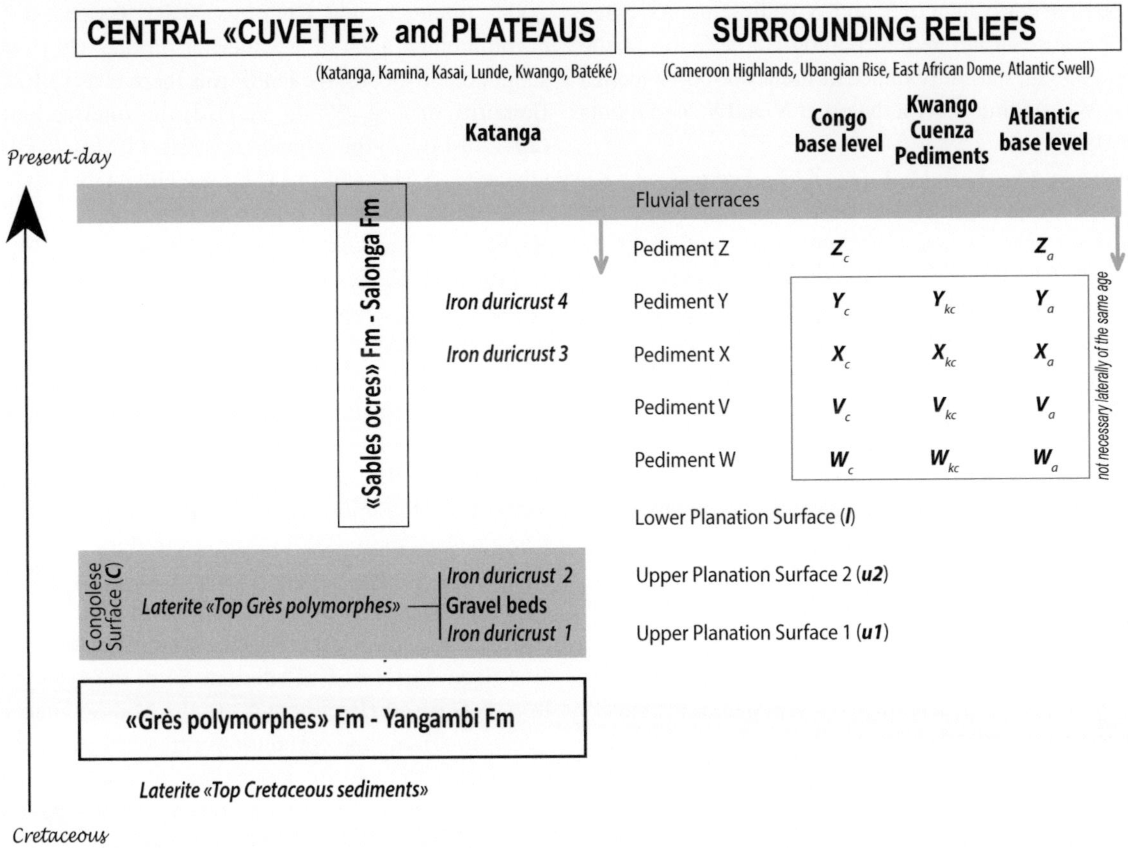

Fig. 14.11 Relative chronology of the sediments and landforms of the CB and surrounding highs

identified with reasonable confidence. It crops out along a crescent from the Cameroon highlands crossing the Ubangian Rise and the Blue Mountains to finish on the southern Itombwe Plateau. Southward, in Katanga and on the Angola Plateau, this surface merges either with Upper planation surface 1 (Katanga) or with the Lower planation surface (Angola). In Katanga, this surface could be equivalent of the uppermost iron duricrust (iron duricrust 2) located above the "Grès Polymorphes" Fm.

- *Congolese Surface (C)* covers the Central CB (dissected planation surface) and the Batéké and Kwango/Lunde/Kasai/Kamina Plateaus (this etchplain is covered by the surficial sediments of the "Sables ocres" Fm).

On the Congolese flank of the Ubangian Rise, the Congolese Surface merges with Upper Surface 2 (*u2*) and with the Lower Surface (*l*). In Katanga, the Congolese Surface corresponds to the iron duricrusts 1 and 2 of Alexandre (2002). The Congolese Surface (*C*) is then contemporaneous with the Upper planation surfaces (*u1* and *u2*) and then represents a long-lasting planation surface.

- *Lower planation surface (l)* is the dominant form in the northern half of the mapped area. It continuously extends from the northern Central African Atlantic Swell (North Gabon) to the Ubangian Rise and the northern East African Dome (Ituri). To the east (Ituri), the Lower planation surface is a pediplain with few preserved weathering profiles that passes to the west (north Gabon–south Cameroon) into a large highly weathered pediplain (with iron duricrusts). Toward the southeast (middle and southern East African Dome and Katanga), this surface is highly degraded by younger pediments (mainly X). Along the central African Atlantic Swell, the Lower planation surface is only preserved on the high relief of Gabon-Congo, the Chaillu Massif. In Angola, because of the absence of spatial continuity with Katanga or Gabon, the summit planation surface may be the Upper planation surface 2 or/and the Lower planation surface.
- *Pediments V (V_a V_c V_{kc}) and W (W_a W_c W_{kc})* record both a narrowing of the pediplains and the emergence of different local base levels and different pedi-valleys drainages, labelled V_a and W_a for the South Atlantic Ocean base level, V_c and W_c for the Congo base level and V_{kc} and W_{kc} for the Kwango–Cuenza large pedi-valleys to pediplains. No V and W pediments were identified along the western flank of the East African Dome, but some forms attributed to the degraded Lower Surface

could be time equivalent of V and W pediments. As stated above, because of the geomorphological isolation of the Angola Plateau, the equivalence of the Kwango—Cuenza V_{kc} and W_{kc} pediment with the other V and W pediments is uncertain.

- *Pediments X (X_a X_c X_{kc}), Y (Y_a Y_c Y_{kc}) and Z (Z_a Z_c)* enhance the trend initiated at time of pediments V and W. The base levels are well differentiated (South Atlantic Ocean, CB, Chad-Sudan) and the pediments are narrower (mainly pedi-valleys, few true pediplains). The labelling is the same than for pediments V and W. No X or Y pediments were identified along the eastern flank of the Central African Atlantic Swell. Again, the equivalence of the Kwango—Cuenza X_{kc} and Y_{kc} pediment with the other X and Y pediments is uncertain.

The only *dates* available for these landforms are from the oldest (*u1*, *C*) and the youngest ones (X, Y). Most of their ages are based on their geometrical relationships with volcanic rocks dated by K-Ar on whole rock (most of the dates) or Ar–Ar on minerals. Dating of the same sample by the two techniques provides an age difference between 1 to 4 Ma.

- *Upper planation surfaces 1 (u1)* are dated (14.5.2.1) between 67 ± 7 Ma and 45.5 ± 1.1 Ma, i.e. Maastrichtian-Danian and Middle Lutetian, with an upper range that could be at 51.8 ± 1.2 Ma (see 14.5.2.1 for discussions).
- The *Congolese Surface (C)* is younger than the "Grès polymorphes" Fm, dated as Paleogene, and probably Eocene.
- *Pediments X (X_a X_c X_{kc})* and *Y (Y_a Y_c Y_{kc})* have different (apparently contradictory) ages.
  - In the Cameroon Highlands (Rumpi Hills, 14.5.2.1)—Atlantic base level—Y_{as} and probably X_{as} are younger than 9.9 ± 0.4 Ma, and may be as young as 5.6 ± 0.2 Ma.
  - Along the East African Dome (Mitumba Mounts, 14.5.2.3)—Congo base level—the upstream incised rivers of the pediment system X_c are younger than 5.9 Ma.
  - In Katanga (Lubumbashi–Kolwezi, 14.5.2.4), the upstream incised rivers of the pediment system X_c may be older than 10.3 Ma and the ones of Y_c (and may be Z_c) between 10.3 and 3.66 Ma.

The apparent differences in age may be due to either a different relative base level or (and) the delayed time response of the upstream incised rivers of the pediment system to the local base level constituted by the pedi-valleys. For these reasons, we suggest a large time range for these pediments, from Late Miocene to Present-Day.

Indirect dating relies on apparent similarities between Congolese landforms and weathering sequences, and those dated in West Africa (Chardon et al. 2006; Beauvais et al. 2008; Beauvais and Chardon 2013). Based on ^{39}Ar-^{40}Ar cryptomelane ages of weathering profiles (Fig. 14.12), the sequence of successive landforms there is as follows: (1) the Bauxitic Surface (59–45 Ma), (2) the Intermediate Surface (29–24 Ma), (3) the High Glacis (18–11 Ma), (4) the Middle Glacis (7–6 Ma) and (5) The Low Glacis (3 Ma).

Correlations were tested in two areas, viz. CAR and Katanga. In the CAR (Ubangian Rise, Table 14.3), Tardy and Roquin (1998), following the interpretations of Boulvert (1985, 1996), suggest that (1) the Lima Bocaranga Surface (here equivalent to the Upper planation surface 1) could be equated to the Bauxitic Surface, (2) the Bouar-Baboua Surface (here equivalent to the Upper planation surface 2) the Intermediate Surface, (2) the Centrafrican Surface (mainly equivalent to the Lower planation surface) the High Glacis and (4) the Piedmonts the Middle Glacis. In Katanga (following Alexandre 2002), the Iron duricrust 1 (possible equivalent to the Upper Surface 1) corresponds to the Fantofa Level (=Bauxitic surface), Iron duricrust 2 (possible equivalent to the Upper Surface 2) to the Intermediate Surface, Iron duricrust 3 (Pediment X) to the High Glacis and Iron duricrust 4 (Pediment Y) to the Middle Glacis.

Summarizing the correlations proposed by Tardy, Roquin and Alexandre, we can deduce than (1) the Bauxitic Surface weathered until 45 Ma corresponds to the Upper planation surface 1 (Katanga), (2) the Intermediate Surface formed between 45 and 24 Ma corresponds to the Upper planation surface 2 (both Katanga and Ubangian Rise), (3) the High Glacis eroded between 24 and 11 Ma is an equivalent of the Lower planation surface (Ubangian Rise) and Pediment X (Katanga).

Our results are consistent with the age of Upper planation surface 1 (here dated between 67 ± 7 Ma and 45.5 ± 1.1), but we disagree with the age of Pediment X, which must be younger (Late Miocene). The 45–29 Ma (Upper Eocene-Early Oligocene) age for the mechanical erosion of the Upper planation surface 2 (if equivalent to the Intermediate Surface) and the 24–18 Ma (Late Miocene) age for the Lower planation surface (if equivalent to the high glacis) have to be tested by other geological means.

One test is to evaluate the balance between the products of the erosion of the planation surfaces against the volume and sedimentation rate of the siliciclastic sediments deposited along the Atlantic passive margin. These data are available for the Congo Delta at low resolution from the Cretaceous (Anka et al. 2010) and at a higher resolution from the uppermost Eocene (Lavier et al. 2001, Fig. 14.12). Both are based on regional seismic lines provided (with ages constraints) by TOTAL (Zaiango Project). Anka et al. (2010) characterized three periods of silicicliclastic sediment supply: Albian-Turonian (200,000 km^3 between

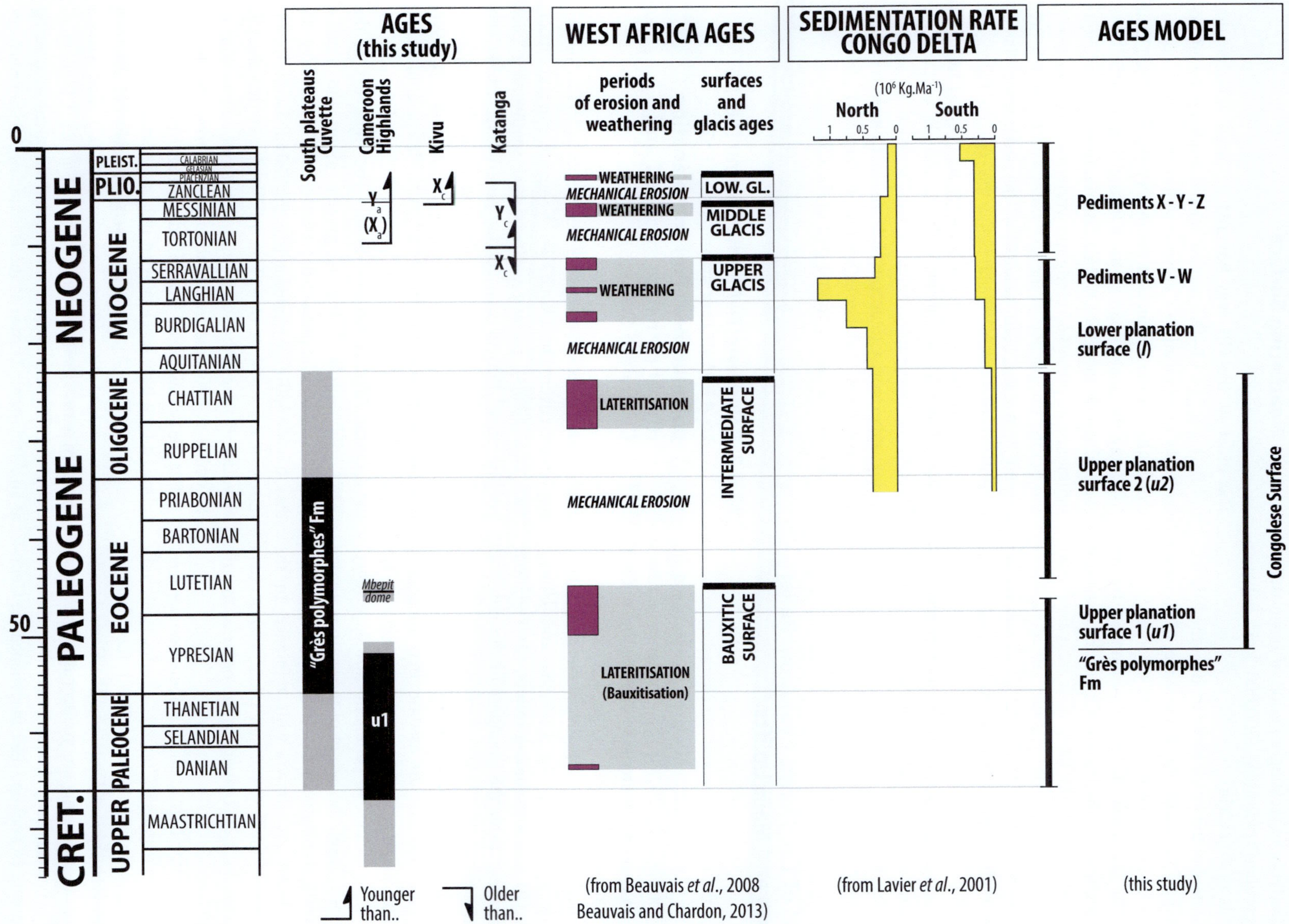

Fig. 14.12 Age model for the sediments and planation surfaces of the CB and surrounding highs. Age of the West Africa planation surfaces and weatherings from Beauvais et al. 2008 and Beauvais and Chardon 2013; Sedimentation rates of the Congo Delta from Lavier et al. 2001

115–90 Ma giving a sedimentation rate of 75 m/Ma); Coniacian-Eocene (around 10,000 km^3 between 90–34 Ma—effectively suggesting a period of basin starvation); and Oligocene-Present-day (700,000 km^3 between 34–0 Ma, at a rate of 80 m/Ma).

The sharp increase of the siliciclastic input at the Eocene-Oligocene boundary confirms the occurrence of a major denudation event onshore that may be the mechanical erosion of the Upper planation surface 2, enhanced by the climate change occurring at the Eocene-Oligocene transition (Séranne 1999). Sedimentation rates also increased from Oligocene to Middle Miocene (Lavier et al. 2001—Fig. 14.12), followed by decreases in the north and further increases in the south (related to delta shifting, Anka et al. 2010). We relate this increase in two steps (Early and Middle Miocene) tentatively to the erosion of the Lower planation surface and of Pediments V and W (for more detail see Linol et al., Chap. 10, this Book).

14.6 Model of Landforms Evolution: Relief Growth and Main Uplifts

The evolution of the topography and the characterisation of the uplifts of the CB and surrounding highs are based on restored cross-sections (Fig. 14.14). The topographic restoration is based on two observations: (1) All pediments and pediplains that frmed subsequent to the deposition of the "Grès polymorphes" Fm, merge together at two base levels, viz. the Atlantic Ocean and the Likouala and Ubangi Swamps for the CB and surrounding reliefs; (2) The slope of pediments increase from the youngest to the oldest in response to uplift.

The principle of topographic restoration is to start from the youngest pediments and to assume that successive pediments have the same profile than the youngest one, i.e. Pediments Z or Y. In the case of pediments dowslope merging to the same base level, the height of the scarp is a good proxy of the vertical displacement. All the restorations shown in Fig. 14.14 are drawn with reference to the present-day sea level; eustasy was not taken into account (see for discussions 14.3.2) compare to the order of magnitude of the uplifts.

The successive restoration of the pediments or the river incisions (rivers of the degraded Congolese Surface), all indicate that: the Central Atlantic African Swell, the Ubangian Rise, the western East African Dome and the Kasai-Lunde-Kwango and Angolese Plateaus are younger than the Upper Surface of Paleocene to Early Eocene age. We therefore propose the following (seven) episodes that shaped the Cenozoic evolution of landscapes in and around the CB:

14.6.1 Late Cretaceous–Early Paleocene (72–66 to 66–56 Ma): Top Cretaceous Sediments Weathering Surface

The crust of the CB was flexured at large wavelengths, uplifted (Fig. 14.13—NS section B) and eroded to be unconformably overlapped by Cenozoic sediments. This is consistent with: (1) the absence of sediments younger than Late Albian in the central part of the CB (e.g. Linol 2013; Linol et al., Chap. 8, this Book) while Santonian to Maastrichtian sediments are preserved along the western CB (Colin 1994); and (2) the occurrence of incised valleys and channels preserved below the "old basalts" of the Adamawa Plateau (Humbel 1966—14.5.2.1) and below the possible Paleogene (Eocene?) sediments of the Yangambi area (Fig. 14.6).

We cannot as yet quantify the amplitude of the Late Cretaceous topography since it was levelled by erosion during uppermost Cretaceous–lowermost Cenozoic (?) and weathered (Top Cretaceous sediments laterites) along the eastern and across the central part of the CB.

14.6.2 Paleocene(?)–Middle Eocene (66–56 to 47 Ma): low subsidence ("Grès polymorphes" Fm) and planation (Upper planation surface 1 and Congolese Surface)

Sedimentation occurred across a long wavelength low subsiding flexure covered by a large lake (Yangambi Fm and subsurface equivalents), passing eastward and southward ("Grès polymorphes" Fm) into an aeolian desert (Fig. 14.7). This environments distribution implies that fresh water derived from the northern very humid equatorial domain (Fig. 14.7) to keep a positive hydrological budget for this perennial lake.

A remaining question relates to the timing of the uplift of the Central African Atlantic Swell. The thickness of the "Grès polymorphes" Fm is greater along the Batéké Plateaus, east of the present-day Gabon Swell. This suggests that this formation extended across the Gabon Swell. Restoration of the E–W section (Fig. 14.14) shows that the CB elevation was near the sea level, but without connections to the ocean as indicated by the absence of marine fossils.

The large lake was first filled during Middle Eocene, followed by a base level fall that induced a second period of weathering (Top "Grès polymorphes" Fm laterite). Limited erosion shaped the Congolese Surface (*C*) that passes upstream across the basement to Upper Surface 1 (a mantled etchplain). These lateritic profiles record chemical weathering at an Africa-scale known as the Bauxitic or African Surfaces (Burke and Gunnell 2008; Beauvais and Chardon 2013).

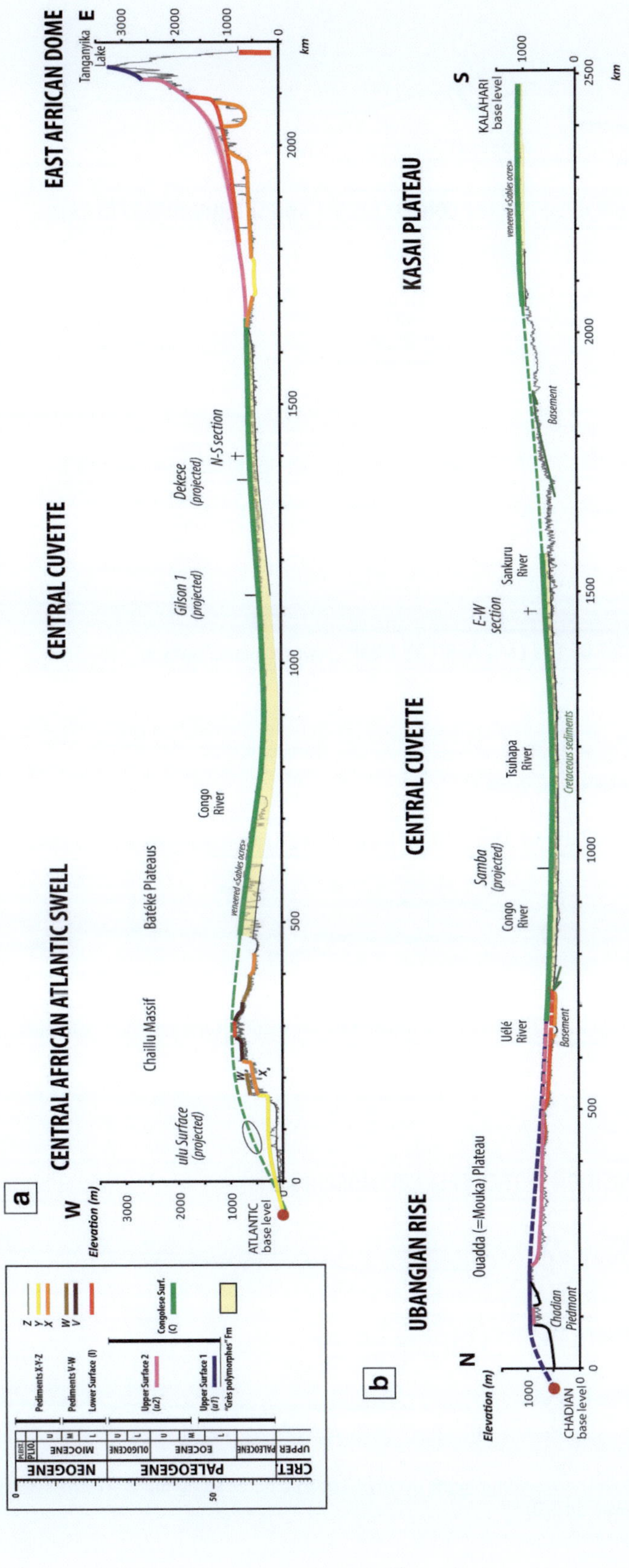

Fig. 14.13 Synthetic cross-sections of the CB and surrounding highs—sections location on Plate 2

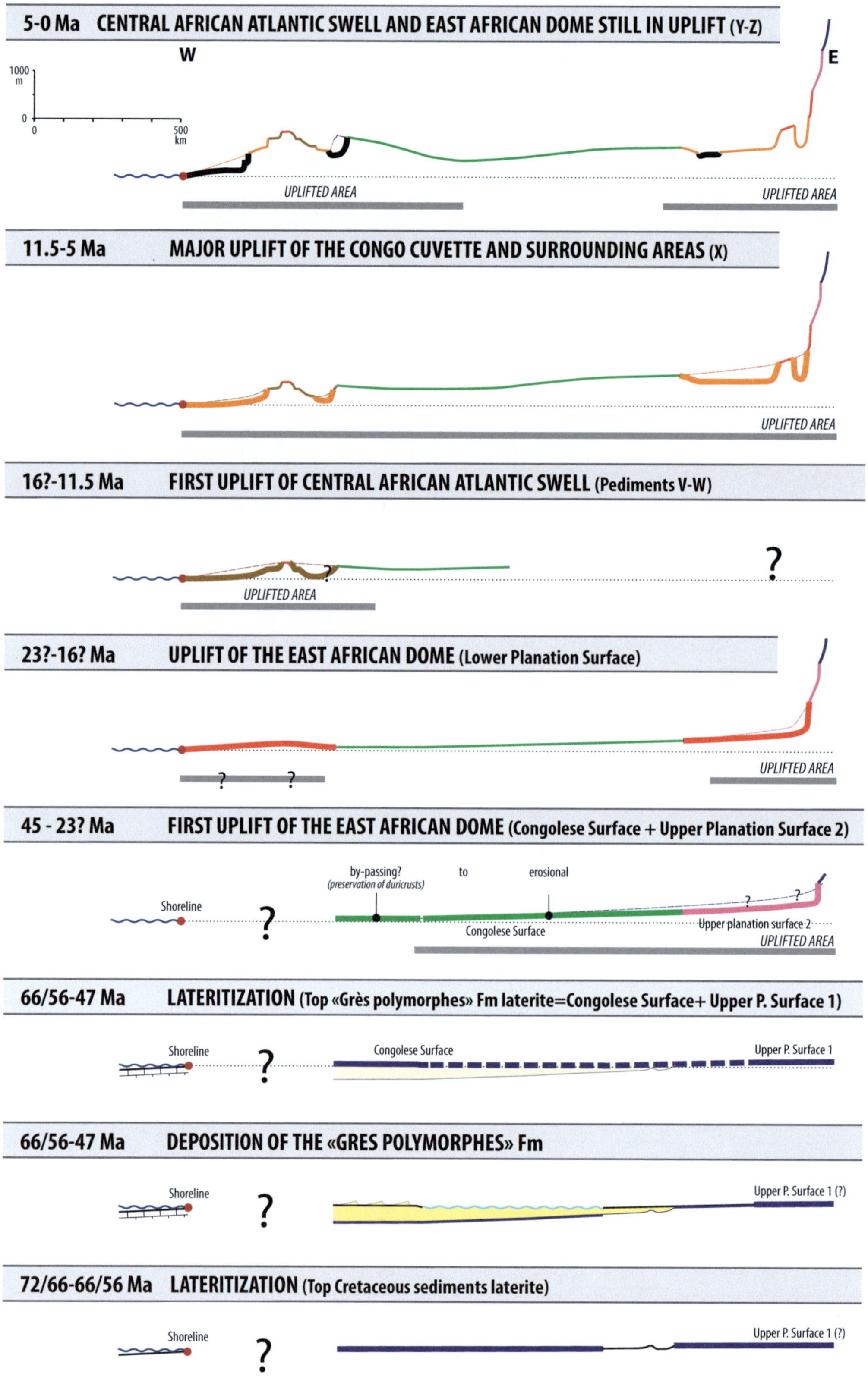

Fig. 14.14 Synthetic evolution of the CB and surrounding highs during Cenozoic times along the E–W cross-section of Fig. 14.13. Sea level is assumed constant (see discussion in text, Sect. 14.3.2)

14.6.3 Middle Eocene–Late Oligocene (?) (45–23? Ma): Upper Planation Surface 2

This is the first occurrence of Cenozoic uplift in the studied area, initiated in the north (Cameroon Highlands and Ubangian Rise) and in the east (western flank of the East African dome). The amplitude of this Middle Eocene–Late Oligocene uplift (taking the height of the scarps as a proxy for the uplift) is greatest along the Cameroon Highlands and Mitumba Mounts (East African Dome), at a magnitude of 100 to more than 500 m, and lower along the Ubangian Rise and the Blue Mountains (East African Dome), between 0 to 200 m.

The restoration of the E–W section (Fig. 14.14) shows that the CB was uplifted along its northern and southern margins (incision of Upper planation surface 2). Newly formed pediments connected to a local base level located in the western and southern parts of the Central CB. The rest of the CB was either affected by limited erosion or was by-passed, creating a long lasting, polygenic surface, known as the Congolese Surface.

A third period of weathering occurred at the end of this period. Upper planation surface 2 stripped most of the lateritic profiles of Upper planation surface 1, leaving a stripped etchplain. We assumed that these weathering processes were contemporaneous with those of the "Intermediate Surface" (Beauvais et al. 2008; Beauvais and Chardon 2013) that formed during Late Oligocene.

14.6.4 Early Miocene (?) (23? to 16? Ma): Lower Planation Surface

Uplift continued along the northern and western flanks of the CB and is more homogeneous in term of amplitude (ca. 0–200 m in the north and 100–300 m in the east).

A fourth period of weathering occurred at the end of this period, best recorded along the Cameroon Highlands and the Ubangian Rise to Ituri. The Lower planation surface is a duricrusted pediplain passing upstream to piedmont pediments (dominant) and pedi-valleys, stripping Upper planation surfaces 1 and 2.

14.6.5 Middle Miocene (?) (16? to 11.5 Ma): Pediments V and W

This is the first clear topographic evidence of the onset of the Central African Atlantic Swell, with maximum uplift of 100–200 m along the Gabon-Congo Swell and up to 650 m along the Kwango large pedi-valley (uplift of the Angolese Plateau). Unfortunately, Pediments V and W were not characterized along the East African dome, and the 100 to 300 m of uplift recorded by the pediments of the Lower planation surface (see above) can probably account for the uplift occurring at time of Pediments V and W.

14.7 Late Miocene (11.5–5 Ma): Pediment X

The entire studied region, including the CB was now further uplifted. Major uplift was located along the Atlantic side, elevating both the Central African Atlantic Swell and the eastern CB, as indicated by the absence of Pediments X on the Congolese side of the Swell. The mean order of uplift is around 100–300 m with higher values (500 m) along the Atlantic side of the Cameroon Highlands. During this period the present-day Ogooué River system crossed the Swell (probably inherited from an alluvial system coming from the CB).

The western part of the CB was the local base level of the pediments of the Ubangian Rise and of the East African Dome, probably initiated at time of Pediments V and W. The base level of these reliefs is not the Atlantic Ocean.

With reference to the CB local base level, the Ubangian Rise, the East African Dome and the Kasai-Lunde-Kwango Plateau were all uplifted between 100–200 m (Ubangian Rise) and 200–500 m (East African Dome). This uplift is difficult to quantify for the Kasai-Lunde-Kwango Plateau, because of the absence of pediments along the incised valleys. But, based on the Katanga and Kwango pediment data, the magnitude of uplift was at least 300 to 500 m. This period records a change in the wavelength of uplift, from × 100–1,000 km (Cameroon Highlands, Ubangian Rise) to >1,000 km of the CB along the Atlantic coast and along the western flank of the East African Dome (e.g. regional curvilinear bending parallel to the Atlantic margin).

14.8 Late Miocene–Present-day (5–0 Ma): Pediments Y and Z

From end of the Miocene onward, uplift mainly affects the Atlantic side of the Central African Atlantic Swell, with higher magnitudes along the Cameroon Highlands (ca. 500 m). This increasing uplift of the Cameroon Highland is recorded by a southward migration of the Ogooué Pediment Y compared to the earlier Pediment X.

In the CB, the uplift was less pronounced. Pediments Y and Z only deepen the previous pediments X, except in the southern plateaus and along large pedi-valleys the the Kwango, where the difference of elevation between X and Y is high, and rivers contemporaneous with deelpment of Y pediments, are well incised.

14.9 Conclusions

14.9.1 Sediment Infill of the Central CB and Surrounding Highs

1. *Age-dating*: Biostratigraphic data (ostracodes and characeans) from the Central CB wells (Gilson 1, Mbandaka 1) indicate that (i) the Central CB deposits are Paleogene (and probably Eocene) in age and that (ii) these sediments are time equivalent of the "Grès polymorphes" Fm, preserved in the western and southern plateaus of the Central CB. The outcrops of the Yangambi Fm (Kisangani area) may be lateral equivalent of the "Grès polymorphes" Fm.
2. *Relative stratigraphy*: From the base to the top, the infilling is as follows: (i) truncated laterite profiles on top of the Cretaceous sediments eroded by the Paleogene deposits; (ii) deposition of the "Grès polymorphes" Fm and its subsurface lateral equivalents; (iii) laterite; and (iv) deposition of the "Sables ocres" Fm. was diachronous from Neogene to the present-day, corresponding to weathered aeolian to alluvial surficial deposits, veneering older planation surfaces.
3. *Paleogene pre- to syn-Middle Eocene environments and palaeogeography*: A hot arid desert with aeolian dunes ("Grès polymorphes" Fm) graded northward into a lake with marshes (northern CB). The "polymorphic" texture of these quarzitic sandstones resulted from a complex diagenetic evolution during a period of base level fall with (i) adjusted ground water circulations and (ii) laterite weathering.

14.9.2 Landforms

1. *Type of landforms*: Two main types of landforms are identified: planation surfaces and incised rivers and channels. According to their size, planation surfaces are etchplains and pediplains (large—$\times 10^4$ to 2×10^6 km^2) or pediments (small—10^3 to $\times 10^4$ km^2).
2. *Relative chronology of landforms*: Nine successive planation surfaces are characterized, from the oldest to the youngest: (i) the Upper planation surface 1 (etchplain); (ii) the Upper planationsurface 2 (etchplain); (iii) the Congolese Surface (long-lasting surface time equivalent of the the two Upper surfaces); (iv) the Lower planation surface and five pediments, labelled from V (the oldest) to Z (the youngest).
3. *Age of formation of landforms*: An age model is proposed, based on: (i) the intersection of the planation surfaces with dated volcanic and plutonic rocks; (ii) the comparison of the landforms chronology and associated weathering with those dated in West Africa and tested: (iii) againt the siliciclastic sediment budget on the Congo passive margin.

The oldest preserved outcropping relief is Paleocene-Early Eocene in age (Upper Surface 1). The only old (Late Cretaceous?) landforms are preserved below the first basalt flows in the Adamawa Basin or below the Paleogene (Eocene?) sediments of the Yangambi area. This implies a major peneplanation of the studied are at the end of the Cretaceous, but no Cretaceous or Jurassic surfaces are preserved (cf. King 1962).

Most of the preserved landforms are Cenozoic, possibly Upper Eocene-Late Oligocene for Upper Surface 2, possibly Early to Middle Miocene for the Lower surface and Pediments V and W and Late Miocene to Pliocene for Pediments X to Z.

14.9.3 Topographic Evolution and Uplifts

1. The CB and most of the surrounding areas are inferred to have been close to sea level during the Paleocene-Middle Eocene times.
2. The Central African Atlantic Swell has been uplifted since the Middle Miocene? (16 Ma) (Pediments V and W); and uplift increased since the Late Miocene (11.5 Ma).
3. The Cameroon Highlands have been uplifted since the Eocene (45 Ma) during two pulses: Middle Eocene-Late Oligocene (Upper Surface 2); and post the Late Miocene.
4. The Ubangian Rise was moderately uplifted from Middle Eocene-Late Oligocene? (45–23 Ma; Upper Surface 2) to the Late Miocene (11.5 Ma).
5. The East African Dome has been uplifted since the Middle Eocene-Late Oligocene (45–23 Ma) (Upper Surface 2) with an increase during Late Miocene (11 Ma) and a decrease subsequently.
6. The Angolese Plateau and the western part of the Kasai-Lunde-Kwango Plateau were uplifted since the Middle Miocene? (16 Ma) (Pediments V and W).
7. The CB was uplifted after the Late Miocene (11 Ma).
8. The present topography of the CB and surrounding areas results from two wavelengths of crustal deformation: long wavelength (100–1,000 km) and very long (>1,000 km). From the Middle Eocene to Middle Miocene, the long wavelength was dominant (Cameroon Highlands, Ubangian Rise); since the Late Miocene, very long wavelength deformation is expressed by the uplift of the CB and parts of the Kalahari Plateau (Kasai-Lunde-Kwango Plateaus).

9. The Cameroon Volcanic Line and the Virunga-Bukavu Volcanic Province represent contemporaneous dome formation (Cameroon Highlands and Mitumba-Virunga Highs) that were initiated during the Middle Eocene (45–47 Ma).
10. The present-day CB results from a differential uplift, low (200–300 m) in the CB and higher in the surrounding areas. From Middle Eocene times onward, the CB has not subsided. These results do not support dynamic topographic models predicting a collapse of the CB after 45 Ma (e.g. Moucha and Forte 2011).

This study is a new modern attempt at mapping the landforms of central Africa. Although we are forced to simplify our interpretation of many areas that in details have much more complex relief patterns. More detailed regional studies are now required to tease out greater complexities and to further improve our knowledge of the landforms and associated weatherings, integrated with more precise ages, with which to test our model of landforms evolution of the CB and surroundings.

Acknowledgments This study was made possible by two projects, "TopoAfrica", funded by the French National Research Agency (ANR) (managed by F. Guillocheau, Géosciences-Rennes) and "! Khure Africa", a South African French Earth System Sciences project, with funding from the South African National Research Foundation and the department of Science and Technology (managed by Maarten J. de Wit, at that time UCT, Cape Town and Vincent Courtillot, IPGP, Paris). We are very grateful to Dominique Chardon for the detailed critical comments that significantly improved an early version of the manuscript.

We would like to thank the colleagues of the Royal Museum of Central Africa (Tervuren, Belgium), Max Fernandez-Alonso, Luc Tack and Damien Delvaux de Fenffe for their help.

We would like to thank Maarten J. de Wit (AEON, NMMU, Port Elizabeth) for his help, his enthusiasm, his friendship and for editing several versions of the manuscript.

References

Alexandre J (2002) Les cuirasses latéritiques et autres formations ferrugineuses tropicales - Exemple du Haut Katanga méridional. Ann Musée roy Afr centrale - Tervuren, Belgique - Sci géol 107:1–118

Alexandre J, Alexandre-Peyre S (1987) La reconstitution à l'aide des cuirasses latéritiques de l'histoire géomorphologique du Haut-Shaba. Z Geomorph NF Suppl-Bd 64:119–131

Alexandre-Pyre S (1971) Le plateau des Biano (Katanga). Acad roy Sci Outre-Mer, Cl Sci Nat Méd, Mém NS Géologie et géomorphologie 18(3):151

Anka Z, Séranne M, di Primio R (2010) Evidence of a large upper-Cretaceous depocentre across the Continent-Ocean boundary of the Congo-Angola basin. Implications for palaeo-drainage and potential ultra-deep source rocks. Mar Petrol Geol 27:601–611

Beauvais A (1999) Geochemical balance of lateritization processes and climatic signatures in weathering profiles overlain by ferricretes in Central Africa. Geochim Cosmochim Acta 63:3939–3957

Beauvais A, Chardon D (2013) Modes, tempo, and spatial variability of Cenozoic cratonic denudation: The West African example. Geochem Geophys Geosyst 40:19. doi:10.1002/ggge.20093

Beauvais A, Mazaltarim D (1988) Etude des cuirasses latéritiques dans la région de Dembia-Zémio en Centrafrique. Pétrographie, minéralogie et géochimie. Sci Géol. Bull (Strasbourg) 41:47–69

Beauvais A, Roquin C (1996) Petrological differentiation patterns on geomorphic distribution of ferricretes in Central Africa. Geoderma 73:63–82

Beauvais A, Ruffet G, Hénocque O, Colin F (2008) Chemical and physical erosion rhythms of the West African Cenozoic morphogenesis: The 39Ar-40Ar dating of supergene K-Mn oxides. J Geophys Res 113(F04007):15. doi:10.1029/2008JF000996

Belinga S (1966) Contribution à l'étude géologique, minéralogique et géochimique des formations bauxitiques de l'Adamaoua (Cameroun). Thèse Doct 3ème cycle, Univ Paris, 155 p

Belinga SE (1972) L'altération des roches basaltiques et le processus de bauxitisation dans l'Adamaoua (Cameroun). Univ Paris VI, Thèse Doct Etat Sci Nat, 571 p

Belinga SE (1973a) Etude des processus métallogénique de la bauxitisation dans l'Adamaoua au Cameroun. C R Acad Sci, Paris Sér D 276:1381–1383

Belinga SE (1973) Nature sédimentaire et autochtone des bauxites de l'Adamaoua au Cameroun. Paper presented at the 3rd International Congress of the International Committe Study Bauxite, Alumina aluminium (ICSOBA), Nice: 209–213

Beugnies A (1950) Contribution à l'étude des sables de l'étage supérieur du Système du Kalahari au plateau de Kamina (Congo belge). Bull Soc belge Géol Pal Hydro 59:93–102

Bigotte G, Bonifas G (1968) Faits nouveaux sur la géologie de la région de Bakouma (Préfecture du M'bomou - République Centrafricaine). Chron Mines Rech Min 36(370):43–46

Boulvert Y (1985) Aplanissements en Afrique centrale. Relations avec le cuirassement, la tectonique, le bioclimat. Problèmes posés. Progrès des connaissances. Bull Assoc Géogr Franç 4:299–309

Boulvert Y (1996) Etude géomorphologique de la République centrafricaine. Carte à 1/1 000 000 en deux feuilles ouest et est. Off Rech Sci Tech Outre-Mer (ORSTOM), Paris, p 258

Boyé M, Fritsch P (1973) Dégagement artificiel d'un dôme cristallin au Sud-Cameroun. Trav Doc Géogr Trop, Centre Et Géogr Trop (CEGET). Talence, Bordeaux 8:31–62

Bremer H (1993) Etchplanation, review and comments of Büdel's model. Z Geomorph NF Suppl-Bd 92:189–200

Breuil H, Janmart J (1950) Les limons et graviers de l'Angola du Nord-Est et leur contenu archéologique. Companhia Diamantes Angola (Diamang). Publ Culturais 5:1–57

Brookfield ME, Silvestro S (2010) Aeolian systems. In: James NP, Dalrymple RW (eds) Facies models 4, GEOtext 6. Geol Assoc Canada, St. John's, Newfoundland, pp 139–166

Büdel J (1957) Sie "Doppelten Einebnungsflächen" in der feuchten Tropen. Z Geomorph NF 1:201–228

Büdel J (1982) Climatic geomorphology. Princeton University Press, Princeton, NJ, 443

Buiter SJH, Steinberger B, Medvedev S, Tetreault JL (2012) Could the mantle have caused subsidence of the Congo Basin? Tectonophys 514–517:62–80

Burke K, Gunnell Y (2008) The African erosion surface: A continental-scale synthesis of geomorphology, tectonics and environmental change over the past 180 million years. Geol Soc Am Mem 201:66

Caen-Vachette M, Tempier P, Kamgang P (1987) Le massif tertiaire du Nkogam (Ouest Cameroun): carcatéristiques principales et géochronologie Rb/Sr sur roches totales. J Afr Earth Sci 6:521–524

Caen-Vachette M, Tempier P, Nana JM (1991) Le granite de Lembo (partie du complexe volcano-plutonique de Bana), témoin du magmatisme Tertiaire du Cameroun. Géochronologie Bull Soc géol Fr 162:497–501

Cahen D (1978) Vers une révision de la nomenclature des industries préhistoriques de l'Afrique centrale. L'Anthropologie (Paris) 82:5–36

Cahen L (1954) Géologie du Congo belge. Vaillant-Caramanne, Liège, 577 p

Cahen L, Ferrand JJ, Haarsma MJF, Lepersonne J, Verbeek T (1959) Résultats scientifiques des missions du syndicat pour l'étude géologique et minière de la Cuvette congolaise et travaux annexes. Géologie. Description du Sondage de Samba. Ann Musée roy Afr centrale - Tervuren, Belgique - Sci géol 29

Cahen L, Ferrand JJ, Haarsma MJF, Lepersonne J, Verbeek T (1960) Résultats scientifiques des missions du syndicat pour l'étude géologique et minière de la Cuvette congolaise et travaux annexes. Géologie. Description du Sondage de Dekese. Ann Musée roy Afr centrale - Tervuren, Belgique - Sci géol 34:32

Cahen L, Lepersonne J (1948) Notes sur la géomorphologie du Congo occidental. Ann Musée roy Afr centrale - Tervuren, Belgique - Tervuren (Belgique), Sér in 8°, Sci Géol 1:95

Cahen L, Lepersonne J (1952) Equivalence entre le Système du Kalahari du Congo belge et les Kalahari Beds d'Afrique australe. Mém Soc belge Géol Pal Hydr 8(4):64

Carvalho GS (1961) Um problema de geomorfologia aplicada. As possibilidades para a prospecçao de minérios de aluminio na Provincia de Angola. Bol Serv Geol Minas Angola 3:19–45

Chardon D, Chevillotte V, Beauvais A, Grandin G, Boulangé B (2006) Planation, bauxite and epeirogeny: one or two paleosurfaces on the West African margin? Geomorphology 82:273–282

Chorowicz J (2005) The East African rift system. J Afr Earth Sci 43:379–410

Claeys E (1947) Première étude de sables du Kalahari du Congo occidental. Bull Soc belge Géol Pal Hydr 56(3):372–382

Colin JP (1994) Mesozoic-Cenozoic lacustrine sediments of the Zaïre interior basin. In: Gierlowski-Kordesch E, Kelts K (eds) Global geological record of lake basins, vol 4, World and regional geology. Cambridge University Press, Cambridge, pp 31–36

Colin JP, Jan du Chêne R (1981) Paleontological study of the Esso/ Texaco wells Gilson-1 and Mbandaka-1. Zaïre, Esso Prod Res European, Tech Serv Rpt, Bordeaux

Crosby AG, Fishwick S, White N (2010) Structure and evolution of the intracratonic Congo Basin. Geochem Geophys Geosyst 11(6), Q06010. doi:10.1029/2009GC003014

de Heinzelin J (1952) Sols, paléosols et désertifications anciennes dans le secteur nord-oriental du bassin du Congo, vol 4. Publ Inst Nat Et Agro Congo Belge (INEAC), Bruxelles, p 168

de Heinzelin J (1962) La formation du Western Rift et de la cuvette angolaise. In: Proc 4th Pan African Congr Prehist, Leopoldville 1959. Ann Musée roy Afr centrale – Tervuren, Belgique - Sér in-8e, Sci Hum 40: 219–243

de Ploey J (1967) Position géomorphologique, genèse et chronologie de certains dépôts superficiels au Congo occidental. Quaternaria 7:131–154

de Ploey J (1969) Report on the Quaternary of the western Congo. Palaeoecol Afr 4:65–68

de Ploey J, Lepersonne J, Stoops G (1968) Sédimentologie et origine des sables de la série des sables ocre et de la série des "grès polymorphes" (Sytème du Kalahari) au Congo occidental. Ann Musée roy Afr centrale - Tervuren, Belgique, Sér in-8°, Sci géol 61:1–72

Decrée S, Deloule E, Ruffet G, Dewaele S, Mees F, Marignac C, Yans J, De Putter T (2010) Geodynamic and climatic controls in the formation of Mio-Pliocene world-class oxidized cobalt and manganese ores in the Katanga province, DR Congo. Mineralium Deposita 45:621–629

Déruelle B, Moreau C, Nkoumbou C, Kambou R, Lissom J, Njonfang E, Ghogomu RT, Nono A (1991) The Cameroon line: a review. In: Kampunzu AB, Lubala RT (eds) Magmatism in extensional structural settings. The Phanerozoic African Plate. Springer, Berlin Heidelberg New York, pp 274–327

Déruelle B, Ngounouno I, Demaiffe D (2007) The "Cameroon Hot Line" (CHL): A unique example of active akaline intraplate structure in both oceanic and continental lithospheres. C R Geosci 339:589–600, 339

Desthieux F (1993) Carte géologique de la République du Congo au 1:1.000.000. Ministère Mines Ener, Dir gén Mines, Rép Congo, Brazzaville, p 27

Dohrenwend JC (1994) Pediments in arid environments. In: Abrahams AD, Parsons AJ (eds) Geomorphology of desert environments. Chapman & Hall, London, pp 321–353

Dohrenwend JC, Parsons AJ (2009) Pediments in arid environments. In: Parsons AJ, Abrahams AD (eds) Geomorphology of desert environments. Springer Science + Business Media B.V, Berlin-Heidelberg, pp 377–411

Downey NJ, Gurnis M (2009) Instantaneous dynamics of the cratonic Congo basin. J Geophys Res 114, B06401. doi:10.1029/ 2008JB006066

Dresch J (1946) Notes de géomorphologie congolaise. Bull Assoc Géograph Franç 181–182:116–123

Dresch J (1950) Sur les pédiments en Afrique méditerranéenne et tropicale. Comptes rendus du Congrès international de Géographie, Union Géographique Internationale, Lisbonne 1949:19–28

Egoroff A, Lombard AL (1961–1962) Présence des couches de Stanleyville dans le sous-sol de Léopoldville. République du Congo - Noté préliminaire Ann Soc géol Belg 85:103–109

Embrechts J, de Dapper M (1990) Morphologie, genèse et sédimentologie des pédiments de versant de la région du Mont-Fébé (Cameroun méridional). In: Lanfranchi R, Schwartz D (eds) Paysages quaternaires de l'Afrique centrale atlantique. Off Rech Sci Tech Outre-Mer (ORSTOM), Paris, pp 138–154

Fair TJD (1948) Hill-slopes and pediments of the semi-arid Karroo. S Afr Geogr J 30:71–79

Feio M (1946) O relêvo de Angola, segundo Jessen. Bol Soc Geol Portugal 5:267–294

Fitton JG, Dunlop HM (1985) The Cameroon line, West Africa, and its bearing on the origin of oceanic and continental alkali basalt. Earth Planet Sci Lett 72:23–38

Fosso J, Ménard JJ, Bardintzeff JM, Wandji P, Tchoua FM, Bellon H (2005) Les laves du mont Bangou : une première manifestation volcanique éocène, à affinité transitionnelle, de la Ligne du Cameroun. C R Geosci 337:315–325

Gilbert GK (1877) Report on the geology of the Henry Mountains. US Geographical and Geological Survey of the Rocky Mountain Region. U.S. Dept Interior, Washington, DC, p 170

Giresse P (1990) Paleoclimatic and structural environment at the end of the Cretaceous along the western flank of the Congo Basin, with application of underground microdiamonds around Brazzaville. J Afr Earth Sci 10:399–408

Giresse P (2005) Mesozoic-Cenozoic history of the Congo Basin. J Afr Earth Sci 43:301–315

Giresse P, K'Vadec JP (1971) Lithologie stratigraphique et paléoclimatologie des formations néogènes et quaternaires de la Cuvette du Congo-Brazzaville. CR Somm Soc géol Fr 8:360–161

Grandin G (1976) Aplanissements cuirassés et enrichissements des gisements de manganèse dans quelques régions d'Afrique de l'Ouest. Mém Office Rech Sci Tech Outre-Mer (ORSTOM) 82:275

Grekoff N (1958) Ostracodes du Bassin du Congo. III. Tertiaire. Ann Musée Roy Congo belge – Tervuren, Belgique - Sér in-8°, Sci géol 22:1–36

Haq BU, Hardenbol J, Vail PR (1987) Chronology of fluctuating sea levels since the Triassic. Science 235:1156–1167

Hartley RW, Allen PA (1994) Interior cratonic basins of Africa: relation to continental break-up and role of mantle convection. Bas Res 6:95–113

Hiéronymus B (1973) Etude minéralogique et géochimique des formations bauxitiques de l'Ouest du Cameroun. Cahiers Office Rech Sci Tech Outre-Mer, sér Géol 5:97–112

Hiéronymus B (1977) Apport de la minéralogie et de la géochimie à la connaissance de la formation et de l'évolution de certaines bauxites du Cameroun. Trav Doc Géogr Trop, Centre Et Géogr Trop (CEGET), Talence, Bordeaux 33:167–184

Hiéronymus B (1983) Etude de l'altération de quelques basaltes dans l'ouest du Cameroun. C R 108e Cong nat Soc sav. Grenoble, Sci fasc 1:373–385

Humbel FX (1966) Contribution pédologique à l'étude géomorphologique de l'Adamaoua. Rapp Office Rech Sci TechOutre-Mer (ORSTOM), Yaoundé: 27 p. (http://horizon.documentation.ird.fr/exl-doc/pleins_textes/divers12-06/01204.pdf)

Jessen O (1936) Reisen und Forschungen in Angola. Verlag von Dietrich Reimer, Berlin, 397 p

Joly F (1950) Pédiments et glacis d'érosion dans le Sud-Est du Maroc. C R Cong inter Géogr, Union Géogr Int, Lisbonne 1949:110–125

Kadima EK, Sebazengi SMN, Lucazeau F (2011) A Proterozoic-rift origin of the structure and the evolution of the cratonic Congo basin. Earth Planet Sci Lett 304:240–250

Kampunzu AB, Bonhomme MG, Kanika M (1998) Geochronology of volcanic rocks and evolution of the Cenozoic Western Branch of the East African Rift System. J Afr Earth Sci 26:441–461

Kaseba MK, Papepe R, Boski T, Herbosch A, Naud J (1997) Genèse et évolution des profils bauxitiques latéritiques sur des surfaces d'aplanissements post-éocènes au Bas-Zaïre (Zaïre). Paper presented at the Colloque international Cornet "Gisements stratiformes de cuivres et minéralisations associées", Mons, 5-9 septembre 1994, Acad Roy Sci Outre-Mer, Bruxelles, pp. 409–442

King L (1949) The pediment landform: some current problems. Geol Mag 86:245–250

King LC (1953) Canons of landscape evolution. Bull Geol Soc Am 64:721–752

King LC (1962) The morphology of the Earth. Oliver & Boyd, London, 699 p

Kuepouo G, Tchouankoue JP, Nagao T, Sato H (2006) Transitional tholeiitic basalts in the Tertiary Bana volcano–plutonic complex, Cameroon Line. J Afr Earth Sci 45:318–332

Lasserre M (1961) Etude géologique de la partie orientale de l'Adamaoua (Cameroun central). Bull Dir Mines Géol, Cameroun, p 130

Lasserre M (1978) Mise au point sur les granitoïdes dits "ultimes" du Cameroun: gisement, pétrographie et géochronologie. Bull Bur Rech Géol Min (2ème sér), Sect IV (2):143–159

Lavier LL, Steckler MS, Brigaud F (2001) Climatic and tectonic control on the Cenozoic evolution of the West African margin. Mar Geol 178:63–80

Le Maréchal A (1966) Contribution à l'étude des plateaux Batékés (Géologie, géomorphologie, hydrogéologie). Office Rech Sci Tech Outre-Mer, Centre Brazzaville, Serv Géol, Rapp: 78 p. (http://horizon.documentation.ird.fr/exl-doc/pleins_textes/divers08-10/00743.pdf)

Leakey LSB (1949) Tentative study of the Pelistocene climatic changes and stone-age culture sequence in north-eastern Angola. Companhia Diamantes Angola (Diamang). Publ Culturais 4:1–82

Lepersonne J (1949) A propos des pénéplaines du Sud-Ouest du bassin du Congo et leurs formations superficielles. Bull Inst Roy Colon Belge 20:664–676

Lepersonne J (1956) Les aplanissements d'érosion du nord-est du Congo belge. Acad roy Sci colon, Cl Sci Nat Méd, Mém in-8°, Nouv sér 4 (7):108 p.

Lepersonne J (1958a) Les surfaces d'érosion des hauts plateaux de l'intérieur de l'Afrique centrale. - Analyse critique d'une étude de M. R.V. Ruhe. Bull Acad roy Sci colon. Nouv Sér 2:596–621

Lepersonne J (1958b) Mise au point concernant la note de M. R.V. Ruhe, intitulée : "Erosion surfaces of the Ituri, Belgian Congo. - Reply to J. Lepersonne". Bull Acad roy Sci colon. Nouv Sér 4:384–407

Lepersonne J (1974) Carte géologique du Zaïre, échelle 1 : 2.000.000 and Notive explicative de la carte géologique du Zaïre au 1/2 000 000. Dir Géol, Dépt Mines, Rép Zaïre, Kinshasa

Lepersonne J (1978) Structure géologique du bassin intérieur du Zaïre. Acad roy Sci Outre-Mer, Cl Sci Nat Méd SN 20(2):3–27

Leriche M (1925) Sur l'âge du calcaire lacustre observé récemment sur le plateau du Kundelungu (Katanga). Ann Soc Géol Belg 48: B128–B130

Leriche M (1926–1927) Les fossiles des "Grès polymorphes" (Couches de Lubilash), aux confins du Congo et de l'Angola. Ann Soc Géol Belg, Publ Relatives Congo Belge 50(2):45–51

Leriche M (1927) Les fossiles des "grès polymorphes" (Couches du Lubilash), aux confins du Congo et de l'Angola. Rev Zool Afr 15:403–409

Letterman M (1984) Etude géologique régionale du bassin d'Anlaoua (Cameroun). Thèse Doct 3ème cycle, Univ Orléans, 194 p.

Linol B (2013) Sedimentology and sequence stratigraphy of the Congo and Kalahari Basins of south-central Africa and their evolution during the formation and break-up of West Gondwana. PhD thesis, Nelson Mandela Metropolitan University, 375p.

Lubala RT, Kampunzu AB, Caron JPH, Vellutini PJ (1984) Minéralogie et pétrologie des basaltes saturés tertiaires du Kahuzi-Biega (Rift du Kivu, Zaïre). Ann Soc Géol Belg 107:125–134

Mabbutt JA (1955) Erosion surfaces in Namaqualand and the ages of surface deposits in the south-western Kalahari. Trans Geol Soc S Afr 58:13–29

Mabbutt JA (1961) Basal surface" or "weathering front. Proc Geologists' Assoc 72:357–358

Mainguet M (1975) Considérations sur les formes superficielles d'enrichissement en fer et le cuirassement des Grès de Mouka-Ouadda (RCA). Trav Doc Géogr Trop, Centre Et Géogr Trop (CEGET), Talence, Bordeaux 22:163–170

Mammerickx J (1959) Géologie et géomorphologie des monts Dipompa (Katanga). Bull Soc belge Géol Pal Hydro 58:411–417

Mammerickx J (1964) Pédiments désertiques et pédiments tropicaux. Acta Geogr Lovaniensia 3:359–370

Marques MM (1992) La cuirasse ferrugineuse conglomératique de Cabatuquila (Malanje - R. P. d'Angola) et sa signification paléogéographique. Ciênc Terra (Univ Nova Lisboa) 11:141–148

Marzoli A, Renne PR, Piccirillo EM, Francesca C, Bellieni G, Melfi AJ, Nyobe JB, N'ni J (1999) Silicic magmas from the continental Cameroon Volcanic Line (Oku, Bambouto and Ngaoudere): ^{40}Ar-^{39}Ar dates, petrology, Sr-Nd-O isotopes and their petrogenetic significance. Contrib Min Petrol 135:133–150

Maufe HB (1928–1929) Observations sur les calcaires silicifiés du Mont Bunza (Kasai) et sur la calcédoine du Kalahari de la Rhodésie du Sud. Ann Soc Géol Belg, Publ Relat Congo Belge 52(3):115–118

Mbuluyo MK (1991) Les principales entités géomorphologiques de l'Ituri oriental et les faciès cuirassés associés (Nord-Est du Zaire). Bull Soc géogr Liège 27:139–148

Miauton JD (1980) Bakouma - genèse et gitologie d'un gisement néoformé continental phosphato-uranifère. Thèse 3ème Cycle, Inst Nat Polytech Lorraine, Ecole Nat Sup Géol appl Prosp min, Nancy, 160 p.

Michel P (1973) Les bassins des fleuves Sénégal et Gambie, étude géomorphologique. Mém Office Rech Sci Tech Outre-Mer (ORSTOM) 63:752 p.

Migon P (2004a) Planation surfaces. In: Goudie AS (ed) Encyclopedia of geomorphology. Routledge, London, pp 788–792

Migon P (2004b) Etching, etchplain and etchplanation. In: Goudie AS (ed) Encyclopedia of geomorphology. Routledge, London, pp 345–347

Miller KG, Kominz MA, Browning JV, Wright JD, Mountain GS, Katz ME, Sugarman PJ, Cramer BS, Christie-Blick N, Pekar SF (2005) The Phanerozoic record of global sea-level change. Science 310:1293–1298

Millot G (1980) Les grands aplanissements des socles continentaux dans les pays subtropicaux, tropicaux et désertiques. In: Livre Jubilaire Soc géol Fr, vol 10. Mém h sér Soc géol Fr, pp 295-305

Millot G (1981) Weathering sequences. "Climatic" planations. Leveled surfaces and paleosurfaces. In: van Olphen H, Veniale F (eds) Proc VII Int Clay Conf, Bologna and Pavia, vol 35. Devt Sedimentology. Elsevier, Amsterdam, pp 585–593

Moeyersons J (1975) Les surfaces d'aplanissement et les cycles géographiques dans le Nord Zaïre. Ann Soc Géol Belg 98:439–448

Mortelmans G (1946) A propos de la présence, au Katanga central, de cailloux éolisés dans le conglomérat de base des "Grès polymorphes". Bull Soc belg Géol. Pal Hydro 55:220–228

Mortelmans G (1947) Préhistoire et Quaternaire du Sud du bassin du Congo. Paper presented at the La géologie des terrains récents dans l'Ouest de l'Europe, Session extraordinaire Soc belges Géol (Bruxelles), Gand, Septembre 1946, pp. 215–244

Mortelmans G (1953) Les antécédents tectoniques du Graben de l'Upemba (Katanga). Bull Volcanol 13:93–98

Moucha R, Forte AM (2011) Changes in African topography driven by mantle convection. Nature Geosci 4:707–712

Moundi A, Wandji P, Bardintzeff JM, Ménard JJ, Okomo Atouba LC, Mouncherou OF, Reusser E, Bellon H, Tchoua F (2007) Les basaltes éocènes à affinité transitionnelle du plateau Bamoun, témoins d'un réservoir mantellique enrichi sous la ligne volcanique du Cameroun. C R Geosci 339:396–406

Mouta F, Dartevelle E (1954) Sur les "Grès polymorphes" fossilifères de la plaine de la Cassanje (Malange, Angola) et leur âge. In: C R 19ème session Cong géol int Alger 1952, 20 (2ème part), 1954. pp 11–24

Mouyoungou J (1990) Les silicifications mésozoïques et cénozoïques de la bordure occidentale du bassin de Paris et de la région de Brazzaville au Congo. Univ Angers, Thèse Doct, 242 p

Nana JM (1988) Le complexe volcano-plutonique de Bana (Ouest Cameroun). Géologie et pétrologie. Thèse Doct, Univ Paris XI, Orsay, 131 p

Njonfang E, Nono A, Kamgang P, Ngako V, Tchoua FM (2011) Cameroon Line alkaline magmatism (central Africa): A reappraisal. In: Beccaluva L, Bianchini G, Wilson M (eds) Volcanism and Evolution of the African Lithosphere. Geol Soc Am, Spec Pap, 478: 173–191

Nkoumbou C (1990) Etude géologiques des Monts Roumpi: un ensemble plutonique et volcanique de la "ligne du Cameroun". Données pétrologiques sur les néphélinites du Mont Etinde (Cameroun). Thèse Doct 3ème cycle, Univ Nancy 1, Nancy, 352 p

Nono A, Déruelle B, Demaiffe D, Kambou R (1994) Tchabal Nganha volcano in Adamawa (Cameroon): petrology of a continental alkaline lava series. J Volcanol Geotherm Res 60:147–178

Novikoff A (1974) L'altération des roches dans le massif du Chaillu (République Populaire du Congo). Formation et évolution des argiles en zone ferrallitique. Thèse Doct 3ème cycle, Univ Strasbourg, 298 p

Oustrière P (1984) Etude géologique et géochimique du bassin lacustre d'Anlaoua (Cameroun). Application à la compréhension de la genèse de la vivianite. Thèse Doct 3ème cycle, Univ Orléans, 344 p

Pasteels P, Villeneuve M, De Paepe P, Klerkx J (1989) Timing of the volcanism of the southern Kivu province: implications for the evolution of the western branch of the East African Rift system. Earth Planet Sci Lett 94:353–363

Petit M (1990) Les grands traits morphologiques de l'Afrique centrale atlantique. In: Lanfranchi R, Schwartz D (eds) Paysages quaternaires de l'Afrique centrale atlantique. Office Rech Sci Tech Outre-Mer (ORSTOM), Paris, pp 20–30

Polinard E (1931–1932) Découverte de gisements fossilifères d'eau douce sur les versants de la Lubudi, au Katanga méridional. Ann Soc Géol Belg, Pub Relat Congo Belge 55:663–681

Polinard E (1954) A propos des couches inférieures du Système du Kalahari au Congo belge. Bull Soc belg Géol Pal Hydro 58:9–16

Rowley DB (2013) Sea level: Earth's dominant elevation—Implications for duration and magnitudes of sea level variations. J Geol 121:445–454

Ruhe RV (1954) Erosion surfaces of Central African interior high plateaus. Publ Inst Nat Et Agro Congo Belge (INEAC), Bruxelles, Sér sci 59:38 p

Ruhe RV (1956) Landscape evolution in the High Ituri Belgian Congo. Publ Inst Nat Et Agro Congo Belge (INEAC), Bruxelles, Sér scient 66:108 p

Ruhe RV (1958) Erosion surfaces of the Ituri, Belgian Congo. Reply to J. Lepersonne. Bull Acad roy Sci colon. Nouv Sér 4:360–383

Ruhe RV (1959) African erosion surfaces. Geol Mag 96:87–88

Schwarz T (1997) Distribution and genesis of bauxite on the Mambilla Plateau, SE Nigeria. Appl Geochem 12:119–131

Ségalen P (1967) Les sols et la géomorphologie du Cameroun. Cah Office Rech Sci Tech Outre-Mer, sér Pédo 2:137–187

Séranne M (1999) Early Oligocene stratigraphic turnover on the west Africa continental margin: a signature of the Tertiary greenhouse-to-icehouse transition? Terra Nova 11:135–140

Séranne M, Bruguier O, Moussavou M (2008) U-Pb single zircon grain dating of Present fluvial and Cenozoic aeolian sediments from Gabon: consequences on sediment provenance, reworking, and erosion processes on the equatorial West African margin. Bull Soc géol Fr 179:29–40

Sluys M (1946) Un gisement de latérite bauxitique pisoolitique sur le substratum granitique (Région de Niapu, Congo-Belge). Bull Soc Géol Belg 69:B218–B220

Tardy Y, Roquin C (1998) Dérive des continents, paléoclimats et altérations tropicales. Ed Bur Rech Géol Min (BRGM), Orléans, p 473

Tator BA (1952) Pediment characteristics and terminology. Ann Am Assoc Geographers 42:295–317

Tator BA (1953) Pediment characteristics and terminology. Part II: Terminology. Ann Am Assoc Geographers 43:47–53

Temdjim R, Njilah IK, Kamgang P, Nkoumbou C (2004) Données nouvelles sur les laves felsiques de Ngaoundere (Adamaoua, Ligne du Cameroun): chronologie K-Ar et pétrologie. Afr J Sci Technol, Sci Eng Ser 5:113–123

Thiéblemont D (2013) Evidence for an aeolian origin of the Holocene lateritic surface cover of Gabon (Central Africa). Quat Int 296:176–197

Thiéblemont D, Flehoc C, Ebang-Obiang M, Rigollet C, Prian JP, Prognon F (2013) Geochronological arguments for a close relationship between surficial formation profiles and environmental crisis (c. 3000–2000 BP) in Gabon (Central Africa). C R Géosci 345:272–283

Thiry M (1999) Diversity of continental silicification features: examples from the Cenozoic deposits in the Paris Basin and neighbouring basement. In: Thiry M, simon-Coinçon R (eds) Palaeoweathering, palaeosurfaces and related continental deposits, vol 27. Spec Publ Int Assoc Sedimentologists, pp 87–127

Thiry M, Maréchal B (2001) Development of tightly cemented sandstone lenses in uncemented sand: Example of the Fontainebleau Sand (Oligocene) in the Paris Basin. J Sedim Res 71:473–483

Thiry M, Ribet I (1999) Groundwater silicification in Paris Basin limestones: fabrics, mechanisms and modeling. J Sed Res 69:171–183

Thomas MF (1989a) The role of etch processes in landform developments: I. Etching concepts and their applications. Z Geomorph NF 33:129–142

Thomas MF (1989b) The role of etch processes in landform developments: II. Etching and the formation of relief. Z Geomorph NF 33:257–274

Tshidibi NB (1986) Le ciment silicieux dans les Grès polymorphes du plateau des Biano (Shaba-Zaïre) provient en partie de la silicification d'une matrice carbonatée. Ann Soc géol Belg 109:579–585

Twidale CR, Bourne JA (2013) Pediments as etch forms: implications for landscape evolution. J Geol 121:607–622

van Kerschaver GC (1975) Répercussions des paléoclimats sur la morphologie de la région de Kimpese. Z Geomorph, NF 19:329–333

van Kerschaver G (1983) Géomorphologie paléoclimatique de la fin du Cénozoïque dans la région de Kimpese (Bas-Zaire). Bull Soc belge Géol 92:301–309

van Kerschaver G (1984a) Géomorphologie paléoclimatique dans le Bas-Zaïre (Afrique centrale). Serv Géol Belg, Prof Pap, Admin Mines 207:52–53

Van Kerschaver G (1984b) Ferruginisations in the region of Kimpese (Lower-Zaire). Geo-Eco-Trop 8:7–24

Vanderstappen R, Cornil C (1955) Note sur les "bauxites" du Congo septentrional. Bull Acad Roy Sci Colon, Nouv Sér 1(4):690–709

Vanderstappen R, Verbeek T (1964) Analcime et minéraux argileux des formations géologiques de la Cuvette congolaise (République du Congo). Ann Musée roy Afr centrale - Tervuren, Belgique, Sér in-8°, Sci géol 47:1–88

Veatch AC (1935) Evolution of the Congo Basin. Geol Soc Am Mem 3: 183

Vogt J (1962) Notes de Géomorphologie gabonaise. Rev Géomorph Dyn 13:161–169

Wandji P, Seuwui DT, Bardintzeff JM, Bellon H, Platevoet B (2008) Rhyolites of the Mbépit Massif in the Cameroon Volcanic Line: an early extrusive volcanic episode of Eocene age. Mineral Petrol 94:271–286

Whitaker CR (1979) The use of the term "pediment" and related terminology. Z Geomorph NF 23:427–439

White K (2004) Pediment. In: Goudie AS (ed) Encyclopedia of geomorphology. Routledge, London, pp 768–771

Yongue-Fouateu R, Ghogomu RT, Penaye J, Ekodeck GE, Stendal H, Colin F (2006) Nickel and cobalt distribution in the laterites of the Lomié region, south-east Cameroon. J Afr Earth Sci 45:33–47

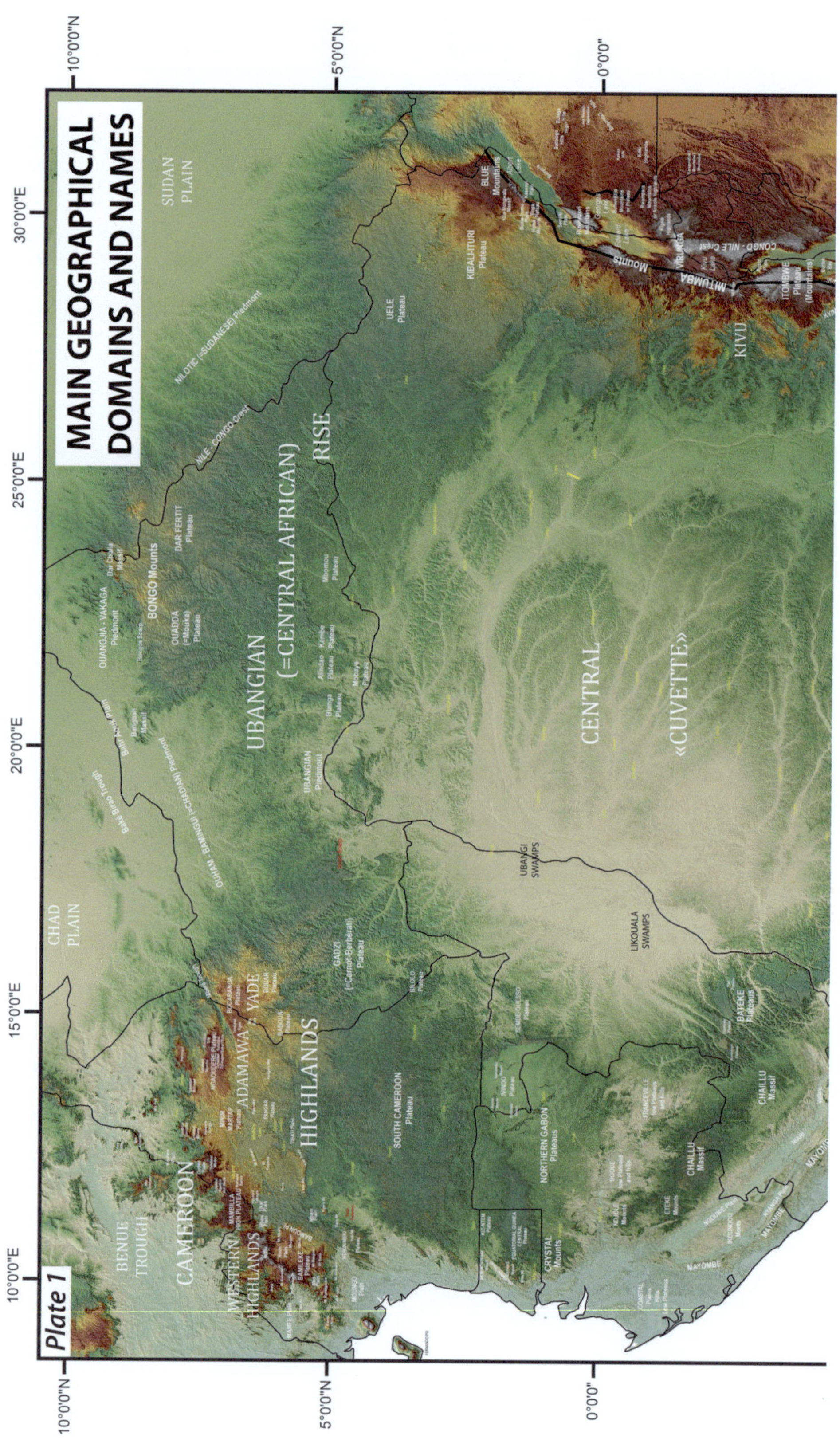
MAIN GEOGRAPHICAL DOMAINS AND NAMES
Plate 1
10°0'0"E
15°0'0"E
20°0'0"E
25°0'0"E
30°0'0"E
10°0'0"N
5°0'0"N
0°0'0"
SUDAN PLAIN
CHAD PLAIN
BENUE TROUGH
CAMEROON
WESTERN HIGHLANDS
ADAMAWA
YADE
HIGHLANDS
UBANGIAN
(=CENTRAL AFRICAN)
RISE
CENTRAL
«CUVETTE»
KIVU
NILOTIC (=SUDANESE) Piedmont
NILE - CONGO Crest
OUANGUIA - VAKAGA Piedmont
BONGO Mounts
DAR FERTIT Plateau
OUADDA (=Mouka) Plateau
UBANGIAN Piedmont
UELE Plateau
KIBALI-ITURI Plateau
BLUE Mountains
MITUMBA Mounts
CONGO - NILE Crest
GADZI (=Carnot-Berbérati) Plateau
SOUTH CAMEROON Plateau
NORTHERN GABON Plateaus
CRYSTAL Mounts
CHAILLU Massif
UBANGI SWAMPS
LIKOUALA SWAMPS
BATEKE Plateaus
MAYOMBE

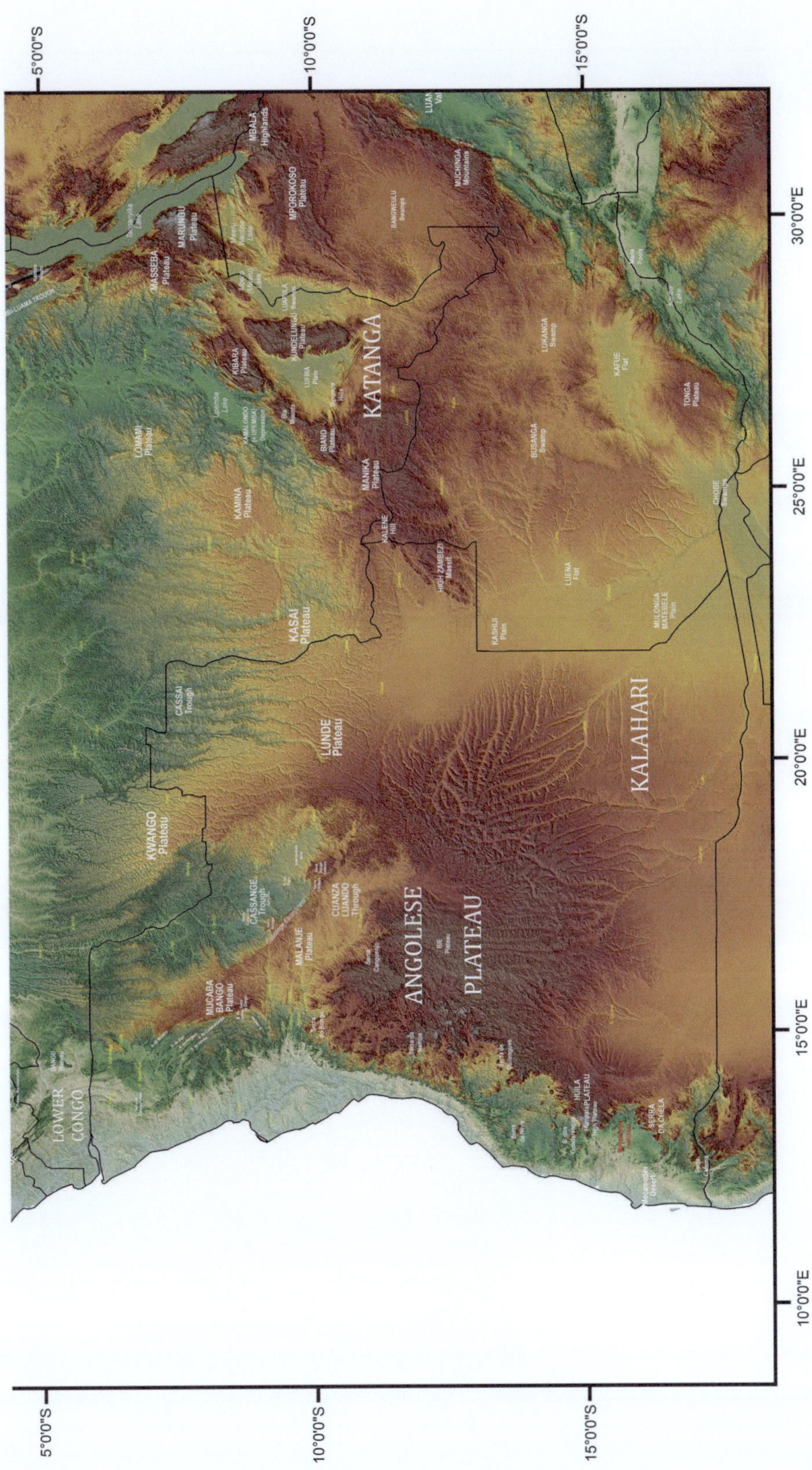
5°0'0"S
10°0'0"S
15°0'0"S
10°0'0"E
15°0'0"E
20°0'0"E
25°0'0"E
30°0'0"E
KATANGA
KALAHARI
ANGOLESE
PLATEAU
LOWER
CONGO
KWANGO
Plateau
CASSAI
Trough
LUNDE
Plateau
KASAI
Plateau
KAMINA
Plateau
LOMAMI
Plateau
MANIKA
Plateau
KUNDELUNGU
Plateau
MPOROKOSO
Plateau
MBALA
Highlands
MARUNGU
Plateau
MASSEBA
Plateau
MUCHINGA
Mountains
HIGH ZAMBEZI
Massif
KASHIJI
Plain
LUENA
Flat
MULONGA
MATEBELE
Plain
BUSANGA
Swamp
LUKANGA
Swamp
KAFUE
Flat
TONGA
Plateau
CHOBE
Swamps
CASSANGE
Trough
CUANZA
LUANDO
Through
MALANJE
Plateau
MUCABA
BANGO
Plateau
HUILA
PLATEAU
SERRA
DA CHELA

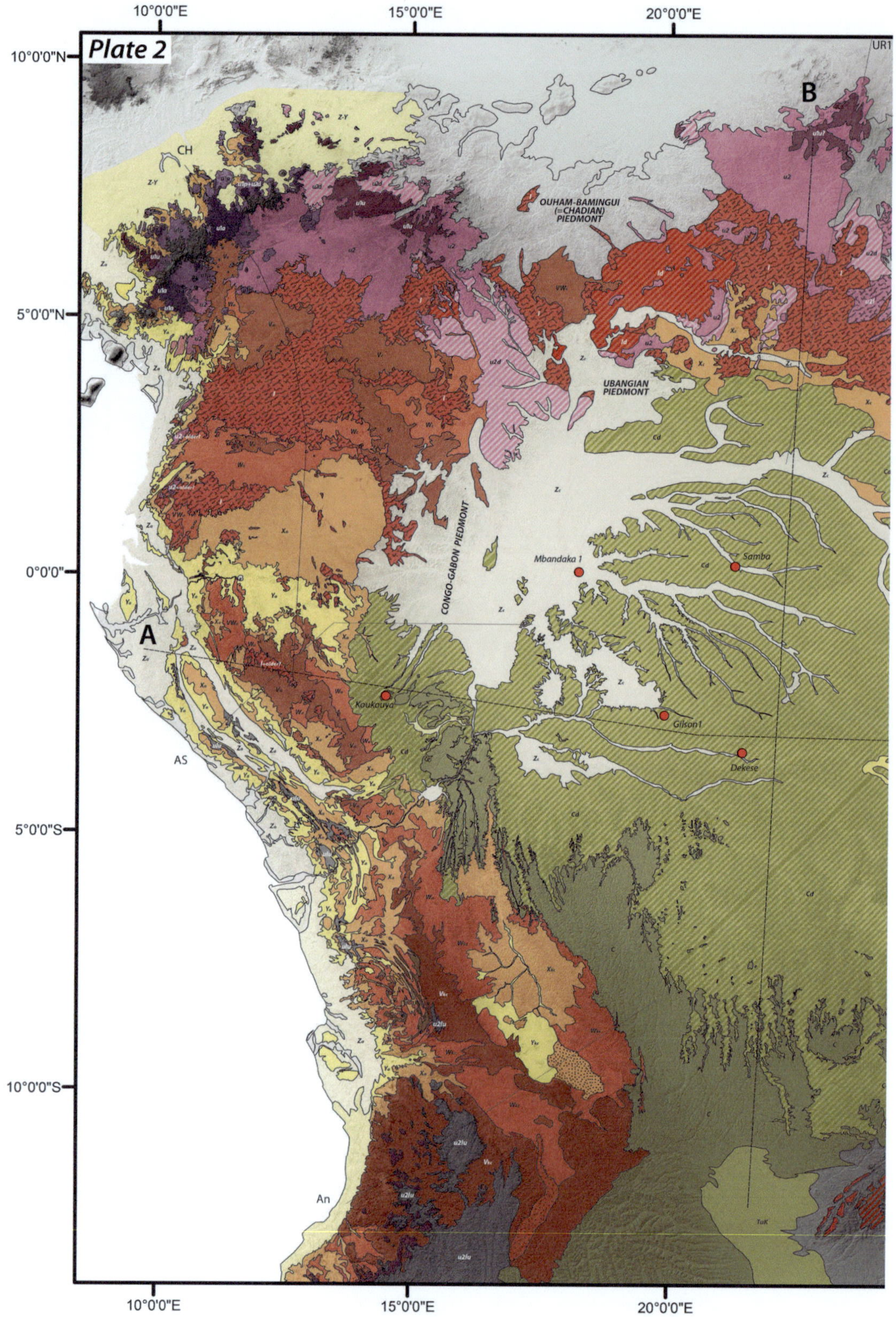
Plate 2
10°0'0"E
15°0'0"E
20°0'0"E
10°0'0"N
5°0'0"N
0°0'0"
5°0'0"S
10°0'0"S
A
B
CH
AS
An
UR1
OUHAM-BAMINGUI
(=CHADIAN)
PIEDMONT
UBANGIAN
PIEDMONT
CONGO-GABON PIEDMONT
Mbandaka 1
Samba
Koukouya
Gilson1
Dekese

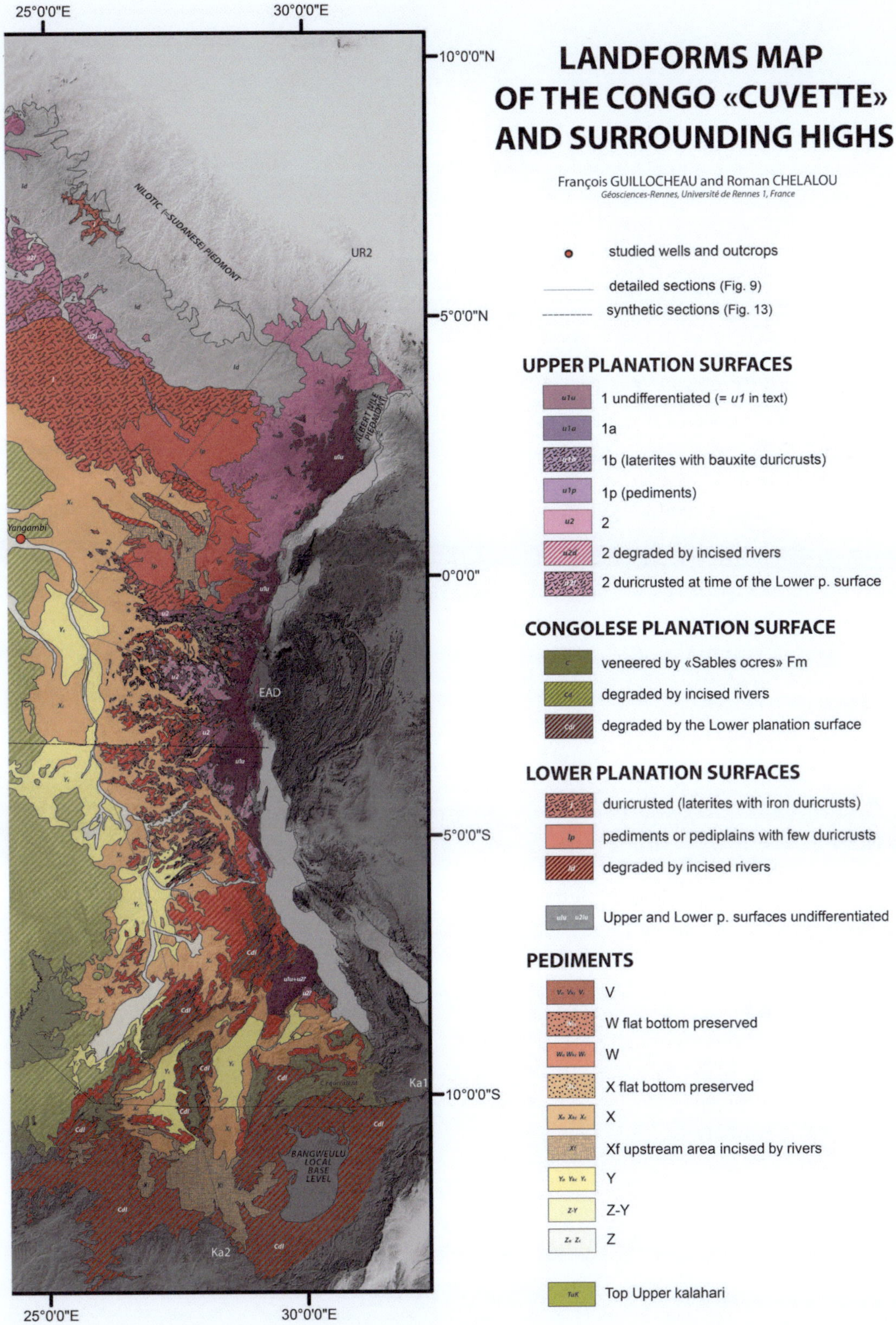
25°0'0"E
30°0'0"E
10°0'0"N
5°0'0"N
0°0'0"
5°0'0"S
10°0'0"S
NILOTIC (=SUDANESE) PIEDMONT
UR2
ALBERT NILE PIEDMONT
Yangambi
EAD
Ka1
Ka2
BANGWEULU LOCAL BASE LEVEL
LANDFORMS MAP
OF THE CONGO «CUVETTE»
AND SURROUNDING HIGHS
François GUILLOCHEAU and Roman CHELALOU
Géosciences-Rennes, Université de Rennes 1, France
studied wells and outcrops
detailed sections (Fig. 9)
synthetic sections (Fig. 13)
UPPER PLANATION SURFACES
1 undifferentiated (= u1 in text)
1a
1b (laterites with bauxite duricrusts)
1p (pediments)
2
2 degraded by incised rivers
2 duricrusted at time of the Lower p. surface
CONGOLESE PLANATION SURFACE
veneered by «Sables ocres» Fm
degraded by incised rivers
degraded by the Lower planation surface
LOWER PLANATION SURFACES
duricrusted (laterites with iron duricrusts)
pediments or pediplains with few duricrusts
degraded by incised rivers
Upper and Lower p. surfaces undifferentiated
PEDIMENTS
V
W flat bottom preserved
W
X flat bottom preserved
X
Xf upstream area incised by rivers
Y
Z-Y
Z
Top Upper kalahari

The Present Day Drainage Patterns of the Congo River System and their Neogene Evolution 15

Tyrel J. Flügel, Frank D. Eckardt, and Fenton P.D. Cotterill

"A river or a drainage basin might best be considered to have a heritage rather than an origin. It is like an organic form, the product of a continuous evolutionary line through time" p. 421 Leopold et al. (1964)

15.1 Introduction

The immense *ca.* 3.67 million km^2 Congo Basin (CB) is an intracontinental basin occupying central Africa. This roughly circular basin stretches across 22° of longitude (from ~12°E to ~34°E) and 21° of latitude (from ~9°N to~13°S) (Figs. 15.1 and 15.2). The basin's unique location straddling the Equator and its size and shape allows the Congo River (CR) to maintain a near constant flow year round: its immense size and exposure to continuous precipitation results in an annual discharge of 1,250 × 10^9 m^3 (Meade 1996). This discharge makes the CR Africa's largest river and the world's second largest river in terms of volume, only being surpassed by the Amazon with a discharge 6,300 × 10^9 m^3 (Meade 1996). The CR is the culmination of a complex and extensive drainage network that exhibits localised drainage patterns and controls. As there appears to be no clear correlation between the present day solid loads of the large river systems and the climatic conditions, Pinet and Souriau (1988) consider that relief is the major controlling factor of the development of the drainage systems of the CB. The drainage network of the Congo River System (CRS) suggests a multi-stage and sometimes interlinked development of the basins river systems (Figs. 15.3 and 15.4). As rivers are one of the major drivers of topographic change, a better understanding of their evolution will provide insights into the evolution of the broader landscape of the CB.

The river systems of the CB are described in terms of their present day drainage pattern and fluvial evolution in order to better understand the geomorphic evolution of the basin during the Neogene. Here, the term CB refers to the hydrographic network of the CRS and its associated landscapes. This network incorporates the geologically delimited central Congo Basin and extends eastward to include the Lake Tanganyika and Malagarasi systems (Figs. 15.1, 15.2 and 15.3). Ultimately, the water from the CRS flows into the Atlantic Ocean via the CR mouth near Monanda (Fig. 15.1c). The discharge of the CR is connected to an active deep sea fan by the 1,135 km long, meandering and deeply incised submarine Congo Canyon (Fig. 15.2; Babonneau et al. 2002). The outflow and sedimentary load of the CR comprises of integrated chemical and lithological inputs from runoff across a diverse range of climatic and tectonic terranes of its catchments (Dupré et al. 1996; Lavier et al. 2001). The CR exports a total of 87 × 10^6 t of matter per a year (Laraque et al. 2009), with a mechanical and chemical erosion rate of 8 t/km^2 yr^{-1} and 5 t/km^2 yr^{-1} respectively, as measured at Pool Malebo (Stanley Pool) (Fig. 15.5; Gaillardet et al. 1995).

The high rainfall associated with the Africa Equatorial region also sustains the world's second large continuous forest covering a *ca.* 2.8 million km^2 area that stretches from ~5°S to ~4°N (Fig. 15.1b). This Equatorial forest supports exceptionally rich biodiversity with the highest species richness in Africa (WWF 2006) and is bisected by the CR. Both the basin's extent and its vast, dense forests have impeded detailed geomorphic investigations. This is especially true of the central basin (*Cuvette Central*) landscapes, and much of the existing data and chronology has been derived from studies closer to the basin's peripheries. The study of the rivers that flow through the

T.J. Flügel (✉) • F.D. Eckardt
Department of Environmental and Geographical Science, Upper Campus, University of Cape Town, Private Bag X3, Rondebosch, Cape Town 7700, South Africa
e-mail: tyrelflugel@gmail.com; frank.eckardt@uct.ac.za

F.P.D. Cotterill
AEON Geoecodynamic Research Hub, Department of Botany and Zoology, Stellenbosch University, Private Bag X1, Matieland 7602, Stellenbosch, South Africa
e-mail: fcotterill@gmail.com

M.J. de Wit et al. (eds.), *Geology and Resource Potential of the Congo Basin*, Regional Geology Reviews,
DOI 10.1007/978-3-642-29482-2_15, © Springer-Verlag Berlin Heidelberg 2015

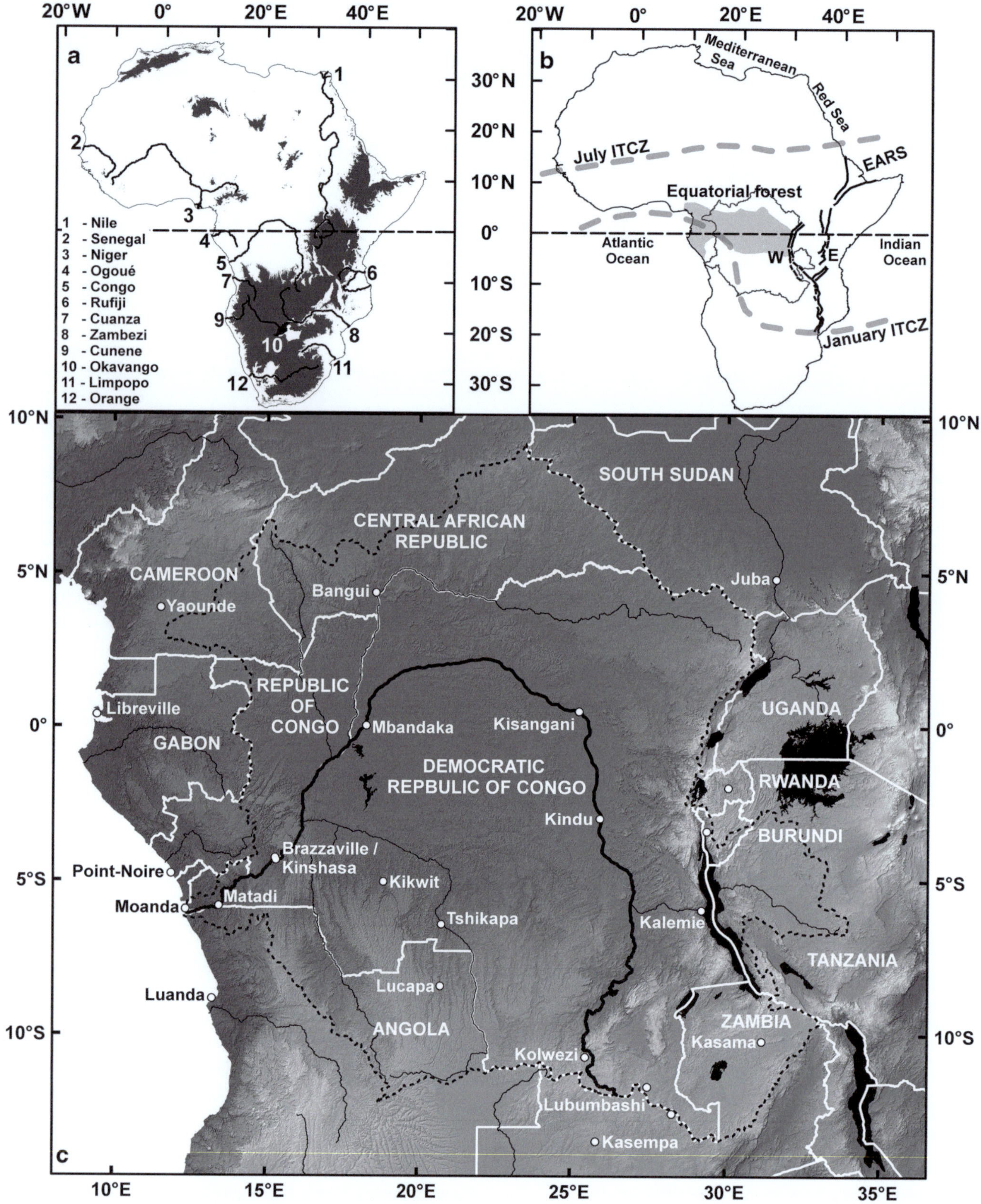

Fig. 15.1 The geographical setting of the Congo River System (CRS). (A) Illustrates the CRS in relation to several other large African rivers (*numbered*). The *grey* shading indicates elevations above 1,000 m.a.s.l., highlighting the biomodal topography of Africa. (B) The extent of the Congo Basin (*dark grey line*) and the Equatorial forest (*light grey* shading) in central Africa. Central Africa is an area of high rainfall, which is related to the movement of the Inter-Tropical Convergence Zone (ITCZ); the maximum extent of which is indicated by the *dashed grey lines*. The East African Rift System (EARS) forms two branches: the Eastern (E) with the Western (W) branch forming part of the eastern boundary of the Congo Basin. (C) The Congo Basin (*dashed black line*) straddles the Equator and measures *ca.* 2,600 km at its longest and *ca.* 2,400 km at its widest. The CB extends beyond the Democratic Republic of Congo into Gabon, Cameroon, Republic of the Congo, Central African Republic, Rwanda, Burundi, Tanzania, Zambia and Angola. Selected town (*white circles*) and countries (*heavy white lines*) are named. Major rivers and waterbodies are shown in solid *black*. Note that the Congo River crosses the Equator twice

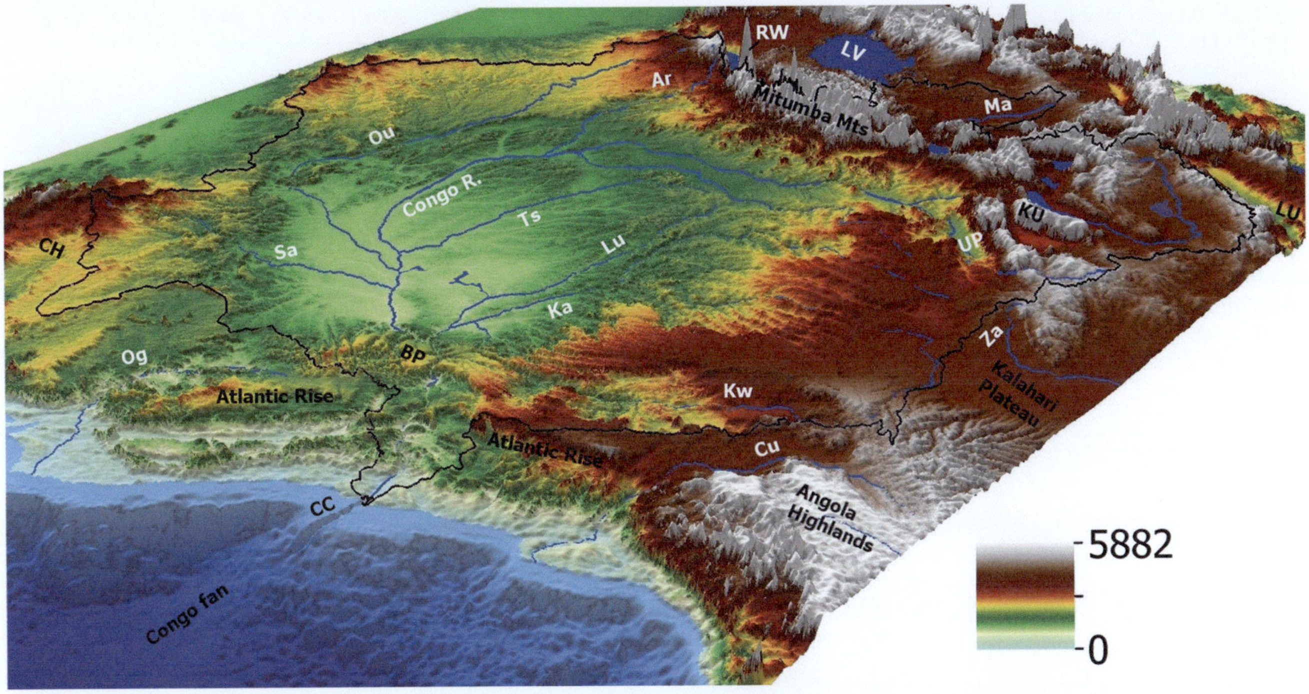

Fig. 15.2 A north-east facing terrain view of central Africa, highlighting major features of the CB. The limit of the hydrographic CB basin is shown (*heavy black line*). Notable rivers are—*Ar* Aruwimi, *Cu* Cuanza, *Ka* Kasai, *Kw* Kwango, *Lu* Lukemie, *Ma* Malagarasi, *Og* Ogoué, *Ou* Oubangui, *Sa* Sangha, *Ts* Tshuapa and Za Zambezi. Major landforms labelled include—CC Congo Canyon, CH Cameroon Highlands, KU Kundelungu Plateau, LU Luangwa Valley, LV Lake Victoria, RW Rwenzori Mountains, UP Upemba trough, BP Batéké Plateau (generated from: ETOPO1 (Amante and Eakins 2009) and SRTM DEM V4 250 m (Reuter et al. 2007; Jarvis et al. 2011)

Equatorial forest, not only the CR and its large tributaries (e.g. Tshuapa) reveal insights into the development of these central basin landscapes. Understanding the interplay between tectonically induced topography and geomorphic responses of fluvial networks can reveal where and when important events occurred in fluvial evolution (Karner and Driscoll 1999; Burbank and Anderson 2001).

Recent advances in remote sensing and availability of digital geospatial datasets makes it feasible to study the CRS in greater detail. In particular the digital surface model (DSM) derived from the Shuttle Radar Topography Mission (SRTM) (Kobrick 2006; Reuter et al. 2007) has enabled the characterisation of the large scale geomorphology of the CB (Figs. 15.2 and 15.3). Utilising this dataset in conjunction with biological evidence (phylogeographic data), we have analysed aspects of the Neogene evolution of the CRS.

15.2 The Congo Basin

15.2.1 Overview

The crescent shaped CR is the only large river in the world to cross the Equator twice (Fig. 15.1), ensuring that some sector of the river always experiences a wet season. The present day climate of the CB is dominantly tropical (hot and humid) around the equatorial region. As the basin straddles the north-south global climatic zone, the seasonal movement of the Inter-Tropical Convergence Zone (ITCZ) is the dominant controlling factor of rainfall in the basin (Fig. 15.1b). The basin thus experiences the northern hemisphere wet season from April to September that maintains the outflow of the southerly flowing tributaries (e.g. the Oubangui River), with northerly flowing tributaries of the southern catchment having high flows during the southern wet season, from October to May. The equatorial basin interior receives 1,600–2,400 mm of rainfall per a year (Nicholson 2000; Runge 2007). A large portion of the northern central and eastern CB north Equator has no dry season, while the rest of the basin experiences two wet and dry season; maximum rainfall occurs over November—December with the rainy season being 7–12 months long (Nicholson 2000). This results in the CR having a double discharge peak flow regime, although there is a limited water level fluctuation in Malebo Pool (Runge 2007). The larger peak occurs during November–January, and a smaller, second peak occurs during April–June (Laraque et al. 2009). The southern highlands (ranging from *ca.* 850 to a maximum of *ca.* 1,400 m.a.s.l) are cooler and drier compared to the central basin, with the highlands (*ca.* 900 m to maximum of *ca.* 3,000 m.a.s.l) in the east being cool and wet. Since the

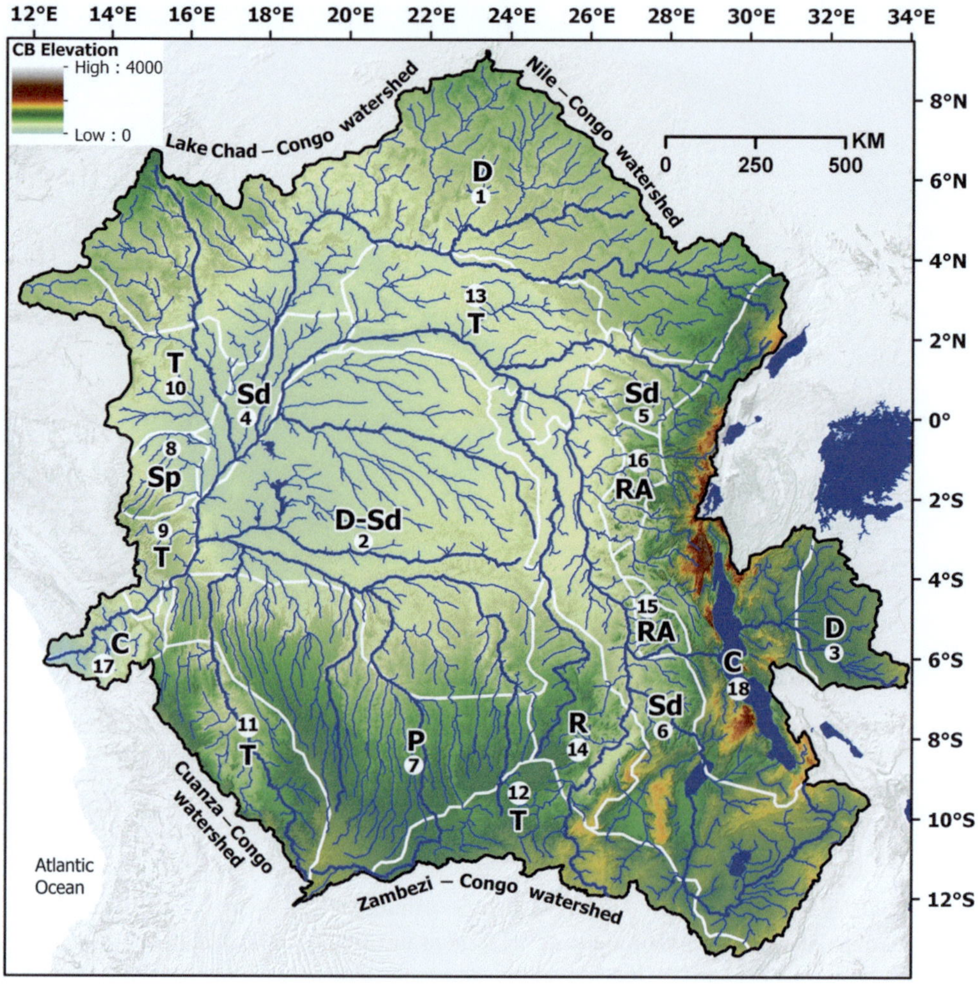

Fig. 15.3 Present day drainage patterns of the CB, elevations and major drainage patterns including *C* contorted, *T* trellis, *D* dendritic, *Sd* sub-dendritic, *P* parallel, *Sp* sub-parallel, *R* rectangular, *RA* rectangular-angulate (*white lines*). Refer to Table 15.1 for pattern descriptions and Table 15.2 for drainage pattern classification. River courses generated from the SRTM DEM V4 250 m (Reuter et al. 2007 and Jarvis et al. 2011)

beginning of the twentieth Century the CR experienced a near constant flow (Laraque et al. 2001). During the 1960s, the CR recorded a discharge exceeding the previous 40 years average, which was followed by a 10 % drop in its interannual discharge in the 1990s which may be related to the high degree of variability of rainfall over central Africa over decadal and centennial time scales (Nicholson 2000; Laraque et al. 2001). Owing to differential denudation between high and low relief of the CB, development in the drainage network will occur different rates throughout the basin (Pinet and Souriau 1988).

The first order geomorphology of the CB can be attributed the interplay between the geodynamics that have led to Africa's bimodal topography (Fig. 15.1a), controls of tectonics (uplift and rifting) and surface processes. The latter have resulted in incision and erosion of valleys around the basin margins, with concomitant deposition of thick lacustrine and fluvial sediments and autogenic river rearrangements, sometimes across drainage divides (Fig. 15.2). For example, during the Palaeogene there was extensive reworking, transport and deposition of sediments within Africa, with limited erosion of the cratonic areas, as evidenced by widely distributed Palaeogene sediments over continental Africa (Seranne et al. 2008). These surface processes, in turn are linked to climatic factors, with past effects of this climatic signal being preserved in the sedimentary record (e.g. Linol et al., this volume, Chap. 11, this Book). The present day CRS preserves evidence of important geomorphic and tectonic events. These events include breaching of pre-existing barriers (pertinently the Atlantic Rise) and expansion of the CRS, especially along its southern margins by drainage capture events. The latter have been most important in significantly increasing the total area of the southern catchment.

Evidence of first order events of the CB's Cenozoic development, including fluvial changes and climatic and geodynamic signals are indicated in the offshore terrigenous sedimentary evidence (Lavier et al. 2001). The Oligocene saw large amounts of sediments transported into the Lower Congo Basin, a consequence of late Cenozoic uplift and formation of the Congo River, forming the Cenozoic Congo deep sea fan (Anka and Séranne 2004; Seranne et al. 2008; Anka et al. 2010; Linol et al., Chap. 10, this book). The Neogene saw an increase of terrigenous sediment deposition in the coastal zone (Lavier et al. 2001) that was likely a result of due to a period of sustained uplift in the Miocene, with sediment supply being further enhanced by a climatic shift (Lavier et al. 2001; Seranne et al. 2008). During the recent past the region was drier during the late Pliocene—Pleistocene and based on sedimentary evidence it is thought that the rivers have remained relatively stable throughout the Quaternary (Guillocheau et al., Chap. 14, this book).

15.2.2 Regional Geomorphological Structure

Since the break-up of Gondwana, Africa's principal drainage systems have undergone substantial rearrangements (Goudie 2005). Details of these rearrangements for the central Africa are poorly known. While much of the Congo Basin's drainage may post-date the break-up of Gondwana, the drainage incorporates many older, inherited structures that have their origins in the assembly of Gondwana. Some of these structures, such as cratons and lithospheric fabrics, continue to maintain a dominant, regional, controlling influence over the geomorphology. This includes consideration of Mesozoic and Cenozoic events, especially where the basin's structural controls persist. However, it appears that events in the Neogene have had the most discernible impacts on the geomorphology of the CB. This is especially true where the eastern Congo Basin has been, and is currently, heavily impacted by the Western branch of the East African Rift System (EARS) and its incipient south-western extension, especially in the Katanga region of south-eastern Democratic Republic of Congo (DRC; Fig. 15.1b, c). Untangling the complex nature of the CRS represented in all the forms and features that comprise the basin, requires an appreciation of their context and interrelationships. Below we summarize the main evidence that allow a synthesis of the Cenozoic evolutionary history of the CB. This evidence comes mainly from the offshore sedimentary record and regional geomorphological structure.

The deep sea Congo Fan, centred on the modern CR outlet, is composed of two units, a Cretaceous and mid- to late-Cenozoic (Oligocene to Present) unit, suggesting two cycles of offshore deposition (Anka et al. 2010). Although the exact palaeo-coastal drainage is unknown, as is the age and nature of the initial canyon, the Congo's outlet has remained fairly stable since Late Cretaceous, although sediment deposition has not been constant (Anka et al. 2010). For example, an early Cenozoic sedimentary hiatus on the Congo Fan was ended by the establishment of a CR-Atlantic Ocean connection during the mid Cenozoic (Anka et al. 2010; see Linol et al., Chap. 10, this book). A steady increase in offshore sediment depositions since the Neogene to the Present suggest an ongoing phase of incision of the Cretaceous–Paleogene sedimentary sequence within the CB and consequent exposure of the cratonic basement (Seranne et al. 2008; Anka et al. 2009). Thus, during the Cenozoic, the CB has undergone a period of reworking followed by a period of increasing incision, a model which fits with the increase of terrigenous sedimentation in the continental passive margin off Gabon-Congo-Angola (equatorial Africa) and the growth of the deep sea fan in the Neogene (Anka and Séranne 2004; Anka et al. 2010). This temporal evolution of erosion-transport-deposition has been correlated with dramatic climatic changes in the Cenozoic that are thought to have resulted in increased mechanical erosion which correlates with increased terrigenous material being transported to the margin observed in the Neogene (Lavier et al. 2001; Seranne et al. 2008).

Presently the CB ranges in elevation, from sea level at the Congo River's outlet near Moanada (downstream of Matadi; Fig. 15.1c), to heights in excess of 3,000 m.a.s.l in the Mitumba Mountains (in the south-east) with the Rwenzori Range, one of Africa's highest mountain ranges (exceeding 5,000 m.a.s.l) lying just outside of the CRS (Fig. 15.2). Overall the Basin has a mean elevation of ~900 m.a.s.l. The *Cuvette Central* and northern areas of the basin are low regions (*ca.* 290–400 m.a.s.l) with subdued relief (Figs. 15.2 and 15.3). They are bounded to the south by dissected watershed, comprised principally of the Angolan Highlands, the northern limit of the Kalahari Plateau in the south and margins of the EARS (Figs. 15.1, 15.2, 15.3 and 15.5). This Western Branch of the EARS runs the entire length (*ca.* 2,100 km) of the eastern margin of the Congo Basin (Fig. 15.1b). This rugged relief consist of a series of elongated, narrow rift valleys and scarps extending from Lake Albert (~3°N) in the north to Lake Malawi (10°S). In the west, Congo basin is bounded by the Atlantic Rise (also known as the Western Escarpment and Monts Crystal), which has been deeply incised by the channel of the lower Congo. The Basin's northern extent is delimited by watersheds shared with the Ogoué River in the west, including parts of the Cameroon Highlands which extends eastward to form the Asande Rise (North Equatorial Plateau) in the north east zone of the CB.

The western and eastern highlands of the Congo Basin, being of tectonic origin (the West Congo Orogen and EARS respectively), exhibit high topographic roughness. The northern watershed incorporates parts of the Central African mobile belt (see de Wit and Linol, Chap. 2 this book). The Basin's highlands are dominated by the plateau, as exemplified by the Congo-Zambezi watershed (Fig. 15.3; Dixey 1943), which corresponds to a change of Africa's bimodal topography, from the low Congo basin (*ca.* 400 m.a.s.l) to the high Kalahari Plateau (*ca.* 1,100 m.a.s.l; Fig. 15.6a–c). The Congo River receive significant inflow in the *Cuvette Centrale* from several large tributary rivers. These systems include the Sanhga and Oubangi from the north and the Kasai (Kasai—Kwango—Lukemie) and

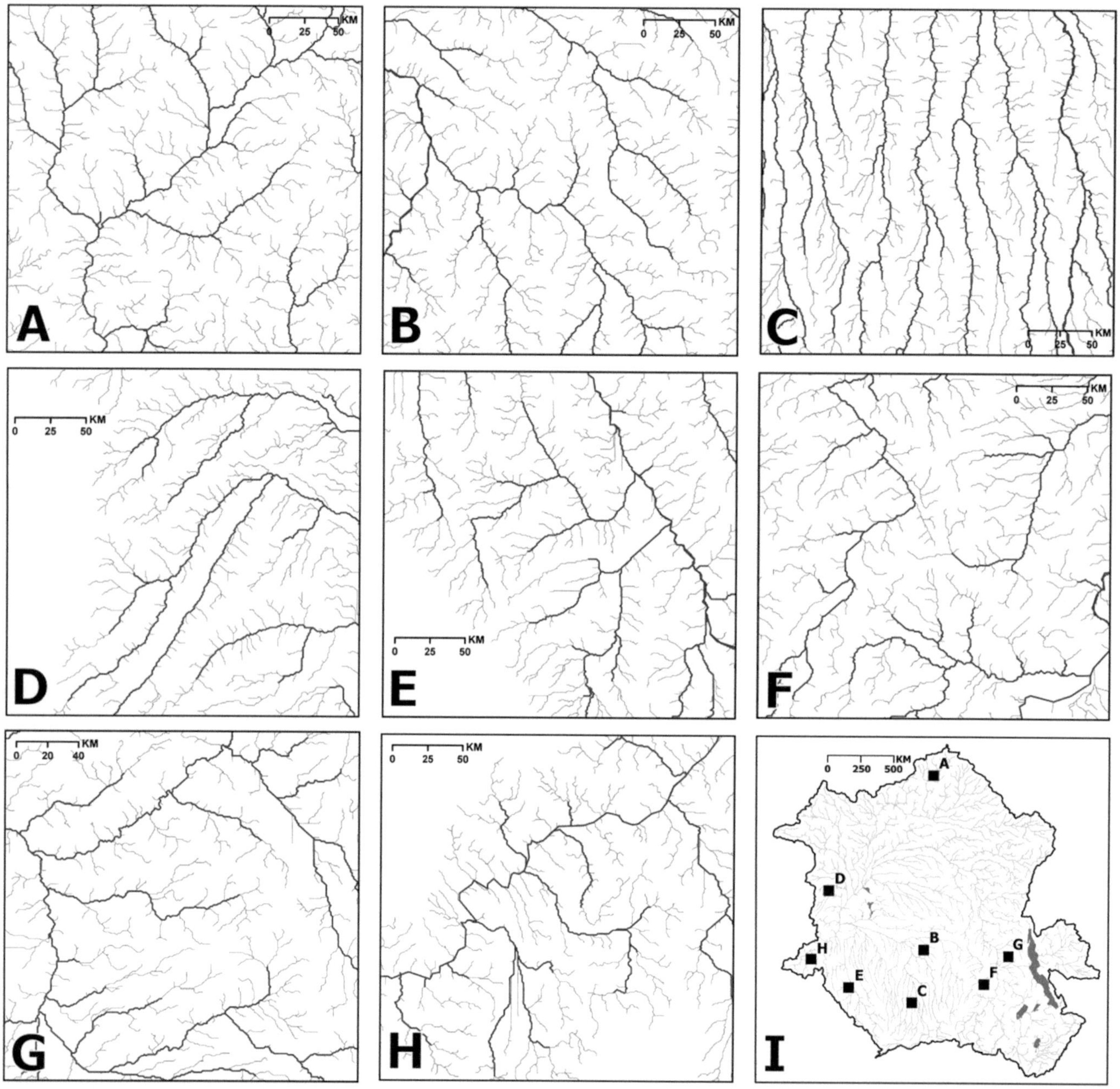

Fig. 15.4 Examples of the drainage types found within the Congo Basin, with stream channels derived from the SRTM data. The corresponding drainage types are : (A) Dendritic, (B) Sub-dendritic, (C) Parallel, (D) Sub-parallel, (E) Trellis, (F) Rectangular, (G) Rectangular-angulate, (H) Contorted. (I) shows the approximate locations of the drainage examples. Refer to Table 15.1 for pattern descriptions and Table 15.2 for drainage pattern classification. Stream channels were generated from SRTM DSM V3 90 m (Reuter et al. 2007; Jarvis et al. 2008)

Lomami and Tshuapa Rivers (Figs. 15.2 and 15.5). Late Neogene and Quaternary alluvium mantling this oval, shallow bowl like depression overlie thick continental sediments (Runge 2007; Guillocheau et al., Chap. 14, this book). These continental sediments are thought to originate from the erosion of the surrounding periphery of the basin and have been accumulating in the basin since the mid Paleozoic (e.g. Linol et al., Chaps. 7 and 11, this book). The subdued surface relief of the *Cuvette Central*, along with thick sedimentary deposits and reworked sediments of the Congo River, has lead several authors to suggest that the central CB has experienced a tendency for recurrent subsidence, with corresponding uplift around its edges (Linol et al., Chap. 11, this book).

The southern reaches of the Congo Basin are dominated by the northward extension of the Kalahari Plateau (Figs. 15.2 and 15.5). It is along this plateau that the watershed of the northward flowing Congo rivers and the

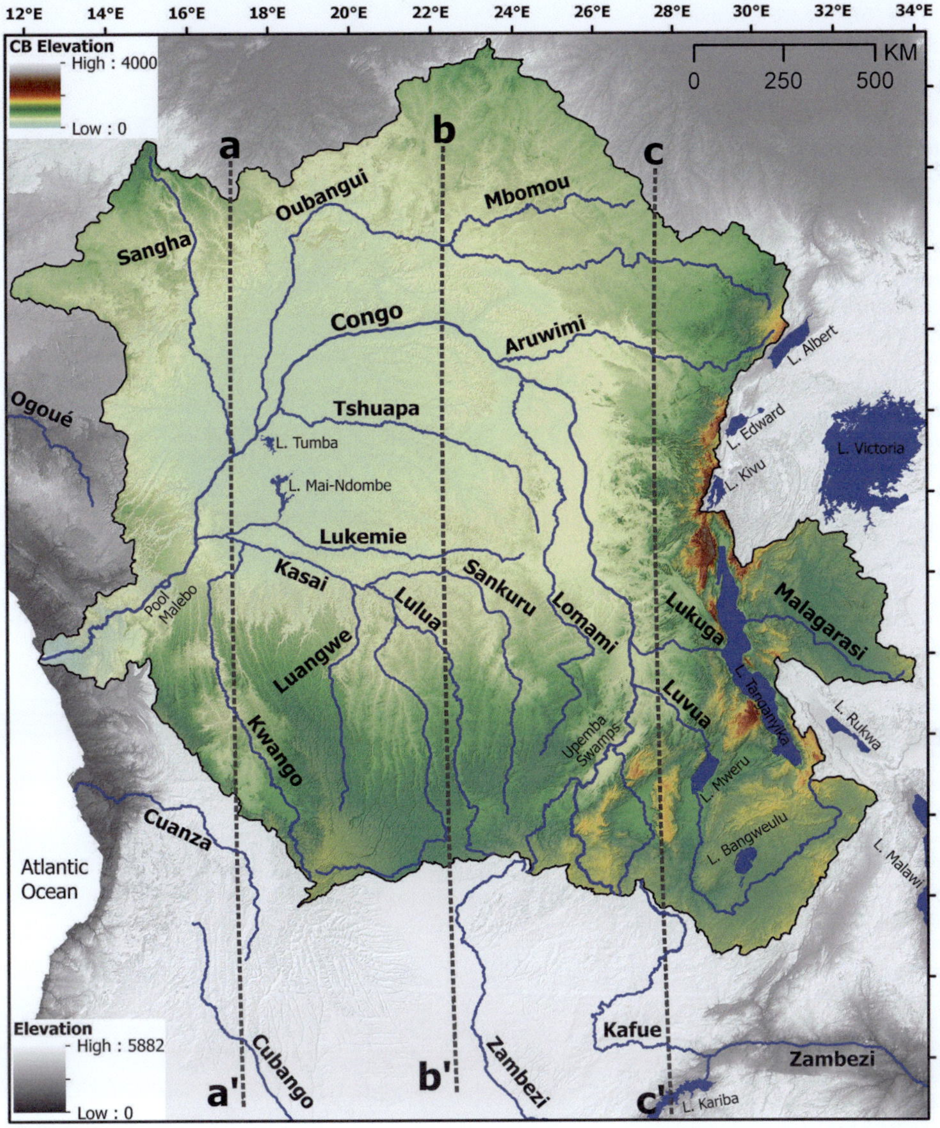

Fig. 15.5 The topography of CB system depicting major rivers and water bodies of the region. The placement of the topographic north-south cross-sections (a, b and c) as seen in Fig. 15.6a, b and c are shown

southward flowing rivers of the Okavango-Zambezi is found as well as that of the coastal draining Cuanza system (Figs. 15.2 and 15.5). In the west this plateau region consists of the dome like Angolan Highlands and its surrounds, further east: including the Bulozi plains and the Kafue Highlands (Moore et al. 2007, Figs. 15.5 and 15.6a). The western region is heavily eroded by the Cuanza River and the Kwango River in the west, with the Kwango River having eroded to the Precambrian basement in parts of its deep valley (Figs. 15.5 and 15.6a). A shallow basin formed by the Sangha, Oubangi and Congo rivers in the north of Fig. 15.6a represents the site of Western Congolian wetland complex. Moving eastward (Fig. 15.6b) one can see that the transition from the southern highland regions to the basin is the most gentle in this zone, relative to the western and eastern sections (Fig. 15.6a, c).

The topographic enhancement contributed by the EARS is illustrated in the rough topography of cross-section C (Fig. 15.6c). The area around the EARS is dominated by rectangular and modified drainage. Several phases of tectonics, rifting and uplift associated with the EARS has resulted in several steep, incised valleys flowing west, directly into the basin (Bauer et al. 2010; Ring 2008; Roller et al. 2010). The relief of the southern portion of the eastern catchments are dominated by the Kundelungu Plateau which has been uplifted to an elevation higher than that of the Congo-Zambezi watershed (Figs. 15.2 and 15.6c). Thus, the mean elevations of these southern Congo tributaries range from 1,000–1,400 m.a.s.l, compared to the elevations of 550–900 m.a.s.l of the northern tributaries (Figs. 15.5 and 15.6c). Of interest is the higher mean elevation of the eastern section of the southern basin (the Chambeshi River region)

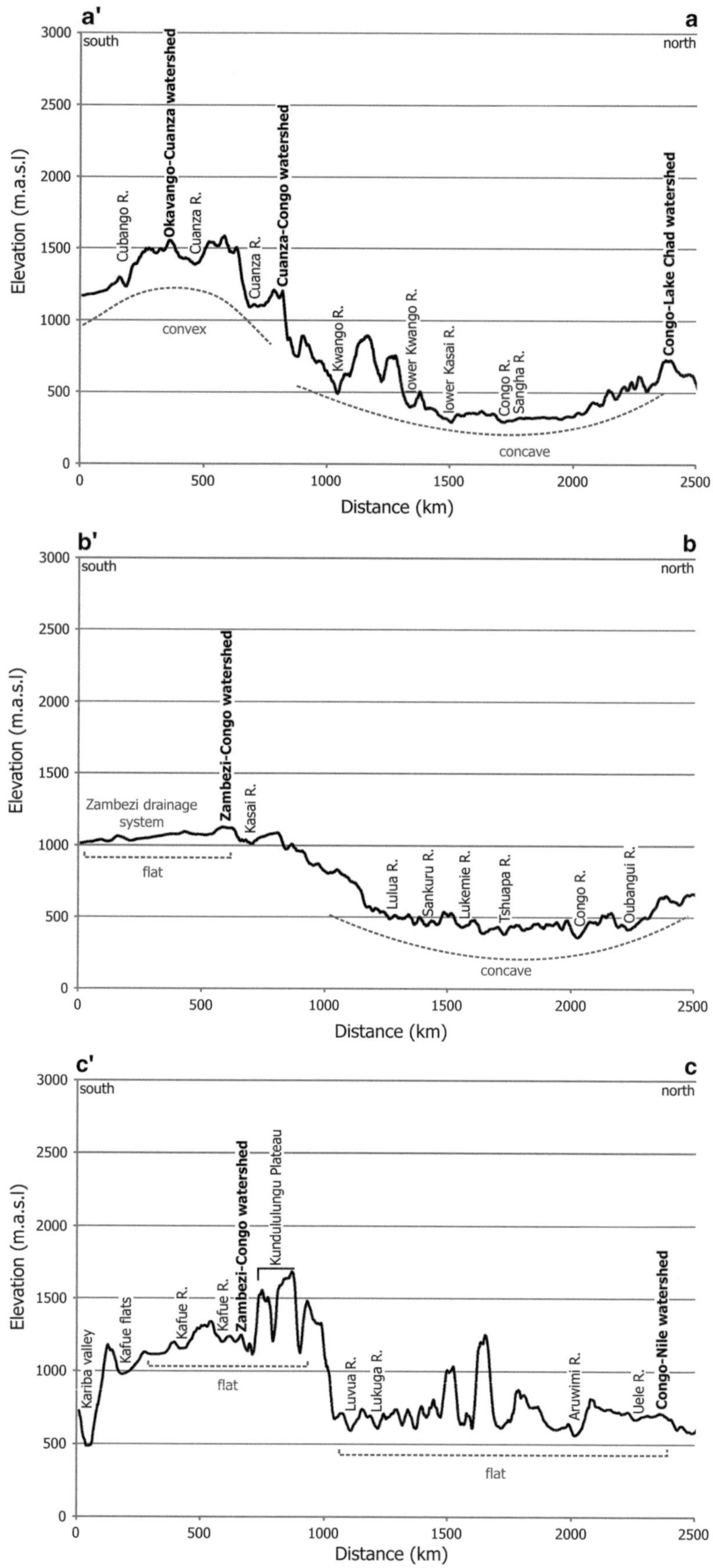

Fig. 15.6 (**a**) The western most cross section from a (north) to a' (south) with the location of major rivers and major watersheds (Okavango–Cuanza, Cuanza–Congo and Congo-Lake Chad) marked in *bold*. Note the convex topography of the Angolan Highlands region to the south and the concave profile to the north, which also includes the lower Kwango River. (**b**) The central north-south cross section depicting the gentle slopes of the CB system to the north and gradual transition towards a convex shape below the Zambezi-Congo watershed in the south. (**c**) Two first order landscapes are visible, a high, flat landscape at *ca.* 1,000 m.a.s.l in the south and a low, flat landscape at *ca.* 500 m. a.s.l. The high, flat landscape is punctuated by the topographic spike of the Kundulungu Plateau. The northern, low, flat landscape has several elevation spikes, representing the Western Branch of the EARS in the form of the eastern highlands of the Congo Basin. The trough of the Gwembe graben (Kariba valley) is conspicuous in the south. The watershed between the Zambezi-Congo and Congo-Nile are shown. The Kundulungu Plateau lies northward of the Zambezi-Congo watershed and is of a greater elevation than the watershed, representing a major deviation from the overall cross–section. The transition from the Zambezi to Congo system does not exhibit the general convex nature that is seen in the profiles further west (Fig. 15.6a, b), with rivers in the region incised to similar elevations

which is dominantly 1,400 m.a.s.l (with a maximum of *ca.* 1,600 m.a.s.l) compared to the rest of southern basin; as are the uplifted plateau separating the lower regions of the Mweru-Lufira-Upemba Complex. To the east of cross-section C, in the NW trending portion of the Western Rift lies the Tanganyika-Rukwa-Malawi (TRM) Rift (Fig. 15.5). The TRM is arranged in a rectilinear en-echelon array that reflects its formation in one of the most seismically active regions of the EARS with several active faults and ongoing seismic activity affecting most of the crust to ~30 km depths (Delvaux et al. 2012). The central zone of the TRM incorporates the well-defined Rukwa basin and the west bounding Ufipa Plateau. The tilted block comprising the Ufipa Plateau was uplifted during the late Cenozoic, largely within the Ufipa terrane (Delvaux et al. 2012). Here alignment of inselbergs and the footwall of faults systems illustrates the multi-stage tectonic history of the region (Delvaux et al. 2012).

The central zone of the CB is likely less erosively active than either of the western and eastern regions, although the convex nature of the transition from the Zambezi basin (high Africa) to the CB (low Africa) implies ongoing erosion. The flat nature of the Zambezi drainage system headwaters suggests limited vertical incision while the concave nature of the CB provides a depositional environment for material eroded from the convex southern headwater regions (Fig. 15.6b). The magnitude of this fluvial action in the central zone appears to be less than that occurring in the western zone, as indicated by the rough topography and deeply incised river valleys of the Kwango and Cuanza in the region of the Angola Highlands (Figs. 15.2 and 15.6a). While the overall concave nature of the central basin suggests a zone of sediment accumulation, the flat nature of the basin north of the Congo and Sangha Rivers suggest horizontal reworking of sediment rather than deposition (Fig. 15.6a). In the east of the CB, the incised river valleys indicate that vertical fluvial incision is dominant, and interestingly many of the rivers have incised to similar channel elevations (Fig. 15.6c). The overall flat nature of the elevated, high Africa region in the south and flat low Africa in the north indicates that these are dominantly sediment transition zones (Fig. 15.6c). The three cross-sections suggest that the eastern basin a zone dominated by vertical river incision and sediment transport, with the central is basin is more balanced between vertical river incision and sediment transport and deposition (Fig. 15.6b, c). In the west the central basin regions provide a sediment sink and horizontal sediment reworking is dominant, while in the south vertical river incision is active in regions (Fig. 15.6a). The implication for drainage pattern development is that, generally the eastern regions are likely to experience more frequent river re-organisation while the central and western regions patterns are, overall more established.

15.3 Present Day Drainage Patterns of the Congo Basin

The complex nature of the Congo Basin's hydrographic network was recognised by pioneering geologists (Veatch 1935; Roberts 1946; Cahen 1954) especially the evidence for the juxtaposition of several, apparently disparate, drainage patterns. It is believed the entire drainage network of the Congo Basin has been active since the late Cretaceous (Roberts 1946; Cahen 1954; Deffontaines and Chorowicz 1991; Goudie 2005). Deffontaines and Chorowicz (1991) have suggested that much of the present day drainage is superimposed, having originated in the early Cenozoic, and Cahen (1954) suggested even older ages for substantial portions of the basin. Our classification of the Congo Basin's drainage patterns (Figs. 15.3 and 15.4, and Tables 15.1 and 15.2) made extensive use of the SRTM DSM (Jarvis et al. 2008) and satellite imagery, namely Landsat Enhanced 7 Thematic Mapper Plus. This DSM was used to generate river networks based on a 90 m pixel resolution. These river networks where than classified into their drainage types, making use of the finer scale satellite imagery (30 m pixel size) in ambiguous areas, such as boundaries and drainage type changes. This allowed for a greater differentiation of pattern types than was previously possible (i.e. Roberts 1946; Deffontaines and Chorowicz 1991). As drainage patterns have a number of controls, such as geology, tectonics and regional topography, they may have convergent and divergent development histories making the interpretation of drainage patterns challenging (Fig. 15.3 and Table 15.1). Furthermore, drainage patterns may represent the cumulative effects of multiple controls. Nevertheless it is possible, and informative, to classify such drainage patterns at a regional scale. Of special interest are the boundary zones between patterns which may be indicative of active geomorphic changes in the basin.

Regionally, the CRS may be divided into eight dominant drainage patterns (see Table 15.1) occurring as 18 discrete zones (Fig. 15.3). This consist of the basic (major) patterns of dendritic, parallel, trellis and rectangular (Zernitz 1932) and contorted (Howard 1967) patterns. Three additional modified patterns, (which are deviations from the basic pattern due to structural or topographic controls, or the transition from one basic pattern to another) were used, namely, sub-dendritic, sub-parallel and a rectangular-angulate compound (Howard 1967). Naturally, more drainage types would be recognised on finer spatial scales. Here modified and palimpsest patterns are of special interest as they may indicate tectonic activity, but as these patterns are more local in nature (Howard 1967), they have not been included in the regional classification. A brief summation of those patterns of interest is shown in Table 15.1. The

Table 15.1 The description of the major drainage types found in the Congo Basin and their most probable controls

Drainage type	Description of drainage pattern	Drainage pattern may indicate the following*
Dendritic	Drainage is irregular, with channels branching in all directions and tributaries joining the main channel at all angles. Drainage is only truly dendritic if the main channels show no slope control. It has the widest modifications of all the basic drainage types	Tectonic stability; regionally uniform uplift or subsidence; low, uniform slopes; geology of uniform resistance; lack or weak structural control; horizontal lithology; superimposition (if occurring on folded or differentially resistant rocks or on a massive igneous body). As this drainage may be the result of several structural conditions, it is of limited use in determining geologic structure compared to other drainage types
Sub-dendritic	Deviation of dendritic	Modification most commonly due to secondary regional controls of structure and topography; transition phase of drainage
Parallel	A number of rivers and streams flow parallel, or near parallel, to one another over a considerable area, or in several successive cases. Parallel drainage may occur as a transition phase of dendritic or trellis drainage	Regional tilting; steep gradients (often on homogenous lithology); parallel topographic features; parallel faulting; transition phase of dendritic or trellis drainage. Parallel drainage may also indicate recent tilting, especially on newly emerged surfaces or may zones of parallel faulting. The parallel pattern suggests homogeneous sediment, with river following the greatest slope
Sub-parallel	Lack regular pattern of parallel drainage. However the distinction between sub-parallel and sub-trellis may be disputed	Slope changes; high relief slopes; mild structural control (i.e. deformed strata of uniform resistance to erosion)
Trellis	A trellis drainage is characterised by the occurrence of secondary tributaries parallel to the major stream or other streams into which primary tributaries enter. The secondary tributaries are often elongated, orientated at right angles to the streams into which they flow. Several modified types of trellis drainage exist, but require relevant data exists to determine their cause. For example fault and joint patterns appear similar. For this study patterns were categorised as trellis although there may be differences in control	Active faulting; the boundary of lithologies of varying resistance (i.e. parallel belts of folded or tilted strata); faulting; lithological jointing
Rectangular	Rectangular drainage is characterised by right-angle bends in both the main channel and its tributaries. It lacks the regular pattern and shorter tributaries that defines trellis drainage	Right-angle channel bends may be due to orthogonal joint or fault systems but lack regularity. Tectonic stability; regionally uniform uplift or subsidence; highly complex fractured systems; ongoing faulting
Rectangular-angulate	A common modified, compound version rectangular drainage but one where the pattern is dominated by non-right angle faults	Modified rectangular drainage where faulting is not dominantly right angled; ongoing faulting
Contorted	Lack a regular pattern, with tributaries exhibiting a multitude of flow directions	Deformed rocks (especially coarsely layered metamorphic rocks) that contain resistant layers (i.e. dykes and migmatised bands); stream and tributary length may indicate general direction of dip, making it possible to distinguish between plunging anticlines and synclines on the local level

Controlling factors are from Zernitz (1932) ; Howard (1967); Deffontaines and Chorowicz (1991) and are given in decreasing order of probability

Table 15.2 Regional drainage pattern classification of the Congo Basin. See Fig. 15.3 for the location of each drainage type

Drainage type	Map code	Location	Example of drainage type from the Congo Basin
Dendritic	D	Dominantly the (D1) Northern Oubangui, (D-Sd2) *Cuvette Central* and (D3) the Upper Malagarasi systems	Figure 15.4A
Sub-dendritic	Sd	Three zones surrounding the Central Basin and parts of the (D-Sd2) *Cuvette Central* ; (Sd4) the Western Congolain wetlands, and in the east the (Sd5) Lindi and (Sd6) Luvua systems	Figure 15.4B
Parallel	P	The (P7) Luangwe System of southern plateau lands of the basin including the middle Kasai	Figure 15.4C
Sub-parallel	Sp	The (Sp8) Northern Batéké Plateau System	Figure 15.4D
Trellis	T	Five systems, three of which are form the western boundary of the basin, *viz.* (T9) southern Batéké Plateau, (T10) Sembe-Ouesso and (T11) Kwango Valley systems. The (T12) Lufupa system forms the extreme southern basin boundary and the (T13) Mid-Congo River zone in the central basin	Figure 15.4E
Rectangular	R	Only the (R14) Buhimay System is the only rectangular pattern, being found in the Katanga region	Figure 15.4F
Rectangular-angulate	RA	The (RA15) Lulwango and (RA16) Ulindi systems are found on the western zone of influence by the EARS	Figure 15.4G
Contorted	C	The (C17) Lower Congo in the extreme west of the basin and the (C18) Lake Tanganyika / EARS in the eastern basin	Figure 15.4H

individual drainage patterns will now be discussed in more detail. Figure 15.3 indicates the extent and location of the various drainage patterns for the CRS and Table 15.2 provides the names and map codes of each drainage type and an example of the drainage pattern from the CB is illustrated in Fig. 15.4.

15.3.1 Dendritic Patterns

In the CB there are three regions exhibiting dendritic patterns and four exhibiting sub-dendritic patterns (see Table 15.2 and Figs. 15.3 and 15.4A):

15.3.1.1 Northern Oubangui System (D1 in Fig. 15.3)

This large, northern area of the basin appears to correspond to the underlying Precambrian basement (Chap. 2, this book). The occurrence of fractured, weathering resistant Proterozoic quartzitic sandstones provides localised structural control on some tributaries (Runge 2007).

15.3.1.2 Cuvette Central (D-Sd2 in Fig. 15.3)

The drainage pattern of the *Cuvette Central* is dominantly dendritic, although several large extents of interleaved sub-dendritic drainage patterns are present. Cahen (1954) suggested a Neogene to early Quaternary age for the large portions the *Cuvette Central* dendritic pattern and that portions of present day channels of the Sangha, Oubangi and Congo Rivers that flow through the Western Congolian wetlands, along with the Lukemie River, are related to the formation of Malebo (Stanley) Pool (Fig. 15.5). These sections thus represent the youngest part of the network (Cahen 1954). It is likely that portions of the *Cuvette Central* have been superimposed, flowing along inherited older trends (Deffontaines and Chorowicz 1991).

15.3.1.3 Upper Malagarasi System (D3 in Fig. 15.3)

The upper Malagarasi system exhibits a dendritic pattern. The dendritic pattern lies predominantly on Tanzanian cratonic region, in the shallow basin region south of Lake Victoria. The tectonic stability and broadly uniform lithology this region would have allowed for the uninterrupted development the Malagarasi system.

15.3.2 Sub-dendritic

15.3.2.1 Cuvette Central (D-Sd3 in Fig. 15.3)

There are several of zones of sub-dendritic drainage that occur within the overall dendritic pattern of the *Cuvette Central*. This sub-dendritic drainage dominantly occurs in the western, eastern and southern zones of the *Cuvette Central* (see Table 15.2 and Figs. 15.3 and 15.4B). Several of these zones are associated with areas of locally outcropping resistant lithologies and structure, in a region that is dominantly alluvial in nature. Abrupt changes in flow direction in zones of transition from one dendritic to sub-dendritic patterns reveal the presence of NE–SW and ESE–WNW trending faults (Deffontaines and Chorowicz 1991). Additionally, to the south, the Lukémié and Kasai Rivers have parallel interfluves that suggest the modification of river patterns by a horst block or tilted block (Deffontaines and Chorowicz 1991). The abrupt change in river direction of the Congo River near Kisangani, the middle section of the Sankuru River and also in the lower Kasai to the south such structural controls, which have experienced vertical movements since the Cenozoic (Figs. 15.1 and 15.5; Deffontaines and Chorowicz 1991). The regular, trending to parallel drainage pattern of the lower Kasai River in the south, may indicate a distinct, major block tilted northwards (Deffontaines and Chorowicz 1991).

15.3.2.2 Western Congolian Wetlands (Sd4 in Fig. 15.3)

The drainage pattern of the rivers flowing south into the CR just before it crosses the Equator for a second time is sub-dendritic. This drainage zone corresponds to a large, inundated region of dominantly Holocene alluvium that forms the Western Congolian wetlands. A combination of low gradients and mosaics of very dense tropical rainforest and swamplands has prevented the establishment of dendritic drainage. This can be seen in the multi-branching meandering form of Likoula aux Herbés River (a tributary of the lower Sangha River, Fig. 15.5) as it flows through the wetlands. Parts of the region have be influenced by neotectonics (namely Lac Télé (Master 2010) which has likely further inhibited the development of a dendritic pattern through network reorganisation.

15.3.2.3 Lindi System (Sd5 in Fig. 15.3)

This small zone of sub-dendritic drainage occurs in the north-east of the basin with trellis and dendritic drainage lying on the downstream side to the west and north, while rectangular angulate and contorted drainage occurs upstream in the southern and eastern sections. The Lindi systems thus appears to be influenced both by the structural influence of the EARS to the west and the trellis network of the downstream mid-Congo. It is possible that the Lindi System was part of the *Cuvette Centrale* and Luvua (sub) dendritic system and has been subsequently isolated by downstream capture and modified by the EARS in its headwater zones.

15.3.2.4 Luvua System (Sd6 in Fig. 15.3)

The Luvua system occurs in the south-western areas of the Congo Basin and extends northward along the CR. The

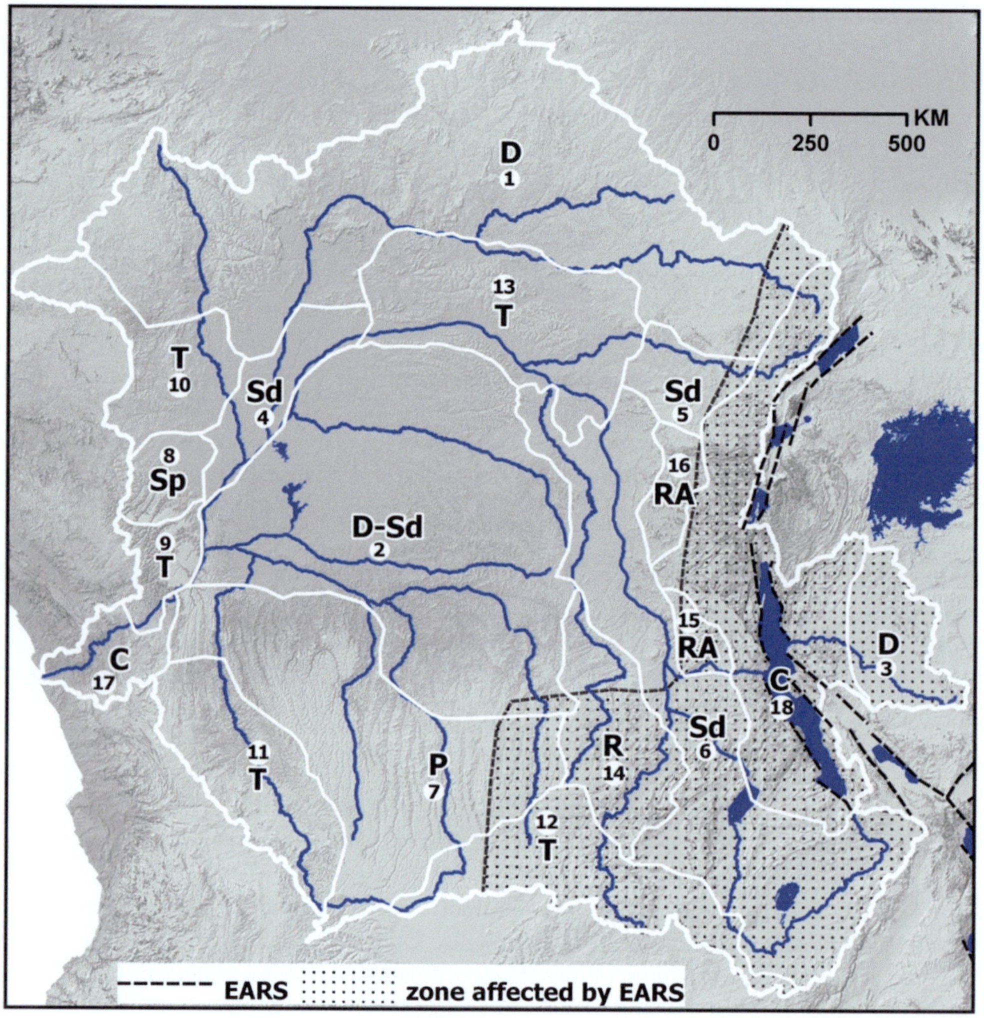

Fig. 15.7 The EARS (*black dashed lines*) and its affect on the CB drainage patterns. The stippling indicates the regions of the CB that been affected by seismic activity linked to the EARS. Drainages close to the rift are most affected by seismic activity, which has a calculated extent of up to 450 km west of the EARS. At distances greater than 450 km, the probability of a seismic event exceeding a peak ground acceleration (PGA) of 0.05 g is less than 10 % in 50 years (Mavonga and Durrheim 2009). This relative higher probability and greater PGA of seismic events closer to the rift could account for the occurrence of Sd5 (Lindi), RA15 (Lulwango), RA16 (Ulindi), Sd6 (Luvua), R14 (Bushimay) and T12 (Lufupa) networks. The limited affects on D3 (Upper Malagarasi) may be related to its proximity to a cratonic region. Interestingly the lower region of the Malagarasi river has become contorted (C18)

southern reaches have a dominant south-west orientation, suggesting some degree of structural control in the region. A large portion of this zone corresponds to the Precambrian basement (Deffontaines and Chorowicz 1991); with evidence of recent tectonic activity along Lake Mweru and a wide alluvial plain upstream of the lake, near Kasenga (Tack et al. 2003). This tectonic activity, associated with the Western Branch of the EARS, is the probable explanation of the sub-dendritic pattern in this region (Fig. 15.7). The similar morphologies of the Ufipa Plateau and Rukwa basins (north of the Chambeshi headwaters, in the Tanganyika-Rukwa-Malawi rift zone) to the Kundelungu Plateau and Lake Mweru provide further evidence of tectonic control of the drainage (Fig. 15.5). While the Rukwa Plateau is generally tilted towards Lake Tanganyika, several of its rivers drain into Lake Rukwa, possibly reflecting a combination of antecedence and tectonic control of these rivers (Delvaux et al. 2012). Additionally there appears to be no river offsets across the scarps (Delvaux et al. 2012). The importance of tectonic events in the region is illustrated by 1910 earthquake in the region that is estimated to have been 7.4 in magnitude (Delvaux et al. 2012). Similarly, localised faulting and uplift appears to have morphed the dendritic network into a sub-dendritic pattern. In summary, much of the sub-dendritic pattern represents the interplay between the inherent structural controls of the Precambrian basement on the one hand, and extensional movements of Neogene to late Holocene age, as is suggested for the Upemba Trough (Fig. 15.5; Deffontaines and Chorowicz 1991). This may be further enhanced by rift associated uplift that promoted erosion and incision of the pre-rift sediment and/or basement, with this rift flank induced river system draining into the basins (i.e. Mweru) in structurally low-lying regions between fault segments.

15.3.3 Parallel

Within the CB, two areas exhibit parallel type drainages: the parallel drainage of the southern plateau region of the basin, as seen in the Luangwe System and the sub-parallel drainage of the Northern Batéké Plateau System of the basin western margin (see Table 15.2 and Figs. 15.3 and 15.4C) .

15.3.3.1 Luangwe System (P7 in Fig. 15.3)

The southern plateau areas, which incorporates the Luangwe basin and associated tributary catchments in the southern Basin has dominantly parallel drainage pattern. The boundary zone varies from parallel to sub-parallel and it covers a region that has been suggested to have remnants of an upper Jurassic drainage network (Cahen 1954). The parallel drainage is associated with the transition zone of the low gradient slopes of the northern extent of the Kalahari Plateau to the basin of *Cuvette Central*. The parallel pattern suggests uniform lithology which allows the river to flow along the greatest slope, as seen to the south of Kasai (Zernitz 1932; Deffontaines and Chorowicz 1991).

15.3.4 Sub-parallel

15.3.4.1 Northern Batéké Plateau System (Sp8 in Fig. 15.3)

The Northern Batéké Plateau System comprises of many near linear rivers, flowing across a NE dipping plateau (see Table 15.2 and Figs. 15.3 and 15.4D). The combination of the a strong NNE flow direction and the linearity of the rivers with connections between parallel rivers being short suggests a degree of fault control. The location of the plateau to the Atlantic margin, lead Karner and Driscoll (1999) to suggest a Gondwana break-up origin of these rivers. These rivers may have begun flowing basinward due to flexural rebound associated with Gondwana break-up, that established numerous, short rivers toward the interior basin (Karner and Driscoll 1999) and these patterns have been maintained to present day.

15.3.5 Trellis

There are five instances where trellis drainage was identified, with this drainage type occurring mainly along the peripheries of the central basin (see Table 15.2 and Figs. 15.3 and 15.4E).

15.3.5.1 Southern Batéké Plateau System (T9 in Fig. 15.3)

The Southern Batéké Plateau System differs from the Northern Batéké Plateau System in that the rivers in the region are shorter, with more right angles and flow directly into CR. The geometry of the rivers and the consistent directionality of many of the river segments, being WWN, reveal a dominant fault control. Additionally much of this zone corresponds to the Precambrian basement (Deffontaines and Chorowicz 1991), which incorporates a portion of the West Congolian orogenic belt making it probable that tectonics, lithological structures and rock differences, have resulted in a trellis drainage forming. Subsequent tectonic activity may have modified the rivers further.

15.3.5.2 Sembe-Ouesso (T10 in Fig. 15.3)

This pattern is controlled by numerous south-north trending faults, with several NNE trending faults, with the rivers flowing on Precambrian basement. Dolerite outcrops further add to the structural control of this trellis pattern. Therefore the Sembe-Ouesso drainage network is predominantly controlled by the directionality of the faults, and the underlying lithology.

15.3.5.3 Kwango Valley (T11 in Fig. 15.3)

The Kwango Valley, found along the southern margin of the CB exhibits a trellis pattern drainage pattern. Here the Kwango River flows in a north-west direction, compared to parallel, north flowing rivers of adjacent Luangwe System to the east. Incision into the Precambrian basement by the Kwango River appears to have caused the diversion and breakup of the longer parallel streams into several shorter rivers. Thus the Kwango River trellis pattern, at least in the eastern region close to the Luangwe System, likely represents a modification of a parallel drainage that existed on the tilted plateau, previous to incision by the Kwango River. Furthermore, its location on the margin of the West Congolian orogenic belt has resulted in the western part of the Kwango Valley developing a strong trellis pattern, shaped by structural control in this region. The Kwango Valley drainage therefore is dominantly a modified parallel network with subsidiary structural control.

15.3.5.4 Lufupa System (T12 in Fig. 15.3)

Forming the southern margin of the Congo Basin the Lufupa trellis drainage system comprises of the easterly flowing Kasai River in the west and the northerly flowing upper Congo (Lualaba) River in the east. The Lufupa system thus forms the main portion of the Congo-Zambezi Watershed. The western part of this trellis network, around the Kasai River, may contain the remnants of late-Cretaceous rivers (Cahen 1954) and thus it inherited to some degree. Whereas, the presence of folded schists and tillites of the Kanga-System (Upper Precambrian) and tectonics (indicated by three major fault scarps) (De Dapper 1988) in the eastern half of the trellis drainage are the likely controlling factors on the drainage pattern.

15.3.5.5 Mid-Congo River (T13 in Fig. 15.3)

The northern bank of the middle CR is dominantly trellised with the southern bank having numerous small rivers join the Congo at right angles. This start of the trellis network closely corresponds to the abrupt and localised braiding of the CR that begins at 0°24°E and continues until 1°N 16°E. The braiding of stream channels suggests an inability to transport the bed load. This may be due to local coarser bedload, and/or loss of velocity due to flattening of gradient (Howard 1967). It is likely that trellis pattern represents a modification of pre-existing dendritic network, as the middle Congo trellis network is bounded between the northern Oubangui and central basin dendritic networks. This modification could be related to the capture of the lower CR. It is probable that this trellis pattern has been enhanced by local structural controls such as exposed outcrops of basement.

15.3.6 Rectangular

The CB contains a single zone of rectangular drainage, this being found in the south-east region of the basin (see Table 15.2 and Figs. 15.3 and 15.4F).

15.3.6.1 Bushimay System (R14 in Fig. 15.3)

The rivers in this zone have a general north-northeast orientation, bounded in the west by a high ridge, which separates the upper CR (also referred to as the Lualaba) from the Kasai system (Cahen 1954). Roberts (1946) suggested that these north-northeast river courses exhibit strong inheritance, where they occupy Karoo palaeovalleys. The persistent seismicity all around the Upemba Trough attests to Neogene to Recent extensional movements (Deffontaines and Chorowicz 1991). Therefore this area is likely a zone of intense faulting.

15.3.7 Rectangular-Angulate

There are two zone of rectangular-angulate drainage, both relatively small and located in the eastern CB in proximity of the Western Branch of the EARS (see Table 15.2 and Figs. 15.3 and 15.4G).

15.3.7.1 Lulwango (RA15 in Fig. 15.3)

The Lulwango system, is a small area, with a mixture of channels with right-angle bends and channel junctions forming acute angles. The combination of NE–SW trending, linear ridges and Basement rock outcrops appears to be the local controls of the drainage. The linear ridges have resulted in channels running near parallel in part, forming right-angles as they cut across the ridges. Additinoally as the sub-dendritic Luvua System (Sd6) lies to the south and west (downstream) and the contorted Lake Tanganyika (C18) to the north and west (upstream) the Luluwango system exhibits both structural and tectonic controls evident in these systems. It is probable that the Lulwango system is an area of ongoing modification, forming a zone of interaction between the Luvua and Lake Tanganyika systems. The multiple phases of deformation and rejuvenation experienced by the Western Branch (Cahen and Snelling 1966) has likely enhanced rectangular-angulate pattern (Fig. 15.7). The interaction of these factors likely accounts for the system having deviated from a topographically controlled (linear ridges) trellis or rectangular pattern.

15.3.7.2 Ulindi (RA16 in Fig. 15.3)

The Ulindi system is surrounded by the sub-dendritic systems of the Lindi (Sd5) and Luvua (Sd6) to the north and west (downstream), with the contorted drainage of Lake Tanganyika (C18) being to the south and north (upstream). The drainage exhibits is more rectangular compared to the Lulwango but lacks perpendicular channel junctions and. It is likely that the outcropping Basement rocks providing a degree of structural control, with the Western Branch having a limited, direct influence on the Ulindi system (Fig. 15.7). Occurring between two dominant drainage types (sub-dendritc downstream and to the north; contorted upstream and to the south) this modified pattern forms a interface zone between the two drainages.

15.3.8 Contorted

In the CB contorted drainages are associated with two zones of deformation, the Atlantic Rise and the EARS (see Table 15.2 and Figs. 15.3 and 15.4H).

15.3.8.1 Lower Congo (C17 in Fig. 15.3)

The contorted drainage of the Lower Congo corresponds to the area where the CR and its associated tributary rivers straddle the Western Escarpment and the coastal region of the CR. Here, the river morphology appears to be controlled dominantly by the exposure of Precambrian rocks; the contorted drainage follows the various zones of the West Congolian orogeny. It is likely that tectonic activity, combination of uplift (thought to be more than 300 m) and warping and/or tilting, associated with the Miocene reactivation of the West African margin, resulted in drainage reorganisation, further enhanced the contorted drainage pattern (Cahen and Snelling 1966; Lavier et al. 2001). It has been suggested that increased sediment loading on the coastal shelf during was the cause of this tectonic reactivation (Deffontaines and Chorowicz 1991; Driscoll and Karner 1994; Anka and Séranne 2004; Anka et al. 2010).

15.3.8.2 Lake Tanganyika/East African Rift System (C18 in Fig. 15.3)

The eastern margin of the CB is dominated by a contorted drainage pattern. The area of contorted pattern corresponds to the Eastern Highlands and thus forms the headwater zone of many of the westerly flowing rivers. The streams and rivers of this zone lack a consistent drainage pattern, with rivers flowing in several different directions and near-straight channelled rivers flow adjacent to rivers. This lack of consistent pattern indicates a dominance of fault and lithology controls in the region. The fact that much of this drainage system corresponds to the Western Branch of the EARS, it is likely that much of the drainage has been influenced by phases of tectonic activity (Fig. 15.7). Indeed several phases of tectonic activity have been identified, resulting in the overprinting of faults by succeeding faults and the formation of valleys and regional blocks (Cahen and Snelling 1966; Ring 2008; Bauer et al. 2010). In the Rwenzori Mountains (Fig. 15.2), the Precambrian basement rocks widely resist fluvial erosion, causing the rivers to predominantly exploit predetermined structures (Bauer et al. 2012). It is thus likely that the contorted drainage pattern is strongly linked to the ongoing tectonic activity associated with development of the Western Branch. For example, the uplift of the Ufipa Plateau severed the drainage linkages to the Tanzanian Plateau in the east, notably those of the greater Rufiji drainage system (Fig. 15.1A). This uplift appears to be Pliocene in age (e.g. Cotterill and de Wit 2011; Goodier et al. 2011).

Therefore a large portion of Lake Tanganyika's drainage is geological young and fault controlled, with the contorted pattern being related to a combination of multiple phases of uplift, subsidence, tectonic activity and exposure of Precambrian rock. This is distinctly seen in the lower Malagarasi (Fig. 15.2) and Lukuga systems that drain into and out of Lake Tanganyika respectively. The Malagarasi systems changes from a dendritic pattern to a contorted pattern in the zone where the river begins to flow off craton and becomes influence by rifting. Previous to the formation of the Western branch there was greater linkage between the drainage of Eastern Africa and the CB. This palaeo-channel is represented in the topological unity of modern Malagarasi River, draining into Lake Tanganyika. These preserved landforms attest to the profound impact of the EARS, in particular its Western branch, on the evolution of the eastern rivers of the CRS (Fig. 15.3).

15.4 Timing of Emplacement of the Modern River Systems of the Congo Basin

15.4.1 Cenozoic Events

The marine sedimentary evidence lead to the suggestion of several scenarios of CB evolution. The three models proposed to account for this marine sedimentary record are : 1) at the end of the Cretaceous there was a progressive change of the depocentre from the Ogoué and Cuanza (Kwanza) Rivers to the Congo as a result of capture of the endorehic Congo Basin (Babonneau et al. 2002; Anka et al. 2010). Today these southern and northern systems exhibit multiple incised valleys and gorges along the coastal areas, often occupied by under fit river channels. 2) There was a net southward migration of the outlet of the CRS from the Cretaceous to the Cenozoic (Karner and Driscoll 1999; Nibbelink and Budihardjo 2002). During the late Cretaceous, the entire CB was drained by an outlet at or near the present day Ogoué River (northern Gabon) (Karner and Driscoll 1999; Nibbelink and Budihardjo 2002). Through the mid-Cenozoic the outlet migrated south, forming a connection near/at the present day Kouilou River, reachingthe present day CR outlet in the Oligocene (Karner and Driscoll 1999; Nibbelink and Budihardjo 2002). Karner and Driscoll (1999) suggest that feedbacks between sediment loading and flexural uplift of the hinterland may have played a role in the migration of the outlet, although the mechanism of migration is unclear. This migration may be evidenced in the misfit of the present Ogoué and Kouilou Rivers (see Fig. 15.1a, c, the Kouilou River is near Point-Noire Fig. 15.1c) that occupy abandoned channels and their valleys, relic drainages and the large estuary near Liberville and the large size of the Late Cretaceous—early Cenozoic delta along the Gabon coast relative to present-day coastal river systems (Karner and Driscoll 1999). This tectonic activity may have lead to the capture of the interior drainage of the Congo Basin by a coastal valley (Cahen 1954).The lower CR crosses an area of lower modelled flexural uplift (Anka et al. 2010) suggesting that the lower Congo River comprises of inherited drainage. This antecedent lower CR is evidenced by the deeply incised gorge and waterfalls downstream of Kinshasa (Anka et al. 2010).

15.4.2 Neogene Events: Evidence from Phylogeographic Studies of Fishes and Mammals

While there is some geologic and geomorphic data regarding the timing of river re-arrangements, this data is often discontinuous or lacking in terms of resolution. This lack of data of river developments can be circumvented by using biological data as proxies to identify changes to drainage topology. These proxies use phylogenetic reconstructions of related species that have become geographically isolated by changes in the landscape in which they live. Over evolutionary time, the isolated populations accumulate changes in their DNA, with the genetic difference between the populations allowing for an estimation of time since separation based on a molecular clock (Avise 2000; DeSalle and Rosenfeld 2013). It is possible to determine the average rate at which a species has accumulated genetic differences (mutation rate) over time. By measuring the number of genetic differences between two related groups and applying the average mutation rate, a time estimate of when the two populations were once freely breeding can be calculated (Avise 2000). This process gives the approximate age of genetic divergence of the two groups from their last shared common population, or most recent common ancestor (MRCA) (Avise 2000). The use of Bayesian methods allows for age estimates to be assigned a measure of statistical error, usually expressed at the 95 % level, which constrains the probability of the true age value (DeSalle and Rosenfeld 2013).

The challenges involved in using phylogenetic data (such as calibration methods and sampling density of species) can be mitigated by using phylogenetic data from several species. The congruence of age estimates across different species provides insights into the spatio-temporal changes in the fluvial geomorphology of central Africa during the late Cenozoic (i.e. Cotterill and de Wit 2011; Goodier et al. 2011; Schwarzer et al. 2011; Hart et al. 2012; Kawamoto et al. 2013). For example, it appears that the CR is the only river that contains the extent of the large primates, namely chimpanzees (*Pan troglodytes*) and gorillas (*Gorilla gorilla*), preventing these primates from occurring south of the CR, although both having successfully dispersed across smaller rivers (see Fig. 15.8a). Fish are the most accessible of these biological indicators and being restricted to water systems their biogeography and evolution are tightly linked to the evolution of fluvial systems (Cotterill and de Wit 2011). This link of biogeography and landform can be seen in the broad congruence that differenct fish assemblages show with regional drainage basins. Although fish assemblages may differ locally, owing to the interplay of numerous abiotic and biotic factors (i.e. water pH, temperature, biological competitors and predators) some fish species occupy definitive niches (stenotypes), and are thus forced to track their local environments. Over evolutionary timescales, these stenotypes either successfully track their specific physical habitats, or become extinct. Therefore stenotopic fish may record the timings of river events, such as disruption or connections of river channels, when water transfer between the river systems ends or begins. The events dates by molecular proxies, record of the start of the current phase of the river drainage structuring (assuming equal ability of dispersal up and down rivers, that is no barriers) as connections may be seasonal (i.e. yearly flood) or event based (i.e. 1 in 10 year floods). The use of primate evidence goes some way in solving problems with the timing of prolonged river channel emplacements as many primate taxa distribution patterns are often contained by major rivers that serve as barriers to movement, both on a genus and species levels (Anthony et al. 2007; Harcourt and Wood 2012).

Yet not all rivers are equally effective barriers, with wider and faster flowing rivers being formidable barriers compared to narrower, smaller tributaries and thus the lower reaches of a river are often more effective barriers compared to their upper reaches; for example, the lower sections of the CR seem to be more effective barriers than its smaller tributaries and headwaters (Harcourt and Wood 2012). As African rivers form the border of distributions for more subspecies rather than genera, Harcourt and Wood (2012) suggest that primate taxonomic level may be a useful, albeit rough, proxy to estimate times of population isolation by river emplacement. Therefore primate divergence estimations (calculated from the MRCA) may serve as proxies of rivers achieving a sufficient size, and continuity of flow, to become a barrier to dispersal. The CR acts a major, regional barrier to primate dispersal, with fewer primate species and subspecies straddling the CR compared to other rivers (Harcourt and Wood 2012). Owing to the relationship between rivers as a barrier and primate size (Harcourt and Wood 2012), it can be argued that timings of divergence of large primates (i.e. bonobos, chimpanzees and gorillas) relate to the development of large rivers (i.e. Congo) with medium and smaller primates serving as proxies for emplacement of medium and smaller rivers. However the inherent heterogeneity of a river in both space and time must be borne in mind when considering rivers as barriers to primate dispersal. This is highlighted by a genetic study of chimpanzees either side of the Malagarasi River, an assumed barrier river (Piel et al. 2013). Although the Malagarsi River is over 100 m wide for extensive stretches, both field observations and genetic analysis of chimpanzees populations either of the river, close to a ford, indicated that the chimpanzees made use of this natural bridge to cross the river (Piel et al. 2013). Thus the Malagarasi River is likely only a seasonal barrier to primate movement in areas where natural rock bridges may have

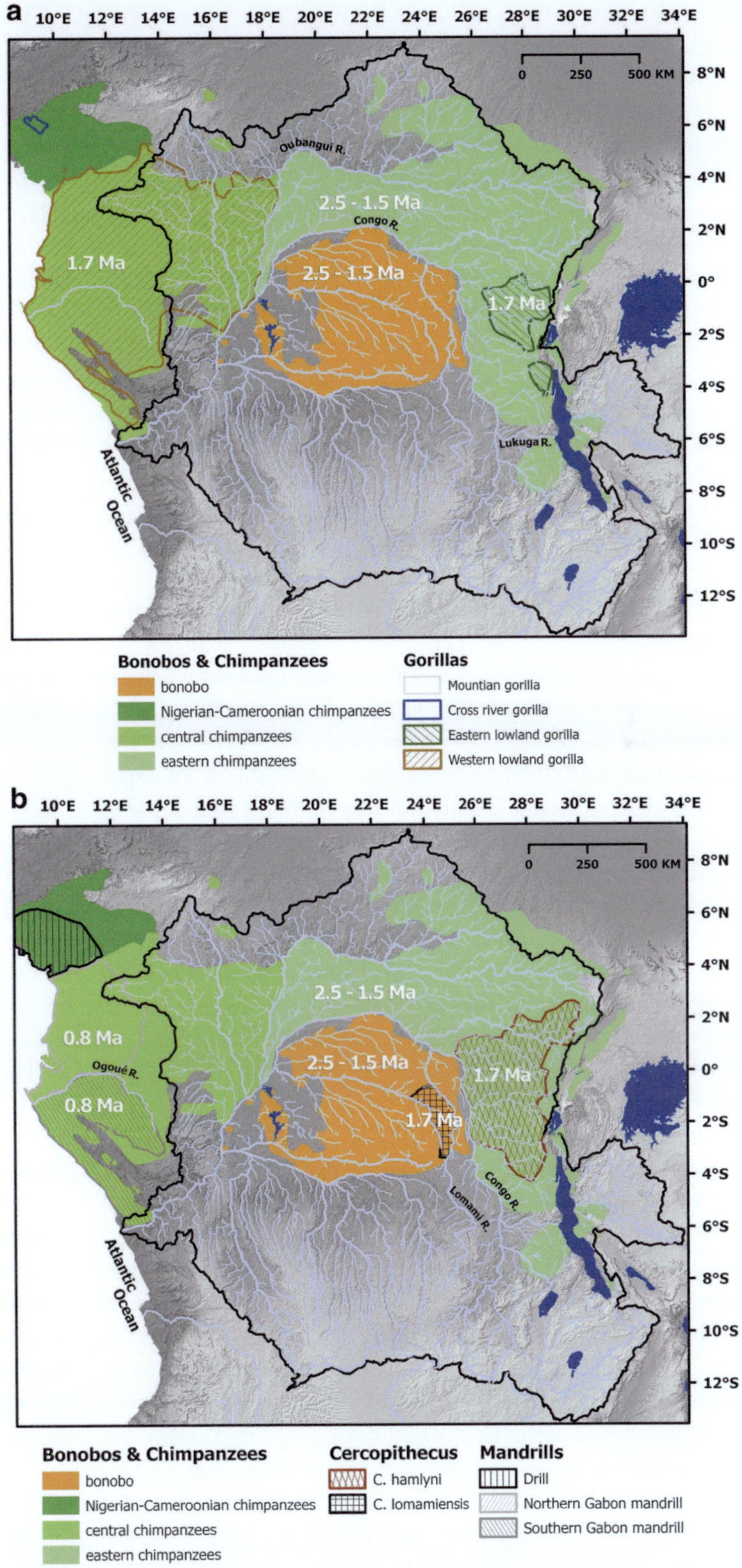

Fig. 15.8 (**a**) The geographical distribution and molecular divergence dates of bonobos, chimpanzees (Genus *Pan*) and gorillas (Genus *Gorilla*), in central and western Africa. The Lukuga River (outflow of Lake Tanganyika) appears to have become a recent barrier to eastern chimpanzees (modified from Prüfer et al. 2012; Scally et al. 2012). Chimpanzees (*green*) and bonobos (*orange*) were separated by the Congo River between 2.5 and 1.5 Ma; and the western and eastern gorillas show an analogous divergence at *ca.* 1.7 Ma. (Bonobo, chimpanzee and gorilla distributions derived from UNEP-WCMC and IUCN International Union for Conservation of Nature 2008a, b, c. (**b**) The geographical distributions of bonobos, chimpanzees, guenons (Genus *Cercopithecus*) and mandrills (Genus *Mandrillus*). See Molecular divergence dates of the southern and northern mandrills of Gabon in the west suggest that the present day Ogoué River became established 0.8 Ma (Telfer et al. 2003). To the east of the CB, divergence dates suggest that the guenon populations were separated by the Lomami – Congo rivers at 1.7 Ma (Hart et al. 2012). See Fig. 15.8a for congruent biogeoraphical patterns in of bonobos and chimpanzees

allowed gene flow (Piel et al. 2013). Similar situations may be found for other rivers of the CRS with more extensive and intensive field research. Ultimately information on paleoenvironmental changes in the CB during the late Noegene and Quaternary are required to fully elucidate the genetic structure of many of the primates in the basin (i.e. Kawamoto et al. 2013). Furthermore, the geographical pattern of a primate grouping may take hundreds of thousands of years to develop (i.e. Kawamoto et al. 2013), thus requiring rivers to be effective barriers during the entire period.

This also means there will be a lag time from when a river becomes a barrier and its effects are seen in the genetic signal.

15.4.2.1 Phylogenetic Age Estimates of Geomorphic Events

In central Africa, a synthesis of phylogenetic studies of primates provide an estimation of river emplacement (Fig. 15.8a, b; Table 15.3) in the Late Pliocene and Early Pleistocene. Evidence from the divergence of bonobos-chimpanzees suggests that the Equatorial arc of the Congo River was an effective barrier to large primate dispersal, with Prüfer et al. (2012) suggesting a 2.5–1.5 Ma, although a different genetic marker provides a 1 Ma age estimate (Kawamoto et al. 2013). The divergence estimate of 1.75 Ma for the western (*Gorilla gorilla*) and eastern (*G. beringei*) gorilla populations (Scally et al. 2012) suggests that the CR (and possibly the Oubangui) formed a barrier to gorilla dispersal at a similar time of the bonobo-chimpanzee divergence (Fig. 15.8a; Table 15.3). Genetic data for the lowland gorillas indicate a subsequent, moderate genetic exchange between the two groups, since the initial split suggesting movement of individuals between the two populations until the late Pleistocene; with the eastern population experiencing a genetic bottleneck that is indicative of low population numbers (Ackermann and Bishop 2010; Scally et al. 2012). Nevertheless both molecular and morphologically studies support the classification of the two species, eastern and western gorillas with further subdivision down to subspecies possible. This is further evidence for genetic structuring within the western and eastern gorilla populations, suggesting a barrier role of rivers smaller than the Congo River (Anthony et al. 2007; Ackermann and Bishop 2010) Yet more extensive sampling of both gorilla populations is needed to allow for better resolution of divergence times (Scally et al. 2012).

While the CR is an effective barrier to bonobos dispersal northward, almost all the dendritic to sub-dendritic drainage of the *Cuvette Central* (D-Sd2 in Fig. 15.3) have had little affect on gene flow between populations (cohorts), apart from the lower Lomami River (Fig. 15.5; Kawamoto et al. 2013). The barrier effect of the Lomami River is seen in the greater genetic distance of the bonobo cohort east of the river (the area between the Congo and Lomami rivers) compared to the central and western cohorts (Kawamoto et al. 2013). Additionally the genetic distances among the central and western bonobos cohorts suggest that in the central region of the basin that rivers only provide a weak barrier (Kawamoto et al. 2013).

The recently discovered species of guenon monkey in the genus *Cercopithecus* (*C. lomamiensis*), based on morphological and molecular evidence, provides proxy evidence of the emplacement of the Lomami River (Hart et al. 2012). *C. lomamiensis* is separated from its nearest congener, *C. Hamlyni*, by both the Congo and the Lomami rivers (Hart et al. 2012) (Figs. 15.5 and 15.8b; Table 15.3). *C. lomamiensis* is endemic to the region confined by the upper Tshuapa River in the west through to the Lomami River in the east, whilst the eastern-most distribution of *C. hamlyni* is bounded by the CR (Hart et al. 2012). Hart et al. (2012) suggest the Lomami and upper Congo rivers are the biogeographic barriers responsible for this speciation event. Molecular divergence dates estimate the time to MRCA of the *C. lomamiensis–C. hamlyni* clade based on two different genetic makers, provide age estimates of *ca.* 1.7 Ma (3.2–0.5 Ma at 95 % confidence level) and *ca.* 2.8 Ma (4.3–0.6 Ma at 95 % confidence level) respectively (Hart et al. 2012).

The congruence of the divergence data for the bonobos-chimpanzees, gorillas and guenon species, supports the emplacement of the modern, crescent shaped CR river by 1.5 Ma. The CR became an effective barrier to gene flow (i.e. larger than 100 m—assumed effective barrier size for chimpanzees see Piel et al. 2013) between 2.5 and 1.5 Ma based on the most recent common ancestor of bonbos and chimpanzees. This places an age constraint on the drainage pattern of the *Cuvette Central* (D-Sd2 in Fig. 15.3), with its relatively young age accounting for the mixture of dendritic and sub-dendritic features. The separation of *Cercopithecus* around 1.7 Ma by the northerly flowing Lomami River strengthens the emplacement age of the modern CR.

The north-western boundary of the CB is formed by the watershed between the Congo and Ogoué rivers is marked by the Batéké Plateau. The eastern portion of the plateau slopes gently toward the CR and is incised by numerous, actively eroding headwaters of tributaries of the Ogoué River (Seranne et al. 2008). Telfer et al. 2003 report a divergence time of the mandrill (*Mandrillus sphinx*) populations north and south of the Ogoué River at *ca.* 800 Ka, constraining the emplacement of the modern Ogoué River to at least 800 Ka. With a sediment load of 19.7×10^6 t a^{-1} the Ogoué River has a comparable sediment to the CR (22.7×10^6 t a^{-1}) although the Ogoué water volume is a magnitude less than the Congo's River, suggesting the Ogoué watershed is undergoing active erosion (Seranne et al. 2008). Thus the headwater regions of the Ogoué are actively incising backward, and have likely captured smaller north-western tributaries of the CR within the last 800 ka.

Table 15.3 The fluvial changes of in the Congo basin during the Neogene and Quaternary. The timing of events is based on age estimates reported by phylogeographic studies.

Period	Geomorphic event and time estimate	Phylogenetic evidence
Neogene		
Early to middle Miocene	(1) Diversion of north eastern Congo of the Aruwimi River headwaters drainage towards the Nile	(1a) "*Haplochromic*" species "Yaekama" is distributed in the north eastern CRS near Kisangani but groups with Lake Victoria superflocks (Schwarzer et al. 2012). (1b) A sistergroup of modern haplochromines found in Lake Kivu occurs in Lake Victoria (Schwarzer et al. 2012)
	(2) Major drainage re-organisations of central CRS drainage flowing off the southern escarpment with: (i) severance of connections between the Fwa, Inkisi and Kiwul Rivers (rivers presently adjacent to one another in south western CB); (ii) disruption of the Sankuru-Lukemie system, with diversion of present day Sankuru flowing into the Kasai system	(2a) Close phylogenetic relationship of "*Haplochromis*" species haplotypes of the Fwa-Inkisi-Kwilu Rivers (Schwarzer et al. 2012) (2b) Genetic markers of the Fwa "*Haplochromis*" closely related to these in the mid-Kasai and mid-Kwango Rivers and; Fwa haplochromides closely relates to "*H.*" cf *bakongo* and "*H.*" snoeki from the lower CR system (Schwarzer et al. 2012)
Late Neogene		
	(3) Partial re-arrangement across the headwaters of the Lufira, Kwango and Cuanza (an Angolan coastal basin) systems; capture of Zamnezian headwaters by the Lufira	(3) Hybrid taxa of northern and southern *Oreochromic torrenticolo* and *Serranochromis* sp. "red scales" that presently occur either side of the Congo-Cuanza and Congo-Zambezi divides (Schwarzer et al. 2012)
	(4) Approximately 4 Ma: Establishment of water flow across the modern day lower rapids of the CR connecting the CRS to the Atlantic Ocean (established by 5 Ma?)	(4) High level of cichlid flocks (*Steatocranus* and *Nanochromis*) diversification in the lower CR; these cichlids are endemic to rapids (Schwarzer et al. 2011)
	(5) 3.1 Ma : Disruption of drainage between Lake Tanganyika and Lake Victoria/East Africa drainage	(5) Phylogenetic dating of the timing of divergence of *Hydrocynus tanzania* and *H. vittatus* (Goodier et al., 2011)
Quaternary		
Pleistocene	(6) 2.5–1.5 Ma: Establishment of the 'modern', sickle shape of CR is established in the region of the Equator	(6a) Divergence date between chimpanzees occurring on the northern and eastern banks of the Congo River with bonobos restricted to an area inside the arc of the CR. Dates of 2.5–1.5 Ma (Prüfer et al. 2012). See Fig. 15.8a (6b) Divergence age between western and eastern gorillas of the CB calculated to have been at 1.75 Ma (Scally et al. 2012). See Fig. 15.8a
	(7) 2.0 Ma: Separation of the Bangweulu and Lufubu drainage systems from the upper-western CR tributaries	(7) Founding of *Hydrocynus* "clade B", and "Clade C" diverges from "Clade D" and *H. vittatus ca.* 2.0 Ma (0.8–3.0 Ma) (Goodier et al. 2011)
	(8) 1.8 Ma: The Lufubu system is isolated from the Bangweulu river system and flows into Lake Tanganyika	(8a) Speciation of *Pseudocranilabrus* "lufubu" (1.4–2.3 Ma) (Koblmüller et al. 2008). (8b) Divergence of *Synodontis nigromaulata 2* from *S. nigromaculata3* (1.0–2.7 Ma) (Day et al. 2009)
	(9) 1.7 Ma: Major south to north flowing tributaries of the *Cuvette Central* region are in place i.e. Lomami River	(9) Two genetic markers of the guenons populations (*Cercopithecus lomamiensis* and *C. hamlyni)* have been estimated to be 1.7 Ma (3.2 Ma–0.5 Ma at 95 % confidence level) and 2.8 Ma (4.3 Ma–0.6 Ma at 95 % confidence). *C. lomamiensis* is separated from *C. hamlyni* by both the Congo and Lomami rivers (Hart et al. 2012). See Fig. 15.8b
	(10) 1.5 Ma: isolation of Lake Taganyika from the CRS, with consequent establishment of the Lukuga River outflow	(10a) Phylogenetic dating of the *Hydrocynus* population in Lake Tanganyika compared populations in the Bangweulu system (Goodier et al. 2011) (10b) Lukuga River appears to from a partial barrier between two chimpanzee population (see Fig. 15.8a)
	(11) 0.8 Ma: Establishment of the modern day Ogoué River sands (Seranne et al. 2008)	(11) Divergence date between southern and northern mandrills (Telfer et al. 2003). See Fig. 15.8b

15.5 The Evolution of the Congo Basin Drainage

Owing to the sensitivity of rivers to vertical tectonic displacements, and their ability to adjust to surface warping through changes in their channel, and subsequently their drainage pattern (Howard 1967; Holbrook and Schumm 1999), the CB rivers patterns provide insight into the evolution of the broader landsurface. Although the time response to change remains difficult to estimate, as it depends on rock strength, stream characteristics and the magnitude of tectonic activity (Holbrook and Schumm 1999; Leturmy et al. 2003; Lucazeau et al. 2003).

Major changes in the fluvial geomorphology of the CB occurred during the late Cenozoic (Table 15.3). Throughout the Neogene, the major rivers of the CRS began to develop their modern day forms. This change was driven by several factors, dominantly structural controls, tectonics and related autogenic fluvial processes. However this ongoing development was not uniform, as can be seen from 18 of the present day drainage patterns, indicating that the development of the CRS has been multi-phase and has several dominant controls.

The rivers of the eastern CB have been highly influenced by the geological young, and ongoing tectonic activity of the EARS, in particular its Western Branch (Fig. 15.7). While the geodynamic effects of the Western Branch have a limited spatial extent in the north-east of the CB, it is extensive in the south-east being and has resulted in the present contorted, trellis and modified rectangular-angulate and sub-dendritic drainages (Table 15.2; Figs. 15.3 and 15.7). These drainages are probably the second youngest of the CB drainages, having undergone significant re-organisation since the development of the EARS. The affects of the EARS on drainage in the region likely began in the Late Miocene with the formation of horst and graben structures, with increasing tectonic activity, including uplift, until the Pliocene-Pleistocene (Ring 2008; Bauer et al. 2010, 2012; Decrée et al. 2010; Roller et al. 2010). This multiple tectonic phases resulted in the contorted pattern of over the majority of the region (Figs. 15.3 and 15.7). This is evident in Fig. 15.6c, where vertical incision is dominant and the river channels in the region have eroded to similar elevations.

According to Pinet and Souriau (1988) there are two phases that characterise regional uplift: the initial 2.5 Myr phase involving constant denudation focused on the uplifted region, resulting in high sediment production and infilling of basin areas. The second phase occurs once the tectonic activity has ceased, with a 25 Myr period of weaker erosion occurring across a greater spatial extent leading to sediment movement throughout the greater basin (Pinet and Souriau 1988). Therefore, in the CB, the uplifting eastern basin margin is likely providing a large sediment supply that is transported westward by the eroding eastern rivers, and along with sediments transported by the northerly flowing rivers of the south CRS, accumulates in the central basin (Fig. 15.6a, b), where it undergoes reworking before being deposited offshore. This accumulation of poorly consolidated sediments has allowed the central drainages (*Cuvette central* and Western Congolain wetlands) to accommodate fluvial changes through horizontal re-working of the sediments. Thus the rivers of the central basin are more likely to have accommodated rejuvenation and adjustment than those on the Precambrian basement, as shown by their sub-dendritic to dendritic pattern. This development of a highly dynamic river network in the region is probably younger than the drainages of the eastern basin; serving as the regional interface between changes in the eastern and western drainages and changes in the CRS base level at the CR mouth. Thus the eastern and central drainages of the CB are likely to be Pliocene in age, with elements being inherited from Middle to Late Miocene.

The peripheral drainage patterns to the south, west and north of the central basin, are dominantly structurally controlled (the trellis drainages of southern Batéké, Sembe-Ouseso, Kwango Valley and mid-Congo; Table 15.2 and Fig. 15.3) and along with parallel drainage in the south (Lunagwe, Table 15.2 and Fig. 15.3) pre-date the central basin drainage (i.e. mid-Cenozoic to Mid-Miocene). This is not to say there have been no changes in these drainages systems but rather that the dominant drainage was already established by the Mid-Miocene.

Biotic evolutionary events provide important lines of evidence to constrain timing of major changes in location of river channels of the Congo drainage net. These include timing of formation of the present day CR and expansion and development of the drainage network. Comparison of river topology at regional scales reveals strong controls of the local geology, as indicated by the juxtaposed drainage patterns.

The south-to-north flowing rivers of the central basin region are likely to precede the east-to-west flowing rivers of the region; with the northerly flowing rivers being emplaced *ca.* 2 Ma. This is evidenced by genetic differences of the guenons and genetic variance of bonobos to the east and west of the Lomami River (Hart et al. 2012; Kawamoto et al. 2013; Table 15.3 and Fig. 15.8a, b). None of the east-to-west flowing rivers (apart from the CR) appear to have been a barrier to bonbos (Kawamoto et al. 2013), these rivers are therefore likely younger and/or of have not been continuously flowing for a sufficient time to form a true barrier. Similarly, it appears that only the north-to-south lower Oubangui River is a barrier dividing central and eastern chimpanzees, with easern chimpanzees moving across the east-to-west flowing rivers on the northern central basin

(Fig. 15.8a). Both these situations in the CB may be analogous to that seen in the east-to-west flowing Malagarasi River (Fig. 15.3) that forms an ephemeral barrier to chimpanzee dispersal (Piel et al. 2013). The smaller east-to-west flowing rivers of the central basin appear to form ephemeral barriers (if they form barriers at all) which suggests that these rivers are dynamic, changing both in size and location.

It is important to consider the relative roles of inheritance and rejuvenation, versus the origin of new features as interacting controls over the overall evolution of the basin. A better understanding of the development of the CRS will be achieved with the increasing merging of geomorphic, geologic (especially direct dates of prominent features, such as large waterfalls, e.g. the Lower Congo) and phylogenetic studies. It is through the combination of the fluvial evidence, data about the landforms over which they flow and the species they host (fishes) and impact (primates) that increasingly detailed and accurate picture of landscape evolution over the Neogene will be elucidated.

Acknowledgements Tyrel Flügel thanks the National Research Foundation (South Africa), John Ellerman Foundation and the ERANDA Foundation for their funding. The authors also thank Maarten J. de Wit for his support and encouragement during the writing of this chapter as well as the two reviewers whose comments improved this document.

References

Ackermann RR, Bishop JM (2010) Morphological and molecular evidence reveals recent hybridization between gorilla taxa. Evolution 64:271–290

Amante C, Eakins BW (2009) ETOPO1 1 Arc-minute global relief model: procedures, data sources and analysis. NOAA technical memorandum NESDIS NGDC-24. National Geophysical Data Center, NOAA. doi: 10.7289/V5C8276M. Available online at: http://www.ngdc.noaa.gov/mgg/global/global.html. Accessed Jun 2011

Anka Z, Séranne M (2004) Reconnaissance study of the ancient Zaire (Congo) deep-sea fan (ZaiAngo Project). Mar Geol 209:223–244

Anka Z, Séranne M, di Primio R (2010) Evidence of a large upper-Cretaceous depocentere across the Continent-Ocean boundary of the Congo-Angola basin. Implications for paleao-drainage and potential ultra-deep source rocks. Mar Petrol Geol 27:601–611

Anka Z, Séranne M, Lopez M, Scheck-Wenderoth M, Savoye B (2009) The long-term evolution of the Congo deep-sea fan: a basin-wide view of the interaction between a giant submarine fan and a mature passive margin (ZaiAngo project). Tectonophysics 470:42–55

Anthony NM, Johnson-Bawe M, Jeffery KJ, Clifford S, Abernethy K, Tutin CEG, Lahm S, White LJT, Utley JW, Wickings EJ, Bruford MW (2007) The role of Pleistocene refugia and rivers in shaping gorilla genetic diversity in central Africa. Proc Natl Acad Sci 104 (51):20432–20436

Avise JC (2000) Phylogeography: the history and formation of species. Harvard University Press, Cambridge, p 447

Babonneau N, Savoye B, Klein B (2002) Morphology and architecture of the present canyon and channel system of the Zaire deep-sea fan. Mar Petrol Geol 19:445–467

Bauer FU, Glasmacher UA, Ring U, Schumann A, Nagudi B (2010) Thermal and exhumation history of the central Rwenzori Mountains, Western Rift of the East African Rift system, Uganda. Int J Earth Sci 99:1575–1597

Bauer FU, Karl M, Glasmacher UA, Nagudi B, Schumann A, Mroszewski L (2012) The Rwenzori Mountains of western Uganda – aspects on the evolution of their remarkable morphology within the Albertine Rift. J Afr Earth Sci 73–74:44–56

Burbank DW, and Anderson RS, (2001) Tectonic Geomorphology. Blackwell Science, Oxford. pp 274

Cahen L (1954) Géologie du Congo Belge. H. Vaillant Carmanne, Liège, p 490

Cahen L, Snelling NJ (1966) The geochronology of equatorial Africa. North-Holland Publishing Company, Amsterdam, Netherlands, p 195

Cotterill FPD, de Wit MJ (2011) Geoecodynamics and the Kalahari epeirogeny: linking its genomic record, tree of life and palimpsest into a unified narrative of landscape evolution. S Afr J Geol 114:489–514

Day JJ, Bills R, Friel JP (2009) Lacustrine radiations in African *Synodontis* catfish. J Evol Biol 22(4):805–817

Decrée S, Deloule É, Ruffet G, Dewaele S, Mees F, Marignac C, De Putter T (2010) Geodynamic and climate controls in the formation of Mio–Pliocene world-class oxidized cobalt and manganese ores in the Katanga province, DR Congo. Mineral Deposita 45:621–629

De Dapper M (1988) Geomorphology of the sand-covered plateaux in southern Shaba, Zaire. In: Dardis GF, Moon BP (eds) Geomorphological studies in Southern Africa. Balkema, Rotterdam, pp 115–135

Delvaux D, Kervyn F, Macheyeki AS, Temu EB (2012) Geodynamic significance of the TRM segment in East African Rift (W-Tanzania): active tectonics and paleostress in the Ufipa plateau and Rukwa basin. J Struct Geol 37:161–180

Deffontaines D, Chorowicz J (1991) Principles of drainage basin analysis from multisource data: application to the structural analysis of the Zaire Basin. Tectonophysics 194:237–263

DeSalle R, Rosenfeld J (2013) Phylogenomics: a primer. Garland Science, New York, p 338

Dixey F (1943) The morphology of the Congo-Zambesi watershed. S Afr Geogr J 25:20–41

Driscoll NW, Karner GD (1994) Flexural deformation due to Amazon fan loading: a feedback mechanism affecting sediment delivery to margins. Geology 22:1015–1018

Dupré B, Gaillardet J, Rosseau D, Allègre CJ (1996) Major and trace elements of river-borne material: The Congo Basin. Geochem Cosmochim Acta 60(8):1301–1321

Gaillardet J, Dupré B, Allèrge CJ (1995) A global geochemical mass budget applied to the Congo Basin rivers: erosion rates and continental crust composition. Geochem Cosmochim Acta 59 (17):3469–3485

Goodier SAM, Cotterill FPD, O'Ryan C, Skelton PH, de Wit MJ (2011) Cryptic diversity of African tigerfish (Genus *Hydrocynus*) reveals palaeogeographic signatures of linked Neogene geotectonic events. PLoS One 6(12):e28775. doi:10.1371/journal.pone.0028775

Goudie AS (2005) The drainage of Africa since the Cretaceous. Geomorphology 67:437–456

Hart JA, Detwiler KM, Gilbert CC, Burrell AS, Fuller JL, Emetshu M, Hart TB, Vosper A, Sargis EJ, Tosi AJ (2012) Lesula: a new species of *Cercopithecus* monkey endemic to the Democratic Republic of Congo and implications for conservation of Congo's Central Basin. PloS One 7(9):e44271

Harcourt AH, Wood MA (2012) Rivers as barriers to primate distributions in Africa. Int J Primatol 33:168–183

Holbrook J, Schumm SA (1999) Geomorphic and sedimentary response of rivers to tectonic deformation: a brief review and critique of a tool for recognizing subtle epeirogenic deformation in modern and ancient settings. Tectonophysics 305:287–306

Howard AD (1967) Drainage analysis in geological interpretation: a summary. AAPG Bull 51:2246–2259

Jarvis A, Reuter HI, Nelson A, Guevara E (2008) Hole-filled SRTM for the globe Version 4, available from the CGIAR-CSI SRTM 90m Database. Available online at: http://srtm.csi.cgiar.org. Accessed August 2010

Jarvis A, Reuter HI, Nelson A, Guevara E (2011) Hole-filled SRTM for the globe Version 4, available from the CGIAR-CSI SRTM 250m Database. Available online at: http://srtm.csi.cgiar.org. Accessed June 2011

Karner GD, Driscoll NW (1999) Tectonic and stratigraphic development of the West African and eastern Brazilian margins: insights from quantitative basin modelling. In: Cameron NR, Bate RH, Clure VS (eds) The oil and gas habitats of the South Atlantic, vol 153. Geological Society, London, pp 11–40, 474 (Special Publications)

Koblmüller S, Schliewen UK, Duftner N, Sefc KM, Katongo C, Sturmbauer C (2008) Age and spread of the haplochromine cichlid fishes in Africa. Mol Phylogenet Evol 49(1):153–169

Kobrick M (2006) On the toes of giants — how SRTM was born. Photogramm Eng Remote Sens 72(3):206–210

Kawamoto Y, Takemoto H, Higuchi S, Sakamaki T, Hart JA, Hart TB, Tokuyama N, Reinartz GE, Guislan P, Dupain J, Cobden AK, Mulavwa MN, Yangozene K, Darroze S, Devos C, Furuichi T (2013) Genetic structure of wild bonobo populations: diversity of mitochondrial DNA and geographical distribution. PLoS One 8(3): e59660. doi:10.1371/journal.pone.0059660

Laraque A, Bricquet JP, Pandi A, Olivry JC (2009) A review of material transport by the Congo River and its tributaries. Hydrol Processes 23:3216–3224

Laraque A, Mahé G, Orange D, Marieu B (2001) Spatiotemporal variations in hydrological regimes within Central Africa during the XXth century. J Hydrol 245:104–117

Lavier LL, Steckler MS, Briguad F (2001) Climate and tectonic control on the Cenozoic evolution of the West African margin. Mar Geol 178:63–80

Leturmy P, Lucazeau F, Briguad F (2003) Dynamice interactions between the Gulf of Guinea passive margin and the Congo River drainage basin: 1. Morphology and mass balance. J Geophys Res 108(B8):2383. doi:10.1029/2002JB001927

Leopold LB, Wolman MG, Miller JP (1964) Fluvial process in geomorphology. W.H. Freeman, San Francisco, 552

Lucazeau F, Brigaud F, Letumy P (2003) Dynamic interactions between the Gulf of Guinea passive margin and the Congo River drainage basin: 2. Isostasy and uplift. J Geophys Res 108(B8):2384. doi:10.1029/2002JB001928

Master S (2010) Lac Télé structure, Republic of Congo: geological setting of a cryptozoological and biodiversity hotspot, and evidence against an impact origin. J Afr Earth Sci 58:667–679

Mavonga T, Durrheim RJ (2009) Probabilistic seismic hazard assessment for the Democratic Republic of Congo and surrounding areas. S Afr J Geol 112:329–342. doi:10.2113/gssajg.112.3-4.329

Meade RH (1996) River-sediment inputs to major deltas. In: Milliman JD, Haq BU (eds) Sea-level rise and coastal subsidence: causes, consequences and strategies. Kluwer, Dordrecht, pp 63–85, 384

Moore AE, Cotterill FPD, Main MPL, Williams HB (2007) The Zambezi River. In: Gutpa A (ed) Large rivers: geomorphology and management. Wiley, New York, pp 311–332

Nicholson SE (2000) The nature of rainfall variability over Africa on times scales of decades to millennia. Global Planet Change 26:137–158

Piel AK, Stwart FA, Pintea L, Li Y, Ramirez MA, Loy DE, Crystal PA, Learn GH, Knapp LA, Sharp PM, Hahn BA (2013) The Malagarasi river does not form an absolute barrier to chimpanzee movement in western Tanzania. PloS One 8(3):e58965. doi:10.1371/journal.pone.0058965

Pinet P, Souriau M (1988) Continental erosion and large-scale relief. Tectonics 7(3):563–582

Prüfer K, Munch K, Hellmann I, Akagi K, Miller JR, Walenz B, Koren S, Sutton G, Kodira C, Winer R, Knight JR, Mullikin JC, Meader SJ, Ponting CP, Lunter G, Higashino S, Hobolth A, Dutheil J, Karakoç E, Alkan C, Sajjadian S, Catacchio CR, Ventura M, Marques-Bonet T, Eichler EE, André C, Atencia R, Mugisha L, Junhold J, Patterson N, Siebauer M, Good JM, Fischer A, Ptak SE, Lachmann M, Symer DE, Mailund T, Schierup MH, Andrés AM, Kelso J, Pääbo S (2012) The bonobo genome compared with the chimpanzee and human genomes. Nature 486:527–531. doi:10.1038/nature11128

Nibbelink K, Budihardjo S (2002) Paleo-Congo River Fan in Northern Gabon, AAPG annual meeting, March, Houston, USA

Robert M (1946) Le Congo Physique (3é edition). Presse Universitaires de France, Liége, p 499

Reuter HI, Nelson A, Jarvis A (2007) An evaluation of void filling interpolation methods for SRTM data. Int J Geogr Inf Sci 21 (9):983–1008

Ring U (2008) Extreme uplift of the Rwenzori Mountains in the East African Rift, Uganda: structural framework and possible role of glaciations. Tectonics 27, TC4018. doi:10.1029/2007TC002176

Roller S, Hornung J, Hinderer M, Ssemmanda I (2010) Middle Miocene to Pleistocene sedimentary record of rift evolution in southern Albert Rift (Uganda). Int J Earth Sci 99:1643–1661. doi:10.1007/s00531-010-0560-z

Runge J (2007) The Congo River, Central Africa. In: Gutpa A (ed) Large rivers: geomorphology and management. Wiley, New York, pp 293–309

Scally A, Dutheil JY, Hillier LW, Jordan GE, Goodhead I, Herrero J, Hobolth A, Lappalainen T, Mailund T, Marques-Bonet T, McCarthy S, Montgomery SH, Schwalie PC, Tang YA, Ward MC, Xue Y, Yngvadottir B, Alkan C, Andersen LN, Ayub O, Ball EV, Beal K, Bradley BJ, Chen Y, Clee CM, Fitzgerald S, Graves TA, Gu Y, Heath P, Heger A, Karakoç E, Kolb-Kokocinski A, Laird GK, Lunter G, Meader S, Mort M, Mullikin JC, Munch K, O'Connor TD, Phillips AD, Prado-Martinez J, Rogers AS, Sajjadian S, Schmidt D, Shaw K, Simpson JT, Stenson PD, Turner DJ, Vigilant L, Vilella AJ, Whitener W, Zhu B, Cooper DN, de Jong P, Dermitzakis ET, Eichler EE, Flicek P, Goldman N, Mundy NI, Ning Z, Odom DT, Ponting CP, Quail MA, Ryder OA, Searle SM, Warren WC, Wilson RK, Schierup MH, Rogers J, Tyler-Smith C, Durbin R (2012) Insights into hominid evolution from the gorilla genome sequence. Nature 483:169–175

Schwarzer J, Misof B, Ifuta SN, Schliewen UK (2011) Time and origin of cichlid colonization of the lower Congo rapids. PLoS One 6: e22380

Schwarzer J, Swartz ER, Vreven E, Snoeks J, Cotterill FPD, Misof B, Schliewen UK (2012) Repeated trans-watershed hybridization among haplochromine cichlids (Cichlidae) triggered by Neogene landscape evolution. Proc R Soc Lond Ser B. Published online, pp 1–11 doi:10.1098/rspb.2012.1667

Seranne M, Bruguier O, Moussavou M (2008) U-Pb single zircon grain dating of present fluvial and Cenozoic aeolian sediments from Gabon: consequences on sediment provenance, reworking, and erosion processes on equatorial West African margin. Bulletin de la Société géologique de France 179(1):29–40

Tack L, Fernandez-Alonso M, Trefois P, Lavreau J (2003) New data raise new questions on the regional geology of the Katanga Province as figured on the 1974 Geological Map (1/2 000 000) of the Democratic Republic of Congo (DRC). In: Cailteux JLH (ed) Proterozoic base metal deposits of Western Gondwana. 3rd IGCP-450 conference and guide book of the field workshop, pp 78–82

Telfer PT, Souquiere S, Clifford SL, Abernethy KA, Bruford MW, Disotell DR, Sterner KN, Roques P, Marx PA, Wickings EJ (2003) Molecular evidence for deep phylogenetic divergence in Mandrillus sphinx. Mol Ecol 12:2019–2024

UNEP-WCMC and IUCN (International Union for Conservation of Nature) (2008a) Gorilla gorilla. In: IUCN 2013. IUCN Red List of Threatened Species. Version 2013.2. http://maps.iucnredlist.org/map.html?id=9404. Accessed 7 Jan 2013

UNEP-WCMC and IUCN (International Union for Conservation of Nature) (2008b) Pan paniscus. In: IUCN 2013. IUCN Red List of Threatened Species. Version 2013.2. http://maps.iucnredlist.org/map.html?id=15932. Accessed 7 Jan 2013

UNEP-WCMC and IUCN (International Union for Conservation of Nature) (2008c) Pan troglodytes. In: IUCN 2013. IUCN Red List of Threatened Species. Version 2013.2. http://maps.iucnredlist.org/map.html?id=15933. Accessed 7 Jan 2013

Veatch AC (1935) Evolution of the Congo Basin. Mem Geol Soc Am 3:183

WWF (2006) A vision for biodiversity conservation in Central Africa: biological priorities for conservation in the Guinean-Congolian Forest and freshwater region. WWF-US/ Central Africa Regional Program Office, Washington, DC

Zernitz ER (1932) Drainage patterns and their significance. J Geol 40 (6):498–521

Part V

Resources: Economic and Potential Deposits

Diamond-Bearing Gravels Along the Lower Kwango River DRC

16

Michiel C.J. de Wit and Edmond Thorose

16.1 Introduction

The Kwango/Cuango River rises in the highlands of Alto Chicapa in the Angolan Province of Lunda Sul, where it is referred to as the Cuango River. It descends from an elevation of 1,395 m amsl and flows for some 1,100 km before it joins the Kasai River in the Democratic Republic of Congo (DRC) at 292 m amsl. It drains a catchment area of some 263,500 km^2 (http://www.wolframalpha.com) (Fig. 16.1) with an average annual discharge of 2,700 m^3/s (en. Wikipedia.org). It is the largest left bank tributary of the Kasai River. The Wamba and Kwilu rivers are the Kwango's main right bank tributaries before the confluence with the Kasai.

In Angola the Cuango River flows in a northerly direction and covers a distance of 455 km before the Angolan/DRC border just south of Tembo, from where it is known as the Kwango River. The river then follows the Angola/DRC border for some 300 km. It then flows for roughly 345 km in the DRC before the Kwango River joins the Kasai near the town of Bandundu.

Diamonds are associated with alluvial sediments within this river system both in Angola and the DRC and this paper covers the exploration results from terrace and flood plain (locally referred to as river flat) deposits associated with the river section that forms the border between the two countries (Fig. 16.1).

M.C.J. de Wit (✉)
Delrand Resources Pty Ltd, Toronto, ON, Canada
e-mail: mdewit@delrand.com

E. Thorose
Vit Gold Corporation, Toronto, ON, Canada
e-mail: ethorose@vitgoldcorp.com

16.2 Historical Background

The first diamond in the DRC was discovered in 1903 in the Katanga Province by *Tanganyika Concessions Ltd (TCL),* which ultimately led to the discovery of the barren Kundulungu kimberlite province in 1908. In 1906 geologists of the newly established company *La Société Internationale Forestière et Minière du Congo (Forminière)* visited the Kwango River mainly in the search for gold. This was followed with prospecting expeditions by Janot and Dunbar in 1909 and 1910 in the search for diamonds but none were found (Fieremans 1977). In 1909 the same company had already discovered the now famous Tshikapa Alluvial diamond fields, 500 km east of the Kwango basin in which Janot was involved (Forminière 1956).

Geological missions of the Lower Kwango were conducted from 1912 to 1914 and of the Wamba River in 1914 by Passau and Asselberghs respectively. The latter person, on behalf of the *Syndicat Général des Recherches Minières du Kasai (SGK),* described some of the local geology but more importantly recognized five terrace levels between Kingushi and the Francois-Joseph waterfalls along the Kwango River (Fig. 16.2) (Asselberghs 1919).

Diamonds were recorded for the first time on the Kwango by Young in 1921 (Fieremans 1977) who recovered five small diamonds (weighing 0.195 cts; average size of 0.039 cts/stn) from two samples just below the confluence of the Kwilu and Kwango Rivers (Young 1921a, b). This was followed with an expedition by Young and Faucett in 1921–1922 who prospected the river from Bandundu to the Lonzo/Kwango confluence (Fig. 16.2) where 10 diamonds were recovered close to this confluence weighing 0.43 cts (average size of 0.043 cts/stn) (Young 1921a, b). Hill and Hoffmann who travelled from the Lovua River, south-west of Tshikapa, to the Francois-Joseph Falls in 1921 and then downstream to Kasongo Lunda reported no diamonds (Fieremans 1977).

Smith and Faucett were instructed by *Forminière* and *SGK* to explore the upper Kwango in the third major

M.J. de Wit et al. (eds.), *Geology and Resource Potential of the Congo Basin*, Regional Geology Reviews,
DOI 10.1007/978-3-642-29482-2_16, © Springer-Verlag Berlin Heidelberg 2015

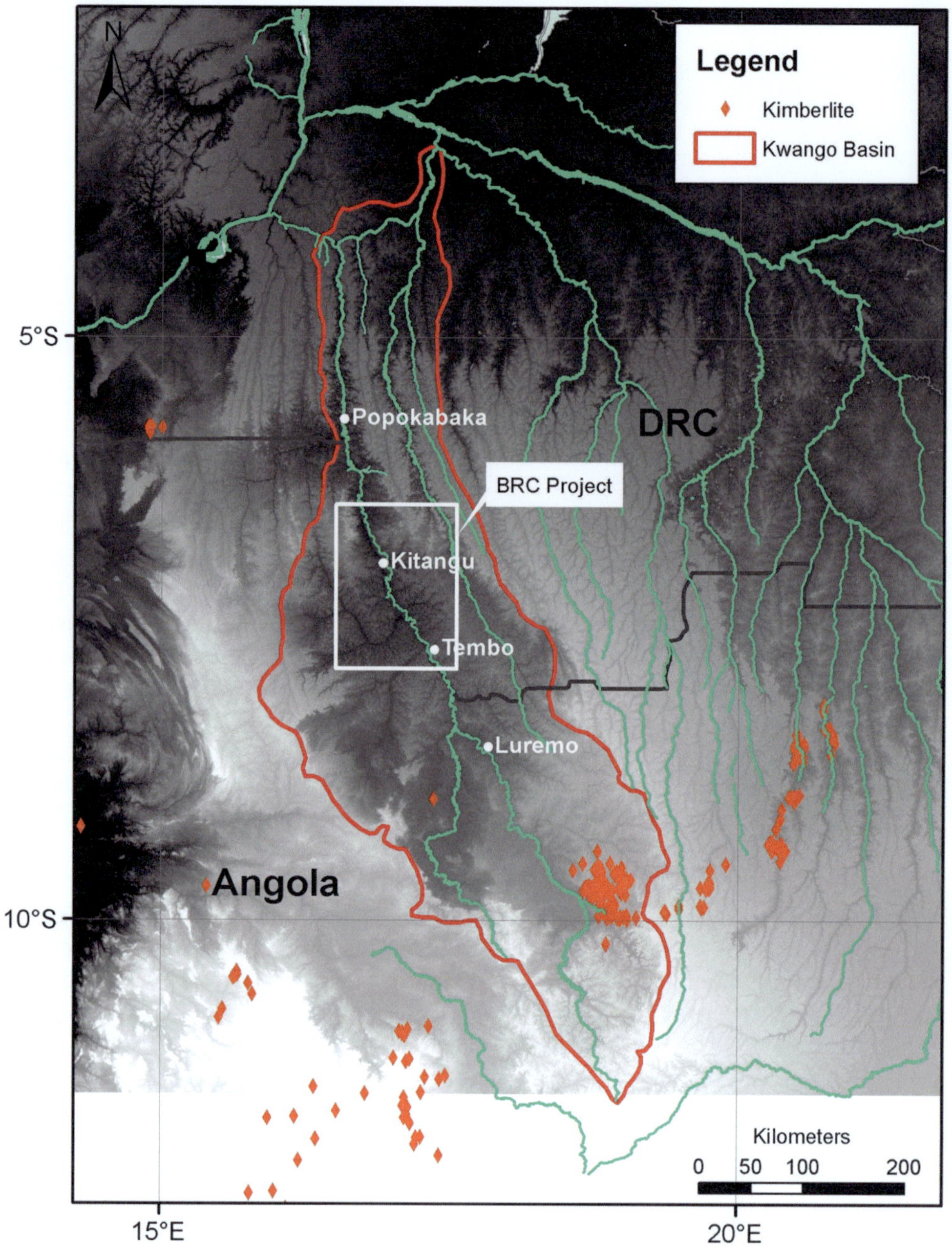

Fig. 16.1 Catchment area (in *red*) of Kwango/Cuango River basin. Main project area is *highlighted* by the *rectangular* block. *Orange diamonds* are kimberlites

prospecting mission in 1922–1923. Prospecting was done by digging pits and drilling holes using Banka 6″ and 4½″ auger. The best recoveries and highest grades were obtained between Popokabaka and the Fufu tributary of the Kwango (Fig. 16.2), where five samples produced 36 diamonds weighing 1.8 cts (average size of 0.05 cts/stn). No diamonds, however, were reported from Kasongo Lunda to the Francois-Joseph Falls further upstream. They believed that this was due to the presence of thick overburden sands and large boulders of silicified sandstone (Smith and Faucett 1922). The diamond grades between Kasongo Lunda and Popokabaka varied from 0.3 to 1.5 cts/m^3. Simultaneously diamonds with grades between 0.04 and 0.5 cts/m^3 were also found in the neighbouring Wamba drainage basin east of the Kwango River. The average diamond size in the Wamba decreased from 0.06 cts/stn (12 diamonds totalling 0.7 cts) at Kabongo to 0.03 cts/stn (12 stones totalling 0.29 cts) at Lukuni (Fig. 16.2) (Smith and Faucett 1922; Grant 1924b).

Young, however, believed that diamonds must increase southward, and this was confirmed by prospectors Johnston and McVey who had found diamonds much larger in size along the Cuango River to the south in Angola (Fieremans 1977). Hence ‘Mission Kwango’ was initiated by *SGK* and between 1924 and 1925 a more detailed sampling programme was conducted along the Kwango River and its tributaries between Popokabaka and Kasongo Lunda.

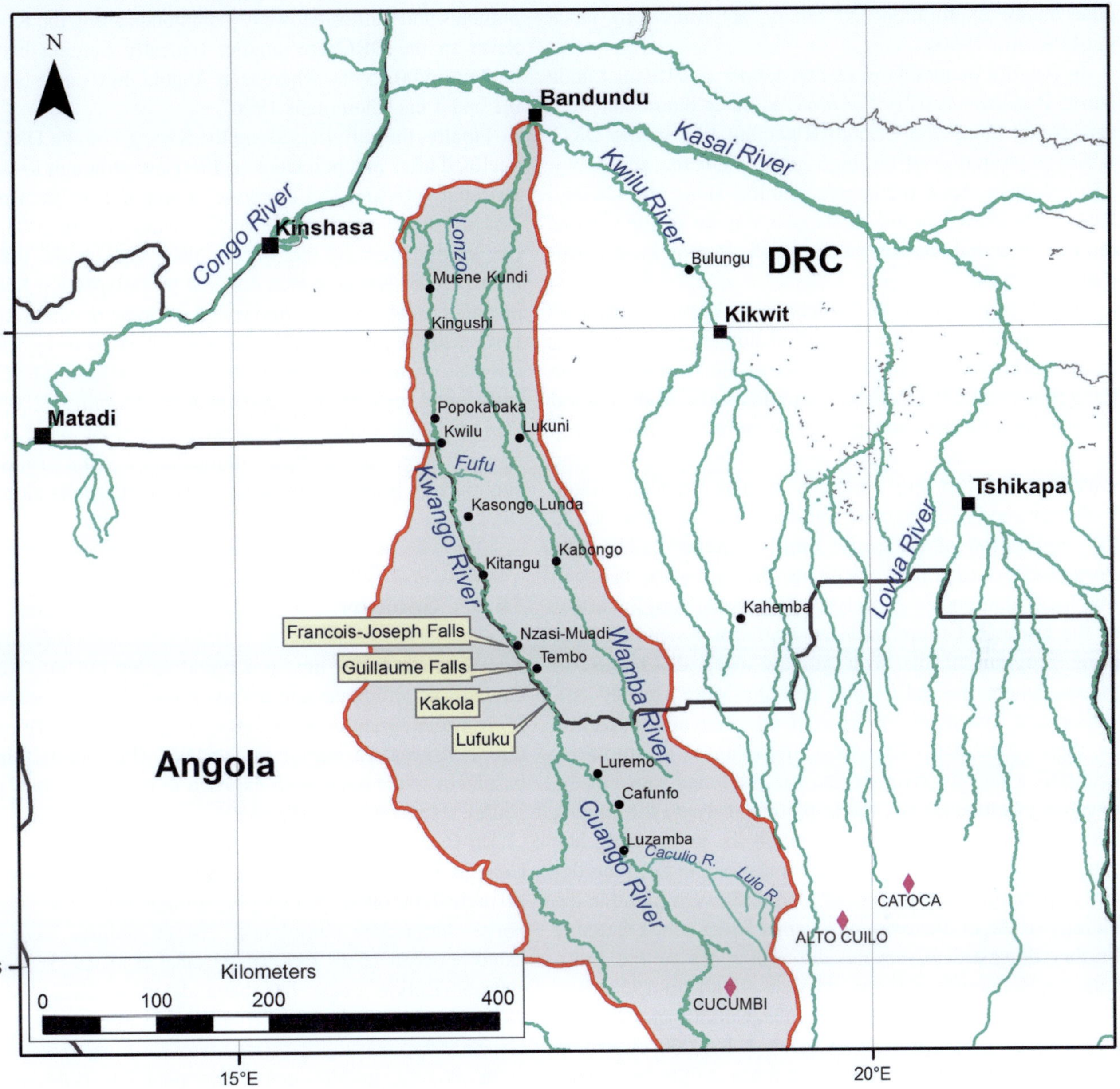

Fig. 16.2 Locality map of the Lower Kwango River Valley with main localities mentioned in the text

Drilling and pitting undertaken by Grant and Faucett in 1924 (Faucett 1924a, b; Grant 1924a) confirmed that diamond grade results were higher between Popokabaka and Kasongo Lunda, and between Kingushi and Muene Kundi (Fig. 16.2). Grades of between 0.01 and 0.39 cts/m^3 were obtained between the Fufu tributary and Popokabaka (the largest stone being 0.87 cts) (Grant 1924a), and between 0.07 and 0.10 cts/m^3 near Kingushi and up to 0.61 cts/m^3 near Muene Kundi but with average stone sizes between 0.04 and 0.06 cts/stn (Faucett 1924a, b; Grant 1924a). They also concluded that the diamonds were restricted to the present river beds bordering the flats, and that no diamonds were present in any of the tributaries (Fieremans 1977). However, after the discovery of the Tshikapa alluvial diamond fields (1909) and then the Mbuji Mayi diamond fields (1918), the search for diamonds in other areas was less important and exploration along the Kwango decreased significantly.

This changed in 1934 when the first two diamonds south of the Francois-Joseph Falls were found (one of which was a 1.52 carat stone, the largest so far recorded) by Verdeyen from *Forminière*, whilst exploring on behalf of the *American Congo Cy*. As a result, that same year, the Kwango River was covered by mining permits allocated to the *Société Minière de Lueta* with *Forminière* responsible for the prospecting and mining operations. Although Aderca spent time exploring the river in 1937 and 1938 (Aderca 1939), no other major exploration and mining activities took place until the mid-1950s.

In Angola diamonds were first found in 1912 in Lunda Norte Province also by *Forminière*. Later diamonds were also found along the Cuango River but like in the DRC, where exploitation of the high-grade Tshikapa and Mbuji Mayi deposits took preference, mining along the Cuango River only started in the 1950s as it was far to the west of the rich diamond deposits of the Lunda Provinces in north-east Angola.

In 1954 *Companhia de Diamantes de Angola (Diamang)* in Angola undertook an investigation along the western bank of the Kwango River in the vicinity of the Francois-Joseph Falls (Fieremans 1977). Prospecting focussed on the natural traps formed by the many gullies, fractures and potholes in the granites around the falls near Tembo where washing gravels, some 25 km upstream of the Francois-Joseph Falls, returned 81 diamonds weighing 9.42 cts. This led to the exploration of the entire upper Cuango Valley. The Portuguese during their early prospecting work believed that local diamonds were derived from basal conglomerates of the Kwango Group. *Diamang* started a systematic sampling program of the river gravels using diversions. In places, where the gullies and potholes were cleaned out, subsequent investigations showed renewed mineralization in these localities. The most important concentrations occurred where the river cut the basement and rocks of the overlying Bembe Group, and in particular where the contact between the two is exposed. By 1958 the prospecting had progressed to about 130–150 km south of the DRC border and important concentrations of diamonds were found in the vicinity of the bridge over the Cuango River near Luzamba. Further downstream, beyond the confluence of the Lulo-Cuango, diamonds of 0.5–1 cts were found. In 1958 river diversions were used around Luzamba, Cafunfo and Luremo and by late 1980s production reached 100,000 cts/month from the Cuango River. Potholes and water falls (e.g Tazua falls) were targeted with record recoveries of up to 50,000 cts from one pothole. By 1983 29 % of Diamang's production came from Cuango Division and by 1991 as much as 80 % of Angola's production came from the Cuango (PAC 2004). The increase in diamond sizes upstream in the Cuango Valley is further illustrated from workings in the Caculio tributary. Here large diamonds (131.5 and 38.3 carat stones) have recently been recovered in alluvial operations (Mining Weekly 2012) that have reported average stone sizes of between 1.35 cts/stn and 2.84 cts/stn (Lonhro Mining 2011).

Based on these results the DRC government ordered another investigation of the region between the Guiliame and Francois-Joseph Falls in 1961. The results confirmed earlier work from 1959 and demonstrated that significant local concentrations of diamonds were to be found in potholes and gullies. However, diamonds along the Kwango River in the DRC are smaller typically ranging between 0.07cts and 0.15 cts, whereas in Angola they range between 0.1 and 1 cts (Fieremans 1996).

Finally, the activities along the Kwango on the DRC side declined after independence in 1960 and although low-level artisanal activities did continue, it was not as intensive as that in Tshikapa and Mbuji Mayi. It was not until 2003, with the introduction of the new mining legislation, that the international Kwango was covered by prospecting licences awarded to *Le Grand Congo sprl* (later to be renamed *Acacia sprl*) with *Midamines* (De Decker 2006) securing ground directly upstream of the Falls. Canadian junior *BRC Pty Ltd* signed an option agreement with *Acacia* in 2005 and started a systematic exploration program of the Kwango alluvial deposits almost immediately. The following summarizes some of the *BRC*'s work and results along the river.

16.3 Geology

The Kwango River lies in a broad valley cut into mainly Cretaceous sediments and the river has exhumed sections of Precambrian granites and gneisses notably around Tembo in the south-western corner of the DRC (Fig. 16.3). Similar basement rocks have been exposed in the southern Wamba Valley some 60 km to the east.

De Carvalho et al. (2000) refer to the basement in the Cuango Valley as the Cuango Shield, which is composed of undifferentiated granite-gneiss-migmatites and gabbro-norite-charnockite complexes. The migmatites have been dated between 2680–2590 Ma (Delhal et al. 1975), and the charnockisation event has been dated at 2822 Ma and 2820 Ma by Delhal et al. (1976) and Cahen et al. (1984) respectively.

In Angola, granite-gneiss basement has been mapped along the Lui River, near the confluence with the Cuango. Neoproterozoic rocks of the West Congolian Belt (Bembe Group) have been described by Forward and Jeffcock (2003) along the Cuango north of Luremo. Interestingly around the town of Cuango in Angola the hills are composed of gneiss and charnokite, whilst the valleys have been developed in the softer 'schisto-calcaire' meta-sediments of the Neoproterozoic Bembe Group (Delhal and Fieremans 1964).

Basement rocks from the Francois-Joseph Falls area were giving a minimum age of 2600 Ma by Mendes (Cahen et al. 1984) based on geochronological results which plot close to the Dibayan Complex gneisses' isochron in the Kasai, approximately 300 km to the east. The Dibayan complex was later dated at 2650 Ma (Delhal 1991). Fieremans (1977) interprets conglomerates and schists that occur close to the Francois-Joseph Falls as being part of the Bembe Group.

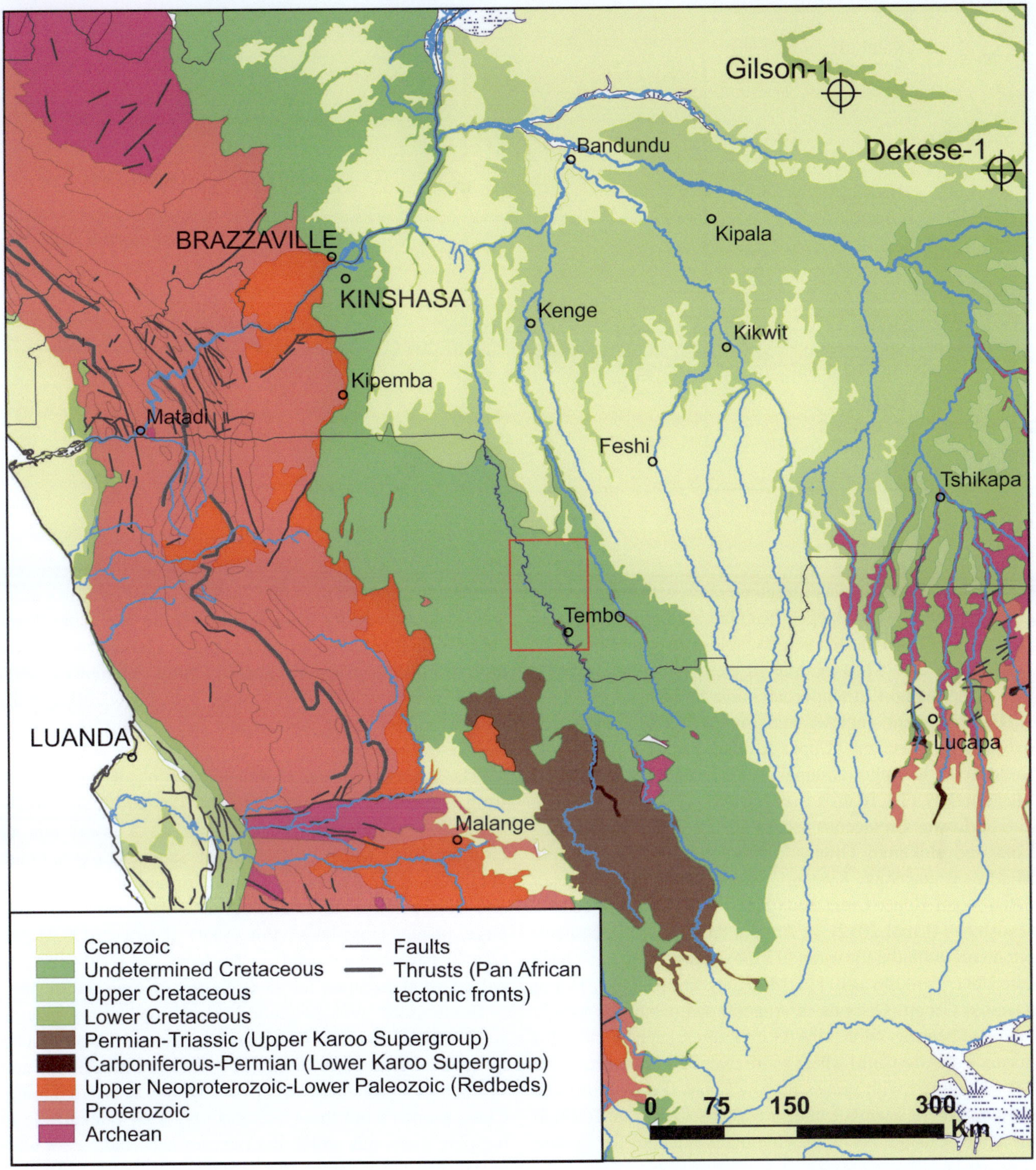

Fig. 16.3 General geology of the Kwango/Cuango river basin in south-western DRC and northern Angola (Linol 2012). *Red rectangle highlights the study area*

Rocha-Campos (1976) describes diamictites just west of Cuango, with boulders of basement granite and gneiss and red arkoses and shales of the 'Schisto-Gréseux Series' of the lower West Congo System. These diamictites are part of the Lutõe Series and in NE Angola contain flora which indicate an age range of Upper Carboniferous to Lower Permian (Rigby 1973). In the Upper Cuango basin these diamictites are overlain by fine-grained sediments with fish fossils considered to be Permian-Triassic in age (Teixeira 1948).

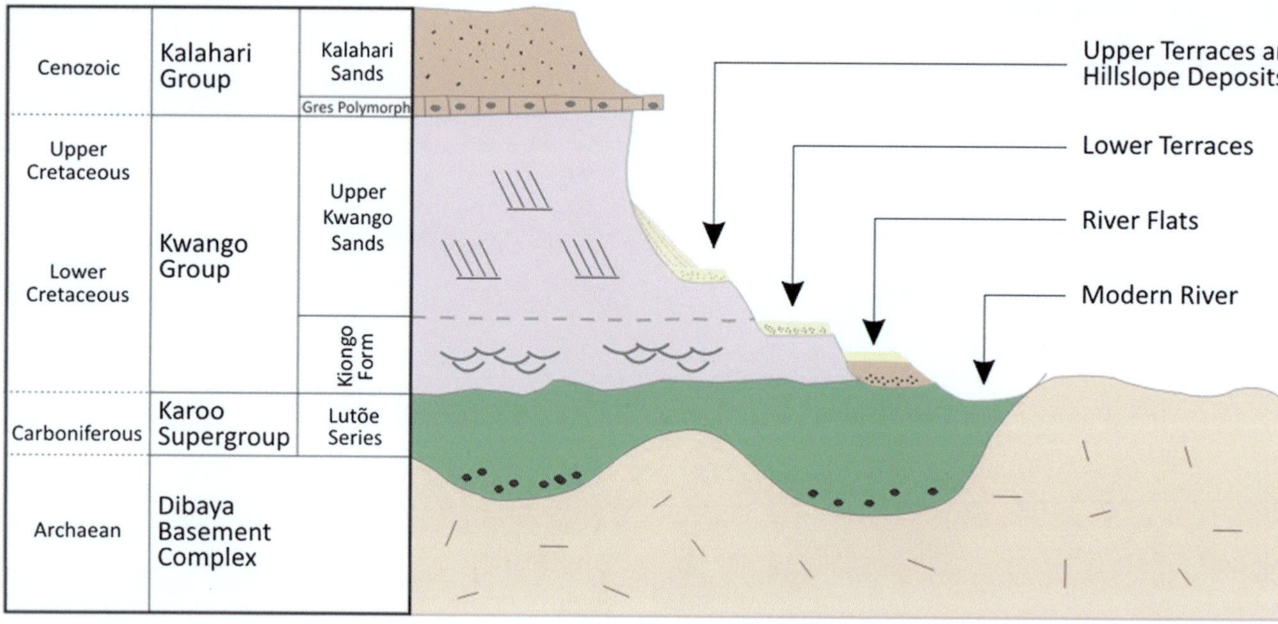

Fig. 16.4 Simplified stratigraphic section

Planar and cross-bedded conglomerates, with red siltstones and dropstones, were mapped from the base of the Francois-Joseph Falls up to the Massemba rapids just upstream of the falls. Linol (2012) has interpreted these sediments as being part of the Permo-Carboniferous Lutõe Series.

The basement gneisses and Proterozoic sediments are overlain by Mesozoic (Upper Jurassic Lualaba Series, and Cretaceous Kwango Group) and Cenozoic sediments as illustrated by Fig. 16.4.

The Lower Cretaceous sediments are represented by the Loia and Bokungu Groups (lacustrine), which pass into Middle Cretaceous Kwango Group sediments (fluvial/deltaic) and Upper Cretaceous Kamina Series cross-bedded sandstones (Linol 2012). In Angola these units are contemporaneous with the intrusion of kimberlites which have been dated between 146 and 111 Ma (Eley et al. 2008). Hence Kwango Group (Calonda in Angola) sediments are locally diamond bearing. Mineralised kimberlites, such as those for instance at Alto Cuilo where more than 73 kimberlites have been identified with grades up to 23 cpht (Pettit 2009) and Cucumbi in the Luangue cluster (Robles-Crus et al. 2009) in the upper Cuango basin, provide diamonds into the system (Fig. 16.1).

The basal unit of the Kwango Group consists of irregular lenses of conglomerate with abundant agates and heavy minerals. These conglomerates have, in places like Tshikapa and along the upper Cuango and Lui Rivers in Angola, been mined for diamonds, but are absent along the 'international' Kwango.

The age of the Kwango Group is believed to be Cenomanian to Turonian for the lower part and Campanian to Maastrichian for the upper part (Gobbo-Rodrigues et al. 2003). This provides a maximum age for the Late Cretaceous Kalamba planation surface that is marked by silcretes and, in turn, by younger Kalahari Group sediments (Cahen and Lepersonne 1952). The latter forms low rolling hills which terminate in a plateau that lies about 150 m above the present river.

The Kwango Group sediments that occur downstream from the Francois-Joseph Falls are composed of aeolian sandstones (Linol 2012), and comprises a lower unit of cross-stratified white, coarse-grained sandstones containing ventifacts referred to as the Kiongo Formation (Linol 2012), and an upper fine- to medium-grained red sandstones with large planar cross-beds (Fig. 16.4) (Linol 2012). Linol (2012) has demonstrated a southerly to south-westerly paleao-wind direction for these Kwango Group sediments.

Silcretes or 'grès polymorphes', which occupy a large section of the southern DRC and Angola, and which contain freshwater gastropod faunas (De Ploey et al. 1968), and sands of the Kalahari Group overly the Kwango Group. Clasts eroded from these silcretes play a significant role in trapping diamonds within the river terraces and recent alluvium that have essentially a Kwango Group sandstone footwall.

Sands of the Kalahari Group, or the 'serie des sables ocre' (ochre sandstones), are homogeneous, reddish to yellowish in colour and between 100 and 300 m thick. Originally thought to have an aeolian origin, De Ploey et al. (1968) suggest that the sands could have been deposited in a fluvial environment, with slow currents, low gradient and constant arid climate and tectonic stability. The grains do indicate

Fig. 16.5 Mid-Pleistocene hand axes made from silcrete or grès polymorph in K33 Hilltop terrace (*Right*)

however, some physical characteristics of a previous aeolian transportation phase.

Mineralogical evidence of kaolinite-iron oxyhydrate clay mineral associations, present in both the Upper and Lower Kalahari sands, indicate deep weathering of the underlying formations, which formed under hot and humid palaeoclimates of the Late Cretaceous and Mid Tertiary pedogenic stages (De Ploery et al. 1968).

Raised terrace deposits and lower level flats are found adjacent to the modern river system. Just below the Francois-Joseph Falls hill-top gravels have been mapped at roughly 40 m above the present river level and these contain artifacts (hand-axes) mostly made from 'grès polymorph' silcrete (Fig. 16.4). These resemble the Mid Pleistocene tools described by Clark (1966) and are characteristic of the Lumpembian industry (Fig. 16.5).

This provides a maximum age for the Kwango terraces at approximately 300,000–500,000 years. Similar tools are found in the terrace gravels farther downstream deposited on Kwango sandstones. The terrain models of some areas highlight the presence of several terraces, of which the most elevated contain the Middle Pleistocene tools. Most of the artisanal activity is however confined to the lower terraces.

16.4 Geomorphology

In Angola the Cuango River rises in Kalahari sands and Upper Cretaceous sediments before it reaches the sandstone and shales of the Karoo Supergroup in the Cassanje Graben, a depression in north-central to north-western Angola. From there the Cuango flows onwards to Luremo where it cuts into basement rocks. It is here that some of the most important diamond concentrations occur within the Kwango/Cuango Valley.

In the DRC crystalline basement also forms the footwall where the Kwango River passes between a narrow, steep-sided valley upstream of Tembo (Fig. 16.6). In these narrow channels, river-bank deposition is limited, but the gravels generally contain higher diamond grades. The migmatite basement forms an irregular floor characterised by gravel-filled troughs and potholes (Fig. 16.6), exhibiting extremely high diamond trapping potential. Extensive indurated, closely-packed diamondiferous gravel was observed in these potholes and gullies.

Between and directly below the Guilaume Falls near Tembo and the Francois-Joseph Falls near Nzasi-Muadi (Fig. 16.7), exposures of Palaeozoic fluvial-glacial sediments have been mapped forming a short but thin package and are overlain by Cretaceous aeolian sediments of the Kwango Group (Linol et al. 2011). These red aeolian sandstones form the footwall of the remainder of the Kwango River downstream of the Francois-Joseph Falls.

Where the Kwango Group sediments form the base to the gravels, the River is wider and has a gentler gradient with more extensive floodplains or flats and terraces. The terraces contain widespread basal gravel layers representing broader, shallower palaeo-channels, dominated by silcrete and calcrete clasts derived from the now eroded Upper Kwango Group sediments. Over the relatively flat Mesozoic base the trapping of diamonds appears to have been controlled by these boulders of silicified sandstone and calcrete. The gravels are generally thin (<30 cm), but appear to be thicker upstream near the basement contact. In all areas the gravels are overlain by younger fluvial quartz sand/silt sediment that is between 1 and 15 m thick. The youngest deposit types are those within the current river bed and these are being exploited for diamonds around Nzasi-Muadi and just downstream from Kitangu.

Based on the geomorphological studies, using air-photography, Landsat imagery, helicopter surveys and field mapping, 24 targets were identified for further investigation. These include 11 terraces (including a hilltop remnant), 11 flats and two splay deposits (Fig. 16.8).

The theoretical influence of the river geomorphology on trap site development, gravel deposition and diamond grades is schematically presented in Fig. 16.9. Geomorphological the river is confined and controlled by bedrock and bedrock

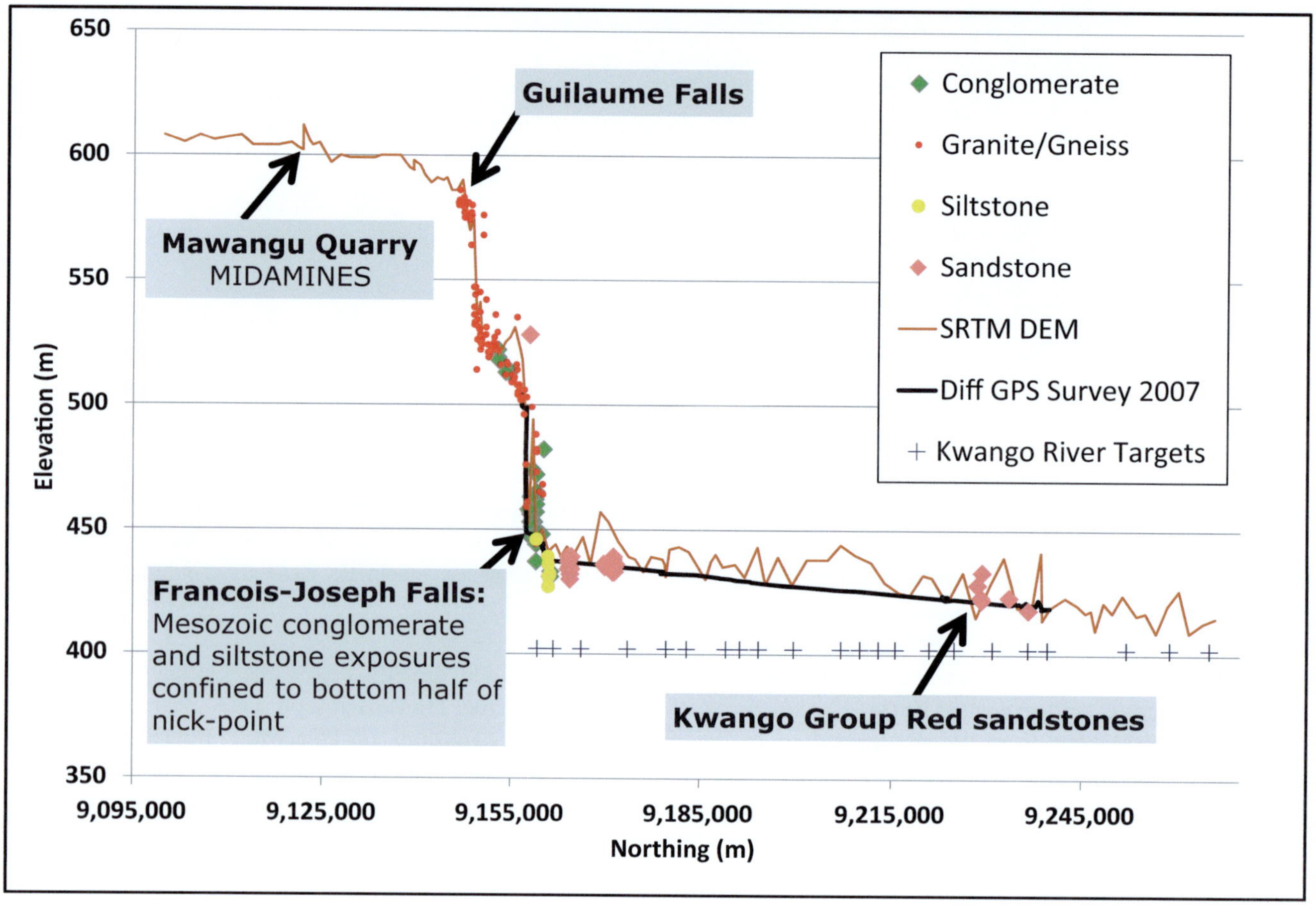

Fig. 16.6 Abandoned channel at *Midamines*' Kakolo Mine. Note straight channels following local basement joint patterns (photo Ward)

structures between targets K24 (just north of Tembo) and K33 (Figs. 16.8 and 16.9). Here the gravel volumes are low, but conversely the grades are high. At K33 downstream from Tembo the first splay develops and it is here where the river enters the Upper Kwango aeolian sediments, depositing its coarse bedload. Basal gravel thicknesses remain limited and although volumes have increased due to lateral expansion, grades have now lowered. Downstream of the first splay the Kwango River Valley widens between the K33 and K48 targets resulting in the formation of wide terraces and flats and deposition of excessive amounts of sand (Flats 3 to 11). Here the gravels are thin and grades are low, virtually no artisanal diamond activity is observed suggesting an absence of any significant deposits. Between K48 and K56 the valley narrows on the soft Kwango Group footwall and narrow terraces have developed (Figs. 16.8 and 16.9). Gravel deposits, dominated by silcrete and calcrete clasts, have increased but the grades remain low to moderate.

16.5 Geophysics

16.5.1 Ground Geophysics

Ground surveys using electromagnetic, gravity and magnetic methods were tested to detect sediment-filled palaeo-channels and/or bedrock irregularities, and whether an airborne equivalent could be used for this purpose. Five areas were surveyed, (K24, K33) near Tembo and (K48, K52 and K56) around Kitangu (Fig. 16.10), where the extent of the gravels could be extrapolated with reasonable confidence from the artisanal workings. The results suggested that an airborne electromagnetic (EM) method would be most effective in locating gravel-filled palaeo-channels, particularly in terraces away from the river.

The electromagnetic profiles across the targets best demonstrate the presence of buried channel features, particular in the

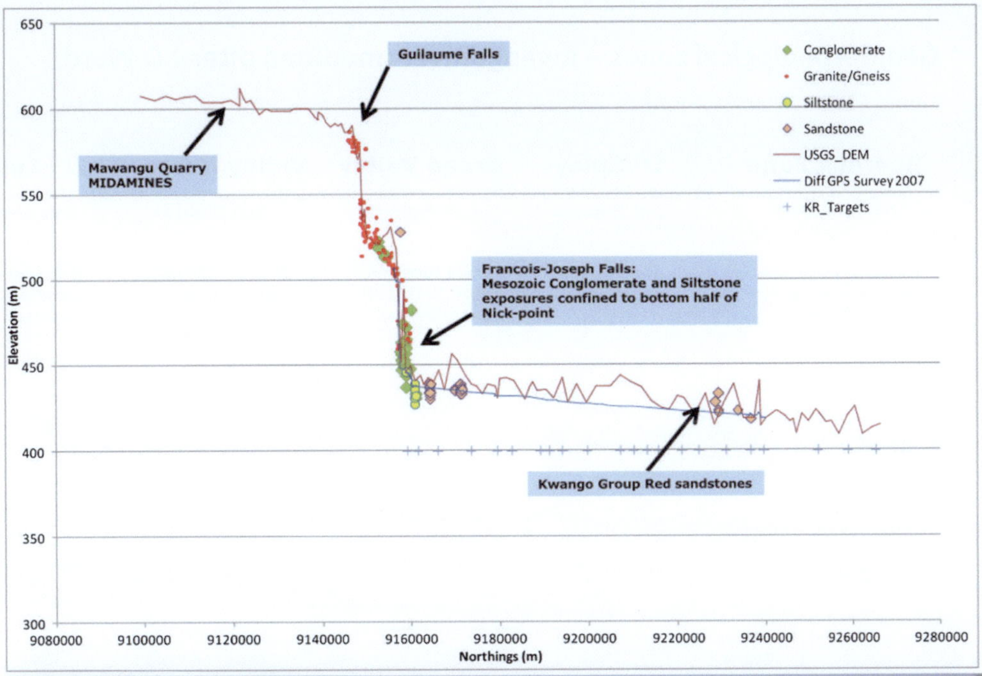

Fig. 16.7 Longitudinal profile of the Kwango River in the Tembo area (*top*) with a photograph of the *lower part* of the Francois-Joseph Falls

south where the highly resistive granite/gneiss bedrock provides better contrast than the sedimentary footwall in the northern areas. Despite the fact that the conductivity contrast is lower on these less competent rocks resistivity-lows on the flats, such as at K52, have been interpreted to be associated with palaeo-channels. These are water-logged clayey material adjacent to the river where pitting in areas coincident with electromagnetic lows has confirmed the presence of this clayey, sandy and gravel material. Although pitting could not reach deeper areas concave resistivity lows within the flats were interpreted as arcuate palaeo-channels, and in many areas the current artisanal workings follow these features.

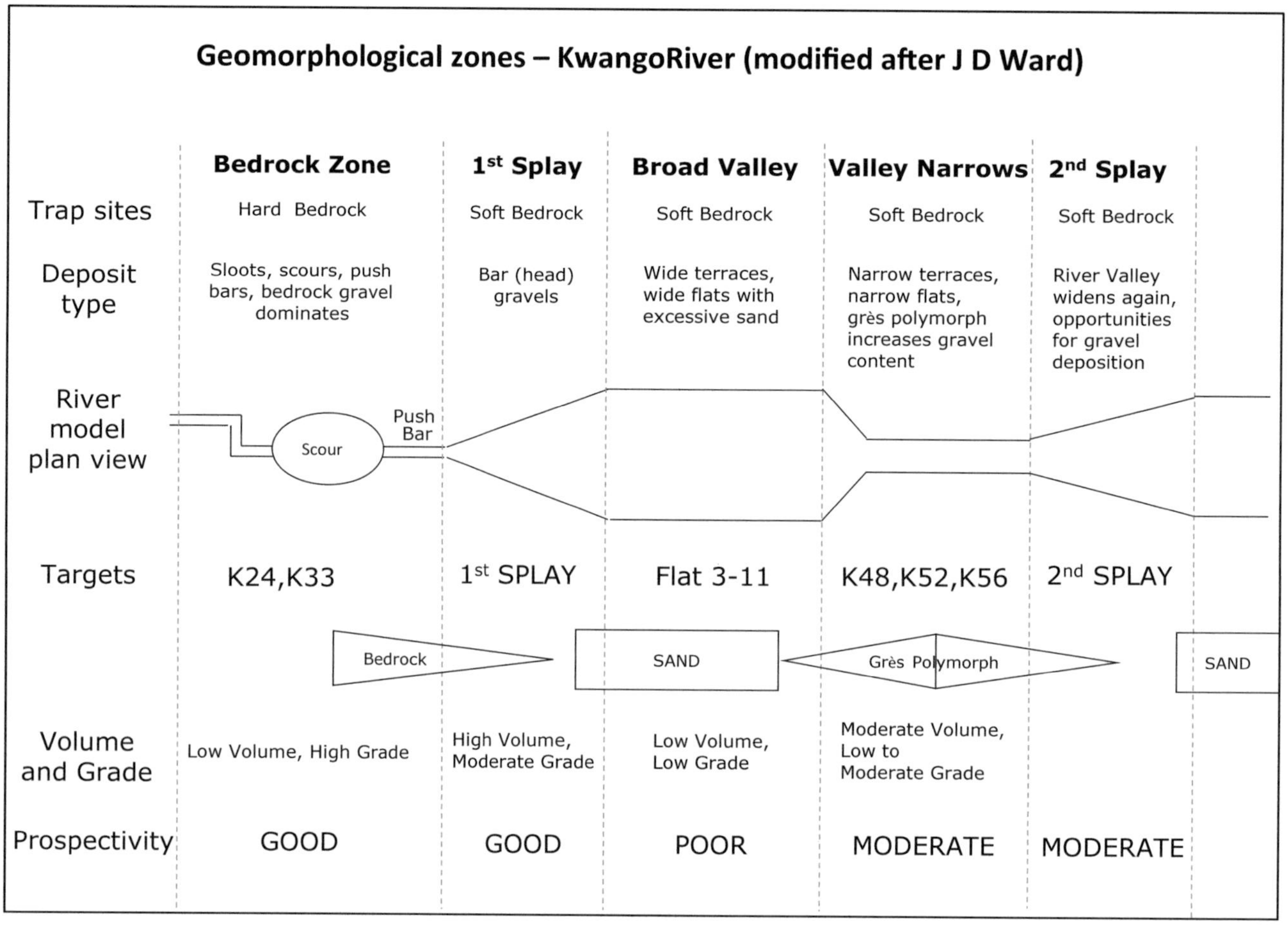

Fig. 16.8 Targets in the form of splays, flats and terraces, along the Kwango River over BRC joint venture ground

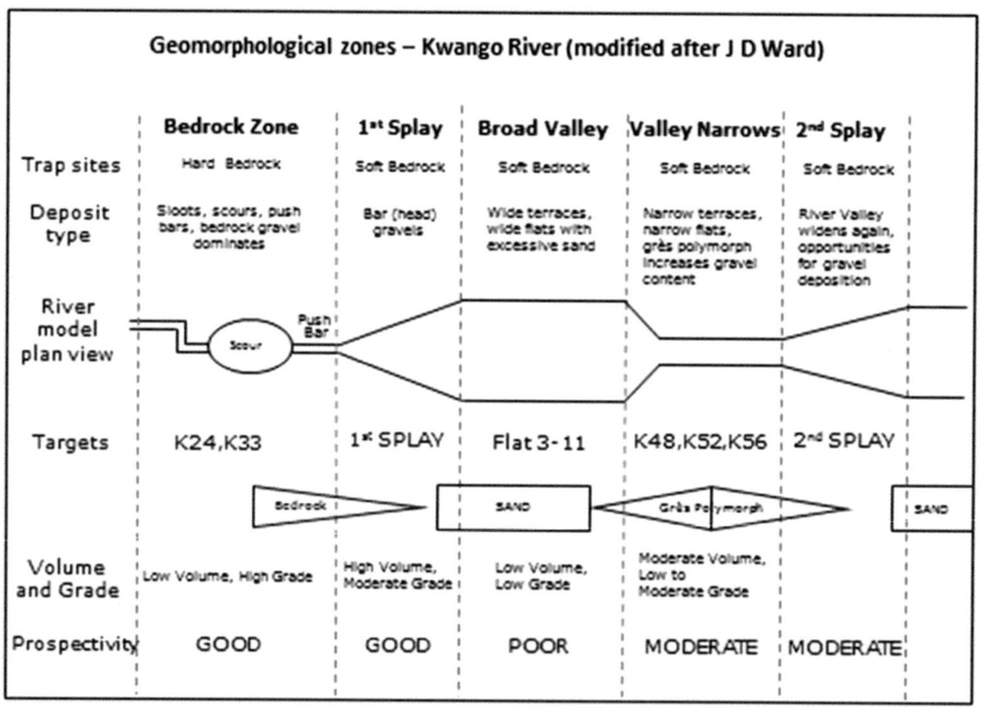

Fig. 16.9 Geomorphological zones in the lower Kwango River with predicted trapsite and gravel development (after Ward)

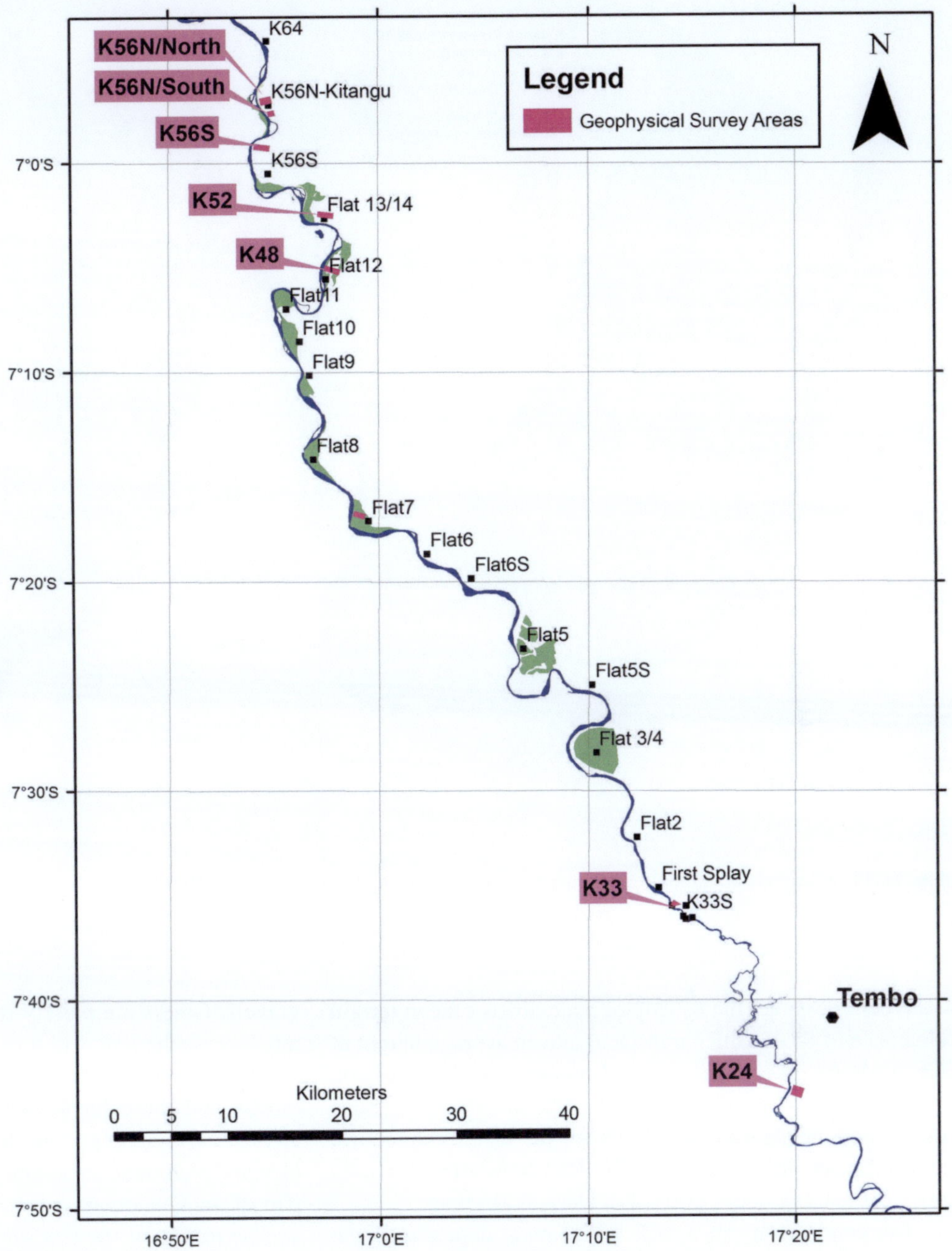

Fig. 16.10 Geophysical areas along the Kwango River for ground survey testing

From the gravity method Bouguer anomalies showed very small variations of density (0.5 milligal) across the area mostly due to possible sediment thicknesses over the basement. In addition, the marshy conditions over most of the valley thwarted any systematic gravity surveys. Finally, several magnetic profiles were undertaken across the project area but provided no advantage in defining sedimentary targets or trap sites that are possibly related to magnetic cross-cutting dykes.

16.5.2 Airborne Survey

Based on the ground work tests described above a helicopter-borne Versatile Time Domain Electromagnetic (VTEM) geophysical survey was flown over eight blocks covering some of the terraces and flats along the eastern side of the Kwango River (Fig. 16.11). Some 3,005 line-km (296.5 km^2) of geophysical data were acquired during the survey (Asiamah 2006). Survey blocks were flown at traverse line spacing of 100 m (Blocks 4, 5, 6, 9 and 10) and 120 m (Blocks 1, 2, and 3) respectively, in a direction 55°N to meet geological target specifications. Tie lines were flown perpendicular to traverse lines.

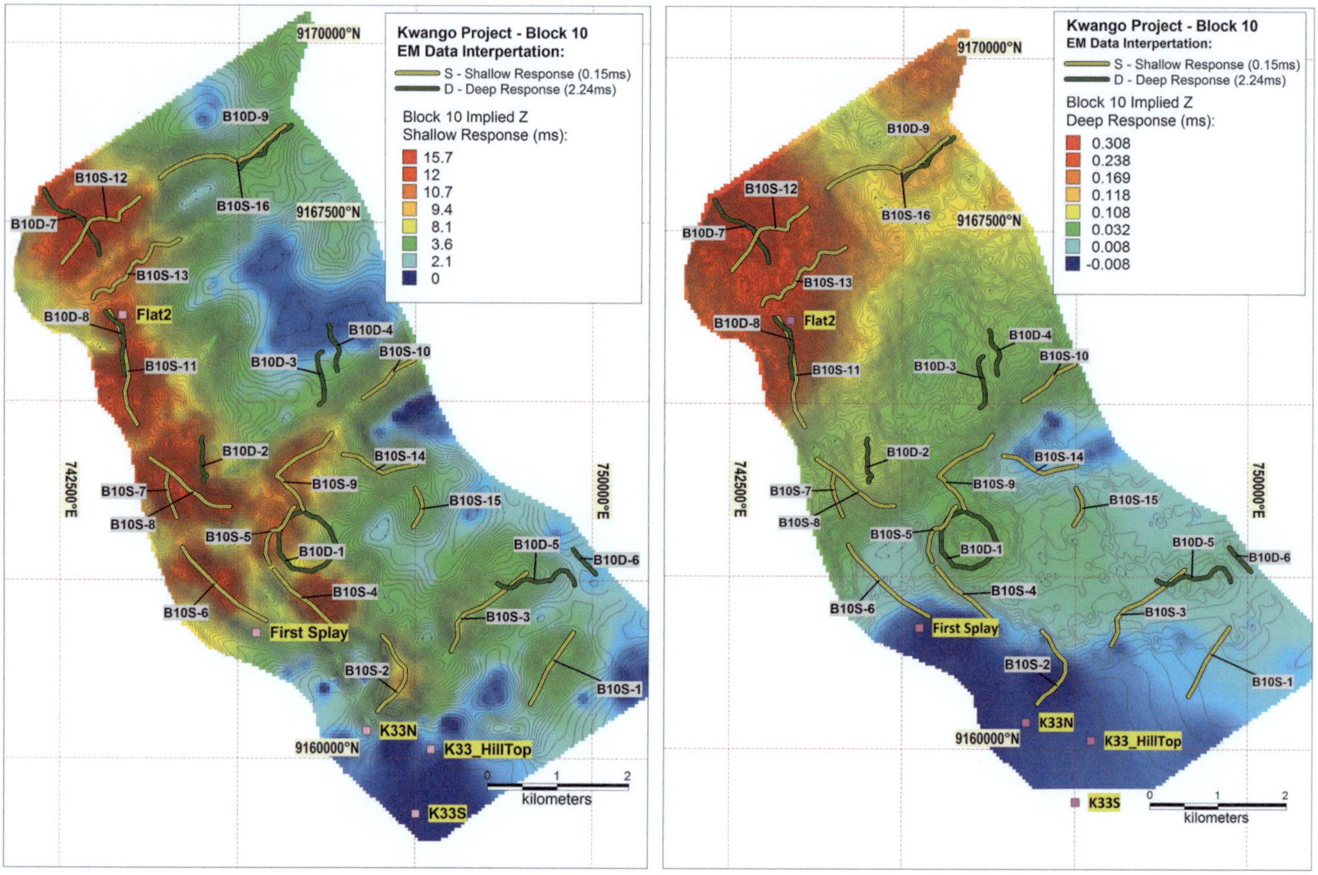

Fig. 16.11 Locations of Airborne electromagnetic survey block along the Kwango River

Where possible, the helicopter maintained a mean terrain clearance of 85 m, which translated into an average height of 33 m above ground for the bird-mounted VTEM system. Nominal survey speed was 80 km/h. The data recording rates of the data acquisition was 0.1 s for the electromagnetics, and 0.2 s for altimeter and GPS. This translates to a geophysical reading about every 2 m along flight track.

The conductivity of a rock or sediment depends on the conductivity of the contained water or fluid, and the amount and manner in which the water is distributed within the rock/sediment. The resistivity varies with porosity, cementation, degree of saturation and water salinity. In general argillaceous sediments, such as clays, can retain water in its structure and are therefore more conductive than sandy or arenaceous sediments.

Block 10 is an example of one of the VTEM surveys (Fig. **16.12**) illustrating the shallow and deep responses. Some anomalies have been interpreted to be possible channels cut into level basement that is footwall to the alluvium and therefore present areas of potentially thicker gravels. Others are clearly related to active tributaries running across the flats and terraces.

Similar interpretations have been done for all of the eight blocks and these have yielded numerous electromagnetic anomalies, including a significant number of topographically elevated sediment packages which are interpreted as being part of the Quaternary cover.

The results of the test surveys have shown that the electromagnetic method is suitable for mapping the shallow geology (less than 50 m) and since there is a high electrical contrast between the palaeo-channel and host rock (Blume and Köstlin 2006) it has shown to be a very effective tool in mapping some of the channels in both the terraces and river flats. Figure **16.12** shows that the shallow responses (yellow lines) and related 150 microsecond time channel (left) versus the deep responses (green lines) picked up with the 2,240 microsecond time channel (right) can be overlapping. The shallow responses are generally associated with channel filled clays-rich sediment whilst the deeper anomalies are more likely to be associated with bedrock structures.

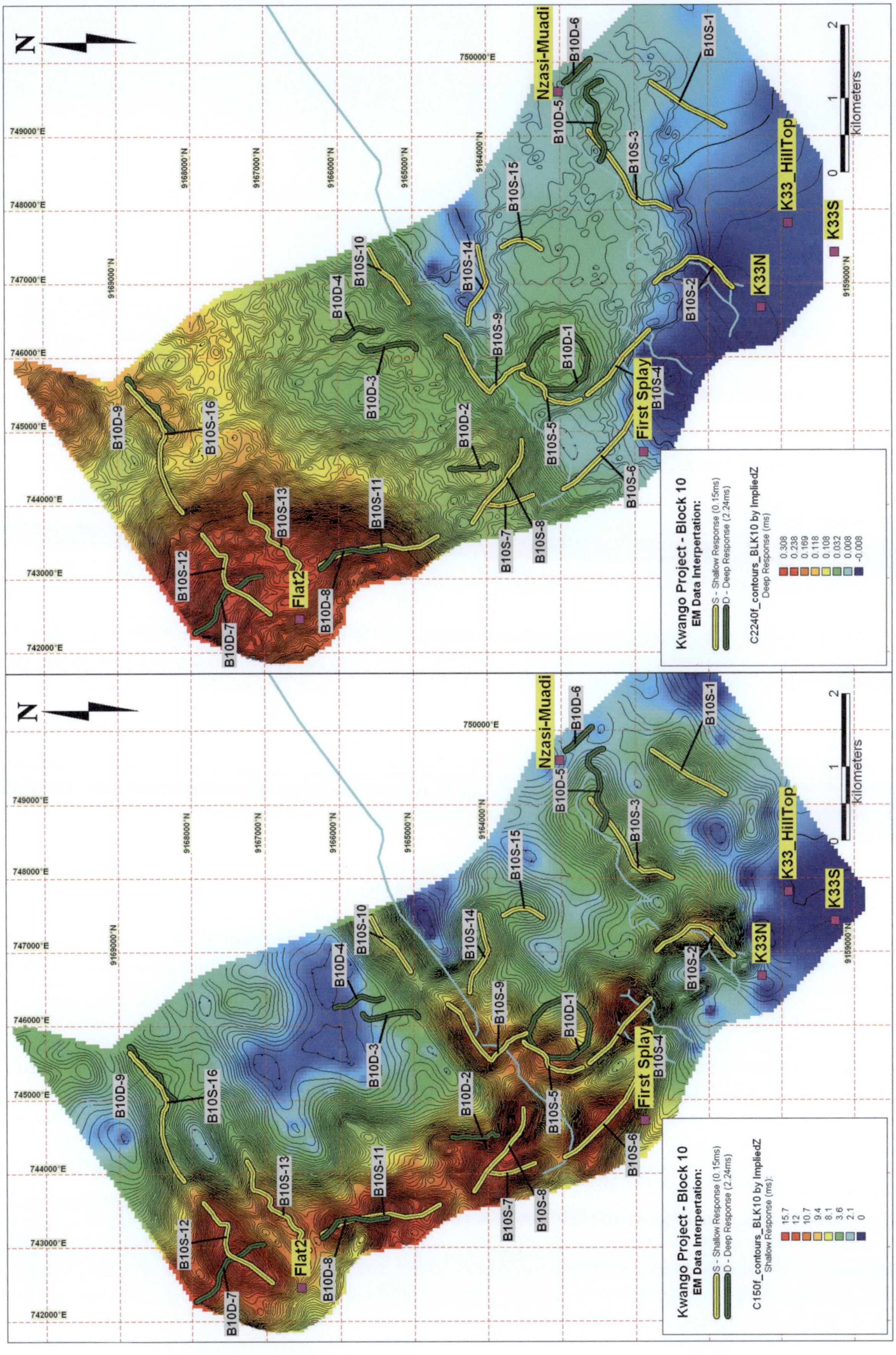

Fig. 16.12 Airborne EM over Block 10. Shallow signals (C150f) are illustrated on the *left* and deep responses (C2240f) on the *right*. Interpreted (*yellow* are shallow; *green* are deep) anomalies are reflected on both. The less resistive materials are reflected by the darker colours

Fig. 16.13 Looking down a Majimba hole dug by hand in order to penetrate terrace deposits

Fig. 16.14 Drilling operation on Kwango Terrace with a trailer-mounted Super Rock 100 (SR100) rig

Table 16.1 Overburden thickness, surface area and gravel tons over the main terraces/flats

Target	First splay	Flat 8	Flat 9	Flat 11	Flat 13/14	K56S	K56N	K64	Second splay	K82	K87	K97	Total
Overburden thickness (m)	4	7	5	3	8	9	11	12	11	11	10	9	N/A
Surface area (km^2)	4.5	4.7	3.4	3.7	7.2	9.2	4.0	6.6	12.1	5.5	9.6	10.6	81.8
Gravel (tons 1,000×)	743	420	310	333	747	1.784	755	63	2,912	443	864	278	9.652

16.5.3 Drilling and Pitting

The primary objective of the pitting and drilling programme was to determine the thicknesses and depths of the various gravel units in the terraces and river flats along the Kwango River. Gravel volumes were calculated on flats and terraces where a sufficiently detailed grid of pitting and drilling data was obtained (typically 200 × 200 m station spacing). Here correlation of the sediments thicknesses with geophysical results was also possible.

Manual pitting, based on the local Majimba digging technique (Fig. 16.13), was found to be a very effective, and if properly managed, safe exploration technique. Some of the skilled local operators, who have for years used this method as a means to rapidly identify diamondiferous gravel horizons, were employed to assist in this program. It also proved to be useful in areas where access with the drill rig was difficult.

The Majimba circular pits measuring 1 m in diameter were dug using pick and shovel (Fig. 16.13). Excavated material would be brought to surface using a bucket tethered to a rope and gathered in mounds for every meter dug. For safety reasons pits were limited to 15 m depth, or less if groundwater was encountered.

In addition to manual pitting, a Super Rock 100 (SR100) drill rig was utilized for ascertaining depth and thickness of gravels. This is a trailer-mounted rotary mud drill rig (Fig. 16.14) which operates by continually re-cycling a mud-mixture fluid. The fluid is pumped down the hollow inner stem of the drill rods (hole diameter 9.5 cm), past the churning drill bits, where loosened material is picked up, then returned to surface through the space between the rods and country rock. This unit was capable of penetrating the loose basal gravels and in most instances also the soft sandstone footwall.

The rock chips and/or sediment derived from the hole were collected using a 0.3 mm screen. These were logged and screened for sedimentological analyses.

During the exploration program 1,248 pits were sunk and 226 holes drilled over 23 of the targets representing approximately 150 km^2 of river terraces and flats. Detailed pitting and drilling was completed over 12 targets, measuring just over 80 km^2 and some 9.6 million tons of gravel were outlined (Table 16.1).

Table 16.1 summarizes overburden thickness as determined by the pitting and drilling program over various targets. Ascertaining depth to gravel was relatively straightforward due to the sharp contact between basal gravels and overlying sands. The results in Table 16.1 highlight the high overburden thicknesses within the Study Area and the challenge faced by miners to access the gravels (Fig. 16.15 and 16.16).

Gravel samples collected during the pitting programme were concentrated and analysed for kimberlitic minerals. Only garnet and ilmenite were recovered. Based on their chemistry these were classified as mantle minerals derived from kimberlites. The level of abrasion of the grains varied

Fig. 16.15 On the *left* overburden sand of some 4.5 m on basal gravel at target K33S—miners at top of gravel layer. On the *right* silicified conglomerate boulders at the base with overburden sands lower down the Kwango at target K56 near Kitangu (photo Ward)

Fig. 16.16 Example of the pitting/drill holes density over a 3D view of the second splay looking south. Average overburden thickness is 11 m. Gravel attained (*Green squares*) and holes abandoned due to ground water problems (*Blue squares*). Distance between pits is 200 m

between slightly to extensively abraded with a higher percentage of extensively abraded grains being recovered from the more downstream targets of the project area (K48, K52 and K56). Since no kimberlitic grains were found in gravel collected from the tributaries flowing into the Kwango it suggests that the source of all the recovered grains are the kimberlites in Angola, transported by the Kwango/Cuango River.

16.5.4 Diamonds

Artisanal workers digging for alluvial diamonds is common in the mineral-rich regions of Central Africa and the Kwango Valley is no exception. A systematic survey of the alluvial diggings on the river terraces by local artisanal workers was conducted. Diamond parcels from artisanal operations on nine targets were screened, using diamond sieve sizes no. 1 (Aperture diameter of 1.09 mm and approximate critical diamond size 0.01 cts/stn) to no. 23 (10.3 mm or 9.2 cts/stn), weighed and photographed. Information pertaining to these parcels, such as diamond size, colour and shape is presented in Tables 16.2 and 16.3.

In total 2,734 stones were examined representing 282.41 cts. The average stone size of the total population is 0.103 cts/stn (Table 16.2).

Maurice Barker, a South African diamond evaluator, classified these data into three broad areas of origin, evaluated these accordingly and presented its size frequency distribution plots (Fig. 16.17). Group 3 (Flats 5 and 8) representing the most upstream areas; Group 2 (Second splay, K56, Flats 13 and 14) being part of the more central areas, and Group 1 (K97 and K105) the most downstream deposits.

Slight variations of the diamond populations between the three groups were noticed and these are believed to be related to different geomorphic settings (Fig. 16.9). For instance Group 3 diamonds are from a wide section of the Kwango Valley where concentration mechanisms are least effective which explains the smallest average stone size. The Group 2 diamonds are from a stretch of river which is relatively narrow and is characterised by many artisanal workings, and the Group 1 diamonds were derived from an area where the valley begins to widen and forms a sediment splay with bigger diamonds being trapped in the proximal part of this splay.

Overall the diamond population is well sorted and variations in stone sizes are generally small ranging between 0.074 to 0.118 cts/stn with an average stone size of 0.103 cts/stn (Fig. 16.18). The largest stone was 4.72 cts and only 14 diamonds of over 1.2 cts in weight (diamond sieve size no. 15) were recovered out of the total population (2,734 diamonds). Eleven of these were recovered in a relatively narrow river section from the Group 2 setting, two from the

Table 16.2 Number and weights of diamonds from nine digging operations along the Kwango River

Group	Parcels	Origin	Stones (n)	Weight (cts)	Carats/stone (cts/stn)
3	40, 41	Flats 5 and 8	286	21.11	0.074
2	13, 22–39, 42	Second splay, K56N, K56S, Flat 13/14	1,338	129.83	0.097
1	1–21 (excl. 13)	K97, K105	1,110	131.47	0.118
Total			2,734	282.41	0.103

Table 16.3 Percentage of diamond colours and shapes for the three Groups in the Lower Kwango diamond groups

Group	No of stones	Colourless	Brown	Yellow	Fragments	Abrasion features	Octa	Dode	Maccle	Cube
3	29	51.7	20.7	10.3	72.4	37.9	6.9	10.3	6.9	3.4
2	469	51.2	26.0	11.9	52.9	35.6	25.6	13.6	2.6	0.0
1	766	61.5	20.6	8.2	35.9	26.2	14.6	3.9	1.3	0.0
Total	1,264	57.4	22.6	9.7	43.0	30.0	18.5	7.7	1.9	0.1

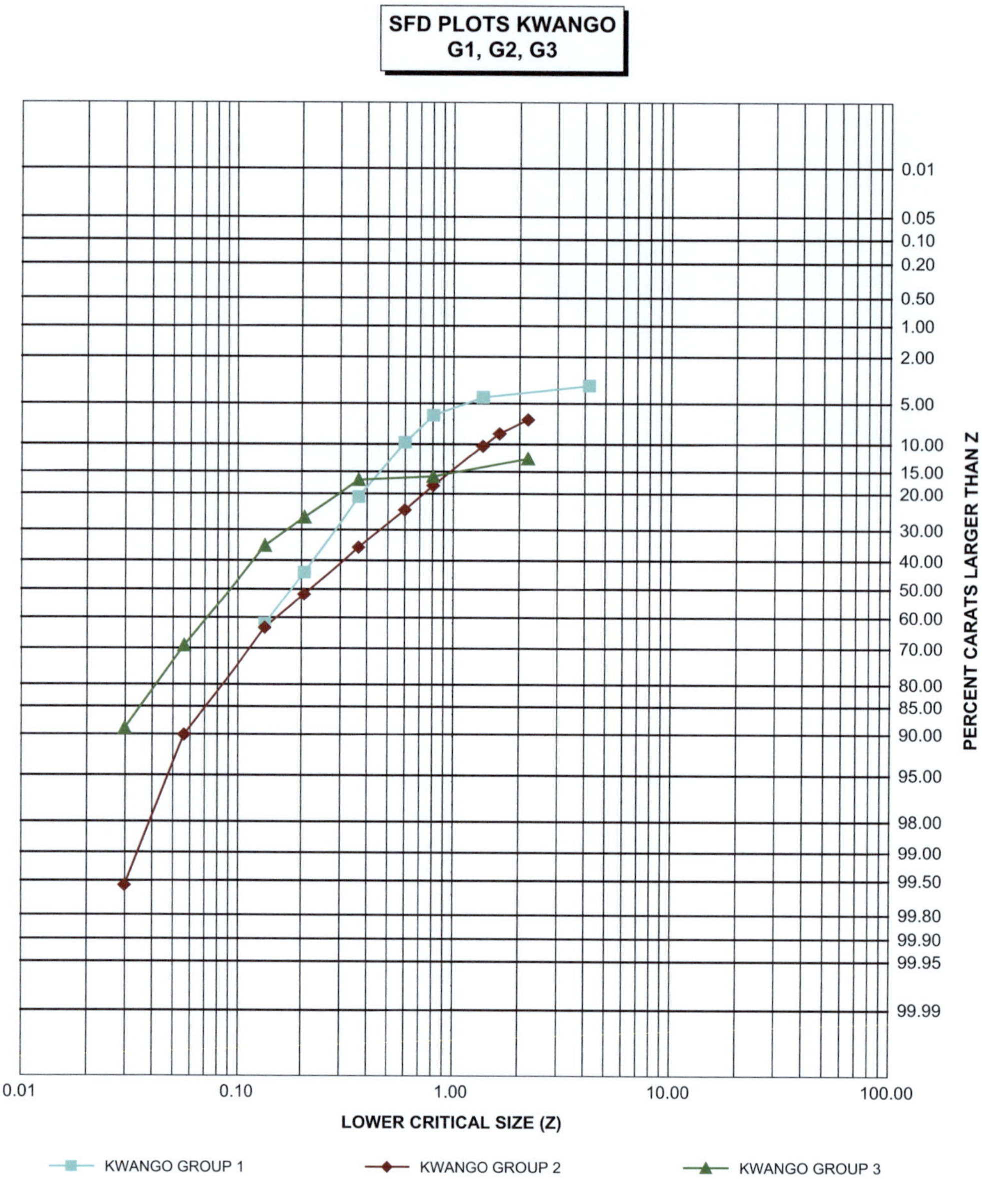

Fig. 16.17 Size frequency distribution plots for the three Groups of diamonds along the 'international' Kwango

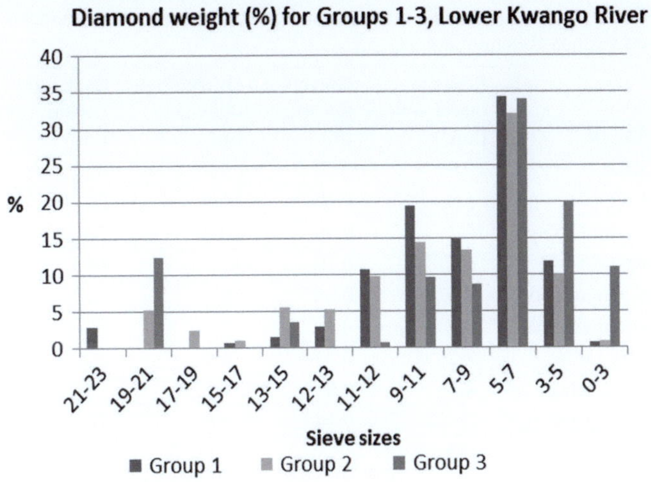

Fig. 16.18 Percentage diamond weight per sieve class for diamonds from lower Kwango River

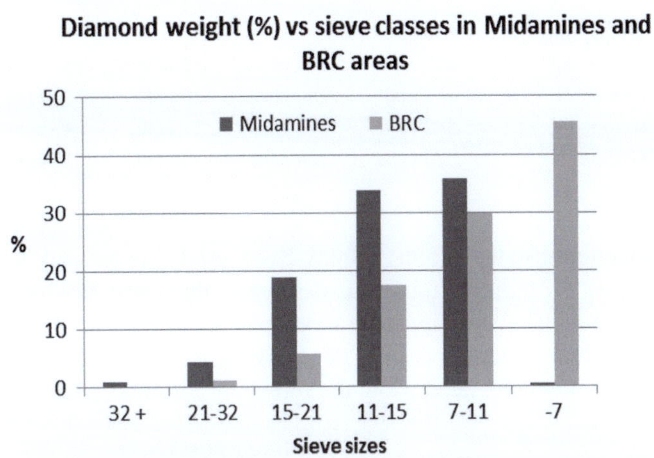

Fig. 16.19 Comparison of the diamond populations between Midamine and Lower Kwango areas

river splays of Group 1 and one from the wide valley section typical of Group 3 (Fig. 16.9).

Bigger diamonds, with an average stone size of 0.25 cts/stn, have been recovered upstream of Tembo at the *Midamines* alluvial diamond mining operations (De Decker 2006) where the Kwango River passes over the basement footwall. In this area *Midamines* were recovering diamonds from Kakolo, an abandoned channel (Fig. 16.18) and from dredging operations within the Kwango River near the Lufutu tributary, some 30 km upstream of the Francois-Joseph Falls (Fig. 16.2).

The diamond population from these areas are also well-sorted with some 24.18 % in the diamond screen size no. 15 (approximate critical diamond size of 1.2 cts/stn) and 0.88 % in the diamond screen size no. 21 (approximate critical diamond size 3.9 cts/stn), compared to 6.93 % in diamond screen size no. 15 and 1.21 % in diamond screen size no. 21 for the lower Kwango area (Fig. 16.19). Both the *BRC* and *Midamines* areas are represented by well-sorted diamond populations but the *Midamines* area showing a distinctly coarser trend compared to those derived from the *BRC* project area. There are three reasons for this. Firstly, the *BRC* study area is located between 30 km and 130 km downstream from the *Midamines* area. This represents the most distal part of the advancing 'front' of the Kwango diamond trail from Angola and will therefore be finer grained. Secondly and more importantly, the footwall of the *Midamines* area is composed of competent basement rock with very favourable concentration sites in the form of gullies and potholes which are able to trap and retain larger diamonds. In contrast, the footwall farther downstream of the Kwango below the Nwazi-Muadi knick-point is composed of relatively soft Cretaceous sediments. This results in a smoother channel floor with only silcrete boulders forming the traps to retain heavy minerals, and is far less effective as a concentrating mechanism.

Finally the diamonds from the *BRC* project area contain a high percentage of fragments and broken stones (42 %), while 30 % of the diamonds show abrasion features (Table 16.3). Since these are all recovered downstream of the two waterfalls many of the bigger diamonds would have been broken in the process of being transported over the Falls, which has a 150 m or more drop over a relatively short distance (Fig. 16.7) on competent basement rock, subjecting the stones to an aggressive process of mechanical abrasion.

As mentioned above the small and well-sorted stone size distribution within the *BRC* areas supports the premise that these diamonds represent the 'head' of a diamond trail originating from Angola, and moving down the Kwango River. Although the Falls are effective in lowering the stone size over a fairly short distance due to mechanical abrasion, the decrease in size continues gradually below the Falls with diamonds reduced to 0.05 cts/stn at Popokabaka, a distance of almost 300 km. The ages of the terraces indicate that this post-Cretaceous distribution is relatively recent, dating from the Early to Mid-Pleistocene. More significantly it also suggests the coarser stone size population has not reached the lower reaches of the Kwango River.

The diamonds which were screened and analysed from the nine targets within the Study Area feature a high proportion of gem quality diamonds, with colourless stones representing between 53 to 68 % of the population (Fig. 16.20). The percentage of brown and yellow diamonds only varies from 10 to 17 %.

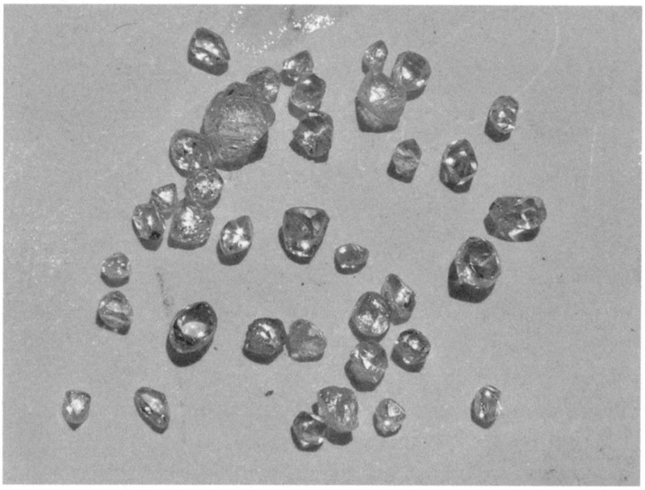

Fig. 16.20 Typical diamonds from the 'international' Kwango below the Falls with a high percentage colourless stones, many fragments and a significant degree of abrasion

16.5.5 Economic Modelling

In order to assess the economics of mining alluvial diamonds along the Kwango, strip ratios, diamond values and grades had to be evaluated against mining costs, which were anticipated to be significantly higher than those for similar sized placer operations in South Africa. This was due to lack of infrastructure and the remoteness of the study area, combined with legal and administrative complexities of operating in the DRC.

Through its pitting and drilling program on the Kwango, BRC was able to quantify with a reasonable level of confidence the strip ratios (i.e. ratio of overburden to gravel thickness) to be expected on various targets within its study area. With the exception of First splay, which occurs at the point where the Kwango River valley widens, strip ratios are greater than 20 for the 8 targets on which sufficient overburden and gravel thickness data was collected (Table **16.4**).

Furthermore, the diamond size frequency distribution analysis carried out by BRC within its Kwango project area suggests stone sizes were generally small, with an average stone size of 0.103 cts/stn (Table 16.2). As a result, it was estimated that diamond parcels from this area would fetch low prices, in the range of US$50/ct to US$75/ct (Maurice Barker, 2007, personal communication).

In order to establish diamond grades on the various targets a trial mining campaign was contemplated by BRC, utilizing a small mobile 5 tph DMS plant. Before proceeding with this campaign, however, a break-even analysis was carried out on the 9 targets for which strip ratios had been sufficiently well established, in an attempt to determine the spectrum of diamond grades required to recoup mining costs, given a range of (i) Mining Costs ($/ton) and (ii) Diamond Values ($/ct). Break-Even grade, expressed in carats per hundred tonnes, is defined as the grade at which revenues from mining operations pay for the costs of the operations, thus generating no profits or losses. Break-even grade was calculated using the following formula:

Table 16.4 Overburden thickness and gravel thickness measurements on various Kwango targets

Target	Primary target area (km^2)	Gravel thickness (m)	Overburden thickness (m)	Strip ratio (waste: ore)
K97	10.6	0.22	9.3	42
K82	5.5	0.32	10.8	34
Second splay	12.1	0.48	10.8	23
K64	6.6	0.18	11.6	64
Kitangu K56N	4.0	0.24	10.7	45
Bomba K56S	9.2	0.18	9.4	52
Flat 13/14	7.2	0.35	6.9	20
First Splay	4.5	0.6	4.1	7
Hilltop Pitting		0.14	3.6	26
Total	150			

Table 16.5 Grade required to Break-Even (cpht) for Second Splay target (strip ratio of 23:1)

Mining cost	Diamond value			
(waste and ore)	$50/ct	$80/ct	$200/ct	$300/ct
$3/ton	141 cpht	88 cpht	35 cpht	24 cpht
$6/ton	282 cpht	176 cpht	71 cpht	47 cpht
$10/ton	470 cpht	294 cpht	118 cpht	78 cpht

$$\text{Break-Even Grade (cpht)} = \frac{\text{Mining cost (\$/t)}}{\text{Diamond Value (\$/ct)}} \times \frac{(\text{Overburden Thickness} + \text{Gravel Thickness})}{\text{Gravel Thickness}} \times 100$$

Table 16.5 and Fig. **16.21** summarize results of the break-even analysis for the Second Splay, with an overall strip ratio of 23:1 (waste to ore).

These results suggest that diamond grades in excess of 150 cpht would be required to break-even on the Second Splay given that:

1. Mining costs for a mechanized placer mining operation on the Kwango in 2007 were anticipated to be in excess of $4/ton.
2. Diamonds were estimated to be valued at $50 to $75/ct.

Had the analysis also taken into account Processing Costs for running a DMS plant (≈$6/ton) and General and Administrative Costs (≈$1/ton), the grades required to break-even would have been even higher, likely in excess of 400 cpht. The probability of attaining such grades, based on historical and contemporary data was deemed to be low.

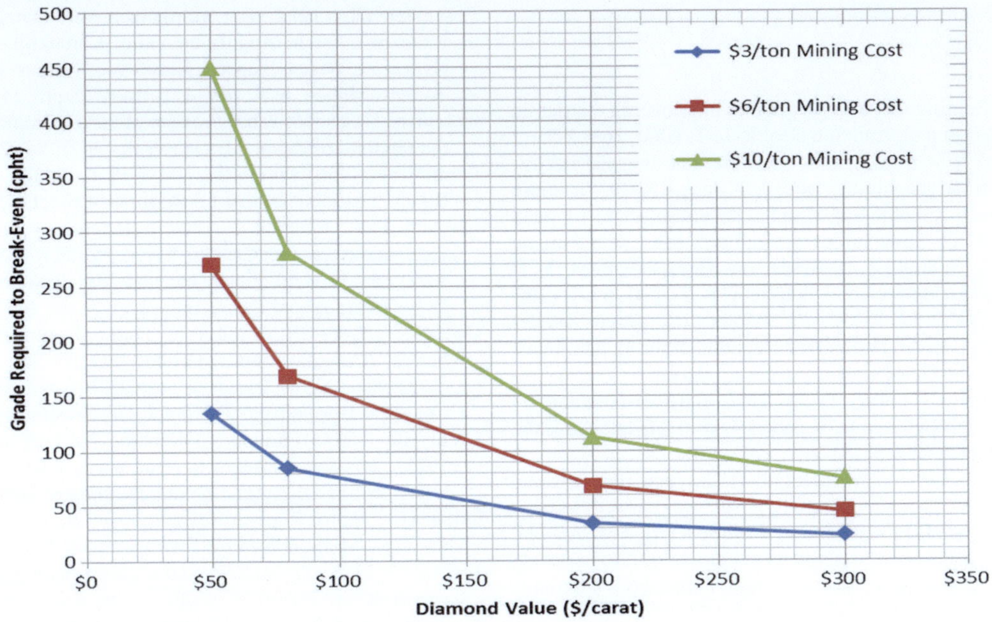

Fig. 16.21 Grade required to Break-Even (cpht) analysis for Second Splay target (strip ratio of 23:1)

For example, anecdotal diamond production figures provided by artisanal diggers at the K56 target, combined with estimates made by BRC staff of gravel volume processed, enabled calculation of a 'back of the envelope' grade figure ranging from 20 cpht to 50 cpht. Notwithstanding concerns with relying on production figures provided by the artisanal diggers these grades coincide well with diamond grades cited in the historical literature (see Historical Background) and grades reported at the Midamines operation which have been estimated to be 25 cpht going up to between 50 and 100 cpht, where gravel has been deposited on basement. Sadly, almost no basement occurs in the BRC project area resulting in less efficient diamond traps. As a result, BRC concluded that a trial mining exercise would not be warranted.

Conclusions

The stone tool artefacts that have been found within the terraces along the lower Kwango indicate that these deposits are younger than Early to Middle Pleistocene. Hence the oldest recorded diamond occurrences introduced into these terraces are relatively recent. It is suggested that the small diamonds along the international Kwango form part of the advancing diamond front stretching from Angola. Stream sediment sampling for kimberlitic heavy minerals at exploration sites along the international Kwango and its main tributaries confirms that no other diamond input points exist at least in the Congo. These kimberlitic minerals also show a high degree of abrasion to support a transported population. In addition, there is a steady decrease in the diamond sizes downstream in the river basin, supporting a transported model.

Importantly the Guiliame and Francois-Joseph Falls near Tembo have a negative impact on the diamond size distribution directly downstream of the Falls. Here a high level of breakages of the diamonds is observed, diamond size is significantly lowered, there is a high proportion of diamond fragments and the diamonds have a high level of abrasion features.

Finally, there is also a significant decrease in diamond concentration in the lower part of the Kwango where the footwall is composed of aeolian facies of the Kwango Group, which does not facilitate diamond concentration compared to the more favourable basement footwall found upstream of the Falls.

These factors coupled with a small average stone size (0.07 to 0.1 cts/stn) that earns a lower carat value and the high stripping ratios and mining costs, presents an unfavourable economic model at today's diamond prices.

Ackowledgements The authors wish to thank the company *BRC Pty Ltd* (now *Delrand Resources Pty Ltd*) for permission to publish these exploration data. Renato Spaggiari and Pat Sheahan are thanked for their comments which significantly improved the manuscript, John Ward for his geological observations on site and Bastien Linol for his help with one of the diagrams. Thanks must also go to the various field personnel that have been involved in the drilling and pitting operations. Reviewers John Bristow, Tania Marshall and Judith Kinnaird are thanked for their valuable input and edits.

References

Aderca B (1939) Contribution à la géographie physique du district du Kwango (Congo Belge). Ann Soc Géol Belg T. LXII, 369–387

Asiamah G (2006) A helicopter-borne versatile time domain electromagnetic (VTEM) geophysical survey, Kwango River. In house report by Geotech Airborne Ltd, Kwango River Project A016, pp38

Asselberghs E (1919). Observations géologiques dans le Bassin du Kwango (partie Sud-Ouest du Bassin du Kasaï). Annales de la Société Géologique de Belgique, Publ.rel.C.B.,T.XLII, p C81–109

Blume J, Köstlin EO (2006) Resistivity imanging and gravity studies along the Kwango River, Democratic Republic of the Congo. Internal report, J Blume Geoelectrical consulting, 15 pp

Cahen L, Lepersonne J (1952) Équivalence entre le Système du Kalahari du Congo Belge et les Kalahari Beds d'Afrique Australe, Mem. Soc Belge de Géol 8(4):1–64

Cahen L, Snelling NJ, Delhal J, Vail JR (1984) The geochronology and evlution of Africa. Clarendon, Oxford, p 512

Clark JD (1966) The distribution of Prehistoric culture in Angola. Subsídios para a história, arqueologia e etnografia dos povos da Lunda, vol 73. Diamang, Museu do Dundo, Publicações Culturais, Angola, p 102

De Carvalho H, Tassinari C, Alves PH, Guimarães F, Simões MC (2000) Geochronological review of the Precambrian in western Angola : links with Brazil. J Afr Earth Sci 31(2):383–402

De Decker RH (2006) Technical report on the alluvial diamond deposits of the Kwango River in the South West of the Democratic Republic of the Congo (DRC). For Rockwell Ventures Incorporated, 67 pp

De Ploey J, Pepersonne J Stoops G (1968) Sédimentologie et origine des sables de la série des 'Grès Polimorphes' (système du Kalahari) au Congo Occidental. Mus. Roy. Afr. Centr., Tervuren, Belgique, Annales du Musée du Congo belge (série in 8°), Sciences Géologiques, Vol. 61, 71 pp

Delhal J (1991) Situation géochronologique 1990 du Précambrian du Sud-Kasai et de l'Ouest-Shaba. Tervuren Museé royal d'Afrique centrale. Annual Report, 1990, pp 119–125

Delhal J, Fieremans C (1964) Extension d'un grand complexe charnockitique en Afrique Centrale, 259th edn. C.R. Acad. Sc, Paris, T, pp 2665–2668

Delhal J, Ledent D, Pastels P (1975) L'âge du Complexe granitique et migmatitique de Dibaya (Région du Kasai, Zaire) par les methodes Rb/Sr et U/Pb. Annales Société Géologique Belgique 98:141–154

Delhal J, Ledent D, Torquato JR (1976) Nouvelles données géochronologiques relatives au complexe gabbro-noritique et charnockitique du bouclier du Kasai et à son prolongement en Angola. Annales Société Géologique Belgique 99:211–226

Eley R, Grütter H, Louw A, Tunguno C, Twidale J (2008) Exploration geology of the Luxinga Kimberlite cluster (Angola) with evidence supporting the presence of kimberlites lava. In: Extended abstracts submitted to the ninth International Kimberlite conference

Faucett FC (1924a) Rapport Techniques No. 1–16 (Jan 1924–April 1925), Mission Kwango, Popokabaka – Kingushi, Internal reports Forminière SGK

Faucett FC (1924b) SGK (Kwango), Mission Grant, Report Faucett 2–16 (Feb1924–April 1925), Dossier No.18, 43 pp

Fieremans CL (1977) Het voorkomen van diamant langsheen de Kwango-rivier in Angola en Zaire. Koninklijke Academie voort Overzeese Wetenschappen, Klasse voor Natuur- en Geneeskundige Wetenschappen, N.R., XX–1, Brussel, 28 pp

Fieremans CL (1996) A brief review of the occurrances of diamonds in the republic of Zaire. Afr Geosci Rev 3(2):247–260

Forminière (1956) Forminière 1906–1956. Editor: L Cuypers, Publisher Les Etablissement Le Berrurier, Bruxelles, 209pp

Forward PW, Jeffcock P (2003) Technical review of diamond concessions on the Cuango River, Northern Angola. Report for American Mineral Fields Inc and Idas Resources NV by ACA Howe International Ltd, 61pp

Gobbo-Rodrigues SR, Coimbre JC, Petri S, Bertini RJ (2003) Kwango Series (Congo), Bauru Group (Brasil) and Neuquén Basin (Argentina) ages, based on ostracodes and vertebrates. XVIII Congreso Brasileiro de Paleontologia, Brasília, pp 152–153

Grant RC (1924a) Rapport Technique No.1-14, (Jan 1924–March 1925), Mission Kwango, Secteur Kwilu-Kwango, Popokabaka – Kasongo-Lunda, Internal reports Forminière SGK

Grant RC (1924b) Rapport Technique No. 9, 10, 11. Secteur Kwilu-Kwango, Mission Wamba, Internal reports Forminière SGK

Linol B (2012) Sedimentology and sequence-stratigraphy of the Congo and Kalahari Basins of south-central Africa and their evolution during the formation and break-up of West Gondwana. Unpubl. PhD thesis, NMMU, 370 pp

Linol B, de Wit MJ, Guilocheau F, de Wit MCJ, Thorose E, Colin JP, Boven A, Fernandez-Alonso M, Tack L (2011) Congo-Kalahari Stratigraphy. Poster at 23rd Colloqium of Africa Geology, Johannesburg

Lonhro Mining Limited (2011) Annual report. 40pp

Mining Weekly (2012) Sparkling find at the Lonrho's Angola Mine. 31 August

PAC (2004) Partnership Africa Canada, Diamond Industry Annual Review, Republic of Angola, pp 1–12

Pettit W (2009) Geophysical signatures of some recently discovered large (>40 ha) kimberlites pipes on the Alto Cuilo concession in northeastern Angola. In: Stephen F et al (eds) Proceedings 9 IKC, Lithos 122S, Special Issue, pp 106–115

Rigby J (1973) Gondwanidium and other similar upper Paleozoic genera and their stratigraphic significance. Geol. Surv. Queensland., Publication No. 350, Paleont. Papers, no. 24

Robles-Crus S, Lomba A, Melgarejo JC, Gali S, Goncalves AO (2009) The Cucumbi kimberlite, NE Angola: Problems to discriminate fertile and barren kimberlites. Revista de la Sociedad española de mineralogia, macla no. 11, 159–160

Rocha-Campos AC (1976) Direction of movement of Late Paleozoic glaciers in Angola (Western Africa), vol 7. Boletim IG, Instituto de Geosciẽcias, USP, pp 39–44

Smith D, Faucett FC (1922) Forminiere (Kwango), Mission Smith and Faucett, Report Smith 1–5 (June-October 1922), Dossier No. 16., 6 pp

Teixeira C (1948) Fõsseis vegetais do Karroo de Angola. Bol Soc Geol Port 7(2):73–76

Young GS (1921a) Extrait du Rapport Young No. 3., Internal report Forminière July 1921

Young GS (1921b) Extrait du Rapport Young No. 4., Internal report Forminière, Aout 1921

A Review of the Kimberlites of the Democratic Republic of Congo

17

Michiel C.J. de Wit and Hielke A. Jelsma

17.1 Introduction

Exploration for diamonds in the DRC, then known as Congo Belge, started in 1900, when King Leopold-II granted the company *Tanganyika Concessions Limited (TCL)* exclusive prospecting rights over 155,000 km^2 in the southern part of the country. Between 1900 and 1909 TCL, a foreign partner of *Union Minière du Haut Katanga (UMHK)* and *Compagnie du Chemin de Fer du Bas-Congo au Katanga (BCK)*, located more than 100 old copper workings in Katanga (then known as Shaba) Province, discovered tin near Bukama in 1904, and discovered the country's first diamond in 1903, in the Mutendele stream in Katanga Province, which is a tributary of the Lualaba River in the headwaters of the Congo River. Further exploration by the company led to the discovery of a large kimberlite field on the Kundelungu plateau in 1908.

The first diamond in the Kasai region was found in 1907 by prospector Narcisse Janot, near Tshikapa in the Kiminina (Tshiminina) River, which is a tributary of the Kasai River and which became part of the alluvial mines around Tshikapa that ultimately produced well over 100 Million carats (Mcts). The discovery encouraged Forminière to continue its exploration program that in 1918 led to the discovery of diamonds near Bakwanga (Mbuji Mayi) Town where geologist George Young recovered 8,840 diamonds in one stream sample in the Bushimaie (Mbujimayi) River. By 1920 this was already known as one of the largest deposits of alluvial diamonds in the world, the production of which in the early 1950s represented some 75 % of world production by volume (Fieremans 1953; Forminière 1956).

The presence of kimberlites in the area was however only realized in 1946 (de Magnée 1946) and the first detailed study of the Bakwanga (Mbuji Mayi) cluster kimberlites was presented by Wasilewski (1950). In 1955 the Tshibua (Tshibwe) cluster kimberlites were found by MIBA. These were the last kimberlites to be found in the country until De Beers started exploring, first between 1970 and 1982 and later between 2004 and 2009. During the first spell the Bas-Congo kimberlites were found south of Kinshasa close to the Angolan border in 1974 and in 2005 two further clusters were discovered near Kabinda, east of Mbuji Mayi. The distribution of the known kimberlite fields is shown in Fig. 17.1.

17.2 Katanga Kundelungu Kimberlite Field

The Kundelungu kimberlites occur as two separate groups on the Kundelungu Plateau in the south-east of the DRC in Katanga Province (Figs. 17.1 and 17.2) and were initially described by Verhoogen (1938). The kimberlites occur to the east of the exposed section of the Bangweulu block, which is separated from the Kasai Craton to the northwest by the Mesoproterozoic Kibaran belt. The Kundelungu Plateau may be underlain by a basement extension of the Bangweulu block.

The kimberlite field comprises an eastern and a western group consisting of 16 and 17 kimberlites, respectively. The eastern group was discovered by TCL in 1908 (Studt 1910) whilst the western group was only found in 1913 by Poulson (Fieremans 1953). The majority of the kimberlites are pipes and were originally evaluated by *Société Belge Industrielle et Minière du Katanga* (SIMKAT) between 1913 and 1923 and later by De Beers between 1977 and 1982. The pipes range in size between less than 1 ha to approximately 50 ha for the Talala pipe. The uppermost rock types comprise both magmatic and volcaniclastic kimberlites suggesting hypabyssal and crater/diatreme facies preserved, respectively.

M.C.J. de Wit (✉)
Department of Geology, University of Pretoria, Private Bag X20, Hatfield 0028, South Africa
e-mail: dewit@icon.co.za

H.A. Jelsma
De Beers Group Services, Exploration, Private Bag X01, Southdale 2135, South Africa
e-mail: hielke.jelsma@debeersgroup.com

M.J. de Wit et al. (eds.), *Geology and Resource Potential of the Congo Basin*, Regional Geology Reviews,
DOI 10.1007/978-3-642-29482-2_17, © Springer-Verlag Berlin Heidelberg 2015

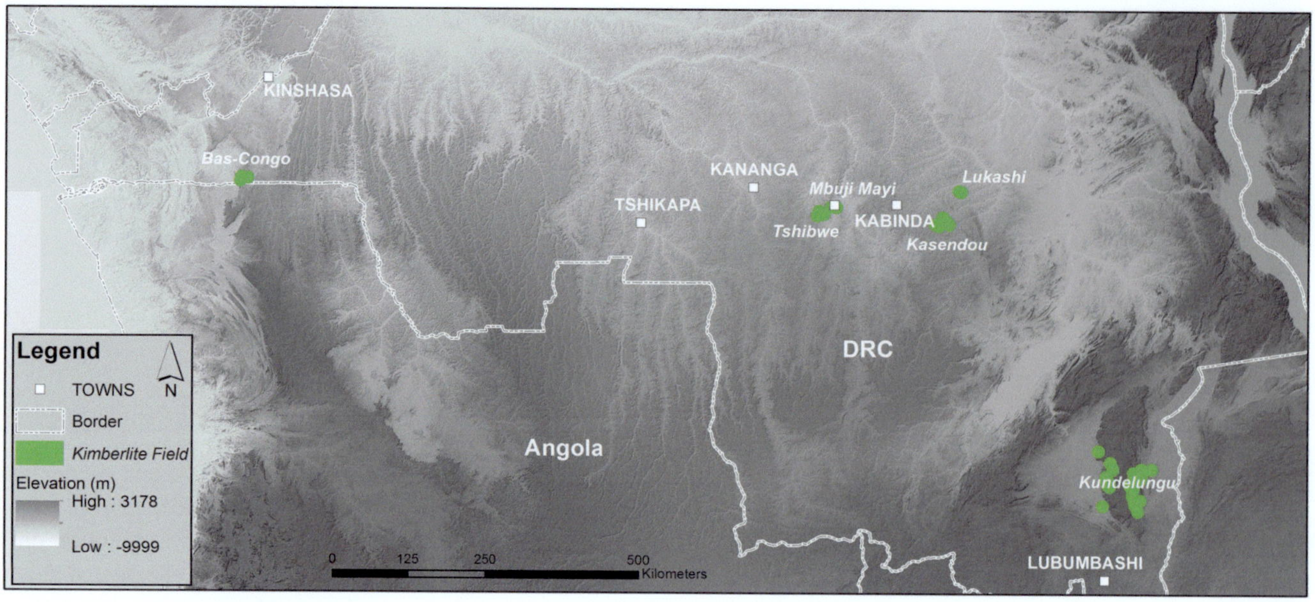

Fig. 17.1 Kimberlite fields of the DRC

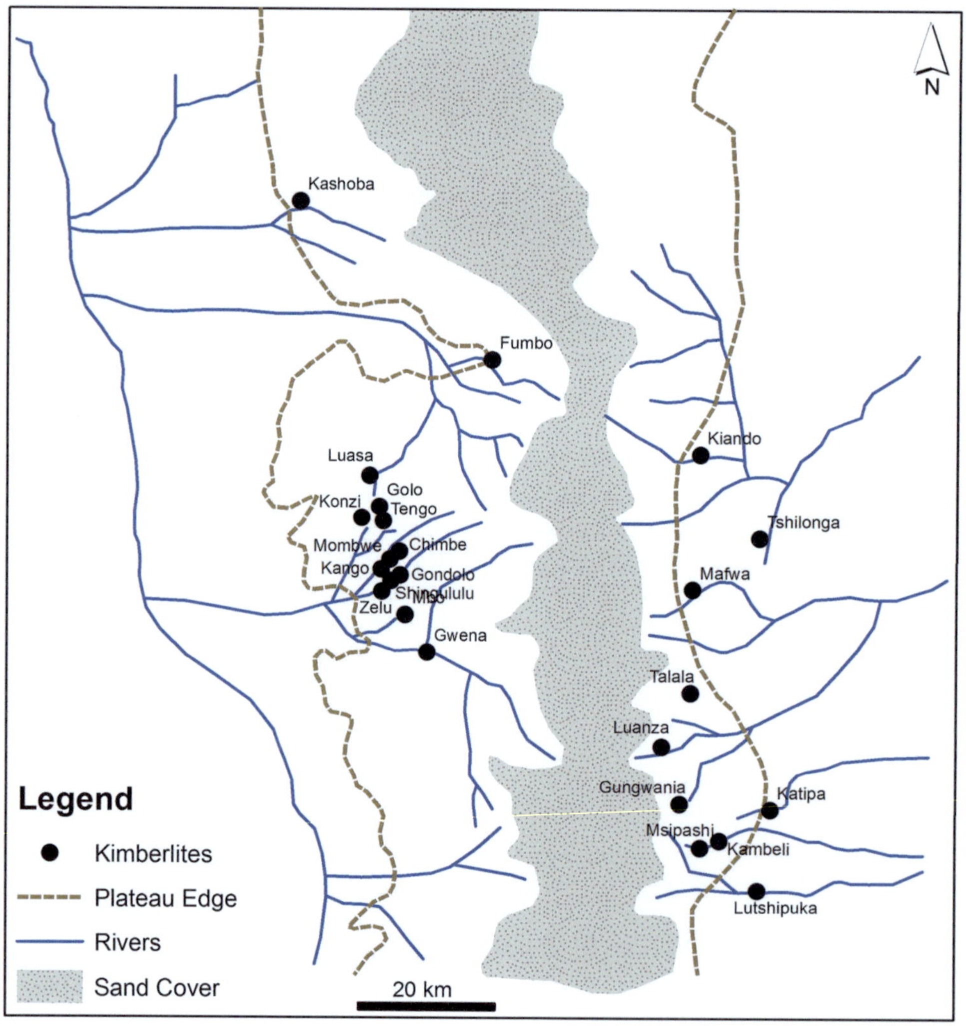

Fig. 17.2 Kimberlites within the Kundelungu Field, Kundelungu Plateau (after Kampata 1993)

Crustal zircons from two of the Kundelungu kimberlites in the eastern group (Gungwania and Talala pipes) have been dated, using U-Pb and Hf-isotope analyses, with results indicating that the basement is Palaeoproterozoic in age (Batumike et al. 2007). The kimberlites have intruded undeformed mudrocks, sandstones and limestones of the Neoproterozoic Biano Subgroup of the upper part of the Katanga Supergroup (Kampunzu and Cailteux 1999).

Palaeontological material from crater sediments from the Cango (Kango) kimberlite (approximately 9.7 ha in size) includes frogs (family *Pipidae,* subfamily *Xenopinae)*, a gastropod (genus *Planorbis*) and plant material in an ostracod-rich matrix (genera *Zonocypris and Ilycypris)*, and were initially interpreted as Late Jurassic to Late Eocene in age (Saxby 1982). However, some of the fossils are similar to those found in Eocene crater sediments in the Mahene kimberlite in Tanzania (Jacobs and Heredeen 2004) and in late Cretaceous crater sediments in the Banke (Haughton 1931) and Stompoor (Smith 1986) kimberlites in South Africa. Based on stratigraphic evidence, Demaiffe et al. (1991) suggested a Cretaceous age for the Kundelungu kimberlites. Recent U-Pb dating of groundmass perovskite of two of the bodies (Kambeli and Msipashi pipes), which occur in the southern end of the eastern group (Fig. 17.2), has provided an age of 32 ± 2 Ma (Batumike et al. 2008).

17.3 Kasai Oriental Kabinda Kimberlite Field

Two clusters of kimberlites were discovered during exploration by the joint venture between the De Beers Group and Bugeco S.A. to the east of the village of Kabinda, some 100 km east of Mbuji Mayi in the eastern part of Kasai Oriental Province (Fig. 17.1). The Kasendou cluster was discovered in 2005 and consists of eight bodies whilst the two known kimberlites of the Lukashi cluster were found in 2007 (www.bugeco.com, Fig. 17.3). The kimberlites include crater-facies volcaniclastic, resedimented volcaniclastic and coherent magmatic varieties and were emplaced through Neoproterozoic dolomitic limestones of the Mbuji Mayi Supergroup with dolerite sills, and overlying Mesozoic siliciclastic sediments (Walker 2011). Micro- and macro-diamonds have been recovered from some of the kimberlites (www.afdiamonds.com). The largest pipe is estimated to be approximately 10 ha in size and the combined total surface area of the kimberlites reaches some 35 ha. In 2008, De Beers withdrew from the joint venture because of the worldwide economic crisis and the small size and low grade of the kimberlites (African Diamonds plc. 2009; USGS 2008).

17.4 Kasai Oriental Mbuji Mayi Kimberlite Field (Mbuji Mayi and Tshibwe Clusters)

17.4.1 Mbuji Mayi Kimberlite Cluster

The first diamonds in Kasai Oriental Province were discovered at Lukelenge by Young in 1918 along the Mbujimayi River (Figs. 17.1, 17.4 and 17.5). It was only in 1946 that Magnée recognized from an electrical survey that the rich eluvial deposits were underlain by kimberlite pipes, locally called "Massifs" (Bardet 1974). The discovery of Massif-1 triggered large-scale systematic prospecting of the high terraces and hills of the Bakwanga (Mbuji Mayi) area and resulted in the discovery of a cluster of ten pipes (M1-M10), six of which are crater- to diatreme-facies pipes and four are interpreted as in-fills of kimberlite in topographic (karst) depressions. These bodies comprise the Northern Group of Mbuji Mayi kimberlites and nine of these occur in the MIBA Mining License known as the "Polygon" (Fig. 17.4).

Follow-up work to determine the source of diamonds in the Mbujimayi River and associated Mudiba and Katsha tributaries resulted in the discovery of six further pipes to the southwest of Mbuji Mayi town, inclusive of the Tshibwe pipe (Fig. 17.5).

The kimberlite pipes of the Mbuji Mayi Cluster occur to the south of Mbuji Mayi town and are primarily aligned along an E-W trend (Figs. 17.5 and 17.6a). They are emplaced through Archean basement, and siliciclastic sediments and stromatolitic, dolomitic limestones of Neoproterozoic age of the Mbuji Mayi Supergroup (Delpomdor et al. 2013; and Chap. 2b), which are overlain by arenaceous sediments of Cretaceous age. Narrow feeder pipes have intruded through the hard limestones and expanded dramatically into the overlying and unconsolidated sediments and resulted in 'champagne glass-shaped' intrusions. The upper parts of the kimberlites show crater facies litho-types, including resedimented volcaniclastic kimberlites (RVK) and pyroclastic kimberlites (PK) (Fig. 17.6b). The RVK shows green and red breccias and tuffs, with variable dilution and which often show graded bedding. The PK is characterized by juvenile-rich micaceous kimberlites that have a massive appearance on the scale of metres to decametres. Country rock clast types are dominated by dolomitic limestone, and subordinate crossbedded sandstone and shale, and basement granite, gabbro, amphibolite and dolerite. Hypabyssal kimberlite is largely absent.

Of note is the presence of sedimentary intervals in Massif-3, separating kimberlite eruptive phases, including a 20 m thick aeolian sandstone horizon and a 50 cm thick bioturbated clay bed (Fieremans, pers. comm. 2006), suggesting that some of eruptive phases are separated in time.

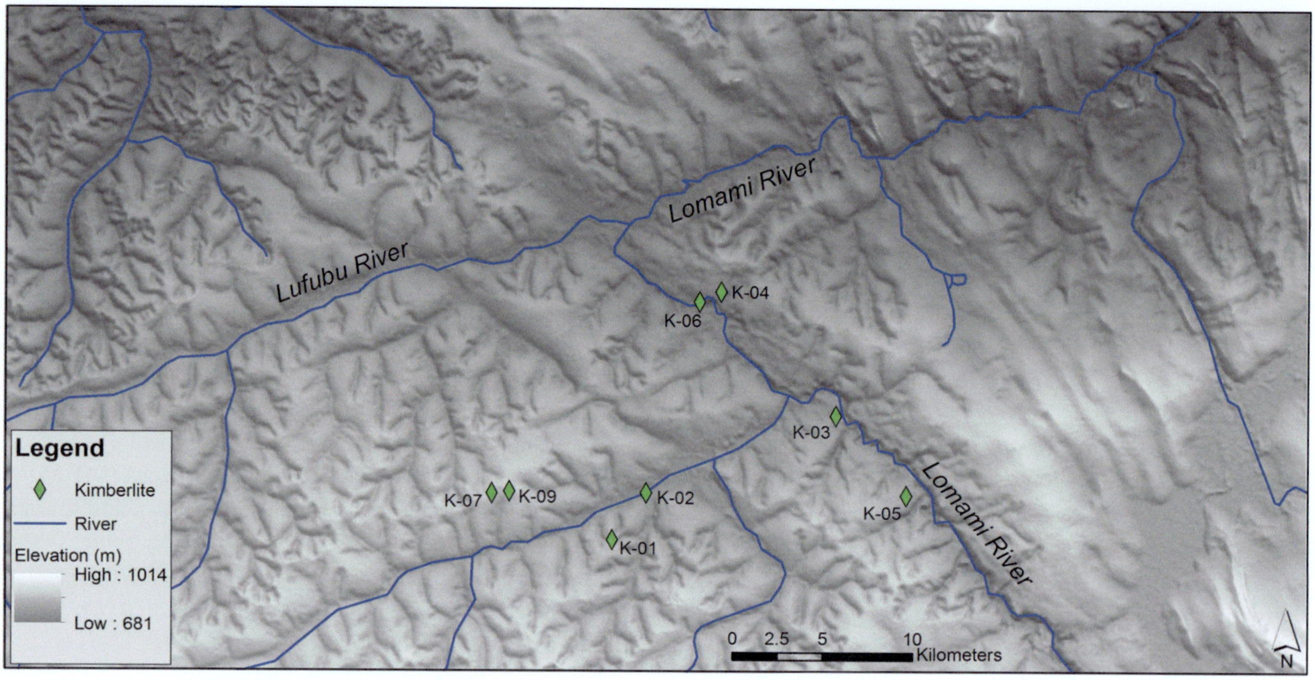

Fig. 17.3 Kimberlites of the Kasendou Cluster within the Kabinda Field (after Walker 2011)

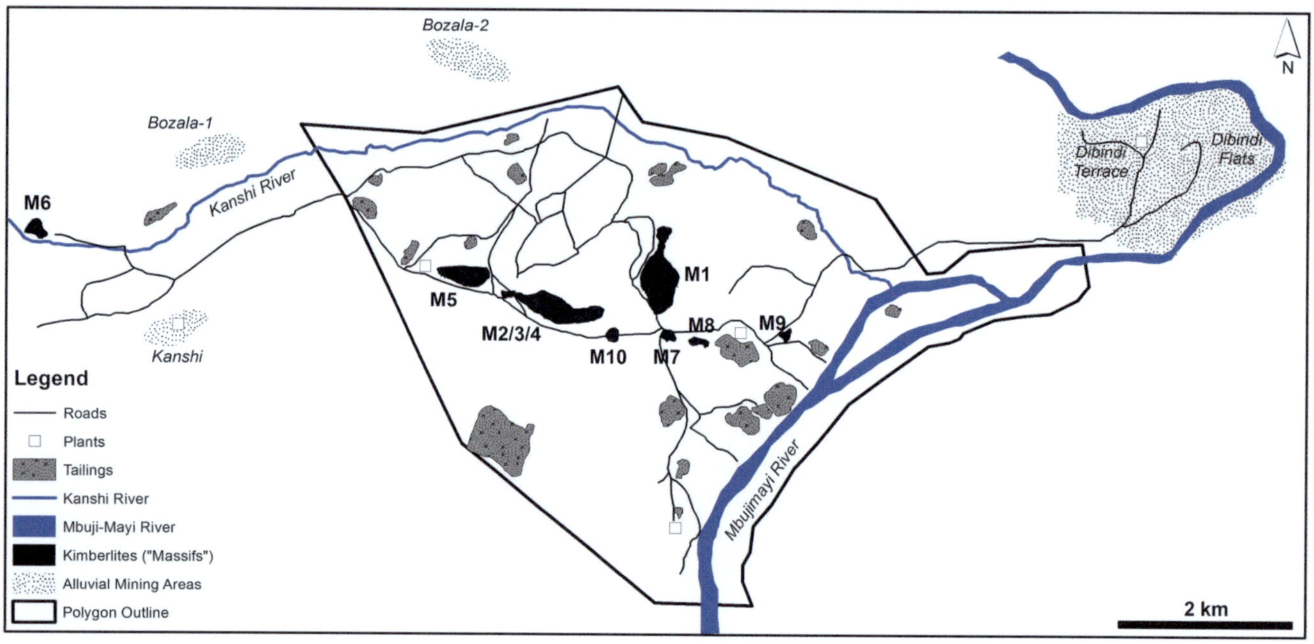

Fig. 17.4 Overview of the mining area and kimberlite "Massifs" at Mbuji Mayi. The location of the "Massifs" within the larger Mbuji Mayi Kimberlite Field is shown in Fig. 17.5

This is also indicated by the interconnection and superimposition of some of the eruptive phases, with magmatism migrating from west to east from Massif-4 to Massif-2 (Fig. 17.7).

The majority of the Mbuji Mayi kimberlites have been evaluated to various degrees of confidence (Table 17.1) and mining has taken place predominantly within weathered ore horizons; Massifs-2 and -3 which share a common feeder

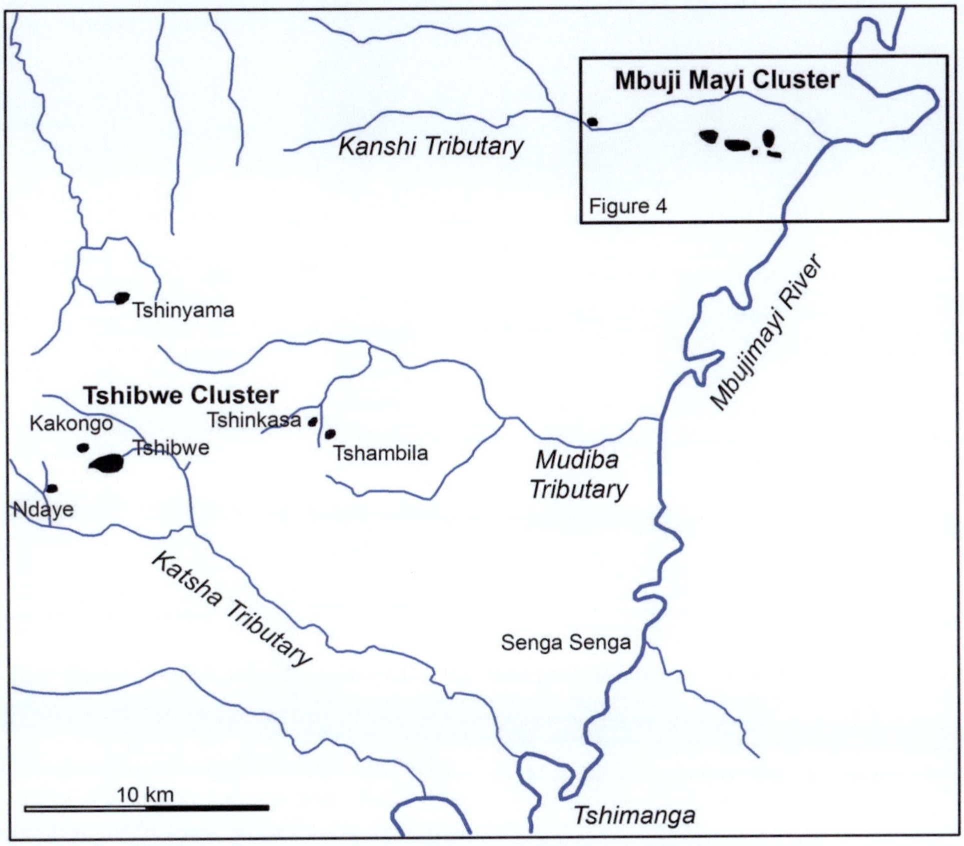

Fig. 17.5 Kimberlite pipes near Mbuji Mayi in East Kasai (after Fieremans and Fieremans 1993)

Fig. 17.6 (**a**) Photograph of Mbuji Mayi kimberlite pipes, with Massif-5 in the foreground, viewing east and Mbuji Mayi town to the north (photo John Ward). (**b**) Different kimberlite phases in Massif-1, including ("epiclastic") red sandy-gritty, highly diluted resedimented volcaniclastic kimberlite (0.5 cts/m^3) and green pyroclastic kimberlite (2 cts/m^3)

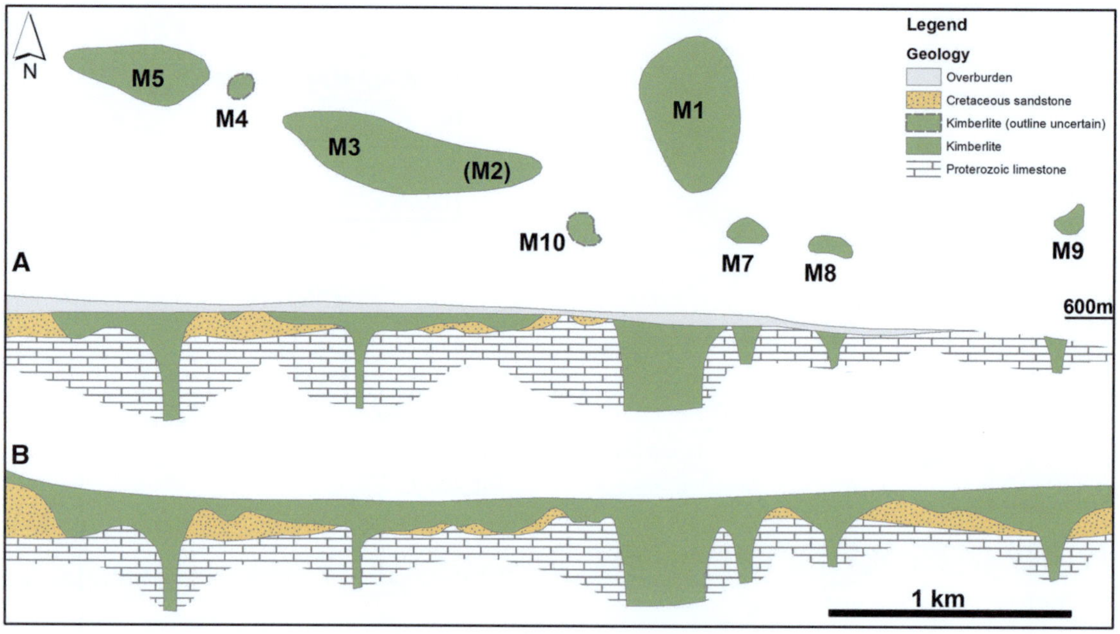

Fig. 17.7 Linear relation between the pipes of the Northern Group (pipe M6 to the west is not shown), after Fieremans and Fieremans (1993). A: longitudinal profile; B: theoretical reconstruction of the profile before erosion to actual level

Table 17.1 Size, grade and revenue data for four of the Mbuji Mayi kimberlites

Kimberlite	Dimension	Surface area (ha)	Grade (cts/m^3)	Revenue (US$/ct)
Massif-1	750 × 450 m	22.6	2.78	12–14
Massif-4	75 × 75 m	0.8	2.57	
Massif-5	650 × 200 m	10.2	2.77	
Massif-6	280 × 160 m	4.4	2.42	

ct carat; *ha* hectare

with Massif-4 have been mined to approximately 60 m below surface while operations have advanced to below 50 m on Massif-5 (Fig. 17.6a). Mining operations are planned on Massif-1 using conventional drill and blast methods although no plant processing capability exists.

Kimberlite grades are variable from 0.5 cts/m^3 in the "epiclastic" kimberlite, to 1.5 cts/m^3 in tuffaceous kimberlite breccias, to 2–3 cts/m^3 in massive pyroclastic kimberlite. The stone value is generally low, averaging at 12 to 14 US$/ct.

The inferred emplacement age of the kimberlites is estimated at ca. 70 Ma (Davis 1977; Shärer et al. 1997), which is consistent with the mean age of a young concordant zircon population recovered from an Upper Cretaceous sandstone of 79 ± 7 Ma (Roberts et al. 2014). $^{40}Ar/^{39}Ar$ dating on diamond inclusions reached consistently older ages for kimberlite emplacement (Phillips et al. 2004) between 115 ± 2 Ma and 801 ± 13 Ma, which can be explained in terms of partial retention of inherited argon, produced between the times of diamond crystallization and kimberlite eruption.

Detrital deposits inclusive of sandy argillaceous deposits with kaolin nodules that overlie the kimberlite pipes run at very high grades as do some of the hill-slope gravels. These pediments or flash flood gravels were deposited onto pre-existing limestone karst topography and are poorly sorted. Further interpretation is complicated by excessive overburden (up to 40 m). However, pre-deposition karst features and potholes have proven to be excellent diamond trap-sites (reported grades of 100 cts/m^3) while post-deposition sinkhole developments exhibit limited diamond potential. Alluvial deposits inclusive of high and low terrace and valley floor deposits occur along the Mbujimayi, Sankuru and Lubi Rivers (and associated tributaries), and riverbed gravel deposits, particularly in the Mbujimayi and Sankuru Rivers.

MIBA mining operations are becoming increasingly marginal as most of the high grade detrital and river bed deposits have been mined to exhaustion and because of declining diamond prices, labour disputes, and power supply problems. Production from the Mbuji Mayi kimberlites ceased in November 2008 and the mine requires significant recapitalisation if mining is to resume.

17.4.2 Tshibwe Kimberlite Cluster

In 1956 a new kimberlite was intersected in extensive eluvial workings in the Katsha-Kabonji-Tshibua area, and is now known as the Tshibwe kimberlite pipe (Figs. 17.5 and 17.8). Following the discovery, a drilling program by MIBA led to the discovery of a further five kimberlites in the area, which

Fig. 17.8 Photograph of Tshibwe kimberlite pipe, viewing northwest (photo John Ward)

are here referred to as the Southern Group of the Mbuji Mayi kimberlite field encompassing Tshibwe, Tshinyama, Kakongo, Tshambila, Ndaye and Tshinkasa pipes (Fig. 17.5).

The Tshibwe kimberlite is located within a broadly east–west orientated valley surrounded by low hills. The Tshibua stream passes across the kimberlite but the flow has been diverted and controlled in order to prevent flooding of the open pit. It joins the Katsha tributary which flows to the southeast to the Tshimanga mining area and there joins the Mbujimayi River (Fig. 17.5). Dredging of gravels downstream of Tshibwe pipe has proven lucrative with the 100 m wide by 20 m deep Senga-Senga pothole in the Mbujimayi River (Fig. 17.4) delivering five million carats from 100,000 m^3 of ore.

Estimates of the extent of the kimberlite sub-outcrop range from 40 to 60 ha, with an E–W axis of 1400 m and an N-S axis of 600 m. In cross-section, the pipe has a flared (albeit asymmetric) champagne-glass shape, with a narrow neck some 200 m in diameter (Fig. 17.9). The facies of the pipe are dominated by volcaniclastic kimberlite, including sandstone-rich, monolithic and heterolithic breccia types (RVK), as well as the more competent, green-grey, massive volcaniclastic (PK) and transitional kimberlite types.

The eluviated and weathered surface horizon ("yellow ground") over Tshibwe pipe was mined by MIBA between 1978 and 1980, with reported grades in excess of 2 cts/m^3. The diamond potential of the hard "blue ground" kimberlite of the pipe was subsequently tested and mined from 670 m to 500 m by MIBA (1981–2000), Sengamines (Senga-Senga SARL, 2000–2005, with MIBA as minority shareholder) and more recently, Societé Anhui-Congo d'Investissement Minier Sprl, or SACIM (2013-present). The Sengamines operations produced 2.03 Mcts at a run-of-mine grade of 0.8 cts/m^3 and revenue of US$ 16/ct (Michaelides 2006).

17.5 Bas-Congo Kimpangu Kimberlite Field

Several kimberlite dykes were found south of Kinshasa close to the border with Angola, intruded into Neoproterozoic sediments of the West Congolian Group (Fig. 17.1). The dykes are up to 3 m wide and trend north-northeast. The kimberlites were discovered by De Beers in 1974 during a reconnaissance sampling program but no diamonds were found associated with these occurrences and the mineral chemistry suggests an off-craton setting (de Wit 2006). One of the kimberlites is referred to as Kimpangu and was explored by SouthernEra (Evans 2004). More recently, Kapapa Eleazard (2011) mentions the presence of a kimberlite pipe at Kimpangu for which little is known—'On signale qu'une pipe kimberlitique existe à Kimpangu mais il est peu étudié' *(A kimberlite pipe is reported at Kimpangu but it has been little studied).*

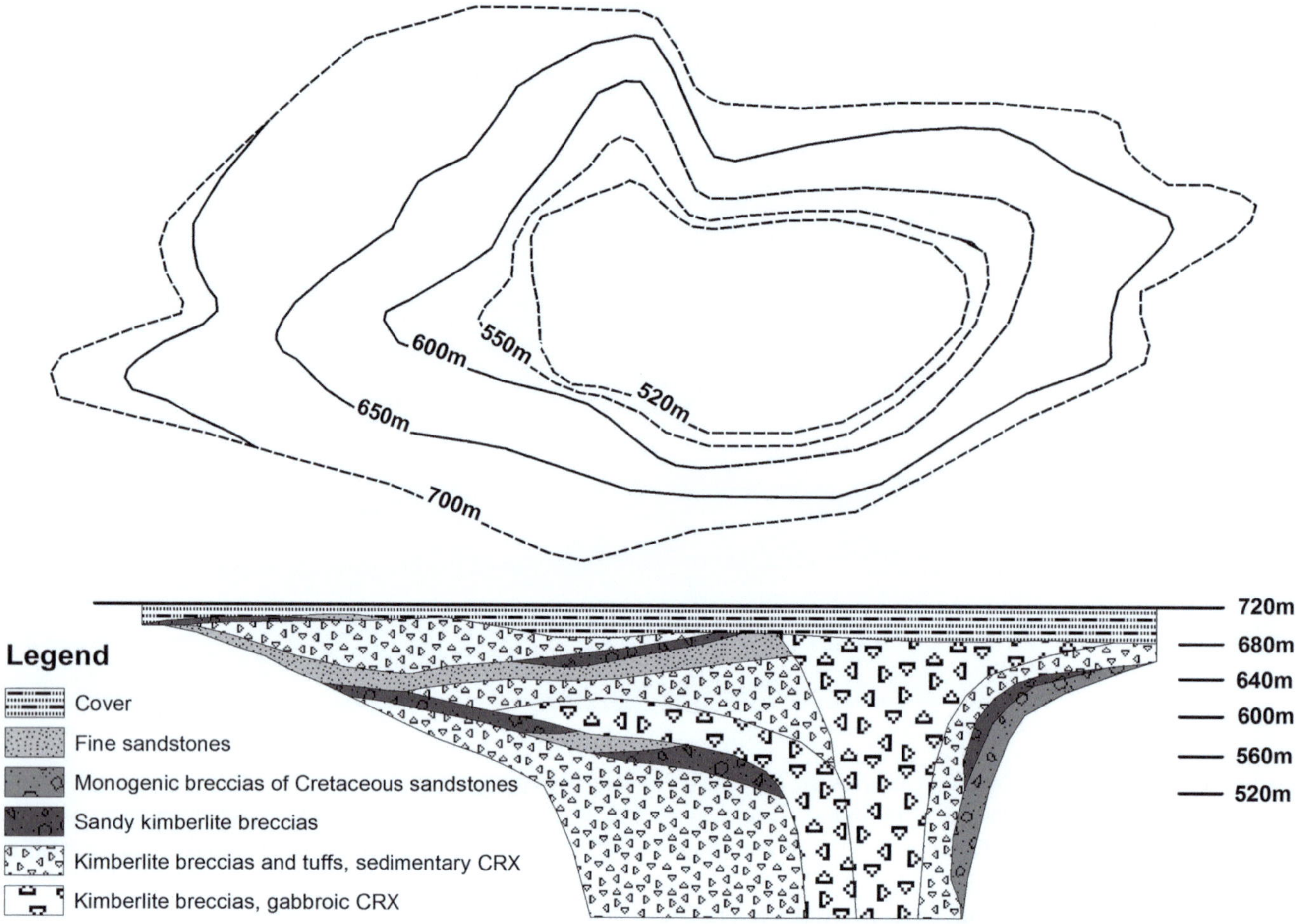

Fig. 17.9 Tshibwe pipe, topographic contour map of crater walls as deduced from drilling, and longitudinal profile (redrawn from Fieremans and Fieremans 1993)

Conclusions

During the past century, the DRC (alternate historic names Congo Belge, Zaire, Congo) has produced over 1,000 Mcts of diamonds, from alluvial and kimberlite mining operations. Sixty-six kimberlites have been discovered since 1946, with economic pipes restricted to the Kasai region, where most of the alluvial mining operations have also taken place. Published age constraints on the kimberlites show Late Cretaceous ages for the Mbuji Mayi kimberlites (~70 Ma) and Eocene ages for the Kundelungu kimberlites (~32 Ma). Emplacement of the Late Cretaceous kimberlites (Mbuji Mayi, Tshibwe, Kasendou and Lukashi) was concomitant with the deposition of Cretaceous sequences. The majority of the pipes show crater-facies preservation and some of the pipes are flared showing so-called 'champagne glass-shaped' morphologies, suggesting emplacement into unconsolidated sediments overlying basement. The age of the Eocene-Oligocene Kundelungu kimberlites corresponds to lithospheric extension associated with the southward propagation of the East African Rift

Acknowledgements We express our gratitude to Marc Fieremans and Paul Kabongo (MIBA) for their hospitality during a mine visit in 2006 and discussions on the geology of the Mbuji Mayi kimberlites. Reviews by John Gurney and Mike Lynn are gratefully acknowledged.

References

African Diamonds plc (2009) Annual report and accounts 2008. London, pp 17–18

Bardet MG (1974) Géologie du Diamant. Deuxième partie: Gisements du diamant d'Afrique. Mémoires du BRGM, 83/3:106-128, Paris

Batumike JM, O'Reilly SY, Griffin WL, Belousova EA (2007) U-Pb and Hf-isotope analysis of zircons from the Kundelungu kimberlites DR Congo: implications for crustal evolution. Precambrian Res 156:195–225

Batumike JM, Griffin WL, Belousova EA, Pearson NJ, O'Reilly SY, Shee SR (2008) LAM-ICPMS U-Pb dating of kimberlitic perovskite: Eocene-Oligocene kimberlites from the Kundelungu Plateau, D.R. Congo. Earth Planet Sci Lett 267:609–619

Davis GL (1977) The ages and uranium content of zircons from kimberlites and associated rocks. Year Book-Carnegie Inst Washington 76: 631–635

De Magnée Y (1946) Présence de kimberlite dans la zone diamantifère de Bakwanga (Kasai, Congo Belge). Bull Soc Belge de Géologie T. LVI, 127–132

De Wit MCJ (2006) Exploring for diamonds in the DRC. Presentation to the Geological Society of South Africa symposium, Johannesburg

Delpomdor F, Linnemann U, Boven A, Gärtner A, Travin A, Blanpied C, Virgone A, Jelsma H, Préat A (2013) Depositional age, provenance, and tectonic and paleoclimatic settings of the late Mesoproterozoic–middle Neoproterozoic Mbuji-Mayi supergroup, Democratic Republic of Congo. Palaeogeogr Palaeoclimatol Palaeoecol 389:4–34

Demaiffe D, Fieremans M, Fieremans C (1991) The kimberlites of Central Africa: a review. In: Kampunzu AB, Lubala RT (eds) Magmatism in extensional structural settings – the Phanerozoic African plate, Chapter 15, pp 537–559

Evans P (2004) Corporate overview. SouthernEra Diamonds Inc.

Fieremans C (1953). Geologie en geochemie der Diamantvelden van Belgisch-Congo. Technisch-Wetenschappelijk Tijdschrift, Jaargang 22(4–5):1–16

Fieremans MR, Fieremans CL (1993) Diamond in its primary rocks with special reference to the diamond deposits of Mbujimayi, East Kasai, Zaire. Bulletin van de Belgische Vereniging voor Geologie 101(1–2):9–39

Forminière (1956) Ouvrage publié par la Société Internationale Forestière et Minière du Congo a l'occassion du 50e anniversaire de sa création. Édition M. L. Cuypers, Bruxelles, 15 juin 1956, 208 p

Haughton SH (1931) On the collection of fossil frogs from the clays at Banke. Trans R Soc S Afr 19:233–249

Jacobs BF, Heredeen PS (2004) Eocene dry climate and woodland vegetation in tropical Africa reconstructed from fossil leaves from northern Tanzania. Palaeogeogr Palaeoclimatol Palaeoecol 213: 115–123

Kampata MD (1993) Minéralogie et géochimie des kimberlites du Haut Plateau de Kundelungu (Shaba, Zaïre). Doc. Thesis, Univ. Cathol. Louvain, 248 pp

Kampunzu AB, Cailteux J (1999) Tectonic evolution of the Lufilian Arc (central African Copperbelt) during the Neoproterozoic Pan-African orogenesis. Gondwana Res 2:401–421

Kapapa Eleazard K (2011) Les formations neoproterozoic du Katanguien et de l'Ouest Congolien: caracteristiques communes et dissemblances. University of Lumbumbashi, Note on geological sciences

Michaelides GN (2006) Compilation of all available geological data, EMIKOR (Entreprise Minière du Kasai Oriental), June 2006

Phillips D, Harris JW, Kiviets GB (2004) Ar40/Ar39 analysis of clinopyroxene inclusions in African diamonds: implications for source ages of detrital diamonds. Geochim Cosmochim Acta 68(1):151–165

Roberts EM, Jelsma HA, Hegna TA (2014) Mesozoic sedimentary cover sequences of the Congo Basin in the Kasai Region, DRC. In: de Wit MJ, Guillocheau F, de Wit MCJ (eds) **The Geology and Resource Potential of the Congo Basin**. Springer, Hiedelberg

Saxby (PB (1982) Report on prospecting in the Kundelungu area between 1977 and 1982. Zairebrit Exploration

Shärer U, Corfu F, Demaiffe D (1997) U-Pb and Lu-Hf isotopes in baddeleyite and zircon megacrysts from the Mbuji-Mayi kimberlites: constraints on the subcontinental mantle. Chem Geol 143: 1–16

Smith RMH (1986) Sedimentation and palaeoenvironment of Late Cretaceous crater-lake deposits in Bushmanland, South Africa. Sedimentology 33:369–386

Studt FE (1910) Report on the Kundelungu pipes, Lubumbashi, 23 p

USGS (2008) Minerals Yearbook Congo Kinshasa

Verhoogen J (1938) Les pipes de Kimberlite du Katanga. Ann. du Service des Mines du CSKT, IX, 3–46

Walker PWE (2011) Competent person's report on a portfolio of diamond exploration properties in Botswana and the Democratic Republic of Congo for Botswana diamonds Plc. and Finncap Ltd. VP3 Geoservices (Pty.) Ltd., 90 pp

Wasilewski I (1950) Note préliminaire sur les gisements de brèche kimberlitique de Bakwanga. Comptes Rendus du Congrès Scientifique, 50ème anniversaire du CSK, Elisabethville, vol. 2, Part 2, pp 291–332

18 Petroleum Potential of the Congo Basin

Damien Delvaux and Max Fernandez-Alonso

18.1 Introduction

The Congo Basin (CB) is a broad and long-lived intracratonic depression in the centre of the African Plate covering most of the Democratic Republic of Congo (DRC, formerly Zaire), the People's Republic of Congo and the Central African Republic (CAR), coinciding with a region of pronounced long-wavelength gravity anomaly (Crosby et al. 2010; see also Raveloson et al., Chap. 1, this Book). The CB has a long (~600 Ma) and complex history of sediment accumulation, tectonic inversion and erosion since the end-Neoproterozoic (Veatch 1935; Cahen 1954; Cahen and Lepersonne 1954; Lepersonne 1974, 1977; Daly et al. 1992; Giresse 2005; Kadima et al. 2011a; see also Kadima et al., Chap. 6, this Book), and it was affected by the break-up of Gondwana during the Mesozoic (Torsvik and Cocks 2011; see also Linol, Chap. 11, this Book). Today, the CB is still tectonically active along the East African Rift margin (Delvaux and Barth 2010).

The CB contains up to ~9 km of Neoproterozoic, Paleozoic and Meso-Cenozoic sedimentary sequences (Kadima et al. 2011a; and Delpomdor et al., Chaps. 3, 4; Linol et al., Chaps. 7–9; Kadima et al., Chap. 6, this Book). The basin is believed to have initiated as a Neoproterozoic rift and a large part of its subsequent subsidence history has been modelled by post-rift thermal relaxation (Kadima et al. 2011b; Buiter et al. 2012; Lucazeau et al., Chap. 12, this Book; but see Linol et al., Chap. 11, this Book, for alternative models). Global climate change, Gondwana motions and break-up, and intraplate stresses are the main factors that controlled the tectonostratigraphic evolution of the CB. Several tectonic inversions occurred during its Phanerozoic development as a response to far-field compressional stresses generated at plate boundaries (e.g. Daly et al. 1992). This caused depositional hiatus, uplift and erosion periods, faulting and folding, leading to the development of stratigraphic and tectonic unconformities (e.g. Kadima et al., Chap. 6, and Linol et al., Chap. 11, this Book).

Although the petroleum potential of this ca. 1.2 million km^2 intra-continental basin has been of interest for decades (Misser 2013), the CB is still one of the largest, least well-understood basin in the world due to relatively limited observational data. Only two stratigraphic wells and two deep petroleum wells have been drilled since exploration began in the 1950s. Many important aspects of its geological knowledge and petroleum systems remain poorly constrained, such as: its regional tectonic framework and internal structure, the stratigraphy of the basin fill, potential source rock levels, reservoirs and seal facies, its burial and thermal history, and so forth. Despite this, the CB is considered by some as a new future petroleum province (e.g. Mello 2008; Pilipili Mawezi (2010). Here, we re-evaluate the petroleum potential of the CB in light of geochemical data from both academic and industrial sources.

18.2 History of Petroleum Investigation

As early as the beginning of the twentieth century, Cornet (1911), then Passau (1923) recognized that the CB (Fig. 18.1) contains organic-rich rocks with potential for oil and gas. The first exploration project was performed between 1952 and 1956 (Mission REMINA—*Syndicat pour l'étude géologique et minière de la cuvette congolaise*), with geological field investigations, combined gravity and magnetic measurements, 600 km of refraction seismic profiles, 131 km of seismic reflection profiles, and two ~ 2,000 m deep fully cored stratigraphic wells (Samba and Dekese). The results have been largely published in the Annales of the Royal Museum for Central Africa (RMCA), in Tervuren, Belgium, where outcrop samples and cores are also archived.

D. Delvaux (✉) • M. Fernandez-Alonso
Geodynamics and Mineral Resources, Royal Museum for Central Africa, Tervuren, Belgium
e-mail: damien.delvaux@africamuseum.be; max.fernandez@africamuseum.be

M.J. de Wit et al. (eds.), *Geology and Resource Potential of the Congo Basin*, Regional Geology Reviews,
DOI 10.1007/978-3-642-29482-2_18, © Springer-Verlag Berlin Heidelberg 2015

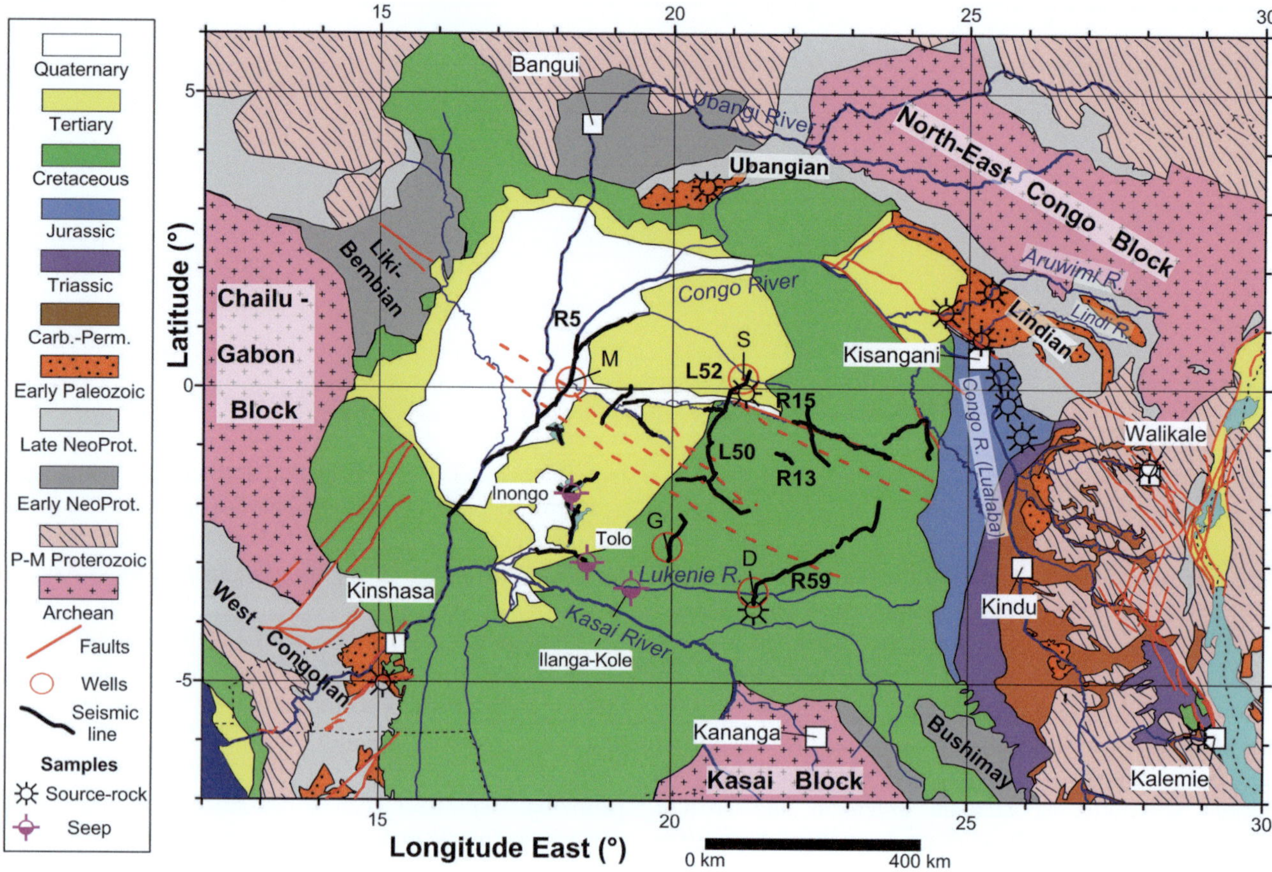

Fig. 18.1 Simplified geological map of the Congo Basin with location of the samples referred to in the text (geology modified from Lepersonne 1974)

A second major exploration phase was conducted between 1974 and 1976 by the Esso-Texaco consortium, which acquired 2,900 km of reflection seismic profiles and drilled two ca. 4,000 m-deep exploration wells (Mbandaka-1 and Gilson-1). The results of this exploration are synthesized in the Hydrocarbon Potential of the Cuvette Centrale Report (ECL 1988) and further discussed in Lawrence and Makazu (1988) and Daly et al. (1992). Unfortunately, most of the detailed results are unpublished. The two exploration wells proved negative with respect to oil potential, but provided valuable stratigraphic and rock information as well as well-logs. Both wells have a dominant siliciclastic lithology, including good potential reservoir rocks (e.g. abundant aeolian sandstone units) but encountered no hydrocarbon shows. A bitumen residue was found in black-grey shales and siltstones of the Mbandaka-1 well between depths of 2,810–3,425 m. This well was drilled almost entirely in red sandstones and conglomerates (ESSO Zaire 1981a). It encountered Neoproterozoic carbonates at 4,133 m deep and bottomed in evaporate and probable halite salt at 4,350 m. Due to its dominantly clastic nature and near absence of microfossil remains, age constraints are scarce (for detailed biostratigraphy, see Colin 1981; Linol 2013, who describes abundant ostracods), and stratigraphic correlations with the other wells and outcrops are difficult. Only a few layers of black shales were encountered, but with negligible source-rock potential. Similarly, the Gilson-1 well encountered no source rock, potential seals or hydrocarbon shows (ESSO Zaire 1981b). It has also a dominant clastic lithology of continental origin and Neoproterozoic carbonates at the base, suggesting restricted marine to lagoon conditions (see Delpomdor et al., Chaps. 3 and 4, this Book for detailed descriptions of the Neoproterozoic carbonates).

The last oil exploration-phase was completed in 1984 by the Japan National Oil Corporation (JNOC), including aeromagnetic and gravity surveys with source rock investigations (JNOC 1984). It concentrated along the eastern rim of the CB, where the Middle Cretaceous Loia and the Upper Jurassic Stanleyville Groups are outcropping. Several source rock horizons have been identified in outcrops and in boreholes. Where sampled, the most promising ones are not within the oil maturation window, but they may have generated hydrocarbons in deeper parts of the basin (Sachse et al. 2012).

Since the JNOC campaign, several reviews of the existing data have been prepared by Oil Search/Pioneer (2007) and HRT Petroleum (Mello 2008), in order to evaluate the hydrocarbon potential of the CB, with the aim of attracting potential investors and promoting hydrocarbon exploration. In parallel, hydrocarbon shows (e.g. oil seeps) have been reported since 2005 along the Lukenie and Tshuapa rivers and along the shores of the Lake Inongo (lumps of bitumen on lake beaches). These were sampled and analysed by HRT Petroleum (Mello 2008), who identified them as remains of black oil generated by prolific and abundant source rocks. He further suggests that giant oil accumulations could have been formed: "*The integration of all geological, geochemical and geophysical data available today suggests that the Cuvette Centrale Basin is an overcharged petroliferous basin that could be considered, today, one of the last provinces in Africa to hold giant to supergiant light oil, condensate and gas accumulations. The presence of oil seeps widespread around most of the Central Basin suggested a light oil/ gas prone system. The size of the structural highs, surpassing more than 1,000 km^{2}and, together with the presence of at least two active oil systems in most of the Basin, indicate that all elements and processes of the petroleum systems are active and works in the Basin.*"

More recently, the basin structure and source rock potential have been re-evaluated by reprocessing existing geophysical data (Kadima et al. 2011a, b) and re-analysing source rock samples stored at the RMCA (Sachse et al. 2012). In the light of these recent studies, Comico SPRL and Centrale Oil and Gas Ltd. asked GhGeochem for an independent review of existing geochemical data on source rocks and seeps (Harriman 2011) and the RMCA to organise in 2011 a new seeps sampling campaign to resample those visited by HRT Petroleum in 2007. The seeps were analysed by the same GhGeochem laboratory and the results compared and re-interpreted in the light of the previous results of HRT Petroleum (Harriman 2012).

18.3 Tectono-Stratigraphic Evolution of the Congo Basin

According to Daly et al. (1992) and Kadima et al. (2011a, b), the evolution of the CB started in the Neoproterozoic (~700 Ma ago), probably in an intracratonic extensional context. They suggested that subsequent subsidence during the Phanerozoic was, at least partly, related to the cooling of the stretched lithosphere and several episodes of basin inversion.

The general stratigraphic evolution has been synthesized by Lepersonne (1977), Lawrence and Makazu (1988) and Daly et al. (1992), based on the results of exploration projects, correlating well-logs with field-based observations along the basin margin (Verbeek 1970; Lepersonne 1977). A synthetic stratigraphic column was presented by Daly et al. (1992), assuming long-distance lateral continuity of the stratigraphic groups. However, this concept appears to be of limited applicability at basin scale, because the stratigraphic units vary laterally both in facies and thickness (Linol 2013; see also Linol et al., Chaps. 7–9 and 11, this Book). A revised and more detailed stratigraphy is presented in Kadima et al. (2011a; and Chap. 6 this Book), taking into account the spatial distribution of the observations (Fig. 18.2).

The development of the CB appears controlled by a series of tectonic events, defining three first-order tectono-stratigraphic units that are separated by prominent seismic reflectors, broadly correlated to the Neoproterozoic, Palaeozoic-Triassic and Jurassic-Cenozoic. Deposition was also controlled by variable climatic factors and changing paleogeographic position of Gondwana/Africa (Scotese 2009; Torsvik and Cocks 2011). The Gondwana continent, of which the Congo Shield formed a central place, was amalgamated at the Neoproterozoic–Palaeozoic transition (550–530 Ma), in a series of events defining the Pan-African orogeny (e.g. De Waele et al. 2008).

Sedimentation in the CB started in the Neoproterozoic during a phase of intracratonic rifting (Daly et al. 1992). This initial rift structure is consistent with aeromagnetic and gravity data (Chorowicz et al. 1990; see also Raveloson et al., Chap. 1, this Book) and might be connected laterally to the Sankuru-Mbuji-Mayi-Lomami-Lovoy failed rift basin that hosts the Meso- and Neoproterozoic Mbuji-Mayi Supergroup (1155 Ma to ca. 800 Ma; Delpomdor et al. 2013; and Delpomdor et al., Chap. 4, this Book). Post-rift subsidence controlled the deposition of a first sedimentary sequence during the Cryogenian and the Ediacaran (seismic unit A of Kadima et al. 2011a; Delpomdor et al., Chap. 4, this Book). This sequence is part of the type Lindian Supergroup described by Henry (1922–1923), Sluys (1945) and Verbeek (1970) in the Lindi-Aruwimi region, north of Kisangani (Fig. 18.1). In this Lindian type region, it comprises ~ 130 m of stromatolitic carbonates at the base (Ituri Group), followed by ~ 470 m of siliciclastics and limestone (Lokoma Group) deposited in an environment interpreted as lagoonal to marine (Verbeek 1970). The Ituri Group contains the Penge arkoses (10–20 m), Lenda carbonates with carbonaceous layers (80–130 m) and Asoso shales and sandstones (50 m). The Lokoma Group comprises the Akwokwo tillites (0–40 m), the Bobwamboli conglomerates and arkoses (50–250 m), the Mamungi grey shales and limestones (200–500 m) and the Kole shales (100 m).

In the Lindi-Aruwimi region (Fig. 18.1), the Ituri Group is unconformably overlain by up to 2,000 m thick siliciclastics (Aruwimi Group: Verbeek 1970; Tait et al. 2011). This possibly forms the base of seismic unit B of Kadima et al. (2011a) in the centre of the CB. In outcrop, it is

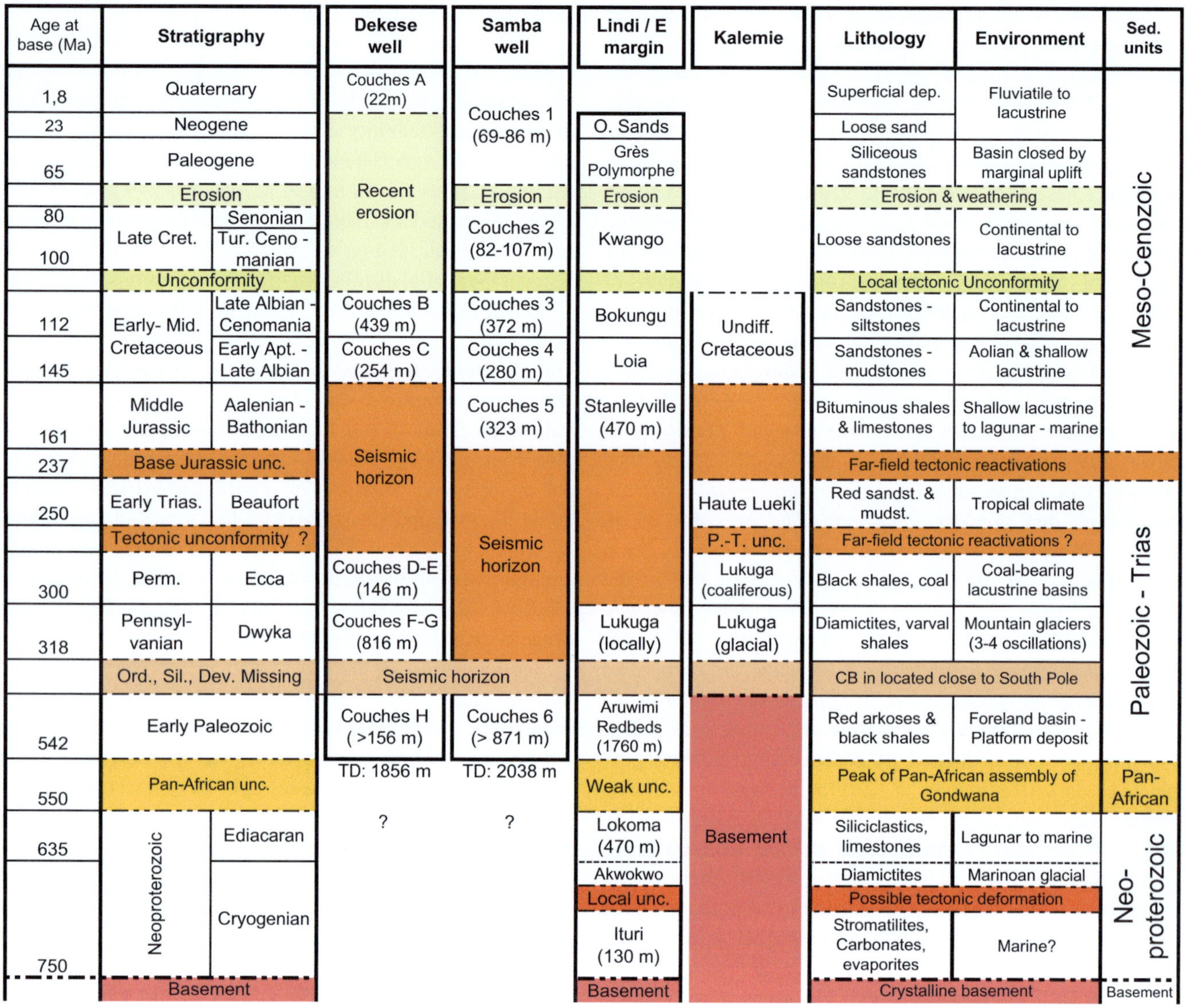

Fig. 18.2 Schematic lithostratigraphy of the Congo Basin (modified from Kadima et al. 2011a), based on the two stratigraphic wells (Samba and Dekese) and outcrop data in in the northeastern and eastern margins of the basin

subdivided successively from base to top into: the Galamboge arkoses and quartzites (100–150 m); the Alolo dark carbonaceous shales (350–400 m); and the Banalia red arkoses (up to 1,200 m), with ~ 100 m thick transitional units between them (Verbeek 1970). The Alolo shales are interpreted to have been deposited in a shallow marine to lagoonal basin (Verbeek 1970). The Aruwimi Group is tentatively correlated to the Inkisi post Pan-African fluvio-deltaic red sandstones in the West Congo Belt (Alvarez et al. 1995; Tait et al. 2011), the Biano red sandstones in the Kundelundu foreland of the Katanga Fold Belt (Kampunzu and Cailteux 1999) and the red sandstones of the lower parts of the Samba and Dekese wells (Cahen et al. 1959, 1960). Using refractions seismic velocities, Evrard (1960) recognized the similarity between the red sandstones in the lower part of the Samba and Dekese wells, but did not exclude that they could be younger than the red sandstones of the Aruwimi, Inkisi and Biano Groups. This correlation is based on lithological similarity (quartzitic red sandstones), taking also into consideration the fact that these series are all post-orogenic relative to the Pan-African deformation. In the Dekese well, the red sandstones are overlain by glacial-interglacial and post-glacial sediments of the Lukuga Group (from the depths of 816 m and 146 m, respectively), dated to be Late Carboniferous-Permianin age (Cahen et al. 1960; Boze and Kar 1978; Cahen and Lepersonne 1978). The age of the Aruwimi Group is therefore constrained only by a regional unconformity at the base, and by the Upper Paleozoic sediments of the Lukuga Group on top, spanning the entire Early–Middle Paleozoic. No prominent discontinuities can be seen between the Aruwimi and the Lukuga Groups, either in the Dekese core or in the seismic profiles. However, a sedimentary hiatus may exist between the two, corresponding to the Ordovician-Early Carboniferous period during which Gondwana drifted across the South Pole (located, respectively, in Algeria in Early Ordovician, in Guinea in Late Ordovician, in Angola in Late Devonian, in northern DRC in Early Carboniferous, in South Africa in Mid Carboniferous, and Antarctica in Late Carboniferous; e. g. Torsvik and Cocks 2011).

The Lukuga Group contains a series of glacial to periglacial massive diamictites and varval dark grey shales in the Dekese core (couches F–G of Cahen et al. 1960), interpreted to be deposited in a large basin adjacent to mountain systems, or as moraine deposits in paleo-glacial valleys outcropping in East DRC. They are overlain by post-glacial brown sandstones and shales in the Dekese core (couches D–E of Cahen et al. 1960), or sandstones with coal seams in the Kalemie region along the western shore of Lake Tanganyika (Fig. 18.1), in the Lukuga coal field (Fourmarier 1914; Jamotte 1931; Cahen and Lepersonne 1978). In the latter locality, fossiliferous reddish sandstones and mudstones of the Lower Triassic Haute Lueki Group overly with a local unconformity Permian sediments (Jamotte 1931; Cahen and Lepersonne 1978). The Triassic has not been formally recognized in the Samba and Dekese wells, but might be present in reduced thickness in the Dekese well above the Lukuga Group (Linol 2013; see also Linol et al., Chap. 7, this Book). Linol (2013) also proposed that the red arkoses of the Samba core could be of Triassic age. In absence of chronological evidence (to date, no spore and pollens or other fossils have been found), we support the views of Cahen et al. (1959) who correlated the red sandstones of the Samba well to the Aruwimi, Inkisi and Biano Groups (Fig. 18.2). Cahen et al. (1959) also mentioned that undetermined organic material was found in one place. This would suggest that the red sandstones are younger than Ordovician. But in their conclusion, Cahen et al. (1959) did not take into account this observation as an age constraint.

A subsequent regional unconformity separates the Paleozoic (and possibly also the Triassic) from the overlying Jurassic-Cretaceous and Cenozoic successions that comprise a third sedimentary unit (Fig. 18.5, seismic unit C of Kadima et al. 2011a; Linol et al., Chaps. 8, 9 and 11 this Book). Its base is defined by the Stanleyville Group, which is exposed along the upper course of the Congo River (locally named Lualaba), south of Kisangani (Fig. 18.1), with up to 470 m of lacustrine sandstones with organic-rich shales and some limestone (Passau 1923; Sluys 1952; Grecoff 1957; Cox 1960; Lombard 1960; Lepersonne 1977; Cahen 1983a; Colin 1994). Towards the basin centre, in the Samba well, 323 m of fluvial-lacustrine red sandstones with thin layers of bituminous shales are also attributed to the Stanleyville Group; here they directly overlie the red arkoses attributed to the Aruwimi Group (Couches 5; Cahen et al. 1959). In contrast, the Stanleyville Group was not recognized in the Dekese well (Cahen et al. 1960) but it occurs in a condensed section in the Kinshasa area (Egoroff and Lombard 1962; Defrétin-Lefranc 1967). The depositional environment of the Stanleyville Group is generally interpreted as lacustrine (Grecoff 1957; Cox 1960; Cahen 1983a; Colin 1994). A limited marine influence was initially suggested by de Sainte Seine and Casier (1962) on the basis of fossil fishes found near the base of the Stanleyville Group. After a paleoichthyological study, Taverne (1975a, b, and personal communication, 2014), however, no clear evidence for a marine influence during deposition of the Stanleyville Group. A Kimmeridgian age was proposed by de Sainte Seine (1955) on the basis of fossil fishes, but Colin (1994), on the basis of palynological and micropaleontological evidence, considered it as Middle Jurassic (Aalenian-Bathonian).

After a long depositional hiatus, the succeeding Middle Cretaceous Loia and Bokungu Groups were deposited across the central CB (Lepersonne 1977; Maheshwari et al. 1977; Cahen 1983b). The Loia Group (Late Aptian–Early Albian; after Colin 1981) was first recognized in the Dekese well by254 m of aeolian sand dunes (Couches C of Cahen et al. 1960; Linol 2013) and in the Samba well, by 280 m of shallow lacustrine sandstones and mudstones with a few thin black shale levels (Cahen et al. 1959, 1960). According to the more recent chronostratigraphic correlations of Linol (2013; Linol et al., Chaps. 8 and 11, this Book) , the aeolian sand unit (the Dekese Formation) is Late Jurassic-Early Cretaceous and not part of the Middle Cretaceous, reducing part of the hiatus between the Stanleyville and the Loia Groups. The Bokungu Group (Late Albian, after Colin 1994), with fluvio-deltaic sandstones and siltstones (372–439 m), is unconformably overlain by unconsolidated siliceous sandstones of the Cenomanian Kwango Group (250–280 m). The Paleogene is represented by the Grès Polymorphe Formation (Eocene, after Colin 1994) comprising silicified sands covering a prominent erosion surface, and in turn by the Neogene 'Ochre Sands' (70–90 m in total for the Cenozoic; Cahen et al. 1959, 1960; Cahen 1983b; Lepersonne 1977).

This above summary demonstrates that the sedimentary sequences display distinct lateral thickness and facies variations at the basin scale. It highlights also some disagreements about their age (e.g. Lower Paleozoic versus Triassic for the red quartzites in Samba). Specifically, locally angular discontinuities between the seismic packages B and C (not shown here) range at least from the Early Paleozoic to the Early Jurassic in the Samba well, and from the Late Triassic to Early Cretaceous in the Dekese well; and is almost absent in the Kalemie area. This loosely constrains an important compressional basin inversion event that Daly et al. (1992) related to far-field effects of the Late Permian—Triassic development of the Cape Fold Belt inSouth Africa (Halbich et al. 1983; Le Roux 1995; Delvaux 2001; Newton et al. 2006; Tankard et al. 2009).

The lateral facies changes across the basin make correlation between the different stratigraphic packages difficult. This is particularly the case for the pre-Jurassic successions (ca.3 km thick) encountered in the Mbandaka-1 and Gilson-1 wells above the Neoproterozoic carbonates. As a consequence, stratigraphic correlations of the Paleozoic to Triassic sequences of the CB (Daly et al. 1992; ECL 1988; ESSO Zaire 1981a, b; Kadima et al. 2011a and other unpublished reports; Linol et al., Chap. 7, this Book) differ each other. This is mainly caused by the lack of chronostratigraphic control for this large part of the basin fill. By contrast, the age of the overlying Jurassic to Neogene sequences and the underlying Neoproterozoic carbonates are better constrained.

18.4 Potential Source Rocks

Several potential source rocks have been recognised in the CB at different stratigraphic levels and locations, but with relatively limited lateral continuity (at the scale of the basin). These are listed below, together with their age estimations:

- Mamungi greyshales of the Lokoma Group (Neoproterozoic)
- Alolo dark shales of the Aruwimi Group in the Aruwimi River region north of Kisangani (Early to Middle Paleozoic)
- Lukuga peri-glacial grey-shales in the Dekese well, and post-glacial coal measures in the Kalemie and Upemba coal fields, respectively, along Lake Tanganyika and in the Kibara Belt in North-Katanga (Late Carboniferous-Permian);
- Stanleyville Group with black shales along the Congo River upstream Kisangani (Lualaba) but red sand and siltstones in the Samba well (Middle or Upper Jurassic);
- Loia Group with seven thin lacustrine black shales levels in the Samba well;
- Kipala black shales south of the Kasai River (Upper Cretaceous, Kwango Group).

The petroleum potential and maturation estimates of these source rocks were initially based on a limited number of organic matter analyses from a selection of samples of core and outcrops, stored at the RMCA. First generation of Rock-Eval instruments (RRI 1988) were used for the analyses. JNOC (1984) also performed a number of analyses on separate outcrop samples, mainly from the Loia and Stanleyville organic-rich shales, but we were not able to re-locate the samples on maps. New petrological and geochemical analyses (Total Organic Carbon (C_{org}), Rock-Eval and Gas Chromatography – Mass Spectrometry) were undertaken by Sachse et al. (2012) on a large number of Palaeozoic and Mesozoic outcrop and core samples from the Dekese and Samba wells. The JNOC Rock-Eval results are reported in Table 18.1. The analysis of the RMCA samples [RWTH data of Sachse et al. (2012) and two data from Kadima (2007)] are reported in Table 18.2. The results, averaged for each stratigraphic unit, are given in Table 18.3.

The type of kerogen in the sediment is characterized using the Rock-Eval results on a Van-Krevelen equivalent diagram (Espitalie et al 1977), by the Hydrogen Index (mg HC equivalents/g C_{org}) and Oxygen Index (mg CO2/g C_{org}). On such diagram (Fig. 18.3) different kerogen types are shown to follow different evolution paths depending on thermal maturation. Progressive decrease in Hydrogen and Oxygen index values reflects loss of hydrocarbon chains and oxygenated functions (Tissot and Welte 1978). Kerogen type I is typical lacustrine organic-matter, rich in aliphatic chains and poor in aromatics, derived from algal lipids, or enriched by microbial activity. Kerogen type II is richer in aromatics, which is usually attributed to marine organic matter deposited in a reducing environment, but which can also correspond to a combination lacustrine algae (type I) and terrestrial (type III) organic matter. Kerogen type III is rich in aromatics and oxygenated functions and is derived from terrestrial higher plants. Organic matter unable to generate oil or gas is qualified as Type IV kerogen. Results of the RMCA samples analyzed by RWTH Aachen (Sachse et al. 2012), with two samples of Kipala shales from Kadima (2007) are shown in Fig. 18.3a, and the JNOC samples (JNOC 1984), in Fig. 18.3b. Sachse et al. (2012) further studied the organic matter by organic petrology and molecular organic geochemistry. Combined with the total organic matter content (C_{org}) and Rock-Eval parameters, this allowed the following characterization of the different source rock horizons (Table 18.3):

- Samples of the Alolo shales from the Aruwimi Group are in general very low in C_{org} (<0.2 % in average), and contain a high amount of degraded organic matter. They are considered barren. Samples from the Yambuya section are slightly richer (0.58 % in average) but few of them are rich enough to allow Rock-Eval analysis; which in any case released very little quantities of hydrocarbon during pyrolysis, but high quantities of CO_2 (very low HI and high OI). They contain a type III/IV kerogen and cannot be considered as a potential source rock. Similarly, an Alolo shale sample taken during the 2011 CoMiCo-RMCA field survey is almost devoid of organic matter (0.08 % $C_{org.}$; Harriman 2012), showing that oxidation of the organic matter during the 60 years storage cannot be invoked to explain these low values.

Table 18.1 Organic matter analytical data (Total Organic Carbon and Rock-Eval) for the JNOC samples (from JNOC 1988)

JNOC Nr	Region	Group	Lithology	Corg (%)	S1 (mg/g)	S2 (mg/g)	S3 (mg/g)	HI (mg/g Corg)	OI (mg/g Corg)	Tmax (°C)	Vr (%)
KS-26	Kisangani	Stanleyville	Oil shales	10.1	0.54	108.1	2	1.067	20	435	
KS-28	Kisangani			12	0.84	125.4	2.24	1.043	19	432	
KS-32	Kisangani			5.55	0.1	58.4	1.77	1.052	32	433	
KS-33	Kisangani			5.02	0.22	59.4	1.1	1.183	22	435	
KS-34	Kisangani			6.21	0.21	72	1.24	1.159	20	437	
KS-35	Kisangani			0.92	0.03	3.97	0.52	432	57	429	
KS-36	Kisangani			1.24	0.06	6.43	0.71	519	57	430	
KS-38	Kisangani			9.23	0.2	118.7	1.48	1.286	16	439	
KS-45	Kisangani			15.5	2.58	172	3.16	1.112	20	422	
KS-46	Kisangani			16.2	1	169	3.44	1.041	21	435	
KS-47	Kisangani			17.6	0.94	169.3	2.18	960	12	437	
KS-48	Kisangani			1.63	0.13	5.59	1.38	343	85	434	0.85
KS-52	Kisangani			9.61	0.22	108.4	2.66	1.128	28	438	
KS-81	Kisangani			5.48	0.51	45.04	2.11	822	39	427	0.69
KS-82	Kisangani			2.8	0.21	25.06	0.98	895	35	429	
KS-83	Kisangani			1.24	0.03	5.45	0.4	440	32	433	
KS-84	Kisangani			8.23	1.82	98.24	1.51	1.194	18	432	
KS-18	Kisangani	Stanleyville	Limestones & calcareous shales	0.38	0.03	0.65	0.21	171	55	423	
KS-23	Kisangani			1.57	0.05	5.41	0.37	345	24	425	
KS-24	Kisangani			2.05	0.07	12.65	0.49	617	24	427	
KS-50	Kisangani			0.56	0.07	0.9	0.74	161	132	429	
KS-53	Kisangani			6.74	0.16	62.68	1.94	930	29	427	
KS-71	Kisangani			3.26	0.12	32.06	0.97	983	30	425	
KS-21	Kisangani	Stanleyville	Mudstones	0.71	0.11	2.45	0.44	345	62	426	
KS-31	Kisangani			0.49	0.02	0.52	0.31	106	63	425	
KS-41	Kisangani			0.44	0.01	0.65	0.34	148	77	432	
KS-72	Kisangani			0.72	0.03	1.75	0.68	243	94	425	
KS-77	Kisangani	Lukuga	Black varval shale	0.95	0.03	0.11	0.57	12	60	411	0.8
KS-78	Kisangani			1.02	0.03	0.14	0.43	14	42	453	0.5
KS-79	Kisangani			0.82	0.01	0.3	0.37	37	45	437	0.64
KS-87	Kisangani			1.42	0.01	0.16	1.18	11	83	434	0.48
KS-88	Kisangani			1.46	0.03	2.31	0.36	158	25	436	0.47
KS-104	Kindu	Lukuga	Black shale	0.61	0.01	0.24	0.41	39	67	435	
KS-105	Kindu			0.88	0.02	0.04	0.8	5	91	443	0.75

Corg: total organic carbon; CaCO3: Carbonates; Rock-Eval pyrolysis parameters; S1: free/absorbed hydrocarbons, S2: hydrocarbons released by the pyrolysis of the kerogen, S3: oxygen-contained volatiles, hydrogen index (HI), oxygen index (OI), Tmax (temperature of the peak of S2 hydrocarbon generation during pyrolysis); vitrinite reflectance data (Vr)

- Samples from the Mamungi shales, Lenda carbonates and Bangu black limestones are all very poor in C_{org}.,and considered barren.
- Permo-Carboniferous sediments from the Lukuga Group (Dekese well and outcrop samples) contain moderate contents of total organic carbon (<=2.4 %C_{org}), with hydrogen-poor type III/IV gas-prone kerogen and a low hydrocarbon generation potential. They may have produced a limited amount of gas, but no oil.
- The Jurassic (Stanleyville Group) and Middle Cretaceous (Loia Group) samples from the Samba well, and outcrops in the north eastern part of the CB, have up to 25 % C_{org}. Sediments from the Stanleyville Group contain type I kerogen of excellent quality; and those of the overlying Loia Group contain type I/II kerogen of good to excellent quality. They are derived from algae deposited in an aquatic anoxic environment, likely lacustrine, and contain small proportions of terrestrial type III material (Sachse et al. 2012).
- The Upper Cretaceous Kipala black shales are also excellent source rock (C_{org}. = 8–9 %, with ca. 750 HI) but are immature (Tmax = 438–439°) and of limited lateral extent.

In summary, the new geochemical data confirm that the Stanleyville Group is the most promising source rock unit in the CB, although it is immature. Samples from this group show a variety of organic matter content and composition, reflecting different depositional environments. In the Samba well, the Stanleyville Group comprises red siltstones and

Table 18.2 Organic matter analytical data (Total Organic Carbon and Rock-Eval) for the RMCA samples (from Sachse et al. 2012), with correspondence of laboratory number (RWTH Nr) with the RMCA RG collection number (RMCA Nr), location, Group, Unit and lithology and coordinates

RWTH Nr	MRAC RG Nr	Location	Group	Unit	Description	Long	Lat	Depth (m)	Corg (%)	CaC03 (%)	S1 (mg/g)	S2 (mg/g)	S3 (mg/g)	HI (mg/g Corg)	OI (mg/g Corg)	Tmax (°C)	VRr (%)
1234	36991	Dekese well	Loia	C	Green shale	21,385	−3.455	703.0	0.09	1.15							
304	36997				Green shale	21,385	−3.455	707.8	0.08	2.47							
305	37001		Lukuga	D	Brown shale	21,385	−3.455	712.0	0.16	52.59							
306	37002				Brown shale	21,385	−3.455	723.0	0.11	7.50							
307	37005				Brown shale	21,385	−3.455	726.0	0.13	5.87							
308	37006				Brown shale	21,385	−3.455	732.0	0.07	6.16							
140	37064-66			F	Dark varval shale	21,385	−3.455	924.6	0.47	0.74							0.97
141	37064-66				Dark varval shale	21,385	−3.455	924.9	0.46	0.94							0.95
151	37067				Dark varval shale	21,385	−3.455	942.2	0.46	1.00							0.93
152	37067				Dark varval shale	21,385	−3.455	942.9	0.53	1.07	0.03	0.10	0.82	19	154	417	0.94
153	37067				Dark varval shale	21,385	−3.455	943.4	0.44	1.06							0.92
154	37067				Dark varval shale	21,385	−3.455	944.4	0.50	1.29							0.97
1235	37074				Dark varval shale	21,385	−3.455	993.5	0.49	0.92	0.04	0.15	0.76	31	156	422	0.84
1237	37076				Dark varval shale	21,385	−3.455	*1019.7*	0.68	1.86	0.09	0.45	1.11	66	162	427	0.77
142	37079				Dark varval shale	21,385	−3.455	1049.2	0.95	0.41	0.05	0.34	0.68	36	72	420	0.96
143	37079				Dark varval shale	21,385	−3.455	1050.2	0.47	0.67							0.97
144	37079				Dark varval shale	21,385	−3.455	1050.5	0.42	0.78							0.99
145	37079				Dark varval shale	21,385	−3.455	1050.9	0.51	0.75	0.05	0.06	0.65	11	128	422	0.94
146	37079				Dark varval shale	21,385	−3.455	1051.2	0.44	0.71							0.95
155	37081-3				Dark varval shale	21,385	−3.455	1067.9	2.30	1.02	0.08	2.27	0.70	99	30	431	0.97
147	37023-81				Dark shale	21,385	−3.455	1068.0	2.40	0.41	0.07	2.48	0.81	103	34	431	0.96
148	37081-83				Dark shale	21,385	−3.455	1068.2	1.83	0.42	0.05	0.65	0.92	36	50	425	0.95
156	37081-83				Dark shale	21,385	−3.455	1068.7	0.43	0.15							1.03
157	37081-83				Dark shale	21,385	−3.455	1069.0	0.37	0.15							
149	37081-83				Dark shale	21,385	−3.455	1069.9	0.63	0.27	0.05	0.08	0.50	13	80	364	

158	37084				Dark varval shale	21,385	−3.455	1095.3	0.64	1.07	0.05	0.18	0.50	27	77	420	0.99
159	37084				Dark varval shale	21,385	−3.455	1096.0	0.61	0.58							1.00
160	37084				Dark varval shale	21,385	−3.455	1098.0	0.73	1.02	0.04	0.24	0.16	33	22	421	0.98
150	37081-83				Dark varval shale	21,385	−3.455	1099.3	0.72	1.18	0.07	0.26	0.43	36	60	420	0.93
161	37085				Dark varval shale	21,385	−3.455	1134.2	0.97	8.52	0.06	0.32	0.67	32	69	424	0.94
162	37085				Dark varval shale	21,385	−3.455	1136.7	0.95	0.75	0.05	0.38	0.95	39	99	423	0.94
163	37085				Dark varval shale	21,385	−3.455	1136.8	0.70	1.27	0.07	0.20	0.71	29	101	421	0.89
164	37085				Dark varval shale	21,385	−3.455	1138.9	0.97	0.97	0.07	0.40	0.78	41	80	421	1.06
1233	37126				Dark varval shale	21,385	−3.455	1189.9	0.56	0.90	0.04	0.21	0.48	38	86	422	0.96
1238	37127				Dark varval shale	21,385	−3.455	1189.9	0.55	1.12	0.04	0.22	0.55	40	100	420	
1236	37195				Grey varval shale	21,385	−3.455	1259.8	0.62	1.27	0.08	0.29	0.40	47	65	428	0.91
1228	37195				Grey varval shale	21,385	−3.455	1261.0	0.58	0.98	0.06	0.29		50		427	1.19
1232	37200				Tillite	21,385	−3.455	1312.3	0.15	2.90							1.22
1231	37202				Tillite	21,385	−3.455	1340.4	0.15	2.43							1.06
1230	37204				Tillite	21,385	−3.455	1379.1	0.14	3.48							
1225	37205				Dark varval shale	21,385	−3.455	1459.0	0.42	0.74	0.04	0.15	0.78	36	188	427	1.05
1224	37234				Dark varval shale	21,385	−3.455	1459.7	0.37	1.66							
1227	37255				Dark shale	21,385	−3.455	1501.5	0.51	0.47	0.08	0.21	1.20	41	235	409	1.11
165	37263-4				Dark shale	21,385	−3.455	1530.0	0.43	0.57							1.09
166	37263-4				Dark shale	21,385	−3.455	1530.8	0.30	0.18							
167	37263-4				Dark shale	21,385	−3.455	1531.4	0.36	0.25							
168	37263-4				Dark shale	21,385	−3.455	1532.2	0.45	0.21							1.13
169	37263-4				Dark shale	21,385	−3.455	1532.5	0.42	0.47							1.17
170	37263-4				Dark shale	21,385	−3.455	1533.2	0.39	0.42							
171	37263-4				Dark shale	21,385	−3.455	1533.5	0.41	1.02							1.19
1226	37268				Tillite	21,385	−3.455	1549.0	0.11	4.54							
1229	37310			G	Tillite	21,385	−3.455	1661.5	0.06	2.94							
299	34505	Samba well	Loia	4	Grey shale	21,202	0.165	567.8	0.38	10.51							
300	35448				Grey shale	21,202	0.165	632.8	0.15	8.56							
301	35457				Bituminous shale	21,202	0.165	654.0	19.38	3.00	3.18	159.20	161.90	821		447	

(continued)

Table 18.2 (continued)

RWTH Nr	MRAC RG Nr	Location	Group	Unit	Description	Long	Lat	Depth (m)	Corg (%)	CaC03 (%)	S1 (mg/g)	S2 (mg/g)	S3 (mg/g)	HI (mg/g Corg)	OI (mg/g Corg)	Tmax (°C)	VRr (%)
292	35464				Grey shale	21,202	0.165	665.2	11.01	17.41	1.98	79.04	3.96	718	36	431	
293	35466				Shale	21,202	0.165	676.2	1.23	5.10	0.36	6.19	1.57	505	128	429	
302	34477				Sandy shale	21,202	0.165	706.4	1.24	0.30	0.08	8.21	0.73	663	59	433	
1202	35495				Bituminous shale	21,202	0.165	734.9	2.91	3.82	1.31	31.89	0.80		27	430	
1203	35495				Bituminous shale	21,202	0.165	734.9	4.51	9.73	2.01	41.55	4.66	922	103	429	
1201	35495				Bituminous shale	21,202	0.165	734.9	8.78	8.71	2.58	72.41	2.68	824	31	433	
1205	35495				Bituminous shale	21,202	0.165	739.7	3.95	12.70	1.59	38.12	1.55	964	39	432	
1204	35495				Bituminous shale	21,202	0.165	739.8	4.67	2.45	2.33	40.79	2.18	874	47	425	
303	35502				Bituminous shale	21,202	0.165	757.0	3.89	6.34	1.24	35.66	2.63	917	68	430	
294	35506				Bituminous shale	21,202	0.165	763.5	1.05	25.96	0.24	6.62	0.86	629	82	432	
295	35514				Bituminous shale	21,202	0.165	781.0	5.85	12.55	2.28	54.81	1.29	936	22	437	
296	35542				Sandy calc. shale	21,202	0.165	825.3	1.60	20.03	0.26	14.31	0.95	893	59	436	
1197	35619		Stanley-ville	5	Grey shale	21,202	0.165	1008.6	0.54	17.52							
1198	35619				Grey shale	21,202	0.165	1008.7	0.20	16.46							
1199	35619				Grey shale	21,202	0.165	1008.8	0.14	21.07							
1200	35619				Grey shale	21,202	0.165	1008.9	1.50	0.95							
297	35619				Grey shale	21,202	0.165	1009.4	0.13	7.58							
298	35625				Sanstone	21,202	0.165	1031.5	0.09	15.10							
180	45196	Aruwimi River Malili-Banalia section	Aruwimi	Alolo	Brown sandy shale	25,398	1.727		0.04	0.46							
181	45194				Brown shale	25,398	1.727		0.09	49.26							
179	45201				Grey shale	25,392	1.722		0.05	0.50							
178	45202				Grey shale	25,387	1.717		0.18	1.00							0.74
177	45205				Grey shale	25,372	1.672		0.18	13.82							1.12
176	45206				Brown shale	25,381	1.626		0.07	0.53							
175	45215				Calcareous shale	25,370	1.607		0.26	24.71				140	89	425	1.44
174	45217				Grey shale	25,369	1.605		0.18	1.55							
173	45219				Grey shale	25,367	1.602		0.17	28.03							1.05
172	45221				Brown shale	25,358	1.595		0.13	26.35							

1243	45577	Aruwimi River Yambuya section	Aruwimi	Alolo	Black shale			0.89	0.91	0.09	0.83		93		439	
1242	4560				Black shale			1.39	0.93	0.08	0.02					
1241	45552				Black calc. shale	24,515	1.266	0.29	68.43							
1247	45547				Black limestone	24,515	1.266	0.36	76.95							
1240	45545				Black limestone	24,515	1.266	0.45	76.71	0.13	0.21	0.00	46		442	
1240	45545				Black limestone	24,515	1.266	0.45	76.71	0.14	0.17	0.51	38	113	450	
309	45576				Black shale	24,515	1.266	1.83	0.92	0.12	0.03	0.30	2	16	416	
1248	45581				Black limestone	24,515	1.266	0.37	63.63							
1249	45582				Black limestone	24,515	1.266	0.30	92.74							
1245	45584				Black calc. shale	24,515	1.266	0.40	72.88	0.11	0.07	1.06	17	264	430	
1245	45584				Black calc. shale	24,515	1.266	0.40	72.88	0.10	0.06	0.81	15	201		
1246	45585				Black shale	24,515	1.266	1.11	0.88	0.01	0.02	3.08	2	277		
310	45595				Black limestone	24,524	1.263	0.19	81.45							
311	45598				Black calc. shale	24,524	1.263	0.28	66.27							
312	45601				Black limestone	24,524	1.263	0.24	79.48							
313	45602				Black limestone	24,524	1.263	0.18	73.07							
1244	45604				Black calc. shale	24,524	1.263	0.25	60.99							
1239	45607				Black shale	24,524	1.263	1.08	0.77	0.06	0.01	0.85	1	78		
1250	45627	Pont de la Lindi	Aruwimi	Alolo	Grey calc. shale	25,207	0.803	0.22	41.94							
1251	45624				Grey calc. shale	25,212	0.811	0.12	53.50							
1252	45622				Grey calc. shale	25,218	0.815	0.27	17.98							
1253	45620				Grey calc. shale	25,222	0.820	0.09	24.91							
182	45065	Aruwimi River Bombwa region	Lokoma	Mamungi	Brown calc. shale	25,715	1.878	0.06	20.47							
183	45067				Brown calc. shale	25,720	1.876	0.13	26.37							
184	45071				Brown calc. shale	25,709	1.887	0.07	1.50							
188	45073				Brown sandy shale	25,708	1.895	0.06	0.71							
187	45075				Grey sandy shale	25,709	1.907	0.07	0.71							0.88
189	45084				Brown calc. shale	25,695	1.891	0.06	21.00							
185	45085				Brown calc. shale	25,688	1.894	0.08	18.21							
186	45086				Grey calc. shale	25,688	1.894	0.08	30.77							

(continued)

Table 18.2 (continued)

RWTH Nr	MRAC RG Nr	Location	Group	Unit	Description	Long	Lat	Depth (m)	Corg (%)	CaC03 (%)	S1 (mg/g)	S2 (mg/g)	S3 (mg/g)	HI (mg/g Corg)	OI (mg/g Corg)	Tmax (°C)	VRr (%)
1261	10669	Ubangui region	Lokoma	Mamungi	Grey-green shale				0.20	30.12							
1255	10731				Dark grey limest	20,786	3.684		0.11	27.92							
1259	10732				Dark grey limest	20,786	3.684		0.19	18.33							
1260	1260				Shale				0.04	−0.18							
1257	22788				Grey shale	22,795	3.673		0.06	0.15							
1258	22883				Grey slate	21,670	4.161		0.05	0.12							
1262	22884				Grey slate	21,670	4.161		0.14	2.13							
1256	22903				Grey calc. shale	22,795	3.673		0.12	19.86							
1254	22904				Grey calc. shale	22,536	3.677		0.05	30.61							
190	2294	Ituri River	Ituri	Lenda	Silic. dol. Limest.	28,028	1.324		0.16	98.50							
191	2272				Silicified dolomite	28,028	1.324		0.15	87.00							
192	2274				Dark-grey limest.	28,028	1.324		0.06	81.64							
193	2293				Dol. Limestone	28,028	1.324		0.08	0.10							
194	2287				Dark limestone	28,028	1.324		0.12	46.98							
195	2303				Grey dol. Limest.	28,028	1.324		0.11	88.91							
1263	119702	Walikale	Lukuga	W3 (Upper Dwyka)	Massive grey shale	28,050	−1.372		0.43	1.53	0.04	0.17	0.47	40	111	446	
1269	119703				Varval sandy shale	28,050	−1.372		0.54	1.15	0.04	0.20	1.39	37	258	442	1.04
1270	119705				Varval sandy shale	28,050	−1.372		0.41	0.87	0.02	0.08	1.08	20	265	449	
1271	119706				Grey shale	28,050	−1.372		0.61	1.47	0.05	0.29	1.26	48	208	436	0.70
1268	119709				Varval sandy shale	28,050	−1.372		0.21	0.63							
1264	119715				Varval black shale	27,877	−1.400		0.65	1.65	0.04	0.34	0.24	52	37	436	1.02
1267	119727				Varval black shale	28,020	−1.210		1.32	2.04	0.09	1.19	0.67	90	51	438	1.01
1265	119729				Varval black shale	27,877	−1.400		0.92	2.17	0.08	0.74	0.00	80	0	439	
1266	119732				Varval black shale	27,877	−1.400		0.30	1.21							

314	92331	Kalemie	Lukuga	Coal	Coal	29,000	−5.830	47.65	3.17	2.60	97.40	18.10	204	38	426
315	92316	Lualaba	Stanley-ville		Lime fine	25,595	−0.200	2.66	85.36	0.80	21.61	*76.16*	811		434
316	92342				Lime fine	25,604	−0.054	4.68	64.82	1.46	36.79	*85.44*	786		428
317	92343				Bituminous shale	25,750	0.270	13.15	20.33	6.27	123.90	2.80	942	21	426
318	92353				Bituminous shale	26,248	0.833	8.49	39.95	3.98	87.30	15.20	1028	179	435
319	92380				Bituminous shale	25,550	0.110	25.43	7.08	5.90	246.40	6.50	969	26	438
320	92393				Bituminous shale	25,730	0.630	10.56	29.99	6.94	96.57	3.84	914	36	424
KP01		Kipala	Kwango	Inzia	Black shale			9.11		1.93	68.02	2.82	747	31	438
KP02			Kwango	Inzia	Black shale			8.21		1.65	61.56	2.30	750	28	439
1165	DRC09-WP048	Bas-Congo	Schisto-Calcaire	Bangu	Black dolomitic limestone	15,123	−5.068	0.08							

Data from the two analyses of the Kipala samples are from Kadima (2007). Parameters as in Table 18.1

Table 18.3 Average Organic matter analytical values, from the data of Tables 18.1 (JNOC samples) and 18.2 (RMCA and Kadima (2007) samples)

Region	Group	Formation	Sample depth (m)/ provenance	n Corg	n RE	n Vr	Corg (%)	CaCo3	Tmax (°C)	HI (mg/g Corg)	OI (mg/g Corg)	Vr (%)	Type	Maturation
Kipala	Kwango	Inzia	Outcrop	2			8.66		438.5	749	30		I	Early mature
Dekese	Loia	Couches C	703–708	2			0.08	1.81					Barren	
Samba	Loia	Couches 4	568–825	22	13		3.52	13.01	432.6	806	58		I	Immature
Kananga	Loia	Red beds	JNOC	31	1		0.15						Barren	
Samba	St'ville	Couches 5	1008–1031	8			0.43	13.11					Barren	
Kisangani	St'ville	Grey shales	Outcrop	6	6		10.8	41.25	430.8	909	66		I	Immature
Kisangani	St'ville	Grey shales	JNOC	17	17	1	7.57		432.8	922	31	0.77	I	Early mature
Kisangani	St'ville	Limestone	JNOC	13	6		1.23		426.0	535	49		II	Immature
Kisangani	St'ville	Mudstones	JNOC	38	8		0.24		426.1	112	83		III	Immature
Kisangani	St'ville	Sandstone	JNOC	6			0.24						Barren	
Lubutu?	Lukuga	Varval clay	JNOC	7	6	5	1.01		433.3	39	46	0.58	III	Transitional
Kindu	Lukuga	Varval clay	JNOC	2	2	1	0.75		439.0	22	79	0.75	III	Transitional
Walikale	Lukuga	Varval clay	Outcrop	10	7	4	0.55	10.99	440.9	52	155	0.94	III	Oil zone
Kalemie	Lukuga	Coal seam	Outcrop	1	1		47.7	3.17	426.0	204	38		III	Early mature
Dekese	Lukuga	Couches D	712–732	4			0.11	18.03					Barren	
Dekese	Lukuga	Couches F	924–1,052	16	5	13	0.54	0.94	421.4	32	134	0.93	III	Oil zone
Dekese	Lukuga	Couches F	1,060–1,100	11	7	8	1.04	0.63	416.0	50	50	0.98	III	Oil zone
Dekese	Lukuga	Couches F	1,130–1,261	9	8	7	0.69	1.99	423.0	40	86	0.98	III	Oil zone
Dekese	Lukuga	Couches F	1,310–1,530	8	2	5	0.33	1.59	418.0	39	211	1.11	III	Gas zone
Dekese	Lukuga	Couches F	1,530–1,550	9		3	0.36	0.96				1.14	III	Gas zone
Malili-Ban.	Aruwimi	Alolo	Outcrop	10			0.13	14.62					Barren	
Yambuya	Aruwimi	Alolo	Outcrop	24	12		0.58	53.70	435.4	27	158		Residual	
Lindi bridge	Aruwimi	Alolo	Outcrop	4			0.18	34.58					Barren	
Bombwa	Lokoma	Mamungi	Outcrop	8			0.08	14.97					Barren	
Ubangi	Lokoma	Mamungi	Outcrop	9			0.11	14.34					Barren	
Ituri	Ituri	Lenda	Outcrop	6			0.11	67.19					Barren	
West-Congo	Schisto-Calcaire	Bangu	Ouctrop	1			0.08						Barren	

For the JNOC samples, only the Rock-Eval data are displayed in Table 18.1. The Corg. Values for the samples for which no Rock-Eval data are available are from JNOC (1988). nCorg, n RE, n Vr: respectively number of Corg, Rock-Eval and vitrinite reflectance analyzed. Parameters as in Tables 18.1 and 18.2. Organic matter type is estimated on the basis of the Van-Krevelen-equivalent diagram (Fig. 18.3) and the maturation, on the basis of the vitrinite reflectance data (Vr) when available, or from the Tmax values otherwise, in a Espiatlie diagram (Fig. 18.4). Note that the maturity estimation from the Lukuga Group samples in the Dekese well is overestimated due to the abundant presence of recycled vitrinite (Fig. 18.5)

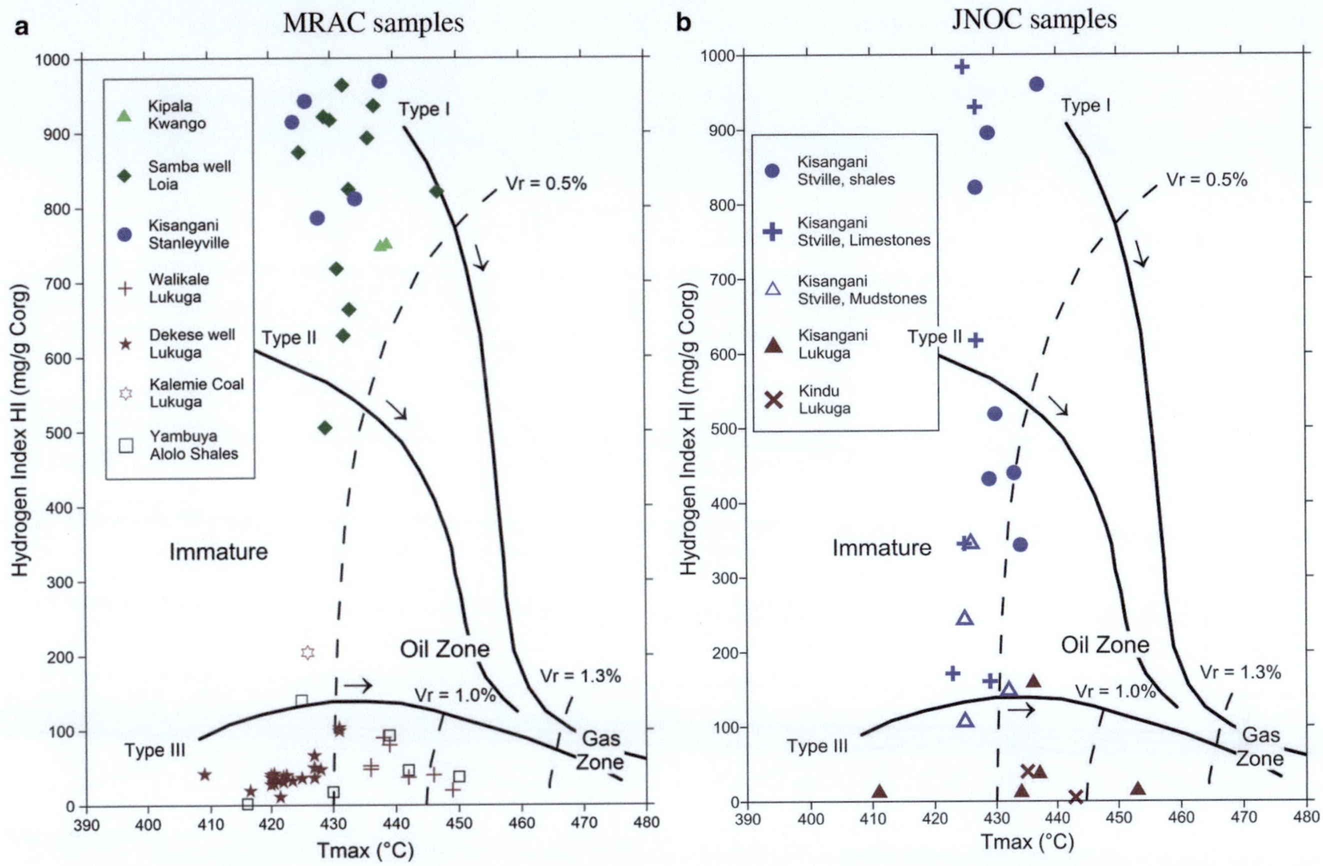

Fig. 18.3 Characterization of kerogen type on a Van-Krevelen equivalent diagram (Espitalie et al. 1977), using Rock-Eval Hydrogen and Oxygen indexes (HI and HO). *Bold lines* show the evolution path of kerogen types with thermal maturation (along the direction shown by the *arrows*). Kerogen type I: rich in aliphatic chains and poor in aromatics, derived from algal lipids or enriched by microbial activity, typical of lacustrine organic-matter; type II: richer in aromatics, usually related to marine organic matter deposited on a reducing environment but also formed by a combination algae (type I) and terrestrial (Type III) organic matter, Type III: aromatics and oxygenated functions, derived from terrestrial higher plants (Tissot and Welte 1978). (**a**) RMCA samples, analyzed by RWTH Aachen (Sachse et al. 2012), with two samples of Kipala shales from Kadima (2007). (**b**) JNOC samples (JNOC 1984). Data from Tables 18.1 and 18.2

sandstones and contain very little quantity of organic matter ($C_{org} = 0.43$ %, Table 18.3). The most organic-rich facies is the brown-grey shales (termed bituminous shales by Passau 1923), found in outcrops and boreholes from the Kisangani-Ubundu portion of the Congo River (Lualaba). Both the RMCA and the JNOC samples from this lithology are equally rich in C_{org}, with ultra-high Hydrogen Index (900–1,000 mg HC equivalent/g C_{org}) and a moderate to low Oxygen Index (30–70 mg CO2/g C_{org}). Some of the JNOC samples have anomalously high HI values (>1,000), which likely reflects an analytical problem in isolating the inorganic from the organic carbon. Limestones are less rich in C_{org} (ca. 1–2 %), but still with relatively high Hydrogen Index (500–600 mg HC equivalent/g C_{org}). Mudstones and sandstones are poor to very poor in C_{org}. On the Van Krevelen-equivalent HI–OI diagram (Fig. 18.3), the samples of the Stanleyville Group display a range of compositions, from type I (algae) to type III (higher plant-derived) kerogen.

Similarly, samples of the Loia Group (Cretaceous) are rich in C_{org}in the Samba well (Couches 4 of Cahen et al. 1959), but very poor in the Dekese well (Couches C of Cahen et al. 1960) and in the region of Kananga (JNOC samples). Those from the Samba well display also high Hydrogen Indexes (~800 mg HC equivalent/g C_{org}) and medium Oxygen Indexes (50–60 mg CO2/gC_{org}).

Sulphur content, microscopic observations, n-alcane and biomarker patterns (Sachse et al. 2012) show that brown-grey shales of the Loia and Stanleyville Groups are composed predominantly of algae-derived aquatic organic matter and small amounts of terrestrial higher plant material. The range of composition of the Stanleyville samples analysed by JNOC (Fig. 18.3b) might therefore reflect various proportions of the end-members (type I algae-derived and type III terrestrial higher plant-derived) rather than the presence of marine type II organic matter.

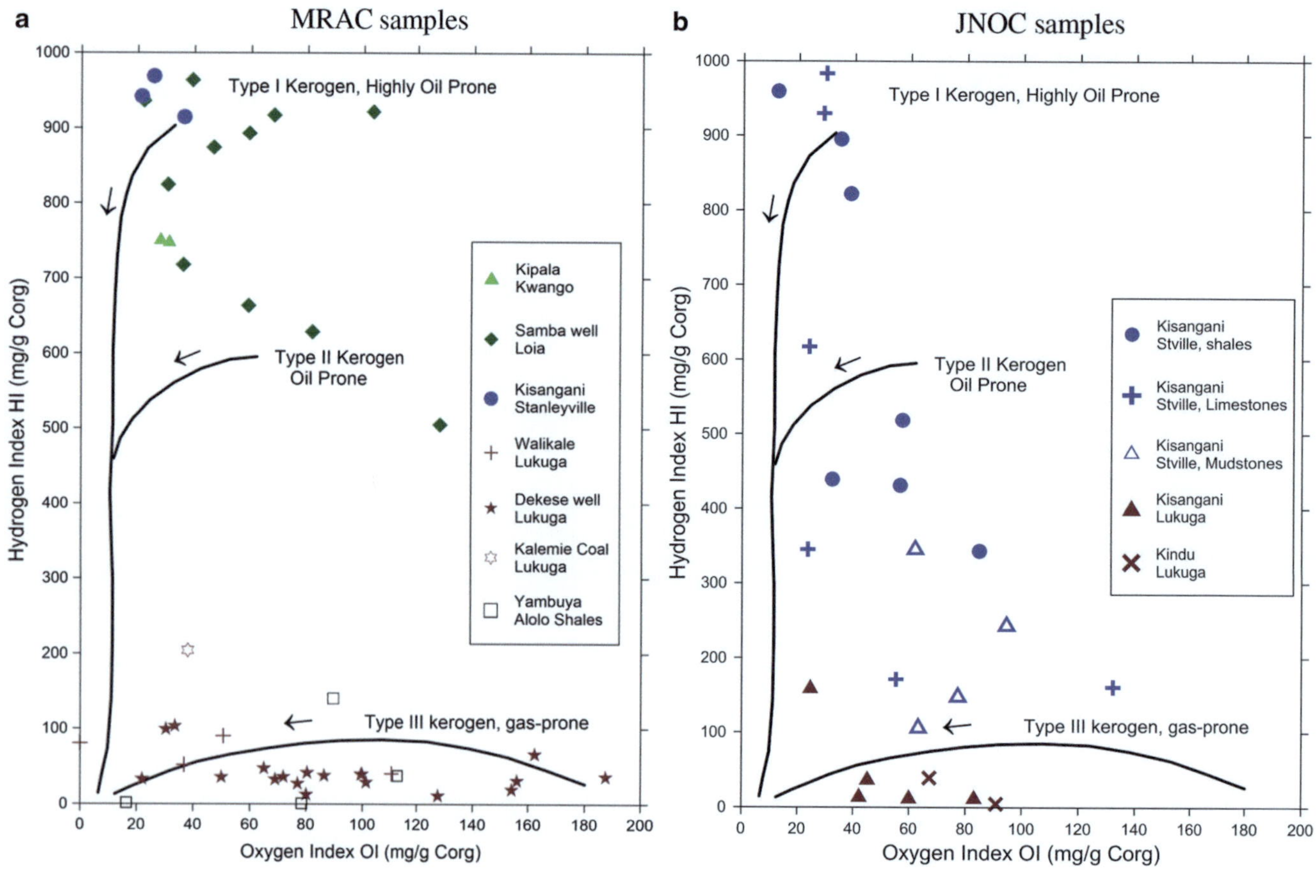

Fig. 18.4 Rock-Eval characterization of kerogen maturation characterized on a Espitalie diagram (Espitalie et al. 1977), by the Hydrogen Index HI and Tmax (temperature of maximum release of hydrocarbons). *Bold lines* show the evolution path of kerogen types with thermal maturation (along the direction shown by the *arrows*). *Dotted lines* show maturation isogrades corresponding to vitrinite (or equivalent) values and separate the maturation zones. (**a**) RMCA samples, analyzed by RWTH Aachen (Sachse et al. 2012), with two samples of Kipala shales from Kadima (2007). (**b**) JNOC samples (JNOC 1984). Data from Tables 18.1 and 18.2

18.5 Thermal Maturation

The thermal evolution of the source rocks is evaluated by the Rock-Eval Tmax (°C) parameter and vitrinite reflectance data (Vr%). Using the Total Organic Carbon (C_{org}) content and Rock-Eval pyrolysis data, kerogen maturity level is expressed in the Van-Krevelen HI-OI (Fig. 18.3) and Espitalie HI-Tmax (Fig. 18.4) diagrams by composition trends in function of thermal maturation for different types of organic matter.

The Rock-Eval Tmax values correspond to temperature of maximal release of hydrocarbons during pyrolysis (summit of the S2 peak; Espitalie et al. 1977). On an Espitalie diagram, the Hydrogen Index (mg HC equivalent/g C_{org}) decreases progressively with maturation, while the Tmax (°C) increases (Fig. 18.4). In the Van-Krevelen diagram (Fig. 18.3), the vitrinite reflectance isolines separate the thermal maturation zones, with Vr = 0.5 % separating the Immature Zone from the Oil generation Zone and Vr = 1.3 % separating the Oil Zone from the Gas Zone. All the samples of the RMCA from the Stanleyville and Loia Groups and from the Kipala Shale fall in the Immature Zone. Some have Tmax values between 430 and 440°, but because they have high HI, they are still immature. Only a few samples of the Stanleyville Group collected by JNOC display HI lower than 450 and are close to the onset of the Oil Zone. The Stanleyville, Loia and Kipala source rocks are all immature, as indicated by their Tmax and Vr values.

Samples from the Lukuga Group present a range of Tmax values, between 410 and 455 °C (Fig. 18.3). The few outcrop samples analysed are in the same range, with a higher maturity for the varval shales of Walikale. Based on vitrinite reflectance (Table 18.2, Fig. 18.5), the gas-prone Lukuga source rocks are generally in the oil window in the Dekese well, except below 1,300 m deep, where they enter the gas window. However, the Tmax values indicate a lower maturity, mostly in the Immature Zone. On the Vr–Tmax diagram (Fig. 18.5), these values display an almost negative correlation, far from

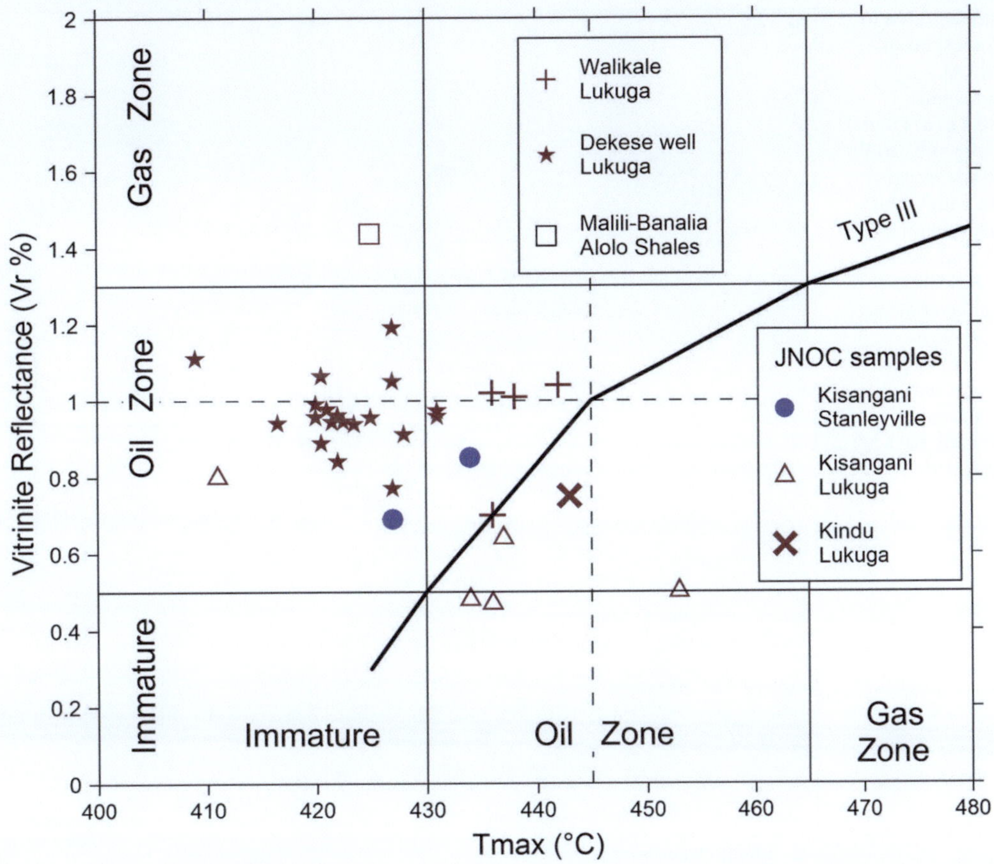

Fig. 18.5 Correlation between Tmax temperature and vitrinite reflectance values. *Bold line* shows the ideal evolution path for type III organic matter (Espitalie et al. 1977; Espitalie 1986). Most of the Lukuga Group samples from the Dekese well appear anomalous, with Vr value stoo high relative to Tmax. Data from Tables 18.1 and 18.2

the ideal evolution path constructed from literature data (Espitalie et al. 1977; Espitalie 1986). With reference to this evolution path, all the Dekese well samples display Vr values too high relative to corresponding Tmax. Such a shift is not seen for the other Lukuga samples (Walikale and Kisangani), or for the Stanleyville samples. This may be explained by contamination of the samples by recycled vitrinite, because petrological observations (Sachse et al. 2012) show that the samples from the Lukuga Group contain large amount of allochtonous vitrinite and vitrinite-like particles, mainly in the Dekese well.

The presence of allochtonous vitrinite in the Lukuga Group samples of the Dekese suggests that they have been recycled, possibly eroded from Lower Carboniferous to Upper Devonian rocks that may have been buried up to 3,000–4,000 m deep. Such rocks are not known in the CB, but do occur farther South, in South Africa and Namibia (e.g. Tankard et al. 2009).

All dark shales and limestones from the Neoproterozoic (Lenda carbonates of the Ituri Group, Mamungi shales of the Lokoma Group, Bangu black limestone of the Schisto-Calcaire Group) and the Lower to Middle Paleozoic sequences (Alolo Shales of the Aruwimi Group) are barren or contain residual organic matter; their maturity cannot be measured therefore with the Rock-Eval Tmax. The Tmax values for those few Alolo shales that are rich enough for a Rock-Eval analysis (from the Yambuya section), are not significant because the small amount of hydrocarbons generated during pyrolysis represent residual organic matter. These samples are also too old (Precambrian) to contain vitrinite particles.

18.6 Reported Oil Seeps and Resource Potential for Petroleum

Over the last 10 years several oil seeps have been reported within the CB (Fig. 18.1), although none of them have been found by the exploration companies that studied the basin in the 1950s and the 1970–1980s. They were re-visited and sampled between 2007 and 2008 by HRT Petroleum for analysing their biomarkers by GC-MS (Gas-Chromatography–Mass spectrometry), GC-IRM and diamontoids (Mello 2008). HRT Petroleum concentrated their investigations near Ilanga-Kole and Tolo along the Lukenie

Fig. 18.6 Field photographs of hydrocarbon seep sites, taken during the 2011 MRAC-COMICO field campaign (location on Fig. 18.1). (**a**) Frontal view of the Tolo seep site. (**b**) Pit drilled in the Tolo site trough a layer of soil mixed with rubbish. (**c**) Ilanga-Kolo seep site in a ditch along an old road, near a bridge. (**d**) Ilanga-Kole site: Impregnated gravel layer at base of superficial and above a compact clay layer. (**e**) Typical shore of Lake Inongo/Mai-Ndombe with lateritic promontories and sandy bays. (**f**) Mpunzuiboko beach with bitumen pellets covering the sand in the upper part of the beach

River and along the beaches of Lake Inongo (Mai-Ndombe). They reported that the biomarkers of the Ilanga-Kole and Tolo seeps suggest derivation of black oil from a marine source rock in the Late Cretaceous, mixed with an older, deeper marine source. The seeps contain a high concentration in Oleanane, a component present in the angiosperms, which appeared in Early Cretaceous and became dominant in the Late Cretaceous-Tertiary. Mello (2008) attribute them to some input from Neogene sediments, but source rocks are not known from the Neogene sequences. Mello (2008) thus predicted the presence of two or three active petroleum systems in the CB, and speculated that it could hold 'giant to supergiant' (size not specified) accumulations of light oil, condensates and gas hydrocarbons, suggesting a promising petroleum province. By contrast, Harriman (2011), disregarded the bitumens found on the beaches of Lake Inongo, and suggested they may be residues from the sealing of local fishing boats, and therefore likely of anthropic origin.

In 2011, we re-visited and re-sampled the Tolo and Ilanga-Kole seeps along the Lukenie River and the bitumen from the Mpunzuiboko beach along Lake Inongo (Figs. 18.1 and 18.6). The Tolo seep site was found along the river bank at the foot of a ~15 m high scarp that separates it from an inhabited plateau occupied by traditional houses. Beige-color clay with a slight odour of light hydrocarbons was found in a 1-m deep pit dug through a superficial layer of soil mixed with rubbish. At Ilanga-Kole, also along the Lukenie River and upstream Tolo, light hydrocarbons were found in a gravel layer on top of clay in a ditch along the old Oshwe-Bandundu road. On the shore of Lake Inongo, we sampled lumps of solid bitumen on the surface of white beach sands. We also sampled a fragment of the bitumen that was used to waterproof our wooden boat.

The collected samples were analysed by GHGeochem laboratories (Harriman 2012), and their anthropogenic origin confirmed for all of them. In summary, none of the hydrocarbon seeps appear to be natural oil seeps. They contain biomarkers that relate them to source rock which are too young, too mature and of marine affinity, for which no source rock are known in the CB.

Conclusions

Based on the evidence presented, we do not share the optimism of HRT Petroleum for large potential oil reserves in the CB (e.g. Mello 2008). Todate no petroleum systems have yet been proven. The samples from the Mamungi shales have negligible organic matter content ($C_{org.} \approx 0.1$ % on average), which do not allow any further organic geochemistry or thermal maturation analysis. Similarly, most of the Alolo shales analyzed also have negligible organic matter content, except a few samples from the Yambuya section, but which, when analyzed with the Rock Eval, appear to have no oil potential. Further, the Mamungi and Alolo shales are only known in outcrop and have not been formally recognized in the 4 wells drilled in the center of theCB.

The Lukuga Group that contains up to 2.4 %C_{org}, is largely a gas-prone peri-glacial or interglacial lacustrine sequence of shales. Lacustrine sediments could have been deposited in more temperate conditions, but OM would be dominated by higher plants. The Stanleyville (Middle or Upper Jurassic) and Loia Groups (Middle Cretaceous) contain the best potential source rocks with some very rich shales (up to 25 %C_{org}). The kerogen is derived from lacustrine algae, and is oil-prone type I (HI up to 950 mg HC equivalent/g C_{org}), but is too immature to have generated significant quantities of oil where they have been found. This Mesozoic petroleum system is, at the present state of knowledge, the best potential target for oil exploration.

In conclusion, there presently are no indications for any active hydrocarbon systems, but as the current knowledge of this large basin is still very limited, there is much room for future exploration. There is a clear need for more, new and modern geophysical and geological work to improve the basic knowledge of this vast region. This should be the backbone for future hydrocarbon exploration surveys. There is currently not enough knowledge on the geological evolution and architecture of the CB to predict the presence of viable hydrocarbon deposits.

Acknowledgements This work benefitted from many informal contacts and discussions that the first author had with individuals interested on the CB. We thank also the editorial efforts of M. de Wit and F. Guillocheau who motivated us to submit this paper.

References

Alvarez P, Maurin J-C, Vicat J-P (1995) La Formation de l'Inkisi (Supergroupe Ouest-congolien) en Afrique centrale (Congo et Bas-Zaïre): un delta d'âge Paléozoïque comblant un bassin en extension. J Afr Earth Sci 20:119–131

Boze MM, Kar RK (1978) Biostratigraphy of the Lukuga Group in Zaïre.Ann. Mus. roy. Afr. cent., Tervuren (Belgique), Série in- 8°, Sciences géologiques, 82, 97–114

Buiter SJH, Stenberger B, Medvedev S, Tereault JL (2012) Could the mantle have caused subsidence of the Congo Basin? Tectonophysics 514–517:62–80

Cahen L (1954) Géologie du Congo belge. Vaillant-Caramanne, Liège, 577 pp

Cahen L (1983a) Le Groupe de Stanleyville (Jurassique supérieur et Wealdien de l'intérieur de la République du Zaïre): Révision des connaissances. Rapp. Ann. 1981-1982, Dép. Géol. Min. Mus. Roy. Afr. Cent., Tervuren (Belgique), 73–91

Cahen L (1983b) Brèves précisions sur l'âge des groupes crétaciques post-Wealdien (Loia, Bokungu, Kwango) du Bassin intérieur du Congo (République du Zaïre). Rapp. Ann., 1981–1982 Dép. Géol. Min. Mus. Roy. Afr. Cent., Tervuren (Belgique), 61–72

Cahen L, Ferrand JJ, Haarsma MJF, Lepersonne J, Verbeek, T. (1959) Description du Sondage de Samba. Ann. Mus. roy. Afr. cent., Tervuren (Belgique), Série in- 8°, Sciences géologiques, 29

Cahen L, Ferrand JJ, Haarsma MJF, Lepersonne J, Verbeek T (1960) Description du Sondage de Dekese. Ann. Mus. roy. Afr. cent., Tervuren (Belgique), Série in-8°, Sciences géologiques, 34

Cahen L, Lepersonne J (1954) État actuel des connaissances relatives aux séries mésozoïques de l'intérieur du Congo. Bull Soc Belge Géol 77:20–37

Cahen L, Lepersonne J (1978) Synthèse des connaissances relatives au Groupe (anciennement Série) de la Lukuga (Permien du Zaire). Ann. Mus. roy. Congo belge, Tervuren (Belgique), série in-8, Sci. géol., 82, 115–152

Chorowicz J, Le Fournier J, MakazuMvumbi M (1990) La Cuvette Centrale du Zaïre : un bassin initié au Protérozoïque supérieur. Contribution de l'analyse du réseau hydrographique. C.R. Acad. Sci. Paris, t.311, Série II, 349–356

Colin JP (1981) Paleontological study of the Esso/Texaco well Gilson-1, Zaire. Unpublishedreport EPR-E.WA19.81

Colin JP (1994) Mesozoic-Cenozoic lacustrine sediments of the Zaïre Interior Basin. In: Gierlowski-Kordesch E, Kelts K (eds) Global geological record of lake basins. Cambridge University Press, Cambridge, pp 31–36

Cornet J (1911) Sur la possibilité de l'existence de gisements de pétrole au Congo. Ann Soc Geol Belg 36(1910–1911):9–15

Cox LH (1960) Further Mollusca from the Lualaba Beds of the Belgian Congo.Ann. Mus. Roy. Afr. Cent., Tervuren (Belgique), série in-8, Sci. géol., 37, 15 p

Crosby AG, Fishwick S, White N (2010) Structure and evolution of the intracratonic Congo Basin. Geochem Geophys Geosyst 11:Q06010

Daly MC, Lawrence SR, Diemu-Thiband K, Matouana B (1992) Tectonic evolution of the Cuvette centrale. Zaire J Geol Soc Lond 149: 539–546

Delvaux D (2001) Tectonic and paleostress evolution of the Tanganyika-Rukwa-Malawi rift segment, East African Rift System. In: PA Ziegler, W Cavazza and AHF Robertson and S Crasquin-Soleau, Eds. Peri-Tethys Memoir 6: PeriTethyan Rift/Wrench Basins and Passive Margins. Mém. Mus. Natn. Hist. nat., 186: 545–567. Paris

Delvaux D, Barth A (2010) African Stress Pattern from formal inversion of focal mechanism data. Implications for rifting dynamics. Tectonophysics 482:105–128

Delpomdor F, Linnemann U, Boven A, Gartner A, Travin A, Blanpied C, Hielsma H, Préat A (2013) Depositional age, provenance, tectonic and palaeoclimatic settings of the late Mesoproterozoic - middle Neoproterozoic Mbuji-Mayi Supergroup, Democratic Republic of Congo. Palaeogeogr Palaeoclimatol Palaeoecol 389: 35–47

de Sainte Seine P (1955) Poissons fossiles de l'étage de Stanleyville (Congo belge). 1ère partie : la faune des argilites et schistes bitumineux. Ann. Mus. royal Afrique centrale, Tervuren (Belgique), série in-8, Sci. géol., 14, 125p

de Sainte Seine P, Casier E (1962) Poissons fossiles de l'étage de Stanleyville (Congo belge). 2ème partie : la faune marine des calcaires de Songa. Ann. Mus. royal Afrique centrale, Tervuren (Belgique), série in-8, Sci. géol., 44, 52p

De Waele B, Johnon SP, Pisarevsky SA (2008) Paleoproterozoic to Neoproterozoic growth and evolution of the eastern Congo Craton: Its role in the Rodinia puzzle. Precambrian Res 160:127–141

Defrétin-Lefranc S (1967) Etude sur les Phyllopodes du Bassin du Congo. Ann. Mus. Roy. Afr. Cent., Tervuren (Belgique), série in-8, Sci. géol., 56, 122 p

ECL (1988) Hydrocarbon potential of cuvette centrale (Republic of Zaire).Exploration Consultants Limited, Cellule Technique Pétrolière, Pétrozaire, unpublished report, 41 p. + figures, tables, appendices and enclosures

Egoroff A, Lombard AL (1962) Présence des couches de Stanleyville dans le sous-sol de Lépoldville, République du Congo (Note préliminaire). Ann Soc Geol Belg 85:103–109

Espitalie J (1986) Use of Tmax as a maturation index for different types of organic matter.Comparison with vitrinite reflectance. In: Burrus J (ed) Thermal modelling in sedimentary basins. Technip, Paris, pp 475–496

Espitalie J, Laporte JL, Madec M, Marquis F, Leplat P, Paulet J, Boutefeu A (1977) Méthode rapide de caractérisation des roches mères, de leur potentiel pétrolier et de leur degré d'évolution. Revue de l'Institut Français du Pétrole 32:23–42

ESSO Zaire SARL (1981a) Geological completion report.Mbandaka-1. Unpublished report

ESSO Zaire SARL (1981b) Geological completion report.Gilson-1. Unpublished report

Evrard P (1960) Sismique. (Résultat scientifique des missions du Syndicat pour l'étude géologique et minière de Cuvette congolaise). Ann. Mus. royal Afrique centrale, Tervuren (Belgique), série in-8, Sci. géol., 33, 87p

Fourmarier P (1914) Le bassin charbonnier d'agePermo-Triassique de la Lukuga. Etude géologique de la région de la Lukuga et de la Lubumba au voisinage du lac Tanganyika. Ann. Soc. géol. Belg., Publ. rel. Congo belge, Liege, 41, C77–227

Halbich W, Fitch FJ, Miller JA (1983) Dating the Cape orogeny, vol 12. Geological society of South Africa, Pretoria, pp 149–164, Special Publication

Harriman G (2011) Geochemical review of the Cuvette Centrale. The hydrocarbon potential of the Cuvette Centrale, DRC. A review of existing geochemical data on source rocks and seeps.GH Geochem Ltd, report GHG 2457, unpublished, 34p

Harriman G (2012) A Geochemical evaluation of outcrop and seep samples from the cuvette centrale, DRC.GH Geochem Ltd, unpublished draft report, 28p

Henry J (1922–1923) Etude géologique du Congo belge dans la region comprise entre Basoko-Stanleyville, à l'Ouest, le lac Albert et la Semliki à l'Est. Ann Soc Geol Belg (Pub Rel Congo belge) 46: C49–313

Giresse P (2005) Mesozoic-Cenozoic history of the Congo Basin. J Afr Earth Sci 43:301–315

Grecoff N (1957) Ostracodes du bassin du Congo. -1 Jurassic supérieur et Crétacé inférieur du Nord du bassin. Ann. Mus. Roy. Afr. Cent., Tervuren (Belgique), série in-8, Sci. géol., 19, 97 p

Jamotte A (1931) Contribution à l'étude géologique du basin charbonnier de la Lukuga. Comité Spécial du Katanga: Annales du Service des Mines, 2, 3–44

JNOC (1984) Rapport des investigations géophysiques et géologiques dans la Cuvette centrale de la République du Zaïre. Japan National Oil Corporation, Unpublished, 205p

Kadima EK (2007) Stratigraphie, structure géologique et prospectivité pétrolière du Bassin de la Cuvette congolaise.DEA dissertation, Department of Geology, University of Lubumbashi, RDC, 126p

Kadima E, Delvaux D, Sebagenzi SN, Tack L, Kabeya M (2011a) Structure and geological history of the Congo Basin: an integrated interpretation of gravity, magnetic and reflection seismic data. Basin Res 23(5):499–527

Kadima EK, Sebagenzi S, Lucazeau F (2011b) A Proterozoic-rift origin for the structure and the evolution of the cratonic Congo Basin. Earth Planet Sci Lett 304:240–250

Kampunzu AB, Cailteux JLH (1999) Tectonic Evolution of the Lufilian Arc (Central Africa Copper Belt) During Neoproterozoic Pan African Orogenesis. Gondwana Res 2(3):401–421

Lawrence S, Makazu MM (1988) Zaire's Central basin.Prospectivity outlook. Oil Gas J 86(38):105–108

Le Roux JP (1995) Heartbeat of a mountain: diagnosing the age of depositional events in the Karoo (Gondwana) basin from the pulse of the Cape Orogen. Geol Rundschau 84:626–635

Lepersonne J (1974) Carte géologique du Zaïre au 1/2.000.000 et notice explicative, République du Zaïre, Direction de la Géologie, Kinshasa & Mus. Roy. Afr. Centr, Tervuren

Lepersonne J (1977) Structure géologique du bassin intérieur du Zaïre. Bull. Acad. Roy. Belg., Cl. Sci., 5e série, 63(12), 941–965, Bruxelles

Linol B (2013) Sedimentology and sequence stratigraphy of the Congo and Kalahari basins of South-Central Africa and their evolution during the formation and breakup of West-Gondwana. Ph.D. thesis Nelson Mandela Metropolitan University, South Africa, 370p

Lombard AL (1960) L'extension méridionale des couches de Stanleyville, Congo belge (Note préliminaire). Bull. Soc. Géol. Belg., 69, 23–27

Maheshwari H, Bose MN, Kumaran KP (1977) Mesozoic sporae dispersae from Zaire. II: The Loia and Bokungu Groups in the Samba borehole. III: Some miospores from the Stanleyville Group. Ann. Mus. roy. Afr. cent., Tervuren (Belgique), Série in- 8°, Sciences géologiques, 80

Mello MR (2008) Field survey, seep collection and HRGT analysis of seeps over CoMiCo exploration areas and samples correlation with Brazil, Africa and Middle-East. High Resolution Technology & Petroleum (HRT), unpublished report, 130p

Misser F (2013) Enjeux et défis d'une province pétrolière en devenir. In : Madrysse S, Omasonbo J (eds) Conjonctures congolaises 2012. Politique, secteur minier et gestion des ressources naturelles en RD Congo. Cahiers Africains 82, 147–177. Musée royal de l'Afrique Centrale et Harmattan, Paris. ISBN 978-2-343-00465-5

Newton AR, Shone RW, Booth PW (2006) The Cape Fold Belt. In: Johnson MR, Anhaeusser CR, Thomas RJ (eds) The geology of South Africa. Geol. Soc. S. Africa. Johannesburg and Council for Geoscience, Pretoria, pp 521–530

Oils Search/Pioneer (2007) Cuvette centrale exploration study oils search limited/pioneer international development. Progress report. 65p

Passau G (1923) La géologie du bassin des schistes bitumineux de Stanleyville (Congo belge). Ann. Soc. géol. Belg., Publ.rel. Congo belge 19 (1921-1922), C91–243

Pilipili Mawezi J (2010) Le pétrole de la République démocratique du Congo. Southern Africa Resource Watch, Johannesburg, South Africa, p 133

RRI (1988) Robertson Research International Ltd, Report 623/Ic.A Geochemical Investigation of the Cuvette Centrale of Zaire

Sachse VF, Delvaux D, Littke R (2012) Petrological and geochemical investigations of potential source rocks of the Central Congo Basin, DRC. AAPG Bull 96(2):277–300

Scotese CR (2009) Late Proterozoic plate tectonics and palaeogeography: a tale of two supercontinents, Rodinia and Pannotia, Special Publications, vol 326. Geological Society, London, pp 67–83

Sluys M (1945) La géologie de l'Ituri. Le Groupe de la Lindi Bull Serv Géol 1:95–182

Sluys M (1952) La géologie des environs de Stanleyville. Bull Inst Roy Col Belge 23(3):870–879

Tait J, Delpomdor F, Preat A, Tack L, Straathof G, Nkula VK (2011) Neoproterozoic sequences of the West Congo and Lindi/Ubangi Supergroups in the Congo Craton, Central Africa. In: Arnaud E, Halverson GP, Shields-Zhou G (eds) The geological record of Neoproterozoic Glaciations, vol 36. Geological Society of London, Memoirs, pp 185–194

Tankard A, Welsink H, Aukes P, Newton P, Settler E (2009) Tectonic evolution of the Cape and Karoo basins of South Africa. Mar Petrol Geol 26:1379–1412

Taverne L (1975a) A propos de trois Téléostéens Salmoniformes du Crétacé inférieur (Wealdien) du Zaïre, précédemment décrits dans les genres *Leptolepis* et *Culpavus* (PiscesTeleostei). Rev Zool Afr 89:481–504

Taverne L (1975b) Etude ostéologique de *Leptolepiscaheni*, Téléostéen fossile du Jurassique supérieur (Kimméridgien) de Kisangani (es-Stanleyville, Zaïre) précédemment décrit dans le genre *Paraclupavus*. Rev Zool Afr 89:821–853

Tissot BP, Welte DH (1978) Petroleum formation and occurrence. Springer, Berlin, p 538

Torsvik TH, Cocks LR (2011) The Palaeozoicpalaeogeography of central Gondwana. In: Van Hinsbergen DJJ, Buiter SJH, Torsvik TH, Gaina C, Webb SJ (eds) The formation and evolution of Africa: a synopsis of 3.8 Ga of Earth history. Geological Society, London, pp 137–166, Special Publications

Veatch AC (1935) Evolution of the Congo Basin. MemGeol Soc Am 3: 184

Verbeek T (1970) Géologie et lithologie du Lindien (Précambrien Supérieur du nord de la République Démocratique du Congo). Ann. Mus. Roy. Afr. Cent., Tervuren (Belgique), série in-8, Sci. géol., 66, 311 p

19 Metallogenic Fingerprints of the Congo Shield with Predictions for Mineral Endowment Beneath the Congo Basin

Christien Thiart and Maarten J. de Wit

19.1 Introduction

The Congo Basin (CB) is richly endowed at surface with alluvial gold, and with alluvial diamond deposits derived from the sedimentary sequences and kimberlites that intrude these sequences, and the basin has potential for conventional oil (Delvaux and Fernandez, Chap. 18, this Book), and unconventional oil and shale gas deposits (Fig. 19.1). Beyond that, there is little robust knowledge about other possible mineral wealth in its sedimentary sequences, or potential ore deposits in the basement rocks beneath this large basin. However, flanking the CB are a large variety of Precambrian mineral-rich basement terranes of the Central African and Congo Shields, ranging between 0.5 and 4.0 Ga in age (e.g. de Wit and Linol, Chap. 2 this Book, and references therein). Some of these Precambrian terranes host world class mineral deposits (Fig. 19.1; Goosens 2009; and personal communications, 2014). The latter include the 0.5–0.6 Ga Copper (-Cobalt) deposits of Katanga that were discovered in 1892 (van Reybrouck 2010) and remain the largest of its kind in the world (Goosens 2009). Similarly gold deposits have been explored in the northeast (in the Ituri region near Bunia, and at Kiva, Moto and Kilo) where they were first discovered as alluvial deposits in 1895 and exploited in 1926 by state mining enterprises during the early days of the 'Congo Free State', and some that were 'rediscovered' in the 1950s, are still being mined from its Archean rocks; and more recently still, since 2012, in Proterozoic metasediments of the Twangiza, Kamituga, Lugushwa and Namoya Gold Belt in South Kivu (Fig. 19.1; Goosens 2009). Tin and tungsten deposits occur across the large scale granitoid province of the Kibaran (1.3–1.4 Ga Karagwe-Ankole Belt) between the CB of the DRC and the eastern Lake region of Rwanda and Burundi and as far south as Katanga. These areas are now sought after also for their unique Coltan deposits (Columbite–Tantalite minerals rich in Niobium and Tantalum) especially in North Kivu (Goosens 2009).

In Katanga, too, uranium and zinc is common; and at Shinkolobwe, Uranium ore rich in Radium was used to build the first atom bomb. Ironically, along the opposite, western margin of the CB, at Okla, rich Uranium placer deposits in the Eburnean-age Francevillian sequences were exploited and later studied because of its 17 unique natural paleo-nuclear reactors at Okla (e.g. Mossman et al. 2005).

Whilst is it not the purpose of this chapter to describe these and the many other deposits around the margins of the CB in detail (for which the reader is referred to other literature, see for example references in de Wit et al. 2004; Schlüter and Trauth 2008; Goosens 2009), we are interested here in determining the mineral resource potential of the region using the known mineral deposits of Central Africa (CA), of which there are about 1600 in our database (Fig. 19.2; Table 19.1), together with the known regional geology. Figures 19.2 and 19.3 shows this simplified Precambrian geology of Central Africa with the location of all the mineral deposits used in this study.

Using the mineral deposits information stored in AEON's G0-GEOID—Gondwana Geoscience Indexing Database (de Wit et al. 2004; http://gondwana.brgm.fr/sig.htm) and the simplified Precambrian geology, we first establish distinct mineral deposit patterns that we term 'metallogenic fingerprints' (c.f. de Wit and Thiart 2005; Thiart and de Wit 2006, 2008) of the Precambrian basement of Central Africa (CA; Fig. 19.2), and of the Central African Shield and the Congo

C. Thiart (✉)
AEON – Africa Earth Observatory Network, Nelson Mandela Metropolitan University, Port Elizabeth, South Africa

Department of Statistical Sciences, University of Cape Town, Private Bag, Rhodes Gift, 7701 Cape Town, South Africa
e-mail: christien.thiart@uct.ac.za

M.J. de Wit
AEON – Africa Earth Observatory Network, Nelson Mandela Metropolitan University, Port Elizabeth, South Africa

ESSRI – Earth Stewardship Science Research Institute, Nelson Mandela Metropolitan University, Port Elizabeth, South Africa
e-mail: Maarten.deWit@nmmu.ac.za

M.J. de Wit et al. (eds.), *Geology and Resource Potential of the Congo Basin*, Regional Geology Reviews,
DOI 10.1007/978-3-642-29482-2_19, © Springer-Verlag Berlin Heidelberg 2015

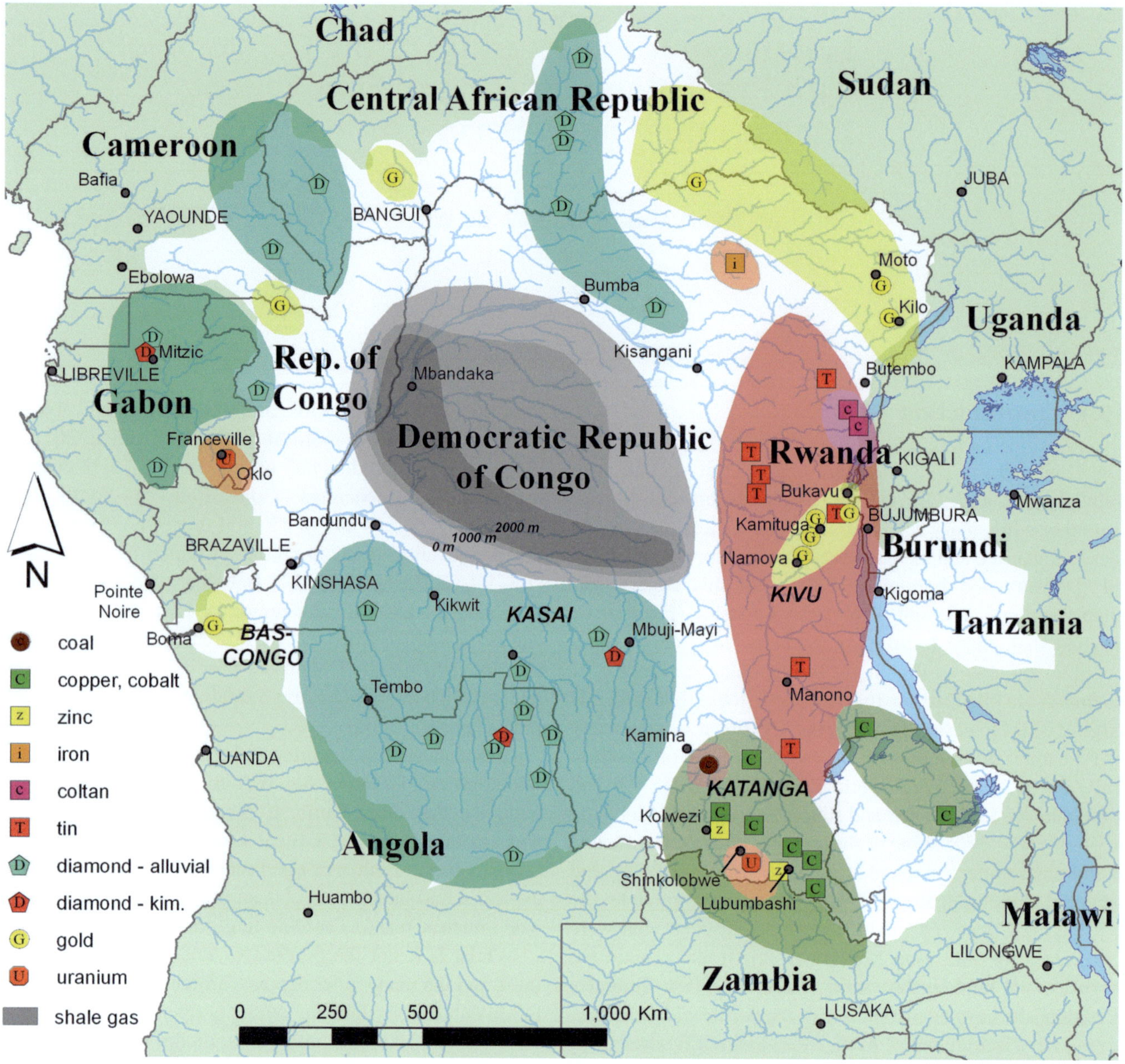

Fig. 19.1 General map of Cental Africa (CA), showing countries with DRC highlighted. The general areas of mineral deposits are shown (modified from the AEON GO-GEOID database; Goosens 2009; and Chaps. 7, 8, 16 and 17, this Book)

Shield (CAS and CS, as defined in de Wit and Linol, Chap. 2, this Book; Fig. 19.3). We then use this data of the CA, CAS and CS to define an "expert weight" fingerprint that reflects an expected mineral endowment in the basement beneath the Phanerozoic cover of the Congo Basin (CB).

19.2 GO-GEOID: The Geology and Mineral Deposit Data Base

The GO-GEOID **GO**ndwana—**GEO**scientific **I**ndexing **D**atabase is a relational database, commonly referred to as the Gondwana database, which consists of three thematic layers: geology, structure/tectonics, and minerals; for simplicity the structure/tectonic layer (e.g. lineaments, faults, foliations etc.) is excluded in this study.

The geology layer of the Gondwana database includes the chronostratigraphy and limited lithostratigraphy modified and updated from the Geological Map of Gondwana (de Wit et al. 1988) from which Central Africa is simplified and displayed in Fig. 19.2. For the purpose here, five different geologic time-domains (e.g. æons, eras) are distinguished (Archean [A], Meso-Paleoproterozoic [MPP]; Neoproterozoic remobilised Archean [NprA]; Neoproterozoic remobilised Proterozoic [NprP]; and Neoproterozoic [Np]; see Thiart and de Wit 2008; Fig. 19.2). Remobilised (r) spacial-domains domains

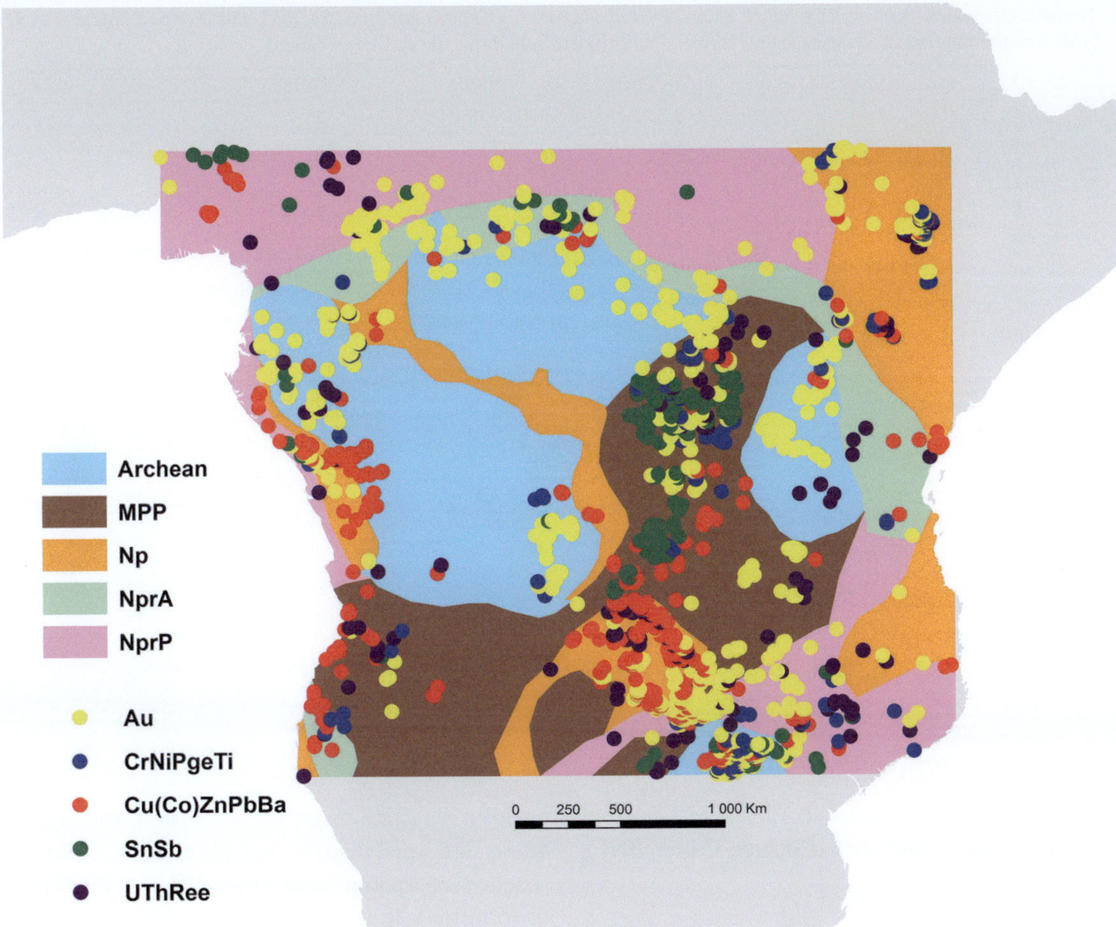

Fig. 19.2 Simplified Precambrian geology of Central Africa with its known mineral deposits. (For details of the geology see Chap. 2, this Book)

comprise predominantly of Precambrian rocks that have been severely deformed and metamorphosed during late Neoproterozoic Pan African thermo-tectonism (ca. 550–650 Ma; for more details see de Wit et al. 2008)

The Gondwana mineral data consists of more than 20 000 deposits covering fifteen commodities (chemical elements) of which 7,571 are in Africa (Thiart and de Wit 2006). In all cases the data were obtained from the literature and/or from geological surveys between 1994 and 2006; and especially in 1999, when our database was merged with that of the BRGM (Bureau de Recherche Geologiques et Minieres, France).

The main features in the mineral layer of the database consist of geographic position (latitude, longitude), deposit-id, deposit-name, commodity and size of commodity. The commodity reported in the database is the major chemical element at that location, whereas size of commodity is reported as an ordinal interval scale. Regular data updates have been well described elsewhere (Wilsher et al. 1993; de Wit et al. 1999; Thiart and de Wit 2006; Thiart and Stein 2013).

For our Metallogenic fingerprint analysis we have selected 11 elements and these are divided into five groups, in part based on their geochemical affinities (e.g. lithophile, chalcophile, etc., where appropriate; Thiart and de Wit 2006). The five groups are cross tabulated with the simplified Precambrian geology for each of the three regions (Table 19.1): within Central Africa (CA; Fig. 19.2 and Table 19.1a): within the Central Africa Shield (CAS) (Fig. 19.3 and Table 19.1b) and the Congo Shield (CS; Fig. 19.3 and Table 19.1c). Note that the numbers reflected for the CA and CAS exclude those of the CS.

19.3 Methodology

We have previously defined 'metallogenic fingerprints' of fragments of continental crust by their spatial association with a combination of element groups, and have applied this at three scales (cratons, continents, and super-continents; de Wit and Thiart 2005). Here we use 3 different local scales as on Table 19.1; viz: (a) Central Africa (CA); (b) Central African Shield (CAS) and (c) Congo Shield (CS).

The measure of spatial association (normalized to area) is referred to as the spatial coefficient (Mihalasky and

Table 19.1 Number of mineral deposits of selected element groups across (a) Central Africa (CA), (b) Central Africa Shield (CAS) and Congo Shield (CS). Note that the number of deposits reflected in (a) and (b) exclude those of the Congo Shield

(a) CA	Au	CrNiPgeTi	CuZPbBa	SnSb	UThRee	Total	Area in km^2
Archean	168	34	27	28	7	264	5.04E+05
MPP	96	31	56	110	33	326	2.13E+06
Np	175	42	133	0	44	394	1.40E+06
NprA	51	8	18	9	8	94	7.25E+05
NprP	195	16	44	20	26	301	2.18E+06
Total	685	131	278	167	118	1379	6.94E+06
(b) CAS							
Archean	66	1	5	0	5	77	3.96E+05
MPP	94	31	56	107	23	311	1.92E+06
Np	10	1	14	0	0	25	1.54E+05
NprA	51	8	18	9	8	94	7.18E+05
NprP	38	1	7	4	2	52	3.66E+05
Total	259	42	100	120	38	559	3.55E+06
(c) CS							
Archean	101	9	43	3	14	170	2.16E+06
MPP	8	2	0	2	5	17	1.26E+05
Np	15	1	9	0	2	27	3.10E+05
NprA	9	1	1	0	1	12	1.03E+05
NprP	0	0	0	0	0	0	4.56E+03
Total	133	13	53	5	22	226	2.70E+06

Bonham-Carter 2001), r_{ij}. The spatial coefficient (r_{ij}) is based on the 'weight of evidence' approach (Bonham-Carter 1996), where

$$r_{ij} = \frac{N(T_i \cap D_j)/N(D_j)}{A(T_i)/A(T_\bullet)},$$

in which A(T_i) is the area of the ith domain (T_i), and A(T.) is the total area of all the domains in the study (A(T.) $= \sum$ A(T_i)). N(D_j) is the total number of deposits in the jth element group (D_j). N($T_i \cap D_j$) represents the number of deposits of group D_j in domain T_i.

The spatial coefficient, r_{ij} represents the proportion of deposits (for example gold—j) of all the jth deposits that occur in the specified geologic domain (i) per unit area of all similar geologic groupings. The spatial coefficient range from 0 to infinity; it is equal to 1 if there is no spatial association between a geological defined domain and an element group (e.g. if the proportion of jth mineral is the same as the proportion of area occupied by the ith). For values of $r_{ij} > 1$ (expected number of deposits is greater than by chance), there is a positive association between mineral j and geologic domain i; $r_{ij} < 1$ (fewer deposits expected than by chance) indicates a negative association. Because all negative associations are compressed in the range from 0 to 1, and all positive associations fall in the range of 1 to infinity, we use the natural log of r_{ij} to eliminate this skewness. ln r_{ij} is now a symmetric value around 0: positive associations are greater than 0, and negative associations are less than 0; values close to zero indicate an absence of spatial associations (i.e. reflecting a random distribution). In a 'weight of evidence' approach for evaluating mineral prospectivity, it can be shown that ln r_{ij} is an approximation for the positive weight W_i^+, and the evaluation is data-driven (Mihalasky and Bonham-Carter 2001, their Appendix). Alternatively, in an expert system for evaluating mineral prospects (e.g. PROSPECTOR, Campbell et al. 1982) W_i^+ is referred to as the sufficiency ratio (LS), and it is assigned by an expert (e.g. the evaluation is knowledge driven by, say, a geologist).

The approximate standard error for ln r_{ij} (Mihalasky and Bonham-Carter (2001, Appendix) and Bonham-Carter (1996, Chap. 9, this Book) are given by

$$s(\ln r_{ij}) = \sqrt{\frac{1}{N(T_i \cap D_j)}}$$

19.4 Results

The natural log of the spatial coefficients (fingerprint) between the element group and each geological group (ln r_{ij}) and its approximate standard errors are given in Table 19.2 and these fingerprints are summarized graphically in Figs. 19.4 (CA), 19.5 (CAS) and 19.6 (CS).

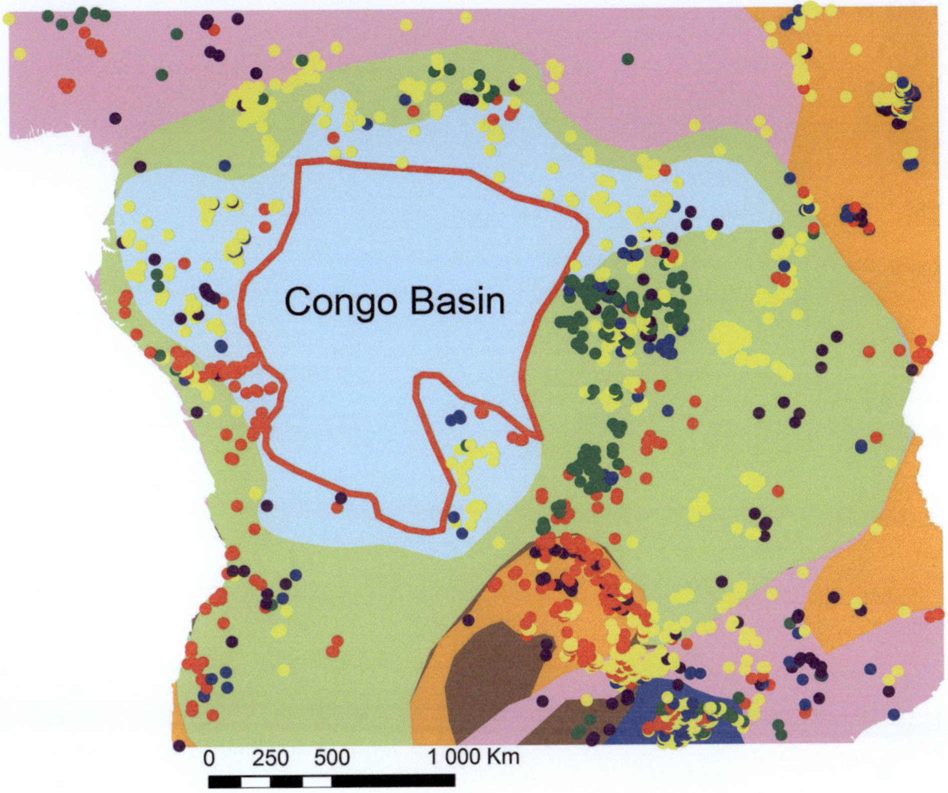

Fig. 19.3 General area of Central Africa (ca. 6.9 million km^2) with its known mineral deposits, showing outlines of the Central African Shield (CAS; *pale green*; ca. 3.6 million km^2); Congo Shield (CS; *pale blue*; ca. 2.7 million km^2) and the approximate outline of the preserved Congo Basin cover (CB; *red line*, ca. 1.3 km^2) for which mineral potential and metalliferous fingerprints are calculated

19.5 Discussion

19.5.1 Fingerprints of the Central Africa, Central African Shield and Congo Shield

First we explore the metallogenic fingerprints of the CA, CAS and the CS. Positive values of the fingerprint (ln r_{ij}) is an indication that there is a positive association between the geology and the mineral deposit distribution within each domain. Negative fingerprints are an indication of negative association between the geology and the mineral deposits within each domain. We also use Standard Errors (in brackets in Table 19.2, and displayed as 2× Standard Deviation in the Figs. 19.4, 19.5, 19.6) to evaluate if a fingerprint has significance: if the error bars include 0, the fingerprint is viewed as displaying random associations.

19.5.1.1 Congo shield

The Congo Shield contains only 226 deposits across a vast area of 2.7 million km^2, resulting in metallogenic fingerprints that are random (Fig. 19.6, Table 19.2c), except for the SnSb fingerprint (2.15) and the UTHREE fingerprint (1.58) in the geologic domain MPP. Note that the NprP domains have no data at all; and the large Standard Deviations in Table 19.1c confirms that we are dealing here with large uncertainties. We therefore exclude the area of the Congo Shield region and its mineral deposits from further analyses of CA and CAS.

19.5.1.2 Central Africa

In Archean areas of Central Africa (Table 19.2a, Fig. 19.4) we observe a strong positive association in three mineral groups (Au (1.22), CrNiPgeTi (1.27), SnSb (0.84); Table 19.2a, line marked Archean). But there is only random distribution for the two mineral groups UThRee and CuZPbBa.

For MPP areas, we observe a positive fingerprint for SnSb (0.77); all other fingerprints are random, except for Au which displays a negative association.

In the Neoproterozoic areas we observe a positive fingerprint for all minerals except SnSb contain (no data). The fingerprint for Au indicates a small (positive) association between Au and Neoproterozoic rocks (e.g. the error bar is just above zero).

In the NprA area we observe only random patterns for all mineral groups, whilst in the NprP areas we observe only a negative associations (e.g. no mineral deposits of any kind are predicted to occur here).

19.5.1.3 Central Africa Shield

In general the fingerprints in the CAS (Table 19.2b, Fig. 19.5) are lower than that of the CA. The most significant contributor to this difference between the metal potential of CA and CAS is the large number of deposits related to the

Table 19.2 Ln r_{ij}, with standard deviations (in brackets)

(a) CA	Au	CrNiPgeTi	CuZnPbBa	SnSb	UThREE	Total
Archean	1.22	1.27	0.29	0.84	−0.20	0.97
	(0.077)	(0.171)	(0.192)	(0.189)	(0.378)	(0.062)
MPP	−0.78	−0.26	−0.42	0.77	−0.09	−0.26
	(0.102)	(0.180)	(0.134)	(0.095)	(0.174)	(0.055)
Np	0.23	0.46	0.86		0.61	0.35
	(0.076)	(0.154)	(0.087)		(0.151)	(0.050)
NprA	−0.34	−0.54	−0.48	−0.66	−0.43	−0.43
	(0.140)	(0.354)	(0.236)	(0.333)	(0.354)	(0.103)
NprP	−0.10	−0.95	−0.69	−0.97	−0.36	−0.37
	(0.072)	(0.250)	(0.151)	(0.224)	(0.196)	(0.058)
(b) CAS	Au	CrNiPgeTi	CuZnPbBa	SnSb	UThREE	Total
Archean	0.83	−1.54	−0.80		0.17	0.21
	(0.123)	(1.000)	(0.447)		(0.447)	(0.114)
MPP	−0.40	0.31	0.04	0.50	0.11	0.03
	(0.103)	(0.180)	(0.134)	(0.097)	(0.209)	(0.057)
Np	−0.12	−0.60	1.17			0.03
	(0.316)	(1.000)	(0.267)			(0.200)
NprA	−0.03	−0.06	−0.12	−0.99	0.04	−0.18
	(0.140)	(0.354)	(0.236)	(0.333)	(0.354)	(0.103)
NprP	0.35	−1.46	−0.39	−1.13	-0.67	−0.10
	(0.162)	(1.000)	(0.378)	(0.500)	(0.707)	(0.139)
(c) CS	Au	CrNiPgeTi	CuZnPbBa	SnSb	UThREE	Total
Archean	−0.05	−0.14	0.02	−0.29	−0.23	−0.06
	(0.100)	(0.333)	(0.152)	(0.577)	(0.267)	(0.077)
MPP	0.25	1.19		2.15	1.58	0.48
	(0.354)	(0.707)		(0.707)	(0.447)	(0.243)
Np	−0.01	−0.40	0.39		−0.23	0.48
	(0.258)	(1.000)	(0.333)		(0.707)	(0.192)
NprA	0.58	0.70	−0.70		0.18	0.33
	(0.333)	(1.000)	(1.000)	–	(1.000)	(0.289)
NprP						

geology of the Lufilian Arc, which are included in the former but not in the latter. Other observations from our data reveal the following.

- Archean domains: potential for gold (positive association of 0.83); for all other minerals groups the fingerprints are random; and for SnSb there is no data.
- MPP domains only display random mineral patterns, but with a slight positive association for SnSb (0.50).
- Neoproterozoic domains display random distribution for all mineral group (no association), except for CuZnPbBa (strong positive signal of 1.17); again for SnSb there is no data.
- For the NprA and NprP, we observe a random distribution for all mineral groups.

19.5.2 Projected Mineral Endowment Beneath the Congo Basin

Central Africa contains the Central Africa Shield, which in turn contains the Congo Shield (CS) that hosts the Congo Basin (CB). Thus, $CA \subset CAS \subset CS \subset CB$, as illustrated in the top right insert of Fig. 19.7. As CB and CS are both contained within the outer regions (CA and CAS) we assume that across the CB and CS, relative to the distribution of their conterminous geological domains in CA and CAS, mineral fingerprints (patterns) will simulate those of the outer regions. Figure 19.7 shows the distribution of the Pre-cambrian areas within CA and CAS. Note that the CS contains no NprP; and that CB contains most Archean (85 %) and Neoproterozoic (14 %) domains (Fig. 19.8).

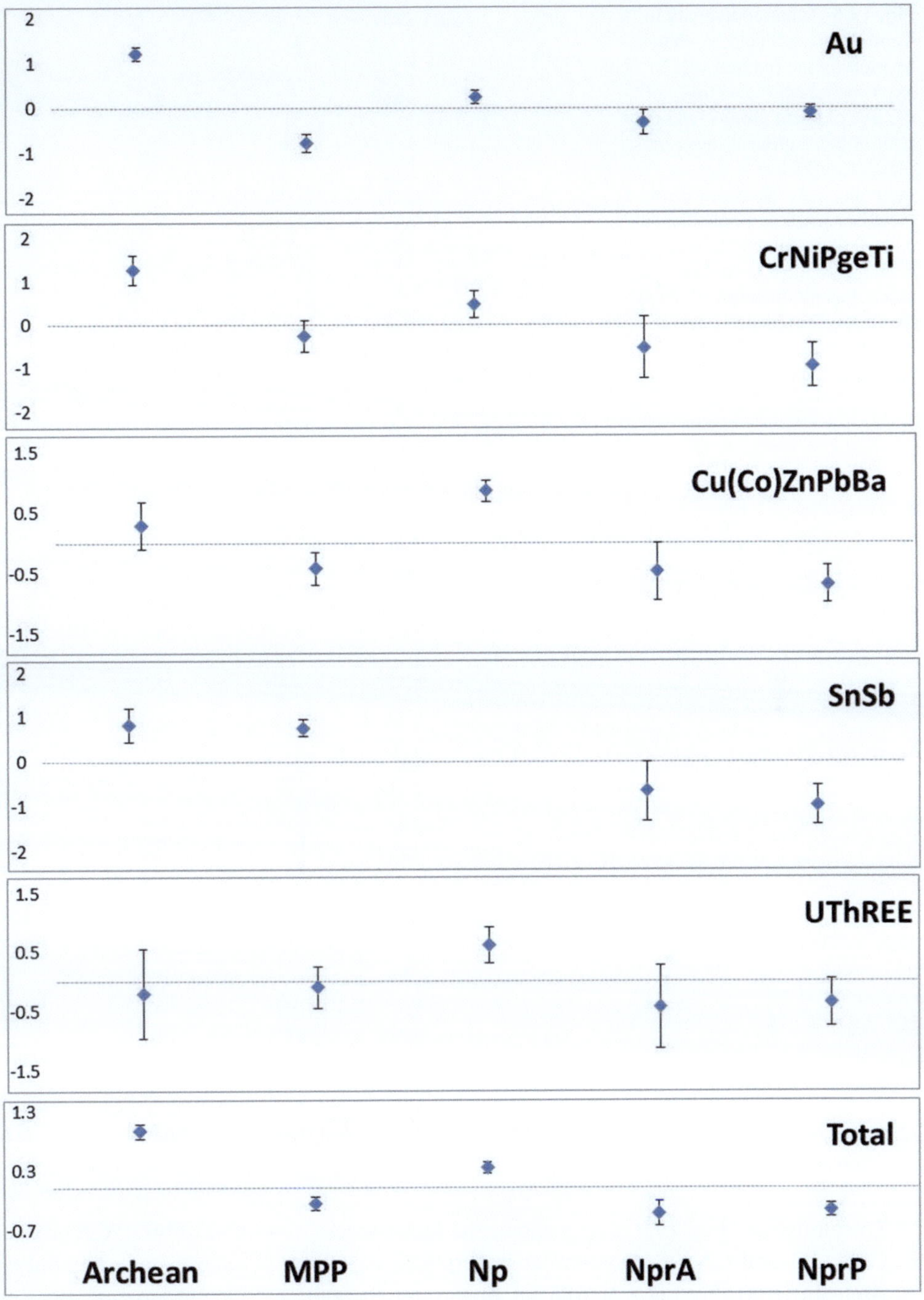

Fig. 19.4 Mineral diversity in Central Africa, as shown by plotting the natural log of the spatial coefficient (ln r_{ij}; Y-axis) of the five element groups (and the total for the combined groups [lower most plot]). Confidence intervals are at 2× Standard Error. Note how each geologic domain has a unique combination of elements that we term their 'Metallogenic Fingerprint'. Note that if the Confidence Interval include 0, the fingerprint is classed as random

We now transfer the expert-weight obtained from CA and CAS across to the two inner areas CS and CB and we treat the two inner regions CS and CB as one. Note that the inner areas do not contain all of the geologic fields represented in the outer areas (CA and CAS). Figure 19.9 compares the data expressed in Figs. 19.4, 19.5, 19.6 at one scale in order to compare each mineral group of all areas (viz. CA, CAS and CB). The circles in the graphs of Fig. 19.9 indicate significant fingerprints, (where significant = fingerprints (bars) whose confidence intervals do not include 0 (e.g. In, in Figs. 19.4, 19.5, 19.6). The significant fingerprints are summarised in Table 19.3 and Fig. 19.9. In Table 19.3 we include the significant fingerprints obtained in the CS, but we exclude them again to derive the 'expert fingerprint' shown in the last column (right hand). Most of the CS patterns appear unstable (large standard deviations), but the (significant) fingerprints in the CS can be analyses further with, for example, greater geologic knowledge of an 'expert'.

Fig. 19.5 Mineral diversity in Central Africa Shield, as shown by plotting the (natural log, Y-axis) spatial coeffient (ln r_{ij}) of the five element groups (and the total of these groups [lower most plot]). Confidence intervals = 2× Standard Error. Note how each geologic domain has a unique combination of elements that we term their 'Metallogenic fingerprint'. Note that if the Confidence Interval include 0, the fingerprint is classed as random

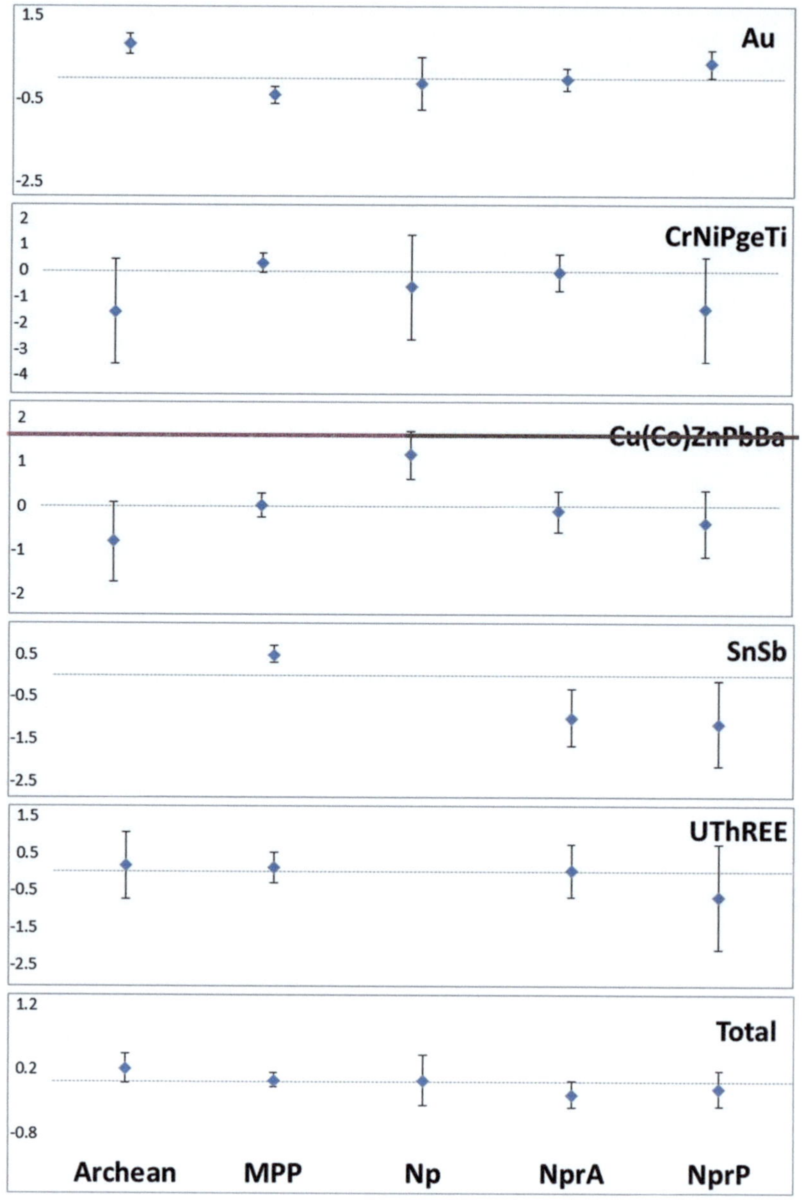

To summarize, Fig. 19.9 clearly shows the following:

- Gold: CA and CAS display similar fingerprints, e.g. for Archean areas there is a significant fingerprint in both regions. For MPP there is a negative association. The data for CS not reliable.
- CrNiPgeTi: CS and CAS have similar fingerprints, while CA displays a different fingerprint. However only for the Archean in the CA do we observe a significant signal.
- CuZnPbBa: CA and CAS display similar fingerprints, except here the only significant signals occur in the Neoproterozoic areas of CAS.
- SnSb: CS shows a significant signal similar to that of the CA; in the CAS there is a lack of data in the Archean domains.
- UTHREE: CS signals are significant in MPP, whilst CA shows significant signals in Neoproterozoic.

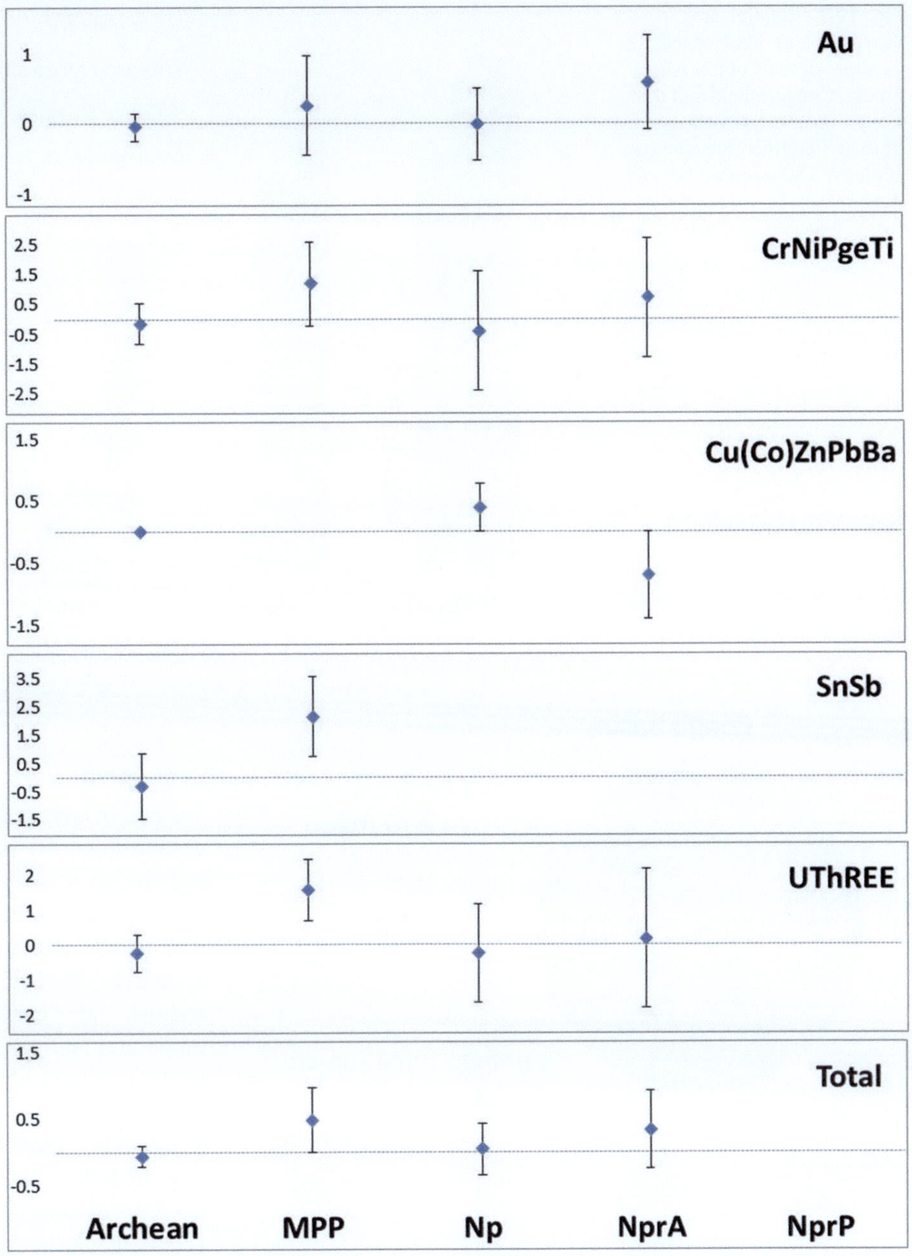

Fig. 19.6 Mineral diversity in Congo Basin, as shown by plotting the (natural log, Y-axis) spatial coeffient (ln r_{ij}) of the five element groups (and the total of these groups [lower most plot]). Confidence intervals = 2× Standard Error. Note how each geologic domain has a unique combination of elements that we term their 'Metallogenic fingerprint'. Note that if the confidence interval include 0, the fingerprint is classed as random

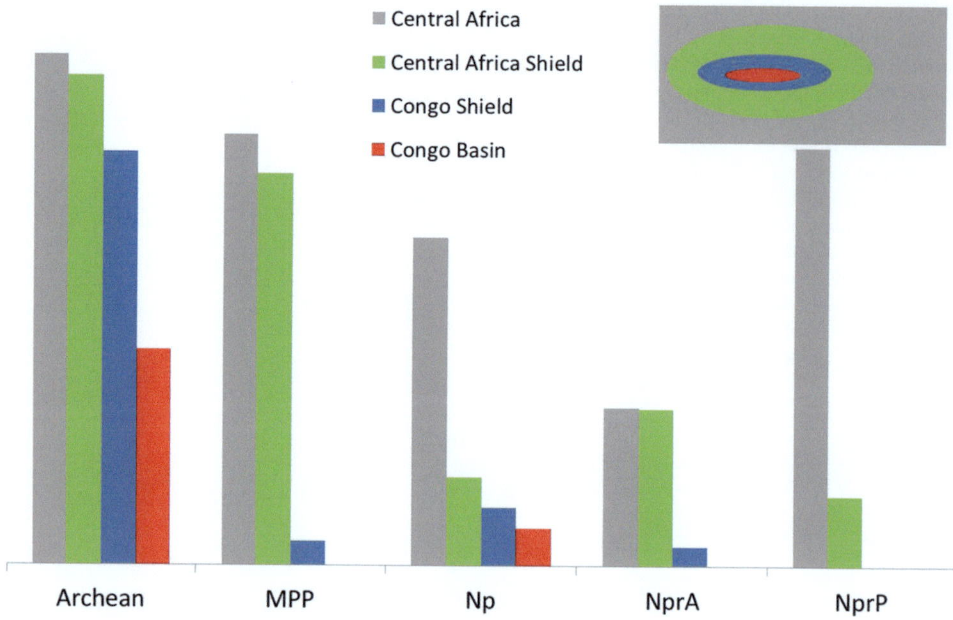

Fig. 19.7 Relative area distribution (Y axis) of the Central Africa, Central Africa Shield, Congo Shield and the Congo Basin expressed in terms of their simplified Precambrian geology. The insert at the *top* illustrates $CA \subset CAS \subset CS \subset CB$

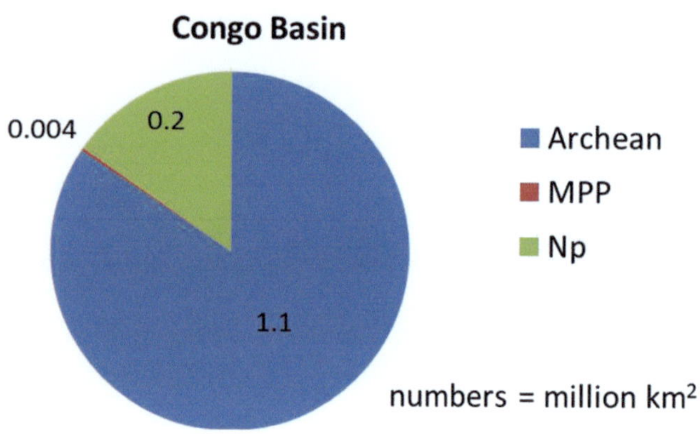

Fig. 19.8 Area distribution of Congo Basin. The values represent the areas in km^2

The final fingerprints that we assign to the CS and CB are given in the last column of Table 19.3. These fingerprints (obtained from the CA and CAS) are based on weights derived from a data-driven 'weight of evidence' (WofE) approach. The results from a data-driven technique can be further enhanced by 'expert knowledge'.

Conclusions

We assign an expert weight to the CS and CB by using the data from known mineral deposits from their surrounding regions, as well as knowledge of the distribution of Precambrian rocks in CA and CAS. Our expert weight is data-driven and it is based on a 'weight of evidence' approach. The value of the weights are judged on an ordinal scale (values between 0 and 0.5 are weak; 0.5–1 moderate; and >1 strong). In general, the Archean areas in the CB (and CS) have a strong affinity for Gold, and CrNiPge; and a moderate affinity for SnSb. This seems reasonable given the general mineral fields shown in Fig. 19.1. The Neoproterozoic areas in the CB (and CS) has a weak affinity for CrNiPgeTi; a moderate affinity for UThREE and a strong affinity for Cu(Co)ZnPbBa. The MPP area beneath the CB contains a moderate affinity for SnSb; however MPP only covers a very small area (less than 1 %). Finally it must be noted that the extrapolation of the basement geology beneath the CB is itself extrapolated from the surrounding regions and has a large uncertainty as is evident from the first two chapters of this Book.

Table 19.3 Summary of metallogenic fingerprints (from Table 19.2) for CA, CAS and CS. Only fingerprints that are significant (as defined in the text) are shown; no numbers reflect random fingerprints. The right-hand column includes the 'expert weight' given to the specified chronostratigraphic domains within the CS and CB

	Observed			Expert weight
	CA	CAS	CB	CB and CS
Au				
Archean	1.22	0.83		0.83–1.22
MPP				
Np				
NprA				
NprP				
CrNiPgeTi				
Archean	1.27			1.27
MPP				
Np	0.46			0.46
NprA				
NprP				
CuZnPbBa				
Archean				
MPP				
Np	0.86	1.17		0.86–1.17
NprA				
NprP				
SnSb				
Archean	0.84			0.84
MPP	0.77	0.50	2.15	0.50–0.77
Np				
NprA				
NprP				
UThREE				
Archean				
MPP			1.58	
Np	0.61			0.61
NprA				
NprP				
Total				
Archean	0.97			0.97
MPP				
Np	0.35			0.35
NprA				
NprP				

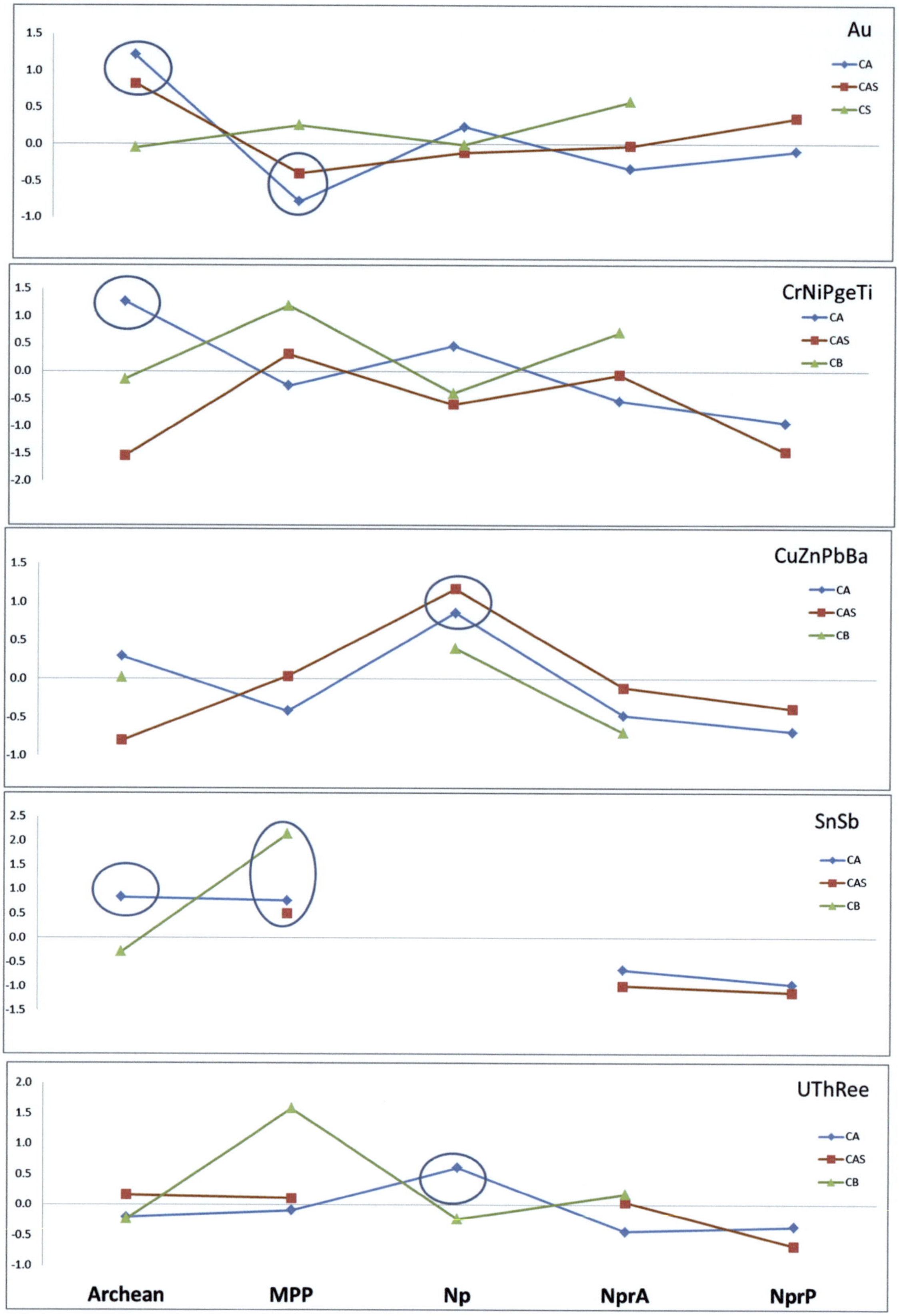

Fig. 19.9 Comparison of the mineral potential CA, CAS and CS. The *circles* in the graphs, indicate significant fingerprints, where significant is defined as the fingerprints whose confidence intervals does not include 0

Acknowledgments We are grateful to Bastien Linol for help with Fig. 19.1, and Fritz Agterberg and Mike de Wit for critical comments on this MS. The research of both authors was partly supported through different grants by the National Research Foundation of South Africa This is AEON contribution number 135.

References

Bonham-Carter GF (1996) Geographic information systems for geoscientists: modelling with GIS. Pergamon/Elsevier Science Publications, Oxford, 398pp

Campbell AN, Hollister VF, Duda RO, Hart PE (1982) Recognition of a hidden mineral deposit by an artificial intelligence program. Science 217:927–929

de Wit MJ, Jeffery M, Bergh H, Nicolaysen L (1988) Geological map of sectors of Gondwana reconstructed to their disposition ~150 Ma, scale 1:10.000.000. American Association Petroleum Geologists, Tulsa, Oklahoma, USA

de Wit MJ, Thiart C, Doucouré CM, Milesi JP, Billa M, Braux C, Nicol N (2004) The GONDWANA Metal-Potential GIS, A geological and metallogenic synthesis of the Gondwana supercontinent at 1:10 million scale, Version 2; using Arcview and Geokiosk. CIGCES/BRGM, Orleans La -Source, France

de Wit MJ, Stankiewicz J, Reeves C (2008) Restoring Pan-African-Brasiliano connections: more Gondwana control, less Trans-Atlantic corruption. In: Pankurst RJ, Trouw RAJ, Brito Neves BB, de Wit MJ (eds) West Gondwana: pre-cenozoic correlations across the South Atlantic region, vol 294. Geological Society, London, pp 399–412, Special Publications

de Wit MJ, Thiart C (2005) Metallogenic fingerprints of Archean cratons. In: McDonald I, Boyce AJ, Butler IB, Herrington RJ, Polya DA (eds) Mineral deposits and earth evolution, vol 248. Geological Society, London, pp 59–70, Special Publications

de Wit MJ, Thiart C, Doucoure CM, Wilsher W (1999) Scent of a supercontinent: Gondwana's ores as chemical tracers- tin, tungsten and the Neoproterozoic Laurentia- Gondwana connection. J Afr Earth Sci 28:35–51

Goosens PJ (2009) Mineral potential of the Democratic Republic of Congo: a geologic scandal? Soc Econ Geol Newsl 77: 1, 13–15

Mihalasky MJ, Bonham-Carter GF (2001) Lithodiversity and its spatial association with metallic mineral sites, Nevada great basin. Nat Resour Res 10:209–226

Mossman DJ, Gauthier-Lafaye F, Jackson SE (2005) Black shales, organic matter, ore genesis and hydrocarbon generation in the Paleoproterozoic Franceville Series, Gabon. Precambrian Res 137: 253–272

Schlüter T, Trauth MH (2008) Geological atlas of Africa: with notes on stratigraphy, tectonics, economic geology, geohazards, geosites and geoscientific education of each country. Springer Science & Business Media, Berlin, 320pp (e-book)

Thiart C, de Wit MJ (2006) Fingerprinting the metal endowment of early continental crust to test for secular changes in global mineralization. In: Kesler SE, Ohmoto H (eds) Evolution of early Earth's atmosphere, vol 198, Hydrosphere and biosphere -constraints from ore deposits. Geological Society of America Memoir., pp 53–66. doi:10.1130/2006.1198(03)

Thiart C, de Wit MJ (2008) African Neoproterozoic mineral deposits and Pan African metallogenesis. Nat Resour Res 17:99–105. doi:10.1007/s11053-008-9063-z

Thiart C, Stein A (2013) Continental – scale kriging of gold – bearing commodities. Spat Stat 6:57–77

Van Reybrouck D (2010) Congo – Een Geschiedenis. De Bezige Bij, Amsterdam. First English translation, 2014. HarperCollins, New York, 649pp

Wilsher W, Herbert R, Wullschleger N, Naicker I, Vitali E, DeWit MJ (1993) Towards intelligent spatial computing for the earth sciences in South Africa. S Afr J Sci 89:315–323

Index

M.J. de Wit et al. (eds.), *Geology and Resource Potential of the Congo Basin*, Regional Geology Reviews,
DOI 10.1007/978-3-642-29482-2, © Springer-Verlag Berlin Heidelberg 2015

Q

R

MIX
Papier aus verantwortungsvollen Quellen
Paper from responsible sources
FSC® C105338

If you have any concerns about our products,
you can contact us on
ProductSafety@springernature.com

In case Publisher is established outside the EU,
the EU authorized representative is:
Springer Nature Customer Service Center GmbH
Europaplatz 3, 69115 Heidelberg, Germany

Printed by Libri Plureos GmbH
in Hamburg, Germany